Physics of the Interstellar and Intergalactic Medium

PRINCETON SERIES IN ASTROPHYSICS

Edited by David N. Spergel

Theory of Rotating Stars, *by Jean-Louis Tassoul*

Theory of Stellar Pulsation, *by John P. Cox*

Galactic Dynamics, *by James Binney and Scott Tremaine*

Dynamical Evolution of Globular Clusters, *by Lyman Spitzer, Jr.*

Supernovae and Nucleosynthesis: An Investigation of the History of Matter, from the Big Bang to the Present, *by David Arnett*

Unsolved Problems in Astrophysics, *edited by John N. Bahcall and Jeremiah P. Ostriker*

Galactic Astronomy, *by James Binney and Michael Merrifield*

Active Galactic Nuclei: From the Central Black Hole to the Galactic Environment, *by Julian H. Krolik*

Plasma Physics for Astrophysics, *by Russell M. Kulsrud*

Electromagnetic Processes, *by Robert J. Gould*

Conversations on Electric and Magnetic Fields in the Cosmos, *by Eugene N. Parker*

High-Energy Astrophysics, *by Fulvio Melia*

Stellar Spectra Classification, *by Richard O. Gray and Christopher J. Corbally*

Exoplanet Atmospheres: Physical Processes, *by Sara Seager*

Physics of the Interstellar and Intergalactic Medium, *by Bruce T. Draine*

Physics of the Interstellar and Intergalactic Medium

Bruce T. Draine

PRINCETON UNIVERSITY PRESS
PRINCETON AND OXFORD

Published by Princeton University Press
41 William Street, Princeton, New Jersey 08540
In the United Kingdom: Princeton University Press, 6 Oxford Street, Woodstock, Oxfordshire OX20 1TW
press.princeton.edu
Cover image courtesy of
NASA/JPL-Caltech/S. Stolovy (Spitzer Science Center/Caltech)

Library of Congress Cataloging-in-Publication Data

Draine, Bruce T., 1947–
Physics of the interstellar and intergalactic medium / Bruce T. Draine.
p. cm. – (Princeton series in astrophysics)
Includes bibliographical references and index.
ISBN 978-0-691-12213-7 (hardback : alk. paper)—ISBN 978-0-691-12214-4 (pbk. : alk. paper) 1. Interstellar matter–Textbooks. 2. Galaxies–Textbooks. 3. Astrophysics–Textbooks. I. Title.
QB790.D73 2011
523.1′135–dc22
2010028285

British Library Cataloging-in-Publication Data is available

The publisher would like to acknowledge the author of this volume for providing the camera-ready copy from which this book was printed.

This book has been composed in Times-Roman

Printed on acid-free paper ∞
Printed in the United States of America

9 10 8

In memory of my parents,
Dolly and Tom

Contents

Preface

This book is intended for use in a graduate-level course on the physics of the interstellar medium within galaxies, and the intergalactic medium between galaxies – diffuse systems dominated by radiative processes and by two-body collisions between electrons, ions, atoms, molecules, and dust grains.

While it is assumed that the reader will have a background in undergraduate-level physics – including some prior exposure to atomic and molecular physics, statistical mechanics, and electromagnetism – the first six chapters of this book include a review of the basic physics that will be used in later chapters, including the notation and nomenclature used for identifying energy levels of atoms, ions, and molecules, and the selection rules for radiative transitions between levels.

In addition to serving as a text for a graduate-level course, it is hoped and intended that this book will also be useful to researchers (myself among them) who need to learn, or reacquaint themselves with, some aspect of interstellar/intergalactic physics. Accordingly, the book contains considerably more material than can realistically be covered in a one-semester course. The table of contents identifies some sections with a $\star$ symbol – these might be skipped in a "first pass" through the book.

The appendices include a list of symbols (Appendix A), values for physical constants (Appendix B), a collection of useful formulae for radiative processes (Appendix C), ionization potentials for atoms and ions up to atomic number 30 (Appendix D), energy-level diagrams for a number of atoms and ions of astrophysical interest (Appendix E), and a compilation of collisional rate coefficients (Appendix F), up-to-date as of 2010.

This book grew out of lecture notes that I started to develop 25 years ago when I began teaching the graduate ISM course at Princeton. It is, I hope, a comprehensive treatment, but inevitably it reflects my own interests and biases. The best part of writing the book was that it forced me to find time to clarify my own understanding of many aspects of interstellar physics.

Errors will no doubt be found. An up-to-date list of errata will be maintained at http://www.astro.princeton.edu/~draine/book/errata_p6.pdf.

The writing of this book has taken more years than I planned. Much work on the text was accomplished during a sabbatical semester at the Institute for Advanced Study in fall 2004; I am deeply grateful to the late John Bahcall for his encouragement. Further work took place during a sabbatical semester in fall 2005 at Osservatorio Arcetri in Florence; the warmth, hospitality, and stimulation of the Arcetri scientists and staff are not forgotten.

I am indebted to many colleagues for permitting me to reproduce figures from their publications, including Rick Arendt, Johannes Blümer, Jim Cordes, Dick

Crutcher, Tom Dame, John Dickey, Don Ellison, Marcello Felli, Doug Finkbeiner, Dale Fixsen, Erika Gibb, JinLin Han, Jelle Kaastra, Peter Kalberla, Ciska Kemper, Tom Kerr, Jin Koda, Richard Larson, Alex Lazarian, Aigen Li, Ron Maddalena, Chris McKee, Jean-Paul Meyer, Jerry Ostriker, Peter Sarre, Steve Snowden, and Joe Weingartner. I especially thank Doug Finkbeiner and Steve Snowden for providing information on local backgrounds, and Ken Dere for providing the radiative cooling function shown in Figures 34.1 – 34.3. I am grateful to Doug Finkbeiner for providing the 100 μm and Hα all-sky maps reproduced in Plates 2 and 3b, to Peter Kalberla for the 21-cm map reproduced in Plate 3a, and to Steve Snowden for the 0.75 keV all-sky map reproduced in Plate 5.

A number of colleagues read early drafts of many of these chapters. I particularly thank Princeton graduate students Gonzalo Aniano, Mike Belyaev, Tim Brandt, Ena Choi, Sudeep Das, Ruobing Dong, Aurelien Fraisse, Josh Green, and Yanfei Jiang for their helpful comments. Doug Finkbeiner and Mark Krumholz field-tested some of the manuscript; they and their students provided valuable feedback. Chris McKee and another reviewer (anonymous) provided thoughtful advice that led to improvements in the manuscript.

I relied on the SM plotting program for creating many of the figures in this book. I thank Robert Lupton for making SM available, and for expert assistance with it; he and Jeremy Goodman provided generous help with LaTeX arcana. The Smithsonian/NASA Astrophysics Data System (http://adsabs.harvard.edu) has been a truly invaluable aid to research, and was used frequently.

I am indebted to all my Princeton colleagues – students, postdocs, and fellow faculty – for creating the most congenial and stimulating research environment I can imagine.

It has been a pleasure to work with Ingrid Gnerlich, Dimitri Karetnikov, Mark Bellis, and Steve Peter at Princeton University Press.

❧ ❧ ❧

My deepest thanks go to my wife, Dina Gutkowicz-Krusin, for her many good suggestions, both scientific and editorial, which greatly improved this work, and for her continuing encouragement, without which it could not have been completed. She didn't expect it to take so long, but it is finally done.

Princeton, October 2010

❧ ❧ ❧

Note added in 2nd printing: R. Allen, B. Catinella, N. Evans, S. Ferraro, G. B. Field, B. Hensley, C. Hill, Xu Huang, E. B. Jenkins, K.-G. Lee, S. Lorenz Martins, B. Ménard, A. Natta, P. Pattarakijwanich, C. Petrovich, J. M. Shull, R. Simons, F. van der Tak, and M. J. Wolfire kindly pointed out various errors in the 1st printing; these have been corrected in the 2nd printing.

Note added in 4th printing: S. Bianchi, N. Evans, B. Jiang, B.-C. Koo, M. Gong, B. Hensley, S. Oh, R. Simons, A.-L. Sun, and W. Vlemmings kindly pointed out various errors in the 2nd printing; these have been corrected in the 4th printing. Some rate coefficients have also been updated.

Note added in 6th printing: M. Gong, J. Greco, E.B. Jenkins, C.D. Kreisch, J. Miralda-Escudé, R. Nakatani, I. Yoon, and G.B. Zhu kindly pointed out various errors remaining in the 4th printing; these have been corrected in the 6th printing.

Chapter One

Introduction

The subject of this book is the most beautiful component of galaxies – the gas and dust between the stars, or **interstellar medium**. The interstellar medium, or **ISM**, is, arguably, also the most important component of galaxies, for it is the ISM that is responsible for forming the stars that are the dominant sources of energy. While it now appears that the mass of most galaxies is primarily in the form of dark matter particles that are collisionless, or nearly so, it is the baryons and electrons (accounting for perhaps $\sim$10% of the total mass) that determine the visible appearance of galaxies, and that are responsible for nearly all of the energy emitted by galaxies, derived from nuclear fusion in stars and the release of gravitational energy in accretion disks around black holes. At early times, the baryonic mass in galaxies was primarily in the gas of the interstellar medium. As galaxies evolve, the interstellar medium is gradually converted to stars, and some part of the interstellar gas may be ejected from the galaxy in the form of galactic winds, or in some cases stripped from the galaxy by the intergalactic medium. Infalling gas from the intergalactic medium may *add* to the mass of the ISM. At the present epoch, the galaxy in which we reside – the Milky Way – has most of its baryons incorporated into stars or stellar remnants. But even today, perhaps 10% of the baryons in the Milky Way are to be found in the ISM. The "mass flow" of the baryons in the Milky Way is illustrated schematically in Figure 1.1.

Our objective is to understand the workings of the ISM – how it is organized and distributed in the Milky Way and other galaxies, what are the conditions (temperature, density, ionization, ...) in different parts of it, and how it dynamically evolves. Eventually, we would like to understand star formation, the process responsible for the very existence of galaxies as luminous objects.

The subject of this book, then, is *everything* in the galaxy that is between the stars – this includes the following constituents:

- **Interstellar gas**: Ions, electrons, atoms, and molecules in the gas phase, with velocity distributions that are very nearly thermal.

- **Interstellar dust**: Small solid particles, mainly less than $\sim 1\,\mu\mathrm{m}$ in size, mixed with the interstellar gas.

- **Cosmic rays**: Ions and electrons with kinetic energies far greater than thermal, often extremely relativistic – energies as high as $10^{21}\,\mathrm{eV}$ have been detected.

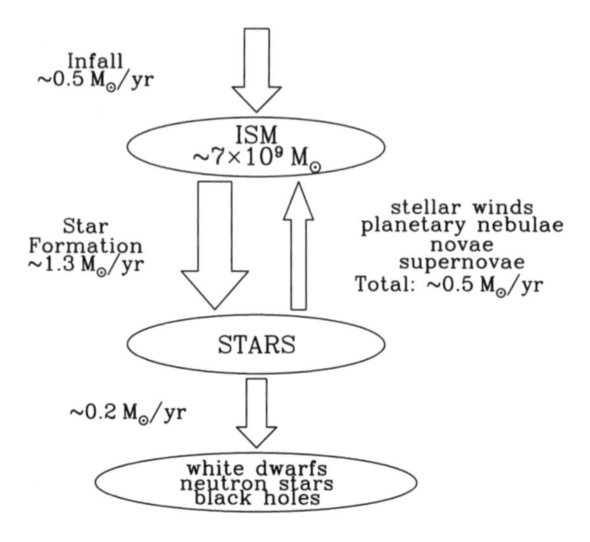

Figure 1.1 Flow of baryons in the Milky Way. See Table 1.2 for the ISM mass budget, and §42.4 for the value of the star formation rate in the Milky Way.

- **Electromagnetic radiation**: Photons from many sources, including the cosmic microwave background (CMB); stellar photospheres (i.e., starlight); radiation emitted by interstellar ions, atoms, and molecules; thermal emission from interstellar grains that have been heated by starlight; free–free emission ("bremsstrahlung") from interstellar plasma; synchrotron radiation from relativistic electrons; and gamma rays emitted in nuclear transitions, π^0 decays, and $e^+ - e^-$ annihilation.

- **Interstellar magnetic field**: The magnetic field resulting from electric currents in the interstellar medium; it guides the cosmic rays, and in some parts of the ISM, the magnetic field is strong enough to be dynamically important.

- **The gravitational field**: This is due to all of the matter in the galaxy – ISM, stars, stellar remnants, and dark matter – but in some regions, the contribution of the ISM to the gravitational potential leads to self-gravitating clouds.

- **The dark matter particles**: To the (currently unknown) extent that these interact nongravitationally with baryons, electrons, or magnetic fields, or either decay or annihilate into particles that interact with baryons, electrons, or magnetic fields, these are properly studied as part of the interstellar medium. The interactions are sufficiently weak that thus far they remain speculative.

There is of course no well-defined boundary to a galaxy, and all of the preceding constituents are inevitably present between galaxies – in the **intergalactic medium** (IGM) – and subject there to the same physical processes that act within the interstellar medium. The purview of this book, therefore, naturally extends to include the intergalactic medium.

Table 1.1 Units

pc	$= 3.086 \times 10^{18}$ cm	parsec
$M_\odot$	$= 1.989 \times 10^{33}$ g	solar mass
$L_\odot$	$= 3.826 \times 10^{33}$ erg s^{-1}	solar luminosity
yr	$= 3.156 \times 10^{7}$ s	sidereal year
Myr	$\equiv 10^{6}$ yr	megayear
AU	$= 1.496 \times 10^{13}$ cm	astronomical unit
Å	$\equiv 10^{-8}$ cm	Ångstrom
nm	$\equiv 10$ Å $\equiv 10^{-7}$ cm	nanometer
μm	$\equiv 10^{-4}$ cm	micron
km s^{-1}	$\equiv 10^{5}$ cm s^{-1}	km per sec
Jy	$\equiv 10^{-23}$ erg s^{-1} cm^{-2} Hz^{-1}	jansky
R	$\equiv (10^6/4\pi)$photons cm^{-2} s^{-1} sr^{-1}	rayleigh
D	$\equiv 10^{-18}$esu cm	debye
eV	$= 1.602 \times 10^{-12}$ erg	electron-volt
G	$= 10^{-4}$ tesla $= 10^{-4}$ weber m^{-2}	gauss

The primary aim of this book is to provide the reader with an exposition of the physics that determines the conditions in, and evolution of, the interstellar medium and the intergalactic medium. We will also emphasize the ways that observational data (e.g., strengths of emission lines or absorption lines) can be used to determine the physical properties of the regions where the emission or absorption is occuring.

We will employ the units of measurement that are currently used routinely by researchers in this field – for the most part, we use cgs units (including for electromagnetism), supplemented by standard astronomical units such as the parsec (pc), solar mass ($M_\odot$), and solar luminosity ($L_\odot$); see Table 1.1.

Historically, astronomers have reported optical wavelengths in Ångstroms (Å). In recent years, much of the physics literature has shifted to nanometers (nm), and consideration was given to doing so here. After weighing pros and cons, I decided to stick with Ångstroms; in practical work, it is necessary to specify optical wavelengths to (at least) four digits to avoid confusion, and it seems easier to remember them without a decimal point. And, after all, conversion from Å to nm is simply division by 10, a rather minor concern in a field that measures distance in pc, brightnesses in magnitudes, and angles in degrees, arcminutes, and arcseconds. So this book will use Ångstroms for wavelengths shorter than 1 μm.

I am, however, departing from established tradition by using wavelengths in vacuo for all transitions. This means that the wavelengths of familiar optical lines are now all shifted by $\sim$1 Å – e.g., the famous [O III] doublet is now 4960, 5008, rather than the wavelengths in air (4959, 5007) that have been entrenched in usage for the past century. This will cause some pain for those who have burned the air wavelengths into their memories, but it is time to abandon this anachronism from days when spectroscopy was done in air at (near) standard temperature and pressure.

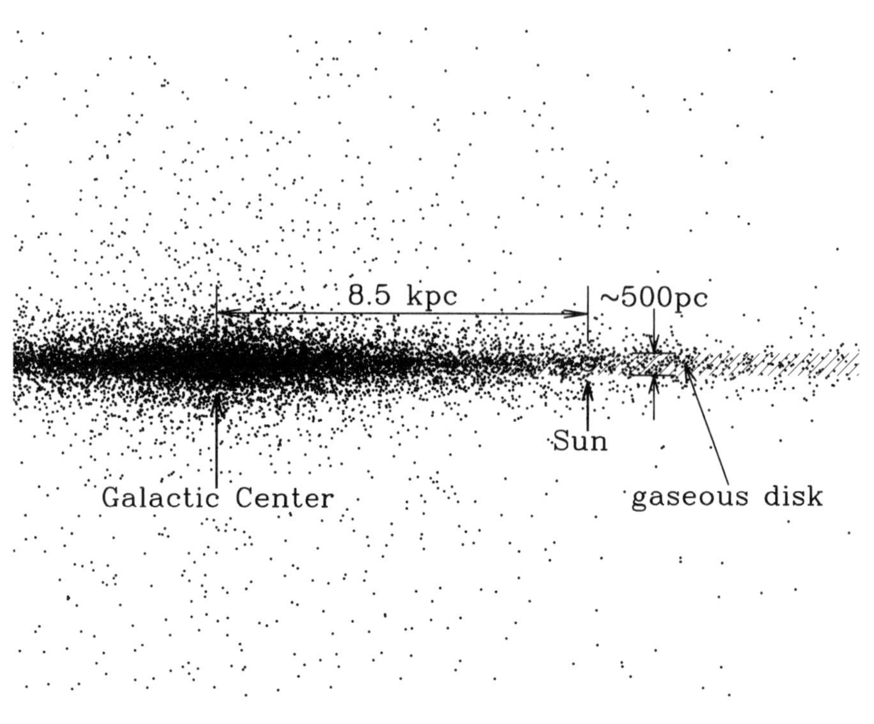

Figure 1.2 Structure of the Milky Way, viewed edge-on. The dots represent a sampling of stars; the volume containing most of the interstellar gas and dust is shaded. Compare with the infrared image of the stars in Plate 1, the dust in Plate 2, and various gas components in Plates 3–5.

1.1 Organization of the ISM: Characteristic Phases

In a spiral galaxy like the Milky Way, most of the dust and gas is to be found within a relatively thin gaseous disk, with a thickness of a few hundred pc (see the diagram in Fig. 1.2 and the images in Plates 1–5), and it is within this disk that nearly all of the star formation takes place. While the ISM extends above and below this disk, much of our attention will concern the behavior of the interstellar matter within a few hundred pc of the disk midplane.

The Sun is located about 8.5 kpc from the center of the Milky Way; as it happens, the Sun is at this time very close to the disk midplane. The total mass of the Milky Way within 15 kpc of the center is approximately $10^{11} M_\odot$; according to current estimates, this includes $\sim 5 \times 10^{10} M_\odot$ of stars, $\sim 5 \times 10^{10} M_\odot$ of dark matter, and $\sim 7 \times 10^9 M_\odot$ of interstellar gas, mostly hydrogen and helium (see Table 1.2). About 60% of the interstellar hydrogen is in the form of H atoms, $\sim$20% is in the form of H_2 molecules, and $\sim$20% is ionized.

The gaseous disk is approximately symmetric about the midplane, but does not have a sharp boundary – it is like an atmosphere. We can define the half-thickness

Table 1.2 Mass of H II, H I, and H_2 in the Milky Way ($R < 20\,\mathrm{kpc}$)

Phase	$M(10^9\,M_\odot)$	fraction	Note
Total H II (not including He)	1.12	23%	see Chapter 11
Total H I (not including He)	2.9	60%	see Chapter 29
Total H_2 (not including He)	0.84	17%	see Chapter 32
Total H II, H I and H_2 (not including He)	4.9		
Total gas (including He)	6.7		

$z_{1/2}$ of the disk to be the distance z above (or below) the plane where the density has dropped to 50% of the midplane value. Observations of radio emission from atomic hydrogen and from the CO molecule indicate that the half-thickness $z_{1/2} \approx 250\,\mathrm{pc}$ in the neighborhood of the Sun. The thickness $2z_{1/2} \approx 500\,\mathrm{pc}$ of the disk is only ~6% of the ~8.5 kpc distance from the Sun to the Galactic center – it is a *thin* disk. The thinness of the distribution of dust and gas is evident from the $100\,\mu\mathrm{m}$ image showing thermal emission from dust in Plate 2, and the H I 21-cm line image in Plate 3.

The baryons in the interstellar medium of the Milky Way are found with a wide range of temperatures and densities; because the interstellar medium is dynamic, all densities and temperatures within these ranges can be found somewhere in the Milky Way. However, it is observed that most of the baryons have temperatures falling close to various characteristic states, or "phases." For purposes of discussion, it is convenient to name these phases. Here we identify seven distinct phases that, between them, account for most of the mass and most of the volume of the interstellar medium. These phases (summarized in Table 1.3) consist of the following:

- **Coronal gas**: Gas that has been shock-heated to temperatures $T \gtrsim 10^{5.5}\,\mathrm{K}$ by blastwaves racing outward from supernova explosions. The gas is collisionally ionized, with ions such as O VI ($\equiv O^{5+}$) present. Most of the coronal gas has low density, filling an appreciable fraction – approximately half – of the volume of the galactic disk. The coronal gas regions may have characteristic dimensions of $\sim 20\,\mathrm{pc}$, and may be connected to other coronal gas volumes. The coronal gas cools on ~Myr time scales. Much of the volume above and below the disk is thought to be pervaded by coronal gas.[1] It is often referred to as the "hot ionized medium," or **HIM**.

- **H II gas**: Gas where the hydrogen has been photoionized by ultraviolet photons from hot stars. Most of this photoionized gas is maintained by radiation from recently formed hot massive O-type stars – the photoionized gas may be dense material from a nearby cloud (in which case the ionized gas is called an **H II region**) or lower density "intercloud" medium (referred to as **diffuse H II**).

[1] This gas is termed "coronal" because its temperature and ionization state is similar to the corona of the Sun.

Table 1.3 Phases of Interstellar Gas

Phase	$T(\mathrm{K})$	$n_{\mathrm{H}}(\mathrm{cm}^{-3})$	Comments
Coronal gas (HIM) $f_V \approx 0.5?$ $\langle n_{\mathrm{H}}\rangle f_V \approx 0.002\,\mathrm{cm}^{-3}$ ($f_V \equiv$ volume filling factor)	$\gtrsim 10^{5.5}$	~ 0.004	Shock-heated Collisionally ionized Either expanding or in pressure equilibrium Cooling by: ◇ Adiabatic expansion ◇ X ray emission Observed by: • UV and x ray emission • Radio synchrotron emission
H II gas $f_V \approx 0.1$ $\langle n_{\mathrm{H}}\rangle f_V \approx 0.02\,\mathrm{cm}^{-3}$	10^4	$0.2 - 10^4$	Heating by photoelectrons from H, He Photoionized Either expanding or in pressure equilibrium Cooling by: ◇ Optical line emission ◇ Free–free emission ◇ Fine-structure line emission Observed by: • Optical line emission • Thermal radio continuum
Warm H I (WNM) $f_V \approx 0.4$ $n_{\mathrm{H}} f_V \approx 0.2\,\mathrm{cm}^{-3}$	~5000	0.6	Heating by photoelectrons from dust Ionization by starlight, cosmic rays Pressure equilibrium Cooling by: ◇ Optical line emission ◇ Fine structure line emission Observed by: • H I 21 cm emission, absorption • Optical, UV absorption lines
Cool H I (CNM) $f_V \approx 0.01$ $n_{\mathrm{H}} f_V \approx 0.3\,\mathrm{cm}^{-3}$	~ 100	30	Heating by photoelectrons from dust Ionization by starlight, cosmic rays Cooling by: ◇ Fine structure line emission Observed by: • H I 21-cm emission, absorption • Optical, UV absorption lines
Diffuse H_2 $f_V \approx 0.001$ $n_{\mathrm{H}} f_V \approx 0.1\,\mathrm{cm}^{-3}$	$\sim 50\,\mathrm{K}$	~ 100	Heating by photoelectrons from dust Ionization by starlight, cosmic rays Cooling by: ◇ Fine structure line emission Observed by: • H I 21-cm emission, absorption • CO 2.6-mm emission • optical, UV absorption lines
Dense H_2 $f_V \approx 10^{-4}$ $\langle n_{\mathrm{H}}\rangle f_V \approx 0.2\,\mathrm{cm}^{-3}$	$10 - 50$	$10^3 - 10^6$	Heating by photoelectrons from dust Ionization and heating by cosmic rays Self-gravitating: $p > p(\text{ambient ISM})$ Cooling by: ◇ CO line emission ◇ C I fine structure line emission Observed by: • CO 2.6-mm emission • dust FIR emission
Cool stellar outflows	$50 - 10^3$	$1 - 10^6$	Observed by: • Optical, UV absorption lines • Dust IR emission • H I, CO, OH radio emission

Bright H II regions, such as the Orion Nebula, have dimensions of a few pc; their lifetimes are essentially those of the ionizing stars, $\sim 3 - 10$ Myr. The extended low-density photoionized regions – often referred to as the **warm ionized medium**, or **WIM** – contain much more total mass than the more visually conspicuous high-density H II regions. According to current estimates, the Galaxy contains $\sim 1.1 \times 10^9\, M_\odot$ of ionized hydrogen; about 50% of this is within 500 pc of the disk midplane (the distribution of the H II is discussed in Chapter 11). In addition to the H II regions, photoionized gas is also found in distinctive structures called **planetary nebulae**[2] – these are created when rapid mass loss during the late stages of evolution of stars with initial mass $0.8 M_\odot < M < 6 M_\odot$ exposes the hot stellar core; the radiation from this core photoionizes the outflowing gas, creating a luminous (and often very beautiful) planetary nebula. Individual planetary nebulae fade away on $\sim 10^4$ yr time scales.

- **Warm H I**: Predominantly atomic gas heated to temperatures $T \approx 10^{3.7}$ K; in the local interstellar medium, this gas is found at densities $n_{\rm H} \approx 0.6\,{\rm cm}^{-3}$. It fills a significant fraction of the volume of the disk – perhaps 40%. Often referred to as the **warm neutral medium**, or **WNM**.

- **Cool H I**: Predominantly atomic gas at temperatures $T \approx 10^2$ K, with densities $n_{\rm H} \approx 30\,{\rm cm}^{-3}$ filling $\sim$1% of the volume of the local interstellar medium. Often referred to as the **cold neutral medium**, or **CNM**.

- **Diffuse molecular gas**: Similar to the cool H I clouds, but with sufficiently large densities and column densities so that H_2 self-shielding (discussed in Chapter 31) allows H_2 molecules to be abundant in the cloud interior.

- **Dense molecular gas**: Gravitationally bound clouds that have achieved $n_{\rm H} \gtrsim 10^3\,{\rm cm}^{-3}$. These clouds are often "dark" – with visual extinction $A_V \gtrsim 3$ mag through their central regions. In these dark clouds, the dust grains are often coated with "mantles" composed of H_2O and other molecular ices. It is within these regions that star formation takes place. It should be noted that the gas pressures in these "dense" clouds would qualify as ultrahigh vacuum in a terrestrial laboratory.

- **Stellar outflows**: Evolved cool stars can have mass loss rates as high as $10^{-4} M_\odot\,{\rm yr}^{-1}$ and low outflow velocities $\lesssim 30\,{\rm km\,s}^{-1}$, leading to relatively high density outflows. Hot stars can have winds that are much faster, although far less dense.

The ISM is dynamic, and the baryons undergo changes of phase for a number of reasons: ionizing photons from stars can convert cold molecular gas to hot H II; radiative cooling can allow hot gas to cool to low temperatures; ions and electrons can recombine to form atoms, and H atoms can recombine to form H_2 molecules.

[2]They are called "planetary" nebulae because of their visual resemblance to planets when viewed through a small telescope.

Table 1.4 Protosolar Abundances of the Elements with $Z \leq 32$ (based on Asplund et al. (2009); see text)

Z	X	$\langle m_X \rangle$/amu	$N_X/N_{\rm H}$	$M_X/M_{\rm H}$	Source
1	H	1.0080	1	1	
2	He	4.0026	$9.55 \times 10^{-2\pm0.01}$	3.82×10^{-1}	Photospheric
3	Li	6.941	$2.00 \times 10^{-9\pm0.05}$	1.38×10^{-8}	Meteoritic
4	Be	9.012	$2.19 \times 10^{-11\pm0.03}$	1.97×10^{-10}	Meteoritic
5	B	10.811	$6.76 \times 10^{-10\pm0.04}$	7.31×10^{-9}	Meteoritic
6	C	12.011	$2.95 \times 10^{-4\pm0.05}$	3.54×10^{-3}	Photospheric
7	N	14.007	$7.41 \times 10^{-5\pm0.05}$	1.04×10^{-3}	Photospheric
8	O	15.999	$5.37 \times 10^{-4\pm0.05}$	8.59×10^{-3}	Photospheric
9	F	18.998	$2.88 \times 10^{-8\pm0.06}$	5.48×10^{-7}	Meteoritic
10	Ne	20.180	$9.33 \times 10^{-5\pm0.10}$	1.88×10^{-3}	Photospheric
11	Na	22.990	$2.04 \times 10^{-6\pm0.02}$	4.69×10^{-5}	Meteoritic
12	Mg	24.305	$4.37 \times 10^{-5\pm0.04}$	1.06×10^{-3}	Photospheric
13	Al	26.982	$2.95 \times 10^{-6\pm0.01}$	8.85×10^{-5}	Meteoritic
14	Si	28.086	$3.55 \times 10^{-5\pm0.04}$	9.07×10^{-4}	Photospheric
15	P	30.974	$2.82 \times 10^{-7\pm0.03}$	8.73×10^{-6}	Photospheric
16	S	32.065	$1.45 \times 10^{-5\pm0.03}$	4.63×10^{-4}	Photospheric
17	Cl	35.453	$1.86 \times 10^{-7\pm0.06}$	6.60×10^{-6}	Meteoritic
18	Ar	39.948	$2.75 \times 10^{-6\pm0.13}$	1.10×10^{-4}	Photospheric
19	K	39.098	$1.32 \times 10^{-7\pm0.02}$	5.15×10^{-6}	Meteoritic
20	Ca	40.078	$2.14 \times 10^{-6\pm0.02}$	8.57×10^{-5}	Meteoritic
21	Sc	44.956	$1.23 \times 10^{-9\pm0.02}$	5.53×10^{-8}	Meteoritic
22	Ti	47.867	$8.91 \times 10^{-8\pm0.03}$	4.27×10^{-6}	Meteoritic
23	V	50.942	$1.00 \times 10^{-8\pm0.02}$	5.09×10^{-7}	Meteoritic
24	Cr	51.996	$4.79 \times 10^{-7\pm0.01}$	2.49×10^{-5}	Meteoritic
25	Mn	54.938	$3.31 \times 10^{-7\pm0.01}$	1.82×10^{-5}	Meteoritic
26	Fe	55.845	$3.47 \times 10^{-5\pm0.04}$	1.94×10^{-3}	Photospheric
27	Co	58.933	$8.13 \times 10^{-8\pm0.01}$	4.79×10^{-6}	Meteoritic
28	Ni	58.693	$1.74 \times 10^{-6\pm0.01}$	1.02×10^{-4}	Meteoritic
29	Cu	63.546	$1.95 \times 10^{-8\pm0.04}$	1.24×10^{-6}	Meteoritic
30	Zn	65.38	$4.68 \times 10^{-8\pm0.04}$	3.06×10^{-6}	Meteoritic
31	Ga	69.723	$1.32 \times 10^{-9\pm0.02}$	9.19×10^{-8}	Meteoritic
32	Ge	72.64	$4.17 \times 10^{-9\pm0.04}$	3.03×10^{-7}	Meteoritic

Asplund et al. (2009) have corrected the measured photospheric abundances of He, C, N, O, Ne, Mg, Si, S, Ar, and Fe to allow for diffusion in the Sun.

As recommended by Asplund et al. (2009), the photospheric abundance of Si, and meteoritic abundances (tied to Si), have been increased by a factor $10^{0.04}$ to allow for diffusion in the Sun. Similarly, the measured photospheric abundance of P has been multiplied by $10^{0.04}$ to allow for diffusion in the Sun.

$M(Z>2)/M_{\rm H} = 0.0199$; $M({\rm total})/M_{\rm H} = 1.402$.

Table 1.5 Energy Densities in the Local ISM

Component		$u(\mathrm{eV\,cm^{-3}})$	Note
Cosmic microwave background	$(T_{\mathrm{CMB}} = 2.725\,\mathrm{K})$	0.265	a
Far-infrared radiation from dust		0.31	b
Starlight ($h\nu < 13.6\,\mathrm{eV}$)		0.54	c
Thermal kinetic energy $(3/2)nkT$		0.49	d
Turbulent kinetic energy $(1/2)\rho v^2$		0.22	e
Magnetic energy $B^2/8\pi$		0.89	f
Cosmic rays		1.39	g

a Fixsen & Mather (2002).
b Chapter 12.
c Chapter 12.
d For $nT = 3800\,\mathrm{cm^{-3}\,K}$ (see §17.7).
e For $n_{\mathrm{H}} = 30\,\mathrm{cm^{-3}}$, $v = 1\,\mathrm{km\,s^{-1}}$, or $\langle n_{\mathrm{H}}\rangle = 1\,\mathrm{cm^{-3}}$, $\langle v^2\rangle^{1/2} = 5.5\,\mathrm{km\,s^{-1}}$.
f For median $B_{\mathrm{tot}} \approx 6.0\,\mu\mathrm{G}$ (Heiles & Crutcher 2005).
g For cosmic ray spectrum X3 in Fig. 13.5.

1.2 Elemental Composition

The interstellar gas is primarily H and He persisting from the Big Bang, with a small reduction in the H fraction, a small increase in the He fraction, and addition of a small amount of heavy elements – from C to U – as the result of the return to the ISM of gas that has been processed in stars and stellar explosions. The abundance of heavy elements in the ISM – e.g., C, O, Mg, Si, and Fe – is a declining function of distance from the Galactic Center, with the abundance near the Sun (galactocentric radius $R \approx 8.5\,\mathrm{kpc}$) being about half the abundance in the Galactic Center region.

The composition of the ISM in the solar neighborhood is not precisely known, but is thought to be similar to the composition of the Sun. The current best estimates of solar abundances for elements with atomic number ≤ 32 (as determined from both observations of the stellar photosphere and studies of primitive carbonaceous chondrite meteorites) are given in Table 1.4. These abundances are intended to be the abundances in the protosun, which differ from photospheric abundances due to diffusion. H and He together account for most of the mass – the elements with $Z \geq 3$ contribute only $\sim 1\%$ of the total mass. Nevertheless, these heavy element "impurities" in many cases determine the chemistry, ionization state, and temperature of the gas, in addition to which they provide valuable observable diagnostics.

1.3 Energy Densities

Energy is present in the ISM in a number of forms: thermal energy $u = (3/2)nkT$, bulk kinetic energy $(1/2)\rho v^2$, cosmic ray energy u_{CR}, magnetic energy $B^2/8\pi$, and energy in photons, which can be subdivided into cosmic microwave back-

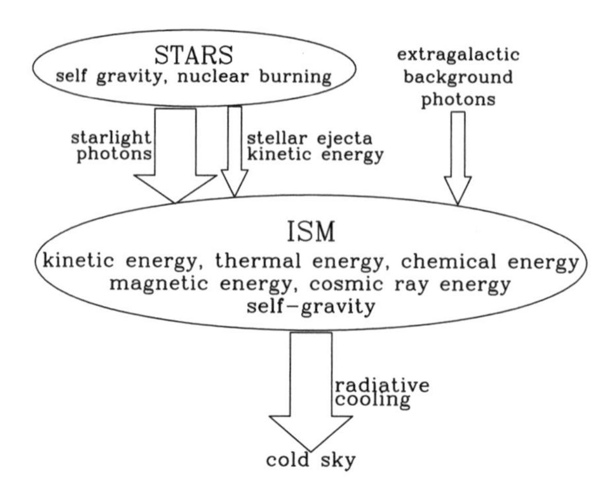

Figure 1.3 Flow of energy in the Milky Way.

ground, far-infrared (FIR) emission from dust, and starlight. It is a remarkable fact that in the local ISM, today, these energy densities all fall within the range $0.2 - 2\,\mathrm{eV\,cm^{-3}}$ – see Table 1.5. This near-equipartition is partly coincidental – the fact that the energy density in the CMB is similar to the other energy densities is surely accidental – but the other six energy densities are in fact coupled: the magnetic energy has been built up by fluid motions, so it is probably not a coincidence that the magnetic energy density $B^2/8\pi$ and the turbulent energy density $(1/2)\rho v^2$ are comparable in magnitude. Similarly, if the cosmic ray energy density were much larger, it would not be possible for the magnetized ISM to confine the cosmic rays, and they would be able to escape freely from the Galaxy – this negative feedback limits the cosmic ray energy density to approximate equipartition with the sum of the turbulent energy density and thermal pressure in the ISM. The fact that the starlight energy density is comparable to the gas pressure may be coincidental. However, if the starlight energy density were much larger (by a factor $\sim 10^2$), radiation pressure acting on dust grains would be able to "levitate" the ISM above and below the Galactic midplane, presumably suppressing star formation; this feedback loop may play a role in regulating the starlight energy density in star-forming galaxies.

The ISM is far from thermodynamic equilibrium, and it is only able to maintain this nonequilibrium state because of the input of "free energy," primarily in the form of ultraviolet radiation emitted by stars, but with a significant and important contribution of kinetic energy from high-velocity gaseous ejecta from supernovae. The overall flow of energy in the ISM is sketched in Figure 1.3. Ultimately, nearly all of the energy injected into the ISM in the form of starlight and kinetic energy of stellar ejecta is lost from the galaxy in the form of emitted photons, departing to the cold extragalactic sky.

Chapter Two

Collisional Processes

Collisions are fundamental to the physics of the interstellar medium (ISM): they allow the gas to (usually) be treated as a fluid; they determine the thermal and electrical conductivity and diffusion coefficients; they produce most of the excitations of ions, atoms, and molecules that result in emission of photons from the ISM; and they are responsible for recombination of electrons and ions, and for chemical reactions.

It is important to become comfortable with the concept of collisional rates – and collisional rate coefficients – and to understand how they depend on temperature. Indeed, the very concept of kinetic temperature depends on elastic scattering rates being fast enough to ensure that the velocity distribution function for particles will be close to a Maxwellian distribution.

In this chapter, we will review the concept of collisional rate coefficients. There are four basic types of collisional interactions that concern us: the long-range $1/r$ Coulomb interaction between ions and ions, ions and electrons, and electrons and electrons; the intermediate range r^{-4} induced-dipole interaction between ions and neutral atoms or molecules; the interaction between electrons and neutrals; and the short-range interaction between neutrals. For the Coulomb interaction, we will estimate collisional ionization rates and scattering rates using the "impact approximation." For the induced-dipole interaction, we will use exact results for scattering by a r^{-4} potential. For electron–neutral scattering, we use experimental data. For neutral–neutral collisions, we use estimates for the effective "hard-sphere" radius.

2.1 Collisional Rate Coefficients

The rate per unit volume of a general two-body collisional process

$$A + B \rightarrow \text{products} \tag{2.1}$$

is written

$$\text{reaction rate per unit volume} = n_A n_B \langle \sigma v \rangle_{AB} \ , \tag{2.2}$$

where the **two-body collisional rate coefficient** for $A + B \rightarrow$ products is

$$\langle \sigma v \rangle_{AB} \equiv \int_0^\infty \sigma_{AB}(v)\, v\, f_v\, dv \quad , \tag{2.3}$$

where $\sigma_{AB}(v)$ is the velocity-dependent reaction cross section for the reaction, and $f_v dv$ is the probability that A and B have relative speed v in dv. In cgs units, a two-body collisional rate coefficient has dimensions of $\mathrm{cm}^3\,\mathrm{s}^{-1}$.

In thermal equilibrium at temperature T, the distribution function for the relative speed in encounters between particles A and B is given by a Maxwellian velocity distribution

$$f_v\,dv = 4\pi\left(\frac{\mu}{2\pi kT}\right)^{3/2} \mathrm{e}^{-\mu v^2/2kT}\, v^2\,dv \quad , \tag{2.4}$$

where $\mu \equiv m_A m_B/(m_A + m_B)$ is the **reduced mass** of the collision partners. It will sometimes be convenient to use the distribution function f_E for the center-of-mass energy $E = \mu v^2/2$. From $f_E dE = f_v dv$, we obtain

$$v f_E = v\frac{dv}{dE} f_v = \left(\frac{8}{\pi\mu kT}\right)^{1/2} \frac{E}{kT}\,\mathrm{e}^{-E/kT} \quad . \tag{2.5}$$

Thus the two-body collisional rate coefficient is

$$\langle\sigma v\rangle_{AB} = \left(\frac{8kT}{\pi\mu}\right)^{1/2} \int_0^\infty \sigma_{AB}(E)\,\frac{E}{kT}\,\mathrm{e}^{-E/kT}\,\frac{dE}{kT} \quad . \tag{2.6}$$

At sufficiently high densities (e.g., the Earth's atmosphere), three-body collisions – where three particles are simultaneously in close proximity and interacting strongly – can become important. The rate per unit volume of a general three-body collisional process $A + B + C \rightarrow$ products is written

$$\text{reaction rate per unit volume} = k_{ABC}\, n_A n_B n_C \ , \tag{2.7}$$

where k_{ABC} is the **three-body collisional rate coefficient**. Even at interstellar densities, three-body processes can be important for some reactions, such as populating the high-n levels of atomic hydrogen (see §3.7). In cgs units, a three-body collisional rate coefficient k_{ABC} has dimensions of $\mathrm{cm}^6\,\mathrm{s}^{-1}$.

2.2 Inverse-Square Law Forces: Elastic Scattering

The classical problem of elastic scattering by an inverse-square force law – Rutherford scattering – can be solved, but the resulting integrals can be tedious to evaluate. Here we will employ a very simple approach – the "impact approximation" – to obtain approximate results with very simple algebra. The impact approximation is accurate when the interaction is weak enough to produce only a small deflection – this often turns out to be the case.

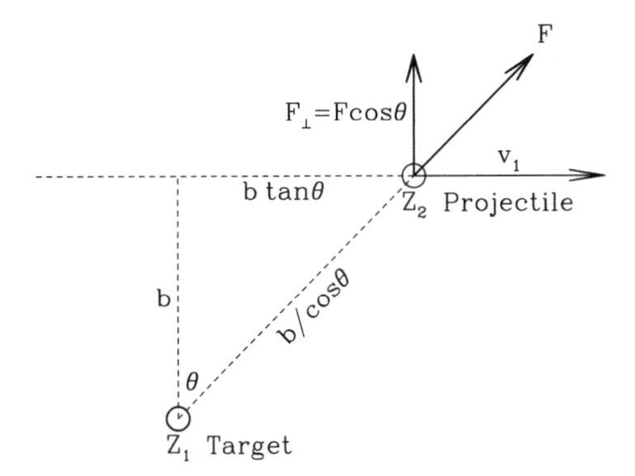

Figure 2.1 Coordinates for impact approximation, in the frame where the target is at rest. The projectile is assumed to travel in a straight line at constant speed, and the target is assumed to remain at rest during the encounter.

2.2.1 The Impact Approximation

The **impact approximation** consists of calculating the exchange of momentum between projectile and target, assuming that the projectile and target velocities remain constant during the encounter – the projectile moves in (almost) a straight line at (almost) constant speed. For the Coulomb interaction between particles with charges Z_1e and Z_2e, the instantaneous force perpendicular to the trajectory is

$$F_\perp = \frac{Z_1Z_2e^2}{(b/\cos\theta)^2}\cos\theta = \frac{Z_1Z_2e^2}{b^2}\cos^3\theta \quad , \tag{2.8}$$

where the "impact parameter" b is the distance of closest approach *if the projectile were to travel undeflected*, and the angle θ is defined in Figure 2.1. If v_1 is the relative velocity, we find

$$dt = \frac{d(b\tan\theta)}{v_1} = \frac{b}{v_1}\frac{d\theta}{\cos^2\theta} \quad , \tag{2.9}$$

and the total momentum transfer is obtained by integrating $F_\perp$ over time[1]:

$$\Delta p_\perp = \int_{-\infty}^{\infty} F_\perp dt = \frac{Z_1Z_2e^2}{bv_1}\int_{-\pi/2}^{\pi/2}\cos\theta d\theta = 2\frac{Z_1Z_2e^2}{bv_1} \quad . \tag{2.10}$$

2.2.2 Example: Collisional Ionization

Consider an atom or ion with a single bound electron. Let I be the energy required to "ionize" it – i.e., to unbind the electron. Suppose that we have a fast-moving

[1] A back-of-the-envelope estimate by taking the product of the force at closest approach ($Z_1Z_2e^2/b^2$) and the characteristic interaction time (b/v_1) differs from Eq. (2.10) by only a factor of 2.

electron with speed $v \gg (2I/m_e)^{1/2}$. We can use the impact approximation to estimate the ionization rate by asking: If $\Delta p_\perp$ is the momentum transfer to the bound electron, for what b is $(\Delta p_\perp)^2 > 2m_e I$? The answer is $b < b_{\rm max}(v) = [2Z_p^2 e^4/m_e v^2 I]^{1/2}$. This gives us an estimate for the ionization cross section:

$$\sigma(v) \approx \pi b_{\rm max}^2 = \frac{2\pi Z_p^2 e^4}{m_e v^2 I} \quad . \tag{2.11}$$

This is of course not quite correct, even classically – we have assumed Δp to be perpendicular to the initial velocity of the bound electron – but it is a good estimate for $(1/2)m_e v^2 \gg I$. If we now assume Eq. (2.11) to apply down to the minimum impact velocity $v_{\rm min} = (2I/m_e)^{1/2}$ for which it is energetically possible to ionize the atom, we can estimate the thermal rate coefficient for collisional ionization:

$$\begin{aligned}
\langle \sigma v \rangle &= \int \sigma(v) \, \times \, v \, \times \, f_v \, dv \\
&= \int_{v_{\rm min}}^{\infty} \frac{2\pi Z_p^2 e^4}{m_e v^2 I} \times v \times 4\pi \left(\frac{m_e}{2\pi kT} \right)^{3/2} v^2 {\rm e}^{-m_e v^2/2kT} dv \\
&= Z_p^2 \left(\frac{8\pi}{m_e kT} \right)^{1/2} \frac{e^4}{I} \, {\rm e}^{-I/kT} \quad .
\end{aligned} \tag{2.12}$$

Now consider a hydrogen atom with principal quantum number n, and ionization energy $I = I_{\rm H}/n^2$, where $I_{\rm H} = 13.602\,{\rm eV}$ is the ionization potential for hydrogen in the ground state. For large n (e.g., $n \approx 100$) and $T \approx 10^4\,{\rm K}$, the ionization energy threshold $I \ll kT$ and $\langle \sigma v \rangle \propto I^{-1} \propto n^2$ – the collisional ionization rate becomes *very* large for hydrogen with large values of the principal quantum number n. We will see in Chapter 3 that such highly excited hydrogen, with $n \gtrsim 10^2$, is in fact present in the interstellar medium, and observable through radio frequency emission lines.

2.2.3 Deflection Time

Consider the special case of a projectile with charge $Z_1 e$ traveling with velocity v_1 through a "field" of stationary "targets," with charge $Z_2 e$ and number density n_2. In the impact approximation, each interaction gives an impulse that is in the plane perpendicular to the direction of motion of the projectile, but is randomly oriented in this plane. Thus the net vector momentum transferred to the projectile undergoes a random walk in this plane, with

$$\begin{aligned}
\langle \frac{d}{dt} [(\Delta p)_\perp]^2 \rangle &= \int_{b_{\rm min}}^{b_{\rm max}} \underbrace{[2\pi b db \, n_2 v_1]}_{d(\text{event rate})} \times \underbrace{\left[\frac{2 Z_1 Z_2 e^2}{b v_1} \right]^2}_{(\Delta p_\perp)^2} \\
&= \frac{8\pi n_2 Z_1^2 Z_2^2 e^4}{v_1} \int_{b_{\rm min}}^{b_{\rm max}} \frac{db}{b} \quad .
\end{aligned} \tag{2.13}$$

The integral is logarithmically divergent at both limits ($b_{\rm min} \to 0$ and $b_{\rm max} \to \infty$), so there must be a physical reason for lower and upper cutoffs, $b_{\rm min}$ and $b_{\rm max}$. At separation $r = Z_1 Z_2 e^2/E$, the interaction energy is equal in magnitude to the initial center-of-mass kinetic energy E and the impact approximation fails, so it is reasonable to take $b_{\rm min} = Z_1 Z_2 e^2/E$.

The upper cutoff $b_{\rm max}$ is more subtle. The integral in Eq. (2.13) assumes the field particles to be randomly located, so that the contribution from each field particle is independent. However, on large scales, the plasma must maintain electrical neutrality, and plasma particles are statistically correlated on length scales larger than the **Debye length** (see Appendix H):

$$L_{\rm D} \equiv \left(\frac{kT}{4\pi n_e e^2}\right)^{1/2} = 690\,{\rm cm}\left(\frac{T}{10^4\,{\rm K}}\right)^{1/2}\left(\frac{{\rm cm}^{-3}}{n_e}\right)^{1/2} . \qquad (2.14)$$

A positive ion will tend to be shielded by a higher-than-average density of electrons, with the shielding being effective on length scales exceeding $L_{\rm D}$. Thus we set $b_{\rm max} = L_{\rm D}$, and find

$$\left\langle \frac{d}{dt}[(\Delta p)_\perp]^2 \right\rangle = \frac{8\pi n_2 Z_1^2 Z_2^2 e^4}{v_1} \ln\Lambda \,, \qquad (2.15)$$

where

$$\Lambda \equiv \frac{b_{\rm max}}{b_{\rm min}} = \frac{E}{kT}\frac{(kT)^{3/2}}{(4\pi n_e)^{1/2} Z_1 Z_2 e^3}$$

$$= 4.13 \times 10^9 \left(\frac{E}{kT}\right)\left(\frac{T}{10^4\,{\rm K}}\right)^{3/2}\left(\frac{{\rm cm}^{-3}}{n_e}\right)^{1/2} \qquad (2.16)$$

$$\ln\Lambda = 22.1 + \ln\left[\left(\frac{E}{kT}\right)\left(\frac{T}{10^4\,{\rm K}}\right)^{3/2}\left(\frac{{\rm cm}^{-3}}{n_e}\right)\right] . \qquad (2.17)$$

We see that Λ is in general a very large number for interstellar conditions. The importance of distant encounters relative to close encounters is given by $\ln\Lambda$. Because $\ln\Lambda \approx 20 - 35$ for interstellar conditions, weak distant encounters dominate, and our simple impact-parameter-based treatment should be quite accurate; uncertainties in exactly how the integral is treated near the upper and lower limits introduce only $\sim$5% uncertainties in the total rate.

It should be noted that Eq. (2.15) is quite general: in the case of species 1 and 2 with velocity distributions, the factor $1/v_1$ in Eq. (2.15) should be replaced by $\langle 1/|v_{12}|\rangle$, where v_{12} is the velocity difference between particles 1 and 2.

Consider now selected cases:

- The energy loss time scale for the projectile is defined by

$$t_{\rm loss} \equiv \frac{E}{\langle (dE/dt)_{\rm loss}\rangle} \,, \qquad (2.18)$$

where $\langle (dE/dt)_{\rm loss} \rangle$ is the mean rate at which the particle gives kinetic energy to the field particles. If the latter have velocity dispersion $\sigma_2 \ll v_1$, then

$$t_{\rm loss} = \frac{m_1 v_1^2}{\langle (d/dt)[(\Delta p)_\perp]^2 \rangle / m_2} = \frac{m_1 m_2 v_1^3}{8\pi n_2 Z_1^2 Z_2^2 e^4 \ln\Lambda} . \tag{2.19}$$

- The time scale for the projectile to be deflected by $\sim 90°$ from its initial trajectory is

$$t_{\rm defl} = \frac{(m_1 v_1)^2}{\langle (d/dt)[(\Delta p)_\perp]^2 \rangle} = \frac{m_1^2 v_1^3}{8\pi n_2 Z_1^2 Z_2^2 e^4 \ln\Lambda} , \tag{2.20}$$

and, for kinetic energy $(3/2)kT$, the mean free path is

$$\mathrm{mfp} = v_1 t_{\rm defl} = \frac{m_1^2 v_1^4}{8\pi n_2 Z_1^2 Z_2^2 e^4 \ln\Lambda} \tag{2.21}$$

$$= 5\times 10^{17}\,\mathrm{cm} \left(\frac{T}{10^6\,\mathrm{K}}\right)^2 \left(\frac{0.01\,\mathrm{cm}^{-3}}{n_e}\right) \left(\frac{25}{\ln\Lambda}\right) . \tag{2.22}$$

For example, consider an electron moving through a field of protons at the root-mean-square (rms) speed $v_1 = (3kT_e/m_e)^{1/2}$. The deflection time is

$$t_{\rm defl}(e \text{ by } p) = \frac{m_e^{1/2}(3kT_e)^{3/2}}{8\pi n_e e^4 \ln\Lambda} \tag{2.23}$$

$$= 7.6\times 10^3\,\mathrm{s} \left(\frac{T_e}{10^4\,\mathrm{K}}\right)^{3/2} \left(\frac{\mathrm{cm}^{-3}}{n_e}\right) \left(\frac{25}{\ln\Lambda}\right) . \tag{2.24}$$

For temperatures $T \lesssim 10^6\,\mathrm{K}$ and densities $n_e \gtrsim 10^{-3}\,\mathrm{cm}^{-3}$, the deflection time is short (compared to astronomical time scales), and we may assume that the velocity distribution will be isotropic. The energy exchange time $t_{\rm loss}$ is larger by the ratio $m_p/m_e = 1836$:

$$t_{\rm loss}(e \text{ to } p) = 1.4\times 10^7\,\mathrm{s} \left(\frac{T_e}{10^4\,\mathrm{K}}\right)^{3/2} \left(\frac{\mathrm{cm}^{-3}}{n_e}\right) \left(\frac{25}{\ln\Lambda}\right) . \tag{2.25}$$

Because we have considered an electron moving with the rms thermal speed, and because the number densities of protons and electrons in ionized gas will be approximately equal, the energy loss time $t_{\rm loss}(e \text{ to } p)$ is the same as the time scales for the electrons and protons to equilibrate at the same temperature, if the electron and proton temperatures initially differed. The energy loss time $t_{\rm loss}$ is generally short by astronomical standards, except in very hot, very low density gas. Accordingly, we may usually (though not always!) assume that the electrons and ions each have Maxwellian velocity distributions, with common temperatures.

2.3 Electron–Ion Inelastic Scattering: Collision Strength $\Omega_{u\ell}$

Inelastic collisions of electrons with ions are responsible for much of the line radiation emitted by hot gas, from H II regions to supernova remnants, as such collisions can leave the ion in an excited state, from which it will decay by emitting a photon.

Earlier we discussed elastic scattering of electrons by ions, finding that the momentum transfer is dominated by "distant" encounters, where the distance of closest approach is much larger than atomic dimensions. However, in those cases where the electron does approach within atomic dimensions of the ion, the electric field of the projectile electron will strongly perturb the wave function of the electrons bound to the ion. Because the projectile electron (attracted by the Coulomb potential) is moving with a speed approaching the velocities of the bound electrons, the perturbation is relatively "sudden," and when the projectile electron recedes, the ion wave function may have made a transition to another energetically allowed state.

Let us consider the case where the ion is initially in an excited state u, with degeneracy g_u. The thermally averaged rate coefficient for deexcitation to a lower energy level ℓ is customarily written in terms of a dimensionless quantity $\Omega_{u\ell}(T)$:

$$\langle \sigma v \rangle_{u \to \ell} \equiv \frac{h^2}{(2\pi m_e)^{3/2}} \frac{1}{(kT)^{1/2}} \frac{\Omega_{u\ell}(T)}{g_u} \tag{2.26}$$

$$= \frac{8.629 \times 10^{-8}}{\sqrt{T_4}} \frac{\Omega_{u\ell}}{g_u} \mathrm{cm^3\,s^{-1}} \quad , \tag{2.27}$$

$$T_4 \equiv \left(\frac{T}{10^4\,\mathrm{K}} \right) \quad . \tag{2.28}$$

Equation Eq. (2.26) serves as the definition of $\Omega_{u\ell}(T)$, called the **collision strength** connecting levels u and ℓ. In principle, the collision strength $\Omega_{u\ell}$ is a function of temperature T, but quantum-mechanical calculations of the inelastic scattering for many ions show that: (1) the $\Omega_{u\ell}$ are approximately independent of temperature T for $T \lesssim 10^4$ K, and (2) the $\Omega_{u\ell}$ typically have values in the range 1 to 10. These quantum-mechanical results can be qualitatively understood in terms of a heuristic classical model (see Appendix I).

2.4 Ion–Neutral Collision Rates

The $1/r$ Coulomb interaction is a long-range interaction, and we found that the elastic scattering process is dominated by distant, weak interactions. We now consider the case of charged particles (e.g., an ion) interacting with neutral particles (atoms or molecules). If the ion and atom are separated by more than a few Å, the principal interaction between the ion and neutral consists of the polarization of the neutral caused by the electric field of the ion. The ion is characterized by its charge Ze, and the neutral by its polarizability α_N: in a uniform static electric field E, the neutral acquires an electric dipole moment $\vec{P} = \alpha_N \vec{E}$. Atomic polarizabilities (see

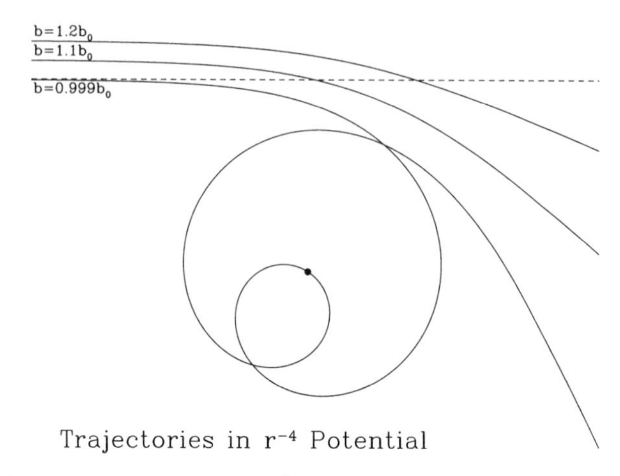

Figure 2.2 Three trajectories in an r^{-4} potential. The trajectory with $b < b_0$ passes through the origin, where b_0 is defined in Eq. (2.30). The total cross section for such "orbiting" trajectories is πb_0^2.

Table 2.1) are of order a few a_0^3, where $a_0 \equiv \hbar^2/m_e e^2 = 5.292 \times 10^{-9}$ cm is the Bohr radius.

In the Coulomb field of the ion, the polarized atom experiences an attractive force $F = P \cdot (dE/dr) = -2\alpha_N Z^2 e^2/r^5$, corresponding to an interaction potential

$$U(r) = -\frac{1}{2}\frac{\alpha_N Z^2 e^2}{r^4} \quad . \tag{2.29}$$

Classical trajectories in this potential have been studied by Wannier (1953) and Osterbrock (1961). Let $E_{\rm cm}$ be the initial center-of-mass kinetic energy. Define

$$b_0 \equiv \left(\frac{2\alpha_N Z^2 e^2}{E_{\rm cm}}\right)^{1/4} \tag{2.30}$$

$$= 6.62 \times 10^{-8}\,{\rm cm}\, Z^{1/2} \left(\frac{\alpha_N}{\alpha_{\rm H}}\right)^{1/4} \left(\frac{0.01\,{\rm eV}}{E_{\rm cm}}\right)^{1/4} \quad , \tag{2.31}$$

where $\alpha_{\rm H}$ is the polarizability of an H atom (see Table 2.1). For $r = b_0$, the interaction energy $U(b_0) = -E_{\rm cm}/4$, so we would expect a trajectory with impact parameter b_0 to be significantly deflected. In fact, for an ideal r^{-4} potential, trajectories with $b < b_0$ are totally overcome by the r^{-4} potential, and actually pass through $r = 0$ – these are referred to as "orbiting" trajectories. An example of one such orbiting trajectory is shown in Figure 2.2. The cross section for orbiting trajectories with $b < b_0$ is

$$\sigma_{\rm orb} = \pi b_0^2 = 2\pi Z e \left(\frac{\alpha_N}{\mu}\right)^{1/2} \frac{1}{v} \quad , \tag{2.32}$$

Table 2.1 Ion–Neutral Scattering Parameters

Neutral	α_N/a_0^3	Ion	$\langle\sigma v\rangle_{\rm orb}$ ($\rm cm^3\,s^{-1}$)[a]	$\langle\sigma v\rangle_{\rm mt}$ ($\rm cm^3\,s^{-1}$)[b]
H	4.500[c]	H^+	2.69×10^{-9}	3.25×10^{-9}
		C^+	1.98×10^{-9}	2.39×10^{-9}
He	1.383[d]	H^+	1.18×10^{-9}	1.42×10^{-9}
H_2	5.315[e]	H^+	2.54×10^{-9}	3.06×10^{-9}
		H_2^+	2.07×10^{-9}	2.49×10^{-9}
		C^+	1.58×10^{-9}	1.91×10^{-9}
O	5.326[f]	H^+	2.14×10^{-9}	2.57×10^{-9}
		C^+	7.95×10^{-10}	9.57×10^{-10}

[a] $\langle\sigma v\rangle_{\rm orb}$ = rate coefficient for orbiting collisions.
[b] $\langle\sigma v\rangle_{\rm mt}$ = momentum transfer rate coefficient.
[c] Landau & Lifshitz (1972).
[d] Thomas & Humbertson (1972).
[e] Marlow (1965).
[f] Kelly (1969).

where $\mu \equiv m_1 m_2/(m_1+m_2)$ is the reduced mass. Note that for this interaction, the cross section $\sigma \propto 1/v$, and the rate coefficient

$$\langle\sigma v\rangle_{\rm orb} = 2\pi Ze\left(\frac{\alpha_N}{\mu}\right)^{1/2} \tag{2.33}$$

$$= 8.980\times10^{-10} Z\left(\frac{\alpha_N}{a_0^3}\right)^{1/2}\left(\frac{m_{\rm H}}{\mu}\right)^{1/2} \rm cm^3\,s^{-1} \tag{2.34}$$

is then independent of temperature! The polarizability α_N is given in Table 2.1 for H, He, H_2, and O, together with $\langle\sigma v\rangle_{\rm orb}$ for orbiting collisions of these neutrals with selected ions.

Obviously, the approximation of an r^{-4} interaction potential fails when the ion and neutral come within a few Å of one another, but it is evident that the orbiting trajectories do bring the ion and neutral into intimate contact. If an outcome is energetically allowed – such as collisional deexcitation of an excited state, or an exothermic charge exchange or chemical exchange reaction – there will be a substantial probability of it happening when the ion and neutral undergoing an orbiting collision with $b < b_0$ come together to form an excited "complex," and the reaction rate coefficient will, therefore, be comparable to the orbiting rate coefficient. Examples of such exothermic reactions include

$$\rm O^+ + H \rightarrow O(^3P_0) + H^+ \quad (charge\ exchange), \tag{2.35}$$

$$\rm H_2^+ + H_2 \rightarrow H_3^+ + H \quad (chemical\ exchange), \tag{2.36}$$

$$\rm CH^+ + H_2 \rightarrow CH_2^+ + H \quad (chemical\ exchange). \tag{2.37}$$

Because their rate coefficients are large even at low temperatures, exothermic ion–neutral reactions play a major role in the chemistry of cool interstellar gas.

Momentum transfer in ion–neutral collisions is important in some astrophysical situations. The momentum transfer cross section is

$$\sigma_{\rm mt} = 2\pi \int_0^\infty (1 - \cos\theta)\, b\, db \,, \tag{2.38}$$

where $\theta(b)$ is the deflection angle. It is reasonable to assume that the short-range interaction between ion and neutral will result in isotropic scattering for collisions with $b < b_0$ (which for an r^{-4} potential would pass through $r = 0$). Adding the contribution from $b > b_0$ gives the momentum transfer rate coefficient:

$$\langle\sigma v\rangle_{\rm mt} = 2.41\pi Ze \left(\frac{\alpha_N}{\mu}\right)^{1/2} = 1.21\langle\sigma v\rangle_{\rm orb} \quad . \tag{2.39}$$

2.5 Electron–Neutral Collision Rates

Elastic scattering of electrons by neutrals can be important in very low ionization regions (such as protoplanetary disks), where electron–neutral scattering limits the electrical conductivity. In regions of very low fractional ionization, the primary collision partner is H_2, followed by He.

Low-energy scattering of electrons by H_2 has been studied theoretically and experimentally. At energies $E < 0.044\,{\rm eV}$ the scattering is purely elastic; for $E > 0.044\,{\rm eV}$, rotational excitation can occur; at $E \gtrsim 0.5\,{\rm eV}$ vibrational excitation is also possible; and for $E > 11\,{\rm eV}$ electronic excitation can take place.

The experimentally measured momentum transfer cross section between 0.01 and 1 eV (Crompton et al. 1969; Ferch et al. 1980) can be approximated by

$$\sigma_{\rm mt} \approx 7.3 \times 10^{-16} (E/0.01\,{\rm eV})^{0.18}\,{\rm cm}^2 \quad . \tag{2.40}$$

The thermal rate coefficient for momentum transfer due to $e - H_2$ scattering is then

$$\langle\sigma v\rangle_{\rm mt} \approx 4.8 \times 10^{-9} \left(\frac{T}{10^2\,{\rm K}}\right)^{0.68} {\rm cm}^3\,{\rm s}^{-1} \,, \tag{2.41}$$

accurate to within $\sim$10% for $50 \lesssim T \lesssim 5000\,{\rm K}$.[2]

The cross section for electron scattering by He (Crompton et al. 1970; Ferch et al. 1980) is somewhat smaller than for H_2.

[2]For $T < 10^4$ K, the rate in Eq. (2.41) is significantly smaller than the classical estimate, Eq. (2.33). The classical estimate is inapplicable because the deBroglie wavelength of the electron, h/p, is larger than the critical impact parameter b_0 for orbiting collisions.

2.6 Neutral–Neutral Collision Rates

The interaction between neutral species is repulsive at small separations and weakly attractive at larger separations due to the van der Waals interaction, which arises because fluctuations in the electron dipole moment of one species induce an electric dipole in the other, resulting in an attractive $U(r) \propto 1/r^6$ interaction. The attractive interaction is sufficiently weak, and the onset of the repulsive interaction sufficiently rapid, that for many purposes the interaction can be approximated by a "hard-sphere" model, with the collision partners each having hard-sphere radii $R_i \approx 1\,\text{Å}$.

Hard-sphere collisions occur for impact parameters $b < R_1 + R_2$, so the collision cross section $\pi(R_1+R_2)^2 \approx 1.2\times10^{-15}\,\text{cm}^2$. The rate coefficient for hard-sphere scattering is

$$\langle\sigma v\rangle = \left(\frac{8kT}{\pi\mu}\right)^{1/2} \pi(R_1+R_2)^2 \tag{2.42}$$

$$= 1.81\times10^{-10}\left(\frac{T}{10^2\,\text{K}}\right)^{1/2}\left(\frac{m_\text{H}}{\mu}\right)^{1/2}\left(\frac{R_1+R_2}{2\,\text{Å}}\right)^2 \text{cm}^3\,\text{s}^{-1}, \tag{2.43}$$

where again μ is the reduced mass. For $T \lesssim 10^2\,\text{K}$, the rate coefficient for neutral–neutral scattering is smaller by more than an order of magnitude than the rate coefficient for ion–neutral scattering.

Chapter Three

Statistical Mechanics and Thermodynamic Equilibrium

The interstellar medium (ISM) and intergalactic medium (IGM) are generally far from thermodynamic equilibrium. Nevertheless, the methods of statistical mechanics and thermodynamics provide powerful tools for understanding the nonequilibrium conditions that prevail, and for relating forward and reverse rates for the processes, such as ionization and recombination, that shape the medium.

This chapter reviews some results from statistical mechanics, and illustrates their use by obtaining the relationship between collisional excitation and deexcitation rate coefficients, and cross sections for forward and reverse reactions. In §§3.7 and 3.8, we estimate the three-body recombination rate, and use this estimate to understand the population of the high-n levels of atomic hydrogen.

3.1 Partition Functions

Consider some physical system (e.g., atoms in a box of volume V) that is able to exchange energy with a "heat reservoir" at temperature T. The theory of statistical mechanics defines the **partition function** Z to be

$$Z(T) \equiv \sum_s \mathrm{e}^{-E(s)/kT} \quad , \tag{3.1}$$

where the sum is over all distinct possible states s of the system, and $E(s)$ is the energy of state s. For dilute gases, we can factor the partition function into "translational" and "internal" partition functions:

$$Z(T) = Z_{\mathrm{tran}}(T) \times z_{\mathrm{int}}(T) \quad . \tag{3.2}$$

For example, let X be a single atom, ion, or molecule. Suppose that the internal energy levels of X are labeled by index $i = 0, 1, 2, ...$ The total energy of X in level i is $E(X_i) = p^2/2M_X + E_i$, where p is the linear momentum, M_X is the mass of X, and E_i is the internal energy of X in level i.

The translational partition function $Z_{\mathrm{tran}}(T)$ is obtained by integrating over six-dimensional phase space and dividing by the "cell size" h^3 (where h = Planck's constant):

$$Z_{\rm tran}(X;T) = \frac{V}{h^3}\int_0^\infty 4\pi p^2 dp\, {\rm e}^{-p^2/2M_X kT} = \frac{(2\pi M_X kT)^{3/2}}{h^3}V \quad , \tag{3.3}$$

and the internal partition function $z_{\rm int}(T)$ is just a sum over the possible internal states of X:

$$z_{\rm int}(X;T) \equiv \sum_i g_i {\rm e}^{-E_i/kT} \quad , \tag{3.4}$$

where the **degeneracy g_i** of level i refers to the number of distinct quantum states that are grouped together and treated as a single energy level. For example, for a free electron, there are just two "internal" quantum states (spin up and spin down), both have $E_i = 0$, and hence $z_{\rm int} = 2$. For dilute gases, the partition function $Z \propto V$, as seen in Eq. (3.3). It is convenient to define the **partition function per unit volume**:

$$f(X;T) \equiv \frac{Z}{V} = \left[\frac{(2\pi M_X kT)^{3/2}}{h^3}\right] z_{\rm int}(X;T) \quad . \tag{3.5}$$

3.2 Detailed Balance: The Law of Mass Action

Suppose that we have a chemical reaction $A + B \leftrightarrow C$. In local thermodynamic equilibrium (LTE), statistical mechanics shows that the number densities of species A, B, C, will satisfy the **law of mass action**:

$$\frac{n_{\rm LTE}(C)}{n_{\rm LTE}(A)n_{\rm LTE}(B)} = \frac{f(C)}{f(A)f(B)} \quad , \tag{3.6}$$

where $f(X)$ is the partition function per unit volume for species X. The LTE abundance of X is proportional to $f(X)$. The law of mass action applies to an arbitrary number of reactants and products. For a general reaction

$$R_1 + R_2 + ... + R_M \leftrightarrow P_1 + P_2 + ... + P_N \quad , \tag{3.7}$$

the law of mass action is

$$\frac{\prod_{j=1}^N n_{\rm LTE}(P_j)}{\prod_{i=1}^M n_{\rm LTE}(R_i)} = \frac{\prod_{j=1}^N f(P_j)}{\prod_{i=1}^M f(R_i)}$$
$$= \left[\frac{(2\pi kT)^{3/2}}{h^3}\right]^{N-M} \left[\frac{\prod_{j=1}^N M(P_j)}{\prod_{i=1}^M M(R_i)}\right]^{3/2} \frac{\prod_{j=1}^N z_{\rm int}(P_j;T)}{\prod_{i=1}^M z_{\rm int}(R_i;T)} \quad . \tag{3.8}$$

Note that the reactants R_i and products P_j in Eq. (3.7) can be species (e.g., H and O, where we sum over internal quantum states), or specific quantum states (e.g., the $^3P_2$ fine structure level of atomic O). In §3.7, we will apply the law of mass action to the states of H with large quantum number n.

We must be careful with our accounting when it comes to internal energies. For each reactant R_i and product P_j, choose "reference" states $R_{i,0}$, and $P_{j,0}$. If we define ΔE by

$$E\left(R_{1,0}+...+R_{M,0}\right)+\Delta E=E\left(P_{1,0}+...+P_{N,0}\right) \quad , \tag{3.9}$$

(so that $\Delta E > 0$ for an *endo*thermic reaction), then

$$\frac{\prod_{j=1}^{N} z_{\rm int}(P_j;T)}{\prod_{i=1}^{M} z_{\rm int}(R_i;T)} = {\rm e}^{-\Delta E/kT} \frac{\prod_{j=1}^{N}\sum_s g(P_{j,s}){\rm e}^{-[E(P_{j,s})-E(P_{j,0})]/kT}}{\prod_{i=1}^{M}\sum_s g(R_{i,s}){\rm e}^{-[E(R_{i,s})-E(R_{i,0})]/kT}} \quad , \tag{3.10}$$

and therefore

$$\frac{\prod_{j=1}^{N} n_{\rm LTE}(P_j)}{\prod_{i=1}^{M} n_{\rm LTE}(R_i)} = \left[\frac{(2\pi kT)^{3/2}}{h^3}\right]^{N-M}\left[\frac{\prod_{j=1}^{N} M(P_j)}{\prod_{i=1}^{M} M(R_i)}\right]^{3/2} \times$$

$${\rm e}^{-\Delta E/kT} \frac{\prod_{j=1}^{N}\sum_s g(P_{j,s}){\rm e}^{-[E(P_{j,s})-E(P_{j,0})]/kT}}{\prod_{i=1}^{M}\sum_s g(R_{i,s}){\rm e}^{-[E(R_{i,s})-E(R_{i,0})]/kT}} \quad . \tag{3.11}$$

This appears daunting, but is in fact fairly straightforward to apply. We will demonstrate the utility of this equation in the following.

3.3 Ionization and Recombination

As a first example, consider the balance between recombination and ionization, where we consider some specific energy level ℓ of species X^{+r}, and some specific energy level u of species X^{+r+1}:

$$e^- + X_u^{+r+1} \leftrightarrow X_\ell^{+r} \quad . \tag{3.12}$$

(If we were to apply this to hydrogen, we would let X = H, and $r = 0$.) Using the law of mass action (3.11), we find the abundance of X_ℓ^{+r} if it is in LTE with $n(X_u^{+r+1})$ and electron density n_e:

$$n_{\rm LTE}(X_\ell^{+r}) = \frac{h^3}{2(2\pi m_e kT)^{3/2}} n(X_u^{+r+1}) n_e \frac{g(X_\ell^{+r})}{g(X_u^{+r+1})} {\rm e}^{-(E_{r,\ell}-E_{r+1,u})/kT} \quad . \tag{3.13}$$

We will apply this result to the high-n energy levels of hydrogen in §§3.7, and 3.8.

3.4 Saha Equation

The overall balance between recombination and ionization

$$e^- + X^{+r+1} \leftrightarrow X^{+r} \tag{3.14}$$

is obtained by summing over energy levels of X^{+r} and X^{+r+1}. Using the law of mass action (3.11), we find

$$\frac{n_{\rm LTE}(X^{+r})}{n_{\rm LTE}(e^-)n_{\rm LTE}(X^{+r+1})} = \frac{h^3}{(2\pi m_e kT)^{3/2}} \frac{\sum_j g_{r,j}{\rm e}^{-E_{r,j}/kT}}{2\sum_j g_{r+1,j}{\rm e}^{-E_{r+1,j}/kT}} \quad . \tag{3.15}$$

We now suppose that we are at a sufficiently low temperature that we can approximate the internal partition functions by retaining only the first term in each sum – the term due to the lowest energy states of X^{+r} and X^{+r+1}. If we let $\Phi_r \equiv E_{r+1,1} - E_{r,1}$ = the ionization energy, we obtain the **Saha equation:**

$$\frac{n_{\rm LTE}(e^-)n_{\rm LTE}(X^{+r+1})}{n_{\rm LTE}(X^{+r})} \approx \frac{2(2\pi m_e kT)^{3/2}}{h^3} \frac{g_{r+1,1}}{g_{r,1}} {\rm e}^{-\Phi_r/kT} \quad . \tag{3.16}$$

If we now apply the Saha equation (3.16) to the specific case of hydrogen, we have $g_{r,0} = g({\rm H}\,1s) = 2 \times 2 = 4$ (the proton can have spin up or spin down, and the electron can have spin up or spin down), $g_{r+1,0} = g({\rm H}^+) = 2$ (the proton can have spin up or spin down), and

$$\frac{n_{\rm LTE}(e^-)n_{\rm LTE}({\rm H}^+)}{n_{\rm LTE}({\rm H}^0)} = \frac{(2\pi m_e kT)^{3/2}}{h^3} {\rm e}^{-I_{\rm H}/kT} \quad , \tag{3.17}$$

where $I_{\rm H} = 13.60\,{\rm eV}$ is the ionization energy of hydrogen.

The law of mass action is a good approximation in a stellar interior, where the radiation is very close to blackbody and the matter has had ample time to come into statistical equilibrium with the radiation field. The Saha equation is a good approximation to the law of mass action, provided the temperature is low enough that retaining only the first term provides a good approximation to the sum over all internal states[1]: $\sum_i g_i {\rm e}^{-E_i/kT} \approx g_0 {\rm e}^{-E_0/kT}$. However, the law of mass action (including the Saha equation) is generally *not* a good approximation in the ISM or IGM, where the electromagnetic radiation field is far from a blackbody.

[1]This is a somewhat delicate point. An isolated atom or positive ion (in an infinite universe) has an infinite number of highly excited "Rydberg states," with the outermost electron in hydrogenic orbits with very large radial quantum number n. Thus, the sum $\Sigma_j g_{r,j} \exp(-E_{r,j}/kT)$ becomes infinite, and the internal partition function $z_{\rm int} \to \infty$. However, at a finite density, the higher Rydberg states can no longer be considered to be bound, as the electron at large radii behaves like a free electron from the plasma rather than as a bound electron.

3.5 Detailed Balance: Ratios of Rate Coefficients

The law of mass action implies very general restrictions on rate coefficients. Suppose that we have a general reaction

$$R_1 + R_2 + ... + R_M \underset{k_r}{\overset{k_f}{\rightleftarrows}} P_1 + P_2 + ... + P_N \quad . \tag{3.18}$$

If the equilibrium abundances of the reactants and products must satisfy the law of mass action, and the forward and reverse rates must balance when in equilibrium (the **principle of detailed balance**), then it follows that the ratio of rate coefficients k_r/k_f must satisfy the condition

$$\frac{k_r}{k_f} = \left[\frac{(2\pi kT)^{3/2}}{h^3}\right]^{M-N} \left[\frac{\prod_{i=1}^{M} M(R_i)}{\prod_{j=1}^{N} M(P_j)}\right]^{3/2} \frac{\prod_{i=1}^{M} z_{\rm int}(R_i;T)}{\prod_{j=1}^{N} z_{\rm int}(P_j;T)} \quad . \tag{3.19}$$

As a simple application of this, consider inelastic scattering

$$X(\ell) + Y \rightarrow X(u) + Y \quad , \tag{3.20}$$

where $X(\ell)$ and $X(u)$ are two different energy levels of species X, with degeneracies g_ℓ and g_u, and Y is some collision partner that does not change internal state during the collision. Let $\langle\sigma v\rangle_{\ell\rightarrow u}$ and $\langle\sigma v\rangle_{u\rightarrow\ell}$ be the "upward" and "downward" rate coefficients (i.e., rate coefficients for excitation $\ell \rightarrow u$ and deexcitation $u \rightarrow \ell$). Equation (3.19) applied to this case yields the **ratio of upward and downward rate coefficients:**

$$\langle\sigma v\rangle_{\ell\rightarrow u} = \frac{g_u}{g_\ell} {\rm e}^{-(E_{u\ell}/kT)} \langle\sigma v\rangle_{u\rightarrow\ell} \ , \tag{3.21}$$

where $E_{u\ell} \equiv E_u - E_\ell$.

3.6 Detailed Balance: Ratios of Cross Sections

3.6.1 Inelastic Scattering

If the ratios of forward and reverse reaction rates are determined by detailed balance considerations, there must be definite conditions on the ratios of forward and reverse cross sections. To abbreviate our notation, let $\sigma_{ij}(E) \equiv \sigma_{i\rightarrow j}(E)$.

For inelastic scattering [Eq. (3.20)], the balancing of forward and reverse rate coefficients implies that, for any temperature T, we must have

$$\int_{E_{u\ell}}^{\infty} E{\rm e}^{-E/kT} \sigma_{\ell u}(E) dE = \frac{g_u}{g_\ell} {\rm e}^{-E_{u\ell}/kT} \int_0^{\infty} E{\rm e}^{-E/kT} \sigma_{u\ell}(E) dE \quad . \tag{3.22}$$

This equation can be true for all T only if the cross sections $\sigma_{\ell u}$ and $\sigma_{u\ell}$ satisfy

$$(E_{ul} + E)\sigma_{\ell u}(E_{ul} + E) = \frac{g_u}{g_\ell} E\sigma_{u\ell}(E) \tag{3.23}$$

for all $E > 0$.

3.6.2 Photoionization and Recombination

These detailed balance considerations also apply to reactions where photons are absorbed and emitted. Consider the balance of photoionization and radiative recombination. Let $\sigma_{\rm pi}(E)$ be the cross section for photoionization from level ℓ of atom X:

$$X_\ell + h\nu \rightarrow X_u^+ + e^- \ , \tag{3.24}$$

where the resulting ion X^+ is in energy level u, and let $\sigma_{\rm rr}(E)$ be the cross section for radiative recombination:

$$X_u^+ + e^- \rightarrow X_\ell + h\nu \quad . \tag{3.25}$$

In LTE, the rate per volume at which photons with energies in $(h\nu, h\nu + hd\nu)$ are removed by photoelectric absorption and the rate at which they are created by radiative recombination must be equal: within an energy interval $dE = hd\nu$, we have

$$n_{\rm LTE}(X_\ell)\frac{4\pi B_\nu d\nu}{h\nu}\sigma_{{\rm pi},\ell u}(h\nu) =$$
$$n_{\rm LTE}(X_u^+)n_{\rm LTE}(e^-)v f_E(h\nu - I_{X,\ell u})hd\nu \ \sigma_{{\rm rr},u\ell}(h\nu - I_{X,\ell u})[1 + (n_\gamma)_{\rm LTE}], \tag{3.26}$$

where $\sigma_{{\rm pi},\ell u}(h\nu)$ is the cross section for photoionization from X_ℓ to create X_u^+, $\sigma_{{\rm rr},u\ell}(E)$ is the cross section for an ion X^+ in quantum state u to capture an electron by radiative recombination to level ℓ of X, and

$$B_\nu(T) = \frac{2h\nu^3}{c^2}\frac{1}{e^{h\nu/kT} - 1} \tag{3.27}$$

is the blackbody radiation intensity [see Eq. (6.6)].

The process of radiative recombination – where the electron drops from a "free" state to a bound state by emission of a photon – can proceed either by spontaneous emission

$$X_u^+ + e^- \rightarrow X_\ell + h\nu \tag{3.28}$$

or by stimulated emission

$$X_u^+ + e^- + h\nu \to X_\ell + 2h\nu \quad . \tag{3.29}$$

The radiative recombination cross section $\sigma_{\mathrm{rr},u\ell}$ applies to the spontaneous process (3.28). Here we anticipate the discussion in Chapter 6, where we will show that the ratio of stimulated emission to spontaneous emission is equal to the photon occupation number [see Eq. (6.11)] in LTE:

$$(n_\gamma)_{\mathrm{LTE}} = \frac{1}{e^{h\nu/kT} - 1} \quad . \tag{3.30}$$

The factor $[1+n_\gamma]$ in Eq. (3.26) therefore allows for the contribution of **stimulated recombination**.

Applying the principle of detailed balance (3.21), we obtain the **Milne relation** between the cross section for photoionization and the cross section for electron capture by spontaneous radiative recombination:

$$\sigma_{\mathrm{rr},u\ell}(E) = \frac{1}{2}\frac{g(X_\ell)}{g(X_u^+)}\frac{(I_{X,\ell u}+E)^2}{Em_ec^2}\sigma_{\mathrm{pi},\ell u}(h\nu = I_{X,\ell u}+E) \quad , \tag{3.31}$$

where $g(X_u^+)$ is the degeneracy of the ion X^+ in level u, and $g(X_\ell)$ is the degeneracy of the ion or atom X in level ℓ.

3.7 Example: Three-Body Recombination

As an example to demonstrate the utility of Eq. (3.8), consider the simple reaction

$$\mathrm{H}^+ + 2e^- \leftrightarrow \mathrm{H}(n) + e^- \quad , \tag{3.32}$$

where $\mathrm{H}(n)$ denotes the hydrogen atom in a level with principal quantum number n. Application of Eq. (3.8) gives

$$\left[\frac{n(\mathrm{H}(n))}{n(\mathrm{H}^+)n_e}\right]_{\mathrm{LTE}} = \left[\frac{h^3}{(2\pi kT)^{3/2}}\right]\left[\frac{m_\mathrm{H}}{m_pm_e}\right]^{3/2}\frac{g[\mathrm{H}(n)]}{g(e^-)g(\mathrm{H}^+)}e^{I_n/kT} \quad , \tag{3.33}$$

where $I_n = I_\mathrm{H}/n^2$ is the energy required to ionize $\mathrm{H}(n) \to \mathrm{H}^+ + e^-$.

The electron and proton are spin 1/2 particles, with $g(e^-) = g(\mathrm{H}^+) = 2$. Hydrogen with principal quantum number n has $g[\mathrm{H}(n)] = 4n^2$: there are two electron spin states, two nuclear spin states, and n^2 distinct (ℓ, m) orbits.[2] Therefore,

$$n_{\mathrm{LTE}}[\mathrm{H}(n)] = n^2\frac{h^3}{(2\pi m_ekT)^{3/2}}n_en(\mathrm{H}^+)e^{I_\mathrm{H}/n^2kT} \quad . \tag{3.34}$$

[2]For example, for $n = 2$ we have four orbits: $\ell = 0$ with $m = 0$, and $\ell = 1$ with $m = -1, 0, 1$.

This is a remarkable result! For $n \gg 1$, we have $n_{\rm LTE}[H(n)] \propto n^2$ – since n could be a large number, this suggests that there could be significant populations in the high quantum levels n, if they are in LTE with the electrons and ions.

Suppose that we would like to estimate the rate for the three-body "collisional recombination" reaction

$$\mathrm{H}^+ + e^- + e^- \rightarrow \mathrm{H}(n) + e^- \quad . \tag{3.35}$$

The rate per volume of this three-body reaction will have the form

$$\frac{d}{dt}\Big(n\,[\mathrm{H}(n)]\Big)_{\mathrm{H}^+ + 2e^-} = \beta_n(T) n(\mathrm{H}^+) n_e^2 \quad , \tag{3.36}$$

where $\beta_n(T)$ is the unknown rate coefficient for three-body recombination to level n. Even if we are willing to use a semiclassical treatment, integration over possible trajectories in three-body collisions is much more complicated and tedious than for two-body reactions. It is therefore pleasing to see that we can use the law of mass action and the principle of detailed balance to relate the three-body rate coefficient to a two-body reaction rate: in LTE, we must have

$$\beta_n(T) n_{\rm LTE}(\mathrm{H}^+)[n_{\rm LTE}(e)]^2 = n_{\rm LTE}\,[\mathrm{H}(n)]\, n_{\rm LTE}(e) \langle \sigma v \rangle_{n\rightarrow c} \quad , \tag{3.37}$$

where $\langle \sigma v \rangle_{n\rightarrow c}$ is the collisional rate coefficient for collisional ionization from $\mathrm{H}(n)$:

$$\mathrm{H}(n) + e^- \rightarrow \mathrm{H}^+ + 2e^- \quad . \tag{3.38}$$

Thus

$$\beta_n(T) = \left[\frac{n[\mathrm{H}(n)]}{n_e n(\mathrm{H}^+)}\right]_{\rm LTE} \langle \sigma v \rangle_{n\rightarrow c} \quad . \tag{3.39}$$

Inserting Eq. (3.34), we obtain

$$\beta_n(T) = n^2 \frac{h^3}{(2\pi m_e kT)^{3/2}} \mathrm{e}^{I_{\rm H}/n^2 kT} \langle \sigma v \rangle_{n\rightarrow c} \quad . \tag{3.40}$$

This is an exact result.

The rate coefficient $\langle \sigma v \rangle_{n\rightarrow c}$ for collisional ionization by electrons was estimated in Eq. (2.12). Applied to ionization from level n, we find

$$\langle \sigma v \rangle_{n\rightarrow c} \approx n^2 \frac{e^4}{I_{\rm H}} \left(\frac{8\pi}{m_e kT}\right)^{1/2} \mathrm{e}^{-I_{\rm H}/n^2 kT} \tag{3.41}$$

$$= 3.5 \times 10^{-3} \left(\frac{n}{100}\right)^2 T_4^{-1/2} \mathrm{e}^{-I_{\rm H}/n^2 kT}\ \mathrm{cm}^3\,\mathrm{s}^{-1} \quad . \tag{3.42}$$

From (3.40), we now obtain the rate coefficient for three-body recombination $\mathrm{H}^+ + 2e^- \rightarrow \mathrm{H}(n) + e^-$:

$$\beta_n(T) \approx n^4 \frac{4a_0^2 h^3 I_{\mathrm{H}}}{\pi (m_e kT)^2} \tag{3.43}$$

$$= 1.4 \times 10^{-20} \left(\frac{n}{100}\right)^4 T_4^{-2} \,\mathrm{cm}^6\,\mathrm{s}^{-1} \quad . \tag{3.44}$$

This three-body rate coefficient appears to be numerically small,[3] but note the dependence on quantum number n: for sufficiently large values of n, the three-body recombination rate can be fast enough to ensure that the high-n levels of H are in collisional equilibrium with the electrons and protons.

3.8 Departure Coefficients

Let us consider the density $n[\mathrm{H}(n)]$ of H with principal quantum number n. When discussing the levels with $n \gtrsim 30$, it is convenient to define the **departure coefficient:**

$$b_n \equiv \frac{n[\mathrm{H}(n)]}{n_{\mathrm{LTE}}[\mathrm{H}(n)]} = \frac{n[\mathrm{H}(n)]}{n_e n(\mathrm{H}^+)} \frac{(2\pi m_e kT)^{3/2}}{n^2 h^3} \mathrm{e}^{-I_{\mathrm{H}}/n^2 kT} \quad , \tag{3.45}$$

where $n_{\mathrm{LTE}}[\mathrm{H}(n)]$ is given by Eq. (3.34). The departure coefficient b_n compares the actual level population $n[\mathrm{H}(n)]$ to the level population that would apply *if* the levels were in LTE with the given electron and proton densities. If only collisional processes (collisional ionization and three-body recombination) were acting, the system would have $b_n = 1$. However, the excited states also can be depopulated by spontaneous emission of a photon, with the rate for this radiative process increasing rapidly as n decreases. The total rate of depopulation by spontaneous radiative decay from level $n \gg 1$ (averaged over the angular momentum states) is approximately [4]

$$A_{n,tot} \approx 7 \times 10^{10} n^{-5} \,\mathrm{s}^{-1} \quad . \tag{3.46}$$

Let n_c be the principal quantum number n for which $A_{n,tot} = n_e \langle \sigma v \rangle_{n \rightarrow c}$. From Eqs. (3.42 and 3.46), we obtain

$$n_c \approx 110 \left(\frac{n_e}{10^3\,\mathrm{cm}^{-3}}\right)^{-1/7} T_4^{1/14} \quad . \tag{3.47}$$

This is only an estimate, but it indicates that we may expect to see appreciable departures from LTE for $n \lesssim 110$ in H II regions where $n_e \approx 10^3\,\mathrm{cm}^{-3}$.

---

[3]Three-body recombination can also be mediated by protons: $2\mathrm{H}^+ + e^- \rightarrow \mathrm{H}(n) + \mathrm{H}^+$. For sufficiently large quantum number n, the three-body rate coefficient for $2\mathrm{H}^+ + e^- \rightarrow \mathrm{H}(n) + \mathrm{H}^+$ is larger than the electron rate (3.43) by a factor $\sqrt{m_p/m_e} = 43$.

[4]Wiese et al. (1966) have transition probabilities for levels up to $n = 20$.

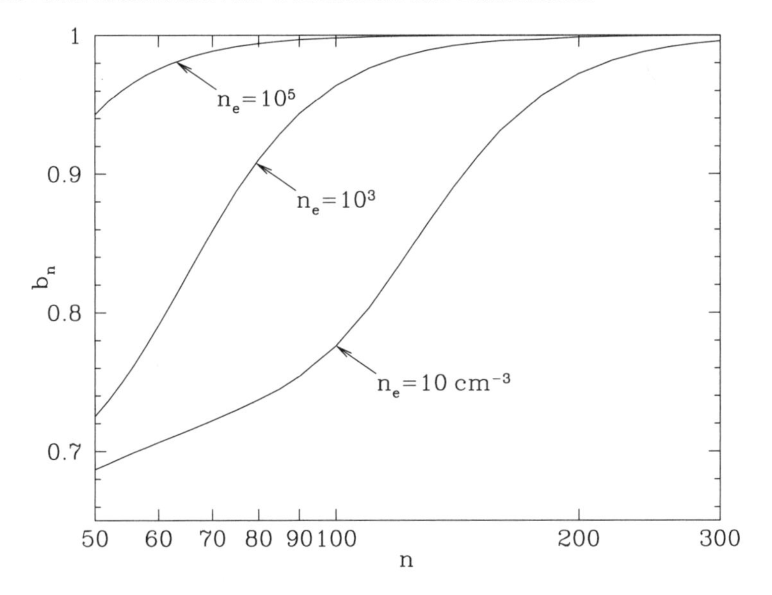

Figure 3.1 Departure coefficient b_n versus principal quantum number n, for H atoms in thermal plasma with $T = 10^{3.9}$ K, for three values of electron density $n_e = 10, 10^3$, and $10^5\,\mathrm{cm}^{-3}$, with no radiation present. Data from Salem & Brocklehurst (1979).

Figure 3.1 shows b_n values calculated by Salem & Brocklehurst (1979). We see that for $n_e = 10^3\,\mathrm{cm}^{-3}$, b_n is noticeably smaller than 1 for $n \lesssim 110$, consistent with our back-of-the-envelope estimate for the value of n for which radiative deexcitation and collisional ionization have comparable rates.

The high-n levels undergo radiative decay via electric dipole transitions to lower levels. Transitions $n + 1 \rightarrow n$ are referred to as $\boldsymbol{n\alpha}$ **transitions** (e.g., Lyman α is the 1α transition). For high n values, the photon emitted is in the radio spectrum; hence these are known as "radio recombination lines." The 166α transition, at $\nu = 1424.7\,\mathrm{MHz}$, falls close to the H I 21-cm line at 1420.4 MHz, and the 159α transition, at $\nu = 1620.7\,\mathrm{MHz}$, falls close to the 1612.2 MHz ($\lambda = 18.6\,\mathrm{cm}$) transition of OH. Because radio observatories often have receivers tuned to work near 21 cm and 18 cm, the 166α and 159α transitions are often selected for observation. When the transition is optically thin, the observed line intensity

$$I_{n\alpha} \;\propto\; A_{n\alpha} h\nu_{n\alpha} \int n[\mathrm{H}(n)] ds \;\propto\; n^{-6} b_n \int n_e n(\mathrm{H}^+) ds. \tag{3.48}$$

By study of the variation of line intensity $I_{n\alpha}$ with n, one can observe the decrease of b_n with decreasing n, and thereby infer the electron density.

When b_n is an increasing function of n, conditions are such that **maser amplification** can occur in the line; such masing, if present, complicates the interpretation of the observations.

Radio recombination lines are discussed further in §10.7.

Chapter Four

Energy Levels of Atoms and Ions

This chapter reviews the energy-level structure of atoms and ions, together with the nomenclature for referring to those levels. It is probably an understatement to say that the material in this chapter is not electrically exciting; it should be regarded as reference material that can be returned to as needed.

Atomic spectroscopists customarily identify the different ionization stages of the elements by roman numerals, with I corresponding to the neutral atom, II to singly ionized, III to doubly ionized, and so on. Thus atomic hydrogen is referred to as H I, ionized hydrogen (H^+) as H II, and five-times ionized oxygen (O^{+5}) as O VI.

We now consider the disposition of the electrons.

4.1 Single-Electron Orbitals

According to the quantum mechanical theory of multielectron atoms, it is a good first approximation to think of the electrons as occupying "single-electron" orbitals characterized by integer quantum numbers n and ℓ: $n = 1, 2, 3, ...$ is the "principal" quantum number (the electron wave function has $n - 1$ radial nodes), and ℓ is the orbital angular momentum in units of $\hbar$. For a given principal quantum number n, the possible values of ℓ are $0 \leq \ell < n$.

The letters s, p, d, f are used to designate orbitals with $\ell = 0, 1, 2, 3$. In addition to the quantum numbers n and ℓ, there is a third quantum number characterizing the orbital: m_z, the projection of the orbital angular momentum/$\hbar$ onto the z axis. Thus m_z can take on $2\ell + 1$ different values: $-\ell, ..., -1, 0, 1, ..., \ell$. If there is no applied magnetic field, the energy of the orbital is independent of m_z.

Electrons are spin $1/2$ particles, and the projection of the electron spin onto the z axis can take on only 2 values: $-\hbar/2$ or $+\hbar/2$. Again, if there is no applied field, these two states are degenerate.

Thus a given pair of quantum numbers $n\ell$ actually refers to $2(2\ell + 1)$ distinct electronic wave functions.

4.2 Configurations

An atom or ion with a single electron can have its electron in any of the allowed orbitals or wave functions. When an atom or ion has more than one electron, the

Pauli exclusion principle forbids two electrons from sharing the same wave function. Therefore, there can be at most $2(2\ell+1)$ electrons in a given **subshell** $n\ell$: s subshells can contain at most 2 electrons, p subshells can contain at most 6 electrons, and d subshells can contain up to 10 electrons.

The orbitals, in order of increasing energy, are $1s$, $2s$, $2p$, $3s$, $3p$, $4s$, $3d$, $4p$, $5s$, and so on. Thus atomic carbon, with 6 electrons, has a ground state **configuration** with 2 electrons in the $1s$ subshell, 2 electrons in the $2s$ subshell, and the remaining 2 electrons in the $2p$ subshell. The number of electrons in each subshell is designated by a superscript: the ground state configuration for neutral carbon is written $1s^2 2s^2 2p^2$. Neutral sodium, with 11 electrons, has ground state configuration $1s^2 2s^2 2p^6 3s$.

4.3 Spectroscopic Terms

Each electron has orbital angular momentum $\ell\hbar$ and spin angular momentum $\hbar/2$. If an orbital has more than one and less than $4\ell+1$ electrons (for the np subshell, this means 2, 3, or 4 electrons), then there is more than one way in which the orbital and spin angular momentum vectors of the electrons in the partially filled shell can add. In the so-called "L-S coupling" approximation, the orbital angular momenta add (vectorially) to give a total orbital angular momentum $L\hbar$, and the individual spin angular momenta similarly add to give a total spin angular momentum $S\hbar$; the wave functions of course must obey the Pauli exclusion principle. Each allowed (L, S) combination is referred to as a **term**.[1] Terms are designated by ${}^{2S+1}\mathcal{L}$, where $\mathcal{L}$ = S, P, D, F for orbital angular momentum $L = 0, 1, 2, 3$.

Different terms (e.g., for an np^2 configuration, the three possible terms ^{3}P, ^{1}D, and ^{1}S) will differ in energy by a significant fraction of the total binding energy of

[1] Determining what terms can be constructed for a given electron configuration can become involved, but it may be helpful to look at one example: two p electrons, i.e., np^2. Each of the p electrons has orbital quantum number $\ell = 1$ and spin quantum number $s = 1/2$. With three possible values of $m_\ell = -1, 0, 1$, and two possible values of $m_s = -1/2, 1/2$, there are $3 \times 2 = 6$ possible one-electron states. The exclusion principle says that both electrons cannot share the same one-electron state, giving $(6 \times 5)/2 = 15$ possible different states for the two indistinguishable electrons:

1. Both electrons could have $m_l = 1$, giving $L_z = 2$, but this would require that one electron be spin up and one spin down, so that $S = 0$. Having $L_z = 2$ requires $L \geq 2$. For two $\ell = 1$ orbitals, the maximum possible value of $L = 2$. Thus it is evident that one of the allowed terms has $S = 0$ and $L = 2$, i.e., ^{1}D. With multiplicity $(2S+1)(2L+1) = 1 \times 5$, this accounts for 5 of the 15 possible quantum states.
2. Both electrons could have $m_s = +1/2$, and $S = 1$. One electron could have $m_\ell = 1$ and one have $m_\ell = 0$, so that $L_z = 1$ is possible, requiring this state to have $L \geq 1$. We have seen earlier that the only way to have $L > 1$ is to have $S = 0$; therefore, this term must have $L = 1$. With degeneracy $(2S+1)(2L+1) = 3 \times 3$, this ^{3}P term accounts for 9 quantum states.
3. We have thus far accounted for 5+9=14 of the 15 quantum states. Therefore there can be only one remaining term, and it must be a singlet, with $S = 0$ and $L = 0$: ^{1}S.

Thus a $2p^2$ configuration gives rise to 3 different terms: ^{1}D, ^{3}P, and ^{1}S. The term with the largest possible values of S and L usually has the lowest Coulomb energy. In this case the ^{3}P term has the lowest energy.

the electrons in the partially filled subshell. Thus for atoms and low-ionization ions, the energy differences between different terms of the ground state configuration will be of order a few eV.

Table 4.1 lists the terms for the ground state configurations of atoms and ions where the outermost subshell is ns or np.

Higher energy states can be constructed by taking one of the electrons out of the ground state configuration and putting it into a higher orbital. For example, in the case of atomic carbon, this can be done by removing one of the $2s$ electrons and promoting it to a $2p$ orbital, giving $1s^2 2s^1 2p^3$ – the electrons in this configuration can also be organized into different terms.

When $L > 0$ and $S > 0$, there is more than one way to add **L** and **S** to get the total angular momentum $\mathbf{J} = \mathbf{L} + \mathbf{S}$. For given L and S, the allowed values of J range from $|L - S|$ to $L + S$. Thus the ^{3}P term can have $J = 0, 1, 2$, with the spin-orbit interaction leading to "fine-structure" splitting between the three different fine-structure levels of the term: $^3\mathrm{P}_0$, $^3\mathrm{P}_1$, and $^3\mathrm{P}_2$.

Because of the possibility of multiple J values for a given L and S, the terms are also referred to as **multiplets**. A term with $L = 0$ or $S = 0$ can have only one possible value of J, and is therefore referred to as **singlet**. Terms with two, three, four, ... possible values of J are referred to as **doublet**, **triplet**, **quartet**, and so on.

4.4 Fine Structure: Spin-Orbit Interaction

As mentioned earlier, when a configuration has $L > 0$ *and* $S > 0$, there are different ways the orbital and spin angular momenta can add to give total angular momentum J. Each will have different value of $\mathbf{L} \cdot \mathbf{S}$, and will differ in energy due to **spin-orbit coupling**. The fractional energy shifts are of order $\sim 10^{-2}$ eV. This splitting of energy levels is referred to as **fine structure**.

4.5 Designation of Energy Levels for Atoms and Ions: Spectroscopic Notation

If

$$L = (\text{total orbital angular momentum})/\hbar \ ,$$
$$S = (\text{total spin angular momentum})/\hbar \ ,$$
$$J = (\text{total electronic angular momentum})/\hbar \ ,$$

then the energy levels (including fine structure splitting) are designated by spectroscopic notation:

$$^{2S+1}\mathcal{L}_J^p \quad ,$$

$$\begin{aligned} &\text{where } \mathcal{L} = S, P, D, F, ... \\ &\quad \text{for } L = 0, 1, 2, 3, ... \\ &\text{and} \quad p = \begin{cases} \textit{blank} & \text{for state of } \textbf{even} \text{ parity} \\ o & \text{for state of } \textbf{odd} \text{ parity} \end{cases} . \end{aligned}$$

The **parity** of an energy level is "even" or "odd" depending on whether the electronic wave function changes sign under reflection of all of the electron positions through the origin. If ℓ_i are the orbital angular momenta of the individual electron orbitals, then

$$\text{parity is } \begin{cases} \text{even} & \text{if } \Pi_i(-1)^{\ell_i} = 1 \quad (\text{i.e., } \sum_i \ell_i \text{ is even}) \\ \text{odd} & \text{if } \Pi_i(-1)^{\ell_i} = -1 \quad (\text{i.e., } \sum_i \ell_i \text{ is odd}) \end{cases} .$$

Note that the adopted notation for designating energy levels overlooks possible hyperfine structure arising from interaction of the electrons with the magnetic moment of the nucleus.

4.5.1 Multiplicity and Degeneracy

Because the total spin **S** and total orbital angular momentum **L** are vectors, they can point in different directions. The **multiplicity** of a term with total spin S and orbital angular momentum L is $g = (2S+1) \times (2L+1)$. Thus the ^{3}P term, with $S = 1$ and $L = 1$, has multiplicity $3 \times 3 = 9$. When spin-orbit coupling is taken into consideration, these states are split into distinct fine-structure levels, each with a definite value of J and a **degeneracy** $g = 2J + 1$: $g = 1$, 3, and 5 for $^3\mathrm{P}_0$, $^3\mathrm{P}_1$, and $^3\mathrm{P}_2$.

4.5.2 Example: Six-electron System

Consider a six-electron system (e.g., C I, N II, O III, F IV, Ne V). The ground configuration $1s^2 2s^2 2p^2$ has *even* parity. The $1s^2$ and $2s^2$ electrons form filled subshells, whereas the $2p$ subshell is only partially filled.

There are three different ways that the two $2p$ electron orbits and spins can be organized into an overall wave function that is antisymmetric under electron exchange, as required by the Pauli exclusion principle – see footnote 1 or, e.g., Bransden & Joachain (2003): ^{3}P (i.e., $L = 1$, $S = 1$), ^{1}D (i.e., $L = 2$, $S = 0$), and ^{1}S (i.e., $L = 0$, $S = 0$). The term with the lowest energy is ^{3}P. With nonzero S and nonzero L, the ^{3}P term splits into 3 fine structure levels: $^3\mathrm{P}_{0,1,2}$. The first excited term is ^{1}D – this is a singlet because it has spin 0, so that the only fine-structure level has $J = L = 2$. The remaining term, ^{1}S, is also a singlet. The energy-level diagram for the ground configuration is shown for N II and O III in Figure 4.1.

Table 4.1 lists the terms corresponding to the ground configuration for atoms or ions where the outermost subshell is either ns or np.

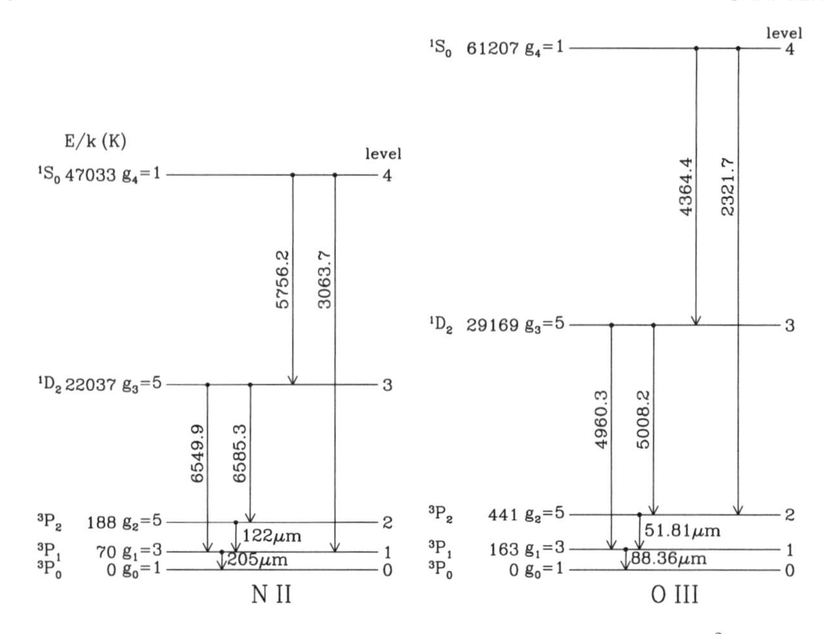

Figure 4.1 Energy-level diagram for the ground configuration of the $2p^2$ ions N II and O III. (Fine-structure splitting is exaggerated for clarity.) Forbidden transitions connecting these levels are shown, with wavelengths in vacuo.

Table 4.1 Terms for ns and np Subshells

Ground configuration	Terms (in order of increasing energy)	Examples
$...ns^1$	$^2S_{1/2}$	H I, He II, C IV, N V, O VI
$...ns^2$	$^1S_0$	He I, C III, N IV, O V
$...np^1$	$^2P^o_{1/2,3/2}$	C II, N III, O IV
$...np^2$	$^3P_{0,1,2}$, $^1D_2$, $^1S_0$	C I, N II, O III, Ne V, S III
$...np^3$	$^4S^o_{3/2}$, $^2D^o_{3/2,5/2}$, $^2P^o_{1/2,3/2}$	N I, O II, Ne IV, S II, Ar IV
$...np^4$	$^3P_{2,1,0}$, $^1D_2$, $^1S_0$	O I, Ne III, Mg V, Ar III
$...np^5$	$^2P^o_{3/2,1/2}$	Ne II, Na III, Mg IV, Ar IV
$...np^6$	$^1S_0$	Ne I, Na II, Mg III, Ar III

4.6 Hyperfine Structure: Interaction with Nuclear Spin

If the nucleus has nonzero spin, it will have a nonzero magnetic moment. If the nucleus has a magnetic moment, then fine-structure levels with nonzero electronic angular momentum can themselves be split due to interaction of the electrons with the magnetic field produced by the nucleus. This "hyperfine" splitting is typically of order 10^{-6} eV. Hyperfine splitting is usually difficult to observe in optical spectra due to Doppler broadening, but it needs to be taken into account if precise modeling of line profiles is required.

It is customary to let

$$\begin{aligned} J &\equiv [\text{electronic angular momentum}]/\hbar\,, \\ I &\equiv [\text{nuclear angular momentum}]/\hbar\,, \text{and} \\ F &\equiv [\text{total angular momentum}]/\hbar\,. \end{aligned}$$

The best-known example of hyperfine splitting is the hydrogen atom, where the ground electronic state $1s\ ^2\mathrm{S}_{1/2}$ has $J = 1/2$ and the proton has $I = 1/2$. The $1s\ ^2\mathrm{S}_{1/2}$ state is split into two levels: The lower level has the electron and proton spins antiparallel, with total angular momentum $F = 0$. The hyperfine excited state has the proton and electron spins parallel, and $F = 1$. The levels are split by $\Delta E = 6.7 \times 10^{-6}$ eV, giving rise to the astronomically important 21-cm transition.

4.7 Zeeman Effect

When a static magnetic field $\mathbf{B}_0$ is applied, each of the fine-structure levels $\mathcal{L}_J$ splits into $2J+1$ energy levels, with energies depending on the value of $\mathbf{J} \cdot \mathbf{B}_0$. The energy splittings are small, of order $\mu_\mathrm{B} B_0 \approx 5.78 \times 10^{-15} (B_0/\mu\mathrm{G})\,\mathrm{eV}$, where $\mu_\mathrm{B} \equiv e\hbar/2m_e c$ is the **Bohr magneton**. Interstellar magnetic field strengths are of order $1 - 100\,\mu$G, and therefore the Zeeman shifts are too small to be measured for transitions in the sub-mm or shortward ($h\nu \gtrsim 10^{-4}\,\mathrm{eV}$).

However, in the case of atomic hydrogen, the hyperfine splitting gives rise to the 21-cm transition, with an energy $h\nu = 5.9 \times 10^{-6}\,\mathrm{eV}$, and, therefore, an applied field of order $10\,\mu$G shifts the frequency by about one part in 10^8. This shift is much smaller than the frequency shift $v/c \sim 10^{-5}$ due to a radial velocity of a few $\mathrm{km\,s}^{-1}$, and it would be nearly impossible to detect, except that it leads to a shift in frequency between the two circular polarization modes. The Zeeman effect in H I 21-cm can therefore be detected by taking the *difference* of the two circular polarization signals. This technique has been used to measure the magnetic field strength in a number of H I regions.

4.8 Further Reading

Bransden & Joachain (2003) provide a comprehensive discussion of the spectroscopy of atoms and ions.

Chapter Five

Energy Levels of Molecules

This chapter reviews the energy-level structure of small molecules, with particular attention to selected molecules of astrophysical interest: H_2, CO, OH, NH_3, and H_2O. Just as for Chapter 4, Chapter 5 should be regarded as reference material – give it a quick once-over now, then return to it when you need to understand observations of some molecule.

5.1 Diatomic Molecules

It is helpful to consider first the hypothetical case where the nuclei are fixed, and only the electrons are free to move – this is known as the **Born-Oppenheimer approximation**. In atoms and atomic ions, the electrons move in a spherically symmetric potential, and the total electronic orbital angular momentum L_e is a good quantum number. In molecules, the electrons move in a Coulomb potential due to two or more nuclei, and spherical symmetry does not apply. However, in the case of diatomic molecules (or, more generally, linear molecules), the Coulomb potential due to the nuclei is symmetric under rotation around the nuclear axis (the line passing through the two nuclei), and L_{ez} =(the projection of the electronic angular momentum onto the internuclear axis)$/\hbar$ is a good quantum number. It is conventional to define $\Lambda \equiv |L_{ez}|$. Because the potential is axially symmetric, the two states $L_{ez} = \pm\Lambda$ have the same energy.

5.1.1 Fine-Structure Splitting

In addition, S_{ez} =(projection of the total electron spin onto the internuclear axis)$/\hbar$ is also a good quantum number; define $\Sigma \equiv |S_{ez}|$.

J_{ez} =(projection of the total electronic angular momentum on the internuclear axis)$/\hbar$ is also a good quantum number. If Λ and Σ are both nonzero, then there are two possible values: $J_{ez} = |\Lambda - \Sigma|$ and $J_{ez} = \Lambda + \Sigma$.

States with different $|J_{ez}|$ will differ in energy due to fine-structure splitting.

5.1.2 Hyperfine Splitting

If one or more nuclei have nonzero nuclear spin *and* $J_{ez} \neq 0$, then there will be an interaction between the nuclear magnetic moment and the magnetic field generated by the electrons, resulting in "hyperfine splitting": the energy will depend on the orientation of the nuclear angular momentum (or angular momenta) relative to the axis. As in atoms, this splitting is small, of order $\sim 10^{-6}$ eV.

5.1.3 Designation of Energy Levels: Term Symbols

Diatomic molecules with identical nuclei (e.g., H_2, N_2, O_2) are referred to as **homonuclear**. Note that the nuclei must be truly *identical* – HD and $^{16}O^{17}O$ are not homonuclear molecules. The energy levels of homonuclear diatomic molecules are designated by **term symbols**

$$^{(2\Sigma+1)}\mathcal{L}_{u,g} \quad ,$$

where

$\mathcal{L} = \Sigma, \Pi, \Delta, ...$ for $\Lambda = 0, 1, 2, ...$, where $\Lambda\hbar$ = projection of the electron orbital angular momentum onto the internuclear axis,

$\Sigma\hbar$ = projection of the electron spin angular momentum onto the internuclear axis.

$$u, g = \begin{cases} g & (\text{"gerade"}) \text{ if symmetric under reflection through the center of mass,} \\ u & (\text{"ungerade"}) \text{ if antisymmetric under reflection through the center of mass.} \end{cases}$$

For the special case of Σ states, a superscript + or – is added to the term symbol:

$$^{(2\Sigma+1)}\Sigma^{\pm}_{u,g} \quad ,$$

where the superscript

$$\pm = \begin{cases} + & \text{if symmetric under reflection through (all) planes containing the nuclei,} \\ - & \text{if antisymmetric under reflection through a plane containing the nuclei.} \end{cases}$$

In the case of a **heteronuclear** diatomic molecule (e.g., HD, OH, or CO), the energy levels are designated

$$^{(2\Sigma+1)}\mathcal{L}_{J_{e,z}}$$

where $\mathcal{L}$ and Σ have the same meaning as for homonuclear diatomic molecules, but now $J_{e,z}$ is indicated as a subscript. As for homonuclear molecules, if the term symbol is Σ, then an additional superscript $\pm$ is applied, specifying the symmetry of the wave function under reflection through planes containing the nuclei.

Because a given molecule may have more than one electronic state with the same term symbol, the electronic states are distinguished by a letter X, A, B, ..., a, b, ...

Table 5.1 Selected Diatomic Molecules[a]

	Ground term	B_0/hc [b] (cm^{-1})	B_0/k [b] (K)	r_0 [d] (Å)	μ [c] (D)	ν_0/c [b] (cm^{-1})	Λ-doubling
H_2	${}^1\Sigma_g^+$	59.335[f]	85.37	0.741	0	4161	–
CH	${}^2\Pi_{1/2,3/2}$	14.190	20.42	1.120[g]	1.406[g]	2733.	$\nu \approx 3.3\,\mathrm{GHz}$
CH^+	${}^1\Sigma_0^+$	13.931	20.04	1.131	1.679[e]	2612.	–
OH	${}^2\Pi_{3/2,1/2}$	18.550	26.69	0.9697	1.6676	3570.	$\nu \approx 1.61\,\mathrm{GHz}$
CN	${}^2\Sigma_{1/2}^+$	1.8910	2.721	1.1718	0.557[i]	2042.	–
CO	${}^1\Sigma_0^+$	1.9225	2.766	1.1283	0.1098	2170.	–
SiO	${}^1\Sigma_0^+$	0.7242	1.042	1.5097	3.088[j]	1230.	–
CS	${}^1\Sigma_0^+$	0.8171	1.175	1.5349	2.001[h]	1272.	–

[a] Data from Huber & Herzberg (1979) unless otherwise noted.
[b] $E(v,J) \approx h\nu_0(v+\frac{1}{2}) + B_0 J(J+1)$ [see Eq. (5.2)].
[c] μ = permanent electric dipole moment.
[d] r_0 = internuclear separation.
[e] Folomeg et al. (1987).
[f] Jennings et al. (1984).
[g] Kalemos et al. (1999).
[h] Maroulis et al. (2000).
[i] Neogrády et al. (2002).
[j] Raymonda et al. (1970).

appearing in front of the term symbol. The letter X is customarily used to designate the electronic ground state. The ground terms for a number of diatomic molecules of astrophysical interest are given in Table 5.1, along with the internuclear separation r_0 and the electric dipole moment μ.

5.1.4 *O*, *P*, *Q*, *R*, and *S* Transitions

A diatomic molecule can vibrate (stretch) along the internuclear axis, and it can rotate around an axis perpendicular to the internuclear axis. The rotational angular momentum adds (vectorially) to the electronic angular momentum.

The rotational levels of diatomic molecules are specified by a single vibrational quantum number v and rotational quantum number J. Transitions will change J by either 0, ±1, or ±2. It is customary to identify transitions by specifying the upper and lower electronic states, upper and lower vibrational states, and one of the following: $O(J_\ell)$, $P(J_\ell)$, $Q(J_\ell)$, $R(J_\ell)$, $S(J_\ell)$, where the usage is given in Table 5.2. Thus, for example, a transition from the $v_\ell = 0$, $J_\ell = 1$ level of the ground electronic state to the $v_u\!=\!5$, $J_u\!=\!2$ level of the first electronic excited state would be written B–X 5–0 $R(1)$.

Table 5.2 Usage of O, P, Q, R, and S

Designation	$(J_u - J_\ell)$	Note
$O(J_\ell)$	-2	Electric quadrupole transition
$P(J_\ell)$	-1	Electric dipole transition
$Q(J_\ell)$	0	Electric dipole or electric quadrupole; $Q(0)$ is forbidden
$R(J_\ell)$	$+1$	Electric dipole transition
$S(J_\ell)$	$+2$	Electric quadrupole transition

5.1.5 H_2

The electronic ground state of H_2 (two electrons) has zero electronic orbital angular momentum ($L_e = 0$), has zero electron spin ($S_e = 0$), is symmetric under reflection through the center of mass (g), and is symmetric under reflection through planes containing the nuclei ($+$). The ground state is $\mathrm{X}\,^1\Sigma_g^+$.

Consider the two nuclei at some fixed separation r_n: one can solve the electron Schrödinger equation for the electrons moving in this potential and obtain the electron eigenfunctions ψ_q and eigenenergies $E_q^{(e)}(r_n)$, where q denotes the quantum numbers that characterize the eigenfunction. If we (slowly) vary the internuclear separation r_n, the electron eigenfunctions ψ_q will change adiabatically, as will the eigenenergies $E_q^{(e)}(r_n)$. Therefore, we can define a function

$$V_q(r_n) \equiv E_q^{(e)}(r_n) + Z_1 Z_2 \frac{e^2}{r_n} \tag{5.1}$$

that is an effective potential governing the internuclear separation. In Figure 5.1, we show the effective internuclear potential $V_q(r_n)$ for the electronic ground state and the first two excited states of H_2.

If we consider only radial, or "vibrational," motions of the two nuclei, the internuclear separation obeys an equation of motion identical to that of a particle with a mass equal to the "reduced mass" $m_r = m_1 m_2/(m_1 + m_2)$, moving in a potential $V_q(r)$. The vibrational energy levels are quantized, with vibrational quantum number $v = 0, 1, 2, ...$ corresponding to the number of nodes in the vibrational wave function. Suppose that $V_q(r)$ has a minimum at nuclear separation r_0. In the neighborhood of r_0, the potential can be approximated $V_q(r) \approx V_q(r_0) + (1/2)k(r - r_0)^2$, corresponding to a "spring constant" k characterizing the curvature of the potential. Classically, for small-amplitude vibrations we would have a harmonic oscillator with angular frequency $\omega_0 = (k/m_r)^{1/2}$. The spring constant $k = d^2V_q/dr^2$ is closely related to the strength of the chemical bond. While k will differ from one chemical bond to another, it varies less than does the reduced mass. Hydrides (i.e., species of chemical formula XH) will have the smallest reduced mass, with H_2 being the extreme limit, with $m_r = m_{\mathrm{H}}/2$. Therefore, the H_2 molecule has an unusually high fundamental vibrational frequency ω_0, corresponding to a wavelength $\lambda \approx 2.1\,\mu\mathrm{m}$.

In addition to vibrational motion, the two nuclei can also undergo rotational motion around their center of mass, with quantized angular momentum $J\hbar$, where $J = 0, 1, 2, ...$ Classically, the rotational kinetic energy of a rigid rotor is $(J\hbar)^2/2I$, where I is the moment of inertia of the molecule. If we consider masses m_1 and m_2 separated by distance r_0, the moment of inertia $I = m_r r_0^2$. Quantum-mechanically, we replace the classical J^2 by $J(J+1)$. Therefore, we expect the rotational kinetic energy $E_{\mathrm{rot}} = J(J+1)\hbar^2/2m_r r_0^2$, and the total vibration-rotation energy when in electronic state q is, in the harmonic-oscillator and rigid-rotor approximation:

$$E_q(v,J) = V_q(r_0) + h\nu_0\left(v + \frac{1}{2}\right) + B_v J(J+1) \quad , \tag{5.2}$$

$$\nu_0 \equiv \frac{\omega_0}{2\pi} \qquad B_v = \frac{\hbar^2}{2m_r r_0^2} \quad . \tag{5.3}$$

The $1/2$ in the $(v+1/2)$ term corresponds to the "zero-point energy" – the quantum vibrator cannot be localized at the potential minimum, and the lowest vibrational level corresponds to an energy $(1/2)\hbar\omega_0$ above $V_q(r_0)$. The constant B_v is referred to as the "rotation constant"; the subscript v is because the moment of inertia depends on the vibrational state. Pure vibrational transitions $v \to v-1$ have energy $h\nu_0$. Pure rotational transitions $J \to J-1$ have energy $h\nu = 2B_v J$

Equation (5.2) is not exact. The potential $V(r)$ is not quadratic, so that the vibrations are not exactly harmonic. In addition, the molecule is not a rigid rotor: the moment of inertia I depends on the state of vibration and also on the state of rotation (in high J states, the molecule gets stretched, resulting in a larger moment of inertia). Note also that r_0 and k depend on which electronic state the molecule

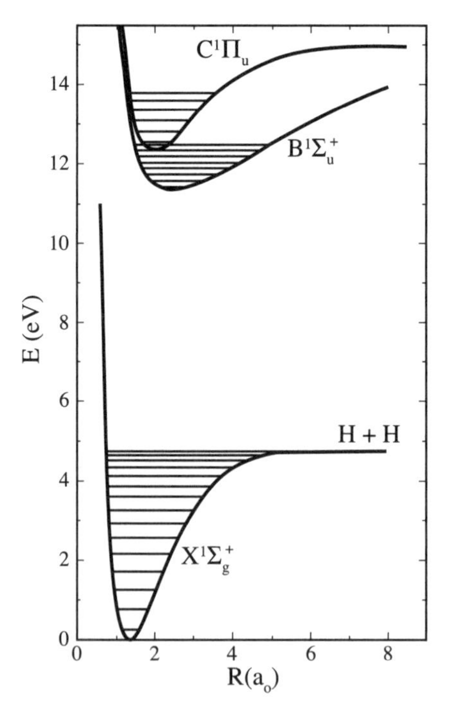

Figure 5.1 Effective internuclear potential for H_2 for the ground state $X^1\Sigma_g^+$ and the first two electronic excited states, $B^1\Sigma_u^+$ and $C^1\Pi_u$

is in: the excited electronic states will have different values of ω_0 and B_v than the ground state.

Each electronic state q therefore supports a vibration–rotation spectrum of energy levels, with energies $E_q(v, J)$. In Figure 5.2, we show the vibration–rotation levels of the ground electronic state of H_2.

5.1.6 Ortho-H_2 and Para-H_2

In the case of H_2, the electronic wave function is required to be antisymmetric under exchange of the two electrons. The two protons, just like the electrons, are identical fermions, and therefore the Pauli exclusion principle antisymmetry requirement also applies to exchange of the two protons. The protons are spin 1/2 particles – the two protons together can have total spin 1 (spins parallel) or total spin 0 (spins antiparallel). Without going into the quantum mechanics, the consequence of the antisymmetry requirement is that if the protons have spin 0, the rotational quantum number J must be even; this is referred to as **para-H_2**, with $J = 0$, 2, 4, ... If the two protons are parallel, with total spin 1, the rotational quantum number J must be odd: this is referred to as **ortho-H_2**, with $J = 1$, 3, 5, ... Because the nuclear spins are only weakly coupled to the electromagnetic field, ortho-H_2 and para-H_2 behave as almost distinct species, with conversion of ortho to para, or para to ortho, happening only very slowly.

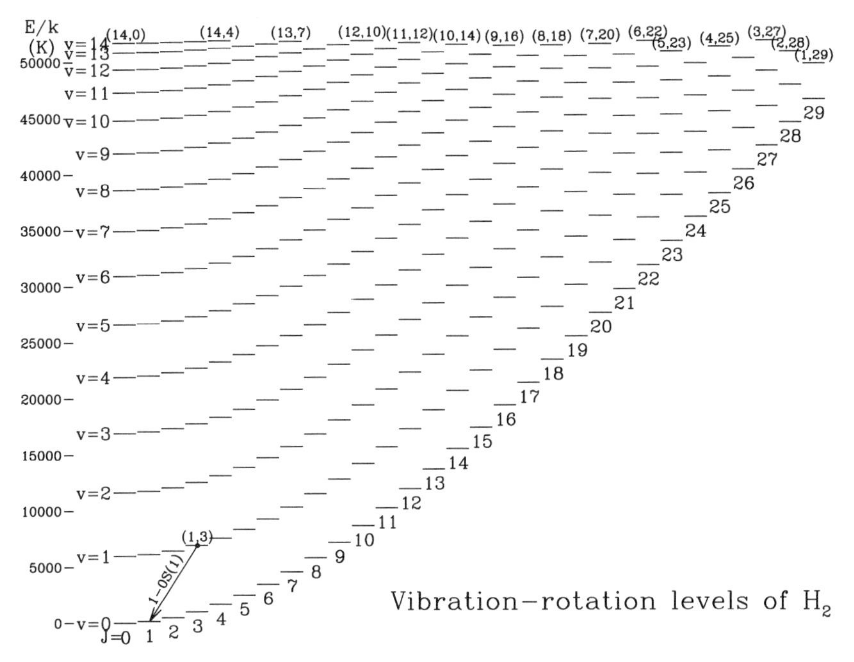

Figure 5.2 Vibration–rotation energy levels of the ground electronic state of H_2 with $J \leq 29$. The $(v, J) = (1, 3)$ level and 1–0S(1) $\lambda = 2.1218\,\mu$m transition are indicated.

Because H_2 has no permanent electric dipole moment, the vibrational states and the rotational states radiate very weakly, via the time-variation of the electric quadrupole moment as the molecule vibrates or rotates. Because the nuclear spin state does not change, the rovibrational radiative transitions of H_2 must have $\Delta J = 0$ or $\Delta J = \pm 2$ – i.e., ortho→ortho or para→para.

The vibration–rotation emission spectrum of H_2 therefore consists of electric quadrupole transitions. The downward transitions are identified by

$$\begin{aligned} &v_u - v_\ell \ \mathrm{S}(J_\ell) \quad \text{if } J_\ell = J_u - 2 \ , \\ &v_u - v_\ell \ \mathrm{Q}(J_\ell) \quad \text{if } J_\ell = J_u \ , \\ &v_u - v_\ell \ \mathrm{O}(J_\ell) \quad \text{if } J_\ell = J_u + 2 \ . \end{aligned}$$

For example, 1–0 S(1) refers to the transition $(v{=}1, J{=}3) \rightarrow (v{=}0, J{=}1)$. This transition is indicated in Fig. 5.2.

5.1.7 CO

CO has 2 p electrons contributed by C and 4 p electrons contributed by O; together, these 6 p electrons fill the $2p$ subshell, and as a result, the ground electronic state of CO has zero electronic angular momentum and zero electronic spin: ${}^1\Sigma_0^+$, just like H_2. The reduced mass of CO is $(12 \times 16/28)\,\mathrm{amu} \approx 6.9\,\mathrm{amu}$. The C=O chemical bond is extremely strong; r_0 is unusually small, the spring constant k is unusually large, and the electric dipole moment (only $\mu = 0.110\,\mathrm{D}$) is unusually small. The fundamental vibrational frequency corresponds to a wavelength $\lambda_0 = c/\nu_0 \approx 4.6\,\mu\mathrm{m}$. (The energy is $\sim 50\%$ of the energy in the H_2 fundamental frequency.) The fundamental rotational frequency $2B_0/h = 115\,\mathrm{GHz}$, and $\hbar^2/Ik \approx 5.5\,\mathrm{K}$ (versus 170 K for H_2). Because the moment of inertia of CO is much larger than that of H_2, the rotational levels of CO are much more closely spaced than those of H_2, and therefore there are many more allowed rotation–vibration levels.

If μ is the permanent electric dipole moment, the Einstein A coefficient for a rotational transition $J \rightarrow J-1$, radiating a photon with energy $\hbar\omega$, is given by

$$A_{J\rightarrow J-1} = \frac{2}{3}\frac{\omega^3}{\hbar c^3}\mu^2 \frac{2J}{2J+1} \tag{5.4}$$

$$= \frac{128\pi^3}{3\hbar}\left(\frac{B_0}{hc}\right)^3 \mu^2 \frac{J^4}{J+\frac{1}{2}}\ \mathrm{s}^{-1} \tag{5.5}$$

$$= 1.07 \times 10^{-7} \frac{J^4}{J+\frac{1}{2}}\ \mathrm{s}^{-1} \tag{5.6}$$

$$= 7.16 \times 10^{-8}\,\mathrm{s}^{-1} \quad \text{for } J = 1 \rightarrow 0\ . \tag{5.7}$$

5.1.8★ OH and Λ-Doubling

OH is an example of a molecule with the ground electronic state having nonzero electronic orbital angular momentum: with seven electrons, the OH ground state has $L_{ez} = 1$ and $S_{ez} = 1/2$, and is therefore designated by $^2\Pi_{1/2,3/2}$. The electron spin and orbital angular momenta can couple to give $J_e = 1/2$ or $3/2$, with energies that are separated due to spin-orbit coupling (i.e., fine-structure splitting in atoms or ions); the $J_e = 3/2$ state has the lower energy.

Now consider either one of these fine-structure states. The projection of the electron angular momentum along the nuclear axis is a constant of the motion, but the vector angular momentum $\mathbf{J}_e$ of the electrons is not. The electric field from the nuclei exerts a torque on the electrons. If the nuclei were held fixed in space, the electron angular momentum vector would precess in a cone centered on the nuclear axis. Now, of course, the nuclei are not held fixed, and if the electron angular momentum $\mathbf{J}_e$ changes, there must be an equal and opposite change in the angular momentum of the nuclei.

For the moment, ignore the nuclear spin – if the nuclear angular momentum is going to change, the nuclei must be undergoing rotation. The implication is that the nuclei undergo rotation even when the OH is in the ground state. Since there is no external torque applied to the OH, the electron angular momentum $\mathbf{J}_e$ and the nuclear angular momentum $\mathbf{J}_n$ both precess around the fixed total angular momentum $\mathbf{J} = \mathbf{J}_n + \mathbf{J}_e$. The magnitude of the total angular momentum J is just equal to the magnitude of the electronic angular momentum that is found when the nuclei are imagined to be held fixed.[1]

If additional angular momentum is given to the nuclei, the rotational kinetic energy will be increased, and each of the fine structure states of OH will have a "rotational ladder": the $J_e = 1/2$ state can have total angular momentum $J = 1/2$, 3/2, 5/2, 7/2, ..., and the $J_e = 3/2$ state can have $J = 3/2$, 5/2, 7/2, ..., and so on. The two rotational ladders are shown in Figure 5.3.

For the moment, let us reexamine the electronic wave functions in the idealization where the nuclei are held fixed, so that the electrons are moving in a potential that is time-independent and symmetric around the nuclear axis. For a linear molecule such as OH, the electronic eigenfunctions are of the form $\psi(r, \theta, \phi) = e^{\pm i\Lambda\phi} f(r, \theta)$, where Λ is the projection of the electronic angular momentum along the nuclear axis, and (r, θ, ϕ) are spherical coordinates with the center of mass as origin and with the polar axis along the internuclear axis: for $\Lambda > 0$, there are two degenerate states. Taking orthogonal linear combinations of these eigenfunctions,

[1] Imagine the nuclei being held fixed, with the electrons orbiting around the nuclear axis. If the nuclei are suddenly released, the total angular momentum will remain unchanged (and equal to the angular momentum of the electrons just prior to the moment of release) but will now be shared by the electrons and the nuclei.

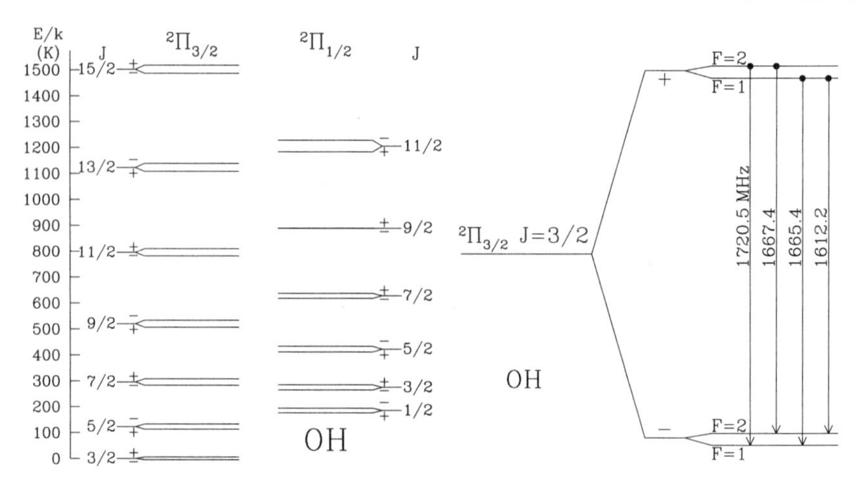

Figure 5.3 Energy levels of OH. Left: The rotational ladders of the $^2\Pi_{3/2}$ and $^2\Pi_{1/2}$ electronic states. The splitting of the levels due to Λ-doubling has been exaggerated. Hyperfine splitting is not shown. Right: Λ-doubling and hyperfine splitting of the $^2\Pi_{3/2}(J=3/2)$ state, showing the four 18-cm lines. In a magnetic field, each of these four lines is further split by the Zeeman effect.

we have

$$\psi_+(r,\theta,\phi) = \frac{e^{i\Lambda\phi} + e^{-i\Lambda\phi}}{\sqrt{2}} f(r,\theta) = \sqrt{2}\cos(\Lambda\phi) f(r,\theta) \ , \tag{5.8}$$

$$\psi_-(r,\theta,\phi) = \frac{e^{i\Lambda\phi} - e^{-i\Lambda\phi}}{i\sqrt{2}} f(r,\theta) = \sqrt{2}\sin(\Lambda\phi) f(r,\theta) \ . \tag{5.9}$$

Now let the nuclear axis be in the $\hat{\mathbf{z}}$ direction, and let $\phi = 0$ correspond to the $\hat{\mathbf{x}}$ direction. The ψ_+ and ψ_- wave functions have different values of $\langle y^2 \rangle$. The moment of inertia of the molecule is dominated by the nuclei, but the electrons make a small contribution. If the rotational angular momentum is oriented along the x axis, the moment of inertia of the molecule, and therefore the rotational kinetic energy, will differ between the ψ_+ and ψ_- states: each of the energy levels is split by this "Λ-doubling" into two states with energies differing by of order $\Delta E_\Lambda \approx [m_e/(m_{AB})] \times [J^2\hbar^2/m_{AB}r_0^2]$, where m_e is the electron rest mass, m_{AB} is the reduced mass of the two nuclei, and r_0 is the internuclear separation. For the OH ground state $^2\Pi_{3/2}(J=3/2)$, the Λ-doubling splitting amounts to $\Delta E_\Lambda = 6.9 \times 10^{-6}$ eV, corresponding to a frequency $\Delta E_\Lambda/h = 1666$ MHz.

5.1.9★ Hyperfine Structure

Many nuclei have nonzero nuclear spin (e.g., ^{1}H, ^{13}C, ^{15}N, and ^{17}O), and, therefore, also have nuclear magnetic moments of magnitude $\sim \mu_N$, where $\mu_N \equiv$

$e^2\hbar/2m_pc$ is the nuclear magneton. The magnetic field from the nuclear magnetic moment couples to the electron motions, and, therefore, the electronic energy depends on the orientation of the nuclear moment and the electron angular momentum. This introduces hyperfine structure in the molecular energy levels. The eigenstates will be states of fixed *total* angular momentum $\mathbf{F} = \mathbf{J} + \mathbf{S}_n$, where $\mathbf{J}$ is the combined angular momentum of the electrons and rotational angular momentum of the nuclei, and S_n is the nuclear spin.

In the case of OH, with $S_n = 1/2$, we have seen that the two rotational ladders have $J = 3/2$, 5/2, 7/2, ..., and $J = 1/2$, 3/2, 5/2, and so on. Each of these states is first split by Λ-doubling into levels denoted $+$ or $-$; each of these levels is in turn split by hyperfine splitting into two levels, with $F = J \pm 1/2$. Therefore, the ground fine-structure level of OH $^2\Pi_{3/2}$ (with $J_e = 3/2$) splits into four sublevels, shown in Figure 5.3.

Radio frequency transitions between the sublevels can be observed in emission and absorption. These levels are important not only because they often are observable as masers (therefore, very bright tracers of the dynamics of molecular gas) but also because the energy levels are subject to Zeeman splitting; because these are radio frequency transitions, the Zeeman splitting produces a frequency shift that is measurable. OH is often used to measure the magnetic field strength in molecular clouds.

5.2★ Energy Levels of Nonlinear Molecules

When we consider nonlinear molecules, the rotational spectrum becomes considerably more complex. Treating the nuclei as point masses at fixed separations (the "rigid rotor" approximation), the moment of inertia tensor has three nonzero eigenvalues and three **principal axes.**

The rotational kinetic energy of a classical rigid rotor can be written

$$E_{\rm rot}^{\rm (class.)} = \frac{(J_A\hbar)^2}{2I_A} + \frac{(J_B\hbar)^2}{2I_B} + \frac{(J_C\hbar)^2}{2I_C} \quad , \tag{5.10}$$

where I_A, I_B, and I_C are the three eigenvalues of the moment of inertia tensor, and $J_A\hbar$, $J_B\hbar$, and $J_C\hbar$ are the (instantaneous) projections of the total angular momentum $\mathbf{J}\hbar$ onto the three principal axes $\hat{A}$, $\hat{B}$, $\hat{C}$ corresponding to the eigenvalues I_A, I_B, I_C. It is conventional to define "rotation constants"

$$A \equiv \frac{\hbar^2}{2I_A} \quad , \quad B \equiv \frac{\hbar^2}{2I_B} \quad , \quad C \equiv \frac{\hbar^2}{2I_C} \quad , \tag{5.11}$$

so that the classical rotational kinetic energy is

$$E_{\rm rot}^{\rm (class.)} = AJ_A^2 + BJ_B^2 + CJ_C^2 \quad . \tag{5.12}$$

Symmetric rotors, also referred to as "symmetric tops," have two degenerate eigenvalues; NH_3 is the primary example of astrophysical interest. Asymmetric rotors (or asymmetric tops) have three nondegenerate eigenvalues – this applies to most polyatomic molecules of astrophysical interest, e.g., H_2O and H_2CO.

5.2.1★ Symmetric Rotor: NH_3

For the classical symmetric rotor, we let $\hat{A}$ be the symmetry axis; then $I_B = I_C$, and we can write

$$\begin{aligned} E_{\rm rot}^{\rm (class.)} &= \frac{(J_A\hbar)^2}{2I_A} + \frac{(J\hbar)^2 - (J_A\hbar)^2}{2I_B} \\ &= \frac{(J\hbar)^2}{2I_B} + (J_A\hbar)^2\left(\frac{1}{2I_A} - \frac{1}{2I_B}\right) \\ &= BJ^2 + (A-B)K^2 \quad , \end{aligned} \tag{5.13}$$

where $K \equiv J_A$ is the projection of **J** onto the symmetry axis. A "prolate" rotor has $I_A < I_B$ ($A > B$); an "oblate" rotor has $I_A > I_B$ ($A < B$). NH_3 is a prolate rotor.

If we now shift from the classical to the quantum treatment, we need only to replace J^2 by the eigenvalue $J(J+1)$:

$$E_{\rm rot} = BJ(J+1) + (A-B)K^2 \quad . \tag{5.14}$$

The quantum symmetric rotor, therefore, has rotational energy levels specified by two quantum numbers, J and K. If J is an integer, then K can take on values $K = 0, 1, ...J-1, J$. The rotational level structure of NH_3 is shown in Figure 5.4.

The electric dipole moment of a symmetric rotor is parallel to axis $\hat{A}$. When $J > K$, $\hat{A}$ precesses around the fixed angular momentum **J**, with a time-varying electric dipole moment. As a result, NH_3 rotational levels with $J > K$ have allowed electric dipole transitions $(J, K) \rightarrow (J-1, K)$, with Einstein A coefficients typically of order 10^{-2} to $10^{-1}\,{\rm s}^{-1}$.

However, when $J = K$, the molecule is spinning around its symmetry axis. Classically, this rotation state has no time-varying electric dipole moment, and would not produce electric dipole radiation. Quantum-mechanically, the levels $J = K$ are metastable, with very long radiative lifetimes. The sequence of levels $J = K$ are referred to as the "rotational backbone"; at interstellar densities, most of the NH_3 will be found occupying these levels, with only very small populations in the levels with $J > K$ above the backbone.

Because NH_3 is a hydride, it has a very small moment of inertia, and the rotational transitions are at relatively high (microwave) frequencies.

However, NH_3 has an additional type of transition that is purely quantum mechanical in nature. Consider the plane defined by the three H atoms. There are two minimum energy positions for the N atom, symmetrically located on either side of the plane. When the problem is treated quantum mechanically, it is found that there are two distinct eigenstates for the NH_3 wave function, separated slightly in

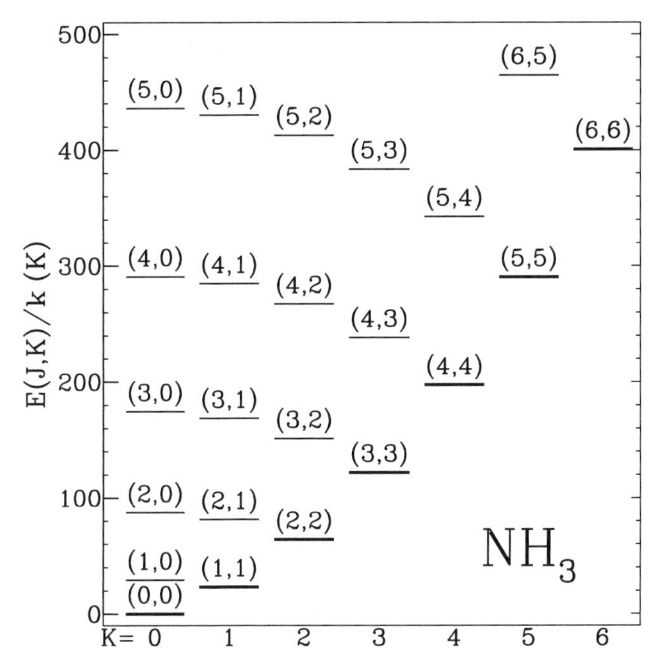

Figure 5.4 Rotational energy levels of NH_3 with $K \leq 6$ and $E/k < 500\,\mathrm{K}$. Each level, labeled by (J, K), is split into two inversion sublevels (not shown). Levels with $J = K$ [e.g., (0,0), (1,1), (2,2), (3,3), ...] are referred to as the "rotational backbone".

energy [9.8×10^{-5} eV in the case of the (1,1) level]. Transitions between these two eigenstates are referred to as **inversion lines**, because the frequency of the transition corresponds to the characteristic frequency for the N atom to tunnel back and forth through the H atom plane from one energy minimum to the other, with associated time-varying electric dipole moment. Transitions between these inversion sublevels are observable, with frequencies near 23 GHz; for the backbone levels $(J, K) = (1, 1)$ and $(2, 2)$, the inversion lines are at 23.694 and 23.723 GHz, respectively.

The inversion sublevels of the (J, K) states are, in turn, further split by interactions with the electric quadrupole moment of the ^{14}N nucleus and the magnetic dipole of the protons. The electric quadrupole splitting breaks the $(1, 1)$ inversion transition into six separate lines, spread over $3.06\,\mathrm{MHz}$ – the splitting is large compared to the thermal line broadening, and therefore these lines are easily resolved. Each of these lines is, in turn, split by hyperfine splitting, but the hyperfine splitting is only on the order of 10 to 40 kHz, corresponding to Doppler shifts of only 0.13 to $0.5\,\mathrm{km\,s^{-1}}$.

NH_3 is not the only molecule with observable inversion transitions. Inversion transitions of H_3O^+ near $181\,\mu\mathrm{m}$ have also been observed (Goicoechea & Cernicharo 2001; Yu et al. 2009).

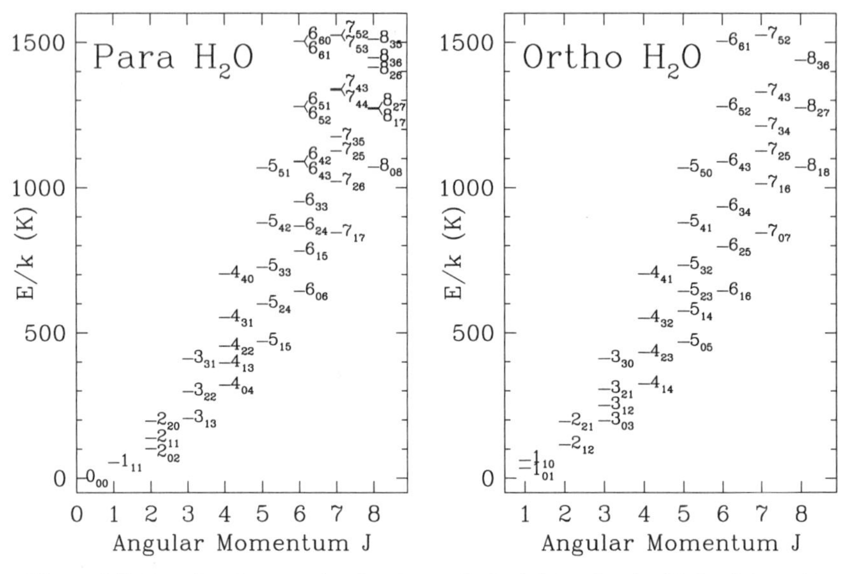

Figure 5.5 Rotational energy levels of para-H_2O (left) and ortho-H_2O (right) with total angular momentum $J \leq 8$ and $E/k < 1500\,\mathrm{K}$ (Chandra et al. 1984).

5.2.2★ Asymmetric Rotor: H_2O

We turn now to the more complicated case of the general asymmetric rotor, with three nondegenerate eigenvalues. We order them $I_A < I_B < I_C$, so that the rotation constants $A > B > C$. The total angular momentum J is of course a constant of the motion, but the projections of J onto individual axes are no longer conserved, except for the special case where $J_A = J$ or $J_C = J$. For the general case where $J_A < J$ and $J_C < J$, the asymmetric rotor undergoes a tumbling motion, with the instantaneous values of J_A, J_B, and J_C all varying while satisfying $J_A^2 + J_B^2 + J_C^2 = J^2$.

For a given value of J, the allowed rotation states are specified by two indices, K_{-1} and K_{+1}; the rotational state is designated $J_{K_{-1},K_{+1}}$. There are seven $J = 3$ states: 3_{03}, 3_{12}, 3_{13}, 3_{21}, 3_{22}, 3_{30}, and 3_{31}.

In the case of H_2O, the two protons can have their nuclear spins either antiparallel or parallel, corresponding to total nuclear spin either 0 or 1. Just as for H_2, the antiparallel spin state is referred to as "para," and the parallel spin state as "ortho." Because the overall wave function must be antisymmetric under proton exchange, it turns out that para-H_2O must have $K_{-1} + K_{+1}$ even while ortho-H_2O must have $K_{-1} + K_{+1}$ odd. Because the nuclear spins do not change in radiative transitions, ortho-H_2O and para-H_2O behave almost like separate species. The rotational levels of para-H_2O and ortho-H_2O are shown in Figure 5.5.

The selection rules for electric dipole radiative transitions in H_2O are $\Delta J = 0, \pm 1$; $\Delta K_{-1} = \pm 1, \pm 3$; and $\Delta K_{+1} = \pm 1, \pm 3$; for less symmetric molecules (e.g., HDO) additional transitions are allowed.

The allowed radiative decays tend to build up the populations in the lowest energy level for a given J – these states are referred to as the **rotational backbone**. All of the rotational backbone levels (except the lowest) have allowed transitions to the next lower backbone state. For most of the backbone states, this is the only allowed downward transition. However, two of the ortho-H_2O backbone states in Fig. 5.5 – 4_{14} and 6_{16} – have permitted downward transitions to nonbackbone levels: $4_{14} \rightarrow 3_{21}$ and $6_{16} \rightarrow 5_{23}$. The nonbackbone levels tend to have very low populations, and, therefore, these two transitions, with the upper level on the backbone and the lower level above the backbone, frequently have population inversions with resulting maser emission in the $4_{14} \rightarrow 3_{21}$ transition at 380 GHz and the $6_{16} \rightarrow 5_{23}$ transition at 22.2 GHz.

The same holds true for the 3_{13} and 5_{15} levels of para-H_2O: the $3_{13} \rightarrow 2_{20}$ 183 GHz and $5_{15} \rightarrow 4_{22}$ 325 GHz transitions are also candidates for masing.

5.3★ Zeeman Splitting

If an external magnetic field $\mathbf{B}$ is present, molecular energy levels with a magnetic moment $\boldsymbol{\mu}$ will be shifted in energy by an amount

$$\Delta E_{\rm mag} = -\boldsymbol{\mu} \cdot \mathbf{B} , \tag{5.15}$$

where $\boldsymbol{\mu}$ is the magnetic moment, with contributions from the orbital and spin angular momenta of the electrons. When both spin and orbital angular momentum are present, neither $\mathbf{S}$ nor $\mathbf{L}$ are fixed in space – both precess around the total angular momentum $\mathbf{F}\hbar = (\mathbf{L} + \mathbf{S})\hbar$. The magnetic moment is antiparallel to $\mathbf{F}$,

$$\boldsymbol{\mu} = -g\mu_{\rm B}\mathbf{F} , \tag{5.16}$$

where $\mu_{\rm B} \equiv e\hbar/2m_e c$ is the **Bohr magneton**, and g is the **Landé g-factor**. Values for the Landé g-factor lie between 0.5 and 1, depending on the values of L and S.

The projection of $\mathbf{F}$ onto the direction of the magnetic field $\mathbf{B}$ is M_F, with $2F+1$ allowed values $M_F = -F, -F+1, ..., F$. The perturbation to the energy level is

$$E_{\rm mag} = (g\mu_{\rm B}B)\, M_F \ . \tag{5.17}$$

For a transition between two levels with different M_F, the energy-level splitting $g\mu_{\rm B}B$ corresponds to a frequency shift

$$(\Delta\nu)_B = \frac{\Delta E_{\rm mag}}{h} = \frac{g\mu_{\rm B}}{h}\, \Delta M_F\, B \tag{5.18}$$

$$= 1.3996\, g\, \Delta M_F \frac{B}{\mu G}\, {\rm Hz} \ . \tag{5.19}$$

For interstellar magnetic fields in the μG– mG range, Eq. (5.19) shows that the frequency shifts are small, typically much smaller than the frequency shift $(\Delta v/c) \times \nu$

due to Doppler broadening.[2] The only hope for detecting this small shift is to (1) use radio frequency transitions where $(\Delta\nu)_B/\nu$ is large enough for the shifts to be measured, and (2) to use the fact that the two circularly polarized components have $\Delta M_F = \pm 1$, so that they are shifted in frequency by $2(\Delta\nu)_B$. If Doppler broadening is assumed to contribute identically to the two circular polarizations, then subtraction of one circular polarization from the other produces a signal that allows the frequency shift $(\Delta\nu)_B$ to be measured, allowing the magnetic field strength to be determined.

Successful detection of the Zeeman splitting requires an atom or molecule with a strong radio frequency transition, and $F > 0$. The hydrogen atom fulfills this requirement, with $F = 1/2$ in the ground state $^2S_{1/2}$, Landé g-factor $g = 0.5$, and the spin-flip transition at $\nu = 1.420$ GHz.

Diatomic molecules with $^2\Pi_{1/2,3/2}$ orbitals have both $F > 0$ (so that Zeeman splitting occurs) and Λ-doubling (providing relatively low frequency radio transitions). The OH transitions at 1.665 and 1.667 GHz are frequently used, both in absorption and in (maser) emission. Higher frequency OH transitions can also be used, as well as CH transitions at 3.33 GHz.

In high-density regions, OH and CH may be unavailable, and it becomes necessary to resort to rotational transitions with much higher frequencies. The CN 1–0 rotational transition at 113 GHz has been used for Zeeman measurements.

Polyatomic molecules also exhibit Zeeman splitting. H_2O does not have an unpaired electron, and hence $F = 0$, but ortho-H_2O, with parallel proton spins, has a small magnetic moment contributed by the protons. Magnetic field measurements have been made using H_2O masers in the $6_{16} \rightarrow 5_{23}$ transition at 22.2 GHz. Further information can be found in reviews by Heiles et al. (1993) and Heiles & Crutcher (2005).

5.4 Further Reading

Bransden & Joachain (2003) has an extensive discussion of molecular quantum mechanics and spectroscopy. The definitive text on microwave spectroscopy is the monograph by Townes & Schawlow (1975).

[2]The only exception appears to be OH – the 1.67 GHz frequency (due to Λ-doubling) is low, and regions hosting OH masers appear to sometimes have large enough magnetic fields to produce frequency shifts exceeding the line width.

Chapter Six

Spontaneous Emission, Stimulated Emission, and Absorption

In this chapter, we review the general principles governing absorption and emission of radiation by absorbers with quantized energy levels. The absorbers in question can be atoms, ions, molecules, dust grains, or *any* objects with energy levels.

6.1 Emission and Absorption of Photons

If an absorber X is in a level ℓ and there is radiation present with photons having an energy equal to $E_u - E_\ell$, where E_ℓ and E_u are the energies of levels ℓ (for "lower") and u (for "upper"), the absorber can absorb a photon and undergo an upward transition:

$$\textbf{absorption}: \qquad \mathrm{X}_\ell + h\nu \rightarrow \mathrm{X}_u \,, \qquad h\nu = E_u - E_\ell \,. \tag{6.1}$$

Suppose that we have number density n_ℓ of absorbers X in level ℓ. The rate per volume at which the absorbers absorb photons will obviously be proportional to both the density of photons of the appropriate energy and the number density n_ℓ, so we can write the rate of change of n_ℓ due to photoabsorption by level ℓ as

$$\underbrace{\left(\frac{dn_u}{dt}\right)_{\ell\rightarrow u}}_{\text{populate level } u} = \underbrace{-\left(\frac{dn_\ell}{dt}\right)_{\ell\rightarrow u}}_{\text{depopulate level } \ell} = n_\ell B_{\ell u} u_\nu \,, \qquad \nu = \frac{E_u - E_\ell}{h} \,, \tag{6.2}$$

where u_ν is the radiation energy density per unit frequency, and the proportionality constant $B_{\ell u}$ is the **Einstein B coefficient**[1] for the transition $\ell \rightarrow u$.

An absorber X in an excited level u can decay to a lower level ℓ with emission of a photon. There are two ways this can happen:

$$\textbf{spontaneous emission}: \qquad \mathrm{X}_u \rightarrow \mathrm{X}_\ell + h\nu \qquad \nu = (E_u - E_\ell)/h \,, \tag{6.3}$$

$$\textbf{stimulated emission}: \qquad \mathrm{X}_u + h\nu \rightarrow \mathrm{X}_\ell + 2h\nu \qquad \nu = (E_u - E_\ell)/h \,. \tag{6.4}$$

[1] Einstein was the first to discuss the statistical mechanics of the interaction of absorbers with the quantized radiation field.

Spontaneous emission is a random process, independent of the presence of a radiation field, with a probability per unit time $A_{u\ell}$ – the **Einstein A coefficient**.

Stimulated emission occurs if photons of the *identical* frequency, polarization, and direction of propagation are already present; the rate of stimulated emission is proportional to the density of these photons. Thus the total rate of depopulation of level u due to emission of photons can be written

$$\left(\frac{dn_\ell}{dt}\right)_{u\rightarrow\ell} = -\left(\frac{dn_u}{dt}\right)_{u\rightarrow\ell} = n_u\left(A_{u\ell} + B_{u\ell}u_\nu\right) \quad , \tag{6.5}$$

where the coefficient $B_{u\ell}$ is the Einstein B coefficient for the downward transition $u \rightarrow \ell$. Thus we now have three coefficients characterizing radiative transitions between levels u and ℓ: $A_{u\ell}$, $B_{u\ell}$ and $B_{\ell u}$. We will now see that they are not independent of one another.

In thermal equilibrium, the radiation field becomes the "blackbody" radiation field, with intensity given by the blackbody spectrum

$$B_\nu = \frac{2h\nu^3}{c^2}\frac{1}{e^{h\nu/kT}-1} \quad , \tag{6.6}$$

with specific energy density

$$(u_\nu)_{\rm LTE} = \frac{4\pi}{c}B_\nu(T) = \frac{8\pi h\nu^3}{c^3}\frac{1}{e^{h\nu/kT}-1} \quad . \tag{6.7}$$

If we place absorbers X into a blackbody radiation field, then the net rate of change of level u is

$$\begin{aligned}\frac{dn_u}{dt} &= \left(\frac{dn_u}{dt}\right)_{\ell\rightarrow u} + \left(\frac{dn_u}{dt}\right)_{u\rightarrow\ell} \\ &= n_\ell B_{\ell u}\frac{8\pi h\nu^3}{c^3}\frac{1}{e^{h\nu/kT}-1} - n_u\left(A_{u\ell} + B_{u\ell}\frac{8\pi h\nu^3}{c^3}\frac{1}{e^{h\nu/kT}-1}\right) . \end{aligned} \tag{6.8}$$

If the absorbers are allowed to come to equilibrium with the radiation field, levels ℓ and u must be populated according to $n_u/n_l = (g_u/g_\ell)e^{(E_\ell-E_u)/kT}$, with $dn_u/dt = 0$. From Eq. (6.8) it is easy to show[2] that $B_{u\ell}$ and $B_{\ell u}$ must be related to $A_{u\ell}$ by

$$B_{u\ell} = \frac{c^3}{8\pi h\nu^3}A_{u\ell} \quad , \tag{6.9}$$

$$B_{\ell u} = \frac{g_u}{g_\ell}B_{u\ell} = \frac{g_u}{g_\ell}\frac{c^3}{8\pi h\nu^3}A_{u\ell} \quad . \tag{6.10}$$

---

[2]Hint: consider the two limits $T \rightarrow 0$ and $T \rightarrow \infty$. Equation (6.8), with $dn_u/dt = 0$, must be valid in both limits.

Thus the strength of stimulated emission ($B_{u\ell}$) and absorption ($B_{\ell u}$) are both determined by $A_{u\ell}$ and the ratio g_u/g_l.

Rather than discussing absorption and stimulated emission in terms of the radiation energy density u_ν, it is helpful to characterize the intensity of the radiation field by a dimensionless quantity, the **photon occupation number** n_γ:

$$n_\gamma \equiv \frac{c^2}{2h\nu^3} I_\nu \quad , \tag{6.11}$$

$$\bar{n}_\gamma \equiv \frac{c^2}{2h\nu^3} \bar{I}_\nu = \frac{c^3}{8\pi h\nu^3} u_\nu \quad , \tag{6.12}$$

where the bar denotes averaging over directions. With this definition of n_γ, we can rewrite Eqs. (6.2 and 6.5) as simply

$$\left(\frac{dn_\ell}{dt}\right)_{u\rightarrow\ell} = n_u \cdot A_{u\ell} \cdot (1 + \bar{n}_\gamma) \quad , \tag{6.13}$$

$$\left(\frac{dn_u}{dt}\right)_{\ell\rightarrow u} = n_\ell \cdot \frac{g_u}{g_\ell} A_{u\ell} \cdot \bar{n}_\gamma \quad . \tag{6.14}$$

If the radiation field depends on frequency in the vicinity of the transition frequency $\nu_{u\ell}$, then n_γ needs to be averaged over the emission profile in (6.13) and over the absorption profile in (6.14).

From Eq. (6.13) we immediately see that the photon occupation number n_γ determines the relative importance of stimulated and spontaneous emission: stimulated emission is unimportant when $\bar{n}_\gamma \ll 1$, but should otherwise be included in analyses of level excitation.

6.2 Absorption Cross Section

Having determined the rate at which photons are absorbed by an absorber exposed to electromagnetic radiation, it is useful to recast this in terms of an absorption cross section. The photon density per unit frequency is just $u_\nu/h\nu$. Let $\sigma_{\ell u}(\nu)$ be the cross section for absorption of photons of frequency ν with resulting $\ell \rightarrow u$ transition. The absorption rate is then

$$\left(\frac{dn_u}{dt}\right)_{\ell\rightarrow u} = n_\ell \int d\nu \, \sigma_{\ell u}(\nu) c \frac{u_\nu}{h\nu} \approx n_\ell u_\nu \frac{c}{h\nu} \int d\nu \, \sigma_{\ell u}(\nu) \quad , \tag{6.15}$$

where we have assumed that u_ν (and $h\nu$) do not vary appreciably over the line profile of $\sigma_{u\ell}$. Thus

$$B_{\ell u} = \frac{c}{h\nu} \int d\nu \, \sigma_{\ell u}(\nu) \quad , \tag{6.16}$$

and, using Eq. (6.10), we obtain the integral over the absorption cross section:

$$\int d\nu\, \sigma_{\ell u}(\nu) = \frac{g_u}{g_\ell} \frac{c^2}{8\pi\nu_{\ell u}^2} A_{u\ell} \quad . \tag{6.17}$$

Thus we may relate the monochromatic absorption cross section $\sigma_{\ell u}(\nu)$ to a normalized line profile ϕ_ν:

$$\sigma_{\ell u}(\nu) = \frac{g_u}{g_\ell} \frac{c^2}{8\pi\nu_{\ell u}^2} A_{u\ell}\phi_\nu \quad \text{with} \quad \int \phi_\nu d\nu = 1 \quad . \tag{6.18}$$

The frequency dependence of the normalized line profile ϕ_ν is discussed in the following.

6.3 Oscillator Strength

Earlier we characterized the strength of radiative transitions by the Einstein A coefficient, $A_{u\ell}$. Equivalently, we can characterize the strength of an absorption transition $\ell \rightarrow u$ by the **oscillator strength** $f_{\ell u}$, defined by the relation

$$f_{\ell u} \equiv \frac{m_e c}{\pi e^2} \int \sigma_{\ell u}(\nu) d\nu \quad . \tag{6.19}$$

From Eqs. (6.17 and 6.19), we see that the Einstein A coefficient for spontaneous decay is related to the absorption oscillator strength of the upward transition by

$$A_{u\ell} = \frac{8\pi^2 e^2 \nu_{\ell u}^2}{m_e c^3} \frac{g_\ell}{g_u} f_{\ell u} = \frac{0.6670\,\mathrm{cm}^2\,\mathrm{s}^{-1}}{\lambda_{\ell u}^2} \frac{g_\ell}{g_u} f_{\ell u} \quad . \tag{6.20}$$

The oscillator strength $f_{u\ell}$ for a downward transition $u \rightarrow \ell$ is negative, and is defined by

$$g_\ell f_{\ell u} = -g_u f_{u\ell} \quad . \tag{6.21}$$

The rate of stimulated emission is proportional to the downward oscillator strength, so it is natural that it should be negative, as it results in depopulation of the upper level. With this definition, the transitions for a one-electron atom in an initial state i obey the Thomas-Reich-Kuhn **sum rule**:

$$\sum_j f_{ij} = 1 \quad , \tag{6.22}$$

where the sum over final states j includes transitions to bound states and also to the continuum (i.e., photoionization). If the initial state i is not the ground state, the sum includes downward transitions with $f_{ij} < 0$. For multielectron atoms or ions, the sum rule (6.22) generalizes to

$$\sum_j f_{ij} = N \quad , \tag{6.23}$$

where N is the number of electrons, and the sum is over *all* transitions out of initial state i. The absorption cross section $\sigma_{\ell u}(\nu)$ is related to the oscillator strength by

$$\sigma_{\ell u}(\nu) = \frac{\pi e^2}{m_e c} f_{\ell u} \phi_\nu \quad \text{with} \quad \int \phi_\nu d\nu = 1 \quad . \tag{6.24}$$

6.4 Intrinsic Line Profile

The intrinsic line profile is characterized by a normalized profile function $\phi_\nu^{\rm intr.}$:

$$\sigma^{\rm intr.}(\nu) = \frac{\pi e^2}{m_e c} f_{\ell u}\, \phi_\nu^{\rm intr.} \qquad \left(\int \phi_\nu^{\rm intr.} d\nu = 1 \right) \quad . \tag{6.25}$$

The intrinsic line profile of an absorption line is normally described by the Lorentz line profile function:

$$\phi_\nu^{\rm intr.} = \frac{4\gamma_{u\ell}}{16\pi^2(\nu - \nu_{u\ell})^2 + \gamma_{u\ell}^2} \quad , \tag{6.26}$$

where $\nu_{u\ell} \equiv (E_u - E_\ell)/h$. The Lorentz profile in Eq. (6.26) provides an accurate (but not exact)[3] approximation to the actual line profile. The Lorentz line profile has a **full width at half maximum (FWHM)**

$$(\Delta\nu)_{\rm FWHM}^{\rm intr.} = \frac{\gamma_{u\ell}}{2\pi} \quad . \tag{6.27}$$

The intrinsic width of the absorption line reflects the uncertainty in the energies of levels u and ℓ due to the finite lifetimes of these levels[4] against transitions to *all* other levels, including both radiative and collisional transitions. If the primary process for depopulating levels u and ℓ is spontaneous decay (as is often the case in the ISM), then

$$\gamma_{u\ell} \equiv \gamma_{\ell u} = \sum_{E_j < E_u} A_{uj} + \sum_{E_j < E_\ell} A_{\ell j} \quad . \tag{6.28}$$

[3]The line profile is more accurately given by the Kramers-Heisenberg formula; Lee (2003) discusses application of this formula to the Lyman α line.

[4]The Heisenberg uncertainty principle $\Delta E \Delta t \geq \hbar$ implies that an energy level u has a width $\Delta E_u \approx \hbar/\tau_u$, where τ_u is the level lifetime.

In the case of a "resonance line," where ℓ is the ground state, the second sum vanishes.

It is convenient to describe line widths in terms of the line-of-sight velocities that would produce Doppler shifts of the same amount. Thus the intrinsic width of an absorption line can be given in terms of velocity:

$$(\Delta v)^{\rm intr.}_{\rm FWHM} = c\frac{(\Delta\nu)^{\rm intr.}_{\rm FWHM}}{\nu_{u\ell}} = \frac{\lambda_{u\ell}\gamma_{u\ell}}{2\pi} = 0.0121\frac{\rm km}{s}\left(\frac{\lambda_{u\ell}\gamma_{u\ell}}{7616\,{\rm cm\,s^{-1}}}\right), \quad (6.29)$$

where $\lambda_{u\ell}\gamma_{u\ell} = 7616\,{\rm cm\,s^{-1}}$ is the value for H Lyman α. The intrinsic line width can also be written in terms of the energy and oscillator strength of the transition:

$$(\Delta v)^{\rm intr}_{\rm FWHM} \geq \alpha^3\left(\frac{h\nu}{I_{\rm H}}\right)\frac{g_\ell}{g_u}f_{\ell u}c = 0.116\left(\frac{h\nu}{I_{\rm H}}\right)\frac{g_\ell}{g_u}f_{\ell u}\,{\rm km\,s^{-1}}, \quad (6.30)$$

where $\alpha \equiv e^2/\hbar c = 1/137.036$ is the fine-structure constant, $I_{\rm H} = (1/2)\alpha^2 m_e c^2 = 13.60\,{\rm eV}$ is the ionization energy of H, and the inequality is because $\gamma_{u\ell} \geq A_{u\ell}$ [see Eq. (6.28)]. From (6.29), we see that optical and ultraviolet absorption lines, for which $h\nu/I_{\rm H} < 1$ and $f_{\ell u} < 1$, will have $(\Delta v)^{\rm intr.}_{\rm FWHM} \lesssim 0.1\,{\rm km\,s^{-1}}$. For example, H Lyman α ($\lambda = 1215.67$ Å) has $h\nu/I_{\rm H} = 3/4$, $f_{lu} = 0.4164$, $g_\ell/g_u = 2/6$, and $(\Delta v)^{\rm intr.}_{\rm FWHM} = 0.0121\,{\rm km\,s^{-1}}$. H Lyman α has a relatively large energy ($0.75 I_{\rm H}$) and relatively large oscillator strength (0.4164); most other optical and ultraviolet permitted lines have even smaller $(\Delta v)^{\rm intr.}_{\rm FWHM}$.

Because $(\Delta v)^{\rm intr.}_{\rm FWHM} \propto h\nu$, X-ray transitions can have considerably larger intrinsic line widths. For example, the 6.68 keV $Fe^{24+}1s2p \rightarrow 1s^2$ line has an intrinsic linewidth $(\Delta v)^{\rm intr.}_{\rm FWHM} = 13.5\,{\rm km\,s^{-1}}$.

6.5 Doppler Broadening: The Voigt Line Profile

Atoms and ions are generally in motion, and the velocity distribution is often approximated by a Gaussian, this being of course the correct form if the velocities are entirely due to thermal motions:

$$p_v = \frac{1}{\sqrt{2\pi}}\frac{1}{\sigma_v}\,{\rm e}^{-(v-v_0)^2/2\sigma_v^2} = \frac{1}{\sqrt{\pi}}\frac{1}{b}\,{\rm e}^{-(v-v_0)^2/b^2}, \quad (6.31)$$

where $p_v dv$ is the probability of the velocity along the line of sight being in the interval $[v, v+dv]$, σ_v is the one-dimensional **velocity dispersion**, and the **"broadening parameter"** $b \equiv \sqrt{2}\sigma_v$.

The width of the velocity distribution is also sometimes specified in terms of the FWHM; for a Gaussian distribution of velocities, this is just

$$(\Delta v)_{\rm FWHM} = \sqrt{8\ln 2}\,\sigma_v = 2\sqrt{\ln 2}\,b\ . \quad (6.32)$$

If the velocity dispersion is entirely due to thermal motion with kinetic temperature $T = 10^4 T_4$ K, then

$$\sigma_v = \left(\frac{kT}{M}\right)^{1/2} = 9.12 \left(\frac{T_4}{M/\text{amu}}\right)^{1/2} \text{km s}^{-1} \quad , \tag{6.33}$$

$$b = \left(\frac{2kT}{M}\right)^{1/2} = 12.90 \left(\frac{T_4}{M/\text{amu}}\right)^{1/2} \text{km s}^{-1} \quad , \tag{6.34}$$

$$(\Delta v)_{\text{FWHM}}^{\text{therm}} = \left[\frac{(8 \ln 2)\, kT}{M}\right]^{1/2} = 21.47 \left(\frac{T_4}{M/\text{amu}}\right)^{1/2} \text{km s}^{-1} \quad . \tag{6.35}$$

Line-of-sight velocity v produces a Doppler shift $\Delta\nu/\nu_{u\ell} = -v/c$. The intrinsic absorption line profile $\phi_\nu^{\text{intr.}}$ must be convolved with the velocity distribution of the absorbers to obtain the line profile

$$\phi_\nu = \int dv\, p_v(v) \frac{4\gamma_{u\ell}}{16\pi^2 \left[\nu - (1 - v/c)\nu_{u\ell}\right]^2 + \gamma_{u\ell}^2} \quad , \tag{6.36}$$

where $p_v dv$ is the probability of the absorber having radial velocity in the interval $(v, v + dv)$. If the absorbers have a Maxwellian (i.e., Gaussian) one-dimensional velocity distribution p_v (Eq. 6.31), then the absorption line will have a so-called **Voigt line profile**:

$$\phi_\nu^{\text{Voigt}} \equiv \frac{1}{\sqrt{2\pi}} \int \frac{dv}{\sigma_v} \mathrm{e}^{-v^2/2\sigma_v^2} \frac{4\gamma_{u\ell}}{16\pi^2 \left[\nu - (1 - v/c)\nu_{u\ell}\right]^2 + \gamma_{u\ell}^2} \quad . \tag{6.37}$$

Unfortunately, the Voigt line profile cannot be obtained analytically except for limiting cases.[5] However, if, as is generally the case, the one-dimensional velocity dispersion $\sigma_v \gg (\Delta v)_{\text{FWHM}}^{\text{intr.}}$, the central core of the line profile is well-approximated by treating the intrinsic line profile as a δ-function, so that the central core of the line has a Maxwellian profile:

$$\phi_\nu \approx \frac{1}{\sqrt{\pi}} \frac{1}{\nu_{ul}} \frac{c}{b} \exp\left(-v^2/b^2\right) \quad , \quad v \equiv -\frac{(\nu - \nu_{u\ell})}{\nu_{u\ell}} c \quad , \quad b \equiv \sqrt{2}\, \sigma_v \, . \tag{6.38}$$

We will discuss the Voigt profile further in Chapter 9.

6.6 Transition from Doppler Core to Damping Wings

Near line-center, the line profile is well-approximated by the Doppler core profile, which for a Gaussian velocity distribution gives

$$\sigma \approx \sqrt{\pi} \frac{e^2}{m_e c} \frac{f_{\ell u} \lambda_{u\ell}}{b} \mathrm{e}^{-v^2/b^2} \quad , \tag{6.39}$$

[5] Accurate approximation formulae have been developed for the Voigt profile – see Armstrong (1967).

where the velocity $v \equiv -(\nu - \nu_{u\ell})c/\nu_{u\ell}$. For very large frequency shifts, the profile can be approximated by just the damping wings:

$$\sigma \approx \sqrt{\pi}\frac{e^2}{m_e c}\frac{f_{\ell u}\lambda_{u\ell}}{b}\left[\frac{1}{4\pi^{3/2}}\frac{\gamma_{u\ell}\lambda_{u\ell}}{b}\frac{b^2}{v^2}\right] . \tag{6.40}$$

For what frequency shift, expressed as a velocity, do we make the transition from the Doppler core to the damping wings? The condition for $z \equiv v/b$ is obtained by equating (6.39) and (6.40):

$$\mathrm{e}^{z^2} = \left[4\pi^{3/2}\frac{b}{\gamma_{u\ell}\lambda_{u\ell}}z^2\right] = 2925\left[\frac{7616\,\mathrm{cm\,s^{-1}}}{\gamma_{u\ell}\lambda_{u\ell}}b_6\right]z^2 , \tag{6.41}$$

where $b_6 \equiv b/10\,\mathrm{km\,s^{-1}}$. The solution to this transcendental equation is

$$z^2 \approx 10.31 + \ln\left[\frac{7616\,\mathrm{cm\,s^{-1}}}{\gamma_{u\ell}\lambda_{u\ell}}b_6\right], \tag{6.42}$$

provided that the quantity in square brackets is not very large or very small. Therefore, for a strong permitted line (such as Lyman α), the damping wings dominate for velocity shifts with $|z| \gtrsim 3.2$, or $|\Delta v| \gtrsim 32 b_6\,\mathrm{km\,s^{-1}}$.

6.7 Selection Rules for Radiative Transitions

Some energy levels are connected by strong radiative transitions; in other cases, radiative transitions between the levels may be extremely slow. The strong transitions always satisfy what are referred to as the **selection rules for electric dipole transitions**. Here, we summarize the selection rules for the strong electric dipole transitions, and we also give the selection rules for **intersystem** and **forbidden** transitions that do not satisfy the electric dipole selection rules but nevertheless are strong enough to be astrophysically important. We will use the ion N II as an example; the first nine energy levels of N II are shown in Fig. 6.1.

6.7.1 Allowed = Electric Dipole Transitions

The strongest transitions are electric dipole transitions. These are transitions satisfying the following selection rules:

1. Parity must change.
2. $\Delta L = 0, \pm 1$.
3. $\Delta J = 0, \pm 1$, but $J = 0 \rightarrow 0$ is forbidden.
4. Only one single-electron wave function $n\ell$ changes, with $\Delta\ell = \pm 1$.

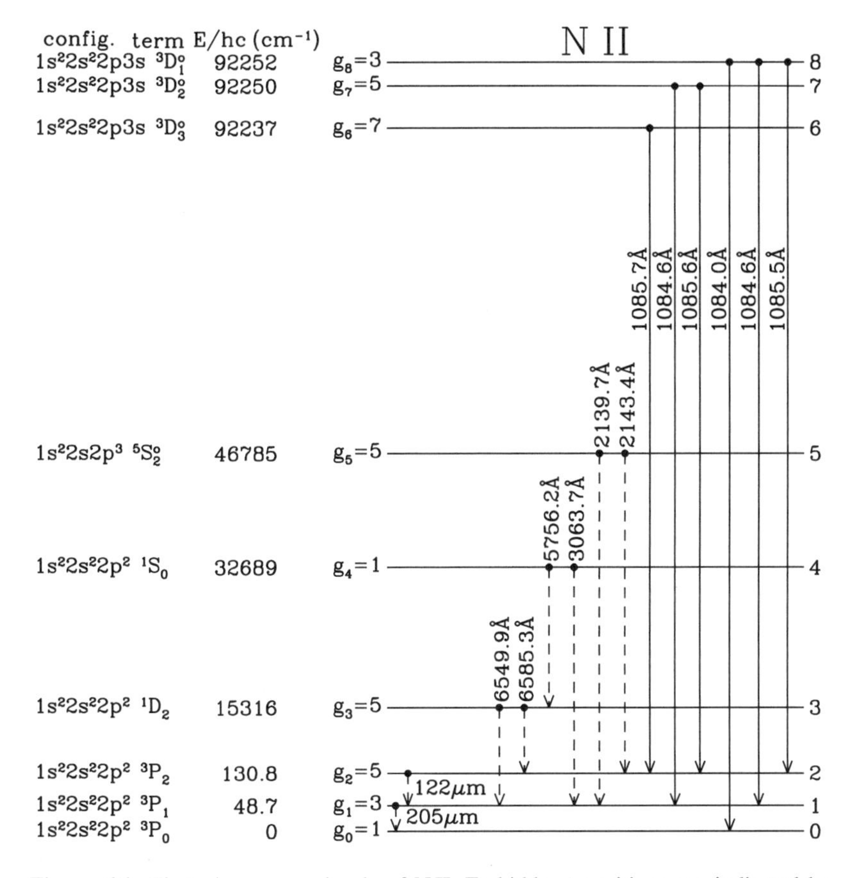

Figure 6.1 First nine energy levels of N II. Forbidden transitions are indicated by broken lines, and allowed transitions by solid lines; forbidden decays are not shown from levels that have permitted decay channels. Fine-structure splitting is not to scale. Hyperfine splitting is not shown.

5. $\Delta S = 0$: Spin does **not** change.

An allowed transition is denoted without square brackets, for example,

$$\text{N II } 1084.0\,\text{Å } ^3\text{P}_0 - {}^3\text{D}_1^\text{o} \quad .$$

This is a transition between the $\ell = 1s^2 2s^2 2p^2\,{}^3\text{P}_0$ and $u = 1s^2 2s^2 2p 3s\,{}^3\text{D}_1^\text{o}$ levels of N II, with a wavelength $\lambda_{u\ell} = 1084.0$ Å. The transition has $A_{u\ell} = 2.18 \times 10^8\,\text{s}^{-1}$. This decay is very fast – the lifetime of the ${}^3\text{D}_1^\text{o}$ level against this decay is only $1/A_{u\ell} = 4.6$ ns!

6.7.2 Spin-Forbidden or Intersystem Transitions

These are transitions that fulfill the electric dipole selection rules 1 to 4 but have $\Delta S \neq 0$. These transitions are considerably weaker than allowed transitions. Such transitions are sometimes referred to as **semiforbidden**, or **intercombination**, or **intersystem transitions**; the latter is the terminology that we will use here. An intersystem transition is denoted with a single right bracket — for example,

$$\mathrm{N\,II]2143.4\,\AA\ {}^3P_2 - {}^5S_2^o} \quad ,$$

a transition between $\ell = 1s^2 2s^2 2p^2\,{}^3\mathrm{P}_2$ and $u = 1s^2 2s 2p^3\,{}^5\mathrm{S}_2^o$, with wavelength $\lambda_{u\ell} = 2143.4\,\text{Å}$ and $A_{u\ell} = 1.27 \times 10^2\,\mathrm{s}^{-1}$.

6.7.3 Forbidden Transitions

Forbidden transitions are those that fail to fulfill at least one of the selection rules 1 to 4. The transition probabilities vary widely, depending on the values of the **electric quadrupole** or **magnetic dipole** matrix elements between the upper and lower states. A forbidden transition is denoted with two square brackets — for example,

$$\mathrm{[N\,II]6549.9\,\AA\ {}^3P_1 - {}^1D_2} \quad ,$$

a transition between $\ell = 1s^2 2s^2 2p^2\,{}^3\mathrm{P}_1$ and $u = 1s^2 2s^2 2p^2\,{}^1\mathrm{D}_2$, with $\lambda_{u\ell} = 6549.9\,\text{Å}$ and $A_{u\ell} = 9.82 \times 10^{-4}\,\mathrm{s}^{-1}$. This fails rule 1 (parity is unchanged) *and* it fails rule 4 (single electron wave functions are unchanged). This is an example of a **magnetic dipole** transition.

Another example of a forbidden transition is the **electric quadrupole** transition

$$\mathrm{[N\,II]5756.2\,\AA\ {}^1D_2 - {}^1S_0} \quad ,$$

between $\ell = 1s^2 2s^2 2p^2\,{}^1\mathrm{D}_2$ and $u = 1s^2 2s^2 2p^2\,{}^1\mathrm{S}_0$, with $\lambda_{u\ell} = 5756.2\,\text{Å}$ and $A_{u\ell} = 1.17\,\mathrm{s}^{-1}$. This fails rules 1 (parity is unchanged) and 4 (single electron wave functions are unchanged) *and* it fails rules 2 and 3 ($\Delta L = -2$ and $\Delta J = -2$), yet its transition probability is three orders of magnitude larger than the magnetic dipole transition [N II]6549.9 Å!

We see then that there is a hierarchy in the transition probabilities: very roughly speaking, intersystem lines are $\sim 10^6$ times weaker than permitted transitions, and forbidden lines are $\sim 10^2 - 10^6$ times weaker than intersystem transitions.

Despite being very "weak," forbidden transitions are important in astrophysics for the simple reason that every atom and ion has excited states that can *only* decay via forbidden transitions. At high densities, such excited states would be depopulated by collisions, but at the very low densities of interstellar space, collisions are sufficiently infrequent that there is time for forbidden radiative transitions to take place.

Chapter Seven

Radiative Transfer

Almost all of what we know about the ISM is based on interpretation of electromagnetic radiation arriving at our telescopes. Chapters 4, 5, and 6 have been devoted to the elementary processes involved in absorption and emission of radiation by atoms, ions, and molecules. The subject of this chapter – **radiative transfer theory** – describes the propagation of radiation through absorbing and emitting media.

7.1 Physical Quantities

There are various ways to describe the strength of the radiation field at location $\vec{\mathbf{r}}$ and at time t:

- **Specific intensity $I_\nu(\nu)$:** The electromagnetic power per unit area, with frequencies in $[\nu, \nu + d\nu]$, propagating in directions $\hat{\mathbf{n}}$ within the solid angle $d\Omega$, is

$$I_\nu(\nu, \hat{\mathbf{n}}, \vec{\mathbf{r}}, t)\ d\nu\ d\Omega \quad . \tag{7.1}$$

  Unless otherwise noted, I_ν includes the power in both polarizations. If the radiation field is in local thermodynamic equilibrium (LTE), the intensity is equal to that of a blackbody:

$$(I_\nu)_{\rm LTE} \rightarrow B_\nu(T) \equiv \frac{2h\nu^3}{c^2} \frac{1}{\exp(h\nu/kT) - 1} \quad . \tag{7.2}$$

- **Photon occupation number $n_\gamma(\nu)$:**

$$n_\gamma(\nu, \hat{\mathbf{n}}, \vec{\mathbf{r}}, t) \equiv \frac{c^2}{2h\nu^3} I_\nu(\nu, \hat{\mathbf{n}}, \vec{\mathbf{r}}, t) \quad . \tag{7.3}$$

  The photon occupation number n_γ is dimensionless, and is simply the number of photons per mode per polarization. If the radiation field is in LTE (i.e., is a blackbody), then

$$(n_\gamma)_{\rm LTE} \rightarrow \frac{1}{\exp(h\nu/kT) - 1} \quad . \tag{7.4}$$

- **Brightness temperature $T_B(\nu)$:** The brightness temperature $T_B(\nu)$ is defined to be the temperature such that a blackbody at that temperature would have specific intensity $B_\nu(T_B) = I_\nu$:

$$T_B(\nu) \equiv \frac{h\nu/k}{\ln[1 + 2h\nu^3/c^2 I_\nu]} \quad . \tag{7.5}$$

In LTE, the brightness temperature T_B is equal to the actual thermodynamic temperature of the emitting and absorbing material.

- **Antenna temperature $T_A(\nu)$:** Brightness temperature has a simple thermodynamic interpretation, but has the disadvantage of being a nonlinear function of intensity I_ν. Radio astronomers have found it convenient to measure intensities in terms of the "antenna temperature:"

$$T_A(\nu) \equiv \frac{c^2}{2k\nu^2} I_\nu \quad . \tag{7.6}$$

The advantage of antenna temperature (as opposed to T_B) is that it is linear in the intensity. In the limit $kT_A \gg h\nu$ (or, equivalently, $n_\gamma \gg 1$), one sees that $T_A \approx T_B$. This is commonly the case at radio frequencies.

- **Specific energy density $u_\nu(\nu)$:**

$$u_\nu(\nu, \vec{\mathbf{r}}) = \frac{1}{c_m} \int I_\nu d\Omega \quad , \tag{7.7}$$

where c_m is the speed of light in the medium (in the ISM, the speed of light in vacuo). Note that as defined here, u_ν has no directional information: it is integrated over all directions and polarizations.

It is clear that four of these – the photon occupation number n_γ, the specific intensity I_ν, the brightness temperature T_B, and the antenna temperature T_A – are entirely equivalent; any one can be obtained from any other of the four. The photon occupation number n_γ has the appeal of being dimensionless, and we have already seen [Eq. (6.13)] that n_γ is equal to the **ratio of stimulated emission to spontaneous emission** from an excited state: the stimulated emission rate is just n_γ times the spontaneous emission rate. Masers and lasers normally have $n_\gamma \gg 1$ at the maser or laser frequency.

The preceding definitions all pertain to the strength of the radiation field. If the radiation field is modified by emission and absorption in a spectral line emitted or absorbed in transitions between levels u and ℓ of some species X, we can characterize the relative importance of emission and absorption using the "excitation temperature":

- **Excitation temperature $T_{\rm exc}$:** The excitation temperature of level u relative to level ℓ is defined by

$$\frac{n_u}{n_\ell} \equiv \frac{g_u}{g_\ell} {\rm e}^{-E_{u\ell}/kT_{{\rm exc},u\ell}} \quad \text{or} \quad T_{{\rm exc},u\ell} \equiv \frac{E_{u\ell}/k}{\ln\left(\frac{n_\ell/g_\ell}{n_u/g_u}\right)} , \tag{7.8}$$

  where n_u, n_ℓ are the populations of the upper and lower levels; g_u, g_ℓ are the degeneracies of the upper and lower levels; and $E_{u\ell} \equiv E_u - E_\ell$ is the difference in energy between the upper and lower levels.

7.2 Equation of Radiative Transfer

Consider now a beam of radiation with intensity I_ν entering a slab of material (see Figure 7.1). Let s measure pathlength along the direction of propagation. Let us neglect scattering and assume that the only processes affecting the intensity are absorption and emission. As the radiation propagates through the medium, the intensity evolves according to the equation of radiative transfer:

$$dI_\nu = -I_\nu \kappa_\nu ds + j_\nu ds \quad . \tag{7.9}$$

The term $-I_\nu \kappa_\nu ds$ is the net change in I_ν due to absorption and stimulated emission (both processes are linear in I_ν), and $j_\nu ds$ is the change in I_ν due to spontaneous emission by the material in the path of the beam:

- κ_ν is the **attenuation coefficient** at frequency ν, with dimensions of 1/length. While κ_ν is normally positive, we will see in §7.5 that it can become negative, in which case amplification takes place (i.e., masers or lasers).
- j_ν is the **emissivity** at frequency ν, with dimensions of power per unit volume per unit frequency per unit solid angle.

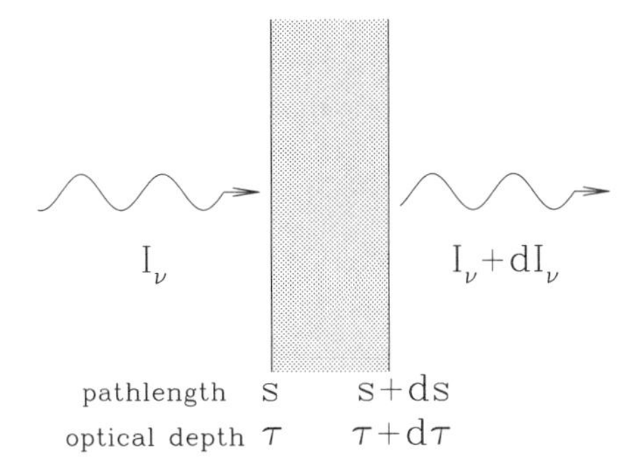

Figure 7.1 Radiative transfer geometry.

7.3 Emission and Absorption Coefficients

Absorption and emission can be due to atoms, ions, molecules, grains, thermal plasma, or a nonthermal population of energetic electrons. Atoms, ions, or molecules with discrete energy levels contribute to the absorption and emission coefficients as follows.

The emissivity j_ν was defined as the power radiated per unit frequency per unit solid angle per unit volume. For randomly oriented emitters:

$$j_\nu = \frac{1}{4\pi} n_u \, A_{u\ell} \, h\nu \, \phi_\nu \quad , \tag{7.10}$$

where $\phi_\nu d\nu$ is the probability that the emitted photon will have frequency in $(\nu, \nu + d\nu)$ (see §6.5). The attenuation coefficient κ_ν is proportional to the *net* absorption – i.e., true absorption minus stimulated emission:

$$\kappa_\nu = n_\ell \sigma_{\ell \to u}(\nu) - n_u \sigma_{u \to \ell}(\nu) \tag{7.11}$$

$$= n_\ell \sigma_{\ell \to u}(\nu) \left[1 - \frac{n_u / n_\ell}{g_u / g_\ell} \right] \tag{7.12}$$

$$= n_\ell \sigma_{\ell \to u}(\nu) \left[1 - \mathrm{e}^{-h\nu / kT_{\mathrm{exc}}} \right] , \tag{7.13}$$

where the absorption cross section $\sigma_{\ell \to u}(\nu)$ is given by Eq. (6.37) in the case where the absorbers have a Maxwellian velocity distribution, and $\sigma_{u \to \ell}(\nu) = (g_\ell / g_u) \sigma_{\ell \to u}(\nu)$ is the cross section for stimulated emission.

7.4 Integration of the Equation of Radiative Transfer

It is convenient to change independent variables from pathlength s to "optical depth" τ_ν, defined by

$$d\tau_\nu \equiv \kappa_\nu ds \, . \tag{7.14}$$

According to this definition,[1] the radiation propagates in the direction of increasing τ_ν (assuming $\kappa_\nu > 0$). The equation of radiative transfer now becomes just

$$dI_\nu = -I_\nu d\tau_\nu + S_\nu d\tau_\nu \quad , \tag{7.15}$$

where

$$S_\nu \equiv \frac{j_\nu}{\kappa_\nu} \tag{7.16}$$

[1]Note that some authors, including Spitzer (1978), adopt the opposite convention of radiation propagating in the direction of decreasing τ.

is referred to as the **source function**. It is important to remember that Eq. (7.15) does not include scattering processes, such as scattering by dust grains or by electrons.

We can formally integrate Eq. (7.15) by moving $I_\nu d\tau_\nu$ to the left hand side, and multipying by the "integrating factor" e^τ:

$$\begin{aligned} e^{\tau_\nu}(dI_\nu + I_\nu d\tau_\nu) &= e^{\tau_\nu} S_\nu d\tau_\nu \\ d(e^{\tau_\nu} I_\nu) &= e^{\tau_\nu} S_\nu d\tau_\nu \end{aligned} \tag{7.17}$$

We now integrate this from some starting point, which we define to be $\tau_\nu = 0$, with initial value $I_\nu(0)$:

$$e^{\tau_\nu} I_\nu - I_\nu(0) = \int_0^{\tau_\nu} e^{\tau'} S_\nu d\tau' \quad . \tag{7.18}$$

Now multiply by $e^{-\tau_\nu}$ to obtain the **equation of radiative transfer in integral form:**

$$I_\nu(\tau_\nu) = I_\nu(0) e^{-\tau_\nu} + \int_0^{\tau_\nu} e^{-(\tau_\nu - \tau')} S_\nu d\tau' \quad . \tag{7.19}$$

This integral equation, with S_ν allowed to be a function of position, is a fully general solution to the equation of radiative transfer (7.15) if scattering is neglected. Equation (7.19) has a simple physical interpretation: the intensity I_ν at optical depth τ_ν is just the initial intensity $I_\nu(0)$ attenuated by a factor $e^{-\tau_\nu}$, plus the integral over the emission $S_\nu d\tau'$ attenuated by the factor $e^{-(\tau_\nu - \tau')}$ due to the effective absorption over the path from the point of emission.

7.4.1 Special Case: Uniform Medium and Kirchhoff's Law

Consider now the special case of a slab of uniform medium with the matter having its energy levels populated according to a single excitation temperature $T_{\rm exc}$. We know that if the slab were infinite in extent, the radiation field within it would be equal to the blackbody radiation field with intensity $I_\nu = B_\nu(T_{\rm exc})$. Therefore, when $I_\nu = B_\nu$, Eq. (7.15) becomes

$$0 = dI_\nu = -B_\nu d\tau_\nu + S_\nu d\tau_\nu \quad , \tag{7.20}$$

which requires that $S_\nu = B_\nu(T_{\rm exc})$. Therefore, the emissivity j_ν and attenuation coefficient κ_ν must satisfy **Kirchhoff's Law**:

$$S_\nu \equiv \frac{j_\nu}{\kappa_\nu} = B_\nu(T_{\rm exc}) \quad . \tag{7.21}$$

Now j_ν and κ_ν depend only on the local properties of the matter: Kirchhoff's law

(7.21) must be true at each point. Without loss of generality, we can therefore rewrite the integral form of the equation of radiative transfer (7.19) as

$$I_\nu = I_\nu(0)\mathrm{e}^{-\tau_\nu} + \int_0^{\tau_\nu} \mathrm{e}^{-(\tau_\nu-\tau')} B_\nu(T_{\rm exc}) d\tau' \quad . \tag{7.22}$$

In the case of a uniform slab with $T_{\rm exc} = constant$, we can immediately integrate Eq. (7.22):

$$I_\nu = I_\nu(0)\mathrm{e}^{-\tau_\nu} + B_\nu(T_{\rm exc})\left(1 - \mathrm{e}^{-\tau_\nu}\right) \quad , \tag{7.23}$$

or, equivalently,

$$n_\gamma(\nu) = n_\gamma^{(0)}(\nu)\mathrm{e}^{-\tau_\nu} + \frac{1 - \mathrm{e}^{-\tau_\nu}}{(n_\ell g_u/n_u g_\ell) - 1} \quad . \tag{7.24}$$

If the matter is in LTE at temperature T, then $T_{\rm exc} = T$. However, under interstellar conditions, the populations of the excited states may depart from a thermal distribution. If emission and absorption at frequency ν are dominated by transitions between levels u and ℓ of a single species, then $T_{\rm exc}$ in Eq. (7.21) is equal to the excitation temperature $T_{{\rm exc},u\ell}$ defined in Eq. (7.8).

Because antenna temperature $T_A \propto I_\nu$, Eq. (7.23) can be rewritten as a transfer equation for T_A:

$$T_A = T_A(0)\mathrm{e}^{-\tau_\nu} + \frac{h\nu/k}{\exp(h\nu/kT_{\rm exc}) - 1}\left(1 - \mathrm{e}^{-\tau_\nu}\right) \tag{7.25}$$

$$\approx T_A(0)\mathrm{e}^{-\tau_\nu} + T_{\rm exc}\left(1 - \mathrm{e}^{-\tau_\nu}\right) \qquad \text{if} \quad \frac{h\nu}{kT_{\rm exc}} \ll 1 \quad . \tag{7.26}$$

7.4.2 Limiting Cases: Radio versus Optical

There are two limiting cases that are often encountered. At radio and sub-mm frequencies, the upper levels are often appreciably populated, and it is important to include both spontaneous emission from the medium [i.e., the term $B_\nu(T_{\rm exc})(1 - \mathrm{e}^{-\tau_\nu})$ in Eq. (7.23)], and stimulated emission [i.e., the term $\mathrm{e}^{-h\nu/kT_{\rm exc}}$ in the attenuation coefficient $\kappa_\nu = n_\ell \sigma_{\ell\to u}(1 - \mathrm{e}^{-h\nu/kT_{\rm exc}})$].

On the other hand, when we consider propagation of optical, ultraviolet, or X-ray radiation through cold interstellar clouds, the upper levels of the atoms and ions usually have negligible populations, and stimulated emission can be neglected: $\kappa_\nu \approx n_\ell \sigma_{\ell\to u}$.

7.5 Maser Lines

Under some conditions, a process may act to "pump" an excited state u by either collisional or radiative excitation of a higher level that then decays to populate level

u. (Radiative pumping will be discussed further in Chapter 20.) If this pumping process is rapid enough (relative to the processes that depopulate u), it may be possible for the *relative* level populations n_u, n_ℓ to satisfy the inequality

$$n_u > \frac{g_u}{g_\ell} n_\ell , \tag{7.27}$$

in which case, the excitation temperature

$$T_{\mathrm{exc},u\ell} < 0 \tag{7.28}$$

[see Eq. (7.8)], and we speak of a **population inversion**. When this occurs, stimulated emission is stronger than absorption, and the radiation is **amplified** as it propagates. The attenuation coefficient

$$\kappa_\nu = \sigma_{\ell \rightarrow u} \left(1 - \frac{n_u/g_u}{n_\ell/g_\ell} \right) < 0 \tag{7.29}$$

[see Eq. (7.12)], and the optical depth $\tau_\nu = \int \kappa_\nu ds < 0$; thus the factor $\mathrm{e}^{-\tau_\nu} > 1$, indicating amplification by stimulated emission rather than attenuation.

Such population inversions in astronomical sources have been observed for microwave transitions of H I, OH, CH, SiO, SiS, H_2O, HCN, NH_3, H_2CO, HC_3N, and CH_3OH, and hence we speak of **maser** (**m**icrowave **a**mplification by **s**timulated **e**mission of **r**adiation) emission. In principle, population inversions can occur for higher frequency transitions as well (e.g., Johansson & Letokhov 2005).

If we assume that $|kT_{\mathrm{exc},u\ell}| \gg h\nu$, then Eq. (7.26) applies:

$$\begin{aligned} T_A &\approx T_A(0)\mathrm{e}^{-\tau_\nu} + T_{\mathrm{exc}} \left(1 - \mathrm{e}^{-\tau_\nu}\right) && (7.30) \\ &= \left[T_A(0) + |T_{\mathrm{exc}}|\right] \mathrm{e}^{|\tau_\nu|} - |T_{\mathrm{exc}}| \quad . && (7.31) \end{aligned}$$

The factor $\mathrm{e}^{|\tau_\nu|}$ is in some cases very large – some OH and H_2O masers have been observed to have $T_A > 10^{11}$ K. Because $\mathrm{e}^{|\tau_\nu|} > 1$:

- $\mathrm{e}^{|\tau_\nu|}$ is more strongly peaked on the sky than $|\tau_\nu|$ — the angular size of the maser is less than the actual transverse dimension of the masing region.
- $\mathrm{e}^{|\tau_\nu|}$ is more strongly peaked in ν than $|\tau_\nu|$ — the maser line is *narrower* than the actual velocity distribution of the masing species.

Some masers can be very bright, allowing the use of interferometry, as well as observations of sources at large distances. This has enabled measurements of proper motions of maser spots in star-forming regions of the Milky Way, as well as in material orbiting a supermassive black hole in the spiral galaxy NGC 4258 (Herrnstein et al. 1999). Maser theory was reviewed by Elitzur (1992).

Chapter Eight

H I 21-cm Emission and Absorption

The ISM is composed primarily of hydrogen, and atomic hydrogen can be conveniently detected and studied via the 21-cm line, originating in the hyperfine splitting of the $1s$ electronic ground state of hydrogen. Studies of this line have allowed mapping of the distribution of H I in the Milky Way and other galaxies, determination of the galactic rotation curve, and measurement of the gas temperature in interstellar clouds. This chapter discusses 21-cm emission and absorption.

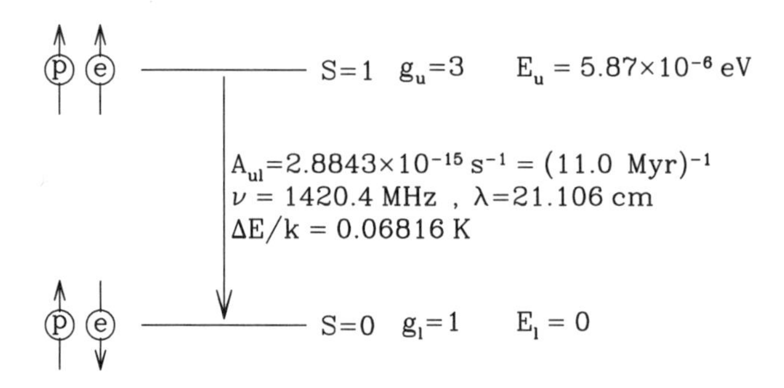

Figure 8.1 Hyperfine splitting of the $1s$ ground state of atomic H (Gould 1994).

8.1 H I Emissivity and Absorption Coefficient

The electron in the electronic ground state ($1s$) of atomic hydrogen can have its spin either parallel or antiparallel to the spin of the proton. The coupling of the electron's magnetic moment to the magnetic field produced by the magnetic moment of the proton results in "hyperfine splitting" of the parallel and antiparallel spin states. The antiparallel spin state (with degeneracy $g = 1$) has the lower energy, which we take to be $E_\ell = 0$. The parallel spin state (with total spin $S = 1$, and degeneracy $g = 2S + 1 = 3$) has an energy $E_u - E_\ell = 5.87 \times 10^{-6}$ eV. When the electron drops to the ground state, the electron spin flips, and a photon is emitted with a wavelength $\lambda = 21.11$ cm.

Because of the small energy splitting between these two spin states, the cosmic microwave background by itself is able to populate the upper level. For all conditions of interest (at least until the expansion factor of the universe increases by a factor 100), we expect the excitation temperature (or "spin temperature") $T_{\rm exc} \equiv T_{\rm spin} \gg .0682\,{\rm K}$, and

$$\frac{n_u}{n_\ell} \equiv \frac{g_u}{g_\ell} {\rm e}^{-h\nu_{u\ell}/kT_{\rm exc}} = 3\,{\rm e}^{-.0682\,{\rm K}/T_{\rm spin}} \approx 3 \quad , \tag{8.1}$$

$$n_u \approx \frac{3}{4} n({\rm H\,I}), \quad n_\ell \approx \frac{1}{4} n({\rm H\,I}) \quad . \tag{8.2}$$

Because the upper level contains $\sim 75\%$ of the H I under all conditions of interest, *the H I 21-cm emissivity is effectively independent of the spin temperature*:

$$j_\nu = n_u \frac{A_{u\ell}}{4\pi} h\nu_{u\ell}\, \phi_\nu \approx \frac{3}{16\pi} A_{u\ell}\, h\nu_{u\ell}\, n({\rm H\,I})\, \phi_\nu \quad , \tag{8.3}$$

where ϕ_ν is the normalized line profile ($\int \phi_\nu d\nu = 1$), determined by the velocity distribution of the H I.

The attenuation coefficient is

$$\kappa_\nu = n_\ell \sigma_{\ell u} - n_u |\sigma_{u\ell}| \tag{8.4}$$

$$= n_\ell \frac{g_u}{g_\ell} \frac{A_{u\ell}}{8\pi} \lambda_{u\ell}^2 \phi_\nu \left[1 - \frac{n_u}{n_\ell}\frac{g_\ell}{g_u}\right] \tag{8.5}$$

$$= n_\ell \frac{g_u}{g_\ell} \frac{A_{u\ell}}{8\pi} \lambda_{u\ell}^2 \phi_\nu \left[1 - {\rm e}^{-h\nu_{u\ell}/kT_{\rm spin}}\right] \quad . \tag{8.6}$$

Because ${\rm e}^{-h\nu_{u\ell}/kT_{\rm spin}} \approx 1$, *the correction for stimulated emission is very important!* Noting that $h\nu_{u\ell}/kT_{\rm spin} \ll 1$ for all conditions of interest, ${\rm e}^{-h\nu_{u\ell}/kT_{\rm spin}} \approx 1 - h\nu_{u\ell}/kT_{\rm spin}$, and

$$\kappa_\nu \approx \frac{3}{32\pi} A_{u\ell} \frac{hc\lambda_{u\ell}}{kT_{\rm spin}} n({\rm H\,I})\, \phi_\nu \quad . \tag{8.7}$$

Thus the attenuation coefficient $\kappa_\nu \propto 1/T_{\rm spin}$.

Suppose that the H I has a Gaussian velocity distribution. Then

$$\phi_\nu = \frac{1}{\sqrt{2\pi}} \frac{c}{\nu_{u\ell}} \frac{1}{\sigma_V} {\rm e}^{-u^2/2\sigma_V^2} \quad , \tag{8.8}$$

and the attenuation coefficient

$$\kappa_\nu = \frac{3}{32\pi} \frac{1}{\sqrt{2\pi}} \frac{A_{u\ell}\lambda_{u\ell}^2}{\sigma_V} \frac{hc}{kT_{\rm spin}} n({\rm H\,I}) {\rm e}^{-u^2/2\sigma_V^2} \tag{8.9}$$

$$= 2.190 \times 10^{-19}\,{\rm cm}^2 n({\rm H\,I}) \left(\frac{\rm K}{T_{\rm spin}}\right) \left(\frac{{\rm km\,s}^{-1}}{\sigma_V}\right) {\rm e}^{-u^2/2\sigma_V^2} \, , \tag{8.10}$$

$$\tau_\nu = 2.190 \left(\frac{N({\rm H\,I})}{10^{21}\,{\rm cm}^{-2}}\right) \left(\frac{100\,{\rm K}}{T_{\rm spin}}\right) \left(\frac{{\rm km\,s}^{-1}}{\sigma_V}\right) {\rm e}^{-u^2/2\sigma_V^2} \, , \tag{8.11}$$

$$N({\rm H\,I}) \equiv \int ds\, n({\rm H\,I}) \, , \tag{8.12}$$

where $N(\mathrm{H\,I})$ is the **column density** of H I. Many sightlines in the ISM have $N(\mathrm{H\,I}) \gtrsim 10^{21}\,\mathrm{cm}^{-2}$, spin temperatures $T_{\rm spin} \approx 10^2\,\mathrm{K}$, and σ_V of order $\mathrm{km\,s}^{-1}$: *self-absorption in the 21-cm line can be important!*

8.2 Optically Thin Cloud

Suppose that we have a cloud with

$$N(\mathrm{H\,I}) \lesssim 10^{20}\,\mathrm{cm}^{-2}\,\frac{T_{\rm spin}}{100\,\mathrm{K}}\,\frac{\sigma_V}{\mathrm{km\,s}^{-1}} \quad , \tag{8.13}$$

so that $\tau_\nu \lesssim 0.2$ even at the center of the line. If we now neglect absorption, then

$$I_\nu = I_\nu(0) + \int j_\nu ds \tag{8.14}$$

$$= I_\nu(0) + \frac{3}{16\pi} A_{u\ell}\, h\nu_{u\ell}\, \phi_\nu\, N(\mathrm{H\,I}) \quad . \tag{8.15}$$

Now suppose that $I_\nu(0)$ is known independently. We can then integrate the intensity over the line:

$$\int [I_\nu - I_\nu(0)]\, d\nu = \frac{3}{16\pi} A_{u\ell}\, h\nu_{u\ell}\, N(\mathrm{H\,I}) \quad . \tag{8.16}$$

This is often expressed in terms of antenna temperature T_A and relative velocity u, where $\nu = \nu_{u\ell}(1 - u/c)$, when u is the radial velocity:

$$\begin{aligned}\int [T_A - T_A(0)]\, du &= \int \frac{c^2}{2k\nu^2} [I_\nu - I_\nu(0)]\, \frac{c}{\nu} d\nu \\ &= \frac{3}{32\pi} \frac{hc\lambda_{u\ell}^2}{k} A_{u\ell}\, N(\mathrm{H\,I}) \\ &= 55.17\,\mathrm{K\,km\,s}^{-1} \frac{N(\mathrm{H\,I})}{10^{20}\,\mathrm{cm}^{-2}} \quad . \end{aligned} \tag{8.17}$$

Thus the intensity integrated over the line profile gives us the total H I column density without need to know $T_{\rm spin}$, provided that self-absorption is not important. Plate 3a is an all-sky map of H I 21-cm integrated line intensity, converted to $N(\mathrm{H\,I})$ assuming self-absorption to be negligible.

With the assumption that the emitting regions are optically thin, the total mass $M_{\rm H\,I}$ of H I in another galaxy can be determined from the observed flux in the 21-cm line, $F_{\rm obs} = \int F_\nu d\nu_{\rm obs}$, where F_ν is the observed **flux density** in the 21-cm line, which may be redshifted to a frequency $\nu_{\rm obs} = \nu_{u\ell}/(1+z)$, where z is the redshift of the emitting galaxy:

$$M_{\rm H\,I} = \frac{16\pi m_{\rm H}}{3A_{ul}h\nu_{u\ell}} D_L^2\, F_{\rm obs} \tag{8.18}$$

$$= 4.945 \times 10^7\, M_\odot \left(\frac{D_L}{\rm Mpc}\right)^2 \left(\frac{F_{\rm obs}}{\rm Jy\,MHz}\right), \tag{8.19}$$

where D_L is the **luminosity distance** to the source.[1] Radio astronomers often report the integrated flux in "Jy km s^{-1}," writing $d\nu_{\rm obs} = [\nu_{u\ell}/(1+z)]dv/c$:

$$M_{\rm H\,I} = \frac{16\pi m_{\rm H}}{3A_{ul}hc} D_L^2 \int F_\nu dv \tag{8.20}$$

$$= 2.343 \times 10^5\, M_\odot \frac{1}{(1+z)} \left(\frac{D_L}{\rm Mpc}\right)^2 \left(\frac{\int F_\nu dv}{\rm Jy\,km\,s^{-1}}\right), \tag{8.21}$$

where z is the source redshift.

8.3 Spin Temperature Determination Using Background Radio Sources

In cases where we have a bright background radio source with a continuum spectrum (a typical radio-loud quasar or an active galactic nucleus, or a radio galaxy), we can study both emission and absorption by the foreground interstellar medium in our galaxy by comparing "on-source" and "off-source" observations. The technique consists of taking on-source observations with the radio telescope pointed at the background radio source, and off-source observations with the telescope pointed so that the background source is out of the beam, and we are observing "blank sky" behind the foreground cloud. We assume that the foreground gas is sufficiently uniform so that the two different sightlines have essentially the same $N({\rm H\,I})$ and $T_{\rm spin}$.

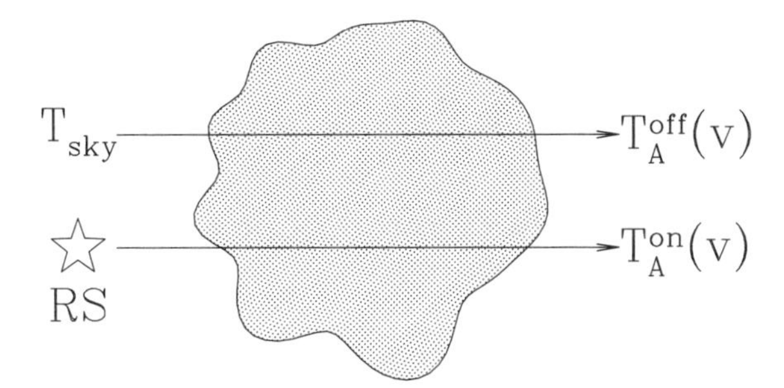

Figure 8.2 On-source and off-source observation geometry. RS is a radio source.

[1]For a nice discussion of distance measures in cosmology, see Hogg (1999).

We measure the on-source and off-source spectra $T_A^{\rm on}(v)$ and $T_A^{\rm off}(v)$, where we continue to report frequency (shifts) in terms of radial velocity. From the measurements at frequencies well above and below the 21-cm emission feature, we determine the antenna temperature $T_{\rm sky}$ of the blank sky, and the antenna temperature $T_{\rm RS}$ of the continuum from the radio source (averaged over the beam of our radio telescope).

At velocity v, the spectrum measured on the blank sky is [see Eq. (7.26)]

$$T_A^{\rm off}(v) = T_{\rm sky}{\rm e}^{-\tau_v} + T_{\rm spin}(1-{\rm e}^{-\tau_v}) \quad , \tag{8.22}$$

while the spectrum on the radio source is

$$T_A^{\rm on}(v) = T_{\rm RS}{\rm e}^{-\tau_v} + T_{\rm spin}(1-{\rm e}^{-\tau_v}) \quad . \tag{8.23}$$

These two equations can be solved for the two unknowns, $\tau(v)$ and $T_{\rm spin}(v)$:

$$\tau(v) = \ln\left[\frac{T_{\rm RS}-T_{\rm sky}}{T_A^{\rm on}(v)-T_A^{\rm off}(v)}\right] \quad , \tag{8.24}$$

$$T_{\rm spin}(v) = \frac{T_A^{\rm off}(v)T_{\rm RS}-T_A^{\rm on}(v)T_{\rm sky}}{(T_{\rm RS}-T_{\rm sky})-[T_A^{\rm on}(v)-T_A^{\rm off}(v)]} \quad . \tag{8.25}$$

When the absorption is strong, $(T_{\rm RS}-T_{\rm sky})$ is measurably larger than $[T_A^{\rm on}(v) - T_A^{\rm off}(v)]$, and both $\tau(v)$ and $T_{\rm spin}$ can be determined from the observations using Eqs. (8.24 and 8.25). When the absorption is weak, however, $(T_{\rm RS}-T_{\rm sky})$ and $[T_A^{\rm on}(v) - T_A^{\rm off}(v)]$ are the same within the errors, and the observations provide only an upper bound on $\tau(v)$, and a lower bound on $T_{\rm spin}(v)$. Surveys employing this technique to determine $T_{\rm spin}$ in H I clouds will be discussed in Chapter 29.

If the gas is optically thin [$\tau(v) \lesssim 0.1$], the H I column density per unit velocity can be directly obtained from the observed intensity:

$$\frac{dN({\rm H}\,I)}{dv} \approx \frac{32\pi}{3\lambda^2}\frac{k}{hcA_{u\ell}}\left[T_A^{\rm off}(v)-T_{\rm sky}\right]$$

$$= 1.813\,\frac{T_A^{\rm off}(v)-T_{\rm sky}}{\rm K} \times \frac{10^{18}\,{\rm cm}^{-2}}{{\rm km\,s}^{-1}} \quad , \tag{8.26}$$

whereas if $\tau(v) \gtrsim 0.1$ and we have measured $\tau(v)$ and $T_{\rm spin}(v)$ using Eqs. (8.24 and 8.25), we can correct for self-absorption:

$$\frac{dN({\rm HI})}{dv} = \frac{32\pi}{3\lambda^2}\frac{k}{hcA_{u\ell}}T_{\rm spin}(v)\tau(v) \tag{8.27}$$

$$= 1.813\,\frac{T_{\rm spin}(v)\tau(v)}{\rm K} \times \frac{10^{18}\,{\rm cm}^{-2}}{{\rm km\,s}^{-1}} \quad . \tag{8.28}$$

Chapter Nine

Absorption Lines: The Curve of Growth

The composition and excitation of interstellar gas can be studied using absorption lines that appear in the spectra of background stars (or other sources of continuum radiation). The interstellar lines are typically narrow compared to spectral features produced by absorption in stellar photospheres, and in practice can be readily distinguished. It is normally possible to detect absorption only by the ground state and perhaps the excited fine-structure levels of the ground electronic state – the populations in the excited electronic states are too small to be detected in absorption.

Observations of interstellar absorption lines allow the radial velocity, elemental composition, and the ionization state of the gas to be determined; if the ground electronic state has fine structure, the populations of the fine-structure levels can constrain the density and temperature.

The widths of absorption lines are usually determined by Doppler broadening, with linewidths of a few $\mathrm{km\,s^{-1}}$ – or $\Delta\lambda/\lambda \approx 10^{-5}$ – often observed in cool clouds. The spectrographs available to astronomers often lack the spectral resolution to resolve the profiles of such narrow lines, but can measure the total amount of "missing power" resulting from a narrow absorption line. The **equivalent width** W is a measure of the strength of an absorption line, in terms of "missing power" in the unresolved absorption line. The **curve of growth** refers to the function $W(N_\ell)$, showing how W depends on the column density N_ℓ of the absorber – in general, we want to invert this function, so that from an observed equivalent width W, we can infer the column density N_ℓ of the absorber.

9.1 Absorption Lines

Suppose that we observe a bright continuum source (e.g., a star) behind a foreground cloud of gas, and we measure the energy **flux density** F_ν as a function of ν. We observe the source using an aperture of solid angle $\Delta\Omega$, and we assume that the properties of the foreground gas are essentially uniform over $\Delta\Omega$. If we integrate Eq. (7.23) over the (small) solid angle $\Delta\Omega$ of our spectrometer aperture, we obtain the flux density at the observer

$$F_\nu = F_\nu(0)\mathrm{e}^{-\tau_\nu} + B_\nu(T_{\mathrm{exc}})\Delta\Omega\left(1 - \mathrm{e}^{-\tau_\nu}\right) \quad , \tag{9.1}$$

where $F_\nu(0)$ is the flux density from the source in the absence of absorption. At optical frequencies, we normally have $n_u/n_\ell \ll 1$, $B_\nu(T_{\rm exc})\Delta\Omega \ll F_\nu$, and we can neglect the emission from the ISM, thus obtaining the very simple result

$$F_\nu \approx F_\nu(0)e^{-\tau_\nu} \quad . \tag{9.2}$$

If absorption is negligible except in a few narrow spectral lines, then, if the background source spectrum is smooth, we can interpolate to estimate the stellar continuum flux $F_\nu(0)$ even at the frequencies where absorption is present, and we can evaluate the **dimensionless equivalent width**

$$W \equiv \int \frac{d\nu}{\nu_0}\left[1 - \frac{F_\nu}{F_\nu(0)}\right] = \int \frac{d\nu}{\nu_0}\left(1 - e^{-\tau_\nu}\right) \quad . \tag{9.3}$$

Some authors instead define the wavelength equivalent width

$$W_\lambda \equiv \int d\lambda \left(1 - e^{-\tau_\nu}\right) \approx \lambda_0 W \quad , \tag{9.4}$$

or the velocity equivalent width $W_v = cW$, but I prefer the dimensionless W.

The integrals in Eqs. (9.3 and 9.4) extend only over the absorption feature, with the integrand $\propto [1 - F_\lambda/F_\lambda(0)] \to 0$ on either side of the feature. It is important to note that W (or W_λ) can be measured even if the absorption line is not resolved: the equivalent width is just proportional to the *total missing power* that has been removed by the absorption line.

The optical depth in an absorption line can be written [see Eq. (6.24)]

$$\tau_\nu = \frac{\pi e^2}{m_e c} f_{\ell u} N_\ell \phi_\nu \left[1 - \frac{N_u/g_u}{N_\ell/g_\ell}\right] \quad , \tag{9.5}$$

where

$$N_\ell \equiv \int n_\ell ds \tag{9.6}$$

is the column density of the absorbers, ϕ_ν is the normalized line profile ($\int \phi_\nu d\nu = 1$), and the term $[1 - (N_u/g_u)/(N_\ell/g_\ell)]$ is the correction for stimulated emission. For interstellar optical absorption lines, $(N_u/g_u)/(N_\ell/g_\ell) \ll 1$, and the correction for stimulated emission can be disregarded.

Suppose that the absorbers have a Gaussian velocity distribution with 1-D velocity dispersion σ_V. Then, from Eq. (6.38), the optical depth near line-center

$$\tau_\nu = \tau_0\, e^{-(u/b)^2} \qquad b \equiv \sqrt{2}\,\sigma_V \ , \tag{9.7}$$

where u is the frequency shift from line-center expressed as a velocity, $u = c(\nu_0 - \nu)/\nu_0$ and the **optical depth at line-center** is

$$\tau_0 = \sqrt{\pi}\frac{e^2}{m_e c}\frac{N_\ell f_{\ell u}\lambda_{\ell u}}{b}\left[1 - \frac{N_u/g_u}{N_\ell/g_\ell}\right] \tag{9.8}$$

$$= 1.497 \times 10^{-2}\frac{\rm cm^2}{\rm s}\frac{N_\ell f_{\ell u}\lambda_{\ell u}}{b}\left[1 - \frac{N_u/g_u}{N_\ell/g_\ell}\right] \quad , \tag{9.9}$$

where the term $(N_u/g_u)/(N_\ell/g_\ell)$ is the correction for stimulated emission, which is usually negligible in the ISM or IGM except for radio frequency transitions. Dropping the correction for stimulated emission, Eq. (9.8) becomes

$$\tau_0 = 0.7580 \left(\frac{N_\ell}{10^{13}\,\mathrm{cm}^{-2}}\right) \left(\frac{f_{\ell u}}{0.4164}\right) \left(\frac{\lambda_{\ell u}}{1215.7\,\text{Å}}\right) \left(\frac{10\,\mathrm{km\,s}^{-1}}{b}\right) \quad , \tag{9.10}$$

where $f_{\ell u} = 0.4164$, $\lambda_{\ell u} = 1215.7$ Å are for H Lyman α.

If we assume that the absorbers have a Gaussian velocity distribution, the line profile shape has three distinct regimes, which we now examine.

9.2 Optically Thin Absorption, $\tau_0 \lesssim 1$

An example of an absorption line in the optically thin regime is shown in the upper panel of Fig. 9.1. If $\tau_\nu \ll 1$, we expand $(1 - \mathrm{e}^{-\tau}) \approx \tau - \tau^2/2 + ...$ and obtain

$$W \approx \sqrt{\pi}\, \frac{b}{c} \tau_0 \left(1 - \frac{\tau_0}{2\sqrt{2}} + ...\right) . \tag{9.11}$$

To obtain a well-behaved approximation formula, we replace $1 - (\tau_0/2\sqrt{2}) + ...$ by $1/[1 + (\tau_0/2\sqrt{2})]$ (which is the same to first order in τ_0) to obtain

$$W \approx \sqrt{\pi}\, \frac{b}{c} \frac{\tau_0}{1 + \tau_0/(2\sqrt{2})} \tag{9.12}$$

$$= \frac{\pi e^2}{m_e c^2} N_\ell f_{\ell u} \lambda_{\ell u} \frac{1}{1 + \tau_0/(2\sqrt{2})} \tag{9.13}$$

$$= 8.853 \times 10^{-13}\,\mathrm{cm}\, N_\ell f_{\ell u} \lambda_{\ell u} \frac{1}{1 + \tau_0/(2\sqrt{2})} \quad . \tag{9.14}$$

The approximation (9.12) is exact for $\tau_0 \to 0$, and is accurate to within 2.6% for $\tau_0 < 1.254$. In the optically thin regime ($\tau_0 \ll 1$), W depends only on the product $N_\ell f_{\ell u} \lambda_{\ell u}$. Hence,

$$N_\ell = 1.130 \times 10^{12}\,\mathrm{cm}^{-1} \frac{W}{f_{\ell u} \lambda_{\ell u}} \qquad \text{if } \tau_0 \ll 1 . \tag{9.15}$$

We see that even if we do not resolve the line profile, measurement of W allows us to determine N_ℓ, provided that the line is actually optically thin.[1]

9.3 Flat Portion of the Curve of Growth, $10 \lesssim \tau_0 \lesssim \tau_{\mathrm{damp}}$

In the linear regime discussed earlier, the quantity $[1 - \mathrm{e}^{-\tau_\nu}]$ has the same Gaussian profile as τ_ν itself. However, as the core of the line becomes saturated, the quantity

[1] If we are not certain whether the line is optically thin, then Eq. (9.15) gives a lower limit on N_ℓ.

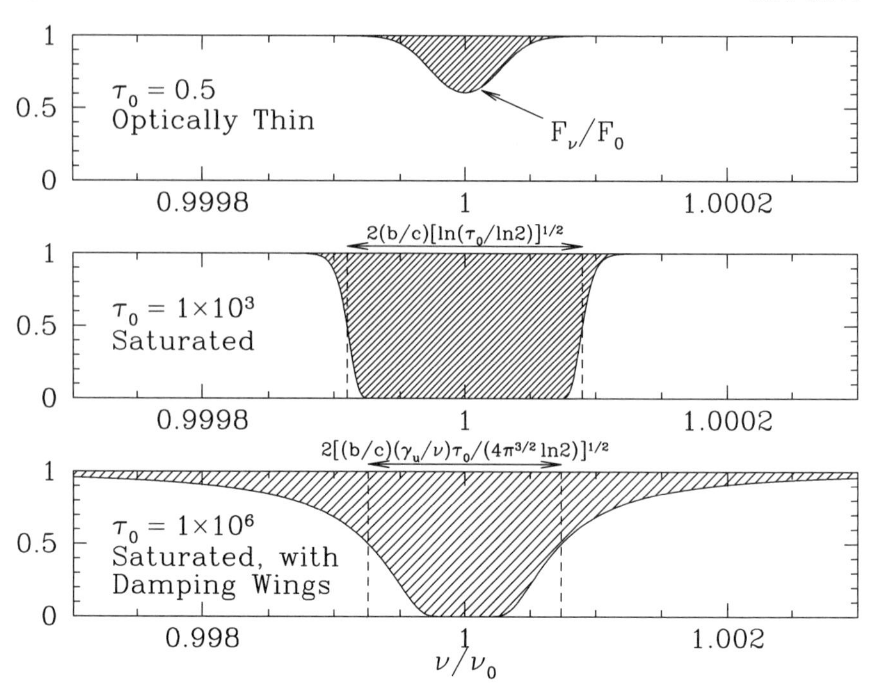

Figure 9.1 Absorption line profiles in three regimes, using as an example H Lyman α with $b = 10\,\mathrm{km\,s^{-1}}$ and $N(\mathrm{H\,I}) = 6.6 \times 10^{12}\,\mathrm{cm^{-2}}$, $1.3 \times 10^{16}\,\mathrm{cm^{-2}}$, and $1.3 \times 10^{19}\,\mathrm{cm^{-2}}$ in the upper, middle, and lower panels. Shaded area is proportional to the missing energy. Note the different abscissa in the lower panel.

$[1 - \mathrm{e}^{-\tau_\nu}]$ becomes increasingly "box-shaped." An example is shown in the middle panel of Fig. 9.1. Treating the opacity as a delta function broadened only by Doppler broadening:

$$\tau(\nu) = \tau_0\, \mathrm{e}^{-(u/b)^2} \quad , \tag{9.16}$$

we approximate W by the fractional full width at half maximum (FWHM):

$$W \approx \frac{(\Delta\nu)_{\mathrm{FWHM}}}{\nu_0} = \frac{(\Delta u)_{\mathrm{FWHM}}}{c} \approx \frac{2b}{c}\sqrt{\ln(\tau_0/\ln 2)} \quad . \tag{9.17}$$

The approximation (9.17) is accurate to within 5% for $1.254 < \tau_0 \lesssim \tau_{\mathrm{damp}}$, where τ_{damp}, the optical depth separating the "flat" from the "damped" regimes, is from Eq. (9.25), shown later.

Since τ_0 depends on $N_\ell f_{\ell u}\lambda_{\ell u}$ and b, we see that W depends on the product $N_\ell f_{\ell u}\lambda_{\ell u}$ and b. Note that W is *very* insensitive to τ_0 (and therefore N_ℓ) in this regime – it varies as the square root of the logarithm of τ_0. Because W increases so slowly with increasing N_ℓ, this is referred to as the **flat portion of the curve of**

growth. If we invert Eq. (9.17) to obtain N_ℓ, we obtain

$$N_\ell \approx \frac{\ln 2}{\sqrt{\pi}} \frac{m_e c}{e^2} \frac{b}{f_{\ell u}\lambda_{\ell u}} \exp\left[\left(\frac{cW}{2b}\right)^2\right] \tag{9.18}$$

$$= 46.29 \left(\frac{b}{f_{\ell u}\lambda_{\ell u}}\right) \exp\left[\left(\frac{cW}{2b}\right)^2\right] \mathrm{cm}^{-2}\,\mathrm{s} \quad . \tag{9.19}$$

Because of the insensitivity of W to N_ℓ — therefore *extreme* sensitivity of N_ℓ to W — Eq. (9.18) should *not* be used to determine N_ℓ unless:

1. We have a *very* accurate measurement of W.
2. We have a *very* accurate value for the Doppler broadening parameter b.
3. We have reason to believe that the velocity profile is *very* accurately described by a single Gaussian.

Unless *all three* conditions are satisfied, column densities estimated from W in this regime should be regarded as highly uncertain.

9.4 Damped Portion of the Curve of Growth, $\tau_0 \gtrsim \tau_{\mathbf{damp}}$

In this regime, the Doppler core of the line is totally saturated, but the "damping wings" of the line provide measurable partial transparency. See the lower panel of Fig. 9.1 for an example with $\tau_0 = 10^6$. In the limiting case, we can entirely neglect the Doppler broadening, and assume that the *wings* of the line are given by a pure Lorentz profile [see Eq. (6.26)]:

$$\tau_\nu \approx \frac{\pi e^2}{m_e c} N_\ell f_{\ell u} \frac{4\gamma_{\ell u}}{16\pi^2(\nu-\nu_0)^2 + \gamma_{\ell u}^2} \quad \text{for } |\nu-\nu_0| \gg \nu_0 \frac{b}{c} \,, \tag{9.20}$$

where $\gamma_{\ell u}$ is given by Eq. (6.28). The full width of the profile at 50% transmission is

$$\frac{(\Delta\lambda)_{\mathrm{FWHM}}}{\lambda} = \frac{(\Delta u)_{\mathrm{FWHM}}}{c} \approx \sqrt{\left(\frac{1}{\pi \ln 2}\right) \frac{e^2}{m_e c^2} N_\ell f_{\ell u}\lambda_{\ell u} \left(\frac{\gamma_{\ell u}}{\nu_{\ell u}}\right)} \,. \tag{9.21}$$

The equivalent width W is larger than the dimensionless FWHM (9.21) by just a factor $(\pi \ln 2)^{1/2} = 1.476$:

$$W = \sqrt{\frac{e^2}{m_e c^2} N_\ell f_{\ell u}\lambda_{\ell u} \left(\frac{\gamma_{\ell u}\lambda_{\ell u}}{c}\right)} = \sqrt{\frac{b}{c}\frac{\tau_0}{\sqrt{\pi}}\frac{\gamma_{\ell u}\lambda_{\ell u}}{c}} \quad . \tag{9.22}$$

In this regime, W depends on the product $N_\ell f_{\ell u}\lambda_{\ell u}$ and on the dimensionless ratio

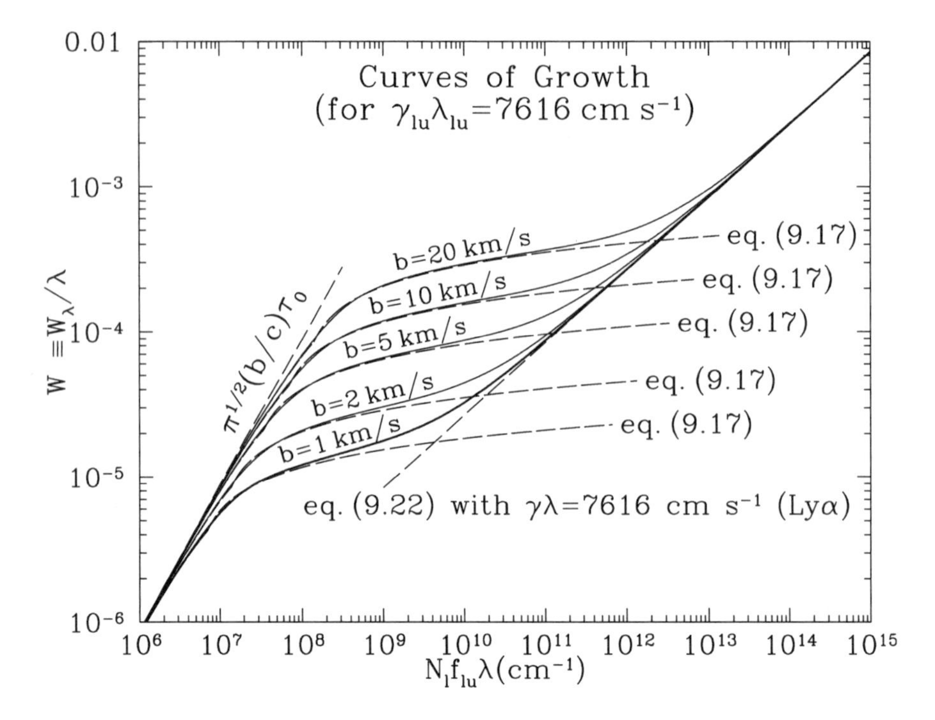

Figure 9.2 Curves of growth for five values of the Doppler broadening parameter b, for a damping constant $\gamma_{\ell u}\lambda_{\ell u} = 7616\,\mathrm{cm\,s^{-1}}$ (appropriate for H Lyman α). Dashed lines show the analytic approximations to the linear, flat, and square-root portions of the curve of growth. Only the flat portion [Eq. (9.17)] depends on b. Only the square-root portion [Eq. (9.22)] depends on $\gamma\lambda$.

$\gamma_{\ell u}\lambda_{\ell u}/c$. In Figure 9.2, we show W as a function of $N_\ell f_{\ell u}\lambda_{\ell u}$, for five values of b, where we have taken $\gamma_{\ell u}\lambda_{\ell u}$ appropriate for the Lyman α transition of atomic H. Because $W \propto \sqrt{N_\ell}$, this regime is often referred to as the **square-root part of the curve of growth**. The column density N_ℓ is given by

$$N_\ell = \frac{m_e c^3}{e^2}\frac{W^2}{f_{\ell u}\gamma_{\ell u}\lambda_{\ell u}^2} \tag{9.23}$$

$$= 2.759\times 10^{24}\,\mathrm{cm}^{-2}\,W^2 \left(\frac{0.4164}{f_{\ell u}}\right)\left(\frac{7616\,\mathrm{cm\,s^{-1}}}{\gamma_{\ell u}\lambda_{\ell u}}\right)\left(\frac{1215.7\,\text{Å}}{\lambda_{\ell u}}\right) \tag{9.24}$$

The transition from the flat to the damped portion of the curve of growth takes place near central optical depth $\tau_{\rm damp}$ obtained by equating[2] Eqs. (9.17) and (9.22):

$$\tau_{\rm damp} \approx 4\sqrt{\pi}\frac{b}{\gamma_{\ell u}\lambda_{\ell u}} \ln\left[\frac{4\sqrt{\pi}}{\ln 2}\frac{b}{\gamma_{\ell u}\lambda_{\ell u}}\right]$$

$$\approx 93\frac{b}{\rm km\,s^{-1}}\left(\frac{7616\,{\rm cm\,s^{-1}}}{\gamma_{\ell u}\lambda_{\ell u}}\right)\ln\left[134\frac{b}{\rm km\,s^{-1}}\frac{7616\,{\rm cm\,s^{-1}}}{\gamma_{\ell u}\lambda_{\ell u}}\right] \quad (9.25)$$

$$[N_\ell]_{\rm damp} \approx \frac{4m_e c}{e^2}\frac{b^2}{f_{\ell u}\gamma_{\ell u}\lambda_{\ell u}^2}\ln\left[\frac{4\sqrt{\pi}}{\ln 2}\frac{b}{\gamma_{\ell u}\lambda_{\ell u}}\right]$$

$$= 1.23\times 10^{14}\,{\rm cm^{-2}}\left(\frac{0.4164}{f_{\ell u}}\right)\left(\frac{b}{\rm km\,s^{-1}}\right)^2\left(\frac{7616\,{\rm cm\,s^{-1}}}{\gamma_{\ell u}\lambda_{\ell u}}\right)\times$$

$$\left(\frac{1215.7\,{\rm \AA}}{\lambda_{\ell u}}\right)\ln\left[134\frac{b}{\rm km\,s^{-1}}\frac{7616\,{\rm cm\,s^{-1}}}{\gamma_{\ell u}\lambda_{\ell u}}\right] \quad . \quad (9.26)$$

9.5 Approximation Formulae for *W*

In general, then, if we assume a Gaussian velocity profile, W depends on three quantities: b, $\gamma_{\ell u}\lambda_{\ell u}$, and $\tau_0 \propto N_\ell f_{\ell u}\lambda_{\ell u}$. Accurate evaluation of W in general requires numerical integration over the Voigt profile (6.37). Rodgers & Williams (1974) give useful approximation formulae for $W(\tau_0, b, \gamma_{\ell u}\lambda_{\ell u})$. Here we provide a simple approximation, based on the discussion in this chapter, that is continuous and accurate to a few percent:

$$W \approx \begin{cases} \sqrt{\pi}\,\frac{b}{c}\,\frac{\tau_0}{1+\tau_0/(2\sqrt{2})} & \text{for } \tau_0 < 1.25393\,, \\ \left[\left(\frac{2b}{c}\right)^2 \ln\left(\frac{\tau_0}{\ln 2}\right) + \frac{b}{c}\frac{\gamma_{\ell u}\lambda_{\ell u}}{c}\frac{(\tau_0-1.25393)}{\sqrt{\pi}}\right]^{1/2} & \text{for } \tau_0 > 1.25393\,, \end{cases} \quad (9.27)$$

where τ_0 is given by Eq. (9.8). This approximation is exact in the limits $\tau_0 \ll 1$ and $\tau_0 \gg b/\gamma_{\ell u}\lambda_{\ell u}$, and is accurate to within 5% for all τ_0 when applied to H I Lyman α with $b = 10\,{\rm km\,s^{-1}}$.

9.6 Doublet Ratio

In some cases, an absorbing level ℓ will have allowed transitions to two different excited states u_1 and u_2. Let λ_2 be the wavelength of the stronger transition (i.e.,

[2]One obtains a transcendental equation of the form $\tau_{\rm damp} = C\ln(\tau_{\rm damp}/\ln 2)$. For $C \gg 1$, $\tau_{\rm damp} \approx C\ln(C/\ln 2)$.

$f_{\ell u_2}\lambda_{\ell u_2} > f_{\ell u_1}\lambda_{\ell u_1}$). In the optically thin limit, the ratio of equivalent widths [from Eq. (9.12)] will be

$$\frac{W_2}{W_1} = \frac{f_{\ell u_2}\lambda_{\ell u_2}}{f_{\ell u_1}\lambda_{\ell u_1}} \quad . \tag{9.28}$$

Now consider the effects of increasing N_ℓ. The stronger line will become optically thick first, and W_2/W_1 will drop below the value in Eq. (9.28). If the measured equivalent width ratio W_2/W_1 is consistent with the optically thin limit (9.28), then one may confidently use (9.15) to calculate the absorbing column density N_ℓ.

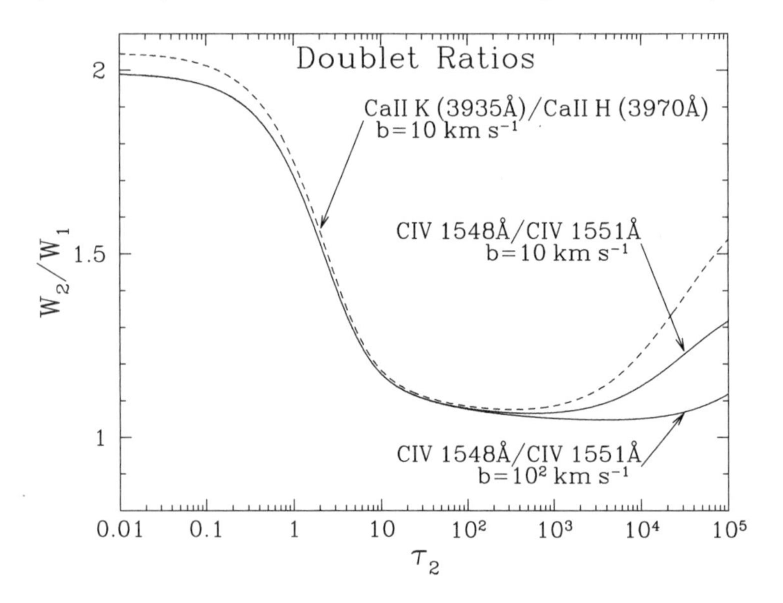

Figure 9.3 Ratio of equivalent widths for the Ca II H (3970Å) and K (3935Å) lines, and for the C IV 1548/1551 doublet, as a function of the line-center optical depth τ_2 for the stronger component of the doublet.

When both lines are on the flat portion of the curve of growth, the ratio will be approximately

$$\frac{W_2}{W_1} \approx \left[1 + \frac{\ln(f_{\ell u_2}\lambda_{\ell u_2}/f_{\ell u_1}\lambda_{\ell u_1})}{\ln(\tau_{\ell u_1}/\ln 2)}\right]^{1/2} \quad . \tag{9.29}$$

With very strong lines, e.g., Lyman α and Lyman β, we may enter the damped portion of the curve of growth, in which case the doublet ratio becomes

$$\frac{W_2}{W_1} \approx \frac{\lambda_{\ell u_2}}{\lambda_{\ell u_1}}\sqrt{\frac{f_{\ell u_2}\gamma_{\ell u_2}}{f_{\ell u_1}\gamma_{\ell u_1}}} \quad . \tag{9.30}$$

Fig. 9.3 shows W_2/W_1 as a function of the optical depth $\tau_{0,2}$ of the stronger transition for two frequently observed doublets – the Ca II H and K lines (3969.6 and 3934.8Å, respectively), and C IV 1548.2,1550.8 .

Table 9.1 Hydrogen Lyman Series

Transition	$\lambda(\text{Å})$	$f_{\ell u}$	$A_{u,tot}(\text{s}^{-1})$
Ly $\alpha(1s-2p)$	1215.67	0.4164	6.265×10^8
Ly $\beta(1s-3p)$	1025.73	0.07912	1.672×10^8
Ly $\gamma(1s-4p)$	972.54	0.02901	6.818×10^7
Ly $\delta(1s-5p)$	949.74	0.01394	3.437×10^7
Ly $\epsilon(1s-6p)$	937.80	0.007799	1.973×10^7
Ly $\zeta(1s-7p)$	930.74	0.004184	1.074×10^7
Ly $\eta(1s-8p)$	926.22	0.003183	8.249×10^6
Ly $\theta(1s-9p)$	923.15	0.002216	5.781×10^6
Ly $\iota(1s-10p)$	920.96	0.001605	4.209×10^6
Ly $(1s-np)$, $n \gg 1$	$\frac{911.75}{1-n^{-2}}$	$1.563n^{-3}$	$4.180 \times 10^9 n^{-3}$

9.7 Lyman Series of Hydrogen: Ly α, Ly β, Ly γ, ...

In the ISM, the atomic hydrogen is found almost entirely in the electronic ground state $1s$. From this level, hydrogen has allowed transitions: $1s \rightarrow 2p$, $1s \rightarrow 3p$, $1s \rightarrow 4p$, and so on. These are referred to as the "Lyman series": Lyman α, Lyman β, Lyman γ, and so on. Lyman α has a rest wavelength $\lambda = 1215.67\,\text{Å}$.

Both the $^2\text{P}^\text{o}_{3/2}$ and $^2\text{P}^\text{o}_{1/2}$ levels have $A_{u\ell} = 6.265 \times 10^8\,\text{s}^{-1}$ for transitions to the ground state $^2\text{S}_{1/2}$, giving an intrinsic linewidth $(\Delta v)^{\text{intr.}}_{\text{FWHM}} = \lambda A_{u\ell}/2\pi = 0.0121\,\text{km}\,\text{s}^{-1}$. This intrinsic linewidth is negligible compared to the FWHM of a thermal velocity distribution

$$(\Delta v)^{\text{therm}}_{\text{FWHM}} = 2\sqrt{2\ln 2}\left(\frac{kT}{M}\right)^{1/2} = 2.15\left(\frac{T/100\,\text{K}}{M/m_\text{H}}\right)^{1/2}\,\text{km}\,\text{s}^{-1}\,, \qquad (9.31)$$

for all gas temperatures of interest.

The $2p$ excited state is actually a doublet $^2\text{P}^\text{o}_{1/2,3/2}$, so there are really two Lyman α lines: $^2\text{S}_{1/2} - {}^2\text{P}^\text{o}_{1/2}$ 1215.674 Å, and $^2\text{S}_{1/2} - {}^2\text{P}^\text{o}_{3/2}$ 1215.668 Å. However, the fine-structure splitting of these levels is extremely small,[3] corresponding to a velocity shift of only $\Delta v = 1.33\,\text{km}\,\text{s}^{-1}$ between the two components of Lyman α. Therefore, in the ISM or IGM, the two lines are always blended by the thermal broadening (9.31). Because the splitting is so small, and because the two excited states have the same $A_{u\ell}$, the two lines can be treated together as a single $1s \rightarrow 2p$ transition.

The Lyman α absorption line is of great importance because it allows us to directly measure the column density of atomic hydrogen. Figure 9.2 shows the curve of growth for Lyman α, for five different values of the Doppler broadening parameter b.

Almost all hydrogen consists of the normal isotope, with atomic weight 1, but of order 20 ppm is deuterium, with atomic weight 2.

[3] $[E(^2\text{P}^\text{o}_{3/2}) - E(^2\text{P}^\text{o}_{1/2})]/hc = 0.366\,\text{cm}^{-1}$.

Deuterium has essentially the same energy levels as the hydrogen atom, the principal difference being that the binding energy of all the energy levels is slightly greater because the reduced mass of the electron-deuteron system is greater than that of the electron–proton system by a factor $(1+m_e/m_p)/(1+m_e/m_d)$. where $m_d \approx 2m_p$ is the mass of the deuteron.[4] The shift in energy levels causes all of the electronic transitions in deuterium to be at slightly higher energy than the corresponding transitions in hydrogen, and the small amount of deuterium in a cloud therefore appears like an equal number of hydrogen atoms but with an apparent blue shift $c(m_e/m_p - m_e/m_d) \approx 82\,\mathrm{km\,s^{-1}}$. Under suitable conditions, it is possible to measure the strength of both the hydrogen lines and the deuterium lines, and thereby determine the D/H ratio in the gas.

9.8 Lyman Limit

As $n \to \infty$, the Lyman series absorption lines converge to a limiting wavelength $\lambda = 911.75\,\text{Å}$ known as the "Lyman limit." The interval between line centers goes to zero as $n \to \infty$. Because the lines have finite width, they will therefore blend together to form a continuum before reaching this limit. The series of lines is essentially indistinguishable from a smooth continuum when

$$\frac{d\lambda_n}{dn} < \frac{v_{\rm FWHM}}{c}\lambda_n \, , \tag{9.32}$$

which is satisfied for quantum number

$$n^{-3} \lesssim \frac{v_{\rm FWHM}}{2c} \quad , \tag{9.33}$$

or

$$n \gtrsim 67 \left(\frac{2\,\mathrm{km\,s^{-1}}}{v_{\rm FWHM}}\right)^{1/3} \quad , \tag{9.34}$$

corresponding to a wavelength

$$\lambda \approx 911.75\,\text{Å} + 0.20 \left(\frac{v_{\rm FWHM}}{2\,\mathrm{km\,s^{-1}}}\right)^{2/3} \text{Å} \, , \tag{9.35}$$

or energy $h\nu/\mathrm{eV} = 13.599 - 0.003(v_{\rm FWHM}/2\,\mathrm{km\,s^{-1}})^2/3$. When the Lyman series absorption lines blend together in this way, the resulting continuum absorption has an absorption cross section per H, $\sigma \approx 6.3 \times 10^{-18}\,\mathrm{cm^2}$, the same as the photoionization cross section for hydrogen at energies just above the photoionization threshold $13.60\,\mathrm{eV}$.

[4]The deuteron has a smaller magnetic moment than a proton. This changes the frequency of the "spin flip" transition from 1420 MHz to 327.4 MHz, but has a negligible effect on the optical and ultraviolet transitions.

Table 9.2 Absorption Lines out of $H_2(v=0, J=1)$ with $\lambda > 970$ Å

Transition	λ(Å)	$f_{\ell u}$	Transition	λ(Å)	$f_{\ell u}$
Ly 0–0 P(1)	1110.07	0.0005775	Ly 7–0 P(1)	1014.33	0.008982
Ly 0–0 R(1)	1108.64	0.001082	Ly 7–0 R(1)	1013.43	0.02045
Ly 1–0 P(1)	1094.05	0.001982	Wer 0–0 Q(1)	1009.77	0.02377
Ly 1–0 R(1)	1092.73	0.003798	Wer 0–0 R(1)	1008.50	0.02058
Ly 2–0 P(1)	1078.92	0.003942	Ly 8–0 P(1)	1003.29	0.008461
Ly 2–0 R(1)	1077.69	0.007716	Ly 8–0 R(1)	1002.45	0.01818
Ly 3–0 P(1)	1064.60	0.005964	Ly 9–0 P(1)	992.81	0.00768
Ly 3–0 R(1)	1063.46	0.01192	Ly 9–0 R(1)	992.01	0.01817
Ly 4–0 P(1)	1051.03	0.007624	Wer 1–0 Q(1)	986.80	0.03645
Ly 4–0 R(1)	1049.96	0.01557	Ly 10–0 P(1)	982.83	0.00679
Ly 5–0 P(1)	1038.16	0.008682	Ly 10–0 R(1)	982.07	0.01358
Ly 5–0 R(1)	1037.15	0.01813	Wer 1–0 R(1)	985.64	0.03313
Ly 6–0 P(1)	1025.93	0.009105	Ly 11–0 P(1)	973.34	0.005875
Ly 6–0 R(1)	1024.99	0.01955	Ly 11–0 R(1)	972.63	0.01395

9.9 H_2: Lyman and Werner Bands

The energy level structure of H_2 was reviewed in Chapter 5. Each vibration–rotation level (v, J) of the electronic ground state $X\,^1\Sigma_g^+$ of the hydrogen molecule H_2 has a large number of permitted electric-dipole transitions to vibration–rotation levels (v_u, J_u) of electronic excited states. The first electronic excited state $B\,^1\Sigma_u^+$ and the second electronic excited state $C\,^1\Pi_u$ have permitted transitions to the lowest (v, J) levels of the electronic ground state at wavelengths $\lambda > 912$ Å, so that these lines can be observed in neutral gas where radiation shortward of the Lyman limit is totally absorbed. Transitions between $X\,^1\Sigma_g^+$ and $B\,^1\Sigma_u^+$ are referred to as **Lyman band transitions**, and transitions between $X\,^1\Sigma_g^+$ and $C\,^1\Pi_u$ are **Werner band transitions.**[5]

Table 9.2 lists the permitted transitions with $\lambda > 970$ Å out of a single vibration–rotation level, $(v=0, J=1)$, of the electronic ground state. The lines have oscillator strengths ranging from 0.00058 for Ly 0-0P(1) to 0.024 for Wer 0-0Q(1) – a factor of 40. The fact that a single lower level has a large number of permitted absorption lines allows the lower-level population to be determined using intrinsically weak transitions if the stronger transitions are on the flat portion of the curve of growth.

Because of the numerous lines and the range of oscillator strengths, observation of the rich absorption spectrum of H_2 can yield the column densities of H_2 from the $J=0$ and $J=1$ rotational levels (that generally account for the bulk of the H_2) to rotational levels as high as $J=5$ or 6 that may contain only $\sim 10^{-4}$ of the H_2. The rotational distribution can tell us much about the physical conditions in the region

[5]The second excited state $C\,^1\Pi_u$ is actually split into two sets of energy states, sometimes denoted C^+ and C^-, with nearly identical energies. The Q branch ($\Delta J = 0$) Werner band transitions are between $X\,^1\Sigma_g^+$ and the C^- levels, and the R and P branch ($\Delta J = \pm 1$) Werner band transitions are between $X\,^1\Sigma_g^+$ and C^+.

where the H_2 is located – see Chapter 31 for further details.

9.10 "Metal" Lines

Following astronomical usage, all elements heavier than He are here referred to as "metals" (including elements such as N, O, and Ne!!). With the exception of the two lightest inert gases He and Ne, all other neutral atoms have **resonance lines** (that is, permitted absorption lines out of the electronic ground state) with $\lambda > 912$ Å, allowing them to be observed on sightlines where neutral hydrogen is present. For most atoms and ions, the resonance lines all fall shortward of ~ 3000 Å. For ground-based observatories, the Earth's atmosphere is effectively opaque at wavelength $\lambda < 3000$ Å, so that $\lambda < 3000$ Å transitions can be observed only from space or in redshifted systems.

A small number of atoms and ions do have resonance lines at wavelengths $\lambda >$ 3000 Å, allowing them to be observed from the ground. Table 9.3 lists selected resonance lines with $\lambda > 3000$ Å.

Some of the lines in Table 9.3 are routinely observed, in particular the Na I D doublet at 5891.6, 5897.6 Å, the K I doublet at 7667.0, 7701.1 Å, and the Ca II doublet at 3934.8, 3969.6 Å.

Because Na I and K I are not the dominant ion stage for these elements, one cannot reliably estimate the total amount of H or H_2 associated with Na I or K I absorption lines, but these lines do serve as evidence that some gas is present at the

Table 9.3 Selected[a] Resonance Lines[b] with $\lambda > 3300$ Å, $f_{\ell u} > 0.015$

	Configurations	ℓ	u	$E_\ell/hc\,(\mathrm{cm}^{-1})$	$\lambda_{\rm vac}$(Å)	$f_{\ell u}$	Note
Na I	$2p^63s - 2p^63p$	$^2S_{1/2}$	$^2P^o_{3/2}$	0	5891.582	0.641	Na D_2
		$^2S_{1/2}$	$^2P^o_{1/2}$	0	5897.558	0.320	Na D_1
Al I	$3s^23p - 3s^24s$	$^2P^o_{1/2}$	$^2S_{1/2}$	0	3945.122	0.115	
		$^2P^o_{3/2}$	$^2S_{3/2}$	112.06	3962.641	0.12	
K I	$3p^64s - 3p^64p$	$^2S_{1/2}$	$^2P^o_{3/2}$	0	7667.01	0.682	
		$^2S_{1/2}$	$^2P^o_{1/2}$	0	7701.08	0.340	
Ca I	$3p^64s^2 - 3p^64s4p$	$^1S_0$	$^1P^o_1$	0	4227.918	1.750	
Ca II	$3p^64s - 3p^64p$	$^2S_{1/2}$	$^2P^o_{3/2}$	0	3934.77	0.682	Ca II K
		$^2S_{1/2}$	$^2P^o_{1/2}$	0	3969.59	0.33	Ca II H
Ti I	$3d^24s^2 - 3d^24s4p$	$^3F_2$	z $^3F^o_2$	0	5175.19	0.0153	
		$^3F_2$	y $^3D^o_1$	0	3949.79	0.0681	
		$^3F_2$	$?^o_2$	0	3930.99	0.0174	
		$^3F_2$	y $^3G^o_3$	0	3636.50	0.223	
		$^3F_2$	w $^3D^o_1$	0	3371.40	0.078	
		$^3F_2$	w $^3D^o_2$	0	3359.24	0.13	
		$^3F_2$	x $^3G^o_3$	0	3342.84	0.15	
		$^3F_3$	y $^3F^o_3$	170.13	3990.88	0.0905	
		$^3F_3$	$?^o_2$	170.13	3957.46	0.0503	
		$^3F_3$	z $^3P^o_2$	170.13	3948.89	0.016	

Table 9.3 contd.

	configurations	ℓ	u	$E_\ell/hc\,(\mathrm{cm}^{-1})$	$\lambda_{\mathrm{vac}}(\mathrm{\AA})$	$f_{\ell u}$	Note
		$^3\mathrm{F}_3$	$?^{\mathrm{o}}_3$	170.13	3925.64	0.0165	
		$^3\mathrm{F}_3$	$\mathrm{x}\,^3\mathrm{F}^{\mathrm{o}}_3$	170.13	3742.12	0.0875	
		$^3\mathrm{F}_3$	$\mathrm{y}\,^3\mathrm{G}^{\mathrm{o}}_4$	170.13	3643.71	0.198	
		$^3\mathrm{F}_3$	$\mathrm{w}\,^3\mathrm{D}^{\mathrm{o}}_2$	170.13	3378.55	0.084	
		$^3\mathrm{F}_3$	$\mathrm{x}\,^3\mathrm{G}^{\mathrm{o}}_4$	170.13	3355.60	0.15	
		$^3\mathrm{F}_4$	$\mathrm{y}\,^3\mathrm{F}^{\mathrm{o}}_4$	386.87	3999.77	0.0979	
		$^3\mathrm{F}_4$	$?^{\mathrm{o}}_3$	386.87	3959.33	0.0740	
		$^3\mathrm{F}_4$	$\mathrm{y}\,^3\mathrm{G}^{\mathrm{o}}_5$	386.87	3654.54	0.185	
		$^3\mathrm{F}_4$	$?^{\mathrm{o}}_3$	386.87	3386.92	0.067	
		$^3\mathrm{F}_4$	$\mathrm{x}\,^3\mathrm{G}^{\mathrm{o}}_5$	386.87	3372.42	0.15	
Ti II	$3d^24s-3d^24p$	$^4\mathrm{F}_{3/2}$	$\mathrm{z}\,^4\mathrm{G}^{\mathrm{o}}_{5/2}$	0	3384.74	0.281	
		$^4\mathrm{F}_{5/2}$	$\mathrm{z}\,^4\mathrm{G}^{\mathrm{o}}_{7/2}$	94.1	3373.77	0.253	
		$^4\mathrm{F}_{7/2}$	$\mathrm{z}\,^4\mathrm{G}^{\mathrm{o}}_{9/2}$	225.73	3362.18	0.23	
Cr I	$3d^54s-3d^54p$	$\mathrm{a}\,^7\mathrm{S}_3$	$\mathrm{z}\,^7\mathrm{P}^{\mathrm{o}}_2$	0	4290.93	0.0623	
		$\mathrm{a}\,^7\mathrm{S}_3$	$\mathrm{z}\,^7\mathrm{P}^{\mathrm{o}}_3$	0	4276.00	0.0842	
		$\mathrm{a}\,^7\mathrm{S}_3$	$\mathrm{z}\,^7\mathrm{P}^{\mathrm{o}}_4$	0	4255.55	0.110	
	$3d^54s-3d^44s4p$	$\mathrm{a}\,^7\mathrm{S}_3$	$\mathrm{y}\,^7\mathrm{P}^{\mathrm{o}}_2$	0	3606.36	0.226	
		$\mathrm{a}\,^7\mathrm{S}_3$	$\mathrm{y}\,^7\mathrm{P}^{\mathrm{o}}_3$	0	3594.52	0.291	
		$\mathrm{a}\,^7\mathrm{S}_3$	$\mathrm{y}\,^7\mathrm{P}^{\mathrm{o}}_4$	0	3579.71	0.366	
Mn I	$3d^54s^2-3d^54s4p$	$\mathrm{a}\,^6\mathrm{S}_{5/2}$	$\mathrm{z}\,^6\mathrm{P}^{\mathrm{o}}_{3/2}$	0	4035.62	0.0257	
		$\mathrm{a}\,^6\mathrm{S}_{5/2}$	$\mathrm{z}\,^6\mathrm{P}^{\mathrm{o}}_{5/2}$	0	4034.21	0.0403	
		$\mathrm{a}\,^6\mathrm{S}_{5/2}$	$\mathrm{z}\,^6\mathrm{P}^{\mathrm{o}}_{7/2}$	0	4031.90	0.055	
Fe I	$3d^64s^2-3d^64s4p$	$^5\mathrm{D}_4$	$\mathrm{z}\,^5\mathrm{D}^{\mathrm{o}}_4$	0	3861.005	0.0217	
	$3d^64s^2-3d^64s4p$	$^5\mathrm{D}_4$	$\mathrm{z}\,^5\mathrm{F}^{\mathrm{o}}_5$	0	3720.993	0.0411	
Co I	$3d^74s^2-3d^74s4p$	$\mathrm{a}^4\mathrm{F}_{9/2}$	$\mathrm{z}^4\mathrm{F}^{\mathrm{o}}_{9/2}$	0	3527.86	0.024	
		$\mathrm{a}^4\mathrm{F}_{9/2}$	$\mathrm{z}^4\mathrm{G}^{\mathrm{o}}_{11/2}$	0	3466.79	0.020	
		$\mathrm{a}^4\mathrm{F}_{9/2}$	$\mathrm{z}^4\mathrm{D}^{\mathrm{o}}_{7/2}$	0	3413.61	0.017	
		$\mathrm{a}^4\mathrm{F}_{9/2}$	$\mathrm{z}^4\mathrm{F}^{\mathrm{o}}_{7/2}$	816.00	3576.38	0.018	
		$\mathrm{a}^4\mathrm{F}_{9/2}$	$\mathrm{z}^4\mathrm{G}^{\mathrm{o}}_{11/2}$	816.00	3514.48	0.018	
Ni I	$3d^84s^2-3d^84s4p$	$^3\mathrm{F}_4$	$?^{\mathrm{o}}_3$	0	3370.54	0.024	
	$3d^94s-3d^94p$	$^3\mathrm{D}_3$	$^3\mathrm{P}^{\mathrm{o}}_2$	204.787	3525.55	0.13	
	$3d^94s-3d^84s4p$	$^3\mathrm{D}_3$	$^3\mathrm{F}^{\mathrm{o}}_4$	204.787	3462.64	0.062	
	$3d^94s-3d^94p$	$^3\mathrm{D}_3$	$^3\mathrm{F}^{\mathrm{o}}_3$	204.787	3434.54	0.030	
		$^3\mathrm{D}_3$	$^3\mathrm{F}^{\mathrm{o}}_4$	204.787	3415.74	0.12	
	$3d^94s-3d^94p$	$^3\mathrm{D}_3$	$?^{\mathrm{o}}_3$	204.787	3393.96	0.041	
	$3d^94s-3d^94p$	$^3\mathrm{D}_2$	$^3\mathrm{F}^{\mathrm{o}}_3$	879.816	3516.06	0.11	
	$3d^94s-3d^94p$	$^3\mathrm{D}_2$	$^3\mathrm{P}^{\mathrm{o}}_1$	879.816	3493.96	0.11	
	$3d^94s-3d^94p$	$^3\mathrm{D}_2$	$?^{\mathrm{o}}_3$	879.816	3473.53	0.03	
	$3d^94s-3d^84s4p$	$^3\mathrm{D}_2$	$^5\mathrm{F}^{\mathrm{o}}_3$	879.816	3453.88	0.025	
	$3d^94s-3d^94p$	$^3\mathrm{D}_2$	$^3\mathrm{D}^{\mathrm{o}}_2$	879.816	3447.25	0.078	

[a] Limited to elements with abundance $(X/\mathrm{H})_\odot > 8\times10^{-8}$

[b] Transition data from NIST Atomic Spectra Database v4.0.0 (Ralchenko et al. 2010)

radial velocity of the Na I or K I lines.

Ca II is a more interesting case, because under some circumstances it may be the dominant ion stage. Ca I, if at all abundant, can be observed with a strong transition at 4228 Å. However, the energy required to photoionize Ca II→Ca III is only 11.87 eV (see Appendix D), and therefore, even in an H I region, some of the

Ca will be in the form of Ca III, which is unobservable. But from the Ca I/Ca II ratio the ionization conditions can be characterized, and the amount of Ca III estimated, allowing the total gas-phase column density of Ca to be estimated. Unfortunately, an unknown (but usually large) fraction of the Ca is generally locked up in dust grains (this will be discussed in Chapter 23), and therefore from the Ca I and Ca II observations alone, one cannot reliably estimate the total amount of H associated with the observed Ca II absorption.

Another interesting case is Ti, where Ti I and Ti II, the two dominant ion stages for Ti in an H I cloud, both have resonance lines in the optical, allowing the total column of gas-phase Ti to be determined from ground-based observations. However, Ti also shares with Ca the problem that a large, but unknown, fraction of the Ti is generally locked up in dust.

Most of the abundant atoms and ions, with a few exceptions (e.g., He, Ne, O II) have permitted absorption lines in the vacuum ultraviolet with wavelengths longward of 912 Å so that they will not photoionize hydrogen. Table 9.4 lists selected resonance lines with 912 Å $< \lambda <$ 3000 Å.

Table 9.4 Selected Resonance Lines[a] with $\lambda <$ 3000 Å

	Configurations	ℓ	u	$E_\ell/hc(\mathrm{cm}^{-1})$	λ_{vac}(Å)	$f_{\ell u}$
C IV	$1s^22s - 1s^22p$	$^2\mathrm{S}_{1/2}$	$^2\mathrm{P}^\mathrm{o}_{1/2}$	0	1550.772	0.0962
		$^2\mathrm{S}_{1/2}$	$^2\mathrm{P}^\mathrm{o}_{3/2}$	0	1548.202	0.190
N V	$1s^22s - 1s^22p$	$^2\mathrm{S}_{1/2}$	$^2\mathrm{P}^\mathrm{o}_{1/2}$	0	1242.804	0.0780
		$^2\mathrm{S}_{1/2}$	$^2\mathrm{P}^\mathrm{o}_{3/2}$	0	1242.821	0.156
O VI	$1s^22s - 1s^22p$	$^2\mathrm{S}_{1/2}$	$^2\mathrm{P}^\mathrm{o}_{1/2}$	0	1037.613	0.066
		$^2\mathrm{S}_{1/2}$	$^2\mathrm{P}^\mathrm{o}_{3/2}$	0	1037.921	0.133
C III	$2s^2 - 2s2p$	$^1\mathrm{S}_0$	$^1\mathrm{P}^\mathrm{o}_1$	0	977.02	0.7586
C II	$2s^22p - 2s2p^2$	$^2\mathrm{P}^\mathrm{o}_{1/2}$	$^2\mathrm{D}_{3/2}$	0	1334.532	0.127
		$^2\mathrm{P}^\mathrm{o}_{3/2}$	$^2\mathrm{D}_{5/2}$	63.42	1335.708	0.114
N III	$2s^22p - 2s2p^2$	$^2\mathrm{P}^\mathrm{o}_{1/2}$	$^2\mathrm{D}_{3/2}$	0	989.790	0.123
		$^2\mathrm{P}^\mathrm{o}_{3/2}$	$^2\mathrm{D}_{5/2}$	174.4	991.577	0.110
C I	$2s^22p^2 - 2s^22p3s$	$^3\mathrm{P}_0$	$^3\mathrm{P}^\mathrm{o}_1$	0	1656.928	0.140
		$^3\mathrm{P}_1$	$^3\mathrm{P}^\mathrm{o}_2$	16.40	1656.267	0.0588
		$^3\mathrm{P}_2$	$^3\mathrm{P}^\mathrm{o}_2$	43.40	1657.008	0.104
N II	$2s^22p^2 - 2s2p^3$	$^3\mathrm{P}_0$	$^3\mathrm{D}^\mathrm{o}_1$	0	1083.990	0.115
		$^3\mathrm{P}_1$	$^3\mathrm{D}^\mathrm{o}_2$	48.7	1084.580	0.0861
		$^3\mathrm{P}_2$	$^3\mathrm{D}^\mathrm{o}_3$	130.8	1085.701	0.0957
N I	$2s^22p^3 - 2s^22p^23s$	$^4\mathrm{S}^\mathrm{o}_{3/2}$	$^4\mathrm{P}_{5/2}$	0	1199.550	0.130
		$^4\mathrm{S}^\mathrm{o}_{3/2}$	$^4\mathrm{P}_{3/2}$	0	1200.223	0.0862
O I	$2s^22p^4 - 2s^22p^33s$	$^3\mathrm{P}_2$	$^3\mathrm{S}^\mathrm{o}_1$	0	1302.168	0.0520
		$^3\mathrm{P}_1$	$^3\mathrm{S}^\mathrm{o}_1$	158.265	1304.858	0.0518
		$^3\mathrm{P}_0$	$^3\mathrm{S}^\mathrm{o}_1$	226.977	1306.029	0.0519
Mg II	$2p^63s - 2p^63p$	$^2\mathrm{S}_{1/2}$	$^2\mathrm{P}^\mathrm{o}_{1/2}$	0	2803.531	0.303
		$^2\mathrm{S}_{1/2}$	$^2\mathrm{P}^\mathrm{o}_{3/2}$	0	2796.352	0.608
Al III	$2p^63s - 2p^63p$	$^2\mathrm{S}_{1/2}$	$^2\mathrm{P}^\mathrm{o}_{1/2}$	0	1862.790	0.277
		$^2\mathrm{S}_{1/2}$	$^2\mathrm{P}^\mathrm{o}_{3/2}$	0	1854.716	0.557

Table 9.4 contd.

	Configurations	ℓ	u	$E_\ell/hc\ (\mathrm{cm}^{-1})$	$\lambda_{\mathrm{vac}}(\mathrm{\AA})$	$f_{\ell u}$
Mg I	$2p^63s^2 - 2p^63s3p$	$^1\mathrm{S}_0$	$^1\mathrm{P}^\mathrm{o}_1$	0	2852.964	1.80
Al II	$2p^63s^2 - 2p^63s3p$	$^1\mathrm{S}_0$	$^1\mathrm{P}^\mathrm{o}_1$	0	1670.787	1.83
Si III	$2p^63s^2 - 2p^63s3p$	$^1\mathrm{S}_0$	$^1\mathrm{P}^\mathrm{o}_1$	0	1206.51	1.67
P IV	$2p^63s^2 - 2p^63s3p$	$^1\mathrm{S}_0$	$^1\mathrm{P}^\mathrm{o}_1$	0	950.655	1.60
Si II	$3s^23p - 3s^24s$	$^2\mathrm{P}^\mathrm{o}_{1/2}$	$^2\mathrm{S}_{1/2}$	0	1526.72	0.133
		$^2\mathrm{P}^\mathrm{o}_{3/2}$	$^2\mathrm{S}_{1/2}$	287.24	1533.45	0.133
P III	$3s^23p - 3s3p^2$	$^2\mathrm{P}^\mathrm{o}_{1/2}$	$^2\mathrm{D}_{3/2}$	0	1334.808	0.029
		$^2\mathrm{P}^\mathrm{o}_{3/2}$	$^2\mathrm{D}_{5/2}$	559.14	1344.327	0.026
Si I	$3s^23p^2 - 3s^23p4s$	$^3\mathrm{P}_0$	$^3\mathrm{P}^\mathrm{o}_1$	0	2515.08	0.17
		$^3\mathrm{P}_1$	$^3\mathrm{P}^\mathrm{o}_2$	77.115	2507.652	0.0732
		$^3\mathrm{P}_2$	$^3\mathrm{P}^\mathrm{o}_2$	223.157	2516.870	0.115
P II	$3s^23p^2 - 3s3p^3$	$^3\mathrm{P}_0$	$^3\mathrm{P}^\mathrm{o}_1$	0	1301.87	0.038
		$^3\mathrm{P}_1$	$^3\mathrm{P}^\mathrm{o}_2$	164.9	1305.48	0.016
		$^3\mathrm{P}_2$	$^3\mathrm{P}^\mathrm{o}_2$	469.12	1310.70	0.115
S III	$3s^23p^2 - 3s3p^3$	$^3\mathrm{P}_0$	$^3\mathrm{D}^\mathrm{o}_1$	0	1190.206	0.61
		$^3\mathrm{P}_1$	$^3\mathrm{D}^\mathrm{o}_2$	298.69	1194.061	0.46
		$^3\mathrm{P}_2$	$^3\mathrm{D}^\mathrm{o}_3$	833.08	1200.07	0.51
Cl IV	$3s^23p^2 - 3s3p^3$	$^3\mathrm{P}_0$	$^3\mathrm{D}^\mathrm{o}_1$	0	973.21	0.55
		$^3\mathrm{P}_1$	$^3\mathrm{D}^\mathrm{o}_2$	492.0	977.56	0.41
		$^3\mathrm{P}_2$	$^3\mathrm{D}^\mathrm{o}_3$	1341.9	984.95	0.47
P I	$3s^23p^3 - 3s^23p^24s$	$^4\mathrm{S}^\mathrm{o}_{3/2}$	$^4\mathrm{P}_{5/2}$	0	1774.951	0.154
S II	$3s^23p^3 - 3s^23p^24s$	$^4\mathrm{S}^\mathrm{o}_{3/2}$	$^4\mathrm{P}_{5/2}$	0	1259.518	0.12
Cl III	$3s^23p^3 - 3s^23p^24s$	$^4\mathrm{S}^\mathrm{o}_{3/2}$	$^4\mathrm{P}_{5/2}$	0	1015.019	0.58
S I	$3s^23p^4 - 3s^23p^34s$	$^3\mathrm{P}_2$	$^3\mathrm{S}^\mathrm{o}_1$	0	1807.311	0.11
		$^3\mathrm{P}_1$	$^3\mathrm{S}^\mathrm{o}_1$	396.055	1820.343	0.11
		$^3\mathrm{P}_0$	$^3\mathrm{S}^\mathrm{o}_1$	573.640	1826.245	0.11
Cl II	$3s^23p^4 - 3s3p^5$	$^3\mathrm{P}_2$	$^3\mathrm{P}^\mathrm{o}_2$	0	1071.036	0.014
		$^3\mathrm{P}_1$	$^3\mathrm{P}^\mathrm{o}_2$	696.00	1079.080	0.00793
		$^3\mathrm{P}_0$	$^3\mathrm{P}^\mathrm{o}_1$	996.47	1075.230	0.019
Cl I	$3s^23p^5 - 3s^23p^44s$	$^2\mathrm{P}^\mathrm{o}_{3/2}$	$^2\mathrm{P}_{3/2}$	0	1347.240	0.114
		$^2\mathrm{P}^\mathrm{o}_{1/2}$	$^2\mathrm{P}_{3/2}$	882.352	1351.657	0.0885
Ar II	$3s^23p^5 - 3s3p^6$	$^2\mathrm{P}^\mathrm{o}_{3/2}$	$^2\mathrm{S}_{1/2}$	0	919.781	0.0089
		$^2\mathrm{P}^\mathrm{o}_{1/2}$	$^2\mathrm{S}_{1/2}$	1431.583	932.054	0.0087
Ar I	$3p^6 - 3p^54s$	$^1\mathrm{S}_0$	$^2[1/2]^\mathrm{o}$	0	1048.220	0.25

[a] Transition data from NIST Atomic Spectra Database v4.0.0 (Ralchenko et al. 2010)

Note that the ultraviolet resonance lines allow detection of many different ionization stages of a given element: a prime example is carbon, which can be detected as C I, C II, C III, or C IV. On the other hand, $\lambda > 912\,\mathrm{\AA}$ ultraviolet absorption spectroscopy cannot detect neon at all, while oxygen can be detected via permitted ultraviolet absorption lines only if either neutral or 5-times ionized.

9.11 Abundances in H I Gas

The gas phase abundances of many elements relative to H have been determined on many different sightlines using interstellar absorption lines. Such studies require that the dominant ionization states be measured. This requirement prevents determination of the abundances of some elements, such as Ne or Na, where the dominant ionization state (Ne^0 or Na^+) lacks resonance lines with $\lambda > 912$ Å.

The observed gas-phase abundances vary from one sightline to another, which is presumed to reflect primarily variations in the amounts of various elements trapped in dust grains. Such removal of elements from the gas is known as **interstellar depletion**. Some elements, like Fe, are extremely underabundant in the gas phase, with gas-phase abundances that are typically only a few percent of the solar abundance, shown in Table 1.4.

Jenkins (2009) analyzed absorption line measurements on many sightlines. Jenkins found that the abundances of different elements can be empirically reproduced to fair accuracy by a fitting formula:

$$\log_{10}(X/\mathrm{H})_{\mathrm{gas}} = \log_{10}(X/\mathrm{H})_{\odot} + C_X + A_X F_\star \quad , \tag{9.36}$$

where C_X and $A_X < 0$ are constants, and $F_\star$ is a parameter that varies from one region to another, scaled so that the highest abundance regions have $F_\star \approx 0$, and low-abundance diffuse regions have $F_\star \approx 1$. Sightlines with $F_\star \gtrsim 0.8$ often have more than trace of amounts of H_2 present.

As was discussed in §1.1, neutral H in the ISM is thought to occur in several distinct "phases," with different levels of depletion of the elements. In Table 9.5, we use the depletion parameters A_X and C_X found by Jenkins to estimate the abundances of selected elements in different phases of the ISM.

Interstellar H I is often classified as "warm neutral medium" (WNM), "cold neutral medium" (CNM), and "diffuse H_2." We estimate that WNM material typically has $F_\star \approx 0.1$, and CNM material typically has $F_\star \approx 0.4$. We include abundance estimates for the "warm ionized medium" (WIM), for which we take $F_\star = -0.1$, and for diffuse H_2, for which we take $F_\star \approx 0.8$.

Table 9.5 lists the gas-phase abundances estimated from Eq. (9.36) using A_X and C_X values obtained from Jenkins (2009), with two adjustments. For carbon, Sofia & Parvathi (2010) find that the oscillator strength for C II] 2325Å – used for many of the published measurements of the dominant ionization state C II – was too low by a factor of $\sim$2; hence we have lowered Jenkins's values for C_{C} by $\log_{10} 2$. For sulfur, we use Jenkins's recommended values for C_{S} and A_{S} to estimate the gas-phase abundances for the CNM and diffuse H_2, but for the WIM and WNM, where Jenkins's fit would predict sulfur abundances in excess of solar, we assume solar abundances.

With the exception of He, Ne, Na, Al, and Ca, Table 9.5 includes abundance estimates for all elements with $(X/H)_\odot > 10^{-6}$, and additionally includes Ti, which has a low abundance but is an example of an element that shows extreme levels of depletion.

Table 9.5 Gas-Phase Abundances (ppm) Relative to H of Selected Elements in H I Regions

Element	Solar[a]	WIM	WNM	CNM	Diffuse H_2
		$F_\star = -0.1$	$F_\star = 0.1$	$F_\star = 0.4$	$F_\star = 0.8$
C [b]	295.	114.	111.	109.	93.
N	74.	62.	62.	62.	62.
O	537.	592.	534.	457.	372.
Na	2.04	(2.)	(2.)	(2.)	(2.)
Mg	43.7	28.1	17.8	8.9	3.6
Al	2.95	(0.54)	(0.27)	(0.097)	(0.025)
Si	35.5	31.6	18.7	8.5	3.0
S	14.5	14.5	14.5	11.8	5.3
Ca	2.14	(0.39)	(0.20)	(0.070)	(0.018)
Ti	0.089	0.013	0.0052	0.0013	0.0002
Fe	34.7	5.2	2.9	1.19	0.36
Ni	1.74	0.32	0.16	0.057	0.015
M^+ [c]	432.	197.	168.	142.	107.

[a] From Table 1.4.
[b] Gas-phase C abundance from Jenkins (2009) reduced by factor 2 (see text).
[c] Photoionizable "metals": $M = \mathrm{C} + \mathrm{Na} + \mathrm{Mg} + \mathrm{Si} + \mathrm{S} + \mathrm{Fe} + 3.9 \times \mathrm{Ni}$.

As we will see in Chapter 16, elements with ionization potentials $I < 13.6\,\mathrm{eV}$ tend to be photoionized in H I clouds, and can make an important contribution to the electron density in cool H I gas. The major electron contributors are C, Na, Mg, Al, Si, S, Ca, Fe, and Ni. The depletion of Na is not well-determined, but circumstantial evidence suggests that it does not strongly deplete (Weingartner & Draine 2001*b*). Jenkins has not determined the depletion parameters for Al and Ca; we provisionally assume that these have the same depletion as Ni (on the well-studied sightline to ζOph, the depletion of Al is similar to Ni.) Thus, we take the gas phase abundance of "metals" with $I_X < I_\mathrm{H}$ to be $M/\mathrm{H} = \mathrm{C/H} + (\mathrm{Na/H})_\odot + \mathrm{Mg/H} + \mathrm{Si/H} + \mathrm{S/H} + \mathrm{Fe/H} + 3.9 \times \mathrm{Ni/H}$, where the factor 3.9 allows for Al and Ca. The resulting estimates for the abundance of photoionizable metals M^+ are given in Table 9.5.

Chapter Ten

Emission and Absorption by a Thermal Plasma

A **thermal plasma** is a partially ionized gas where the particle velocity distributions are very close to Maxwellian distributions. Such plasmas are ubiquitous in interstellar and intergalactic space, with temperatures ranging from $\sim 10^3$ K to $\sim 10^8$ K. There are three important emission processes in thermal plasmas:

1. **Free–free** transitions, where the emitting electron is inelastically scattered from one free state to another, with emission of a continuum photon.

2. **Free–bound** transitions, where the emitting electron is initially free, but is captured into a bound state, with emission of a continuum photon.

3. **Bound-bound** transitions, where the emitting electron makes a transition from one bound state to another, with the difference in energy carried away by one or two photons.

In this chapter, we describe the first two of these processes. Free–free emission and free–bound emission together produce a characteristic spectrum that, when measured, can be used to estimate the temperature of the plasma. Bound-bound transitions will be discussed in Chapters 14 and 17.

10.1 Free–Free Emission (Bremsstrahlung)

The electrons and ions in a thermal plasma scatter off one another. In classical electromagnetism, an accelerating charge radiates electromagnetic energy; therefore, we expect these scattering events to produce electromagnetic radiation. The electrons, with their much smaller mass, undergo much larger accelerations and therefore dominate the radiation.[1]

The emission is a continuum extending from very low (radio) frequencies, up to frequencies where the emitted photon energies are comparable to the thermal energy kT. A classical analysis of the power radiated by electrons scattered by ions with charge $Z_i e$ leads one to write the **free–free emissivity** (power radiated

[1]In classical electromagnetism, the power radiated by an accelerated charge q is proportional to $|q\mathbf{a}|^2$, where $\mathbf{a}$ is the instantaneous acceleration; hence, in an electron–proton plasma, the power radiated by the protons is smaller by a factor $(m_e/m_p)^2 = 3 \times 10^{-7}$ than that radiated by the electrons.

per unit frequency, per unit volume, per steradian) as

$$j_{\mathrm{ff},\nu} = \frac{8}{3}\left(\frac{2\pi}{3}\right)^{1/2} g_{\mathrm{ff},i}\,\frac{e^6}{m_e^2c^3}\left(\frac{m_e}{kT}\right)^{1/2} \mathrm{e}^{-h\nu/kT} n_e Z_i^2 n_i \tag{10.1}$$

$$= 5.444\times10^{-41} g_{\mathrm{ff}} T_4^{-1/2} \mathrm{e}^{-h\nu/kT} Z_i^2 n_i n_e \,\mathrm{erg\,cm^3\,s^{-1}\,sr^{-1}\,Hz^{-1}}, \tag{10.2}$$

where the dimensionless factor $g_{\mathrm{ff},i}(\nu, T)$ is called the **Gaunt factor for free–free transitions**. A classical treatment (Kramers 1923) gives $g_{\mathrm{ff}} = 1$. The actual Gaunt factor $g_{\mathrm{ff}}(\nu, T)$ differs from unity because of quantum effects. Note that if the Gaunt factor were a constant, the emissivity $j_{\mathrm{ff},\nu}$ [Eq. (10.1)] would be independent of frequency ν – i.e., "flat" – at frequencies $\nu \ll kT/h$.

The power radiated per volume by free–free emission is

$$\Lambda_{\mathrm{ff}} = 4\pi\int_0^\infty j_{\mathrm{ff},\nu}d\nu = \frac{32\pi}{3}\left(\frac{2\pi}{3}\right)^{1/2}\frac{e^6}{m_e^2hc^3}(m_ekT)^{1/2}\langle g_{\mathrm{ff}}\rangle_T Z_i^2 n_i n_e\,, \tag{10.3}$$

where

$$\langle g_{\mathrm{ff}}\rangle_T \equiv \int_0^\infty \frac{dh\nu}{kT}\mathrm{e}^{-h\nu/kT} g_{\mathrm{ff}}(\nu, T) \tag{10.4}$$

is the frequency-averaged Gaunt factor. Because $\langle g_{\mathrm{ff}}\rangle_T$ is almost independent of T, the radiated power $\Lambda_{\mathrm{ff}} \propto n_i n_e T^{0.5}$.

10.2★ Gaunt Factor

The Gaunt factor takes into account quantum-mechanical effects. At very low frequencies, approaching the "plasma frequency" $\nu_p = 8.98(n_e/\,\mathrm{cm}^{-3})^{1/2}\,\mathrm{kHz}$ (see §11.1), collective effects in the plasma become important, but for frequencies $\nu \gg \nu_p$, we have just a two-body Coulomb scattering problem. For frequencies $\nu_p \ll \nu \ll kT/h$, the Gaunt factor from quantum-mechanical calculations is approximately

$$g_{\mathrm{ff}} \approx \frac{\sqrt{3}}{\pi}\left[\ln\frac{(2kT)^{3/2}}{\pi Z_i e^2 m_e^{1/2}\nu} - \frac{5\gamma}{2}\right] \tag{10.5}$$

$$= 4.691\left[1 - 0.118\ln\left(\frac{Z_i\nu_9/10}{T_4^{3/2}}\right)\right] \tag{10.6}$$

$$\approx 6.155\,(Z_i\nu_9)^{-0.118}\,T_4^{0.177}\quad, \tag{10.7}$$

where $\gamma = 0.577216\ldots$ is Euler's constant, $\nu_9 \equiv \nu/\,\mathrm{GHz} = 30\,\mathrm{cm}/\lambda$, and $T_4 \equiv$

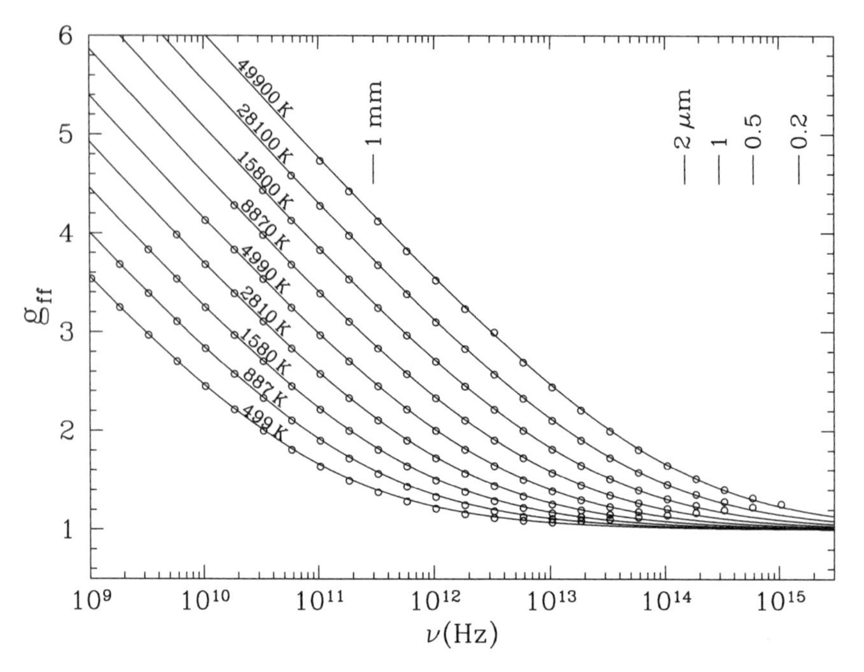

Figure 10.1 Gaunt factor $g_{\rm ff}(\nu)$ for electron-proton free–free transitions. Curves (labeled by temperature T) are Eq. (10.9). The points are numerical results from Hummer (1988), showing that Eq. (10.9) is accurate to $\sim 1\%$.

$T/10^4$ K. The power-law approximation (10.7) is accurate to $\pm 10\%$ for $0.14 < Z_i \nu_9 / T_4^{3/2} < 250$; for frequencies ν satisfying this condition [and the conditions $\nu_p \ll \nu \ll kT/h$ required by approximation (10.5)], the free–free emissivity

$$j_{{\rm ff},\nu} \approx 3.35 \times 10^{-40} Z_i^{1.882} n_i n_e \nu_9^{-0.118} T_4^{-0.323} \, {\rm erg\, cm^3\, s^{-1}\, sr^{-1}\, Hz^{-1}} \quad . \quad (10.8)$$

Thus the radio and microwave free–free emission spectrum is almost flat, declining with increasing frequency as $\sim \nu^{-0.12}$.

At higher frequencies, the Gaunt factor has been calculated numerically; values of $g_{\rm ff}(\nu)$ calculated by Hummer (1988) are shown in Figure 10.1 for nine values of T. The numerical results can be approximated by

$$g_{\rm ff}(\nu, T) \approx \ln \left\{ \exp \left[5.960 - \frac{\sqrt{3}}{\pi} \ln \left(Z_i \nu_9 T_4^{-3/2} \right) \right] + {\rm e} \right\} \quad , \qquad (10.9)$$

where e $= 2.71828....$ This analytic approximation, shown in Figure 10.1, is accurate to within $\sim 1\%$ over a wide range of temperatures and frequencies.

For $T_4 \approx 1$, the numerical results give $g_{\rm ff} \approx 1$ for $\nu \gtrsim 10^{14}$ Hz. With the emissivity $j_\nu \propto g_{\rm ff}(\nu) {\rm e}^{-h\nu/kT}$, it follows that most of the free–free power is near $h\nu \approx kT$.

10.3 Frequency-Averaged Gaunt Factor

The frequency-averaged Gaunt factor (10.4) has been calculated numerically by Karzas & Latter (1961) for $10^{4.2}\,\mathrm{K} < T/Z^2 < 10^{8.2}\,\mathrm{K}$. In the vicinity of $10^4\,\mathrm{K}$,

$$\langle g_{\rm ff}\rangle_T \approx 1.34\left(T_4/Z_i^2\right)^{0.05} \tag{10.10}$$

to within $\sim$2% for $0.1 < T_4/Z_i^2 < 3$. Over the full temperature range $10^{4.2}\,\mathrm{K} \leq (T/Z_i^2) \leq 10^{8.2}\,\mathrm{K}$, their results are reproduced to within 2% by the fitting function

$$\langle g_{\rm ff}\rangle_T \approx 1 + \frac{0.44}{1+0.058[\ln(T/10^{5.4}Z_i^2\,\mathrm{K})]^2} \ , \tag{10.11}$$

and the fitting function should remain applicable up to $\sim 10^9\,\mathrm{K}$. Inserting Eq. (10.11) into Eq. (10.3), we obtain the **free–free power per volume** for $T \geq 10^{4.2}\,\mathrm{K}$:

$$\Lambda_{\rm ff}(T) \approx 1.422\times10^{-25}\left\{1+\frac{0.44}{1+0.058[\ln(T/10^{5.4}Z_i^2\,\mathrm{K})]^2}\right\}T_4^{1/2}Z_i^2 n_i n_e \frac{\mathrm{erg\,cm^3}}{\mathrm{s}} . \tag{10.12}$$

10.4 Free–Free Absorption

As we have seen in §7.4.1, absorption and emission are intimately related: if the energy levels involved in the absorption and emission processes[2] are populated according to a thermal distribution with temperature T, the attenuation coefficient κ_ν and the emissivity j_ν must satisfy Kirchhoff's law: $\kappa_\nu = j_\nu/B_\nu(T)$, where B_ν is the Planck function. Thus, the **attenuation coefficient due to free–free absorption** is obtained from the emissivity in Eq. (10.1):

$$\kappa_{{\rm ff},\nu} = \frac{4}{3}\left(\frac{2\pi}{3}\right)^{1/2}\frac{e^6}{m_e^{3/2}(kT)^{1/2}hc\,\nu^3}\left[1-\mathrm{e}^{-h\nu/kT}\right]Z_i^2 n_i n_e g_{\rm ff} \ , \tag{10.13}$$

where $g_{\rm ff}(\nu,T)$ is the Gaunt factor for free–free emission. Note that Eq. (10.13) includes the correction for stimulated emission through the factor $\left[1-\mathrm{e}^{-h\nu/kT}\right]$,

[2]The energy levels involved are of course just the translational motions of the electrons – the emission process involves a loss of kinetic energy equal to $h\nu$. Electron–ion and electron–electron elastic scattering will ensure that these kinetic energy states are populated according to a thermal distribution.

In the "radio" limit $h\nu \ll kT$, we expand the exponential and obtain

$$\frac{\kappa_{\mathrm{ff},\nu}}{n_i n_e} \approx \frac{4}{3}\left(\frac{2\pi}{3}\right)^{1/2} \frac{e^6}{(m_e kT)^{3/2} c \nu^2} Z_i^2 g_{\mathrm{ff}} \quad (h\nu \ll kT) \tag{10.14}$$

$$= 1.772 \times 10^{-26} T_4^{-1.5} \nu_9^{-2} Z_i^2 g_{\mathrm{ff}} \text{ cm}^5 \tag{10.15}$$

$$\approx 1.091 \times 10^{-25} Z_i^{1.882} T_4^{-1.323} \nu_9^{-2.118} \text{ cm}^5 \quad , \tag{10.16}$$

where we have used eq (10.7). Equation (10.16) is valid for $0.14 \lesssim \nu_9/T_4^{1.5} \lesssim 250$.

We see from Eq. (10.16) that free–free absorption becomes strong at low frequencies.

10.5 Emission Measure

The intensity I_ν due to free–free emission from an ionized region is obtained by integrating the equation of radiative transfer:

$$I_\nu(s) = I_\nu(0)\mathrm{e}^{-\tau_\nu} + \int_0^s ds' \, j_{\mathrm{ff},\nu} \, \mathrm{e}^{-[\tau_\nu(s)-\tau_\nu(s')]} \tag{10.17}$$

$$= I_\nu(0)\mathrm{e}^{-\tau_\nu} + \int_0^{\tau_\nu} d\tau' \left[\frac{j_{\mathrm{ff},\nu}}{\kappa_{\mathrm{ff},\nu}}\right] \mathrm{e}^{-(\tau-\tau')} \, . \tag{10.18}$$

If the ionized region has a uniform temperature T, then

$$I_\nu = I_\nu(0)\mathrm{e}^{-\tau_\nu} + \left[\frac{j_{\mathrm{ff},\nu}}{\kappa_{\mathrm{ff},\nu}}\right]_T \left(1 - \mathrm{e}^{-\tau_\nu}\right) \tag{10.19}$$

$$= I_\nu(0)\mathrm{e}^{-\tau_\nu} + \frac{(1-\mathrm{e}^{-\tau_\nu})}{\tau_\nu} \left[\frac{j_{\mathrm{ff},\nu}}{n_e n_p}\right]_T EM \quad , \tag{10.20}$$

where the **emission measure** EM is defined by

$$EM \equiv \int n_e n_p ds = \left[\frac{n_e n_p}{\kappa_{\mathrm{ff},\nu}}\right]_T \tau_\nu \quad . \tag{10.21}$$

For $\tau \ll 1$, the factor $[(1-\mathrm{e}^{-\tau})/\tau] \approx 1$, and I_ν increases linearly with EM. The optical depth τ_ν is (setting $Z_i = 1$)

$$\tau_\nu = 1.772 \times 10^{-26} T_4^{-1.5} \nu_9^{-2} \, g_{\mathrm{ff}} \left(\frac{n_i}{n_p}\right) EM \text{ cm}^5 \tag{10.22}$$

$$\approx 1.091 T_4^{-1.323} \nu_9^{-2.118} \left(\frac{n_i}{n_p}\right) \frac{EM}{10^{25}\text{ cm}^{-5}} \quad . \tag{10.23}$$

Bright H II regions may have $n_e \approx n_p \approx 10^3 \text{ cm}^{-3}$ over pathlengths $\Delta s \approx 1 \text{ pc}$, resulting in emission measures $EM \approx 10^6\,\mathrm{cm}^{-6}\,\mathrm{pc} \approx 10^{24.5}\,\mathrm{cm}^{-5}$, with a wide range around this value.[3] Giant H II regions or very dense H II regions can have $EM \approx 10^{26}\,\mathrm{cm}^{-5}$ or even larger. Attenuation by free–free absorption is generally negligible in the ISM for $\nu \gtrsim 10\,\mathrm{GHz}$, but H II regions can become optically thick for $\nu \lesssim 1\,\mathrm{GHz}$.

10.6 Free–Bound Transitions: Recombination Continuum

Free–free emission involves a transition of an electron from one "free" kinetic energy state to another. It is also possible for the electron to be captured by an ion, making a transition from a free state to a specific bound energy level, with emission of a photon: this is referred to as a "free → bound transition," or "radiative recombination," and associated with this process is a specific and distinctive emission spectrum.

The inverse of radiative recombination is photoionization, and we can use Kirchhoff's law to relate the bound-free emissivity to the cross section for photoionization:

$$j_{\mathrm{fb},\nu} = \kappa_{\mathrm{bf},\nu} B_\nu(T) = [n_{\mathrm{b}}]_{\mathrm{LTE}} \sigma_{\mathrm{b,pi}}(\nu) \left[1 - \mathrm{e}^{-h\nu/kT}\right] B_\nu(T) \,, \qquad (10.24)$$

where $[n_{\mathrm{b}}]_{\mathrm{LTE}}$ is the number density of atoms in bound state b if in LTE at temperature T with electron and ion densities n_i and n_e, and $\sigma_{\mathrm{b,pi}}(\nu)$ is the photoionization cross section for bound state b. Statistical mechanics tells us how to relate $[n_{\mathrm{b}}]_{\mathrm{LTE}}$ to the number densities of electrons and ions – see Eq. (3.33). Using this result, we obtain the **free–bound emissivity**

$$j_{\mathrm{fb},\nu} = \frac{g_{\mathrm{b}}}{g_e g_i} \frac{h^4 \nu^3}{(2\pi m_e kT)^{3/2} c^2} \, \mathrm{e}^{(I_{\mathrm{b}} - h\nu)/kT} \, \sigma_{\mathrm{b,pi}}(\nu) n_e n_i \quad , \qquad (10.25)$$

where g_{b} is the degeneracy of bound state b, g_i is the degeneracy of the ion (normally in its electronic ground state), $g_e = 2$ is the degeneracy of the free electron state, and I_{b} is the energy required to ionize from bound state b. Not surprisingly, the free–bound emissivity $j_{\mathrm{fb},\nu}$ is proportional to the product $n_e n_i$. Each bound state b contributes a recombination continuum beginning at $h\nu = I_b$, cut off at high frequencies by the factor $\exp(-(h\nu - I_b)/kT)$. For hydrogen, $I_b = I_{\mathrm{H}}/n^2$ and $g_b/g_i = 2n^2$ for bound states of principal quantum number n. The spectrum of this emission for a hydrogen plasma is illustrated in Figure 10.2.

10.7 Radio Recombination Lines

In a hydrogen plasma, collisional processes will maintain a population of H atoms in very high quantum states with principal quantum number $n \gtrsim 100$ – referred

[3] We will see in Chapter 28 that the Orion Nebula has an rms electron density $n_e \approx 3200\,\mathrm{cm}^{-3}$ and a diameter $2R \approx 0.5\,\mathrm{pc}$, corresponding to a maximum emission measure $EM \approx 5 \times 10^6\,\mathrm{cm}^{-6}\,\mathrm{pc} \approx 1.5 \times 10^{25}\,\mathrm{cm}^{-5}$.

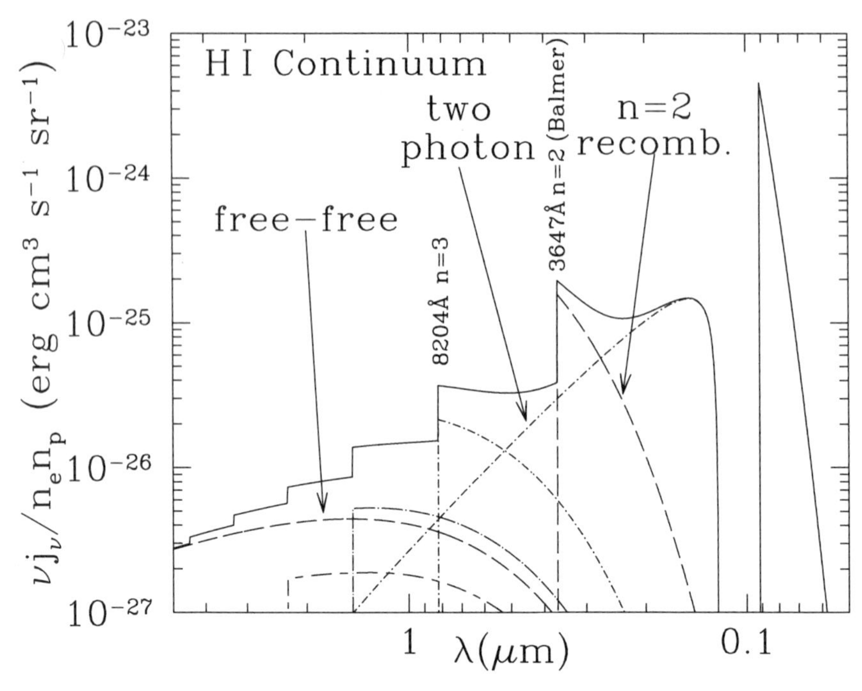

Figure 10.2 Solid line: The continuous emission spectrum of a $T = 8000\,\mathrm{K}$ hydrogen plasma, including free–free emission, recombination continuum emission, and two-photon emission from H $2s$ (see Chapter 14). Separate contributions of free–free and 2-photon emission are labeled. The two-photon component is suppressed at densities $n_e \gtrsim 10^3\,\mathrm{cm}^{-3}$ (see Chapter 14). Emission lines are not shown.

to as **Rydberg states**. These levels are populated mainly by three-body collisional processes (see §3.7). For a given principal quantum number n, collisions maintain the different angular momentum states ℓ populated very nearly in proportion to their degeneracies $2\ell + 1$.

These Rydberg levels can undergo spontaneous decay to lower levels. The $n + 1 \rightarrow n$ transitions, referred to as $n\alpha$ transitions, are at frequencies

$$\nu_{n\alpha} = \frac{2n+1}{[n(n+1)]^2}\frac{I_{\mathrm{H}}}{h} \approx 6.479 \left(\frac{100.5}{n+0.5}\right)^3 \mathrm{GHz} \quad . \tag{10.26}$$

The 166α transition is frequently observed, because its 1.425-GHz frequency is conveniently close to the 1.420-GHz ($\lambda = 21.1\,\mathrm{cm}$) transition of atomic hydrogen. The $n\alpha$ transitions with $n \gg 1$ have Einstein coefficients (averaged over the ℓ states)[4]

$$A_{n+1\rightarrow n} \approx 6.130 \times 10^9 (n + 0.7)^{-5}\,\mathrm{s}^{-1} \quad . \tag{10.27}$$

[4]Equation (10.27), obtained from an asymptotic formula by Menzel (1968), reproduces numerical values $A_{n+1\rightarrow n}$ from Wiese et al. (1966) to within 1% for $n \geq 4$.

The attenuation coefficient κ for a radio recombination line is

$$\kappa(\nu) = n_\ell \frac{\lambda^2}{8\pi} \frac{g_u}{g_\ell} A_{u\ell} \phi_\nu \left(1 - \frac{n_u/g_u}{n_\ell/g_\ell}\right) \quad , \tag{10.28}$$

where ϕ_ν is the normalized line profile ($\int \phi_\nu d\nu = 1$), and the negative term is the correction for stimulated emission. Recall the definition of departure coefficient b_n in Eq. (3.45): for a level with principal quantum number n, $n_n/g_n \propto b_n e^{I_\mathrm{H}/n^2kT}$. For $n\alpha$ transitions (levels ℓ and u have principal quantum numbers n and $n+1$, respectively),

$$\left(1 - \frac{n_u/g_u}{n_\ell/g_\ell}\right) = 1 - \frac{b_{n+1}}{b_n} \exp\left[-\frac{(2n+1)}{n^2(n+1)^2} \frac{I_\mathrm{H}}{kT}\right] \quad . \tag{10.29}$$

Salem & Brocklehurst (1979) define

$$\beta_n \equiv \frac{1 - \frac{n_{n+1}/g_{n+1}}{n_n/g_n}}{1 - e^{-h\nu/kT}} \quad , \tag{10.30}$$

so that the attenuation coefficient for radiation in the $n\alpha$ transition becomes

$$\kappa_\nu = n_n \frac{\lambda^2_{n+1,n}}{8\pi} \left(\frac{n+1}{n}\right)^2 A_{n+1\rightarrow n} \phi_\nu \beta_n \left(1 - e^{-h\nu/kT}\right) \quad , \tag{10.31}$$

where $h\nu = E_{n+1} - E_n = (2n+1)I_\mathrm{H}/[n(n+1)]^2$. In LTE, the level populations would have $\beta_n = 1$. If $\beta_n < 0$, there is a population inversion, and maser amplification will take place.

Figure 10.3 shows β_n for three different electron densities. For the $n_e = 10\,\mathrm{cm}^{-3}$ case, $\beta_n < 0$ for $n \lesssim 400$: the competition between radiative decay and collisions results in a population inversion for these levels – stimulated emission exceeds absorption, and maser amplification of the line radiation will take place. For the $n_e = 10^3\,\mathrm{cm}^{-3}$ case, a population inversion is present for $n \lesssim 300$: collisions are able to keep the $n > 300$ levels close enough to thermal so that the attenuation coefficient remains positive. For the $n_e = 10^5\,\mathrm{cm}^{-3}$ case, corresponding to a very dense H II region, collisions are rapid enough that masing occurs only for $n \lesssim 130$.

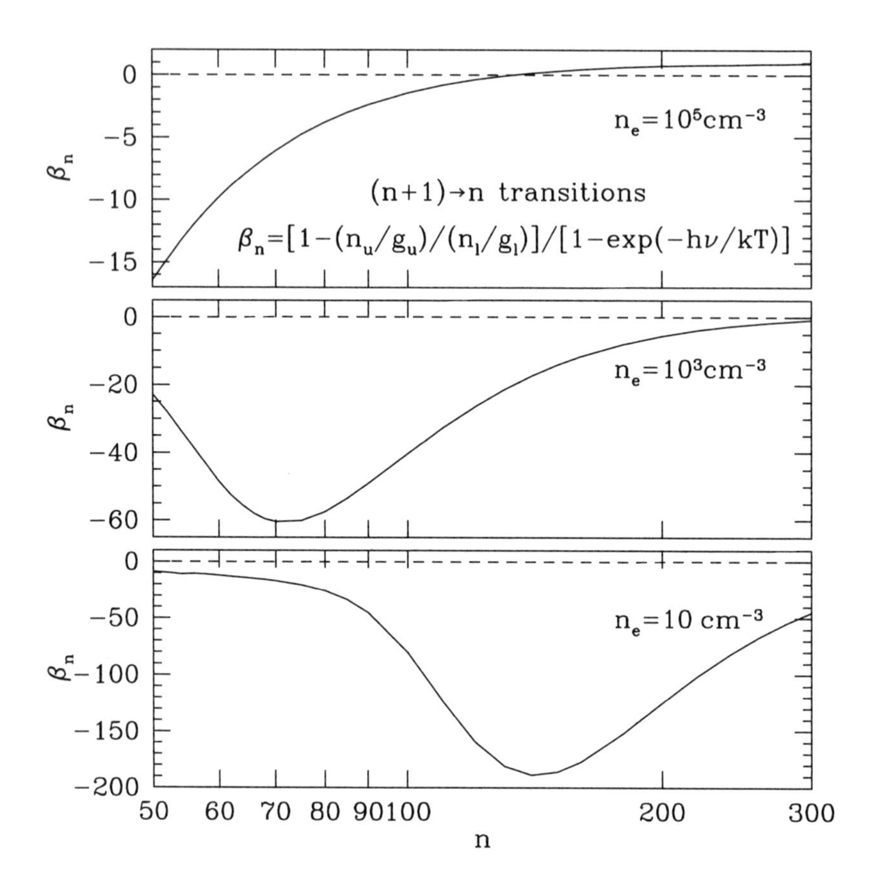

Figure 10.3 Stimulated emission correction factor β for $n\alpha$ radio recombination lines (Salem & Brocklehurst 1979). When $\beta_n < 0$, line intensities will be affected by maser amplification. See Fig. 3.1 for the departure coefficients for these three cases.

Chapter Eleven

Propagation of Radio Waves through the ISM

Radio waves propagating through a plasma interact with the plasma particles, and, as a result, the velocity of propagation of the wave differs from the speed of light in vacuo. The interstellar electron density n_e is tiny compared to the densities of laboratory plasmas, so these effects are small, but because of the long pathlengths, the small change in propagation speed can be measurable, thus providing information about the electron density n_e. In addition, the presence of a magnetic field leads to a difference in the phase velocity for right- and left-circularly polarized radio waves; this small difference results in measurable **Faraday rotation** of the plane of polarization of linearly polarized radio waves, thus providing us with an important diagnostic of the interstellar magnetic field.

Radio waves can also be refracted by inhomogeneities in the plasma, leading to the phenomenon of **interstellar scintillation**; observations of this scintillation provide valuable – and in some cases, perplexing – information about the density structure of the ISM, with implications for interstellar turbulence.

11.1 Dispersion Relation for Cold Plasmas

In a cold plasma with electron density n_e, electromagnetic waves propagating with $\mathbf{E} \propto e^{ikx-i\omega t}$ must satisfy the **dispersion relation** (Kulsrud 2005)

$$k^2c^2 = \omega^2 - \omega_p^2 \quad , \tag{11.1}$$

where

$$\omega_p \equiv \left(\frac{4\pi n_e e^2}{m_e}\right)^{1/2} = 5.641 \times 10^4 \left(\frac{n_e}{\mathrm{cm}^{-3}}\right)^{1/2} \mathrm{s}^{-1} \tag{11.2}$$

is known as the **plasma frequency**. From this equation, it is evident that there are no propagating modes with frequencies below the plasma frequency ω_p. Radio astronomical observations are in general conducted at frequencies $\nu \gtrsim 10^8$ Hz, far above the plasma frequency $\nu_p = \omega_p/2\pi = 8.979 \times 10^3 (n_e/\,\mathrm{cm}^{-3})^{1/2}$ Hz, so we can in general make the approximation $\omega_p/\omega \ll 1$ when discussing propagation of electromagnetic waves through the ISM.

The **phase velocity** – the speed of propagation of a **surface of constant phase** for a monochromatic wave – is

$$v_{\mathrm{phase}} \equiv \frac{\omega}{k} = \frac{c}{\sqrt{1-(\omega_p/\omega)^2}} > c \quad , \tag{11.3}$$

and the plasma has a refractive index

$$m(\omega) \equiv \frac{c}{v_{\mathrm{phase}}} = \left(1 - \frac{\omega_p^2}{\omega^2}\right)^{1/2} < 1 \quad . \tag{11.4}$$

11.2 Dispersion

A "wave packet" in a dispersive medium will propagate with the **group velocity**

$$v_g(\omega) = \frac{d\omega}{dk} \quad . \tag{11.5}$$

In a plasma, the group velocity is

$$v_g = c\left(1 - \frac{\omega_p^2}{\omega^2}\right)^{1/2} \quad . \tag{11.6}$$

This is the speed at which information can be transmitted; note that $v_g < c$.

Suppose that an astronomical object, e.g., a pulsar, emits a pulse of radiation at $t = 0$. If the distance to the pulsar is L, the time of arrival of energy at frequency $\nu = \omega/2\pi$ is

$$t_{\mathrm{arrival}} = \int_0^L \frac{dL}{v_g(\omega)} \tag{11.7}$$

$$\approx \int_0^L \frac{dL}{c}\left(1 + \frac{1}{2}\frac{\omega_p^2}{\omega^2}\right) \tag{11.8}$$

$$= \frac{L}{c} + \frac{1}{2c\omega^2}\int_0^L \omega_p^2 \, dL \tag{11.9}$$

$$= \frac{L}{c} + \frac{e^2}{2\pi m_e c}\frac{1}{\nu^2} DM \tag{11.10}$$

$$= \frac{L}{c} + 4.146 \times 10^{-3}\left(\frac{\nu}{\mathrm{GHz}}\right)^{-2} \frac{DM}{\mathrm{cm^{-3}\,pc}}\,\mathrm{s} \quad , \tag{11.11}$$

where the time delay has been written in terms of the **dispersion measure**:

$$DM \equiv \int_0^L n_e dL \quad . \tag{11.12}$$

A typical dispersion measure to a pulsar $\sim 3\,\mathrm{kpc}$ away might be $DM \approx 10^2\,\mathrm{cm^{-3}\,pc}$; for this value of DM, the pulse arrival time at $1\,\mathrm{GHz}$ would be delayed by $\sim 0.4\,\mathrm{s}$ after traveling $\sim 10^4$ years.

Because of the dependence on ν^{-2}, low frequencies arrive later, and we speak of the pulse as being "dispersed." The actual pulse travel time t_{arrival} is unknown, but from measurement of the pulse arrival time at different frequencies, we can observationally determine $dt_{\mathrm{arrival}}/d\nu$. Differentiating Eq. (11.10), we obtain

$$DM = -\frac{\pi m_e c}{e^2}\,\nu^3\,\frac{dt_{\mathrm{arrival}}}{d\nu} \quad . \tag{11.13}$$

This method has been used to determine the dispersion measure DM to many hundreds of pulsars (Taylor & Cordes 1993; Cordes & Lazio 2003). If the distance

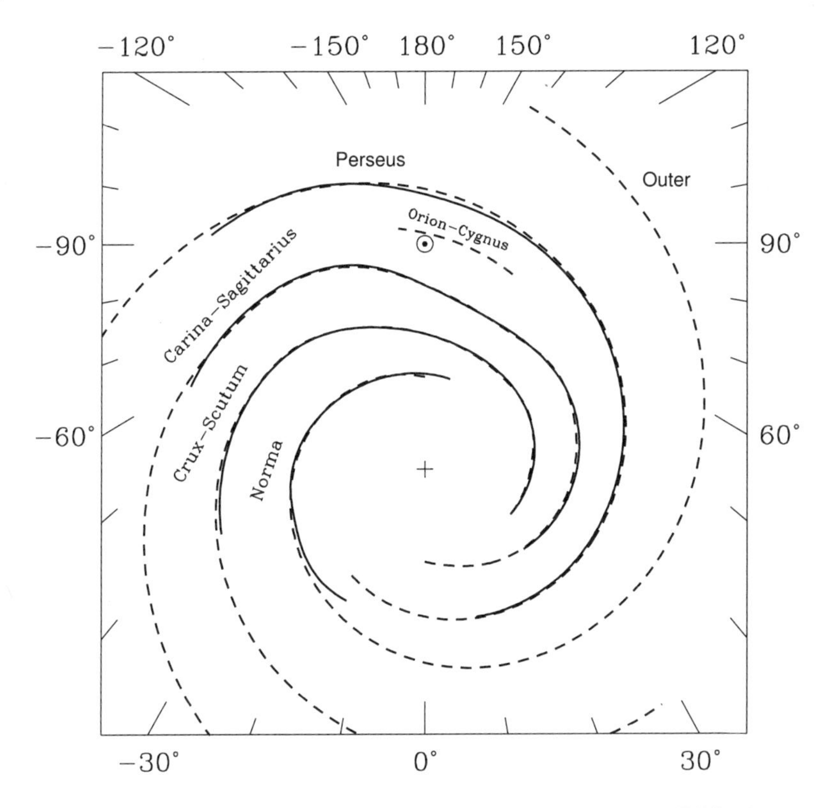

Figure 11.1 The spiral pattern for the Milky Way as estimated from pulsar DM, where spiral arms are assumed to coincide with regions of enhanced electron density n_e. Dashed line: Pattern favored by Cordes & Lazio (2003). Solid line: Pattern obtained by Taylor & Cordes (1993). A similar pattern was inferred by Georgelin & Georgelin (1976) based on the locations of bright H II regions. Figure from Cordes & Lazio (2003).

L to the pulsar can be estimated from other considerations, the measured DM provides the mean value of the electron density along this line of sight: $\langle n_e \rangle = DM/L$. A three-dimensional model for the electron density $n_e(x, y, z)$ can be constrained by the DM measurements for many hundreds of pulsars. The electron density model includes enhanced electron density in a four-arm logarithmic spiral pattern, plus an overall density enhancement in a broad ring with radius $\sim 3.5\,\mathrm{kpc}$ (for an assumed Galactic Center distance of $R_{\rm GC} = 8.5\,\mathrm{kpc}$). Figure 11.1 shows the spiral pattern adopted by Cordes & Lazio (2003), and their estimate for the electron density along a line passing through the Sun and the Galactic Center is shown in Figure 11.2.

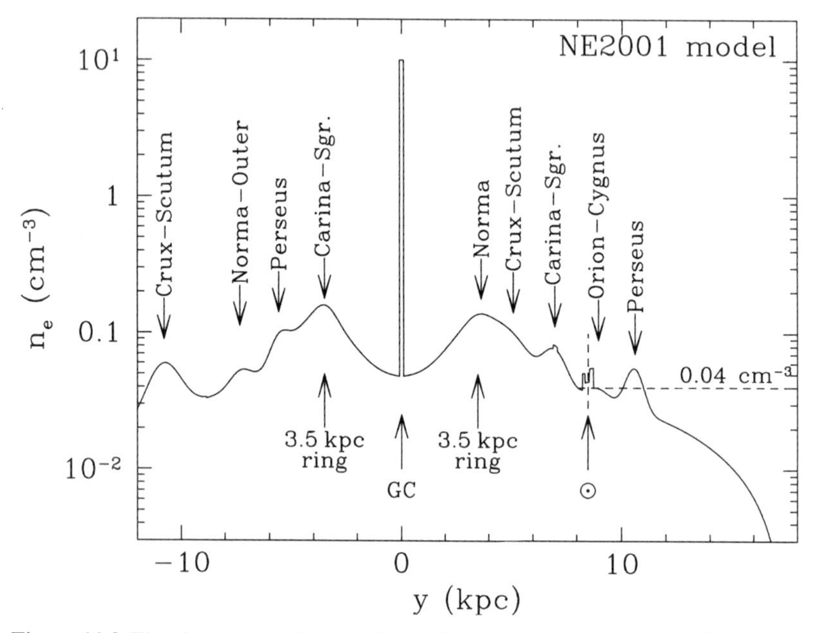

Figure 11.2 The electron density n_e at the midplane ($z = 0$), on a line passing through the Galactic Center (GC) and the Sun, as estimated by the NE2001 model of Cordes & Lazio (2003). Spiral arm locations are indicated. The Sun is at $y = 8.5\,\mathrm{kpc}$.

Because of our greater knowledge of the solar neighborhood, the 3-D electron density model of Cordes & Lazio (2003) includes structure near the Sun, including enhanced electron density in a shell referred to as "Loop I" (thought to be material that has been shock-compressed as the result of one or more supernova explosions), an underdensity of electrons in the "Local Hot Bubble" with $n_e \approx 0.005\,\mathrm{cm}^{-3}$ (also the result of supernovae), and enhanced electron density in the neighborhood of the Vela supernova remnant and in the Gum Nebula. The Galactic Center region has enhanced electron density, represented by Cordes & Lazio as $n_e \approx 10\,\mathrm{cm}^{-3}$ in a region of radius $\sim 145\,\mathrm{pc}$ in the Galactic plane, with a vertical scale height of $\sim 26\,\mathrm{pc}$.

The vertical structure of the electron distribution is shown at four locations in

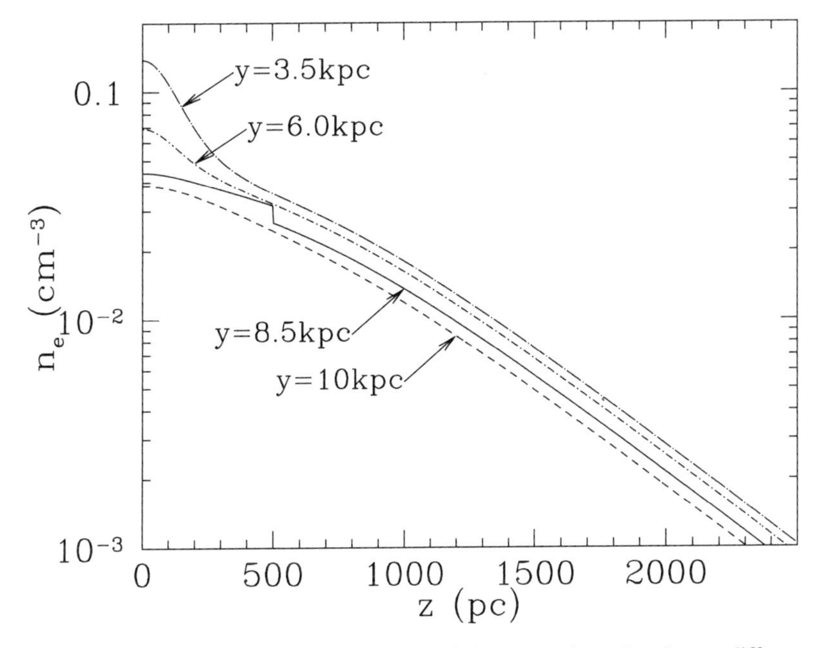

Figure 11.3 Electron density n_e as a function of distance z above the plane at different distances y from the Galactic Center, for the NE2001 model of Cordes & Lazio (2003). The Sun is at $y = 8.5\,\mathrm{kpc}$.

Fig. 11.3; the curve for $y = 8.5\,\mathrm{kpc}$ is for the location of the Sun. In the solar neighborhood, the electron density is $\sim 0.045\,\mathrm{cm}^{-3}$ near the midplane. The electron density n_e generally declines with increasing distance from the plane, reaching $\sim 0.002\,\mathrm{cm}^{-3}$ by $z = 2\,\mathrm{kpc}$.

Figure 11.4 shows the azimuthally averaged surface density of H II, projected onto the disk, as a function of galactocentric radius. The upper curve shows all of the H II mass within $\pm 10\,\mathrm{kpc}$ of the disk; the lower curve is limited to $\pm 500\,\mathrm{pc}$. Integrated over the entire galaxy, the H II mass is $M(\mathrm{H\,II}) = 1.1 \times 10^9\,M_\odot$, or about 20% of the total diffuse H. Approximately 50% of the free electrons are located more than 500 pc from the plane.

11.3 Faraday Rotation

Now consider a static magnetic field B_0 to be present in the plasma, and consider the propagation of electromagnetic waves with angular frequency ω. In cold plasma, the dispersion relation for circularly polarized waves is

$$k^2 c^2 = \omega^2 - \frac{\omega_p^2}{1 \pm \frac{\omega_B}{\omega}} \ , \tag{11.14}$$

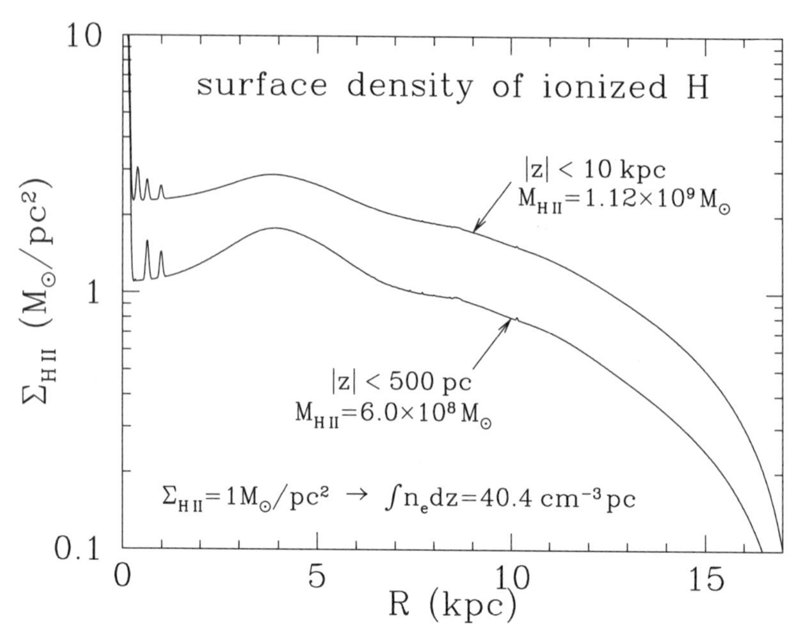

Figure 11.4 Surface density of H II, as a function of galactocentric radius, for the NE2001 electron density distribution (Cordes & Lazio 2003). The Sun is at $R = 8.5\,\mathrm{kpc}$.

where

$$\omega_B \equiv \frac{eB_\parallel}{m_e c} \tag{11.15}$$

is the cyclotron frequency calculated for the component of the magnetic field that is parallel to the direction of propagation. Since

$$\frac{\omega_B}{2\pi} = 2.80 \frac{B_\parallel}{\mu\mathrm{G}}\,\mathrm{Hz} \quad , \tag{11.16}$$

and interstellar magnetic field strengths are typically in the μG range, the validity condition $\omega_B/\omega \ll 1$ will be satisfied for all interstellar conditions of interest. We now calculate the velocity of propagation of surfaces of constant phase, or **phase velocity**:

$$v_{\mathrm{phase}}(\omega) \equiv \frac{\omega}{k(\omega)} \approx c\left[1 + \frac{1}{2}\frac{\omega_p^2}{\omega^2} \mp \frac{1}{2}\frac{\omega_p^2\omega_B}{\omega^3}\right] \quad , \tag{11.17}$$

where the plus/minus sign in Eq. (11.17) applies to right/left circular polarization.[1]

[1]We follow the conventions of optical physicists, e.g., Born & Wolf (1999) or Bohren & Huffman (1983), and use "right-handed" to refer to the mode where the $\mathbf{E}$ vector rotates clockwise as viewed by the observer.

The two circular polarization modes differ in phase velocity and k:

$$v_{\rm phase,L} - v_{\rm phase,R} = c\frac{\omega_p^2\omega_B}{\omega^3} = \frac{4\pi e^3}{m_e^2}\frac{n_e B_\parallel}{\omega^3} \tag{11.18}$$

$$\Delta k \equiv k_{\rm R} - k_{\rm L} = \frac{\omega}{c}\frac{\omega_p^2\omega_B}{\omega^3} = \frac{4\pi e^3}{m_e^2 c^2}\frac{n_e B_\parallel}{\omega^2} , \tag{11.19}$$

where, for $\omega_B > 0$, the left-handed mode has larger $v_{\rm phase}$. Now consider the case where the source (e.g., pulsar or active galactic nucleus, or AGN) emits linearly polarized radiation. Pure linear polarization can be decomposed into equal amounts of left- and right-circularly polarized waves with an initial phase relationship determined by the direction of the linear polarization at the source. In a vacuum, the two circular polarization modes would each propagate at the speed of light, maintaining the same phase relationship and therefore the same direction of linear polarization. In a magnetized plasma, the two circular polarization modes have different phase velocities; after propagating a distance L, the two waves will differ in phase by $L\Delta k$. The linear polarization mode obtained by adding the two circular polarization modes together will be rotated counterclockwise relative to the linear polarization of the source by a rotation angle

$$\Psi = \frac{1}{2}\int_0^L \Delta k \, dL = \int_0^L \frac{\omega_p^2\omega_B}{2c\omega^2} \, dL \tag{11.20}$$

$$= \frac{e^3}{2\pi m_e^2 c^2}\frac{1}{\nu^2}\int_0^L n_e B_\parallel \, dL \tag{11.21}$$

$$= RM \, \lambda^2 , \tag{11.22}$$

where the **rotation measure** is defined to be

$$RM \equiv \frac{1}{2\pi}\frac{e^3}{m_e^2 c^4}\int_0^L n_e B_\parallel dL \tag{11.23}$$

$$= 8.120 \times 10^{-5} \frac{\int_0^L n_e B_\parallel dL}{{\rm cm}^{-3}\,\mu{\rm G\,pc}} {\rm rad\,cm}^{-2} . \tag{11.24}$$

In general, we do not know, a priori, the direction of the linear polarization at the source, but we can determine the RM if we measure the difference between the linear polarization angles Ψ_1 and Ψ_2 at two different wavelengths λ_1 and λ_2:

$$RM = \frac{\Psi_2 - \Psi_1}{\lambda_2^2 - \lambda_1^2} . \tag{11.25}$$

If the dispersion measure and rotation measure can both be measured, we can determine the electron-density-weighted mean value of the line-of-sight magnetic field:

$$\langle B_{\parallel} \rangle = \frac{2\pi m_e^2 c^4}{e^3} \frac{RM}{DM} , \tag{11.26}$$

$$\frac{\langle B_{\parallel} \rangle}{\mu\mathrm{G}} = \frac{RM}{8.120 \times 10^{-5}\,\mathrm{rad\,cm^{-2}}} \times \frac{\mathrm{cm^{-3}\,pc}}{DM} . \tag{11.27}$$

Pulsars and AGNs are, in general, strongly linearly polarized (because they emit synchrotron radiation), and can be used to determine the RM for many sightlines through the Galactic ISM. Simultaneous measures of DM and RM for pulsars that are nearby on the sky but at differing distances allows the line-of-sight component of the magnetic field $B_{\parallel}$ to be determined:

$$B_{\parallel} = \frac{2\pi m_e^2 c^4}{e^3} \frac{\Delta RM}{\Delta DM} . \tag{11.28}$$

Using this method, Han et al. (2006) conclude that the magnetic field in the disk generally follows the spiral structure, but with reversals of magnetic field direction from spiral arms to interarm regions. The spiral arm magnetic field strength

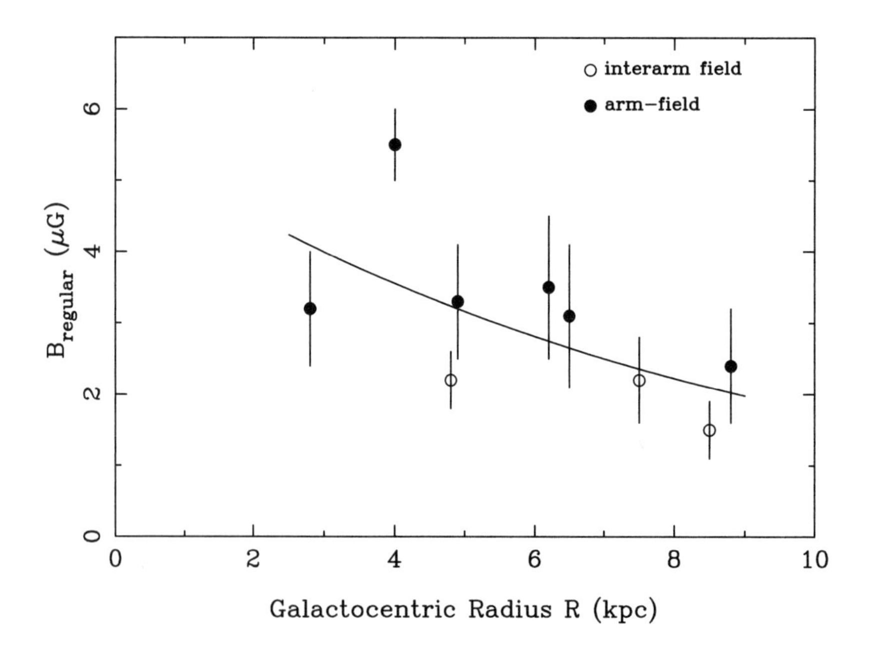

Figure 11.5 Large-scale ordered magnetic field in spiral arms and in interarm regions, as estimated by Han et al. (2006). The spiral arm magnetic fields are all counterclockwise (as viewed from the North Galactic pole), while the interarm fields are clockwise. Han et al. assumed a galactocentric distance of $R = 8.5\,\mathrm{kpc}$. Figure from Han et al. (2006), reproduced by permission of the AAS.

appears to be greater than the interarm value by a factor of $\sim$1.5 . The estimated field strengths are shown in Figure 11.5. Nota & Katgert (2010) used published rotation measures to reexamine the Galactic field in the fourth quadrant. They also find counterclockwise fields in the spiral arms and clockwise fields in the interarm regions, but their estimate for the strength of the ordered field is somewhat smaller than the results of Han et al. (2006).

11.4★ Refraction

The group velocity determines the speed of propagation of pulses (i.e., wave packets), but discussion of propagation (i.e., the effects of refraction) is best approached through consideration of steady monochromatic waves, since the direction of propagation is normal to surfaces of constant phase.

The ISM is inhomogeneous. Variations in the electron density n_e cause radio waves propagating through it to be refracted. Let us assume that we have a wavefront that has propagated a distance D through the clumpy ISM. Suppose that the variations in n_e have a characteristic length scale L, with characteristic density variations $(\Delta n_e)_{L,\mathrm{rms}}$ (deviations from the average density) and variations in phase velocity

$$(\Delta v_{\mathrm{phase}})_{L,\mathrm{rms}} = \frac{c}{2}\frac{(\Delta\omega_p^2)_{L,\mathrm{rms}}}{\omega^2} = \frac{2\pi e^2 c(\Delta n_e)_{L,\mathrm{rms}}}{m_e\omega^2} \quad , \tag{11.29}$$

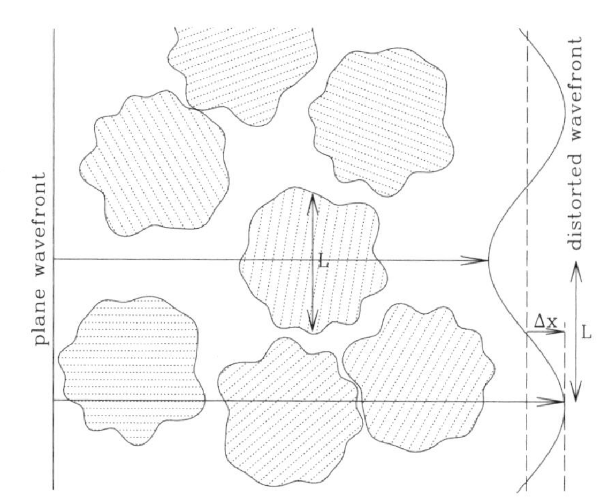

Figure 11.6 Refractive distortion of an electromagnetic wave propagating through the ISM with electron density variations on a length scale $\sim L$. The characteristic scattering angle is $\theta \approx \Delta x/L$ (see text).

or variations in refractive index m with

$$(\Delta m)_{L,\mathrm{rms}} = \frac{(\Delta v_{\mathrm{phase}})_{L,\mathrm{rms}}}{c} = \frac{2\pi e^2 (\Delta n_e)_{L,\mathrm{rms}}}{m_e \omega^2} \quad . \tag{11.30}$$

If we take a snapshot after the wavefront has traversed a single clump of thickness L, the wavefront is advanced (relative to where it would be in the absence of the density perturbation) by a distance

$$\Delta x = -L\Delta m \quad . \tag{11.31}$$

After traveling a distance D (see Fig. 11.6), the wavefront has crossed (D/L) independent clumps, and the rms advance or retardation of the wavefront will be $(D/L)^{1/2} L(\Delta m)_{L,\mathrm{rms}}$. Two rays separated by a transverse distance L will traverse independent clumps, and the r.m.s. difference in advance of the wavefront will be

$$\Delta x \sim \sqrt{2}(D/L)^{1/2} L(\Delta m)_{L,\mathrm{rms}} = (2DL)^{1/2}(\Delta m)_{L,\mathrm{rms}}. \tag{11.32}$$

The wavefront surface will be tilted, and therefore the ray will be refracted, through a characteristic refraction angle (see Fig. 11.6)

$$\theta_s \approx \frac{\Delta x}{L} \approx \frac{(2DL)^{1/2}(\Delta m)_{L,\mathrm{rms}}}{L} = \left(\frac{2D}{L}\right)^{1/2} \frac{2\pi e^2}{m_e \omega^2} (\Delta n_e)_{L,\mathrm{rms}} \tag{11.33}$$

$$= 0.653'' \left(\frac{D/\,\mathrm{kpc}}{L/10^{14}\,\mathrm{cm}}\right)^{1/2} \frac{(\Delta n_e)_{L,\mathrm{rms}}}{10^{-3}\,\mathrm{cm}^{-3}} \nu_9^{-2} \quad . \tag{11.34}$$

The source image would have angular size $\sim \theta_s$. This image broadening can in some cases be observed with long baseline interferometry.

We have been discussing the wave propagation using the conceptual framework of "ray optics." This will be valid provided that the refraction angle $\sim (\Delta m)_{L,\mathrm{rms}}$ for a single clump is large compared to the diffraction angle $\theta_{\mathrm{diff}} \approx \lambda/L$ for an object of size L: the ray optics treatment will be valid only for

$$\nu \ll \frac{e^2 (\Delta n_e)_{L,\mathrm{rms}}}{2\pi m_e c} (2LD)^{1/2} = 1\times 10^3\,\mathrm{GHz} \frac{(\Delta n_e)_{L,\mathrm{rms}}}{10^{-3}\,\mathrm{cm}^{-3}} \left(\frac{L}{10^{14}\,\mathrm{cm}} \frac{D}{\mathrm{kpc}}\right)^{1/2} . \tag{11.35}$$

If this condition is not satisfied, the analysis should be extended to include the diffractive effects of the inhomogeneities.[2]

[2]It is curious to observe that the diffractive effects become important at *high* frequencies. This happens because of the $1/\nu^2$ dependence of the dielectric function — for a given structure, the refraction angle varies as $1/\nu^2$, while the diffraction angle varies only as $1/\nu$.

11.5★ Scintillation

Because of the refractive effects mentioned earlier, there may be more than one ray path from the source to the observer. The wavefronts arriving on different paths will interfere, perhaps destructively. If the medium between the observer and source is changing (due to motion of the source, medium, or observer), the interference conditions can vary, leading to variations in the intensity of the radiation reaching the observer – this is referred to as **scintillation**. For example, scintillation resulting from turbulence is responsible for the twinkling of stars when viewed through the Earth's atmosphere.

A complete description of scintillation due to the ISM would entail using a three-dimensional representation of the inhomogeneities in electron density in the region between source and observer, and modeling the distortion of a wavefront as it propagates through the medium. A simpler approach is to approximate the scattering by the ISM as being produced by a two-dimensional "scattering screen" approximately halfway between the source and the observer. The scattering can be due to refraction of the wave, with a characteristic scattering angle as estimated in Eq. (11.34), or, at very high frequencies, it can be dominated by the diffractive effects of the inhomogeneities, with characteristic scattering angles $\theta_s \approx \lambda/L$.

When the scattering angles are small, the slight curvature in the wavefront as it reaches the observer will produce small changes in the flux from the source (i.e., variations in apparent brightness). When the fractional changes are small, this is referred to as **weak scattering**.

However, when the scattering angles are sufficiently large, it is possible for rays to reach the observer along multiple paths. For characteristic scattering angle θ_s, Fig. 11.7 shows two ray paths reaching an observer at location A. The pathlength difference between these two paths is

$$\Delta s = 2\sqrt{(D/2)^2 + (\theta_s D/4)^2} - D \approx D\theta_s^2/8 \quad , \tag{11.36}$$

and the phase difference between these two paths to location A is

$$\Delta\phi = \frac{2\pi\Delta s}{\lambda} = \frac{\pi D\theta_s^2}{4\lambda} \quad . \tag{11.37}$$

If $\Delta\phi \ll \pi$, the arriving waves will interfere constructively. Because the path geometry is determined by the structure of a turbulent medium, the apparent brightness of the source will vary as the medium moves across the line of sight, but the brightness variations will not be extreme.

If, however, $\Delta\phi > \pi$, the phase differences are large enough for either constructive or destructive interference to occur, and we will have **scintillation**. For refractive scattering, with θ_s given by Eq. (11.34), the condition for scintillation is simply $\theta_s^2 > 4\lambda/D$, or

$$\nu < \left(\frac{e^4 D^2}{8\pi^2 m_e^2 cL}\right)^{1/3} (\Delta n_e)_{L,\mathrm{rms}}^{2/3} \quad , \tag{11.38}$$

$$\nu < 2.0\,\mathrm{GHz} \left(\frac{D}{\mathrm{kpc}}\right)^{2/3} \left(\frac{10^{14}\,\mathrm{cm}}{L}\right)^{1/3} \left(\frac{(\Delta n_e)_{L,\mathrm{rms}}}{10^{-3}\,\mathrm{cm}^{-3}}\right)^{2/3} \quad . \tag{11.39}$$

Therefore, scintillation effects will disappear at high frequencies. And, at a fixed frequency ν, scintillation will not occur for sources that are close to us (small D).

Let us now inquire into the characteristics of the scintillation when it does occur. Consider a surface located a distance D from the source – call this the "detection plane". The observer's detector is located at one point on the detection plane. The source, observer, and interstellar medium all have transverse motions, and, therefore, the ISM along the line from source to observer is varying with time. Let the observer be moving with a transverse motion $v_\perp$ in a coordinate frame where the source and phase screen are both fixed.

Each point on the detection plane will be illuminated by a region of radius $\sim(\theta_s/2)(D/2) = \theta_s D/4$ on the scattering screen, containing $\sim\pi(\theta_s D/4)^2/L^2$ independent patches.

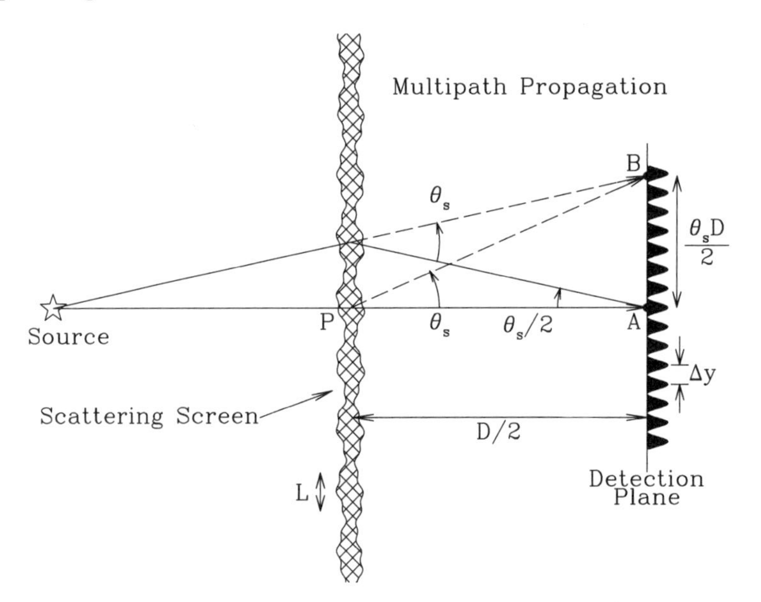

Figure 11.7 Scattering screen and multipath propagation.

Consider one position P on the phase screen, and two different points, A and B, on the detection plane (see Figure 11.7), with point B a distance $\theta_s D/2$ from point A. The direct propagation distance to B will exceed that to A by $D\theta_s^2/8$, corresponding to a phase shift $\phi_B - \phi_A = \pi\theta_s^2 D/4\lambda$. The wave from patch P will interfere with the wave from other patches; as we move from A to B on the detection plane, we may expect to encounter $\sim\theta_s^2 D/4\lambda$ maxima, separated by distance $\sim (\theta_s D/2)/(\theta_s^2 D/4\lambda) = 2\lambda/\theta_s$. The power detected by the moving observer will vary over a scintillation time scale:

$$(\Delta t)_{\rm scint.} \approx \frac{2\lambda}{\theta_s v_\perp} \tag{11.40}$$

$$\approx 2\frac{m_e c}{e^2}\left(\frac{2L}{D}\right)^{1/2} \frac{1}{(\Delta n_e)_{L,\rm rms}}\frac{1}{v_\perp}\nu \tag{11.41}$$

$$\approx 2\times 10^4\,{\rm s}\left(\frac{L/10^{14}\,{\rm cm}}{D/\,{\rm kpc}}\right)^{1/2} \frac{10^{-3}\,{\rm cm}^{-3}}{(\Delta n_e)_{L,\rm rms}}\frac{30\,{\rm km\,s}^{-1}}{v_\perp}\nu_9 \ . \tag{11.42}$$

Returning to our discussion of the propagation distance to position B, we can ask by how much the frequency must change for the phase shift due to this extra pathlength to change by π radians. The condition

$$(\Delta\nu)_{\rm scin}\,\frac{d}{d\nu}(\phi_B - \phi_A) = \pi \tag{11.43}$$

yields the scintillation bandwidth:

$$(\Delta\nu)_{\rm scin} = \frac{4c}{D\theta_s^2} \tag{11.44}$$

$$= 8\pi^2\frac{m_e^2 c}{e^4}\frac{L}{D^2}\frac{\nu^4}{(\Delta n_e)^2_{L,\rm rms}} \tag{11.45}$$

$$= 390\,{\rm MHz}\frac{L/10^{14}\,{\rm cm}}{(D/\,{\rm kpc})^2}\left(\frac{10^{-3}\,{\rm cm}^{-3}}{(\Delta n_e)_{L,\rm rms}}\right)^2 \nu_9^4 \ . \tag{11.46}$$

To observe strong scintillation, it is necessary to use a bandwidth smaller than $(\Delta\nu)_{\rm scin}$ – otherwise, the scintillations are washed out. In fact, by measuring the scintillation bandwidth $(\Delta\nu)_{\rm scin}$, one can determine $(\Delta n_e)^2_{L,\rm rms}D^2/L$.

11.6★ Interstellar Electron Density Power Spectrum

The preceding discussion of scintillation has been framed in terms of a characteristic length scale L for electron density fluctuations. The real, turbulent interstellar medium actually contains electron density fluctuations over a wide range of length scales. It is natural to describe the electron density fluctuations in terms of the **electron density power spectrum** $P_{n_e}(\vec{k})$, with normalization so that

$$\langle(\Delta n_e)^2\rangle = \int d^3k\, P_{n_e}(\vec{k}) \ . \tag{11.47}$$

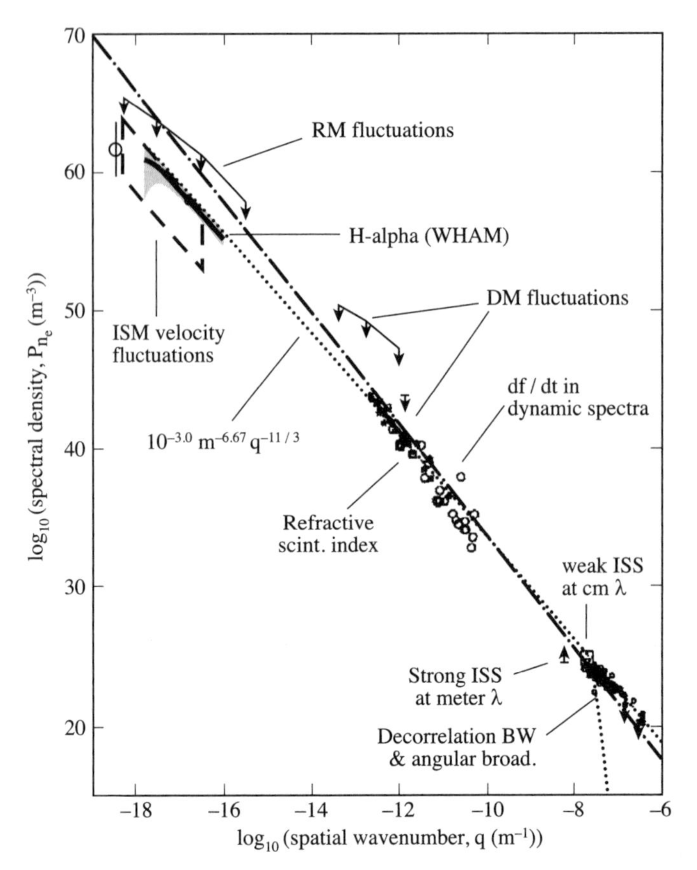

Figure 11.8 Electron density power spectrum P_{n_e} [see Eq. (11.48)], measured by a number of different methods. The dotted line is eq. (11.48) with $C_n^2 \approx 5 \times 10^{-17}\,\mathrm{cm}^{-20/3}$. From Chepurnov & Lazarian (2010). See Armstrong et al. (1995) and Chepurnov & Lazarian (2010) for details.

Observations of scintillation in the nearby interstellar medium within $\sim 1\,\mathrm{kpc}$ (see Fig. 11.8) appear to be consistent with a power-law spectrum for the fluctuations in electron density over 11 decades: $10^{-18}\,\mathrm{cm}^{-1} \lesssim k \lesssim 10^{-6.5}\,\mathrm{cm}^{-1}$ (or $10^{6.5}\,\mathrm{cm} \lesssim L/2\pi \lesssim 10^{18}\,\mathrm{cm}$):

$$P_{n_e}(\vec{k}) \approx C_n^2\, k^{-11/3} \quad . \tag{11.48}$$

Measurements for a single sightline determine an average C_n^2 for that sightline. By comparing different sightlines, it is found that C_n^2 varies considerably with location, as shown in Fig. 11.9. Within a few kpc of the Sun, it appears that the electron

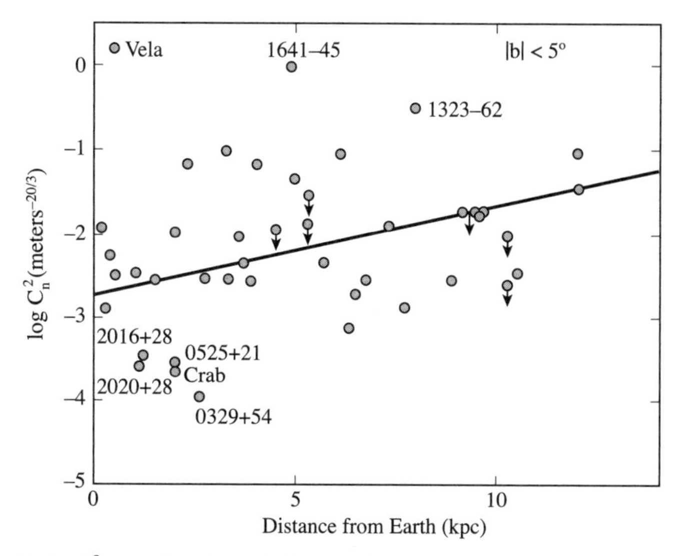

Figure 11.9 C_n^2 as a function of distance from Earth, from Cordes et al. (1985). Points are sightlines to individual pulsars. Note: $C_n^2 = 5 \times 10^{-17}\,\mathrm{cm}^{-20/3} = 1.1 \times 10^{-3}\,\mathrm{m}^{-20/3}$.

density fluctuations have

$$C_n^2 \approx 1 \times 10^{-3}\,\mathrm{m}^{-20/3} \approx 5 \times 10^{-17}\,\mathrm{cm}^{-20/3} \tag{11.49}$$

(Armstrong et al. 1995), although with large sightline-to-sightline variations (see Fig. 11.9). The power-law dependence $P_{n_e} \propto k^{-11/3}$ corresponds to what would be expected for incompressible turbulence – so-called Kolmogorov turbulence – if the ionization fraction behaves like a passive tracer, with ionization fluctuations "injected" on large scales and removed on very small scales.

With P_{n_e} from Eq. (11.48), we can integrate Eq. (11.47) from the largest length scale $L_{\rm max} = 2\pi/k_{\rm min} = 6 \times 10^{18}\,\mathrm{cm}$ down to the smallest scales, to obtain total rms electron density fluctuations $\langle(\Delta n_e)^2\rangle^{1/2} = 0.03\,\mathrm{cm}^{-3}$, comparable to the mean electron density $\langle n_e \rangle \approx 0.04\,\mathrm{cm}^{-3}$ obtained from measurements of the dispersion measure DM toward pulsars (see §11.2). Evidently, the electron density has variations of order unity on length scales $2\pi/k_{\rm min} \approx 6 \times 10^{18}\,\mathrm{cm}$ and longer. To find $(\Delta n_e)^2_{L,\rm rms}$, we integrate from $2\pi/L\sqrt{e}$ to $2\pi\sqrt{e}/L$ (one e-folding):

$$\langle(\Delta n_e^2)_L\rangle^{1/2} = \frac{6.4 \times 10^{-4}}{\mathrm{cm}^3}\left(\frac{C_n^2}{5 \times 10^{-17}\,\mathrm{cm}^{-20/3}}\right)^{1/2}\left(\frac{L}{10^{14}\,\mathrm{cm}}\right)^{1/3} . \tag{11.50}$$

With $(\Delta n_e)_{L,\rm rms} \propto L^{1/3}$ (for $L \lesssim 6 \times 10^{18}$ cm), we see that electron density fluctuations are largest on large scales. Nevertheless, small scales dominate some phenomena, such as image broadening (11.34). A proper analysis of image broadening and scintillation should take into account the full spectrum of density fluctuations, but that is beyond the scope of this text.

There are at least two remarkable aspects to Eq. (11.48):

- Why is the spectrum consistent with the simple Kolmogorov power-law? There are two issues: (1) The Kolmogorov spectrum is expected for small-amplitude, incompressible, hydrodynamic turbulence, but the ISM is *compressible*, *magnetized*, and the turbulent fluctuations are *large in amplitude when k is small.* (2) The Kolmogorov spectrum is expected only for scales smaller than the smallest length scale on which the medium is being actively stirred – yet the electron density fluctuations appear to obey the Kolmogorov spectrum from 2 pc length scales on down, seemingly implying that the important "stirring" takes place only on scales larger than $\sim 2\,\mathrm{pc}$.

- The power spectrum appears to continue down to length scales as short as $\sim 10^7$ cm – this is shorter than the collisional mean free path for electrons or ions for the plasma densities that are present. This must imply that the density variations on the shortest length scales are occurring perpendicular to the magnetic field direction, which is possible as long as the scale lengths are larger than the ion gyroradius:

$$r_{\mathrm{g}} = \frac{c}{eB}\left(m_p kT\right)^{1/2} = 1.9\times 10^7\,\mathrm{cm}\left(\frac{5\,\mu\mathrm{G}}{B}\right) T_4^{1/2} \quad . \tag{11.51}$$

The electron density power spectrum amplitude $C_n^2 \approx 5\times 10^{-17}\,\mathrm{cm}^{-20/3}$ appears to be representative of the interstellar medium within $\sim 1\,\mathrm{kpc}$ of the Sun. However, on some sightlines (e.g., to the Vela pulsar, to the pulsar 1641-45, or to Sgr A* at the Galactic Center), the interstellar scintillation is stronger – by up to two orders of magnitude – than expected for electron density fluctuations corresponding to Eq. (11.49). Figure 11.9 shows C_n^2 for sightlines to sources at various distances from the Earth, showing strong enhancements in C_n^2 along some sightlines. It is not known what is responsible for these "enhanced scattering regions."

In addition to scintillation, the electron density structure can be probed by observing temporal variation in the dispersion measure as structure in the interstellar medium drifts in the direction perpendicular to the line of sight to a pulsar. The density structure can also be probed by observing pulse broadening due to arrival of the pulse along more then one path (e.g., Bhat et al. 2004).

11.7$\star$ Extreme Scattering Events

The scattering phenomena described above appear to be consistent with an ISM with a power-law spectrum of electron density fluctuations, presumably resulting from interstellar turbulence. However, in the course of monitoring the fluxes of compact extragalactic radio sources, Fiedler et al. (1987) discovered that one source (0954+658), which had been steady for years, underwent remarkable changes in apparent brightness over the course of ~3 months. At 2.7 GHz, the source brightness declined by a factor ~2 for ~0.15 yr, but with brightness peaks just before and just after the minimum (see Figure 11.10). The 8.1 GHz light curve differed from the 2.7 GHz light curve, with an interval of strong brightness variations that seemed to be of shorter duration than at 2.7 GHz.

It seems unlikely that the brightness variations are intrinsic to the source, and most likely that they are instead the result of refraction as the waves propagate through the ISM of the Milky Way – some plasma structure in the ISM acts as a lens, either magnifying or demagnifying the source.

The plasma lens producing the refraction must be small – if it is moving across the LOS at $\lesssim 250\,\mathrm{km\,s^{-1}}$, then the transverse size of the lensing structure must be $L \lesssim (250\,\mathrm{km\,s^{-1}} \times 0.2\,\mathrm{yr}) \approx 10\,\mathrm{AU}$. If the lens is at a distance of $d \approx 10\,\mathrm{kpc}$, then the angular size is only $\theta \equiv L/d \approx 1$ milli-arc-sec. Order-unity changes in brightness require that the blob be able to bend rays by an angle $\gtrsim \theta/2$, which requires that $(|\Delta m|\, L)/(L/2) \gtrsim \theta/2 = L/2d$. Thus $|\Delta m| \gtrsim L/4d \approx 10\,\mathrm{AU}/4d$. Evaluating this condition at 2 GHz, using Eq. (11.30) for Δm, we find

$$n_e = \frac{m_e \omega^2}{2\pi e^2} |\Delta m| = 1 \times 10^3 \left(\frac{5\,\mathrm{kpc}}{d} \right) \mathrm{cm}^{-3} . \tag{11.52}$$

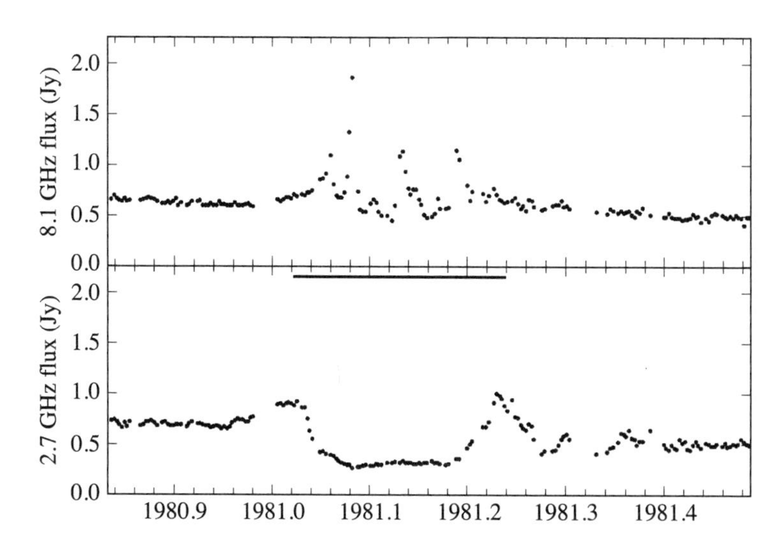

Figure 11.10 The so-called "Fiedler event" reported by Fiedler et al. (1987), showing remarkable changes in the brightnes of the the extragalactic radio source 0954+658 at 2.7 GHz and 8.1 GHz. The large brightness variations may be due to refraction by a small but dense blob of ionized gas moving across the line-of-sight (see text). Figure from Lazio et al. (2001).

Ionized gas normally has $T \gtrsim 10^4$ K, with sound speed $\gtrsim 10\,\mathrm{km\,s^{-1}}$, and therefore the pressure in a blob with $n_e \approx 10^3\,\mathrm{cm^{-3}}$ would far exceed the typical thermal pressure $p/k \approx 3800\,\mathrm{cm^{-3}}\,\mathrm{K}$ near the Galactic plane; well above and below the plane the ambient pressures are even smaller. The overpressured plasma would therefore expand supersonically, and would disperse on an expansion time scale $L/10\,\mathrm{km\,s^{-1}} \approx 5\,\mathrm{yr}$. Such blobs are therefore not expected to be present in the ISM, except perhaps as rare transient events following a sudden strong compression of ionized gas, or resulting from sudden ionization of cool dense gas by some very strong source of ionization.

Monitoring of radio sources has turned up a total of 10 possible lensing events – now referred to as **extreme scattering events (ESEs)** – in 594 source-years of monitoring (Fiedler et al. 1994). The sources being monitored appeared to be undergoing lensing events fully $\sim$0.1% of the time, implying that these plasma blobs cover $\sim$0.1% of the sky! However, *there are no known interstellar phenomena that are expected to have such ionized structures associated with them.* The lensing events may be associated with prominent "loops" of synchrotron emission (Fiedler et al. 1994), suggesting that interstellar blast waves viewed edge-on might provide the plasma structures required to account for the ESEs. However, no satisfactory models have been put forward that reproduce observed ESE light curves.

Chapter Twelve

Interstellar Radiation Fields

The physical state of interstellar gas is determined in large part by interaction of the gas and dust with the radiation field. The chemical and ionization state of the gas depends on the rates for photoionization and photodissociation. The thermal state is strongly affected by the heating effects of photolectrons ejected from both atoms and dust grains. Dust grains are heated by the radiation field and reradiate the energy at longer wavelengths; the dust grain emission spectrum is determined by the spectrum and intensity of the radiation field to which the dust is exposed. The radiation field can also have dynamical consequences: radiation pressure from anisotropic radiation can accelerate dust and gas, sometimes to high velocities.

The interstellar radiation field in the solar neighborhood is dominated by six components:

- Galactic synchrotron radiation from relativistic electrons.
- The cosmic microwave background radiation.
- Far-infrared (FIR) and infrared (IR) emission from dust grains heated by starlight.
- Emission from $\sim 10^4$ K plasma – free–free, free–bound, and bound–bound transitions.
- Starlight – photons from stellar photospheres.
- X-ray emission from hot (10^5 to 10^8 K) plasma.

The contribution of each of these to the interstellar radiation field will be briefly discussed in this Chapter. Our objective is to gain an estimate for the specific energy density u_ν to enable calculation of rates for photoexcitation, photoionization, and photodissociation for atoms, ions, and molecules of interest; heating of dust grains; and photoelectric emission from dust. Components will be discussed in order of increasing frequency.

12.1 Galactic Synchrotron Radiation

The interstellar medium (ISM) contains relativistic electrons that emit synchrotron radiation in the galactic magnetic field; this synchrotron radiation dominates the sky brightness at frequencies $\nu \lesssim 1\,\mathrm{GHz}$. The synchrotron emissivity is spatially variable, with enhancements near supernova remnants due in part to electron acceleration associated with the supernova blastwave and in part to increased magnetic field strengths in the shocked gas.

The all-sky 408-MHz map by Haslam et al. (1982) – see Plate 4 – has a mean intensity in the sychrotron component corresponding to an antenna temperature $\langle T_A \rangle = 31.3\,\mathrm{K}$ (Finkbeiner 2004, private communication). The spectrum is approximated by a power-law, $(u_\nu)_{\rm synch} \propto \nu^\beta$. La Porta et al. (2008) find $\beta = -0.95 \pm 0.15$ between 0.408 and 1.42 GHz for $|b| > 5°$. Thus the sky-averaged Galactic synchrotron background has

$$\nu u_\nu \approx 2.86 \times 10^{-19} \nu_9^{0.05}\,\mathrm{erg\,cm^{-3}} \quad , \quad \text{where } \nu_9 \equiv \nu/\,\mathrm{GHz}, \tag{12.1}$$

between 408 MHz and 1.42 GHz. At higher frequencies, the synchrotron spectrum appears to steepen. Finkbeiner (2004) finds $u_\nu \propto \nu^{-1.04}$ between 0.4 and 23 GHz for the bright synchrotron-emitting structure known as "Loop I," and assumes this spectrum to be applicable throughout the high-latitude sky. The high-frequency behavior depends on the processes responsible for accelerating the ultrarelativistic electron population. In Figure 12.1, the synchrotron spectrum is approximated by

$$(\nu u_\nu)_{\rm synch} \approx 3.05 \times 10^{-19} \frac{\nu_9^{0.1}}{1 + 0.04\nu_9}\,\mathrm{erg\,cm^{-3}} \quad ; \tag{12.2}$$

this is consistent with what is known about the synchrotron background at $\nu \leq 23\,\mathrm{GHz}$.[1] The sky-averaged antenna temperature due to synchrotron radiation is

$$\langle \Delta T_A \rangle_{\rm synch} \approx 2.37 \frac{\nu_9^{-2.9}}{1 + 0.04\nu_9}\,\mathrm{K} \quad , \tag{12.3}$$

which gives $\langle \Delta T_A \rangle_{\rm synch} \approx 705\,\mathrm{K}$ at $\nu = 140\,\mathrm{MHz}$ – the radio sky is *very* bright! From Eq. (12.3), it is apparent that the sky-averaged Galactic synchrotron intensity equals the intensity of the $T = 2.7\,\mathrm{K}$ cosmic microwave background at $\sim 1\,\mathrm{GHz}$. The total energy density in synchrotron radiation obtained by integrating Eq. (12.2) over frequency is small (see Table 12.1).

12.2 Cosmic Microwave Background Radiation

The cosmic microwave background (CMB) is very close to blackbody radiation with a temperature $T_{\rm CMB} = 2.7255 \pm 0.0006\,\mathrm{K}$ (Fixsen 2009). The radiation is essentially isotropic, with the primary departure from isotropy consisting of a dipole perturbation due to motion of the Sun relative to the CMB rest frame with a velocity $v = 372{\pm}1\,\mathrm{km\,s^{-1}}$ toward $\ell = (264.31{\pm}0.15)\,\mathrm{deg}$, $b = (48.05{\pm}0.10)\,\mathrm{deg}$. The CMB exceeds the Galactic synchrotron background at frequencies $\nu \gtrsim 1\,\mathrm{GHz}$.

[1] The functional form assumed in Eq. (12.2) to describe the steepening for $\nu \gg 1\,\mathrm{GHz}$ is rather arbitrary, but of little consequence, as other radiation components dominate for $\nu \gtrsim 1\,\mathrm{GHz}$.

Table 12.1 Interstellar Radiation Field (ISRF) Components

Component		$u_{\rm rad}$ ($\rm erg\,cm^{-3}$)
Radio synchrotron [Eq. (12.2)]		2.7×10^{-18}
CMB, $T = 2.725\,\rm K$		4.17×10^{-13}
Dust emission		5.0×10^{-13}
Free–free,free–bound,two-photon		4.5×10^{-15}
Starlight: $T_1 = 3000\,\rm K$, $W_1 = 7 \times 10^{-13}$	4.29×10^{-13}	
$T_2 = 4000\,\rm K$, $W_2 = 1.65 \times 10^{-13}$	3.20×10^{-13}	
$T_3 = 7500\,\rm K$, $W_3 = 1 \times 10^{-14}$	2.39×10^{-13}	
$\lambda < 2460$ Å UV (Eq. 12.7)	7.11×10^{-14}	
Starlight total		1.06×10^{-12}
Hα		8×10^{-16}
Other $\lambda \geq 3648$ Å H lines = $1.1 \times$ Hα:		9×10^{-16}
$0.1 - 2\,\rm keV$ x rays		1×10^{-17}
ISRF total		1.98×10^{-12}

12.3 Free–Free Emission and Recombination Continuum

The thermal plasma radiates free–free emission and recombination continuum, as discussed in Chapter 10. Plate 3b shows an all-sky map of Hα emission (Finkbeiner 2003). The Hα intensity exceeds 100 R in many bright H II regions (see Plate 3b). The all-sky-averaged Hα intensity is 8.04 R (Finkbeiner 2005, private communication), corresponding to a sky-averaged emission measure $\langle EM \rangle \approx 18\,\rm cm^{-6}\,pc$, if the H$\alpha$ is emitted from $T \approx 8000\,\rm K$ gas. In Figure 12.1, we show the spectrum calculated for plasma at $T = 8000\,\rm K$ and an angle-averaged emission measure $\langle EM \rangle = 18\,\rm cm^{-6}\,pc$.[2]

12.4 Infrared Emission from Dust

Infrared emission from dust dominates the sky brightness between $\nu \approx 500\,\rm GHz$ ($\lambda = 600\,\mu\rm m$) and $\sim 6 \times 10^{13}\,\rm Hz$ ($\lambda = 5\,\mu\rm m$). Plate 2 is an all-sky map of the emission from dust at $100\,\mu\rm m$, produced by Schlegel et al. (1998) from observations by the Infrared Astronomy Satellite (IRAS) and the Diffuse Infrared Background Experiment (DIRBE) on the Cosmology Background Explorer (COBE) satellite. About $\frac{2}{3}$ of the power radiated by the dust is at $\lambda > 50\,\mu\rm m$; this portion of the emission spectrum can be approximated as thermal emission from dust grains at a temperature $T_{\rm dust} \approx 17\,\rm K$, with an angle-averaged optical depth $\tau(\lambda) \approx 1.5 \times 10^{-3} (100\,\mu{\rm m}/\lambda)^{1.7}$. (The optical properties of dust grains in the FIR will be

[2]The correlation of Hα and 43 GHz free–free emission at $|b| \gtrsim 10^\circ$ indicates a lower ratio of 43 GHz free-free to Hα than expected for $T \approx 8000\,\rm K$ H II (Dobler et al. 2009), apparently because some of the high-latitude Hα is actually reflected light, and some of the ionized gas is cooling and recombining at temperatures less than $\sim$8000 K (Dong & Draine 2011).

discussed in Chapter 24).

However, about $\frac{1}{3}$ of the power radiated by dust is at $\lambda < 50\,\mu\text{m}$, and much of this power is radiated in vibrational emission bands at 3.3, 6.2, 7.7, 8.6, 11.3, and 12.7 μm that are thought to be emitted by very small polycyclic aromatic hydrocarbon (PAH) particles that have undergone single-photon heating, as will be discussed in Chapter 24.

Figure 12.1 shows the infrared emission spectrum calculated for the "DL07" dust model (Draine & Li 2007) consisting of carbonaceous grains (including PAHs) and amorphous silicate grains illuminated by a range of starlight intensities. The all-sky average intensity is approximately reproduced by the dust emission from a medium with hydrogen column density $N_{\rm H} = 2.7 \times 10^{21}\,\text{cm}^{-2}$, with the dust grains illuminated by a range of radiation intensities, but with an average intensity equal to 1.2 times the radiation field estimated by Mathis et al. (1983).

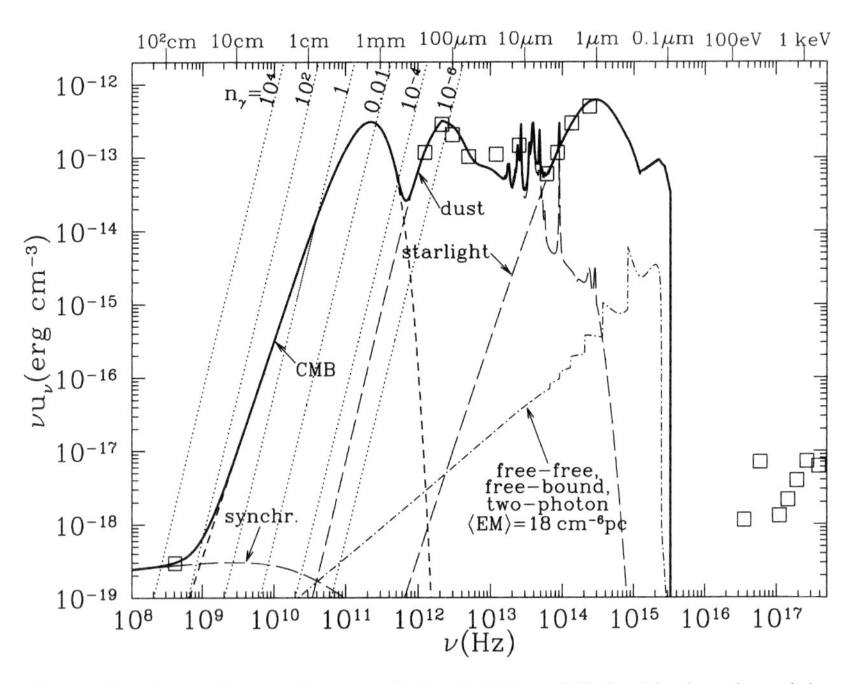

Figure 12.1 Interstellar continuum radiation field in an H I cloud in the solar neighborhood (see text). Spectral lines are not included. Solid line is the sum of all components for $h\nu \leq 13.6\,\text{eV}$. Squares show the measured sky brightness at 408 MHz (Haslam et al. 1982), the all-sky measurements by COBE-DIRBE in 10 broad bands from 240 μm to 1.25 μm (Arendt et al. 1998), and all-sky measurements by ROSAT between 150 eV and 2 keV (Snowden 2005, private communication). Dotted lines are contours of constant photon occupation number n_γ.

12.5 Starlight in an H I Region

Within an H I region, there is very little radiation at energies between the hydrogen ionization energy of 13.6 eV and $\sim 10^2$ eV, because photons in this energy range are very strongly absorbed by neutral H and He. In the energy range 1 to 13.6 eV, most of the photons are starlight. Mathis et al. (1983) have approximated the local starlight background as a sum of three dilute[3] blackbodies:

$$\nu u_\nu = \sum_{j=1}^{3} \frac{8\pi h\nu^4}{c^3} \frac{W_j}{e^{h\nu/kT_j} - 1} \qquad \text{for } \lambda > 2450\,\text{Å}, \tag{12.4}$$

and a piecewise power-law approximation for $912\,\text{Å} < \lambda < 2450\,\text{Å}$ [Eq. (12.7) below]. To improve agreement with the COBE-DIRBE photometry, we have increased the dilution factor W_1 by 75%, from $W_1 = 4\times10^{-13}$ to 7×10^{-13}, and raised W_2 from 1.0×10^{-13} to 1.65×10^{-13}. This estimate is plotted in Figure 12.1, and the blackbody parameters T_j, W_j are given in Table 12.1.

Far-ultraviolet radiation is of considerable importance in the neutral ISM, because it can (1) photoexcite and photodissociate H_2 and other molecules, (2) photoionize many heavy elements, and (3) eject photoelectrons from dust grains. The intensity of this ultraviolet radiation will be spatially variable, because the O and B stars that are the primary emitters are neither numerous nor randomly distributed, and because of the strong attenuating effects of interstellar dust in the far ultraviolet. Figure 12.2 shows various estimates for the intensity of far-ultraviolet radiation in the solar neighborhood.

Habing (1968) made an early estimate of the intensity of the ultraviolet radiation field, $\nu u_\nu \approx 4\times10^{-14}\,\text{erg cm}^{-3}$ at $\lambda = 1000\,\text{Å}$, i.e., $h\nu = 12.4\,\text{eV}$. It is convenient to reference other estimates of the intensity near 1000 Å to this canonical value, so we define the dimensionless parameter

$$\chi \equiv \frac{(\nu u_\nu)_{1000\,\text{Å}}}{4\times10^{-14}\,\text{erg cm}^{-3}} \quad . \tag{12.5}$$

χ is a good parameter for characterizing the radiation field when considering photoexcitation of H_2, or photoionization of species with ionization potentials I between 10 and 13.6 eV. For some purposes, a wider range of UV is of interest. Habing's UV spectrum, if integrated between 6.0 and 13.6 eV, gives an energy density $u_{\text{Hab}}(6 - 13.6\,\text{eV}) = 5.29\times10^{-14}\,\text{erg cm}^{-3}$. We define a parameter

$$G_0 \equiv \frac{u(6 - 13.6\,\text{eV})}{5.29\times10^{-14}\,\text{erg cm}^{-3}} , \tag{12.6}$$

giving the overall 6 to 13.6 eV intensity relative to Habing's estimate.

The ultraviolet background was estimated by Draine (1978) from published observations, and fitted by a polynomial. This estimate had $G_0 = 1.69$ and $\chi = 1.71$.

---

[3]A "dilute blackbody" is defined to be a spectrum equal to a blackbody multiplied by a "dilution factor" $W < 1$.

Mathis et al. (1983, henceforth MMP83) represented the local ultraviolet background between 2460Å and 912Å by three power-law segments:

$$\nu u_\nu = \begin{cases} 2.373 \times 10^{-14} (\lambda/\,\mu\mathrm{m})^{-0.6678}\,\mathrm{erg\,cm^{-3}} & 1340 - 2460\,\text{Å} \\ 6.825 \times 10^{-13} (\lambda/\,\mu\mathrm{m})\,\mathrm{erg\,cm^{-3}} & 1100 - 1340\,\text{Å} \\ 1.287 \times 10^{-9} (\lambda/\,\mu\mathrm{m})^{4.4172}\,\mathrm{erg\,cm^{-3}} & 912 - 1100\,\text{Å} \end{cases} . \quad (12.7)$$

The MMP83 estimate for the ISRF has $\chi = 1.23$ and $G_0 = 1.14$.

Henry et al. (1980) measured the UV ISRF over 1/3 of the sky, and from this estimated the full-sky radiation field in the UV; their results are shown in Fig. 12.2. Gondhalekar et al. (1980) measured direct starlight in the UV; because they did not measure diffuse radiation, this gives a lower bound[4] on the ISRF. The available evidence indicates that the MMP83 radiation field is a good estimate for the ISRF in the solar neighborhood.[5]

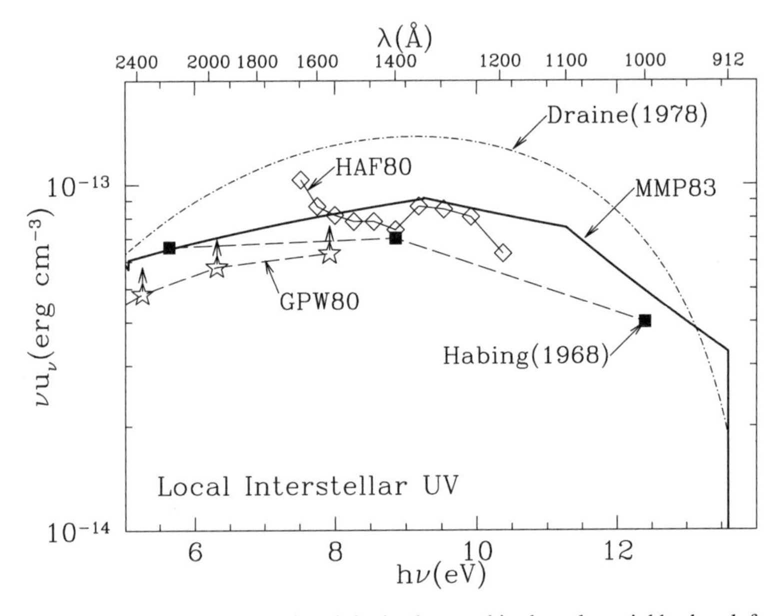

Figure 12.2 Estimates for the ultraviolet background in the solar neighborhood, from Habing (1968), Draine (1978), and Mathis et al. (1983, MMP83). The observational determination by Henry et al. (1980, HAF80), and the observational lower bound from Gondhalekar et al. (1980, GPW80) are also shown.

[4]Diffuse UV (mainly light scattered by dust plus fluorescent emission from H_2) may be comparable in energy density to the direct starlight.

[5]A very recent measurement of the 1370–1720Å radiation field by Seon et al. (2011) obtains intensities approximately a factor 2 below previous measurements by HAF80 and GPW80.

12.6 X Rays from Hot Plasma

Supernovae are estimated to inject energy into the ISM at a rate of $\sim 10^{51}\,\mathrm{erg}/100\,\mathrm{yr}$, producing hot plasma as well as accelerating nuclei and electrons to ultrarelativistic energies (see Chapters 39 and 40).

Most of the supernova kinetic energy goes into thermal plasma, and is eventually radiated as X-rays and extreme ultraviolet (EUV), with a broad range of energies. The lower energy (EUV and soft X-ray) photons can be absorbed by small amounts of neutral gas; as a result, the soft X-ray background is highly variable within the galaxy. We can measure it directly only at the location of the Sun.

The X-ray sky has been mapped by ROSAT in seven broad energy bands (Snowden et al. 1994). Plate 5 is an all-sky map of the 0.5–1.0 keV emission obtained by the ROSAT satellite (Snowden et al. 1995). Snowden (2005, private communication) has integrated over the all-sky maps to determine the density of 0.1 to 2 keV photons originating outside the solar system; the resulting energy density is plotted in Figure 12.1. It is evident that the energy density in interstellar X-rays is very small, of order $\sim 10^{-6}$ of the total radiation field. Nevertheless, these x rays are of some importance because of their ionizing properties.

In galaxies with active galactic nuclei, the X-ray intensities can be much higher than in the Milky Way.

12.7 Radiation Field in a Photodissociation Region near a Hot Star

Consider a luminous massive star of spectral type O (accompanied by additional lower luminosity stars) exciting an H II region adjacent to a molecular cloud – the Orion Nebula (M42) is a nearby example. As will be discussed in Chapter 15, the radiation from the star at energies between the hydrogen ionization threshold of 13.6 eV and $\sim 100\,\mathrm{eV}$ will be absorbed within the H II region surrounding the star. Radiation below 13.6 eV will arrive at the boundary of the H II region attenuated only by absorption by dust. After crossing the ionization front separating the ionized and neutral regions, the photons will enter what is referred to as a **photodissociation region**, or PDR (the structure of PDRs is discussed in §31.7). Fig. 12.3 shows the spectrum of the radiation field within the PDR.

The PDR is illuminated by the same CMB and galactic synchrotron radiation as a diffuse H I cloud. In addition, the radiation field includes:

- $h\nu < 13.6\,\mathrm{eV}$ radiation from the nearby O star. (The photons with $h\nu > 13.6\,\mathrm{eV}$ are absorbed within the H II region.)
- Free–free radiation from the H II region.
- Line emission from the H II region.
- Emission from the warm dust in the PDR.

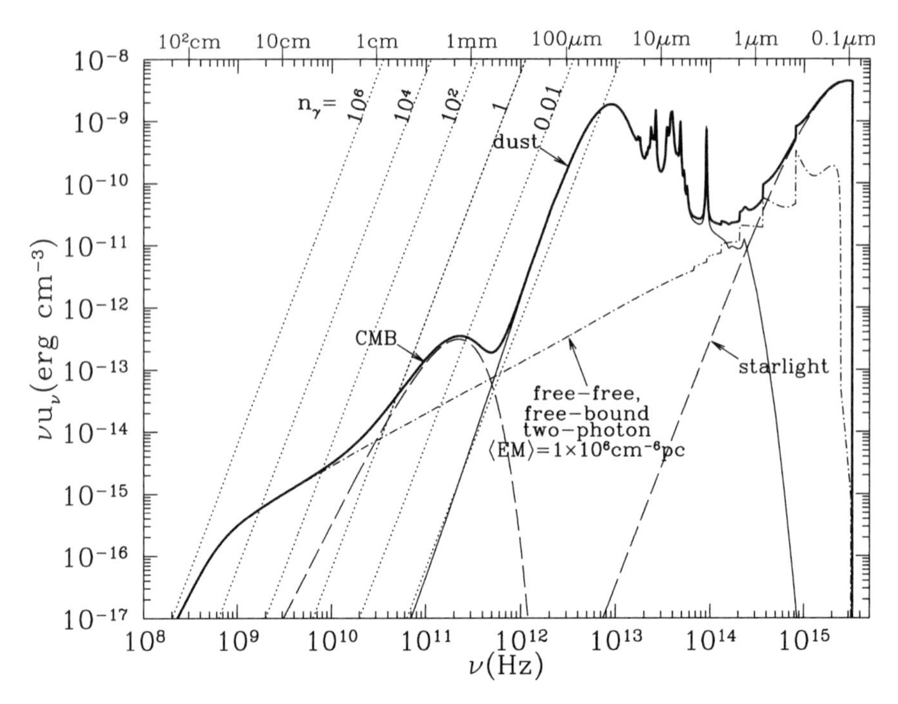

Figure 12.3 Radiation field in neutral gas adjacent to an H II region with peak EM $\approx 4 \times 10^6\,\mathrm{cm}^{-6}\,\mathrm{pc}$. Starlight is from an O star with $T = 3.5 \times 10^4$ K and dilution factor $W = 2 \times 10^{-13}$ (e.g., at a distance $d \approx 6.4 \times 10^{17}$ cm from a star of spectral type O8V). Spectral lines, e.g., Hα, are not shown. The infrared emission from dust is calculated for the DL07 dust model with a starlight heating rate 3000 times the ISRF shown in Fig. 12.1. Dotted lines are contours of constant photon occupation number n_γ.

Bright H II regions, such as the Orion Nebula, have peak emission measures $EM \approx 10^6 - 10^7\,\mathrm{cm}^{-6}\,\mathrm{pc}$. From the perspective of a point in the PDR, the H II region will cover ~50% of the sky, and the angle-averaged EM will be $\sim \frac{1}{4}$ of the peak value. In Figure 12.3, we have assumed an angle-averaged $EM \approx 1 \times 10^6\,\mathrm{cm}^{-6}\,\mathrm{pc}$.

The $h\nu < 13.6\,\mathrm{eV}$ starlight entering the PDR mostly ends up absorbed by dust, with the energy reradiated in the IR. The energy density of this radiation will be similar to the energy density of starlight entering the PDR. A simple estimate for the IR radiation field is shown in Fig. 12.3.[6]

[6]The dust emission spectrum shown in Fig. 12.3 is for the DL07 dust model illuminated by starlight with an intensity $U = 3000$ relative to the ISRF, with an angle-averaged column density $N_{\mathrm{H}} = 5 \times 10^{21}\,\mathrm{cm}^{-2}$.

PLATE 1. All-sky image of $\sim 5 \times 10^8$ stars detected by the 2MASS survey: Blue = 1.2 μm, green = 1.65 μm, and red = 2.2 μm. This Hammer equal-area projection has the Galactic Center $[(\ell, b) = (0, 0)]$ at the center of the image, $\ell = 180°$ at left and right, with ℓ increasing from right to left. The stars are in a disk-like geometry, and in a stellar "bulge" concentrated around the center. The ISM (seen by the obscuration and reddening produced by dust) is primarily in a disk-like geometry. The LMC and SMC are visible below and to the right of center. This mosaic image was obtained as part of the Two Micron All Sky Survey (2MASS), a joint project of the Univ. of Massachusetts and IPAC/Caltech, funded by NASA and the NSF. Credit: 2MASS/J. Carpenter, T.H. Jarrett, & R. Hurt.

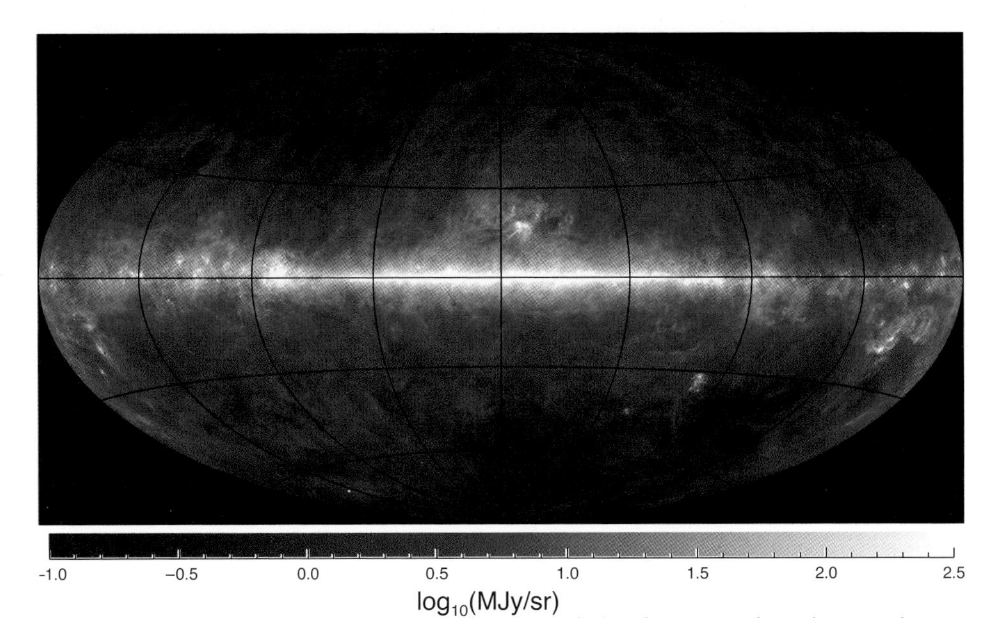

PLATE 2. The 100 μm sky, after subtracting the emission from warm interplanetary dust particles within the Solar system. The LMC and SMC are visible at $(\ell, b) = (280°, -33°)$ and $(303°, -44°)$. The bright emission near $\ell = 80°$ (in Cygnus) corresponds to dust in the Perseus spiral arm (see Fig. 11.1) and the Cygnus OB2 association, at a distance of ~ 1.45 kpc. Based on observations with the IRAS and COBE satellites. Image courtesy of D. Finkbeiner.

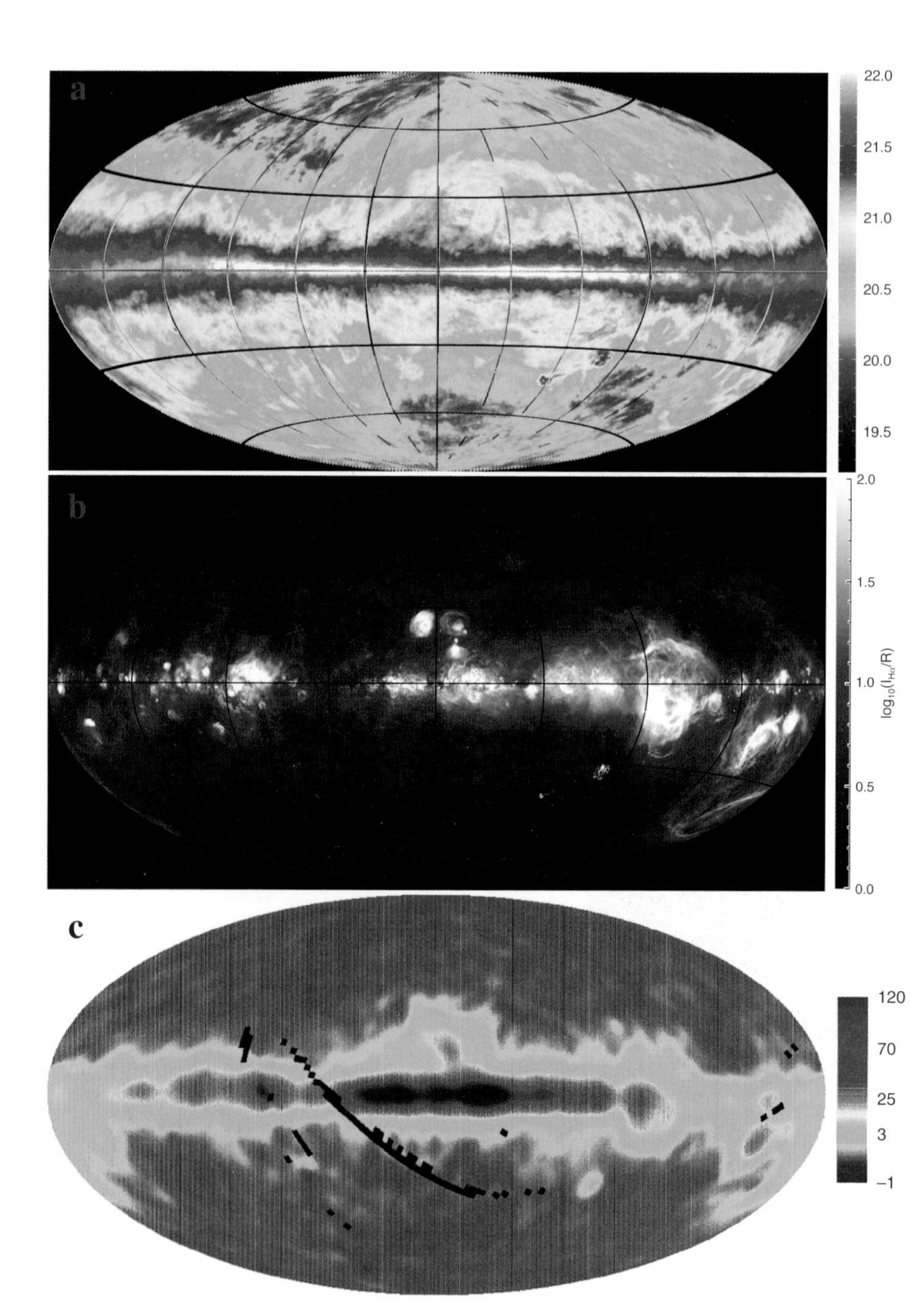

PLATE 3. a: All-sky map of H I 21 cm line intensity from the LAB survey (Kalberla et al. 2005), with angular resolution $\sim 0.6^\circ$. Scale gives $\log_{10} N(\text{H I})(\text{cm}^{-2})$. The LMC and SMC are visible (cf. Plate 1), with a connecting H I "bridge." Image courtesy of P. Kalberla. **b:** Hα from Finkbeiner (2003). Many bubble-like ionized structures are apparent, including the Orion Nebula and Barnard's Loop near $(\ell, b) \approx (209^\circ, -19^\circ)$; the Gum nebula $(256^\circ, -9^\circ)$; and the ζ Oph H II region $(6^\circ, 23^\circ)$. Image courtesy of D. Finkbeiner. **c:** $[\text{C II}]158\,\mu\text{m}$ from Fixsen et al. (1999), reproduced by permission of the AAS. The map is smoothed to 10° resolution. Black stripes are unobserved regions. Scalebar shows $[\text{C II}]158\,\mu\text{m}$ intensity in $\text{nW m}^{-2}\,\text{sr}^{-1}$.

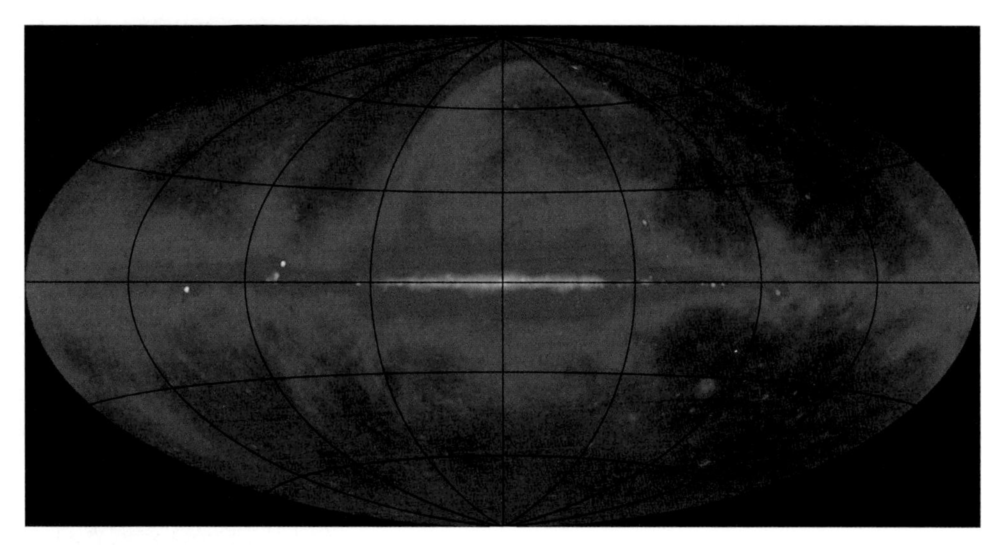

PLATE 4. All-sky map of 408 MHz continuum emission (Haslam et al. 1982), synchrotron radiation from highly relativistic electrons. In addition to bright emission along the Galactic plane, there are conspicuous extensions above and below the plane, in particular the so-called North polar spur, starting from the Galactic plane at $\ell \approx 15^\circ$ and extending upward in a loop-like geometry. Bright sources in the Galactic plane include: the Vela/Puppis SNR near $(261^\circ, -3^\circ)$, Sgr A at the Galactic Center; the Cygnus superbubble near $(85^\circ, 0)$, and Cas A at $(112^\circ, -2^\circ)$. The LMC and SMC are visible. Image produced by NASA SkyView, using data from Max-Planck Institüt für Radioastronomie.

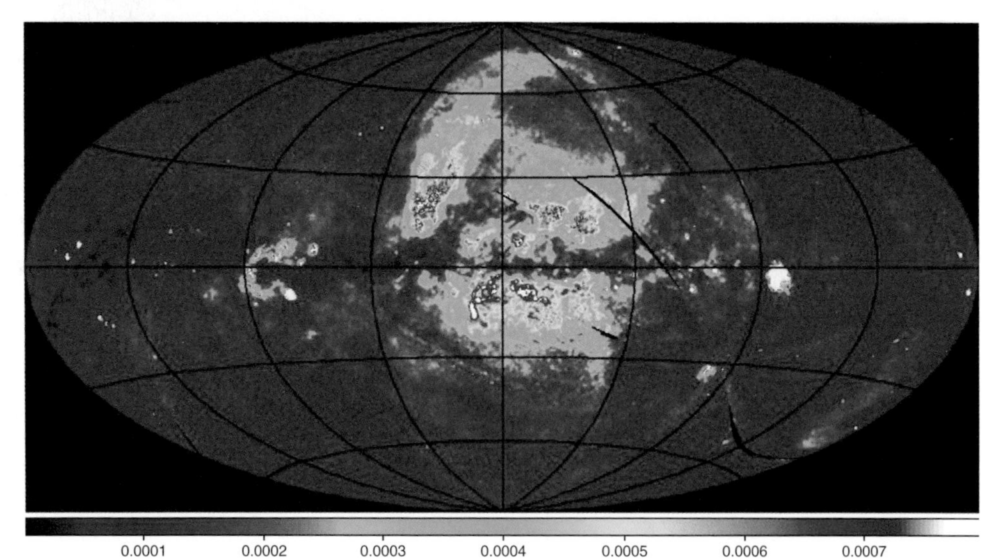

PLATE 5. ROSAT PSPC all-sky map (Snowden et al. 1995) in the 0.5-1.0 keV band. Color bar shows ROSAT count rate in cts s^{-1} arcmin^{-2}. Note the X-ray emission from the North Polar Spur, from the Cygnus superbubble, and from the Vela/Puppis SNR, all with associated radio synchrotron emission seen in Plate 6. Much of the emission seen toward the inner Galaxy above and below the Galactic plane, including the North Polar Spur, is thought to be plasma within a few hundred pc. Absorption is evident along the Galactic plane. Courtesy of Steve Snowden (NASA/GSFC), based on data from the ROSAT All-Sky Survey (MPE/Garching).

PLATE 6. The Andromeda galaxy (M31), and its small companions M32 and NGC 205. Image width = 3.5° = 47 kpc / 770 kpc; N is 50.0° left of vertical. Upper image: blue = 3.6 μm (mainly starlight), red = 8 μm (mainly emission from PAHs). NGC 205 (above the center of M31) and M32 (below and to the right of center) are easily seen in starlight, but have very little 8 μm emission. Lower left image is 3.6 μm only, showing the relatively smooth distribution of stars. Lower right image is 8 μm only, showing the concentration of the ISM into rings and arms. Image courtesy: NASA / JPL-Caltech / P. Barmby (Harvard-Smithsonian CfA).

PLATE 7. The nearby starburst galaxy M82. Blue = 1.5 keV X-rays, green = starlight, red = 8 μm (PAH emission). Image width = 10.7′ = 10.9 kpc / 3.5 Mpc. Energy released by stars (stellar winds, SNe) appears to be driving a hot outflow (seen in X-rays) out of the stellar disk. There is also extended 8 μm emission from PAHs as much as ~4 kpc above the disk, presumably located in cooler gas, bounding the X-ray outflow, that has somehow been lifted out of the disk. Credit: NASA / JPL-Caltech / STScI / CXC/ Univ. of Arizona.

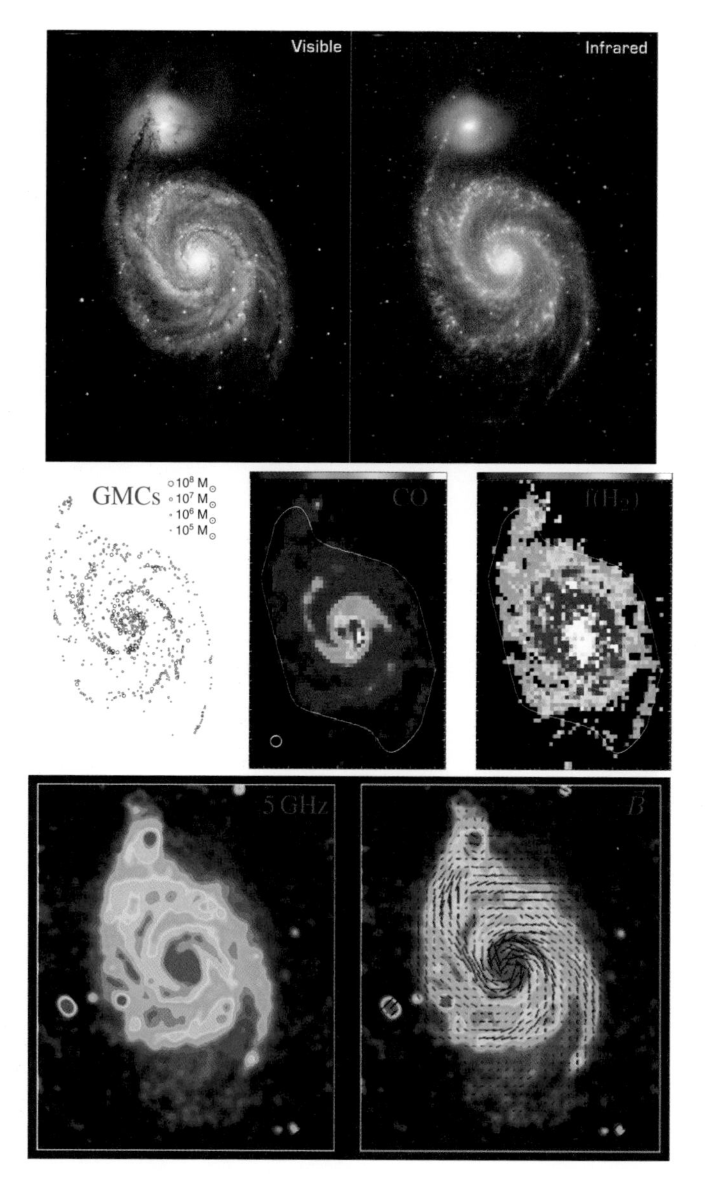

PLATE 8. M51a (NGC 5194, the Whirlpool) and its companion M51b (NGC 5195).
Vis.: Visible-light image, with conspicuous dust lanes located along the spiral arms.
IR: blue = 3.6 μm, red = 8 μm. The dust lanes, dark in the visible-light image, are glowing in 8 μm PAH emission. Note the numerous "holes" where PAH emission is weak. The companion galaxy, M51b, is bright in starlight, but has very little PAH emission.
GMCs: Locations of GMCs with $M > 4 \times 10^5\ M_\odot$ (detected in CO 1–0).
CO: CO 1-0 line intensity, smoothed over a $22''$ beam (circle in lower left corner); violet = $5\ \mathrm{K\,km\,s^{-1}}$, red = $50\ \mathrm{K\,km\,s^{-1}}$.
f(H$_2$): H$_2$ fraction $2N(\mathrm{H_2})/[2N(\mathrm{H_2})+N(\mathrm{H\,I})]$: green = 0.5, white = 1.
5 GHz: 5 GHz emission (mainly synchrotron radiation from relativistic electrons).
$\vec{B}$: B-polarization of 5 GHz emission. The magnetic field is aligned with the spiral arms.
Credits: Vis. and IR: NASA/JPL-Caltech/R. Kennicutt (Univ. of Arizona)/DSS; **GMCs, CO, and f(H$_2$):** from Koda et al. (2009), reproduced by permission of the AAS; **5 GHz and $\vec{\mathrm{B}}$:** R. Beck (MPIfR Bonn), C. Horellou (Onsala Space Observatory). Image courtesy of NRAO/AUI/NSF.

PLATE 9. The Orion Nebula (M42) (Henney et al. 2007). North is up. Image width = 30$'$ = 3.6 pc / 414 pc; red = Hα + [N II]6584 + [S III]9071 + scattered starlight, green = [O III]5008 + scattered starlight, blue = Hγ + Hδ + scattered starlight. The smaller (and more obscured) M43 H II region is visible 8$'$ to the NNE. Credit: NASA, ESA, M. Robberto (Space Telescope Science Institute/ESA) and the Hubble Space Telescope Orion Treasury Project Team.

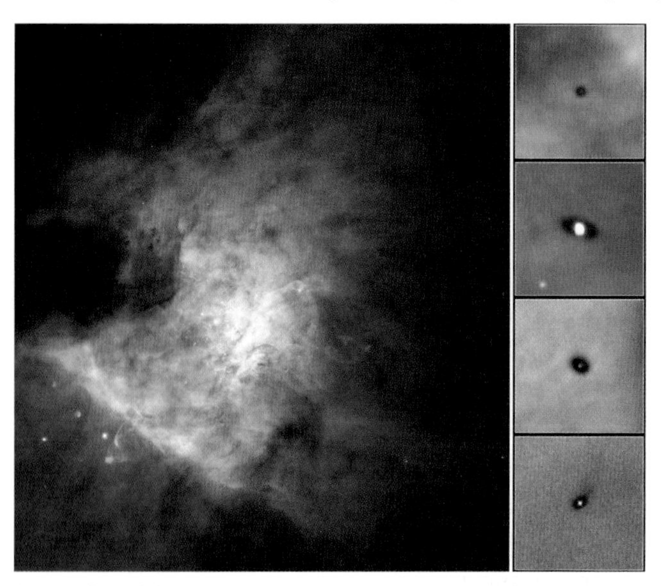

PLATE 10. Center of the Orion Nebula, imaged by Hubble Space Telescope. The bright Orion Bar is an edge-on ionization front separating neutral gas at the lower left from the H II region. Red = [N II]6584, green = Hα, blue = [O III]5008. The 4 Trapezium stars are visible near the image center. See Fig. 28.1 for a radio free-free image. At right are images of 4 protoplanetary disks silhouetted against the H II region; each image is $\sim 4''$ wide (1700 AU / 414 pc). Credit: NASA, C.R. O'Dell and S.K. Wong (Rice University), and M.J. McCaughrean (MPIA),

Plate 11. The Trifid Nebula (M20). Image width = $17.3' = 8.5\,\mathrm{pc} / 1.68\,\mathrm{kpc}$. **Left:** optical image, blue = 4400Å, green = 5500Å, red = 7000Å (incl. Hα). The Trifid Nebula is an H II region powered by a single O7 star. The H II region is bright in Hα (red); the blue haze to the North is starlight scattered by dust. **Right:** IR image (Rho et al. 2006) with blue = $3.6\,\mu\mathrm{m} + 4.5\,\mu\mathrm{m}$ (mainly starlight), green = $6 + 8\,\mu\mathrm{m}$ (mainly PAH emission), red = $24\,\mu\mathrm{m}$ (hot dust). The dust lanes across the H II region are dark in the optical image, but glow brightly in the IR image. Note the dark filaments to the NNW, that are opaque at 8 and $24\,\mu\mathrm{m}$. Credit: NOAO / NASA / JPL-Caltech / J. Rho (SSC/Caltech).

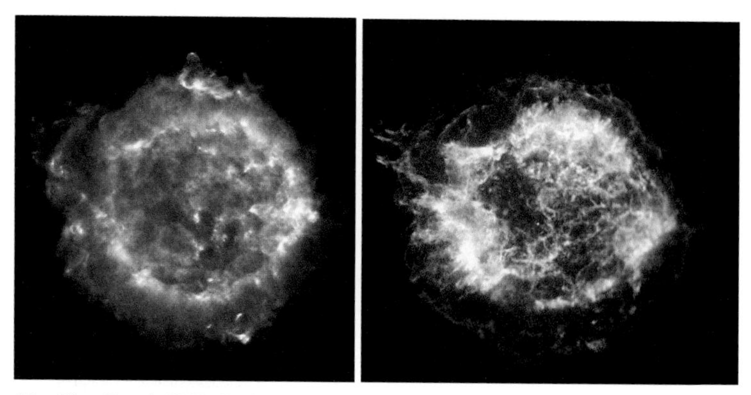

Plate 12. The Cas A SNR. **Left:** Radio synchrotron image at 1.4, 5.0, and 8.4GHz. Image width = $6.0' = 5.9\,\mathrm{pc} / 3.4\,\mathrm{kpc}$. Image courtesy of NRAO/AUI. **Right:** X-ray image, red = 0.5–1.5 keV, green = 1.5–2.5 keV, blue = 4.0–6.0 keV. From the observed expansion the explosion date is estimated to be 1681 ± 19 (Fesen et al. 2006). The SNR has a jet-like extension toward the NE. The hard X-ray emission at the outer edge is thought to be synchrotron emission from relativistic electrons. Credit: NASA / CXC / MIT / UMass Amherst / M.D. Stage et al.

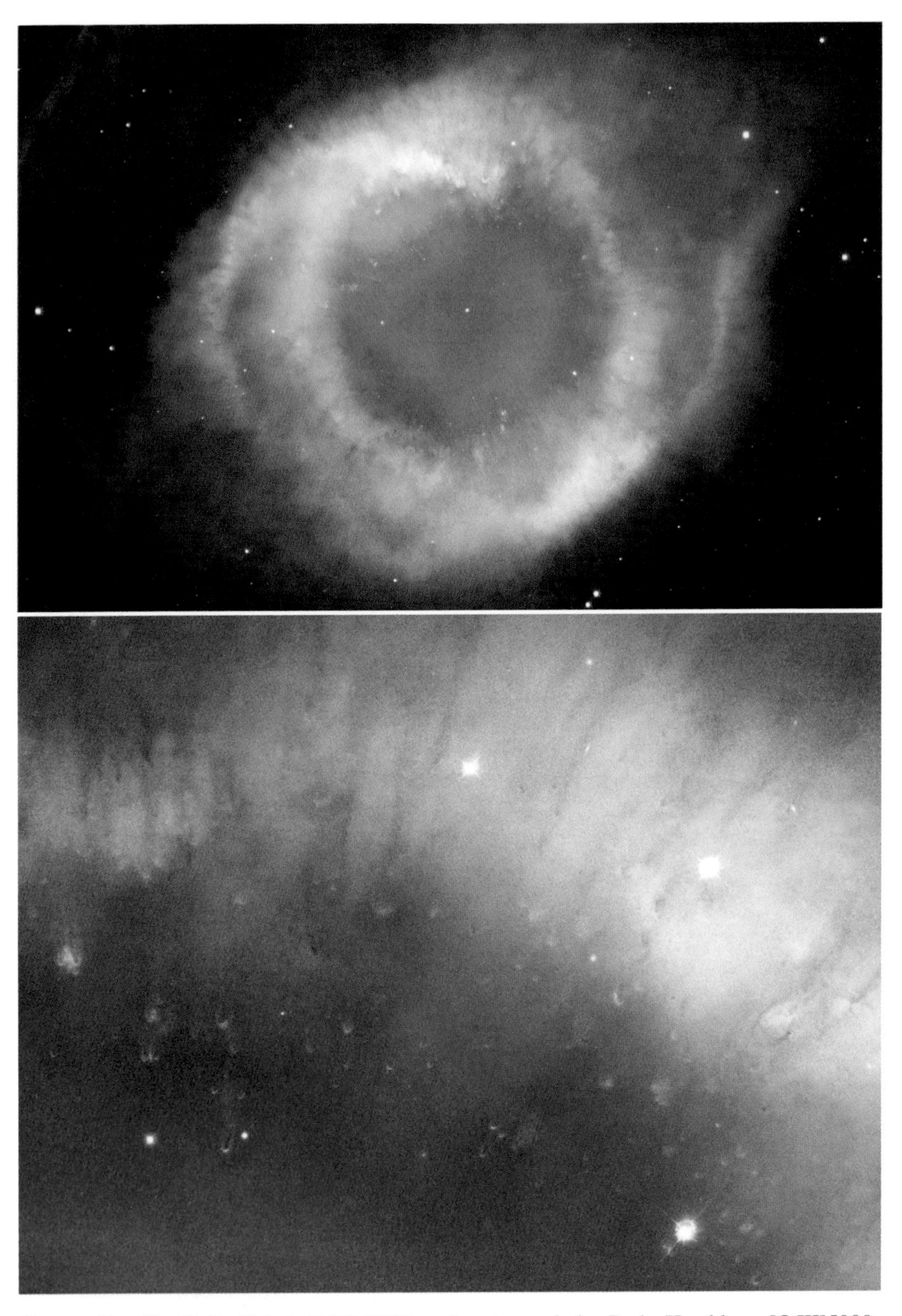

PLATE 13. The Helix Nebula NGC 7293, a planetary nebula. Red = Hα, blue = [O III]5008. **Top:** Image width = 28.7′ = 1.67 pc / 200 pc; North is up. The central star temperature is $\sim 1.0 \times 10^5$ K. The central zone, bright in [O III]5008, has a diameter $\sim$0.4 pc, and is surrounded by a ring bright in Hα. **Bottom:** Image width = 4.7′ = 0.28 pc / 200 pc; North is 71° L of vertical. Note the numerous dusty cometary filaments located in the inner edge of the gaseous ring. Each cometary filament has a head with a bright ionization/dissociation front on the side facing the central star. Credit: NASA, NOAO, ESA, the Hubble Helix Nebula Team, M. Meixner (STScI), and T.A. Rector (NRAO).

Chapter Thirteen

Ionization Processes

The ionization state of the gas varies greatly from one region to another in both the ISM and the IGM:

- In dense molecular clouds, the material is almost entirely neutral, with $x_e \equiv n_e/n_{\rm H} \lesssim 10^{-6}$.
- In H I gas, elements such as C are photoionized by starlight, and the hydrogen is partially ionized by cosmic rays, with resulting ionization fractions in the range $10^{-3} \lesssim n_e/n_{\rm H} \lesssim 10^{-1}$, depending on the density, temperature, and cosmic ray ionization rate.
- In an H II region around an early O star (e.g., spectral type O6), the hydrogen may be mostly ionized, the helium may be mostly singly ionized, and elements like oxygen or neon mainly doubly ionized (O III and Ne III).
- In a "Lyman α cloud" with $N({\rm H\,I}) \lesssim 10^{17}\,{\rm cm}^{-2}$ in the IGM, the hydrogen and helium may be mostly ionized (H II, He III), with C triply ionized (C IV).
- In a supernova remnant, elements up through carbon may be fully ionized, and elements like oxygen or neon may retain only electrons in the innermost $1s$ shell.

These variations in ionization arise primarily from the regional differences in the rates for ionizing processes. The ionization state of gas is obviously critical to a number of processes – the rate of radiative cooling, the rate at which the gas absorbs ionizing photons from stars or active galactic nuclei, and the chemical processes that can proceed within the gas. Furthermore, we are often in the situation where we cannot observe the dominant ionization states directly and need to estimate how much unseen material is present – an example is observation of H I Lyman α absorption from a region where the hydrogen is mostly ionized. This chapter is devoted to ionizing processes.

Consider some atom, molecule, or ion X. There are many different processes acting to change its ionization state by removing one or more electrons:

- **Photoelectric absorption:** $X + h\nu \rightarrow X^+ + e^-$.
- **Photoelectric absorption followed by the Auger effect:** $X + h\nu \rightarrow (X^+)^* + e^- \rightarrow X^{+n} + ne^- \ (n \geq 2)$.

- **Collisional ionization:** $X + e^- \rightarrow X^+ + 2e^-$.
- **Cosmic ray ionization:** $X + CR \rightarrow X^+ + e^- + CR$.
- **Charge exchange:** $X + Y^+ \rightarrow X^+ + Y$.

The first four processes will be discussed in this chapter; charge exchange will be discussed in Chapter 14.

13.1 Photoionization

The nonrelativistic quantum mechanics of hydrogen and one-electron (i.e., "hydrogenic") ions is simple enough that the ground-state photoelectric cross section for photons with energy $h\nu > Z^2 I_{\rm H}$ is given by an analytic expression:

$$\sigma_{\rm pi}(\nu) = \sigma_0 \left(\frac{Z^2 I_{\rm H}}{h\nu}\right)^4 \frac{{\rm e}^{4-4\arctan(x)/x}}{1-{\rm e}^{-2\pi/x}} \quad , \quad x \equiv \sqrt{\frac{h\nu}{Z^2 I_{\rm H}} - 1} \quad , \tag{13.1}$$

where Z is the atomic number of the nucleus, and σ_0, the cross section at threshold, is given by

$$\sigma_0 \equiv \frac{2^9 \pi}{3{\rm e}^4} Z^{-2} \alpha \pi a_0^2 = 6.304 \times 10^{-18} Z^{-2}\,{\rm cm}^2 \quad , \tag{13.2}$$

where ${\rm e} = 2.71828...$ in Eqs. (13.1 and 13.2). Over the energy range $Z^2 I_{\rm H} < h\nu < 10^2 Z^2 I_{\rm H}$, the exact (nonrelativistic) result (13.1) is reproduced reasonably well by a simple power-law approximation:

$$\sigma_{\rm pi}(\nu) \approx \sigma_0 \left(\frac{h\nu}{Z^2 I_{\rm H}}\right)^{-3} \quad \text{for } Z^2 I_{\rm H} < h\nu \lesssim 10^2 Z^2 I_{\rm H} \quad . \tag{13.3}$$

This approximation for H is shown in Figure 13.1a. At high energies, the asymptotic behavior of Eq. (13.1) is

$$\sigma_{\rm pi} \rightarrow \frac{2^8}{3Z^2} \alpha \pi a_0^2 \left(\frac{h\nu}{Z^2 I_{\rm H}}\right)^{-3.5} \quad \text{for } h\nu \gg Z^2 I_{\rm H} \quad . \tag{13.4}$$

The hydrogen photoelectric cross section becomes equal to the Compton scattering cross section for $h\nu \approx 2.5\,{\rm keV}$; above this energy photoionization of H is dominated by Compton scattering rather than photoelectric absorption.

The photoionization cross section for H_2 is also shown in Figure 13.1a. The $15.4\,{\rm eV}$ ionization threshold for H_2 is 15% above that for atomic H. Note that for $h\nu \gtrsim 20\,{\rm eV}$, the H_2 photoionization cross section is approximately twice that for H. The limiting value at high energy is $\sigma_{\rm pi}({\rm H}_2) = 2.8\sigma_{\rm pi}({\rm H})$ (Yan et al. 1998).

For atoms with three or more electrons, the energy dependence of the photoionization cross section can be considerably more complicated because there is more

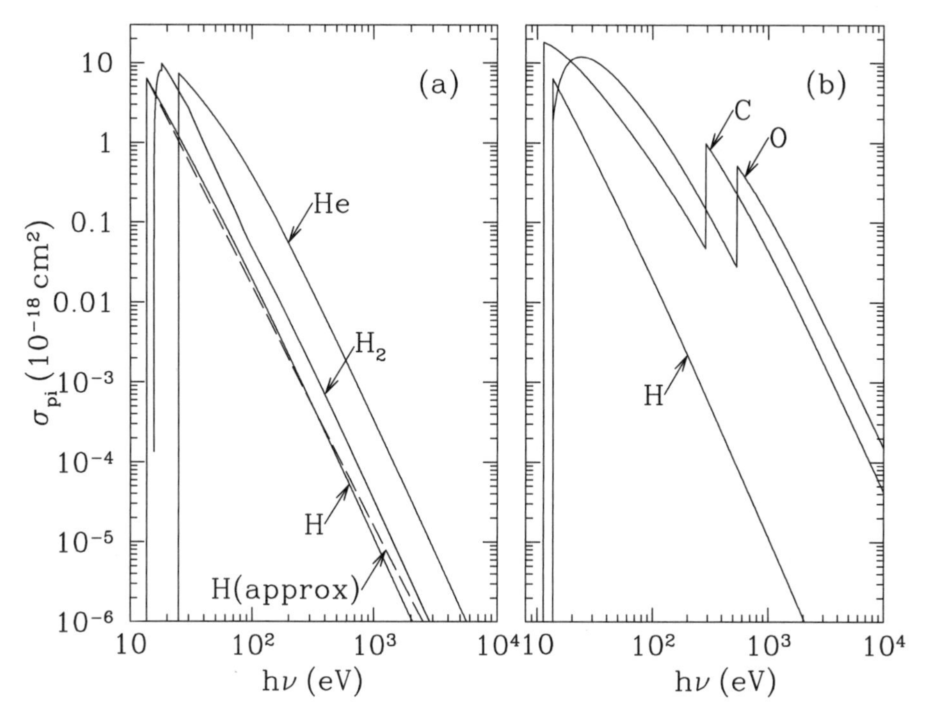

Figure 13.1 Photoionization cross sections for H, H_2, He, C, and O. The dashed line in (a) shows the power-law approximation (13.3) for H.

than one available state for the resulting ion – for example, in the case of photoionization from the O I ground state $^3P_2$ ($1s^2 2s^2 2p^4$), the electron being photoejected could come from the $1s$, $2s$ or $2p$ levels. If a $2p$ electron is removed, and $13.6\,\mathrm{eV} < h\nu < 16.9\,\mathrm{eV}$, the resulting O II ion will be in the $1s^2 2s^2 2p^3\ ^4S^o_{3/2}$ state, but if $h\nu > 16.9\,\mathrm{eV}$, the ion could also be left in the $1s^2 2s^2 2p^3\ ^2D^o_{3/2,5/2}$ state. The availability of multiple channels leads to complex structure in the photoionization cross section. For ionization by continuum radiation, this detailed structure can be smoothed and averaged over. Convenient analytic fits to the contribution of individual shells to photoionization cross sections are given by Verner & Yakovlev (1995) and Verner et al. (1996). Figure 13.1b shows the photoionization cross sections for C and O, each of which has a conspicuous **absorption edge** at the minimum photon energy for photoionization from the K shell ("K shell" $\equiv 1s$ shell). At high energies (i.e., above the K-edge), the $1s^2$ electrons provide a photoionization cross section that is $\sim 10^4$ times larger than the cross section for an H atom. Thus at high energies, the heavy elements can dominate the total photoionization cross section, even though the total abundance of heavy elements is only $\sim 10^{-3}$ that of H. Figure 13.2 shows photoionization cross sections $\sigma_{\rm pi}$ for O, Ne, Mg, and Si.

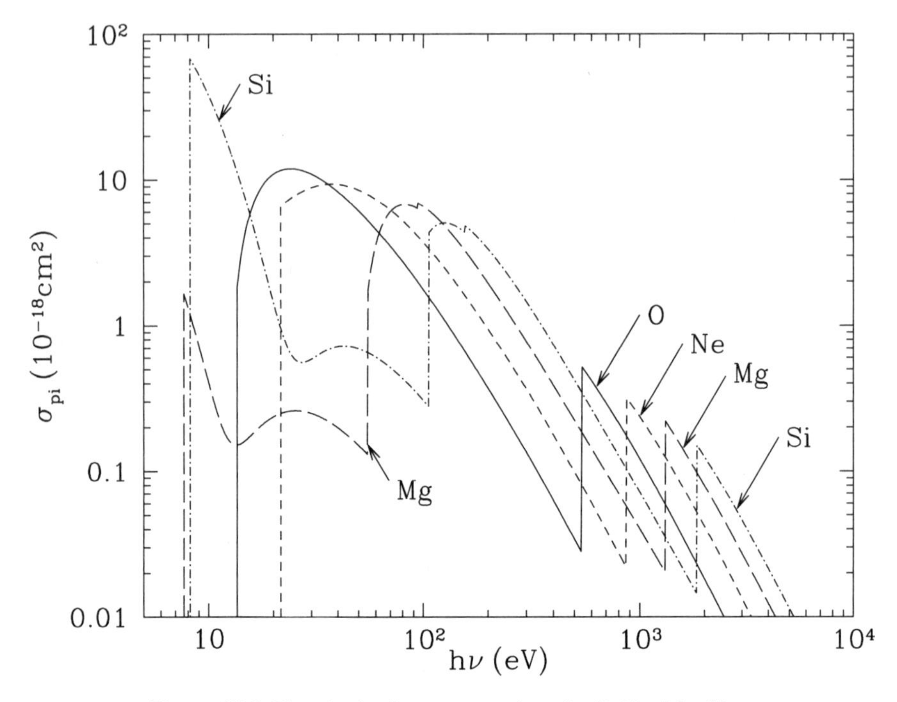

Figure 13.2 Photoionization cross sections for O, Ne, Mg, Si.

The photoionization rate (the probability per unit time of photoionization) is

$$\zeta_{\rm pi} = \int_{\nu_1}^{\infty} \sigma_{\rm pi}(\nu)\, c\, \frac{u_\nu}{h\nu}\, d\nu \quad , \tag{13.5}$$

where $u_\nu d\nu$ is the energy density of radiation in the frequency interval $(\nu, \nu + d\nu)$. Photoionization rates have been calculated for selected atoms and ions, for two estimates of the local interstellar radiation field (Draine 1978; Mathis et al. 1983). The resulting photoionization rates are given in Table 13.1.

The rates for neutral atoms cover a wide range, from $7 \times 10^{-12}\,{\rm s}^{-1}$ for K to $3 \times 10^{-9}\,{\rm s}^{-1}$ for Si – this wide variation comes from the large differences in $h\nu <$ 13.6 eV photoionization cross sections. (The difference in $\sigma_{\rm pi}$ between Mg and Si in Figure 13.2 accounts for the factor of ~ 50 difference in photoionization rates for Mg and Si.) It is curious that the two elements with the lowest ionization potentials – Na and K – have relatively low photoionization rates.

Note that of the elements in Table 13.1, two – Ca and Ti – have two ion states with ionization potentials $I < I_{\rm H}$. These elements can therefore be found in *three* ionization states in H I regions: Ca I, Ca II, and Ca III; and Ti I, Ti II, and Ti III. However, ionization of Ti II→Ti III can only be accomplished by the very few photons in the range 13.576–13.598eV, so the rate is very small (see Table 13.1).

Table 13.1 Photoionization Rates[a] for Elements with Abundance $X/\mathrm{H} > 1 \times 10^{-8}$ and Ionization Potential (IP) $< 13.60\,\mathrm{eV}$, for Two Estimates of the ISRF

Z	X	IP (eV)	ISRF from MMP83[b] $\zeta_{\rm pi}(\mathrm{s}^{-1})$	ISRF from D78[c] $\zeta_{\rm pi}(\mathrm{s}^{-1})$
6	C	11.2603	2.58×10^{-10}	3.43×10^{-10}
11	Na	5.1391	7.59×10^{-12}	1.13×10^{-11}
12	Mg	7.6462	5.39×10^{-11}	8.37×10^{-11}
13	Al	5.9858	1.05×10^{-9}	1.63×10^{-9}
14	Si	8.1517	2.77×10^{-9}	4.29×10^{-9}
15	P	10.4867	7.93×10^{-10}	1.14×10^{-9}
16	S	10.3600	9.25×10^{-10}	1.29×10^{-9}
17	Cl	12.9676	3.59×10^{-10}	3.17×10^{-10}
19	K	4.3407	6.85×10^{-12}	1.04×10^{-11}
20	Ca	6.1132	1.21×10^{-10}	1.88×10^{-10}
"	Ca II	11.872	4.64×10^{-12}	5.77×10^{-12}
22	Ti	6.8281	1.45×10^{-10}	2.12×10^{-10}
"	Ti II	13.576	1.13×10^{-14}	5.12×10^{-15}
23	V	6.7462	3.64×10^{-11}	4.59×10^{-11}
24	Cr	6.7665	4.67×10^{-10}	6.93×10^{-10}
25	Mn	7.4340	2.41×10^{-11}	3.77×10^{-11}
26	Fe	7.9024	1.92×10^{-10}	2.91×10^{-10}
27	Co	7.8810	3.96×10^{-11}	6.19×10^{-11}
28	Ni	7.6398	7.24×10^{-11}	1.13×10^{-10}
29	Cu	7.7264	1.45×10^{-10}	2.04×10^{-10}
30	Zn	9.3942	2.94×10^{-11}	4.49×10^{-11}

[a] $\sigma_{\rm pi}$ from Verner & Yakovlev (1995) and Verner et al. (1996).
[b] Mathis et al. (1983) radiation field [Eq. (12.7)], with $\chi = 1.231$, and $G_0 = 1.137$ [see Eq. (12.5 and 12.6) for definitions of χ and G_0].
[c] Draine (1978) radiation field ($\chi = 1.71$, $G_0 = 1.69$).

13.2 Auger Ionization and X-Ray Fluorescence

When photoionization ejects an electron from an inner shell, it leaves the resulting ion with an inner-shell vacancy. This highly excited state can, in principle, relax radiatively, but it can instead undergo a radiationless two-electron transition, with one electron dropping to fill the vacancy, and a second electron promoted to an excited level. If the excited level is unbound, the second electron will escape: this is referred to as the **Auger effect**.

Figure 13.3 shows the total number of electrons ejected following one photoelectric absorption by the K-, L_1-, or M_1-shell in initially neutral atoms.[1] For neutral C, N, O, or Ne, photoionization from the K shell is followed by ejection of one Auger electron (Kaastra & Mewe 1993). For atoms with larger atomic numbers, there can be a sequence of such transitions, with a number of Auger electrons

[1] K-shell = $1s$ subshell; L_1-shell = $2s$ subshell; M_1-shell = $3s$ subshell.

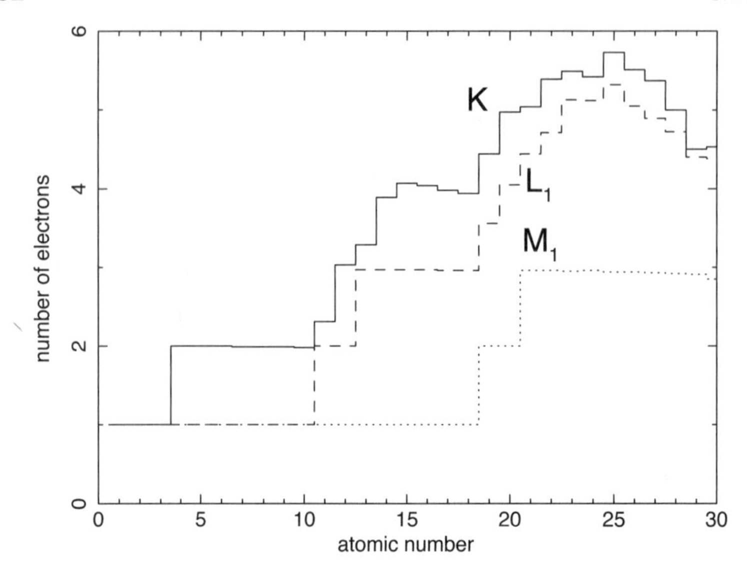

Figure 13.3 Average number of electrons (including the photoelectron) emitted following photoelectric absorption by the K, L_1, and M_1 subshells of neutral atoms. From Kaastra & Mewe (1993).

ejected following a single photoionization. For example, photoionization from the K shell of a neutral Fe atom will, on the average, be followed by ejection of $\sim$4.7 Auger electrons (Kaastra & Mewe 1993). Auger ionization plays an important role in the ionization of heavy elements by x rays. Gorczyca et al. (2003) discuss the accuracy of the estimated Auger yields in Fig. 13.3.

It is also possible for a K-shell vacancy to be filled by a radiative transition as an electron from the L-shell drops to the K-shell, with emission of a $K\alpha$ photon. Similarly, an electron from the M-shell can drop to the K-shell, with emission of a $K\beta$ photon. This is known as "$K\alpha$ and $K\beta$ fluorescence," and is thought to be an important source of Fe Kα and Kβ line emission from gas near x ray sources, such as accretion disks around AGN. The energies of the $K\alpha$ and $K\beta$ lines, and the $K\alpha/K\beta$ intensity ratio, depend on the ionization state of the Fe ion that was originally photoionized. Figure 13.4 shows estimates of Kaastra & Mewe (1993) for the $K\alpha$ fluorescence yield for Fe in ion stages running from Fe^0 to Fe^{+22}; see also Palmeri et al. (2003).

13.3 Secondary Ionizations

The initial kinetic energy of a photoelectron ejected from shell s of an atom or ion is just $E_{\mathrm{pe}} = h\nu - I_s$, where I_s is the threshold energy for photoionization from shell s. In the case of photoionization by x rays, the photoelectron energy can easily be 10^2 to 10^3 eV.

In a fully ionized plasma, an energetic photoelectron slows down by Coulomb

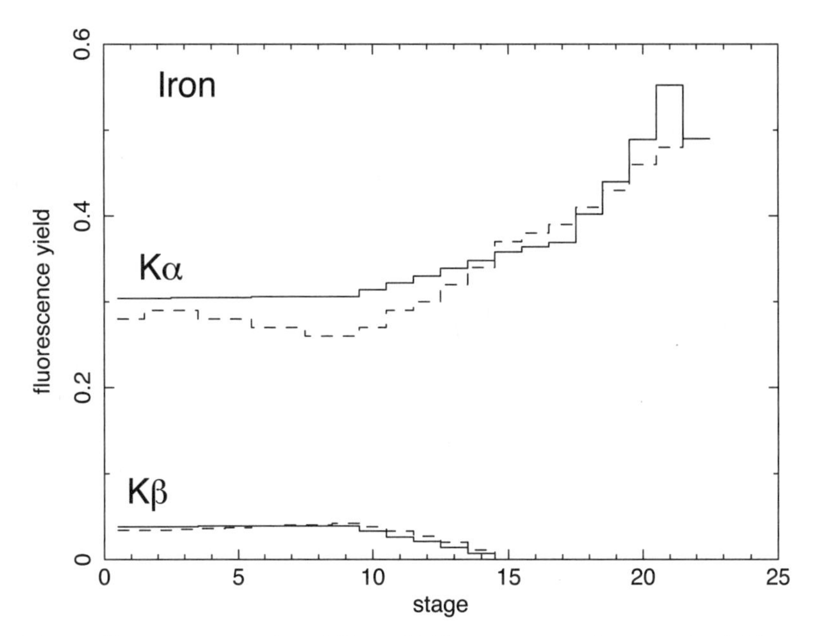

Figure 13.4 Average number of $K\alpha$ and $K\beta$ fluorescence photons emitted following creation of a K-shell vacancy, for all ions of Fe (stage 1 = Fe^0; stage 22 = Fe^{+21}). Solid curves show results of Kaastra & Mewe (1993); dashed lines are results of Jacobs & Rozsnyai (1986). From Kaastra & Mewe (1993).

scattering off thermal electrons, with all of its initial kinetic energy eventually converted to heat. However, if there are bound electrons present (e.g., H atoms), the photoelectron can also excite electrons to excited bound states (e.g., $1s \rightarrow 2p$) by collisional excitation, or to free electron states (e.g., $\mathrm{H} \rightarrow \mathrm{H}^+ + e^-$) by collisional ionization. If molecules are present (e.g., H_2), an energetic electron can also excite rovibrational excited states of the molecule. To understand the effects of energetic electrons in a partially ionized gas, it is necessary to carry out a statistical study of slowing-down process.

The collisional ionizations by both the photoelectron and energetic secondary electrons are referred to as **secondary ionizations**. The average number of secondary ionizations per photoelectron $\equiv \phi_s(E_{\mathrm{pe}}, x_e)$ depends on the photoelectron kinetic energy E_{pe} and on the ionization state of the gas.

Numerical studies of ionization by $E \gtrsim 50\,\mathrm{eV}$ electrons in partially ionized gas (Dalgarno & McCray 1972; Voit 1991) are approximately reproduced by

$$\phi_s(E, x_e) \approx \left(1 - \frac{x_e}{1.2}\right) \left(\frac{E - 15\,\mathrm{eV}}{35\,\mathrm{eV}}\right) \frac{1}{1 + 18x_e^{0.8}/\ln(E/35\,\mathrm{eV})} , \tag{13.6}$$

where $x_e \equiv n_e/n_{\mathrm{H}}$, the number of free electrons per H nucleon. According to (13.6), in neutral gas each ionization costs about 35 eV. Energy loss to the thermal electrons suppresses ϕ_s by a factor > 2 for $x_e \gtrsim 0.027[\ln(E/35\,\mathrm{eV})]^{1.25}$.

13.4 Collisional Ionization

Let the cross section for collisional ionization by an electron be $\sigma_{\rm ci}(E)$. The rate coefficient for collisional ionization by the thermal electrons in a plasma is

$$k_{\rm ci} = \int_I^\infty \sigma_{\rm ci}(E)\, v f_E dE \tag{13.7}$$

$$= \left(\frac{8kT}{\pi m_e}\right)^{1/2} \int_I^\infty \sigma_{\rm ci}(E) \frac{E}{kT} {\rm e}^{-E/kT} \frac{dE}{kT} \quad . \tag{13.8}$$

From phase-space considerations, it can be seen that the cross section must rise smoothly from zero at the threshold energy I. For moderate energies $I < E \lesssim 3I$, the actual cross section can often be approximated by the simple form

$$\sigma_{\rm ci}(E) \approx C\pi a_0^2 \left(1 - \frac{I}{E}\right) , \tag{13.9}$$

where C is a constant of order unity. For this simple form, the thermal rate coefficient is readily calculated and is given by is

$$k_{\rm ci} = C\pi a_0^2 \left(\frac{8kT}{\pi m_e}\right)^{1/2} {\rm e}^{-I/kT} \tag{13.10}$$

$$= 5.466 \times 10^{-9} C T_4^{1/2} {\rm e}^{-I/kT} \,{\rm cm}^3\,{\rm s}^{-1} \quad . \tag{13.11}$$

Collisional ionization cross sections are given by Bell et al. (1983) and Lennon et al. (1988). The collisional ionization cross section for H is approximated by Eq. (13.9) with $C = 1.07$ and $I = I_{\rm H}$.

13.5 Cosmic Ray Ionization

Most electrons and ions in interstellar space have velocities drawn from the local thermal distribution, but a small fraction of the particles have energies that are much larger than thermal – these "nonthermal" electrons and ions are referred to as **cosmic rays**. The cosmic ray population in the solar neighborhood is observed to extend to ultrarelativistic energies. Let $\zeta_{\rm CR}$ be the rate of ionization of a hydrogen atom exposed to the cosmic rays. Cosmic ray ionization produces electrons with a spectrum of energies. The mean kinetic energy $\langle E \rangle \approx 35\,{\rm eV}$ is essentially independent of the energy of the ionizing particle, as long as the particle is moving fast compared to the "Bohr velocity" $c/137$. Just like photoelectrons produced by X-ray ionization, the electrons produced by cosmic ray ionization may cause secondary ionizations. In neutral gas, the number of secondary ionizations is ~ 0.67 (Dalgarno & McCray 1972), with the number of secondary ionizations decreasing with increasing fractional ionization approximately as

$$\phi_s(\mathrm{CR}) \approx \left(1 - \frac{x_e}{1.2}\right) \frac{0.67}{1 + (x_e/0.05)} \quad . \tag{13.12}$$

The total ionization rate per volume resulting from cosmic rays is

$$\left(\frac{dn_e}{dt}\right)_{\mathrm{CR}} \approx 1.1 n_{\mathrm{H}} \zeta_{\mathrm{CR}} \left(1 - \frac{x_e}{1.2}\right) [1 + \phi_s(\mathrm{CR})] \quad . \tag{13.13}$$

The cosmic ray ionization rate ζ_{CR} is uncertain. The density of H and He cosmic rays can be measured near the Earth down to $\sim$ 20 MeV/nucleon (Wang et al. 2002), but below $\sim$ 2 GeV the observed flux (see Fig. 13.5a) must be corrected for the effects of the solar wind. Webber & Yushak (1983) used data from the *Voyager* probe to calibrate the modulating effects of the solar wind, and estimated the interstellar proton flux down to 60 MeV. As seen in Figure 13.5, the (uncertain) corrections for the solar wind are a factor of $\sim$3 at 1 GeV, and substantially larger at $E \lesssim 0.3$ GeV.

For an assumed flux per unit solid angle $F(E)$ of particles with energy less than E, the primary ionization rate ζ_{CR} due to particles with $E > E_{\mathrm{min}}$ is

$$\zeta_{\mathrm{CR}} = 4\pi \int_{E_{\mathrm{min}}}^{\infty} \sigma_{\mathrm{ci}}(E) E \frac{dF}{dE} \cdot \frac{dE}{E} \quad , \tag{13.14}$$

where the collisional ionization cross section σ_{ci} for a particle with charge Ze and velocity βc is given by (Bethe 1933):

$$\sigma_{\mathrm{ci}} = 0.285 \frac{2\pi e^4 Z^2}{m_e c^2 I_{\mathrm{H}} \beta^2} \left\{ \ln \left[\frac{2 m_e c^2 \beta^2}{I_{\mathrm{H}}(1 - \beta^2)} \right] + 3.04 - \beta^2 \right\} \quad , \tag{13.15}$$

for $\beta \gtrsim .025$. Smoothly extrapolating the Webber & Yushak (1983) estimate for the proton flux (curve WY83 in Fig. 13.5a) down to 1 MeV, the primary ionization rate for an H atom due to $E > 1$ MeV protons is calculated to be $\zeta_{\mathrm{CR},p} = 6.6 \times 10^{-18}\,\mathrm{s}^{-1}$. Allowance for the effects of helium and heavier ions raises the total primary ionization rate to $\zeta_{\mathrm{CR}} \approx 2\zeta_{\mathrm{CR},p}$.[2] Primary ionizations will be followed by secondary ionizations, with the number of secondary ionizations of hydrogen approximately given by Eq. (13.12). When the fractional ionization is low, $x_e \lesssim 0.03$ (as in a cool H I cloud), the total ionization rate per H (including ionization by cosmic rays heavier than H) will be $(1 + \phi_s)\zeta_{\mathrm{CR}} \approx 3\zeta_{\mathrm{CR},p}$.

From Figure 13.5b, it can be seen that $\zeta_{\mathrm{CR},p}$ is primarily due to energies $E < 1$ GeV. Because of the uncertainty in correcting for the effects of the solar wind

[2] At fixed kinetic energy per nucleon, the ionization cross section (13.15) scales as Z^2. For solar abundances, He nuclei would contribute an ionization rate equal to $0.38\zeta_{\mathrm{CR},p}$, and Fe nuclei would contribute $0.023\zeta_{\mathrm{CR},p}$. In fact, the cosmic-ray abundances relative to H of nuclei such as Si and Fe are super-solar by as much as a factor of 20 (see Fig. 40.3). As a result, the total cosmic ray primary ionization rate $\zeta_{\mathrm{CR}} \approx 2\zeta_{\mathrm{CR},p}$.

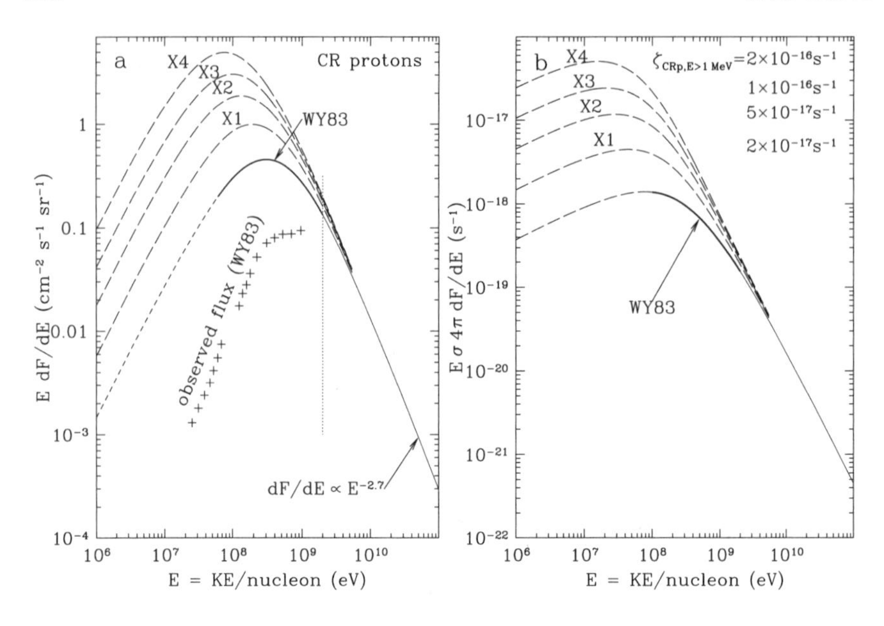

Figure 13.5 (a) Symbols show *Voyager* measurements of cosmic ray protons (Webber & Yushak 1983). Solid curve labeled WY83 is the estimate of Webber & Yushak (1983) and Webber & Lockwood (2001) for the local interstellar cosmic ray protons. Curves $X1 - X4$ are hypothetical spectra that match the observed high-energy proton flux, but with larger numbers of low-energy particles than the Webber & Yushak (1983) estimate. (b) H primary ionization rate per logarithmic interval in cosmic ray energy E calculated for spectra $X1 - X4$ in (a).

at low energies, as well as the possibility that the low-energy cosmic ray spectrum may vary from point to point in the ISM, Fig. 13.5a also shows four hypothetical proton spectra, labeled $X1$–$X4$, where the maximum in $E \cdot dF/dE = dF/d\ln E$ has been arbitrarily shifted to energies lower than estimated by Webber & Yushak (1983) for the local interstellar cosmic rays. The primary ionization rates for the proton spectra $X1$–$X4$ are $\zeta_{\mathrm{CR},p} = 2 \times 10^{-17}\,\mathrm{s}^{-1}$, $5 \times 10^{-17}\,\mathrm{s}^{-1}$, $1 \times 10^{-16}\,\mathrm{s}^{-1}$, and $2 \times 10^{-16}\,\mathrm{s}^{-1}$. Note that models $X1$–$X4$ all have nearly identical cosmic ray fluxes at $E > 2\,\mathrm{GeV}$, where the interstellar flux can be measured reliably by spacecraft within the Solar System – they differ only at energies $E < 1\,\mathrm{GeV}$, where we do not have reliable measurements.

As will be discussed further in Chapter 16, studies of the products of ion–neutral chemistry inside molecular clouds (Black & van Dishoeck 1991; Lepp 1992; McCall et al. 2003; Indriolo et al. 2007) can be used to estimate cosmic ray ionization rates. In particular, the observed abundances of H_3^+ in diffuse molecular clouds appear to require a cosmic ray *primary* ionization rate $\zeta_{\mathrm{CR}} = (0.5 - 3) \times 10^{-16}\,\mathrm{s}^{-1}$, with an average value $2 \times 10^{-16}\,\mathrm{s}^{-1}$ (Indriolo et al. 2007). If ∼50% of this is due to protons, with the balance due to heavier nuclei, it is an indication that the interstellar cosmic ray flux, in at least some clouds, may be similar to spectrum X3 in Fig. 13.5, with $dF/d\ln E$ peaking near $\sim 100\,\mathrm{MeV}$.

Chapter Fourteen

Recombination of Ions with Electrons

Consider some atom or molecule X. If X is ionized, there are many different processes that act to neutralize it, by "recombining" the ion with an electron:

- **Radiative recombination:** $X^+ + e^- \rightarrow X + h\nu$.
- **Dielectronic recombination:** $X^+ + e^- \rightarrow X^{**} \rightarrow X + h\nu$.
- **Three-body recombination:** $X^+ + e^- + e^- \rightarrow X + e^-$.
- **Charge exchange:** $X^+ + Y \rightarrow X + Y^+$.
- **Dissociative recombination:** $AB^+ + e^- \rightarrow A + B$.
- **Neutralization by grain:** $X^+ + \text{grain} \rightarrow X + \text{grain}^+$.

The relative importance of these different channels depends on the ion and on the physical conditions. Each of these pathways is discussed in the following sections.

14.1 Radiative Recombination

Consider an ion with its electrons in some configuration that we will refer to as the "core". In a low-density thermal plasma, free electrons can undergo transitions to bound states by emission of a photon:

$$X^+(\text{core}) + e^- \rightarrow X(\text{core} + n\ell) + h\nu \ , \tag{14.1}$$

where the electron is captured into some specific state $n\ell$ that was initially unoccupied. The cross section for electron capture via this "radiative recombination" process is $\sigma_{\text{rr},n\ell}(E)$. As discussed in Chapter 3, if the photoionization cross section $\sigma_{\text{pi,n}\ell}(h\nu)$ is either measured or calculated for level $X(\text{core} + n\ell)$, then the Milne relation (3.31) can be used to obtain the electron capture cross section $\sigma_{\text{rr},n\ell}(E)$.

The thermal rate coefficient $\alpha_{n\ell}$ for electron capture directly to level $n\ell$, with emission of a photon of energy $h\nu = I_{n\ell} + E$ (where $I_{n\ell}$ is the energy required for ionization from level $n\ell$), is

$$\alpha_{n\ell}(T) = \left(\frac{8kT}{\pi m_e}\right)^{1/2} \int_0^\infty \sigma_{\text{rr},n\ell}(E) \frac{E}{kT} \mathrm{e}^{-E/kT} \frac{dE}{kT} \quad . \tag{14.2}$$

Table 14.1 Recombination Coefficients $\alpha_{n\ell}$ ($\mathrm{cm^3\,s^{-1}}$) for H.[a] The approximation formulae are valid for $0.3 \lesssim T_4 \lesssim 3$. For a broader range of T, see Eq. (14.5,14.6).

	Temperature T			
$\alpha_n(^2L)$	5×10^3 K	1×10^4 K	2×10^4 K	approximation
α_{1s}	2.28×10^{-13}	1.58×10^{-13}	1.08×10^{-13}	$1.58\times10^{-13}T_4^{-0.540-0.017\ln T_4}$
α_{2s}	3.37×10^{-14}	2.34×10^{-14}	1.60×10^{-14}	$2.34\times10^{-14}T_4^{-0.537-0.019\ln T_4}$
α_{2p}	8.33×10^{-14}	5.36×10^{-14}	3.24×10^{-14}	$5.36\times10^{-14}T_4^{-0.681-0.061\ln T_4}$
α_2	1.17×10^{-13}	7.70×10^{-14}	4.84×10^{-14}	$7.70\times10^{-14}T_4^{-0.636-0.046\ln T_4}$
α_{3s}	1.13×10^{-14}	7.82×10^{-15}	5.29×10^{-15}	$7.82\times10^{-15}T_4^{-0.547-0.024\ln T_4}$
α_{3p}	3.17×10^{-14}	2.04×10^{-14}	1.23×10^{-14}	$2.04\times10^{-15}T_4^{-0.683-0.062\ln T_4}$
α_{3d}	3.03×10^{-14}	1.73×10^{-14}	9.09×10^{-15}	$1.73\times10^{-14}T_4^{-0.868-0.093\ln T_4}$
α_3	7.33×10^{-14}	4.55×10^{-14}	2.67×10^{-14}	$4.55\times10^{-14}T_4^{-0.729-0.060\ln T_4}$
α_{4s}	5.23×10^{-15}	3.59×10^{-15}	2.40×10^{-15}	$3.59\times10^{-15}T_4^{-0.562-0.026\ln T_4}$
α_{4p}	1.51×10^{-14}	9.66×10^{-15}	5.80×10^{-15}	$9.66\times10^{-15}T_4^{-0.691-0.064\ln T_4}$
α_{4d}	1.90×10^{-14}	1.08×10^{-14}	5.67×10^{-15}	$1.08\times10^{-14}T_4^{-0.870-0.094\ln T_4}$
α_{4f}	1.09×10^{-14}	5.54×10^{-15}	2.57×10^{-15}	$5.54\times10^{-15}T_4^{-1.041-0.100\ln T_4}$
α_4	5.02×10^{-14}	2.96×10^{-14}	1.64×10^{-14}	$2.96\times10^{-14}T_4^{-0.805-0.065\ln T_4}$
α_A	6.81×10^{-13}	4.17×10^{-13}	2.51×10^{-13}	$4.17\times10^{-13}T_4^{-0.721-0.018\ln T_4}$
α_B	4.53×10^{-13}	2.59×10^{-13}	1.43×10^{-13}	$2.59\times10^{-13}T_4^{-0.833-0.035\ln T_4}$

[a] $\alpha_{n\ell}$ from Burgess (1965); α_B from Hummer & Storey (1987) (for $n_e = 10^3\,\mathrm{cm^{-3}}$)

14.2 Radiative Recombination of Hydrogen

The most important element in astrophysics is, of course, hydrogen. Cross sections $\sigma_{\mathrm{rr},n\ell}(E)$ for radiative recombination of hydrogen to level $n\ell$ have been calculated, and the thermal rate coefficients for radiative recombination to level $n\ell$ have been obtained for every level $n\ell$ using Eq. (14.2). Table 14.1 presents rate coefficients $\alpha_{n\ell}$ for direct recombination to level $n\ell$, for $n = 1-4$.

An electron of kinetic energy E can undergo radiative recombination into any level $n\ell$ of the hydrogen atom, with emission of a photon of energy $h\nu = E + I_{n\ell}$, where $I_{n\ell} = I_{\mathrm{H}}/n^2$ is the binding energy of an electron in level $n\ell$. Therefore, when recombination takes place directly to the ground state ($n = 1$), a photon will be emitted that can ionize hydrogen.

If the region under consideration has a significant amount of neutral hydrogen present, the emitted photon will have a high probability of being absorbed by another hydrogen atom very near the point of emission, with creation of a hydrogen ion. Therefore, aside from the transport of the ionization energy a short distance, a recombination directly to the ground state under these circumstances has virtually no effect on the ionization state of the gas. This was first pointed out by Baker &

Menzel (1938), who proposed distinguishing between two limits:

- **Case A:** Optically thin to ionizing radiation, so that every ionizing photon emitted during the recombination process escapes. For this case, we sum the radiative capture rate coefficient $\alpha_{n\ell}$ over all levels $n\ell$.

- **Case B:** Optically thick to radiation just above $I_{\rm H} = 13.60\,{\rm eV}$, so that ionizing photons emitted during recombination are immediately reabsorbed, creating another ion and free electron by photoionization. In this case, the recombinations directly to $n = 1$ do not reduce the ionization of the gas: only recombinations to $n \geq 2$ act to reduce the ionization.

These two limiting cases are in fact realistic approximations for two physical situations. Photoionized nebulae around O and B stars – H II regions – usually have large enough densities of neutral H so that free–bound photons emitted in recombinations to level $n = 1$ have a very small probability of escaping from the nebula – for this situation, case B is an excellent approximation.

Regions where the hydrogen is collisionally ionized, on the other hand, are typically very hot ($T \gtrsim 10^6$ K) and contain a very small density of neutral hydrogen. Because of the high T, the free-bound emission is also relatively hard ($h\nu \gg I_{\rm H}$), so that the photoionization cross sections are well below the threshold value. As a result, only a negligible fraction of photons emitted in recombinations to $n = 1$ will photoionize H within the nebula – for these shock-heated regions, case A is an excellent approximation.

The effective radiative recombination rates for hydrogen for these two limiting cases are

$$\alpha_A(T) \equiv \sum_{n=1}^{\infty} \sum_{\ell=0}^{n-1} \alpha_{n\ell}(T) \quad , \tag{14.3}$$

$$\alpha_B(T) \equiv \sum_{n=2}^{\infty} \sum_{\ell=0}^{n-1} \alpha_{n\ell}(T) = \alpha_A(T) - \alpha_{1s}(T) \quad . \tag{14.4}$$

Good approximations to the radiative recombination rates for hydrogenic ions (e.g., $\rm H^+$ or $\rm He^{++}$) are provided over the temperature range $30\,{\rm K} < T/Z^2 < 3 \times 10^4$ K by the fitting formulae

$$\alpha_A(T) \approx 4.13 \times 10^{-13}\, Z\, (T_4/Z^2)^{-0.7131-0.0115\ln(T_4/Z^2)}\,{\rm cm^3\,s^{-1}} \ , \tag{14.5}$$

$$\alpha_B(T) \approx 2.54 \times 10^{-13}\, Z\, (T_4/Z^2)^{-0.8163-0.0208\ln(T_4/Z^2)}\,{\rm cm^3\,s^{-1}} \ . \tag{14.6}$$

These are plotted in Figure 14.1.

14.2.1 Hydrogen Recombination Emission Spectrum

The hydrogenic energy levels are designated by quantum numbers $n = 1, 2, 3, 4, ...$ and $\ell = 0, 1, 2, 3, ..., n-1$ (or $s, p, d, f, ...$). With the exception of the $2s$ level,

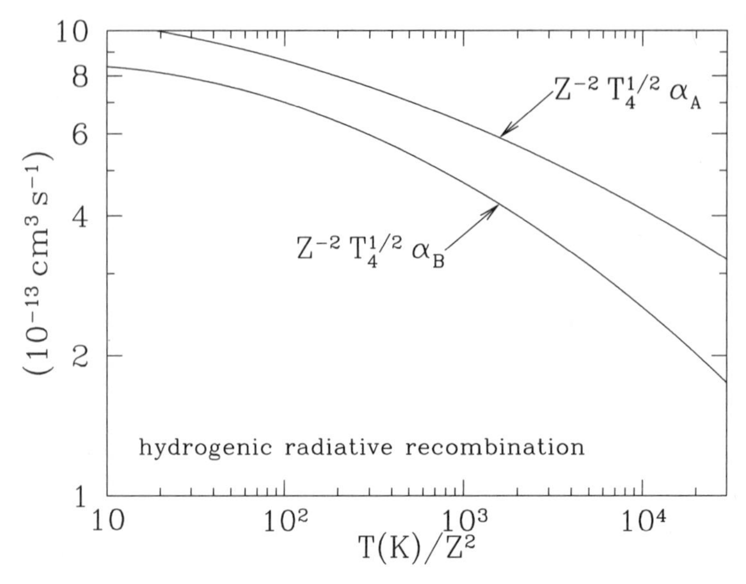

Figure 14.1 Case A and Case B rate coefficients α_A and α_B for radiative recombination of hydrogen, multiplied by $T_4^{1/2}$ (equations 14.5,14.6). Note that no single power-law fit can reproduce the T-dependence over a wide range in T.

every excited state $n\ell$ of hydrogen has allowed radiative decays to one or more lower levels $n'\ell'$, satisfying the selection rules $n > n'$ and $\ell - \ell' = \pm 1$. Figure 14.2a shows the three different $4\ell \rightarrow 2\ell'$ transitions that contribute to the Balmer series $n = 4 \rightarrow 2$ transition, known as "H β," with a wavelength $911.77\,\text{Å}/(\frac{1}{4} - \frac{1}{16}) = 4862.7\,\text{Å}$. Figure 14.2b shows all of the allowed transitions open to an atom in the $4p$ state. Therefore, as radiative recombination takes place to excited levels $n\ell$, these (except for $2s$, which we discuss further in the following) promptly decay to populate lower levels. The radiative decay process produces a distinctive emission spectrum characteristic of hydrogen recombination.

The hydrogen recombination spectrum depends on temperature T, and therefore measured line ratios can be used to estimate T. The power radiated in the recombination lines can be used to determine the total rate of hydrogen recombination in the ionized region. Measurements of the relative intensities of recombination lines with different wavelengths can be used to estimate the reddening by dust between us and the emitting region.

14.2.2 Case A Recombination Spectrum

In the optically thin limit where no emitted photons are reabsorbed – so-called Case A – the emission spectrum of recombining hydrogen can be calculated. The power

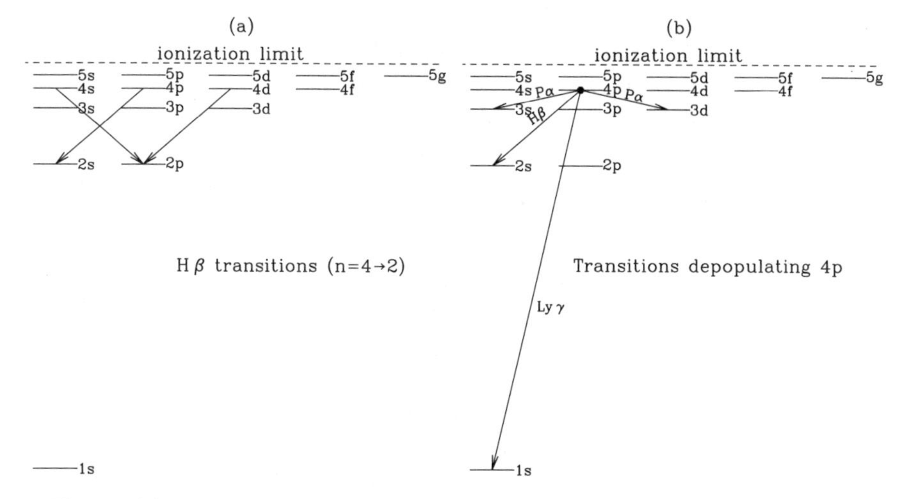

Figure 14.2 (a) The three transitions contributing to Hβ; (b) Allowed radiative transitions depopulating the $4p$ level.

radiated per volume in the transition $n\ell \to n'\ell'$ is just

$$4\pi j(n\ell \to n'\ell') = n_e n(\mathrm{H}^+) \frac{A(n\ell \to n'\ell') h\nu_{u\ell}}{\sum_{n''\ell''} A(n\ell \to n''\ell'')} \times \left[\alpha(n\ell) + \sum_{n''\ell'', n''>n} \alpha(n''\ell'') P_A(n''\ell'', n\ell)\right] , \qquad (14.7)$$

where $P_A(n''\ell'', n\ell)$ is the Case A probability that an atom in level $n''\ell''$ will follow a decay path that takes it through level $n\ell$. The probabilities $P_A(n''\ell'', n\ell)$ are readily calculated from the known transition probabilities $A(n\ell \to n'\ell')$ using straightforward branching probability arguments.

14.2.3 Case B Recombination Spectrum

All np levels have allowed transitions to the ground state $1s$, with large transition rates $A_{np,1s}$ for $n \lesssim 10$ (see Table 9.1). In the ideal Case A limit, these photons are all assumed to escape freely.

The resonant absorption cross section for Lyman α, Lyman β, and so on, are much larger than photoionization cross sections. Any nebula that is optically thick to Lyman continuum radiation – the defining condition for Case B – will therefore be *very* optically thick to Lyman series $n \to 1$ photons for small values of n. While the cross sections for absorption in the $n \to 1$ transitions is a declining function of n, as $n \to \infty$ this cross section becomes equal to the photoionization cross section 6.3×10^{-18} cm^2 near threshold; thus, for Case B conditions, *all* of the Lyman series transitions are optically thick.

Therefore, Lyman series photons that are emitted will be resonantly absorbed by other hydrogen atoms in the ground state, immediately exciting the absorber to the same quantum state np as characterized the emitting atom.

Because the resonant absorption cross section is large, under Case B conditions, a Lyman α, β, γ, . . . photon will travel only a short distance before being reabsorbed. If we disregard the displacement – this is known as the **on-the-spot approximation** – it is helpful to think about the radiative decay and resonant reabsorption process as though the photon were reabsorbed by the same atom as emitted it. We will use such language in the following discussion.

Consider a hydrogen atom in level np, with $n \geq 3$. It has allowed decays to levels $n's$ where $n' < n$. If $n \geq 4$, (see Fig. 14.2b), there are also allowed decays to levels $n'd$ with $3 \leq n' < n$. Therefore, if the upper state np decays by emission of a Lyman series photon ($np \rightarrow 1s$), the photon will immediately be resonantly absorbed, returning the excitation back to the np state. After returning to the np state, the atom will again decay along one of its allowed decay paths, based on its branching probabilities. The atom may emit another Lyman series photon, which will again be absorbed. This process will repeat until eventually a non-Lyman decay occurs, taking the atom to level $n' > 1$, with the emitted photon escaping freely.[1]

Therefore, under Case B conditions, insofar as the spontaneous emission of photons from levels $n \geq 3$ is concerned, it is just as though the Lyman series transitions did not even occur – every time a Lyman series photon depopulates a level np, the level is almost instantly repopulated by reabsorption of the photon, and, because the levels np with $n \geq 3$ always have at least one non-Lyman decay channel available, eventually the np level will depopulate by emission of a non-Lyman series photon (e.g., a Balmer series photon). The case B emissivities $j_B(n\ell \rightarrow n'\ell')$ can therefore be calculated using Eq. (14.7), but with the $A_{np\rightarrow 1s}$ rates replaced by zero, and with the probabilities $P_A(n\ell, n'\ell')$ replaced by probabilities $P_B(n\ell, n'\ell')$ calculated with the Lyman series transition probabilities set to 0. Case B emissivities for selected lines are given in Table 14.2 for $T = 5000$, 10,000, and 20,000 K. The two strongest lines are Hα 6564.6 Å and Hβ 4862.7 Å. The rate coefficients for recombinations that result in emission of these lines can be approximated by

$$\alpha_{\text{eff,H}\alpha} \approx 1.17 \times 10^{-13} T_4^{-0.942-0.031 \ln T_4} \,\text{cm}^3\,\text{s}^{-1} \quad , \tag{14.8}$$

$$\alpha_{\text{eff,H}\beta} \approx 3.03 \times 10^{-14} T_4^{-0.874-0.058 \ln T_4} \,\text{cm}^3\,\text{s}^{-1} \quad ; \tag{14.9}$$

these approximations are accurate to within $\sim 2\%$ for $0.1 < T_4 < 3$. Because collisions can affect the high-n and $2s$ levels, the rate coefficients for emission of recombination lines are weakly dependent on density. Equations (14.8 and 14.9) are for $n_e \approx 10^3\,\text{cm}^{-3}$, but are valid to within a few percent for $n_e \lesssim 10^6\,\text{cm}^{-3}$.

[1]This argument assumes that the population of hydrogen in excited states is too small to present significant opacity, even for allowed transitions, such as Hα. As will be discussed later, this is not always true, but it is usually an excellent approximation.

Table 14.2 Case B Hydrogen Recombination Spectrum[a] for $n_e = 10^3\,\mathrm{cm}^{-3}$

	T(K) 5000	10,000	20,000
$\alpha_B(\mathrm{cm}^3\,\mathrm{s}^{-1})$	4.53×10^{-13}	2.59×10^{-13}	1.43×10^{-13}
$\alpha_{\mathrm{eff},2s}/\alpha_B$	0.305	0.325	0.356
$\alpha_{\mathrm{eff,H}\alpha}(\mathrm{cm}^3\,\mathrm{s}^{-1})$	2.20×10^{-13}	1.17×10^{-13}	5.96×10^{-14}
$\alpha_{\mathrm{eff,H}\beta}(\mathrm{cm}^3\,\mathrm{s}^{-1})$	5.40×10^{-14}	3.03×10^{-14}	1.61×10^{-14}
$4\pi j_{\mathrm{H}\beta}/n_e n_p(\mathrm{erg\,cm}^3\,\mathrm{s}^{-1})$	2.21×10^{-25}	1.24×10^{-25}	6.58×10^{-26}
Balmer-line intensities relative to Hβ 0.48627 μm			
$j_{\mathrm{H}\alpha\,0.65646}/j_{\mathrm{H}\beta}$	3.03	2.86	2.74
$j_{\mathrm{H}\beta\,0.48627}/j_{\mathrm{H}\beta}$	1.	1.	1.
$j_{\mathrm{H}\gamma\,0.43418}/j_{\mathrm{H}\beta}$	0.459	0.469	0.475
$j_{\mathrm{H}\delta\,0.41030}/j_{\mathrm{H}\beta}$	0.252	0.259	0.264
$j_{\mathrm{H}\epsilon\,0.39713}/j_{\mathrm{H}\beta}$	0.154	0.159	0.163
$j_{\mathrm{H8}\,0.38902}/j_{\mathrm{H}\beta}$	0.102	0.105	0.106
$j_{\mathrm{H9}\,0.38365}/j_{\mathrm{H}\beta}$	0.0711	0.0732	0.0746
$j_{\mathrm{H10}\,0.37990}/j_{\mathrm{H}\beta}$	0.0517	0.0531	0.0540
Paschen ($n\rightarrow3$) line intensities relative to corresponding Balmer lines			
$j_{\mathrm{P}\alpha\,1.8756}/j_{\mathrm{H}\beta}$	0.405	0.336	0.283
$j_{\mathrm{P}\beta\,1.2821}/j_{\mathrm{H}\gamma\,0.43418}$	0.399	0.347	0.305
$j_{\mathrm{P}\gamma\,1.0941}/j_{\mathrm{H}\delta\,0.41030}$	0.391	0.348	0.311
$j_{\mathrm{P}\delta\,1.0052}/j_{\mathrm{H}\epsilon\,0.39713}$	0.386	0.348	0.314
$j_{\mathrm{P}\epsilon\,0.95487}/j_{\mathrm{H8}\,0.38902}$	0.382	0.348	0.316
$j_{\mathrm{P9}\,0.92317}/j_{\mathrm{H9}\,0.38365}$	0.380	0.347	0.317
$j_{\mathrm{P10}\,0.90175}/j_{\mathrm{H10}\,0.37990}$	0.380	0.347	0.317
Brackett ($n\rightarrow4$) line intensities relative to corresponding Balmer lines			
$j_{\mathrm{Br}\alpha\,4.0523}/j_{\mathrm{H}\gamma\,0.43418}$	0.223	0.169	0.131
$j_{\mathrm{Br}\beta\,2.6259}/j_{\mathrm{H}\delta\,0.41030}$	0.219	0.174	0.141
$j_{\mathrm{Br}\gamma\,2.1661}/j_{\mathrm{H}\epsilon\,0.39713}$	0.212	0.174	0.144
$j_{\mathrm{Br}\delta\,1.9451}/j_{\mathrm{H8}\,0.38902}$	0.208	0.173	0.145
$j_{\mathrm{Br}\epsilon\,1.8179}/j_{\mathrm{H9}\,0.38365}$	0.204	0.173	0.146
$j_{\mathrm{Br10}\,1.7367}/j_{\mathrm{H10}\,0.37990}$	0.202	0.172	0.146
Pfund ($n\rightarrow5$) line intensities relative to corresponding Balmer lines			
$j_{65\,7.4599}/j_{\mathrm{H}\delta\,0.41030}$	0.134	0.0969	0.0719
$j_{75\,4.6538}/j_{\mathrm{H}\epsilon\,0.39713}$	0.134	0.101	0.0774
$j_{85\,3.7406}/j_{\mathrm{H8}\,0.38902}$	0.130	0.101	0.0790
$j_{95\,3.2970}/j_{\mathrm{H9}\,0.38365}$	0.127	0.100	0.0797
$j_{10\,5\,3.0392}/j_{\mathrm{H10}\,0.37990}$	0.125	0.0997	0.0801
Humphreys ($n\rightarrow6$) line intensities relative to corresponding Balmer lines			
$j_{76\,12.372}/j_{\mathrm{H}\epsilon\,0.39713}$	0.0855	0.0601	0.0435
$j_{86\,7.5026}/j_{\mathrm{H8}\,0.38902}$	0.0867	0.0632	0.0471
$j_{96\,5.9083}/j_{\mathrm{H9}\,0.38365}$	0.0850	0.0634	0.0481
$j_{10\,6\,5.1287}/j_{\mathrm{H10}\,0.37990}$	0.0833	0.0632	0.0486

[a] Emissivities from Hummer & Storey (1987)

14.2.4 Emission of Lyman α

Let $\alpha_{\rm eff2s}$ and $\alpha_{\rm eff2p}$ be the effective rate coefficients for populating the $2s$ and $2p$ states. It is clear that the case B radiative recombination process must eventually take the atom to either the $2s$ level or the $2p$ level – thus $\alpha_{\rm eff2s} + \alpha_{\rm eff2p} = \alpha_B$. The fractions $f(2s) = \alpha_{\rm eff2s}/\alpha_B \approx \frac{1}{3}$ and $f(2p) = \alpha_{\rm eff2p}/\alpha_B \approx \frac{2}{3}$ of the case B recombinations that populate the $2s$ and $2p$ states are given in Table 14.3. What happens after the electron enters either the $2s$ or $2p$ levels?

The only possible radiative decay path for the $2s$ state is $2s \to 1s$, and this is strongly forbidden. The transition proceeds by two-photon decay, with $A_{2s\to 1s} = 8.23\,{\rm s}^{-1}$ (Drake 1986), emitting a continuous spectrum extending from $\nu = 0$ to $\nu_{L\alpha} = 3I_H/4h$. Let $P_\nu^{(2s)} d\nu$ be the probability that one of the emitted photons will be in $(\nu, \nu + d\nu)$. Energy conservation requires that $P_\nu^{(2s)}(\nu) = P_\nu^{(2s)}(\nu_{L\alpha} - \nu)$: $P_\nu^{(2s)}$ must be symmetric around $\nu = \nu_{L\alpha}/2$. In the low-density limit, where every transition to $2s$ is followed by two-photon decay, the two-photon emissivity is

$$j_\nu(2s \to 1s) = n_e n_p \alpha_{{\rm eff},2s} \left[\frac{h\nu}{4\pi} P_\nu^{(2s)}\right] \quad . \tag{14.10}$$

The function $P_\nu^{(2s)}$ peaks at $\nu = \nu_{L\alpha}/2$. The product $h\nu P_\nu^{(2s)}$ appearing in Eq. (14.10) can be found, e.g., in Osterbrock & Ferland (2006) (Table 4.12). The two-photon decay spectrum is shown in Figure 10.2.

Because the radiative lifetime of the $2s$ level is long ($0.122\,{\rm s}$), it is possible for collisions with electrons or protons to depopulate the $2s$ level before a spontaneous decay occurs. Collisional deexcitation to the ground state $1s$ is possible, but the collisional rates for $2s \to 2p$ transitions are much larger, involving only a change in orbital angular momentum with essentially no change in energy. The rate coefficients $q_{e,2s\to 2p}$ and $q_{p,2s\to 2p}$ are given in Table 14.4. Collisions with protons dominate; since $n_e \approx n_p$, the critical density at which deexcitation by electron and proton collisions is equal to the radiative decay rate is

$$n_{e,{\rm crit}} = \frac{A_{2s\to 1s}}{q_{p,2s\to 2p} + q_{e,2s\to 2p}} \approx 1.55\times 10^4\,{\rm cm}^{-3} \quad . \tag{14.11}$$

When this process is taken into account, the emissivity contributed by two-photon decay becomes

Table 14.3 Fraction[a] of Case B Recombinations of Hydrogen that Populate $2s\ ^2{\rm S}_{1/2}$ and $2p\ ^2{\rm P}^{\rm o}_{1/2,3/2}$.

T(K)	$f(2s)$	$f(2p)$
4000	0.285	0.715
10000	0.325	0.675
20000	0.356	0.644

[a] From Brown & Mathews (1970)

Table 14.4 Rate Coefficients[a] q for $2s \rightarrow 2p$ transitions in H due to electron and proton collisions at $T = 10^4$ K.

$q_{p,2s\rightarrow 2^2\mathrm{P}^\mathrm{o}_{1/2}}$	$2.51 \times 10^{-4}\,\mathrm{cm}^3\,\mathrm{s}^{-1}$
$q_{p,2s\rightarrow 2^2\mathrm{P}^\mathrm{o}_{3/2}}$	$2.23 \times 10^{-4}\,\mathrm{cm}^3\,\mathrm{s}^{-1}$
$q_{e,2s\rightarrow 2^2\mathrm{P}^\mathrm{o}_{1/2}}$	$0.22 \times 10^{-4}\,\mathrm{cm}^3\,\mathrm{s}^{-1}$
$q_{e,2s\rightarrow 2^2\mathrm{P}^\mathrm{o}_{3/2}}$	$0.35 \times 10^{-4}\,\mathrm{cm}^3\,\mathrm{s}^{-1}$
$q_{p,2s\rightarrow 2p} + q_{e,2s\rightarrow 2p}$	$5.31 \times 10^{-4}\,\mathrm{cm}^3\,\mathrm{s}^{-1}$

[a] From Osterbrock (1974); Osterbrock & Ferland (2006).

$$j_\nu = \frac{n_e n_p \alpha_{\mathrm{eff}2s}}{(1 + n_e/n_{e,\mathrm{crit}})} \frac{h\nu}{4\pi} P_\nu^{(2s)} \quad \text{for } 0 < \nu < \nu(\mathrm{Ly}\alpha) \quad . \tag{14.12}$$

Therefore, in bright H II regions (e.g., the Orion Nebula), the two-photon continuum is somewhat suppressed by collisional processes: the fraction of radiative recombinations that produce two-photon emission is a decreasing function of density n_e. Thus, measurement of the ratio of two-photon continuum to Hα, Hβ, or other recombination line from levels $n \geq 3$ allows one to determine the electron density in an ionized nebula.

An atom entering the $2p$ level (whether $^2\mathrm{P}^\mathrm{o}_{1/2}$ or $^2\mathrm{P}^\mathrm{o}_{3/2}$) has a radiative lifetime of only $1/A_{2p\rightarrow 1s} = 1.59$ ns. At interstellar densities, the probability of collisional depopulation of the $2p$ state is negligible.

The Lyman α optical depth at line center can be written

$$\tau_0(\mathrm{Ly}\alpha) = 8.02 \times 10^4 \left(\frac{15\,\mathrm{km\,s}^{-1}}{b}\right) \tau(\mathrm{Ly\ cont}) , \tag{14.13}$$

where $\tau(\mathrm{Ly\ cont}) = 6.30 \times 10^{-18}\,\mathrm{cm}^2 N(\mathrm{H})$ is the optical depth due to photoionization just above the H ionization edge. Case B conditions presume that the Lyman continuum optical depth $\tau(\mathrm{Ly\ cont}) > 1$, so it is apparent that the Lyman α optical depth is very large, $\tau(\mathrm{Ly}\alpha) \gtrsim 10^5$. The Lyman α photons will therefore be scattered (i.e., reabsorbed and reemitted in a different direction) many times. The scattering is coherent: the frequency is essentially unchanged in the center-of-mass frame of the incident photon and the H atom doing the scattering. However, because the H atoms have velocities $\sim 10\,\mathrm{km\,s}^{-1}$, each scattered photon undergoes a random walk in frequency, which tends to take the photon away from line center and into the wings of the profile, where the optical depth is smaller. The combined effects of spatial diffusion and diffusion in frequency space will eventually allow the photon to escape the nebula, if it is not first absorbed by some absorber other than H — for example, dust — and if collisional deexcitation does not occur during the ~ 1.59 ns intervals between absorption and reemission. This is discussed further in §15.7.

14.3★ Radiative Recombination: Helium

14.3.1 $\mathrm{He}^{++} + e^- \rightarrow \mathrm{He}^+$

Massive stars with He in their atmospheres produce very little emission at $h\nu > 54.4\,\mathrm{eV} = I(\mathrm{He}^+ \rightarrow \mathrm{He}^{++})$, and therefore He^{++} is not abundant in normal H II regions. However, He^{++} can be abundant in the ionized gas around Wolf-Rayet stars, in planetary nebulae, and around accreting compact objects (e.g., AGNs). Radiative recombination of He^{++} to He^+ at temperature T follows a radiative cascade with the same branching probabilities as for H recombination at temperature $T/4$. However, all of the He II Lyman series $np \rightarrow 1s$ are capable of ionizing H. The resonance photon may therefore photoionize an H atom before undergoing resonance absorption by He II, but because the resonance cross sections are very large, this affects only a negligible fraction of the radiative cascades. Therefore, when case B conditions apply for atomic H, the radiative recombination cascade following recombination of He^{++} will proceed as though the $3p \rightarrow 1s$, $4p \rightarrow 1s$, $5p \rightarrow 1s$, ... transitions were forbidden, producing a recombination line spectrum just as for H I but with all frequencies larger by a factor 4 [assuming, of course, that $n(\mathrm{He}^+)/n_{\mathrm{He}}$ is not extremely small].

All of the He^+ ions produced by radiative recombination of He^{++} will end up in either the $2s$ or $2p$ states. The $2s$ state has $A(2s \rightarrow 1s) = 526.7\,\mathrm{s}^{-1}$ (Drake 1986), which is fast enough so that collisional depopulation is generally not important at nebular densities, and each recombination populating $2s$ will contribute a two-photon continuum with $h\nu_1 + h\nu_2 = 40.80\,\mathrm{eV}$. Decay of the $2s$ level produces at least one, and usually two, photons that are capable of ionizing H.

The $2p$ state will immediately decay to produce He II Lyα at $\lambda = 304\,\text{Å}$, which will undergo some number of resonant scatterings but will eventually photoionize either H or He^0. Recombination of He^{++} to He^+ $2s$ and $2p$ produces Balmer continuum radiation with $h\nu > I_{\mathrm{H}}$. In fact, in a nebula with $n(\mathrm{He}^{++})/n(\mathrm{H}^+) = 0.1$ and $T = 1 \times 10^4\,\mathrm{K}$, the production rate per volume of these H-ionizing photons is equal to 80% of the H recombination rate per volume (Osterbrock 1989) – these photons play an important role in the ionization of H in regions where He is doubly ionized.

14.3.2 $\mathrm{He}^+ + e^- \rightarrow \mathrm{He}^0$

Recombination of He^+ to He^0 is more complex. The rate coefficients for recombination directly to the ground state, or recombination to any level *except* the ground state, are approximated by

$$\alpha_{1s^2}(\mathrm{He}) = 1.54 \times 10^{-13} T_4^{-0.486}\,\mathrm{cm}^3\,\mathrm{s}^{-1} \tag{14.14}$$

$$\alpha_B(\mathrm{He}) = 2.72 \times 10^{-13} T_4^{-0.789}\,\mathrm{cm}^3\,\mathrm{s}^{-1} \tag{14.15}$$

for $0.5 < T_4 < 2$ (Benjamin et al. 1999). In Case B, the recombinations directly to the ground state $1s^2\,{}^1\mathrm{S}_0$ generate $h\nu > 24.60\,\mathrm{eV}$ photons that can ionize either

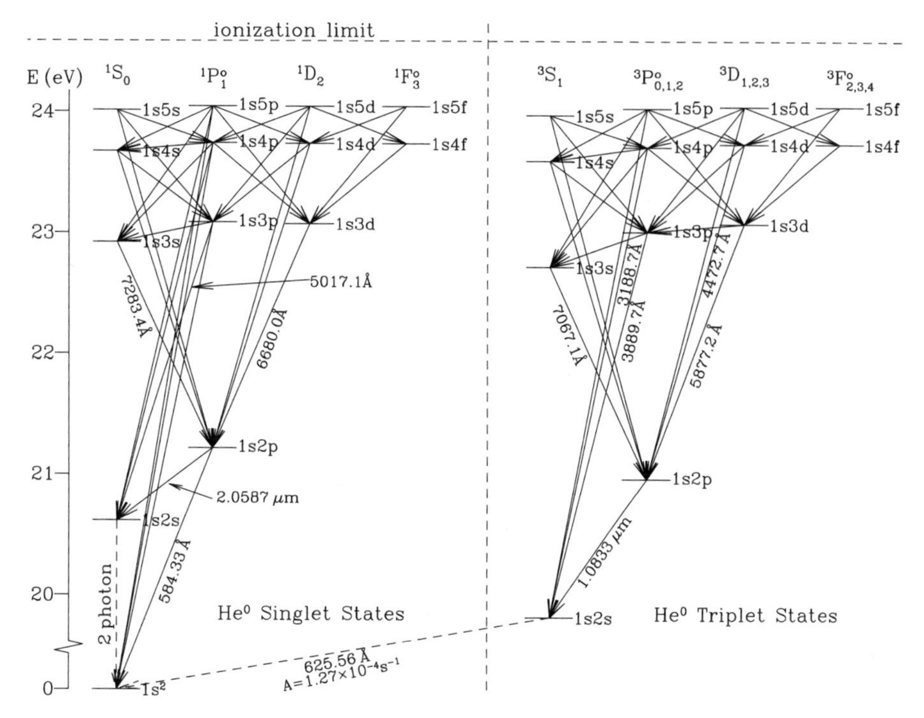

Figure 14.3 Radiative decay pathways for He^0 (see text). Selected lines are labeled by vacuum wavelength.

H^0 or He^0; the fraction y of these that ionize H is just

$$y \approx \frac{n(\mathrm{H}^0)\sigma_{\mathrm{pi,H}^0}(24.60\,\mathrm{eV} + kT)}{n(\mathrm{H}^0)\sigma_{\mathrm{pi,H}^0}(24.60\,\mathrm{eV} + kT) + n(\mathrm{He}^0)\sigma_{\mathrm{pi,He}^0}(24.60\,\mathrm{eV} + kT)}$$

$$= \left[1 + \frac{n(\mathrm{He}^0)}{n(\mathrm{H}^0)}\frac{\sigma_{\mathrm{pi,He}^0}(24.60\,\mathrm{eV} + kT)}{\sigma_{\mathrm{pi,H}^0}(24.60\,\mathrm{eV} + kT)}\right]^{-1} \quad . \tag{14.16}$$

For $kT > 0$, we have $\sigma_{\mathrm{pi,He}^0}(24.6\,\mathrm{eV} + kT)/\sigma_{\mathrm{pi,H}^0}(24.6\,\mathrm{eV} + kT) > 6.0$. Thus if $n(\mathrm{He}^0) > 0.16 n(\mathrm{H}^0)$, we will have $y < 0.5$.

The effective recombination rate for $He^+ \to He^0$ is then

$$\alpha_{\mathrm{eff}}(\mathrm{He}) = \alpha_B(\mathrm{He}) + y\alpha_{1s^2}(\mathrm{He}) \quad , \tag{14.17}$$

where α_B is the recombination rate to all states except the ground state, and α_{1s^2} is the recombination rate to the ground state $1s^2\ ^1\mathrm{S}_0$.

Consider now the recombinations to *excited* states of He^0. Approximately 25% of these will be to states with total spin $S = 0$ – i.e., **singlet** states. Recombinations to the singlet excited states of He^0 (see Figure 14.3) are followed by a radiative

Table 14.5 $\mathrm{He}^+ \rightarrow \mathrm{He}^0$ Recombination Spectrum[a] for $n_e = 10^2\,\mathrm{cm}^{-3}$ and $10^4\,\mathrm{cm}^{-3}$. Lines with $j > 0.7 j_{4472.73}$ are shown in bold.

	$n_e = 10^2\,\mathrm{cm}^{-3}$			$n_e = 10^4\,\mathrm{cm}^{-3}$		
$T(\mathrm{K})$	5000	10,000	20,000	5000	10,000	20,000
$\alpha_{\mathrm{eff},4472.73}\,(10^{-14}\,\mathrm{cm}^3\,\mathrm{s}^{-1})$	2.66	1.39	0.700	2.66	1.47	1.14
$4\pi j_{4472.73}/n_e n(\mathrm{He}^+)(10^{-26}\,\mathrm{erg\,cm}^3\,\mathrm{s}^{-1})$	11.8	6.16	3.11	11.8	6.52	5.07
$j_{2945.97}/j_{4472.73}$	0.356	0.438	0.540	0.357	0.419	0.359
$j_{\mathbf{3188.66}}/j_{4472.73}$	**0.745**	**0.913**	**1.136**	**0.748**	**0.931**	**0.997**
$j_{3614.67}/j_{4472.73}$	0.094	0.110	0.127	0.094	0.104	0.078
$j_{\mathbf{3889.75}}/j_{4472.73}$	**1.864**	**2.231**	**2.742**	**1.891**	**2.555**	**2.953**
$j_{3965.85}/j_{4472.73}$	0.193	0.226	0.261	0.194	0.219	0.187
$j_{4027.33}/j_{4472.73}$	0.450	0.464	0.463	0.450	0.440	0.284
$j_{4121.98}/j_{4472.73}$	0.028	0.040	0.059	0.028	0.037	0.036
$j_{4389.16}/j_{4472.73}$	0.120	0.123	0.121	0.120	0.116	0.074
$j_{4438.80}/j_{4472.73}$	0.013	0.017	0.023	0.013	0.016	0.014
$j_{\mathbf{4472.73}}/j_{4472.73}$	**1.000**	**1.000**	**1.000**	**1.000**	**1.000**	**1.000**
$j_{4714.5}/j_{4472.73}$	0.075	0.105	0.164	0.076	0.146	0.291
$j_{4923.31}/j_{4472.73}$	0.269	0.266	0.257	0.269	0.259	0.196
$j_{5017.08}/j_{4472.73}$	0.496	0.567	0.646	0.498	0.572	0.503
$j_{5049.15}/j_{4472.73}$	0.031	0.041	0.057	0.031	0.045	0.059
$j_{\mathbf{5877.25,5877.60}}/j_{4472.73}$	**2.989**	**2.743**	**2.591**	**2.966**	**2.933**	**3.118**
$j_{\mathbf{6679.99}}/j_{4472.73}$	**0.857**	**0.778**	**0.702**	**0.849**	**0.775**	**0.554**
$j_{7067.14,7067.66}/j_{4472.73}$	0.358	0.481	**0.713**	0.398	**0.943**	**1.525**
$j_{7283.36}/j_{4472.73}$	0.114	0.146	0.197	0.118	0.196	0.236
$j_{9466.21}/j_{4472.73}$	0.020	0.024	0.030	0.020	0.023	0.020
$j_{\mathbf{10832.1,10833.2,10833.3}}/j_{4472.73}$	**4.298**	**5.515**	**7.895**	**12.91**	**31.39**	**40.74**
$j_{11972.4}/j_{4472.73}$	0.045	0.047	0.047	0.045	0.044	0.029
$j_{12531.0}/j_{4472.73}$	0.024	0.029	0.036	0.024	0.030	0.032
$j_{12788.5}/j_{4472.73}$	0.174	0.152	0.129	0.173	0.143	0.079
$j_{12794.1}/j_{4472.73}$	0.058	0.051	0.043	0.058	0.048	0.026
$j_{12972.0}/j_{4472.73}$	0.015	0.016	0.015	0.015	0.015	0.009
$j_{15087.8}/j_{4472.73}$	0.010	0.012	0.014	0.010	0.012	0.010
$j_{17007.1}/j_{4472.73}$	0.066	0.066	0.066	0.066	0.066	0.066
$j_{18690.4}/j_{4472.73}$	0.440	0.360	0.295	0.429	0.347	0.239
$j_{18702.3}/j_{4472.73}$	0.146	0.120	0.097	0.143	0.113	0.064
$j_{19094.6}/j_{4472.73}$	0.025	0.025	0.024	0.025	0.024	0.018
$j_{19548.4}/j_{4472.73}$	0.014	0.017	0.021	0.014	0.017	0.019
$j_{\mathbf{20586.9}}/j_{4472.73}$	0.629	0.670	**0.723**	**0.758**	**1.036**	**0.899**
$j_{21126}/j_{4472.73}$	0.011	0.016	0.025	0.012	0.022	0.044

[a] Emissivities from Benjamin et al. (1999)

cascade down to the first two singlet excited states: approximately $\frac{1}{3}$ end up in $1s2s\,^1\mathrm{S}_0$, and approximately $\frac{2}{3}$ in $1s2p\,^1\mathrm{P}_1^{\mathrm{o}}$. The $1s2p\,^1\mathrm{P}_1^{\mathrm{o}}$ state has allowed decays to the ground state ($\lambda = 584.33$ Å) or to the $1s2s\,^1\mathrm{S}_0$ state ($\lambda = 2.0587\,\mu$m), with branching ratio

$$\frac{A_{1s2p\,^1\mathrm{P}_1^{\mathrm{o}} \rightarrow 1s2s\,^1\mathrm{S}_0\;2.0587\,\mu\mathrm{m}}}{A_{1s2p\,^1\mathrm{P}_1^{\mathrm{o}} \rightarrow 1s^2\,^1\mathrm{S}_0\;584.33\,\mathrm{\mathring{A}}}} = \frac{1.976 \times 10^6\,\mathrm{s}^{-1}}{1.799 \times 10^9\,\mathrm{s}^{-1}} = 1.098 \times 10^{-3} \quad . \qquad (14.18)$$

Therefore, 99.9% of the decays of $1s2p\,^1\mathrm{P}_1^{\mathrm{o}}$ will be by emission of a 584.33Å photon; under Case B conditions, this photon will either ionize a hydrogen atom or will be absorbed by a nearby He^0, exciting it to $1s2p\,^1\mathrm{P}_1^{\mathrm{o}}$, which will then undergo another decay. The cycle will be repeated until the photon is absorbed by dust, it produces a hydrogen ionization, or a spontaneous decay to $1s2s$ occurs.

Transitions leading to $1s2s\ ^1\mathrm{S}_0$ will be followed by two-photon decay with $A = 51.0\,\mathrm{s}^{-1}$ (Drake 1986) and a total photon energy 20.62 eV; 56% of these two-photon decays produce a photon with $h\nu > 13.60\,\mathrm{eV}$.

Approximately 75% of the recombinations to excited states will be to states with spin $S = 1$ – i.e., the **triplet** states. As seen in Figure 14.3, recombinations to triplet states of He^0 will be followed by a rapid cascade down to the lowest triplet state $1s2s\ ^3\mathrm{S}_1$. A large fraction of these cascades pass through the $1s2p\ ^3\mathrm{P}$ state, resulting in emission of the He I 10833 Å triplet (the components are at 10832.1, 10833.2, 10833.3Å), with the strongest observable line[2] produced by recombination of He^+ to He^0.

The $1s2s\ ^3\mathrm{S}_1$ level is metastable – it can decay by spontaneous emission of a single 19.82 eV photon, but the transition is highly forbidden, with $A_{1s2s\ ^3\mathrm{S}_1 \rightarrow 1s^2\ ^1\mathrm{S}_0} = 1.27 \times 10^{-4}\,\mathrm{s}^{-1}$. Spin-changing collisions with electrons can collisionally excite $1s2s\ ^3\mathrm{S}_1$ to $1s2s\ ^1\mathrm{S}_0$, requiring an energy $20.62 - 19.82 = 0.80\,\mathrm{eV}$; at interstellar densities, this will almost always be followed by two-photon decay to the ground state with $A(1s2s\ ^1\mathrm{S}_0 \rightarrow 1s^2\ ^1\mathrm{S}_0) = 51.0\,\mathrm{s}^{-1}$; 56% of these decays produce a photon with $h\nu > 13.60\,\mathrm{eV}$. About 10% of the spin-changing collisions will collisionally excite $1s2p\,^1\mathrm{P}_1^\mathrm{o}$, which immediately decays with emission of a 21.2 eV photon. The critical density for triplet-to-singlet conversion is

$$n_{\mathrm{crit},e}(1s2s\ ^3\mathrm{S}_1) = \frac{A_{1s2s\ ^3\mathrm{S} \rightarrow 1s^2\ ^1\mathrm{S}}}{q_{1s2s\ ^3\mathrm{S} \rightarrow 1s2s\ ^1\mathrm{S}} + q_{1s2s\ ^3\mathrm{S} \rightarrow 1s2p\ ^1\mathrm{P}}} \tag{14.19}$$

$$\approx 1100\,\mathrm{e}^{1.2/T_4} T_4^{0.5}\,\mathrm{cm}^{-3} \quad . \tag{14.20}$$

For $n_e \gg n_{\mathrm{crit},e}(1s2s\ ^3\mathrm{S}_1)$, nearly all recombinations to the triplet state are collisionally converted to the singlet states before emission of a 625.6Å (19.82 eV) photon can take place.

Non-spin-changing collisions can excite $1s2s\ ^3\mathrm{S}_1 \rightarrow 1s2p\ ^3\mathrm{P}^\mathrm{o}_{0,1,2}$ (the energy required is only 1.14 eV), which will be immediately followed by emission of He I 10833 Å. Even at low densities, He I 10833 Å is a strong emission line, but at high densities the line strength can be increased by this collisional excitation process. For example, Table 14.5 shows that at $T = 10^4\,\mathrm{K}$, the ratio of $j_{10833}/j_{4472.7}$ increases by a factor of 5.7 as the density is increased from $10^2\,\mathrm{cm}^{-3}$ to $10^4\,\mathrm{cm}^{-3}$. Because it is enhanced by collisional excitation, He I 10833 is also temperature-sensitive: at $n_e = 10^4\,\mathrm{cm}^{-3}$, the ratio $j_{10833}/j_{4472.7}$ increases by a factor of 3.2 as T varies from $5 \times 10^3\,\mathrm{K}$ to $2 \times 10^4\,\mathrm{K}$.

Collisional excitation from the metastable $1s2s\ ^3\mathrm{S}_1$ level can also enhance the intensity of the 7067.1 Å and 3889.7 Å lines; inspection of Table 14.5 shows that these lines are indeed enhanced by increased density and increased temperature, although not to the same degree as 10833 Å itself.

[2]Certain H-ionizing lines emitted in permitted transitions to the ground state, such as He I 584.33 Å, are more powerful, but are unobservable.

Table 14.6 Selected recombination rate coefficients[a] $\alpha_{rr}(\mathrm{cm^3\,s^{-1}})$ for H I regions ($T \leq 10^4$ K)

Reactants	$T = 30$ K	$T = 100$ K	$T = 10^3$ K
H II + e^- → H I (case B)	1.44×10^{-11}	7.01×10^{-12}	1.49×10^{-12}
He II + e^- → He I	1.75×10^{-11}	8.42×10^{-12}	1.00×10^{-12}
C II + e^- → C I	1.77×10^{-11}	8.63×10^{-12}	2.18×10^{-12}
Na II + e^- → Na I	1.44×10^{-11}	6.77×10^{-12}	1.31×10^{-12}
Mg II + e^- → Mg I	2.01×10^{-11}	7.18×10^{-12}	1.00×10^{-12}
Al II + e^- → Al I	4.20×10^{-11}	1.60×10^{-11}	2.52×10^{-12}
Si II + e^- → Si I	1.94×10^{-11}	9.39×10^{-12}	2.35×10^{-12}
P II + e^- → P I	3.60×10^{-11}	1.70×10^{-11}	4.08×10^{-12}
S II + e^- → S I	1.59×10^{-11}	7.46×10^{-12}	1.75×10^{-12}
Cl II + e^- → Cl I	7.35×10^{-11}	3.02×10^{-11}	5.53×10^{-12}
K II + e^- → K I	2.92×10^{-11}	1.11×10^{-11}	1.75×10^{-12}
Ca II + e^- → Ca I	2.09×10^{-11}	7.07×10^{-12}	8.90×10^{-13}
Ca III + e^- → Ca II	7.07×10^{-11}	2.70×10^{-11}	4.28×10^{-12}
Ti II + e^- → Ti I	2.24×10^{-11}	7.59×10^{-12}	9.62×10^{-13}
Ti III + e^- → Ti II	8.06×10^{-10}	2.85×10^{-10}	3.90×10^{-11}
Cr II + e^- → Cr I	2.38×10^{-11}	8.10×10^{-12}	1.03×10^{-12}
Mn II + e^- → Mn I	2.44×10^{-11}	8.33×10^{-12}	1.07×10^{-12}
Fe II + e^- → Fe I	2.46×10^{-11}	8.52×10^{-12}	1.10×10^{-12}
Co II + e^- → Co I	2.54×10^{-11}	9.77×10^{-12}	1.57×10^{-12}
Ni II + e^- → Ni I	2.10×10^{-11}	9.04×10^{-12}	1.80×10^{-12}
Zn II + e^- → Zn I	2.10×10^{-11}	9.04×10^{-12}	1.80×10^{-12}

[a] From Verner (1999)

14.4 Radiative Recombination: Heavy Elements

Radiative recombination of elements heavier than H or He proceeds similarly, with the important exception that we do not concern ourselves with the possibility that photons emitted from recombination to the ground state could be reabsorbed locally by another atom of the "recombined" species. That is, we assume Case A conditions when studying the recombination of heavy elements. If recombination radiation is going to produce ionization, the photons from recombination of H and He will likely be of much greater importance than photons from recombination of the heavier elements.

In regions where the gas temperature is $\lesssim 10^3$ K, the H and He will generally be predominantly neutral, and most ions that are present will be either singly or doubly ionized. Table 14.6 provides radiative recombination rates for H II, He II, and ions that can be produced by photoionization by $h\nu < 13.6$ eV photons, for elements with solar abundances $X/\mathrm{H} > 4 \times 10^{-8}$.

In H II regions, higher ionization stages can be produced by photoionization. Table 14.7 lists radiative recombination rates for ions with ionization potentials < 100 eV; the table is limited to the elements with solar abundances greater than or equal to sulfur $[(\mathrm{S/H})_\odot = 1.8 \times 10^{-5}]$.

Radiative recombination of elements such as O and Ne is accompanied by emission of characteristic recombination lines – the recombining electrons are captured into excited states, which then emit a cascade of line radiation. For example, radiative recombination of O III sometimes populates the excited state $2s2p^2(^4\mathrm{P})3d$, resulting in O II 4462.86 Å $[2s2p^2(^4\mathrm{P})3p - 2s2p^2(^4\mathrm{P})3d]$ and O II 4073.79, 4075.13 Å $[2s2p^2(^4\mathrm{P})3s - 2s2p^2(^4\mathrm{P})3p]$ emission, in addition to other lines.

In H II regions and planetary nebulae, these recombination lines will be faint compared to the recombination lines of H, simply because of the greatly reduced abundance of the heavy elements compared to hydrogen, but can nevertheless be measured. The line strengths allow the abundance of the recombining ion stage to be inferred. Because the radiative recombination cross sections to different excited states will have different temperature dependence, the ratios of recombination lines can be used to determine the electron temperature. When the recombining ion (e.g., O III) has fine structure splitting of the ground state, the recombination line strengths will depend on the relative populations of the different fine structure states of the recombining ion, and the line ratios will therefore also be density dependent. It is important to choose recombination lines for observation that are unlikely to be contaminated by emission resulting from starlight excitation in the nebula.

The abundances obtained from recombination lines should, in principle, agree with the abundances derived from the much stronger collisionally excited lines, such as [O III]4960,5008 (to be discussed in §18.5). Interestingly, recombination lines in the Orion Nebula give abundances that are *larger* than deduced from collisionally excited lines of these species, by factors of $10^{0.14\pm0.02}$ for O III, $10^{0.39\pm0.20}$ for O II, $10^{0.39\pm0.15}$ for C III, and $10^{0.26\pm0.10}$ for Ne III (Esteban et al. 2004).

More extreme results are found in some planetary nebulae, with elemental abundances derived from recombination lines exceeding the abundances based on collisionally excited lines by factors ranging from 1.5 to 12 in a sample of 23 Galactic planetary nebulae (Wesson et al. 2005). The optical recombination lines tend to give abundance estimates that are significantly higher than those based on collisionally excited lines (e.g., Wesson et al. 2005). The reason for these discrepancies is uncertain. It has been interpreted as evidence for temperature variations within the emitting gas (Esteban et al. 2004); the extreme discrepancies seen in some planetary nebulae suggest that they may contain cool ionized gas where recombination line emission is enhanced and collisionally excited emission is weak. This is a puzzle that is yet to be resolved.

14.5 Dielectronic Recombination

For an electron that is initially free to be captured to a bound state of an atom or ion, the electron must lose energy. Radiative recombination is relatively slow because it is necessary to create a photon to remove this energy as part of the capture process, and this can take place only during the brief time that the free electron is appreciably

Table 14.7 Selected Recombination Rate Coefficients at $T = 10^4$ K

Reactants	$\alpha_{\rm rr}(\rm cm^3\,s^{-1})$	$\alpha_{\rm diel}(\rm cm^3\,s^{-1})$	$\alpha_{\rm diel}/\alpha_{\rm rr}$	Reference
H II + e^- → H I (case A)	4.18×10^{-13}	0	0	*d*
H II + e^- → H I (case B)	2.54×10^{-13}	0	0	*d*
He II + e^- → He I (case A)	4.26×10^{-13}	$< 10^{-23}$	0	*d*
He II + e^- → He I (case B)	2.72×10^{-13}	$< 10^{-23}$	0	*d*
He III + e^- → He II	2.19×10^{-12}	0	0	*f*
C II + e^- → C I	5.48×10^{-13}	–	–	*e*
C III + e^- → C II	2.48×10^{-12}	6.06×10^{-12}	2.44	*e,b*
C IV + e^- → C III	5.12×10^{-12}	1.31×10^{-11}	2.56	*f, b*
N II + e^- → N I	3.87×10^{-13}	–	–	*e*
N III + e^- → N II	2.36×10^{-12}	2.04×10^{-12}	0.86	*e, b*
N IV + e^- → N III	5.23×10^{-12}	2.16×10^{-11}	4.13	*e, b*
N V + e^- → N IV	9.84×10^{-12}	1.54×10^{-11}	1.56	*e, b*
O II + e^- → O I	3.25×10^{-13}	–	–	*e*
O III + e^- → O II	1.99×10^{-12}	1.66×10^{-12}	0.83	*e, b*
O IV + e^- → O III	5.76×10^{-12}	1.14×10^{-11}	1.98	*e, b*
O V + e^- → O IV	1.00×10^{-11}	3.45×10^{-11}	3.45	*e, b*
Ne II + e^- → Ne I	2.14×10^{-13}	–	–	*e*
Ne III + e^- → Ne II	1.53×10^{-12}	–	–	*e*
Ne IV + e^- → Ne III	5.82×10^{-12}	–	–	*e*
Ne V + e^- → Ne IV	9.84×10^{-12}	–	–	*e*
Mg II + e^- → Mg I	1.40×10^{-13}	7.06×10^{-13}	5.04	*e, c*
Mg III + e^- → Mg II	1.22×10^{-12}	–	–	*e*
Mg IV + e^- → Mg III	3.50×10^{-12}	–	–	*e*
Si II + e^- → Si I	5.90×10^{-13}	4.19×10^{-13}	0.71	*e, c*
Si III + e^- → Si II	1.00×10^{-12}	3.87×10^{-12}	3.87	*e, c*
Si IV + e^- → Si III	3.70×10^{-12}	1.47×10^{-11}	3.97	*e, c*
S II + e^- → S I	4.10×10^{-13}	6.04×10^{-15}	0.01	*e*
S III + e^- → S II	1.80×10^{-12}	–	–	*e*
S IV + e^- → S III	2.70×10^{-12}	–	–	*e*
Fe II + e^- → Fe I	1.42×10^{-13}	5.79×10^{-13}	4.07	*e, a*
Fe III + e^- → Fe II	1.02×10^{-12}	1.29×10^{-17}	1.3×10^{-5}	*e*
Fe IV + e^- → Fe III	3.32×10^{-12}	–	–	*e*
Fe V + e^- → Fe IV	7.79×10^{-12}	–	–	*e*

a Mazzotta et al. (1998)
b Nussbaumer & Storey (1983)
c Nussbaumer & Storey (1986)
d Osterbrock & Ferland (2006)
e Verner (1999)
f Verner & Ferland (1996)

accelerated by the electric field of the ion.

However, if an ion has at least one bound electron to begin with, then it is possible to have a capture process where the incoming electron transfers energy to a bound electron, promoting the bound electron to an excited state, and removing enough energy from the first electron that it too can be captured in an excited state: the ion now has **two** electrons in excited states. As long as the energy has not been radiated away, the process is reversible: the doubly excited electronic state is capable of **autoionizing**. However, either of the electrons can undergo a spontaneous radiative decay and remove enough energy from the system that it can no longer autoionize. Because two electrons are involved, the process is referred to as **dielectronic recombination**.

The rate coefficient for dielectronic recombination depends on detailed atomic physics: the initial cross section for capture to a doubly excited state, and the probabilities per unit time of autoionization and spontaneous radiative decay.

Dielectronic recombination is important in high-temperature plasmas, where it often exceeds the radiative recombination rate. At low temperatures, dielectronic recombination is generally suppressed because of the need for the electron to have sufficient energy to produce a doubly excited state. Nevertheless, for selected ions – including recombination of Mg II, and C III – dielectronic recombination is important at the $\sim 10^4$ K temperatures of H II regions.

Dielectronic recombination can also proceed via excitation of fine-structure levels of the target ion, with the recombining electron captured into a high-lying "Rydberg state" – this channel can be important in low-temperature gas (Bryans et al. 2009), both in H I regions (with $T \approx 10^2$ K) and in molecular clouds (with $T \approx 20$ K). Examples of species where this could be important include recombination of C II $\rightarrow$ C I, Si II $\rightarrow$ Si I, and O III $\rightarrow$ O II.

The dielectronic recombination process populates specific energy levels, which then undergo radiative decay, with predicted line ratios that differ from those resulting from pure radiative recombination (Nussbaumer & Storey 1983, 1984, 1986). For example, the C III 2296Å line in planetary nebulae is produced by dielectronic recombination of C IV (Storey 1981).

There has been considerable recent work on dielectronic recombination: see Colgan et al. (2003), Zatsarinny et al. (2003), Altun et al. (2004), Colgan et al. (2004), Gu (2004), Mitnik & Badnell (2004), Zatsarinny et al. (2004*b*,*a*), and Badnell (2006).

14.6 Dissociative Recombination

When a molecular ion AB^+ captures an electron, the electronic wave function of the resulting molecule AB must of course be different from the wave function of AB^+, because an additional electron is present. Because ionization potentials generally exceed chemical binding energies, the newly recombined molecule will have enough energy to dissociate, and normally will find itself in an electronic state where the electronic energy can be reduced by increasing the separation of the nuclei – i.e., a repulsive state. When this happens, A and B will begin to move apart, with the nuclear motion accelerating on the vibrational time scale, $\sim 10^{-13}$ s. The excited state can of course reduce its electronic energy by emitting a photon, but the Einstein A coefficient for the transition will typically be $\lesssim 10^8\,\mathrm{s}^{-1}$ – the time scale for radiative decay $1/A \approx 10^{-8}$ s. Because the vibrational time scale is only $\sim 10^{-5}$ of the time scale for spontaneous emission, photon emission is highly improbable, and A and B will fly apart, with the difference between the ionization energy I_{AB} and the chemical binding energy of AB appearing as kinetic energy of A and B.

The cross section for the molecular ion to capture an electron is similar to the cross section for an excited ion to be deexcited by a thermal electron – as discussed in §2.3, the rate coefficient for this process is typically $k_d \approx 10^{-7} T_2^{-1/2}\,\mathrm{cm}^3\,\mathrm{s}^{-1}$, albeit with significant differences from one reaction to another. Table 14.8 shows some selected dissociative recombination rates.

Table 14.8 Dissociative Recombination for Selected Molecular Ions

Reaction	$k_d(\mathrm{cm}^3\,\mathrm{s}^{-1})$	Reference
$\mathrm{OH}^+ + e^- \to \mathrm{O} + \mathrm{H}$	$6.50 \times 10^{-8} T_2^{-0.50}$	Le Teuff et al. (2000)
$\mathrm{H_2}^+ + e^- \to \mathrm{H} + \mathrm{H}$	$1.1 \times 10^{-8} T_2^{-0.89}$	Schneider et al. (1994)
$\mathrm{CO}^+ + e^- \to \mathrm{C} + \mathrm{O}$	$3.39 \times 10^{-7} T_2^{-0.48}$	Le Teuff et al. (2000)
$\mathrm{H}_3^+ + e^- \to \mathrm{H} + \mathrm{H} + \mathrm{H}$	$8.9 \times 10^{-8} T_2^{-0.48}$	McCall et al. (2004)
$\mathrm{H}_3^+ + e^- \to \mathrm{H}_2 + \mathrm{H}$	$5.0 \times 10^{-8} T_2^{-0.48}$	McCall et al. (2004)
$\mathrm{CH}^+ + e^- \to \mathrm{C} + \mathrm{H}$	$2.38 \times 10^{-7} T_2^{-0.42}$	Le Teuff et al. (2000)

14.7 Charge Exchange

In a collision between an ion A^+ and a neutral B, sometimes the ion can seize one of the electrons from the neutral: $A^+ + B \to A + B^+ + \Delta E$. The energy release $\Delta E = I(A) - I(B)$, where $I(A)$ and $I(B)$ are the ionization potentials for A and B. (In the case of A, this should be the ionization potential for whatever electronic state the electron is captured into – this need not be the ground state of A.)

A necessary (but not sufficient) condition is that the reaction either be exothermic ($\Delta E > 0$) or, if endothermic, that the energy required ($-\Delta E$) not be large compared to kT: $-\Delta E \lesssim kT$.

A second necessary condition is that there be "level crossing." Let $V_{A^+ + B}(r)$ and $V_{A+B^+}(r)$ be potential energy functions obtained by solving the Schrödinger equation for the electrons for the nuclei at fixed separation r, where $V_{A^+ + B}$ is the energy of the eigenfunction that, for increasing r, evolves adiabatically into

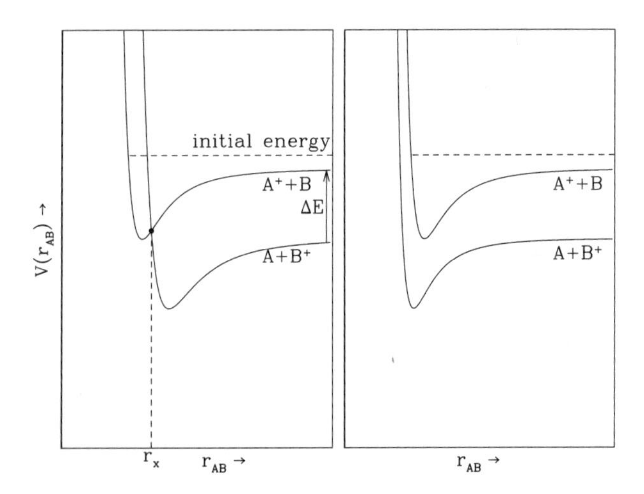

Figure 14.4 Diagrams showing the effective potential, as a function of internuclear separation r_{AB}, for the $A^+ + B$ system and the $A + B^+$ system. Left: The energy levels cross at $r = r_x$; charge exchange will occur in a large fraction of orbiting encounters as the internuclear separation passes through r_x. Right: The energy levels do not cross, and charge exchange, even though exothermic, will not proceed.

the incoming state $A^+ + B$, and V_{A+B^+} is the energy of the eigenfunction that corresponds to $A + B^+$ at large separation. In order for charge exchange to occur, these two potential energy functions must cross, so that the electronic wave function can make a transition from one state to another when the internuclear separation is just right. Even though the reaction may be exothermic, level crossing is not guaranteed: Figure 14.4 shows an example where level crossing does occur, and another example where it does not. Furthermore, the level crossing must take place at an energy that is within a few kT of the energy of the incoming electronic state $A^+ + B$, so that the approaching nuclei will have sufficient kinetic energy to achieve the required internuclear separation. This condition tends to rule out charge exchange reactions where the incoming species are both charged, or the outgoing species are both charged, because the Coulomb repulsion prevents small internuclear separations to be achieved.

If conditions are favorable, the electron transfer will occur whenever A^+ and B approach within $\sim 10^{-8}$ cm; the rate coefficient will therefore be approximately equal to that for ion–neutral "orbiting" collisions, with $k \gtrsim 10^{-9}\,\mathrm{cm^3\,s^{-1}}$ (see the discussion in §2.4).

14.7.1 Important Special Case: $O^+ + H \leftrightarrow O + H^+$

An important example is charge exchange between oxygen and hydrogen, $H^0 + O^+ \leftrightarrow H^+ + O^0$. Atomic oxygen and hydrogen have almost identical ionization potentials ($I_H = 13.5984\,\mathrm{eV}$, $I_{O(^3P_2)} = 13.6181\,\mathrm{eV}$), differing by only 0.0197 eV. There are three different fine-structure levels of O I that could be produced when an electron is exchanged; we must therefore consider the three separate channels:

$$\mathrm{H}(^1\mathrm{S}_{1/2}) + \mathrm{O}^+(^4\mathrm{S}^o_{3/2}) \rightarrow \mathrm{H}^+ + \mathrm{O}(^3\mathrm{P}_2) + .0197\,\mathrm{eV} \tag{14.21}$$

$$\rightarrow \mathrm{H}^+ + \mathrm{O}(^3\mathrm{P}_1) + .0001\,\mathrm{eV} \tag{14.22}$$

$$\rightarrow \mathrm{H}^+ + \mathrm{O}(^3\mathrm{P}_0) - .0084\,\mathrm{eV}\,. \tag{14.23}$$

The first two reactions are exothermic, and proceed rapidly; the third is slightly endothermic. The rate coefficients given by Stancil et al. (1999) are fit by

$$k_0 \approx 1.14 \times 10^{-9} T_4^{0.400+0.018\ln T_4}\,\mathrm{cm^3\,s^{-1}}\,, \tag{14.24}$$

$$k_1 \approx 3.44 \times 10^{-10} T_4^{0.451+0.036\ln T_4}\,\mathrm{cm^3\,s^{-1}}\,, \tag{14.25}$$

$$k_2 \approx 5.33 \times 10^{-10} T_4^{0.384+0.024\ln T_4} e^{-97\,\mathrm{K}/T}\,\mathrm{cm^3\,s^{-1}}\,. \tag{14.26}$$

The fits are accurate to within $\sim 10\%$ for $50 < T < 10^4$ K. By detailed balance, the rate coefficient for the reverse reaction $\mathrm{H}^+ + \mathrm{O}(^3\mathrm{P}_2) \rightarrow \mathrm{H}^0(^2\mathrm{S}_{1/2}) + \mathrm{O}^+(^4\mathrm{S}^o_{3/2})$ is

$$k_{0r} = \frac{g(^2\mathrm{S}_{1/2})g(^4\mathrm{S}^o_{3/2})}{g(\mathrm{H}^+)g(^3\mathrm{P}_2)}\, k_0\, e^{-.0197\,\mathrm{eV}/kT} \tag{14.27}$$

$$= \frac{8}{5}\, k_0\, e^{-229\,\mathrm{K}/T}\,. \tag{14.28}$$

Similarly, the rates for the reactions $\mathrm{H^+ + O(^3P_1) \rightarrow H + O^+}$ and $\mathrm{H^+ + O(^3P_0) \rightarrow H + O^+}$ are

$$k_{1r} = \frac{8}{3}\, k_1\, \mathrm{e}^{-1\,\mathrm{K}/T} \tag{14.29}$$

$$k_{2r} = \frac{8}{1}\, k_2\, \mathrm{e}^{+97\,\mathrm{K}/T} \quad . \tag{14.30}$$

Consider now a region where the hydrogen is partially ionized. At low densities, the excited fine-structure levels of O will decay radiatively to populate the ground state $\mathrm{O(^3P_2)}$ before collisions with $\mathrm{H^+}$ take place. The oxygen ionization balance will then be the steady state solution to

$$n(\mathrm{H^0})n(\mathrm{O^+})(k_0 + k_1 + k_2) = n(\mathrm{H^+})n(\mathrm{O^0})k_{0r} \quad , \tag{14.31}$$

or

$$\frac{n(\mathrm{O^+})}{n(\mathrm{O^0})} = \left(\frac{k_{0r}}{k_0 + k_1 + k_2}\right)\frac{n(\mathrm{H^+})}{n(\mathrm{H^0})} \quad . \tag{14.32}$$

The excited fine-structure levels are efficiently depopulated by spontaneous decay for $n_\mathrm{H} < n_\mathrm{crit}$, where

$$n_\mathrm{crit} = \frac{A(^3\mathrm{P}_1 \rightarrow ^3\mathrm{P}_2)}{(n(\mathrm{H})/n_\mathrm{H})k_{10}(\mathrm{H}) + (n_e/n_\mathrm{H})[k_{10}(M^+) + k_{10}(e^-)]} \tag{14.33}$$

$$\approx \frac{2.5 \times 10^5\,\mathrm{cm}^{-3}}{(n(\mathrm{H})/n_\mathrm{H})T_2^{0.40} + (n_e/n_\mathrm{H})(5.6 + 4.1T_2^{0.19})} . \tag{14.34}$$

Under most conditions of interest, we will be in the low-density limit $n_\mathrm{H} \ll n_\mathrm{crit}$.

At high densities $n \gg n_\mathrm{crit}$, collisions will ensure that the $\mathrm{O^0}$ fine structure levels $^3\mathrm{P}_{2,1,0}$ are populated according to a thermal distribution. In the high density limit, the oxygen ionization balance in steady state will then satisfy

$$n(\mathrm{O^+})\, n(\mathrm{H^0})\, (k_0 + k_1 + k_2)$$

$$= n(\mathrm{H^+}) \left[k_{0r} n(\mathrm{O^0}\,{}^3\mathrm{P}_2) + k_{1r} n(\mathrm{O^0}\,{}^3\mathrm{P}_1) + k_{2r} n(\mathrm{O^0}\,{}^3\mathrm{P}_0)\right]$$

$$= n(\mathrm{H^+})n(\mathrm{O^0}) \frac{\left[5k_{0r} + 3k_{1r}\mathrm{e}^{-228\,\mathrm{K}/T} + k_{2r}\mathrm{e}^{-326\,\mathrm{K}/T}\right]}{5 + 3\mathrm{e}^{-228\,\mathrm{K}/T} + \mathrm{e}^{-326\,\mathrm{K}/T}} . \tag{14.35}$$

The ratio $\left[n(\mathrm{O^+})/n(\mathrm{O^0})\right] / \left[n(\mathrm{H^+})/n(\mathrm{H^0})\right]$ is plotted in Figure 14.5 for the low- and high-density limits.

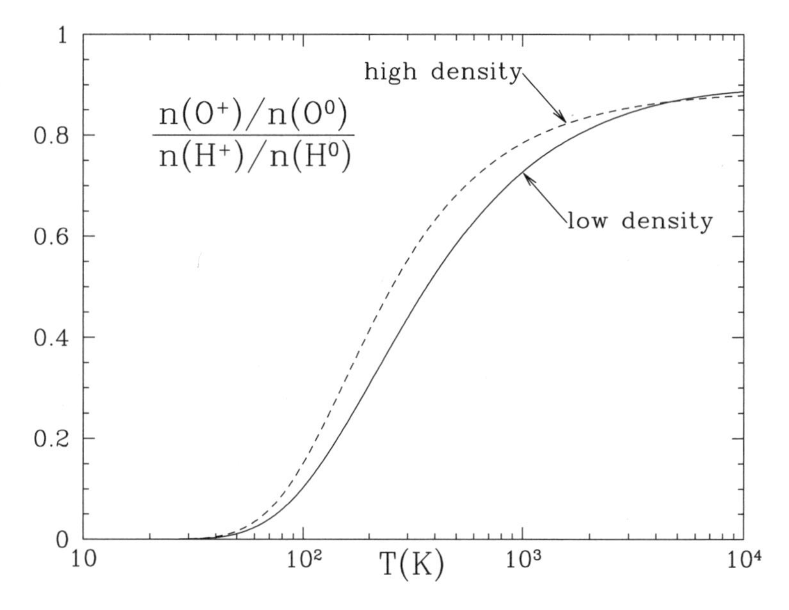

Figure 14.5 Dependence of oxygen ionization fraction on hydrogen ionization fraction due to charge exchange. The low-density limit applies for $n_{\rm H} \lesssim 10^4\,{\rm cm}^{-3}$.

At low temperatures, $T \lesssim 300\,{\rm K}$, ionized oxygen is suppressed by charge transfer reactions with $\rm H^0$. However, for $T \gtrsim 10^3\,{\rm K}$, the oxygen ionization fraction will be close to the hydrogen ionization fraction – this applies in an H II region provided that the rate of photoionization of O is slow compared to the rate of charge exchange $(k_{0r} + k_{1r} + k_{2r})n({\rm H}^+)$.

14.8 Ion Neutralization by Dust Grains

Astrophysical gas is commonly mixed with dust, and an ion A^+ colliding with a dust grain may capture an electron from the grain if the ionization potential $I(A)$ exceeds the "work function" W for the grain material – W is, essentially, the energy required to ionize the grain material. The rate for this process will depend on the charge state of the dust grains: positive ions will be repelled by positively charged grains, but will have large collision rates with negatively charged grains due to Coulomb focusing. Collision rates with neutral grains will also be enhanced by the same induced-dipole interaction that leads to large ion-neutral collision rates.

This process has been discussed by Weingartner & Draine (2001*b*), who have calculated the grain charge distribution for realistic dust models that include a population of very small polycyclic aromatic hydrocarbon (PAH) grains.

The rate per volume for recombination of some ion X^+ via electron capture from

Table 14.9 Parameters[a] for Fit (14.37) to Grain Recombination Rate coefficients $\alpha_{\rm gr}(X^+)$ for Selected ions.

Ion	C_0	C_1	C_2	C_3	C_4	C_5	C_6
H^+	12.25	8.074×10^{-6}	1.378	5.087×10^{2}	1.586×10^{-2}	0.4723	1.102×10^{-5}
He^+	5.572	3.185×10^{-7}	1.512	5.115×10^{3}	3.903×10^{-7}	0.4956	5.494×10^{-7}
C^+	45.58	6.089×10^{-3}	1.128	4.331×10^{2}	4.845×10^{-2}	0.8120	1.333×10^{-4}
Mg^+	2.510	8.116×10^{-8}	1.864	6.170×10^{4}	2.169×10^{-6}	0.9605	7.232×10^{-5}
S^+	3.064	7.769×10^{-5}	1.319	1.087×10^{2}	3.475×10^{-1}	0.4790	4.689×10^{-2}
Ca^+	1.636	8.208×10^{-9}	2.289	1.254×10^{5}	1.349×10^{-9}	1.1506	7.204×10^{-4}

[a] From Weingartner & Draine (2001*b*)

grains can be written

$$\frac{d}{dt}n(X^+) = -n_{\rm H}\alpha_{\rm gr}(X^+, G_0/n_e, T)n(X^+)\,, \tag{14.36}$$

where $\alpha_{\rm gr}$, the effective rate coefficient, depends on n_e/G_0 and T. Weingartner & Draine (2001*b*) fit $\alpha_{\rm gr}$ with

$$\alpha_{\rm gr}(X^+, G_0/n_e, T) = \frac{10^{-14}C_0\,{\rm cm}^3\,{\rm s}^{-1}}{1 + C_1\psi^{C_2}(1 + C_3T^{C_4}\psi^{-C_5-C_6\ln T})}\,, \tag{14.37}$$

where the charge state of the grains is determined by T and a dimensionless "charging parameter"

$$\psi \equiv \frac{G_0\sqrt{T/{\rm K}}}{n_e/\,{\rm cm}^{-3}}\,. \tag{14.38}$$

Here, G_0 is the starlight intensity relative to Habing (1968) [see Eq. (12.6)]. We will see in Chapter 25 why the grain charge depends on ψ. Table 14.9 gives the fit parameters for selected ions, and Figure 14.6 shows the effective rate coefficient $\alpha_{\rm gr}$ for recombination of H^+, C^+, and S^+ as functions of the charging parameter ψ, for three different temperatures.

The importance of grain-assisted recombination is illustrated in Chapter 16 (see Figure 16.1), where it is found to lead to significant reductions in the ionization in diffuse clouds. The rate coefficients $\alpha_{\rm gr}$ depend on both the physics of grain charging (see Weingartner & Draine 2001*b*) and on the assumed abundances of grains of different sizes. The parameters in Table 14.9 and the rates shown in Figure 14.6 were calculated using a model for Milky Way dust that includes a substantial population of polycyclic aromatic hydrocarbons (PAHs). Most of the recombination takes place on the smallest grains, as these dominate the total grain surface area, and in addition are often neutral or even negatively charged. In low metallicity galaxies, where the dust mass per H is reduced, grain-assisted recombination will be less important than in the Milky Way.

In dense clouds, changes in the ultraviolet extinction curve indicate that the abundances of the smallest grains are suppressed, presumably because of coagulation with larger grains. This should reduce the effective rate coefficient for grain-assisted recombination; the values of $\alpha_{\rm gr}$ in dense clouds are therefore expected to be smaller than the values in Fig. 14.6.

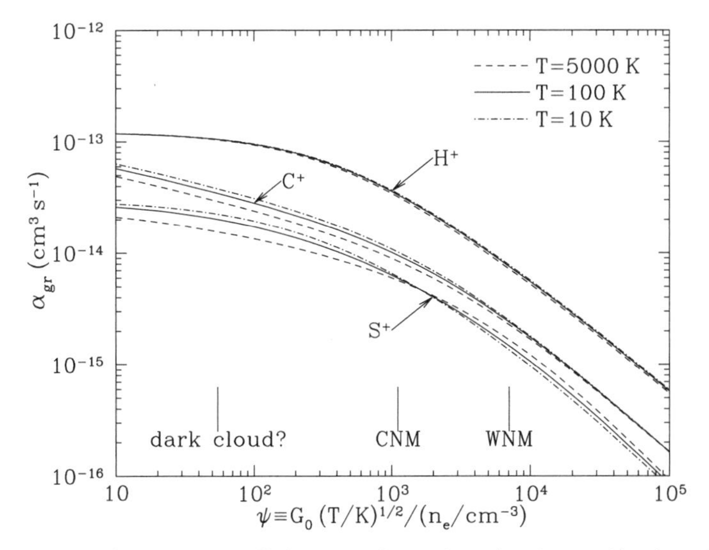

Figure 14.6 Effective rate coefficients $\alpha_{\rm gr}$ for grain-assisted recombination of H^+, C^+, and S^+ on dust grains in diffuse clouds, as a function of the grain charging parameter ψ, and for three gas temperatures. The values of ψ for nominal CNM and WNM conditions are shown. ψ varies over a large range within a single dark cloud; one possible value is indicated. Note that the values shown for $\alpha_{\rm gr}$ were calculated for a model for dust in diffuse clouds; because of grain coagulation, $\alpha_{\rm gr}$ in dark clouds is likely to be smaller than the values shown here.

14.9 Ionization Balance in Collisionally Ionized Gas

Consider hot gas where the ionization is dominated by collisional ionization. The steady state balance between collisional ionization and radiative recombination is given by

$$n_e \langle \sigma v \rangle_{\rm ci} n(X^{n+}) = n_e \langle \sigma v \rangle_{\rm rr} n(X^{(n+1)+}) \quad . \qquad (14.39)$$

Above the threshold energy I, the collisional ionization cross section $\sigma_{\rm ci} \approx C\pi a_0^2(1-I/E)$, where C is a constant of order unity (see §13.4). The collisional ionization rate is given by Eq. (13.11):

$$\langle \sigma v \rangle_{\rm ci} \approx C\pi a_0^2 \left(\frac{8kT}{\pi m_e}\right)^{1/2} {\rm e}^{-I/kT} \quad . \qquad (14.40)$$

To estimate the radiative recombination rate, recall that the Milne relation (3.31) relates the radiative recombination cross section $\sigma_{\rm rr}$ to the photoionization cross section $\sigma_{\rm pi}$:

$$\sigma_{\rm rr}(E) = \frac{g_\ell}{2g_u} \frac{(I+E)^2}{Em_ec^2} \sigma_{\rm pi}(h\nu = I+E) \quad . \qquad (14.41)$$

If $f_{\rm pi}$ is the integrated oscillator strength for photoionization from the ground state, then $\int_I^\infty \sigma_{\rm pi} h\, d\nu = (\pi e^2/m_e c) h f_{\rm pi}$. If we assume that $\sigma_{\rm pi} \propto (h\nu)^{-3}$ (c.f. Eq. 13.3) then

$$\sigma_{\rm pi}(h\nu = I) \approx \frac{2\pi e^2}{m_e c} f_{\rm pi} \frac{h}{I} \quad . \tag{14.42}$$

With these two estimates, we can show that

$$\frac{\langle\sigma v\rangle_{\rm rr}}{\langle\sigma v\rangle_{\rm ci}} \approx 2\pi\alpha^3 \frac{f_{\rm pi}}{C} \frac{I}{kT} \, {\rm e}^{I/kT} \quad , \tag{14.43}$$

where $\alpha \equiv e^2/\hbar c = 1/137.04...$ is the fine-structure constant. The radiative recombination cross section is small because the fine-structure constant $\alpha \ll 1$.

For what temperature T will we have equal amounts of the two ionization stages, $n(X^{+n}) = n(X^{+(n+1)})$? This will take place when $\langle\sigma v\rangle_{\rm rr} = \langle\sigma v\rangle_{\rm ci}$. For this to be the case, the ratio I/kT must satisfy

$$\frac{I}{kT} \, {\rm e}^{I/kT} = \frac{C}{2\pi f_{\rm pi}} \frac{1}{\alpha^3} \quad . \tag{14.44}$$

If $C \approx 1$ and $f_{\rm pi} \approx 1$, this has solution $I/kT \approx 10.6$. This is a good rule-of-thumb: Balancing collisional ionization against radiative recombination leads to a 50/50 balance of ion stages when $kT \approx I/10$. If dielectronic recombination is strong, it can raise the temperature at which 50/50 ion/neutral balance occurs.

If we apply this to hydrogen, for which $I/k = 157{,}800\,{\rm K}$, we estimate $n({\rm H}^0) = n({\rm H}^+)$ when $T = 15{,}000\,{\rm K}$. This is close to the value found using detailed collisional ionization and radiative recombination rates.

For elements other than hydrogen, dielectronic recombination must be included, so that the condition for the steady state "collisional ionization equilibrium" is

$$\langle\sigma v\rangle_{\rm ci} n_e n(X^{+j}) = \left(\langle\sigma v\rangle_{\rm rr} + \langle\sigma v\rangle_{\rm diel}\right) n_e n(X^{+j+1}) \quad , \tag{14.45}$$

where it is assumed that photoionization can be neglected. The ion ratios

$$\frac{n(X^{+j+1})}{n(X^{+j})} = \frac{\langle\sigma v\rangle_{\rm ci}}{\langle\sigma v\rangle_{\rm rr} + \langle\sigma v\rangle_{\rm diel}} \tag{14.46}$$

depend on temperature T but not on the electron density n_e (in the low-density limit where all species are assumed to have radiatively relaxed to the ground electronic state before collisions).

Solutions for collisional ionization equilibrium have been calculated by many authors. Figure 14.7 shows results calculated by Mazzotta et al. (1998) for C, O, Ne, Si, and Fe. Low atomic number species such as C are readily stripped of their electrons, with fully ionized C VII prevailing for $T > 1.5 \times 10^6\,{\rm K}$. Iron, on the other hand, is not fully stripped until $T > 2 \times 10^8\,{\rm K}$.

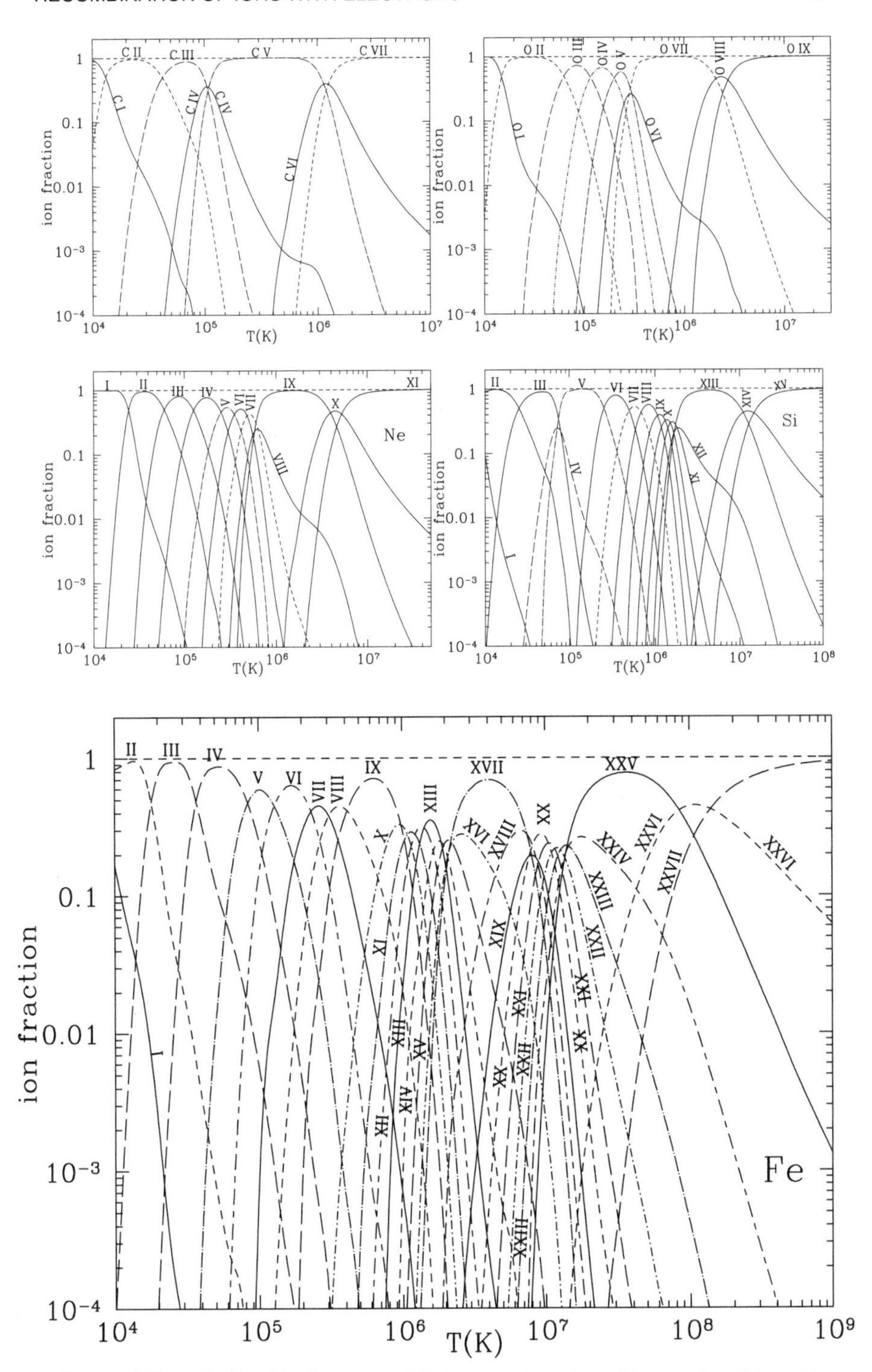

Figure 14.7 Collisional ionization equilibrium, as a function of temperature T, for C, O, Ne, Si, and Fe in a thermal plasma. Data from Mazzotta et al. (1998).

Chapter Fifteen

Photoionized Gas

Galaxies are powered primarily by the release of nuclear energy in stars, with part of this energy transferred to the ISM in the form of starlight photons and mechanical energy in stellar ejecta, including stellar winds and supernova explosions. A $T = 35{,}000\,\mathrm{K}$ blackbody – corresponding to stellar type O8 V – emits 32% of its power in $h\nu > I_{\mathrm{H}}$ photons, and these photons can ionize a substantial fraction of the surrounding ISM.

The photoionized gas surrounding a hot, luminous star is referred to as an **H II region**. H II regions – the Orion Nebula (Plates 9 and 10) and the Trifid Nebula (Plate 11) are beautiful examples – are conspicuous in telescopic observations of the night sky, and are responsible for most of the optical emission lines present in the spectra of star-forming galaxies. They have gas temperatures in the range $7000 - 15{,}000\,\mathrm{K}$ (depending on the metallicity of the gas and the temperature of the exciting star). The brightest regions in the all-sky Hα map in Plate 3b are well-known H II regions, including the Orion Nebula, the Gum Nebula, and the H II region around the nearby star ζ Oph.

15.1 H II Regions as Strömgren Spheres

15.1.1 Hydrogen

Because it is the most abundant element, we are primarily concerned with ionization of hydrogen. Following Strömgren (1939), we consider first the idealized problem of a fully ionized, spherical region of uniform density – now known as a **Strömgren sphere** – with the ionization assumed to be maintained by absorption of the ionizing photons radiated by a central hot star. For the moment, assume the gas to be pure hydrogen. We seek the steady state solution, where hydrogen recombination is balanced by photoionization.

Let Q_0 be the rate of emission of hydrogen-ionizing photons, i.e., $h\nu > I_{\mathrm{H}} = 13.6\,\mathrm{eV}$. The electrons and protons will be undergoing radiative recombination, but it is assumed that every radiative recombination $\mathrm{H}^+ + e^- \rightarrow \mathrm{H} + h\nu$ is balanced by a photoionization $\mathrm{H} + h\nu \rightarrow \mathrm{H}^+ + e^-$. Let the ionized hydrogen extend to radius R_S, and the hydrogen density be $n_{\mathrm{H}} = n(\mathrm{H}^+)$. Equating the rates of photoionization and radiative recombination (using the Case B recombination coefficient α_B

discussed in §14.2) gives the steady state condition for ionization balance:

$$Q_0 = \frac{4\pi}{3} {R_{S0}}^3 \, \alpha_B \, n(\mathrm{H}^+) n_e \quad . \tag{15.1}$$

Since $n(\mathrm{H}^+) = n_e = n_\mathrm{H}$, we can solve for the **Strömgren radius**:

$$R_{S0} \equiv \left(\frac{3\, Q_0}{4\pi {n_\mathrm{H}}^2 \alpha_B} \right)^{1/3} \tag{15.2}$$

$$= 9.77 \times 10^{18} \, Q_{0,49}^{1/3} \, {n_2}^{-2/3} \, T_4^{0.28} \, \mathrm{cm} \quad , \tag{15.3}$$

where $Q_{0,49} \equiv Q_0/10^{49}\,\mathrm{s}^{-1}$, $n_2 \equiv n_\mathrm{H}/10^2\,\mathrm{cm}^{-3}$, and we take $\alpha_B \approx 2.56 \times 10^{-13} T_4^{-0.83}\,\mathrm{cm}^3\,\mathrm{s}^{-1}$. At fixed density n_H, the volume of the Strömgren sphere is proportional to Q_0, hence $R_{S0} \propto Q_0^{1/3}$; at fixed ionizing input Q_0, the volume is proportional to ${n_\mathrm{H}}^{-2}$, hence $R_{S0} \propto {n_\mathrm{H}}^{-2/3}$. When observed from distance $D \gg R_{S0}$, the mean emission measure [averaged over the solid angle $\pi(R_{S0}/D)^2$] is

$$EM = \frac{4}{3} {n_\mathrm{H}}^2 R_S \approx 4.22 \times 10^4 \, Q_{0,49}^{1/3} \, n_2^{4/3} \, T_4^{0.28} \, \mathrm{cm}^{-6}\,\mathrm{pc} \quad . \tag{15.4}$$

For stellar temperatures $T < 50{,}000\,\mathrm{K}$, most of the photons with $h\nu > 13.6\,\mathrm{eV}$ will have $h\nu < 20\,\mathrm{eV}$ – let us take 18 eV as the "typical" energy of $h\nu > 13.6\,\mathrm{eV}$ photons.[1] In chapter 13 we saw that atomic hydrogen has a photoionization cross section $\sigma_\mathrm{pi}(h\nu = 18\,\mathrm{eV}) = 2.95 \times 10^{-18}\,\mathrm{cm}^2$.

The mean free path (mfp) of an 18 eV photon in neutral hydrogen is

$$\mathrm{mfp} = \frac{1}{n(\mathrm{H}^0)\sigma_\mathrm{pi}} = 3.39 \times 10^{17} \left(\frac{\mathrm{cm}^{-3}}{n(\mathrm{H}^0)} \right) \mathrm{cm} \quad . \tag{15.5}$$

For the values of Q_0 appropriate to O stars (see Table 15.1), we see that $\mathrm{mfp} \ll R_{S0}$ in neutral gas of density $n_\mathrm{H} \gtrsim 1\,\mathrm{cm}^{-3}$. This tells us that the transition from ionized gas to neutral gas at the boundary of the H II region will occur over a distance that is small compared to R_{S0}, and it is therefore reasonable to idealize the boundary as a discontinous transition from fully ionized to neutral – this was Strömgren's insight.

15.1.2 Helium Ionization

Now, what about helium? At $T \approx 10^4\,\mathrm{K}$, $\alpha_\mathrm{eff}(\mathrm{He})/\alpha_B(\mathrm{H}) \approx 1.1 - 1.7$, depending on the fraction y of $h\nu > 24.6\,\mathrm{eV}$ photons that are absorbed by H. Since the abundance ratio $n_\mathrm{He}/n_\mathrm{H} \approx 0.096$, it follows that in regions where the H is fully

[1] For a 35,000 K blackbody, $h\nu > 13.6\,\mathrm{eV}$ photons have mean energy $\langle h\nu \rangle_i = 17.9\,\mathrm{eV}$.

Table 15.1 Radiative Properties of Massive Stars[a].

SpTp	$M/M_\odot$	$T_{\rm eff}(\rm K)$	$\log_{10}(Q_0/\rm s^{-1})$ [b]	Q_1/Q_0 [c]	$\log_{10}(L/L_\odot)$ [d]
O3V	58.0	44850	49.64	0.251	5.84
O4V	46.9	42860	49.44	0.224	5.67
O5V	38.1	40860	49.22	0.209	5.49
O5.5V	34.4	39870	49.10	0.204	5.41
O6V	31.0	38870	48.99	0.186	5.32
O6.5V	28.0	37870	48.88	0.162	5.23
O7V	25.3	36870	48.75	0.135	5.14
O7.5V	22.9	35870	48.61	0.107	5.05
O8V	20.8	34880	48.44	0.072	4.96
O8.5V	18.8	33880	48.27	0.0347	4.86
O9V	17.1	32830	48.06	0.0145	4.77
O9.5V	15.6	31880	47.88	0.0083	4.68
O3III	56.0	44540	49.77	0.234	5.96
O4III	47.4	42420	49.64	0.204	5.85
O5III	40.4	40310	49.48	0.186	5.73
O5.5III	37.4	39250	49.40	0.170	5.67
O6III	34.5	38190	49.32	0.158	5.61
O6.5III	32.0	37130	49.23	0.141	5.54
O7III	29.6	36080	49.13	0.129	5.48
O7.5III	27.5	35020	49.01	0.105	5.42
O8III	25.5	33960	48.88	0.072	5.35
O8.5III	23.7	32900	48.75	0.0417	5.28
O9III	22.0	31850	48.65	0.0257	5.21
O9.5III	20.6	30790	48.42	0.0129	5.15
O3I	67.5	42230	49.78	0.204	5.99
O4I	58.5	40420	49.70	0.182	5.93
O5I	50.7	38610	49.62	0.158	5.87
O5.5I	47.3	37710	49.58	0.151	5.84
O6I	44.1	36800	49.52	0.141	5.81
O6.5I	41.2	35900	49.46	0.132	5.78
O7I	38.4	34990	49.41	0.115	5.75
O7.5I	36.0	34080	49.31	0.100	5.72
O8I	33.7	33180	49.25	0.079	5.68
O8.5I	31.5	32270	49.19	0.065	5.65
O9I	29.6	31370	49.11	0.0363	5.61
O9.5I	27.8	30460	49.00	0.0224	5.57

[a] After Martins et al. (2005).
[b] Q_0 = rate of emission of $h\nu > 13.6\,\rm eV$ photons.
[c] Q_1 = rate of emission of $h\nu > 24.6\,\rm eV$ photons.
[d] L = total electromagnetic luminosity.

ionized and the He is singly ionized, the rate per volume of helium recombinations will be $\sim 14\%$ of the hydrogen recombination rate.

Let Q_1 be the rate of emission of photons with $h\nu > I_{\rm He} = 24.6\,{\rm eV}$. The radiative cascade following recombination of $\rm He^+$ in most cases[2] produces a photon that can ionize H (see Figure 14.3). To a first approximation, we can assume that every stellar photon with $h\nu > 24.6\,{\rm eV}$ will result in one He ionization and one H ionization. Then, if $Q_1 \lesssim 0.14 Q_0$, the He ionization zone will be smaller than the H ionization zone. If $Q_1/Q_0 \gtrsim 0.14$, then the He will be singly ionized throughout the region where the H is fully ionized.

From Table 15.1 we see that the hotter O stars – spectral types O6.9 V and earlier, O6.5 III and earlier, and O6 I and earlier – have $Q_1/Q_0 \gtrsim 0.14$. For these stars, we expect helium to be ionized throughout the Strömgren sphere.

We will revisit the question of He ionization in §15.5; a more refined treatment of the ionizing effects of He recombination photons gives a slightly higher estimate for the minimum value of Q_1/Q_0 required for He to be ionized throughout the H II region.

15.2 Time Scales

The Strömgren sphere analysis assumes a steady state solution. Is this a reasonable idealization? What is the time scale for approach to the steady state?

Suppose that we start with a neutral region, and the ionizing source is suddenly turned on. How long will it take to ionize the region? The total number of ions to be created is $(4/3)\pi R_{s0}^3 n_{\rm H}$; the time to supply this many ionizations is

$$\tau_{\rm ioniz.} \equiv \frac{(4/3)\pi {R_{\rm S0}}^3 n_{\rm H}}{Q_0} = \frac{1}{\alpha_B n_{\rm H}} = \frac{1.22\times 10^3\,{\rm yr}}{n_2} \quad . \tag{15.6}$$

This is the time required for the ionization state of the gas to respond to an increase in the supply of ionizing photons from the source.

Suppose that the ionizing source suddenly turns off. The ionized region will recombine on the recombination time scale

$$\tau_{\rm rec} = \frac{1}{\alpha_B n_{\rm H}} = \frac{1.22\times 10^3\,{\rm yr}}{n_2} \quad . \tag{15.7}$$

Note that the recombination time scale $\tau_{\rm rec}$ is *identical* to the ionization time scale

[2]The only exception is the path involving two-photon decays from $1s2s\ ^1{\rm S}_0$; 56% of the two-photon decays from $^1{\rm S}_0$ will produce a photon with $h\nu > 13.6\,{\rm eV}$. For electron densities $n_e \ll 4000\,{\rm cm}^{-3}$, only about 10% of the radiative decays populate the $1s2s\ ^1{\rm S}_0$ state; thus about 95% of the decays produce hydrogen-ionizing photons. The resonance-line photons (e.g., $1s2p\ ^1{\rm P}_1^o \rightarrow 1s^2\ ^1{\rm S}_1$) can be scattered by $\rm He^0$ (with a small probability that the $1s2p$ level will decay to the $1s2s\ ^1{\rm S}_0$ level that decays by two-photon emission), but the bottom line is that a large fraction of radiative recombinations to $\rm He^0$ will produce a photon that will ionize H.

$\tau_{\rm ioniz.}$! For densities $n_{\rm H} > 0.03\,{\rm cm}^{-3}$, the ionization/recombination time scale is shorter than the main-sequence lifetime $\gtrsim 5\,{\rm Myr}$ for a massive star.

The other relevant time scale is the sound crossing time $\tau_{\rm sound}$. The ionized region will likely be overpressured relative to its surroundings, in which case it will expand on the sound crossing time. The isothermal sound speed in ionized hydrogen is $c_s = (2kT/m_{\rm H})^{1/2} = 13T_4^{1/2}\,{\rm km\,s}^{-1}$; the time for a pressure wave to propagate a distance equal to Strömgren radius is therefore

$$\tau_{\rm sound} = \frac{R_{\rm S0}}{c_s} \approx 2.39 \times 10^5\,{\rm yr}\frac{Q_{0,49}^{1/3}}{n_2^{2/3}} \quad . \tag{15.8}$$

For $Q_{0,49} n_2 > 1.3 \times 10^{-7}$, we have $\tau_{\rm sound} > \tau_{\rm rec}$: the ionization responds on a time scale short compared to the hydrodynamic time. But we also see that for densities $n_{\rm H} \lesssim 1\,{\rm cm}^{-3} Q_{0,49}^{1/2}$, the sound-crossing time will be comparable to the lifetime of an O star.

We will discuss the propagation of ionization fronts and the expansion of H II regions in Chapter 37.

15.3 Neutral Fraction within an H II Region

The Strömgren sphere approximation assumes the hydrogen to be fully ionized within the radius $R_{\rm S0}$. What, in fact, is the fractional ionization within the H II region?

Consider a pure hydrogen nebula for simplicity. We use the Case B approximation (i.e., we neglect direct recombinations to level $n = 1$). Let us further assume that the ionizing radiation from the star has a median photon energy $h\nu$, and that $\sigma_{\rm pi}$ is the photoionization cross section for this median photon energy.

Let $Q(r)$ be the rate at which ionizing photons cross a spherical surface of radius r. Clearly, $Q(0) = Q_0$, where Q_0 is the rate at which H-ionizing photons are emitted from the star. (We assume that the radius of the star is negligible compared to the radius of the H II region.) In a steady state,

$$Q(r) = Q_0 - \int_0^r {n_{\rm H}}^2 \alpha_B x^2 4\pi (r')^2 dr' = Q_0 \left[1 - 3\int_0^{r/R_{\rm S0}} x^2 y^2 dy\right] \quad , \tag{15.9}$$

where $x \equiv n({\rm H}^+)/n_{\rm H} = n_e/n_{\rm H}$, $y \equiv r/R_{\rm S0}$, and $n_{\rm H}$ is assumed to be uniform. At each point, the rate of Case B recombinations per volume must be balanced by the rate of photoionization per volume:

$${n_{\rm H}}^2 \alpha_B x^2 = \frac{Q(r)}{4\pi r^2} n_{\rm H}(1-x)\sigma_{\rm pi} \quad , \tag{15.10}$$

which we rewrite as

$$\frac{x^2}{1-x} = \frac{Q(r)}{Q(0)} \frac{n_{\rm H}\sigma_{\rm pi} R_{\rm S0}}{3y^2} = \frac{Q(r)}{Q(0)} \frac{\tau_{\rm S0}}{3y^2} \quad , \tag{15.11}$$

where we define

$$\tau_{S0} \equiv n_H \sigma_{pi} R_{S0} \tag{15.12}$$

$$= 2880 \, (Q_{0,49})^{1/3} n_2^{1/3} T_4^{0.28} \quad , \tag{15.13}$$

where we have taken $\sigma_{pi} = 2.95 \times 10^{-18}\,\mathrm{cm}^2$. Because $\tau_{S0} \gg 1$, it follows from (15.11) that $1 - x \ll 1$. The neutral fraction is small:

$$1 - x \approx \frac{Q(0)}{Q(r)} \frac{3y^2}{\tau_{S0}} \ll 1 \quad . \tag{15.14}$$

If we now assume $x \approx 1$ in Eq. (15.9), then

$$\frac{Q(r)}{Q_0} \approx 1 - y^3 = 1 - \left(\frac{r}{R_{S0}}\right)^3 \quad , \tag{15.15}$$

and

$$\frac{x^2}{1-x} \approx \frac{1-y^3}{3y^2} \tau_{S0} \quad . \tag{15.16}$$

At the center of the H II region, $y \to 0$, and we see that $x^2/(1-x) \to \infty$, or $x \to 1$. At the boundary of the H II region, $y \to 1$, and we see that $x^2/(1-x) \to 0$, or $x \to 0$.

What is the "typical" fractional ionization in the H II region? The median fractional ionization x_m is found at the "half-mass" or "half-volume" radius, $y_m = (1/2)^{1/3} = 0.794$. At this point, the neutral fraction

$$(1 - x_m) = \frac{3(0.5)^{2/3}}{0.5} \frac{x_m^2}{\tau_{S0}} = \frac{3.78}{\tau_{S0}} x_m^2 \quad . \tag{15.17}$$

Usually $\tau_{S0} \gg 1$ [see Eq. (15.12)], and the neutral fraction $(1-x_m) \approx 3.78/\tau_{S0} \ll 1$.

If we solve Eqs. (15.9) and (15.11) self-consistently, we would find that $Q(r)$ does not go to zero at $y = 1$, but decays exponentially at large radii. It is straightforward to integrate Eq. (15.9) numerically to obtain $Q(r)$, but we have neglected an important physical effect – absorption by dust – which we now bring into the analysis.

15.4 Dusty H II Regions with Radiation Pressure

Thus far, the discussion has neglected the effects of dust. Dust is important in two ways: (1) dust will absorb some of the $h\nu > 13.6\,\mathrm{eV}$ photons emitted by the star, thus reducing the rate of photoionization of hydrogen and reducing the size of the H II region; and (2) radiation pressure acting on the dust will also affect the density structure of the H II region.

Petrosian et al. (1972) discussed the first effect – the absorption of ionizing photons by the dust and resulting reduction in size of the ionized region. Petrosian et al. (1972) assumed that the density of the ionized gas was uniform, as in the original treatment by Strömgren (1939).

However, radiation pressure acting on the dust (and gas) in the H II region will result in nonuniform density, because gradients in the gas pressure will be required to counteract the effects of radiation pressure. This will result in enhancement of the ionized gas density near the edge of the H II region, and lowering of the gas density near the star.

The theory of dusty H II regions, including the effects of dust and radiation pressure, has been discussed by Draine (2011). We follow this treatment here.

Consider a star of luminosity $L = L_n + L_i = L_{39} 10^{39}\,\mathrm{erg\,s^{-1}}$, where L_n and L_i are the luminosities in $h\nu < 13.6\,\mathrm{eV}$ and $h\nu > 13.6\,\mathrm{eV}$ photons, respectively. The mean energy of the ionizing photons is $\langle h\nu \rangle_i \equiv L_i/Q_0$. Ignore He, and assume the H to be nearly fully ionized, with photoionizations balancing Case B recombinations in the "on-the-spot" approximation. If the gas is in dynamical equilibrium, then the force per unit volume due to photon absorptions must be balanced by the gradient of the gas pressure:

$$n_{\mathrm{H}}\sigma_{\mathrm{dust}} \frac{L_n e^{-\tau} + L_i \phi(r)}{4\pi r^2 c} + \alpha_B {n_{\mathrm{H}}}^2 \frac{\langle h\nu \rangle_i}{c} - \frac{d}{dr}\left(2 n_{\mathrm{H}} kT\right) = 0 \quad , \tag{15.18}$$

where $n_{\mathrm{H}}(r)$ is the proton density, $L_i\phi(r)$ is the power in $h\nu > 13.6\,\mathrm{eV}$ photons crossing a sphere of radius r, and σ_{dust} and $\tau(r)$ are the UV attenuation cross section per H nucleon and attenuation optical depth due to dust, here assumed to be independent of frequency. The functions $\phi(r)$ and $\tau(r)$ are determined by

$$\frac{d\phi}{dr} = \frac{-1}{Q_0} \alpha_B {n_{\mathrm{H}}}^2 4\pi r^2 - n_{\mathrm{H}}\sigma_{\mathrm{dust}}\phi \quad , \tag{15.19}$$

$$\frac{d\tau}{dr} = n_{\mathrm{H}}\sigma_{\mathrm{dust}} \quad , \tag{15.20}$$

where we have assumed the fractional ionization $x \approx 1$ in Eq. (15.19). The boundary conditions are $\phi(0) = 1$ and $\tau(0) = 0$. Defining

$$n_0 \equiv \frac{4\pi\alpha_B}{Q_0}\left(\frac{2ckT}{\alpha_B \langle h\nu \rangle_i}\right)^3 = 4.54 \times 10^5 \frac{T_4^{4.66}}{Q_{0,49}}\left(\frac{18\,\mathrm{eV}}{\langle h\nu \rangle_i}\right)^3 \mathrm{cm}^{-3} \quad , \tag{15.21}$$

$$\lambda_0 \equiv \frac{Q_0}{4\pi\alpha_B}\left(\frac{\alpha_B \langle h\nu \rangle_i}{2ckT}\right)^2 = 2.47 \times 10^{16} \frac{Q_{0,49}}{T_4^{2.83}}\left(\frac{\langle h\nu \rangle_i}{18\,\mathrm{eV}}\right)^2 \mathrm{cm} \quad , \tag{15.22}$$

we switch to the dimensionless variables $y \equiv r/\lambda_0$, $u \equiv n_0/n$. The governing equations then become

$$\frac{du}{dy} = -1 - \gamma \left(\beta e^{-\tau} + \phi\right) \frac{u}{y^2} \quad , \tag{15.23}$$

$$\frac{d\phi}{dy} = -\frac{y^2}{u^2} - \gamma \frac{\phi}{u} \quad , \tag{15.24}$$

$$\frac{d\tau}{dy} = \frac{\gamma}{u} \quad , \tag{15.25}$$

with initial conditions $\phi(0) = 1$ and $\tau(0) = 0$ and dimensionless parameters

$$\beta \equiv \frac{L_n}{L_i} = \frac{L}{L_i} - 1 = 3.47 \frac{(L/10^{39}\,\mathrm{erg\,s^{-1}})}{Q_{0,49}} \left(\frac{18\,\mathrm{eV}}{\langle h\nu \rangle_i}\right) - 1 \quad , \tag{15.26}$$

$$\gamma \equiv \left(\frac{2ckT}{\alpha_B \langle h\nu \rangle_i}\right) \sigma_{\mathrm{dust}} = 11.2\, T_4^{1.83} \left(\frac{18\,\mathrm{eV}}{\langle h\nu \rangle_i}\right) \left(\frac{\sigma_{\mathrm{dust}}}{10^{-21}\,\mathrm{cm}^2}\right) \quad . \tag{15.27}$$

The parameter β depends only on the stellar spectrum. We take $\beta = 3$ as our standard value, corresponding to the spectrum of a $T_{bb} = 32{,}000\,\mathrm{K}$ blackbody, but we also consider $\beta = 2$ ($T_{bb} = 45{,}000\,\mathrm{K}$) and $\beta = 5$ ($T_{bb} = 28{,}000\,\mathrm{K}$); the latter value may apply to a cluster of O and B stars.

The parameter γ is proportional to σ_{dust}, which is proportional to the dust-to-gas mass ratio, but also depends on the distribution of grain sizes. γ also depends on the gas temperature T and on the mean ionizing photon energy $\langle h\nu \rangle_i$, but these are not likely to vary much for H II regions around OB stars. Models for dust in the diffuse interstellar medium (e.g., Weingartner & Draine 2001*a*) have $\sigma_{\mathrm{dust}} \approx 2 \times 10^{-21}\,\mathrm{cm}^2$ from 13.6 to $\sim 18\,\mathrm{eV}$, then declining to $\sim 0.7 \times 10^{-21}\,\mathrm{cm}^2$ at 50 eV. Dust within H II regions has not been well-characterized. We will take $\gamma = 10$ as a standard value, but will also consider $\gamma = 5$ and $\gamma = 20$. Low-metallicity systems would be characterized by small values of γ.

The solutions are defined for $0 < y < y_{\mathrm{max}}$, where y_{max} is determined by the boundary condition $\phi(y_{\mathrm{max}}) = 0$. For each solution $u(y)$, the root-mean-square density is

$$n_{\mathrm{rms}} \equiv n_0 \left[\frac{3}{y_{\mathrm{max}}^3} \int_0^{y_{\mathrm{max}}} \frac{y^2}{u^2} dy\right]^{1/2} \quad . \tag{15.28}$$

The radius is $R = y_{\mathrm{max}} \lambda_0$. The fraction of the $h\nu > 13.6\,\mathrm{eV}$ photons that are absorbed by H is

$$f_{\mathrm{ion}} = \left(\frac{R}{R_{s0}}\right)^3 \quad , \quad R_{s0} \equiv \left(\frac{3Q_0}{4\pi n_{\mathrm{rms}}^2 \alpha_B}\right)^{1/3} \quad , \tag{15.29}$$

where R_{s0} is the radius of a dustless Strömgren sphere.

For a given β and γ, varying the initial conditions on the function $u(y)$ generates a family of solutions, each with a different value of n_{rms}/n_0. The solutions are

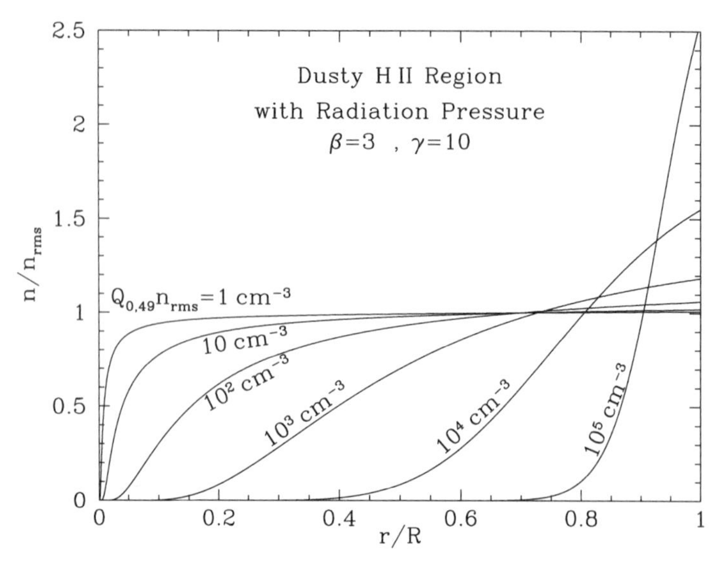

Figure 15.1 Density profiles for static equilibrium H II regions with radiation pressure, for $\beta = 3$, $\gamma = 10$, and selected values of $Q_{0,49}n_{\rm rms}$. From Draine (2011).

parameterized by β, γ, and a third parameter, which may be taken to be the dust optical depth along a path with density $n_{\rm rms}$ and length R_{s0}:

$$\tau_{d0} \equiv n_{\rm rms} R_{s0} \sigma_{\rm dust} = 2.10 \left(Q_{0,49} n_{\rm rms,3}\right)^{1/3} T_4^{0.28} \sigma_{d,-21} \quad . \tag{15.30}$$

where $\sigma_{d,-21} \equiv \sigma_{\rm dust}/10^{-21}\,{\rm cm}^2$. Examples of solutions for $\beta = 3, \gamma = 10$ are shown in Figure 15.1. The solution with $\tau_{d0} = 0.21$ has nearly uniform density, except for a small cavity near the center, where radiation pressure on the dust has forced the gas away from the star. As τ_{d0} is increased, the fractional size of the central cavity increases, and as the central cavity enlarges, the gradient in the gas density in the outer regions becomes stronger.

The change in the density structure produced by radiation pressure leads to changes in the fraction of the ionizing photons that are absorbed by dust. Figure 15.2 shows $f_{\rm ion}$, the fraction of the H-ionizing photons that produce photoionizations (i.e., that escape absorption by dust). Results are shown for $\gamma = 10$ [i.e., $\sigma_{d,-21} = 0.89 T_4^{-0.83} (\langle h\nu\rangle_i / 18\,{\rm eV})$]. For small τ_{d0}, the value of $f_{\rm ion}$ is nearly the same as was calculated assuming a uniform density H II region (see Fig. 15.2). However, for $\tau_{\rm d0} \gtrsim 2$, the central cavity results in significantly higher values of $f_{\rm ion}$ compared to what Petrosian et al. (1972) found assuming uniform density. The center-to-edge optical depth $\tau(R)$, shown in Fig. 15.3, is also much lower than was found assuming uniform density. A significant fraction ($e^{-\tau(R)}$) of the non-ionizing radiation from the star escapes absorption within the H II region.

The emission measure profile

$$EM(b) = 2\int_b^R [n(r)]^2 \frac{rdr}{\sqrt{r^2-b^2}} \quad , \tag{15.31}$$

where b is the transverse distance from the central star, has been computed for these solutions; examples are shown in Fig. 15.4. For $Q_{0,49}n_{\rm rms} \gtrsim 10^3\,{\rm cm}^{-3}$, the emissivity profile becomes conspicuously ring-like rather than centrally peaked as in a uniform density H II region.

Observed H II regions often have central cavities where the density is much lower than the average value, resulting in a ring-like appearance. The H II region designated N49 (Watson et al. 2008) is a very nice example of ring-like morphology seen in free–free emission. However, the actual cavity appears to be larger than would be expected from radiation pressure alone (Draine 2011). It is likely that the wind from the central star contributes to formation of the N49 bubble. Stellar wind bubbles are discussed in Chapter 38.

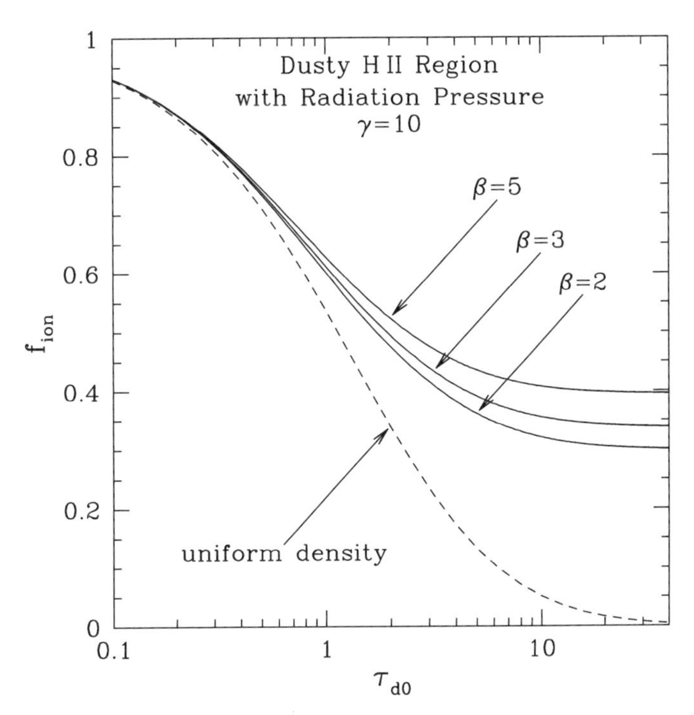

Figure 15.2 $f_{\rm ion}$ = fraction of $h\nu > 13.6\,{\rm eV}$ photons that escape absorption by dust, as a function of the parameter τ_{d0} defined in Eq. (15.30). Results are shown for static equilibrium H II regions with radiation pressure, and also (dashed line) for uniform density H II regions.

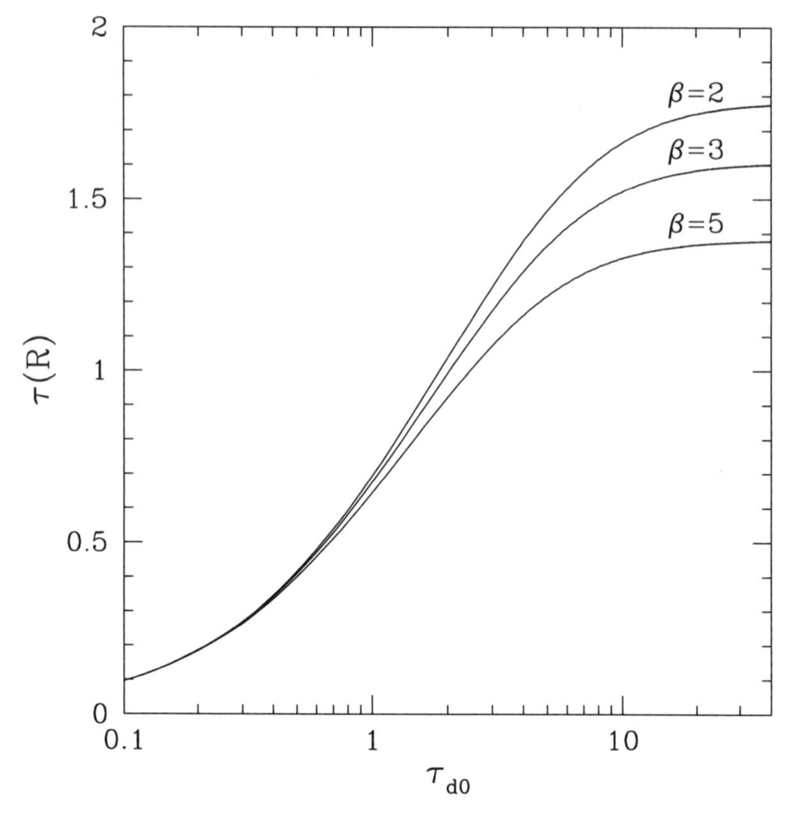

Figure 15.3 Dust optical depth from center to edge of dusty H II regions, as a function of τ_{d0} defined in Eq. (15.30).

15.5 Ionization of Helium and Other Elements

Because helium is abundant, it strongly affects both the spectrum of emission from O stars and the ionization structure of H II regions. The He^0 opacity in stellar atmospheres causes the emission from B stars and late O-type stars to have very little emission above the He ionization edge at $24.59\,\mathrm{eV}$. Figure 15.5 shows the fraction of H-ionizing photons that are also capable of ionizing He for stars with spectral types between B0 and O4.

For low values of Q_1/Q_0, the radius $R(\mathrm{He}^+)$ of the region where He will be ionized will be smaller than the radius $R(\mathrm{H}^+)$ of the H II region. Let y be the fraction of He-ionizing photons that are absorbed by H rather than He, and let z be the fraction of the He recombinations to excited levels that produce a photon with $h\nu > 13.6\,\mathrm{eV}$. The number of H^+ and He^+ ions can be estimated by balancing

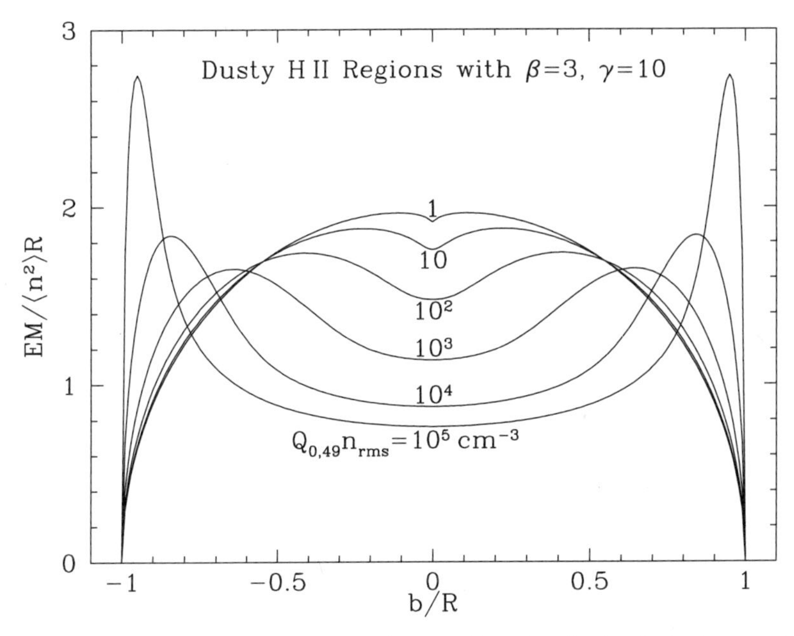

Figure 15.4 Emission measure profiles for the H II regions in Fig. 15.1. From Draine (2011).

recombinations with photoionizations:

$$N(\mathrm{He}^+)n_e\left[\alpha_B(\mathrm{He}) + y\alpha_{1s^2}(\mathrm{He})\right] = (1-y)Q_1 \quad , \tag{15.32}$$

$$N(\mathrm{H}^+)n_e\alpha_B(\mathrm{H}) = (Q_0 - Q_1) + yQ_1 + N(\mathrm{He}^+)n_e\left[z\alpha_B(\mathrm{He}) + y\alpha_{1s^2}(\mathrm{He})\right] \; , \tag{15.33}$$

which can be combined to obtain

$$\frac{N(\mathrm{He}^+)}{N(\mathrm{H}^+)} = \frac{(1-y)\alpha_B(\mathrm{H})(Q_1/Q_0)}{\alpha_B(\mathrm{He}) + y\alpha_{1s^2}(\mathrm{He}) - (1-y)(1-z)(Q_1/Q_0)\alpha_B(\mathrm{He})} \; . \tag{15.34}$$

Both y and z depend on conditions in the H II region. We expect H to be more highly ionized than He even near the center of the H II region, and therefore $y \lesssim 0.6$ [see Eq. (14.16)]. In an accurate calculation, y would be obtained from the actual ratio $n(\mathrm{He}^0)/n(\mathrm{H}^0)$ using Eq. 14.16. Here we will assume $y \approx 0.2$ to be a reasonable estimate for the fraction of the $h\nu > 24.6\,\mathrm{eV}$ photons that are absorbed by H rather than He. The fraction z of He Case B recombinations that produce an H-ionizing photon varies from $z \approx 0.96$ at low densities $n_e \ll n_{e,\mathrm{crit}}(\mathrm{He}^0\, 2\,^3\mathrm{S}_1) \approx 1100 e^{1.2/T_4} T_4^{0.5}\,\mathrm{cm}^{-3}$ to $z \approx 0.67$ at high densities $n_e \gg n_{e,\mathrm{crit}}(\mathrm{He}^0\, 2\,^3\mathrm{S}_1)$. Taking an intermediate value $z \approx 0.8$, setting $T = 8000\,\mathrm{K}$ and $y = 0.2$, we obtain

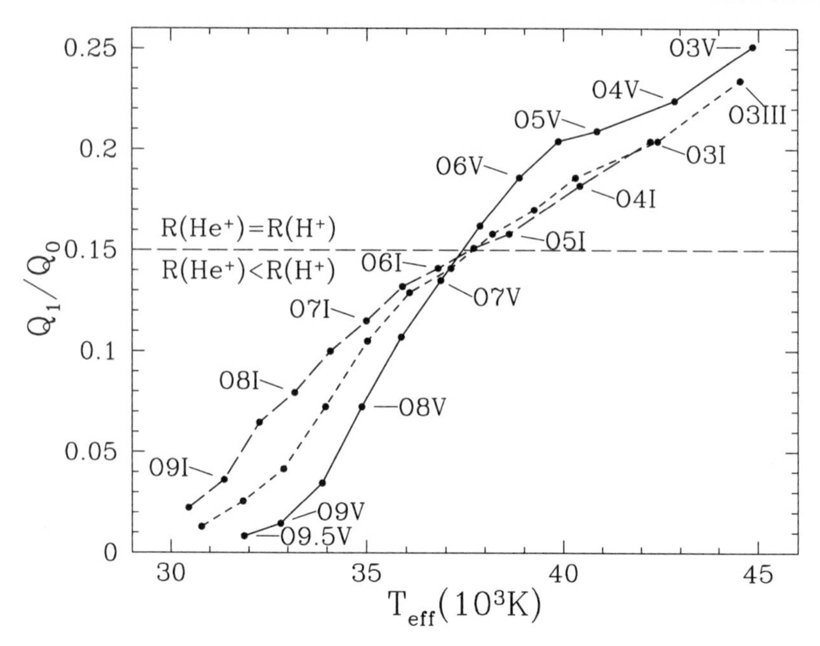

Figure 15.5 Q_1/Q_0 = ratio of rate of emission of $h\nu > 24.6\,\mathrm{eV}$ photons to rate of emission of $h\nu > 13.6\,\mathrm{eV}$ photons, as a function of $T_{\rm eff}$, for luminosity classes V, III, and I (see Table 15.1). $Q_1/Q_0 \gtrsim 0.15$ is required for He to be ionized throughout the H II region, corresponding to $T_{\rm eff} \gtrsim 37{,}000\,\mathrm{K}$.

$$\frac{N(\mathrm{He}^+)}{N(\mathrm{H}^+)} \approx \frac{0.68(Q_1/Q_0)}{1-0.17(Q_1/Q_0)} \quad . \tag{15.35}$$

Full ionization of the He is achieved when $N(\mathrm{He}^+)/N(\mathrm{H}^+) = n_{\mathrm{He}}/n_{\mathrm{H}} = 0.096$, which is attained for $Q_1/Q_0 \approx 0.15$.

For $N(\mathrm{He}^+)/N(\mathrm{H}^+) < n_{\mathrm{H}}/n_{\mathrm{He}}$, the radius of the He^+ zone is smaller than the H^+ zone:

$$\frac{R(\mathrm{He}^+)}{R(\mathrm{H}^+)} \approx \left[\frac{(n_{\mathrm{H}}/n_{\mathrm{He}})(1-y)\alpha_B(\mathrm{H})(Q_1/Q_0)}{\alpha_B(\mathrm{He}) + y\alpha_{1s^2}(\mathrm{He}) - (1-y)(1-z)(Q_1/Q_0)\alpha_B(\mathrm{He})}\right]^{1/3} . \tag{15.36}$$

Ions that require $h\nu > 24.6\,\mathrm{eV}$ for their formation will be present only in the He^+ zone. A list of these ions is given in Table 15.2.

Table 15.2 Abundant Ions in H II Regions[a]

	H II and He I zone[b]		H II and He II zone[c]	
Element	Ion	$h\nu$ (eV)[d]	Ion	$h\nu$ (eV)[d]
H	H II	13.60	H II	13.60
He	He I	0	He II	24.59
C	C II	11.26	C III [e]	24.38
			C IV	47.88
N	N II	14.53	N III	29.60
			N IV	47.45
O	O II	13.62	O III	35.12
Ne	Ne II	21.56	Ne III	40.96
Na	(Na II)[f]	5.14	(Na II)[f]	5.14
			Na III	47.29
Mg	Mg II	7.65	(Mg III)[f]	15.04
	(Mg III)[f]	15.04		
Al	Al III	18.83	(Al IV)[f]	28.45
Si	Si III	16.35	Si IV	33.49
			(Si V)[f]	45.14
S	S II	10.36	S III	23.33
	S III	23.33	S IV	34.83
Ar	Ar II	15.76	Ar III	27.63
			Ar IV	40.74
Ca	Ca III	11.87	Ca IV	50.91
Fe	Fe III	16.16	Fe IV	30.65
Ni	Ni III	18.17	Ni IV	35.17

[a] Limited to elements X with $N_X/N_H > 10^{-6}$.
[b] Ions that can be created by radiation with $13.60 < h\nu < 24.59$ eV.
[c] Ions that can be created by radiation with $24.59 < h\nu < 54.42$ eV.
[d] Photon energy required to create ion.
[e] Ionization potential is just below 24.59 eV.
[f] Closed shell, with no excited states below 13.6 eV.

15.6 Planetary Nebulae

Planetary nebulae are a special class of H II regions, where the ionized gas is a dense stellar wind that is ionized when the central star evolves quickly from being a cool red giant to an extremely hot "planetary nebula nucleus."

There are two ways in which planetary nebulae can differ from classical H II regions produced by O or B0 stars.

- First, the dense gas being ionized is limited to the outflowing wind. A typical planetary nebula precursor can have a mass loss rate $\dot{M}_w \approx 10^{-4} M_\odot/\,\mathrm{yr}$ for $\sim 2000\,\mathrm{yr}$, with a wind velocity $v_w \approx 15\,\mathrm{km\,s^{-1}}$. A spherically symmetric steady outflow with these properties would have a r^{-2} density profile, but in fact the density profiles in planetary nebulae are much more complicated, with a visually beautiful morphology that often has bipolar symmetry. The complex morphology of the Helix nebula (see color Plate 13) includes many dusty cometary filaments.

- Second, the central stars are generally much hotter than O stars, with effective temperatures $T_\star$ often exceeding 10^5 K. For example, the hot white dwarf at the center of the Helix Nebula (NGC 7293) has $T_\star \approx 1.0 \times 10^5$ K (Mendez et al. 1988; Napiwotzki 1999), and the central star in NGC 7027 has $T_\star \approx 2.2 \times 10^5$ K (Zhang et al. 2005). As a result, helium is generally fully ionized (to He^{++}), and elements such as C or O can be ionized to high stages (such as O IV, Ne IV, or Ne V) that are not present in classical H II regions ionized by O stars. The harder photons also produce greater photoelectric heating per photoionization event, and gas temperatures in planetary nebulae are generally higher (up to $\sim$16,000 K – see, e.g., Zhang et al. 2004) than in classical H II regions (e.g., 7,000 to 10,000 K).

15.7★ Escape of Lyman α

15.7.1 Escape from the H II Region

The recombining hydrogen in the H II region generates Lyman α radiation. Under Case B conditions at densities $n_e \lesssim 10^3\,\mathrm{cm}^{-3}$, the rate of production of Lyman α photons is equal to $f(2p)f_{\mathrm{ion}}Q_0$, where $f(2p) \approx \frac{2}{3}$ is the fraction of Case B recombinations that populate the $2p$ level (see §14.2.4). We have seen earlier that the gas in the H II region is highly ionized, but the Lyman α photons can be resonantly scattered by the small amount of neutral hydrogen present. This inhibits escape of the Lyman α photons from the ionized region. We can think of the Lyman α photon as bouncing around inside the nebula, until it finally escapes or is destroyed.

Scattering of Lyman α radiation is a resonant process, involving absorption and reemission of the photon. Prior to the scattering, the H atom is in the ground state. The photon is absorbed, and then reemitted, with the H atom returning to the ground state. Except for the possibility that the initial and final states may be different hyperfine levels, the energy of the photon in the center of mass frame of the photon and H atom is unchanged. However, the H atom producing the scattering will in general be moving relative to the gas, and the emitted photon will be radiated in a direction differing from the incident direction; the net result is that the photon undergoes a shift in frequency. Therefore, a photon involved in a sequence of scattering events will undergo a random walk in frequency space. At the same time, it also does a random walk in position space. These two random walks, together, will tend to take the photon away from the initial frequency, and away from the initial location.

The initial photon tends to be emitted in the "Doppler core" of the line profile. As it migrates away from the Doppler core into the wings of the line profile, the cross section for scattering goes down, and the mean-free-path between scatterings goes up. After a sufficient number of scatterings, the photon will have travelled sufficiently far from its point of emission that we will count it as having "escaped."

The problem of resonant scattering has been considered by a number of authors, both numerically (e.g., Bonilha et al. 1979) and analytically (e.g., Neufeld 1990). Here we present a very simple analysis that captures the essential features. For the moment, let us consider a dustless H II region.

First, let us consider the diffusion in frequency space. If an H atom moving with velocity v produces the scattering, then the mean-square frequency shift for the scattered photon will be $\langle(\Delta\nu)^2\rangle = \frac{2}{3}\nu^2(v/c)^2$, if the direction of the absorbed photon and emitted photon are uncorrelated with one another,[3] and with the velocity vector of the H atom.[4] Averaging over the velocity distribution, $\langle v^2\rangle = \frac{3}{2}b^2$, the frequency shift $\Delta\nu$ from each scattering event has

$$\langle(\Delta\nu)^2\rangle = \nu^2(b/c)^2 \quad . \tag{15.37}$$

Earlier, we assumed that the direction of propagation of the absorbed photon is uncorrelated with the vector velocity of the H atom. This is not the case for photons in the "core" of the line, but it is a good approximation in the "wings," and it is the scattering in the wings that concerns us here. Let

$$z \equiv \frac{(\nu - \nu_0)/\nu_0}{b/c} \tag{15.38}$$

measure the displacement from line center in terms of the Doppler broadening parameter b. Then, from Eq. (15.37), after N_s scatterings,

$$\langle z^2\rangle = N_s + 1/2 \quad . \tag{15.39}$$

The originally "injected" photons tend to be in the Doppler core of the line (with $\langle z^2\rangle = \frac{1}{2}$), but after a few scatterings the photons move out to frequencies where the Voigt profile for the thermal H atoms is approximated by the Lorentzian wings [see Eq. (6.40)]:

$$\sigma(z) \approx \frac{\sigma_w}{z^2} \quad , \tag{15.40}$$

$$\sigma_w \equiv \frac{g_u}{g_\ell}\frac{\lambda^2 A_{u\ell}}{32\pi^3}\frac{\gamma\lambda_{u\ell}}{b^2} = 2.59\times10^{-17}\,\mathrm{cm}^2\,b_6^{-2}, \tag{15.41}$$

$$b_6 \equiv \frac{b}{10\,\mathrm{km\,s^{-1}}} = 1.29\,T_4^{1/2} \quad . \tag{15.42}$$

[3]Consider an atom with velocity $\mathbf{v}$. If $\hat{n}_a$, $\hat{n}_s$ = the direction of propagation of the absorbed and emitted photons, then $\Delta\nu = (\nu_0/c)(\hat{\mathbf{n}}_s - \hat{\mathbf{n}}_a)\cdot\mathbf{v}$, $(\Delta\nu)^2 = (\nu_0/c)^2[n_{s\|}^2 - 2n_{s\|}n_{a\|} + n_{a\|}^2]v^2$, where $n_{s\|}v \equiv \hat{n}_s\cdot\mathbf{v}$, $n_{a\|}v \equiv \hat{n}_a\cdot\mathbf{v}$. The scattering process has $\langle n_{s\|}n_{a\|}\rangle = 0$. In the damping wings, the scattering cross section is nearly independent of $\mathbf{v}$; hence $\langle n_{s\|}^2\rangle = \langle n_{a\|}^2\rangle \approx \frac{1}{3}$, and $\langle(\delta\nu)^2\rangle = (\nu_0/c)^2(\frac{1}{3}+\frac{1}{3})\langle v^2\rangle$.

[4]While not true for frequencies in the Doppler core of the line, the scattering cross section in the damping wings is nearly independent of the random motions of the atoms.

Now consider the random walk away from the point of emission. In the line wings, the photon mean-free path is

$$\mathrm{mfp} = L_0 z^2 \quad , \tag{15.43}$$

$$L_0 \equiv \frac{1}{n(\mathrm{H}^0)\sigma_w} \quad . \tag{15.44}$$

If r is the displacement from the point of emission, we can approximate[5]

$$\frac{d\langle r^2\rangle}{dN_s} = 2\left(L_0 z^2\right)^2 \approx 2L_0^2 \left(N_s + \frac{1}{2}\right)^2 \quad . \tag{15.45}$$

Integration yields

$$\langle r^2\rangle = \frac{2}{3}L_0^2 \left[\left(N_s + \frac{1}{2}\right)^3 - \frac{1}{8}\right] \quad . \tag{15.46}$$

In the case of an H II region, a reasonable criterion for "escape" would be when $\langle r\rangle^2 \approx R_{\mathrm{S0}}^2$. The number of scatterings required to accomplish this is

$$N_s \approx \left(\frac{3R_{\mathrm{S0}}^2}{2L_0^2}\right)^{1/3} \quad . \tag{15.47}$$

Because the mfp increases as z^2, most of the displacement is accomplished during the last portion of the scattering sequence, when the photon is relatively far from line-center. The number of scatterings depends on the ratio

$$\frac{R_{\mathrm{S0}}}{L_0} \approx \frac{\tau_{\mathrm{S0}}}{n_{\mathrm{H}}\sigma_{\mathrm{pi}}}(1 - x_m)n_{\mathrm{H}}\sigma_w \approx 33.2 b_6^{-2} \quad , \tag{15.48}$$

where we have used Eq. (15.17), and taken $\sigma_{\mathrm{pi}} = 2.95 \times 10^{-18}\,\mathrm{cm}^2$, appropriate for $h\nu = 18\,\mathrm{eV}$. Thus,

$$N_s \approx 11.8\, b_6^{-4/3} \quad . \tag{15.49}$$

This is a modest number of scatterings, but sufficient to frequency-shift a typical photon out to $z^2 \approx 12$, at which point the mfp, evaluated using the neutral fraction $(1 - x_m)$, is $\sim 0.37 R_{\mathrm{S0}}$. Note also that when $z^2 \approx 12$, we are (barely) far enough from line-center for the line profile to be well-approximated by just the damping wings [see Eq. (6.42)], as has been assumed in Eq. (15.43).

To estimate the effects of dust absorption, we first estimate the total pathlength s traveled by an escaping photon. We can write

$$\frac{ds}{dN_s} \approx L_0 z^2 \approx L_0 \left(N_s + \frac{1}{2}\right) \quad . \tag{15.50}$$

[5] The mean step size $=$ mfp, but the mean (step size)$^2 = 2(\mathrm{mfp})^2$.

Integrating this, we find the pathlength s:

$$s \approx \frac{1}{2} L_0 \left[\left(N_s + \frac{1}{2} \right)^2 - \frac{1}{4} \right] \quad , \tag{15.51}$$

$$\frac{s}{R_{S0}} \approx \frac{L_0}{R_{S0}} \frac{N_s^2}{2} \approx 2.10 b_6^{-2/3} \quad . \tag{15.52}$$

In order to "escape," the Lyman α photon needs to escape being absorbed by dust. The escaping fraction is

$$f_{\rm esc} \approx \exp(-n_{\rm H} \sigma_{\rm dust}\, s) = \exp(-2.1 b_6^{-2/3} \tau_{d0}). \tag{15.53}$$

Dust can therefore suppress the escape of Lyman α photons even in H II regions with only moderate dust optical depths τ_{d0}. However, it must also be noted that Eq. (15.53) assumes that the Lyman α photon must achieve a displacement $\sim R_{S0}$ from the point of emission to escape; in fact, photons that are created near the edge of the H II region can escape with a smaller displacement (if they happen to random-walk in the outward direction), and thus (15.53) underestimates the overall Lyman α escape fraction.

15.7.2 Lyman α Photons in an H I Region

The escape of Lyman α from the H II region is facilitated by the low neutral fraction in the H II region, resulting in a relatively large mfp. Now suppose that the photon enters a surrounding cloud of atomic H. We repeat the preceding analysis, but now consider a photon to be injected at the center of a slab with total column density $N(\mathrm{H}^0)$. We will use superscript (HI) to denote the Doppler broadening parameter characterizing the H I cloud: $b^{(\mathrm{HI})} \equiv b_5^{(\mathrm{HI})}\,\mathrm{km\,s^{-1}}$.

As before, let z measure the frequency shift $\Delta\nu$ in units of $\nu_0 b^{(\mathrm{HI})}/c$. If x is the coordinate normal to the slab, then the criterion for escape from the slab is that $\Delta x = \pm 0.5 N(\mathrm{H}^0)/n_{\mathrm{H}}$, or

$$\langle r^2 \rangle \approx 3 \left[0.5 N(\mathrm{H}^0)/n_{\mathrm{H}} \right]^2 \quad . \tag{15.54}$$

The number of scatterings required is [from Eq. (15.51)]

$$N_s \approx \frac{1}{2} \left[3 N(\mathrm{H}^0) \sigma_w^{(\mathrm{HI})} \right]^{2/3} \tag{15.55}$$

$$\approx 2.0 \times 10^4 \left[\frac{N(\mathrm{H}^0)}{10^{21}\,\mathrm{cm}^{-2}} \right]^{2/3} \left(b_5^{(\mathrm{HI})} \right)^{-4/3} \quad . \tag{15.56}$$

The total pathlength s traveled before reaching the cloud boundary is

$$s \approx \frac{1}{2} L_0 N_s^2 \approx 7.4 \times 10^{22}\,\mathrm{cm} \left[\frac{\mathrm{cm}^{-3}}{n(\mathrm{H}^0)} \right] \left[\frac{N(\mathrm{H}^0)}{10^{21}\,\mathrm{cm}^{-2}} \right]^{4/3} \left(b_5^{(\mathrm{HI})} \right)^{-4/3} \quad . \tag{15.57}$$

The probability of escaping absorption by dust is

$$f_{\rm esc} \approx \exp\left\{-74\sigma_{d,21}\left[\frac{N({\rm H}^0)}{10^{21}\,{\rm cm}^{-2}}\right]^{4/3}\left(b_5^{({\rm HI})}\right)^{-4/3}\right\} \quad . \tag{15.58}$$

Therefore, we see that for slabs with $N(\mathrm{H}^0)/b_5^{(\mathrm{HI})} \gtrsim 10^{20}\,\mathrm{cm}^{-2}$ and dust abundances characteristic of the Milky Way ($\sigma_{d,21} \approx 1$), the probability of photon escape is small. Most of the Lyman α photons generated within a typical H I cloud will, therefore, be absorbed by dust within the cloud

The Lyman α photons that enter from an adjacent H II region (which we imagine to be surrounded by a spherical shell of H I) begin their random walk in the H I with $z^2 \approx 12(b^{(\mathrm{HII})}/b^{(\mathrm{HI})})^2 \approx 10^2$ (for $b^{(\mathrm{HI})} \approx 3\,\mathrm{km\,s^{-1}}$). This is small compared to N_s estimated from Eq. (15.55), and, therefore, $f_{\rm esc}$ will not be substantially increased by the fact that the Lyman α photons from the H II region "begin" with a headstart $z^2 \approx 10^2$: the Lyman α photons from the H II region will likely be absorbed in the surrounding H I.

Escape of Lyman α photons from galaxies therefore requires either very low dust abundances or "breakout" of the ionization front so that, in at least some directions, the H II is not bounded by dusty H I gas.

15.8★ Ionization by Power-Law Spectra

Although energy release in stellar flares, or radiation-driven shocks in fast stellar winds, may result in X-ray emission in addition to the thermal emission from the stellar photosphere, the bulk of the power radiated by normal stars can be approximated by a thermal (blackbody) spectrum. An active galactic nucleus (AGN), on the other hand, emits a large fraction of its luminosity via synchrotron emission from a population of relativistic electrons. The resulting spectrum is often approximated by a power-law $L_\nu \propto \nu^{-\alpha}$, with $\alpha \approx 1.2$. Gas that is irradiated by such a power-law spectrum is heated and ionized, and the X-ray component of the spectrum can produce highly ionized species that are not present in normal H II regions.

The state of gas exposed to a power-law spectrum will depend primarily on the ratio of photoionization rates (proportional to the energy density u_ν) to radiative and dielectronic recombination rates (proportional to electron density n_e). It is customary to define the dimensionless **ionization parameter**

$$U \equiv \frac{1}{n_{\rm H}}\int_{\nu_0}^{\infty}\frac{u_\nu d\nu}{h\nu} \qquad \left(\text{where} \quad \nu_0 \equiv \frac{I_{\rm H}}{h}\right) \quad , \tag{15.59}$$

which is simply the ratio of the number density of photons with $h\nu > I_{\rm H}$ to the number density of H nuclei. At densities that are low enough that collisional deexcitation is not important, the steady state temperature and ionization of the gas will be determined by U, by the shape of the ionizing spectrum (the spectral index α), and by the composition of the gas (the abundances of coolants). Note that some authors (e.g., Krolik et al. 1981) use a different definition of "ionization parameter":

$$\Xi \equiv \frac{u(I_{\rm H} < h\nu < 10^3 I_{\rm H})}{p} \quad , \tag{15.60}$$

where u is the radiation energy density between $I_{\rm H}$ and $10^3 I_{\rm H}$, and $p \approx 2.3 n_{\rm H} kT$ is the gas pressure. The two parameters U and Ξ are directly related. If $\langle h\nu \rangle \equiv \int_{I_{\rm H}/h} u_\nu d\nu / \int_{I_{\rm H}/h} (u_\nu / h\nu) d\nu$ is the mean energy of $h\nu > I_{\rm H}$ photons, and the energy above $10^3 I_{\rm H}$ is negligible, then

$$\Xi \equiv \frac{n_{\rm H}}{n} \frac{\langle h\nu \rangle}{kT} U \approx 5 \frac{\langle h\nu \rangle}{100\,{\rm eV}} \frac{U}{T_4} \quad . \tag{15.61}$$

The equilibrium gas temperature rises monotonically with increasing U. For $10^{-4.5} \lesssim U \lesssim 10^{-2}$ (or $10^{-3.5} \lesssim \Xi \lesssim 10^{-1}$), photoionization heats the gas to $1 \lesssim T_4 \lesssim 1.5$, but as U is increased further, T rises as the gas becomes highly ionized, reducing its ability to cool by collisional excitation of bound electrons. For large values of U or Ξ, T depends on the spectrum of the ionizing source. For a standard quasar spectrum, Krolik et al. (1981) found a limiting $T \approx 1.5 \times 10^8$ K for $\Xi \gtrsim 10$, or $U \gtrsim 3 \times 10^4$. The ionizing continuum from the quasar can evidently sustain a $T \gtrsim 10^8$ K plasma, which can pressure-confine higher-density gas that is heated to $T \approx 10^4$ K and accounts for the observed optical and UV emission lines.

Chapter Sixteen

Ionization in Predominantly Neutral Regions

In this chapter, we discuss the degree of ionization expected in predominantly neutral interstellar clouds. The abundance of free electrons in the gas determines the ionization balance of various species (e.g., Ca^+/Ca^0). Collisions with free electrons play a role in determining the charge state for interstellar grains. Electrons also play a key role in interstellar chemistry and, under some conditions, to cooling of the gas via collisional excitation.

There are three quite distinct regimes:

- Diffuse H I regions, where the metals are photoionized by starlight. Cosmic rays create small amounts of H^+ and He^+. The gas may be "cold," with $T \approx 10^2$ K (the **cold neutral medium** or **CNM**) or "warm," with $T \approx 5000$ K (the **warm neutral medium**, or **WNM**).

- Diffuse molecular clouds (visual extinction $0.3 \lesssim A_V \lesssim 2$ mag), where most of the hydrogen is in H_2, and the metals are still predominantly photoionized by starlight. Cosmic rays produce ${H_2}^+$, which leads to the formation of H_3^+. The cosmic ray ionization rate ζ_{CR} can be determined through measurements of the H_3^+ abundance in these clouds. Observations toward the bright star ζ Per indicate a cosmic ray ionization rate $\zeta_{CR} \approx 1 \times 10^{-16}\,\mathrm{s}^{-1}$.

- Dark molecular clouds (visual extinction $A_V \gtrsim 3$ mag), where there is insufficient ultraviolet radiation to photoionize elements such as C and S. Cosmic rays can maintain only a very small fractional ionization $x_e \equiv n_e/n_H \approx 10^{-7}(10^4\,\mathrm{cm}^{-3}/n_H)^{1/2}$. The low fractional ionization in dense molecular clouds has implications for coupling of the magnetic field to the gas and damping of MHD waves; these will be discussed in later chapters.

16.1 H I Regions: Ionization of Metals

Carbon is the fourth most abundant element, following hydrogen, helium, and oxygen. The abundance of C in the local ISM is approximately equal to the solar value, $\sim 2.5 \times 10^{-4}$. Of this, approximately 60% appears to be in solid grains, leaving a gas phase abundance $n_C/n_H \approx 1 \times 10^{-4}$ (see Table 9.5).

With an ionization threshold of 11.26 eV, carbon can be photoionized by starlight that can penetrate into neutral H I clouds. Let $\zeta(C^0)$ be the probability per unit time of ionization of a carbon atom; this will be primarily due to photoionization by starlight but, in principle, also includes photoionization by x rays and collisional ionization by cosmic rays. Pulsar dispersion measures indicate a mean electron density $\langle n_e \rangle \approx 0.04\,\mathrm{cm}^{-3}$ near the midplane of the disk, and within about 1 kpc of the Sun (see Fig. 11.2).

Consider the ionization state of carbon in a region with the mean electron density $n_e \approx 0.04\,\mathrm{cm}^{-3}$. Let n_C be the density of C nuclei in the gas phase, with neutral fraction $x(\mathrm{C}^0) \equiv n(\mathrm{C}^0)/n_\mathrm{C}$. Under steady state conditions, ionization and recombination must balance; if the dominant recombination process is radiative recombination, with rate coefficient α_rr, we have

$$\alpha_\mathrm{rr}(\mathrm{C}^+)n_e\left[1 - x(\mathrm{C}^0)\right] = \zeta(\mathrm{C}^0)x(\mathrm{C}^0) \quad , \tag{16.1}$$

giving a neutral fraction

$$x(\mathrm{C}^0) = \frac{\alpha_\mathrm{rr}(\mathrm{C}^+)n_e}{\alpha_\mathrm{rr}(\mathrm{C}^+)n_e + \zeta(\mathrm{C}^0)} \quad . \tag{16.2}$$

Inserting $\zeta(\mathrm{C}^0) = 2.58 \times 10^{-10}\,\mathrm{s}^{-1}$ (see Table 13.1) and $\alpha_\mathrm{rr}(\mathrm{C}^+) = 8.63 \times 10^{-12}\,\mathrm{cm}^3\,\mathrm{s}^{-1}$ for $T = 100\,\mathrm{K}$ (see Table 14.6), we find $x(\mathrm{C}^0) = 1.3 \times 10^{-3}$. According to this estimate, then, $\sim$99.9% of the gas phase carbon is ionized – photoionization of C is much faster than radiative recombination of C^+ for the densities of electrons and ultraviolet photons found in the diffuse ISM. A similar conclusion is found for other heavy elements with ionization potentials $I < 13\,\mathrm{eV}$.

16.1.1 Grain-Assisted Recombination

The process of grain-assisted recombination, discussed in Chapter 14, can under some circumstances be faster than radiative recombination. Let us examine the importance of grain recombination for our example of carbon ionization.

We write the grain-assisted recombination rate per volume as $n_\mathrm{H}\alpha_\mathrm{gr}n(\mathrm{C}^+)$, where α_gr, the effective rate coefficient, depends on the ion (in this case, C^+), the UV intensity $\propto G_0$ [see Eq. (12.6)], the electron density n_e, and the gas temperature T. Using the parameters in Table 14.9 and assuming $G_0 \approx 1$ and $T \approx 100\,\mathrm{K}$, we calculate the charging parameter $\psi \approx 250$ and obtain an effective rate coefficient $\alpha_\mathrm{gr} \approx 2.83 \times 10^{-14}\,\mathrm{cm}^3\,\mathrm{s}^{-1}$. Grain-assisted recombination is faster than radiative recombination by a factor $n_\mathrm{H}\alpha_\mathrm{gr}/n_e\alpha_\mathrm{rr} \approx 2.5$. With grain-assisted recombination included, we would now estimate a larger neutral fraction $x(\mathrm{C}^0) \approx .0045$, but this is still $\ll 1$. Therefore, even with grain-assisted recombination included, 99.5% of the gas-phase carbon would be photoionized in our hypothetical region with $n_e = 0.04\,\mathrm{cm}^{-3}$ and $T = 100\,\mathrm{K}$. Hence we can assume that nearly all of the carbon and other gas-phase species (e.g., S) with ionization potential $< 13.6\,\mathrm{eV}$ will be photoionized in diffuse H I regions.

Let $n(M^+)$ be the number density of ions of species with ionization potentials $I_X < 13.6\,\mathrm{eV}$. Table 9.5 gives estimates for $x_M \equiv n(M^+)/n_\mathrm{H}$ in different types of H I gas, with

$$x_M \approx \begin{cases} 1.68 \times 10^{-4} & \text{in the WNM.} \\ 1.42 \times 10^{-4} & \text{in the CNM.} \\ 1.07 \times 10^{-4} & \text{in diffuse H}_2. \end{cases} \quad (16.3)$$

16.2 Cool H I Regions: Ionization of Hydrogen

Typical cool H I regions, the so-called cold neutral medium (CNM), may have a density $n_{\rm H} \approx 30\,{\rm cm}^{-3}$ and half-thickness $\approx 1\,{\rm pc}$, giving a column density $N({\rm H\,I}) \approx 10^{20}\,{\rm cm}^{-2}$. Starlight will not ionize H or He within an H I region, because the $13.6 < h\nu < 54.4\,{\rm eV}$ ionizing photons cannot penetrate into the cloud (see Fig. 13.1), but x rays of sufficiently high energy can reach the cloud interior. A photon with energy $h\nu = 150\,{\rm eV}$ has an absorption cross section per H of $1.8 \times 10^{-20}\,{\rm cm}^2$ (mostly due to He); x rays with $h\nu > 150\,{\rm eV}$ can therefore affect the ionization within the interiors of H I clouds.

Let $\zeta_{\rm CR}$ be the rate of primary ionizations per H atom by either cosmic rays or x rays, and let ϕ_s be the average number of secondary ionizations per primary ionization of H. The extremely rapid rate of photoionization of C and S ensures that they will be almost fully ionized in diffuse H I, contributing an electron density $\sim n_{\rm H} x_M$. In a steady state, the hydrogen ionization fraction $x({\rm H}^+) \equiv n({\rm H}^+)/n_{\rm H}$ must obey

$$\begin{aligned} &\zeta_{\rm CR}\,(1+\phi_s) n_{\rm H} \left[1 - x({\rm H}^+)\right] \\ &\quad = \alpha_{\rm rr}({\rm H}^+) n_{\rm H}{}^2 \left[x({\rm H}^+) + x_M\right] x({\rm H}^+) + \alpha_{\rm gr}({\rm H}^+) n_{\rm H}{}^2 x({\rm H}^+)\,, \end{aligned} \quad (16.4)$$

where $\alpha_{\rm rr}$ is the rate coefficient for radiative recombination of $\rm H^+$, and $\alpha_{\rm gr}$ is the effective rate coefficient for grain-assisted recombination of $\rm H^+$ (see §14.8). This is a quadratic equation for $x({\rm H}^+)$ with solution

$$x({\rm H}^+) = \frac{\left[(\beta + \gamma + x_M)^2 + 4\beta\right]^{1/2} - (\beta + \gamma + x_M)}{2}\,, \quad (16.5)$$

$$\beta \equiv \frac{\zeta_{\rm CR}(1 + \phi_s)}{\alpha_{\rm rr} n_{\rm H}}\,, \quad (16.6)$$

$$\gamma \equiv \frac{\alpha_{\rm gr}}{\alpha_{\rm rr}}\,. \quad (16.7)$$

The rate coefficient $\alpha_{\rm gr}$ depends on the charge state of the grains, which in turn depends on the electron density n_e (through the "charging parameter" ψ [defined in Eq. (14.38)]. We therefore iterate: assume some n_e to calculate ψ, with this ψ calculate $\alpha_{\rm gr}$, with this $\alpha_{\rm gr}$ solve Eq. (16.5), with this n_e recalculate ψ, and iterate to converge on the self-consistent steady state solution for n_e. Figure 16.1

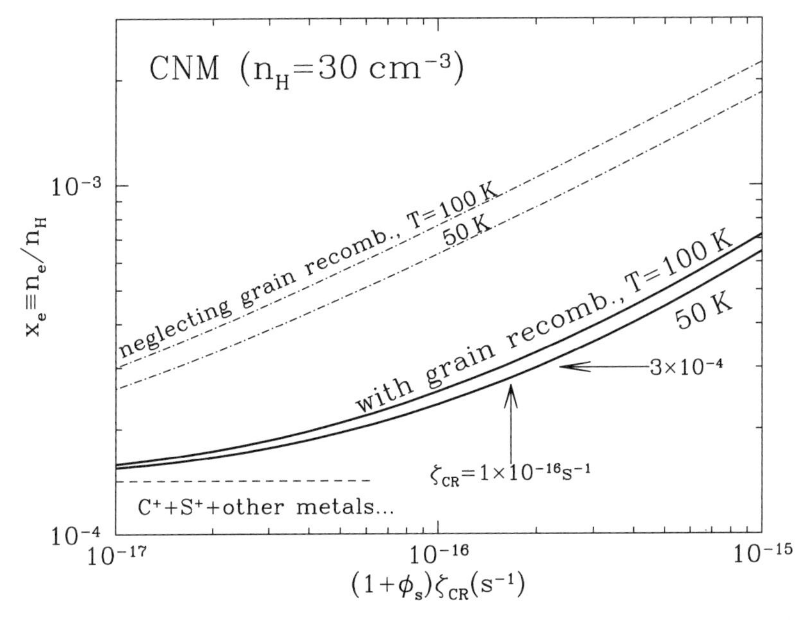

Figure 16.1 Solid lines show fractional ionization x_e in the CNM as a function of cosmic ray ionization rate $\zeta_{\rm CR}$, for two gas temperatures, $T = 50$ and 100 K (see text). Dashed-dotted curves show n_e calculated if grain-assisted recombination is neglected. For $\zeta_{\rm CR} \approx 1 \times 10^{-16}\,{\rm s}^{-1}$ determined from observations of diffuse molecular clouds (see §16.4), the CNM electron density $n_e \approx 0.010\,{\rm cm}^{-3}$.

shows the electron density in the CNM with $n_{\rm H} \approx 30\,{\rm cm}^{-3}$, given by Eq. (16.5). For a given cosmic ray ionization rate, grain-assisted recombination significantly lowers the electron density. It is therefore important to include this process to obtain realistic estimates of the electron density in diffuse H I clouds. If the cosmic ray ionization rate $\zeta_{\rm CR}(1+\phi_s) \lesssim 2\times10^{-16}\,{\rm s}^{-1}$, the electron density $n_e \lesssim 0.01\,{\rm cm}^{-3}$. At this time, there is considerable uncertainty regarding the ionization rate $\zeta_{\rm CR}$, but observations of diffuse molecular clouds (see §16.4) provide some constraints.

16.3 Warm H I Regions

A considerable fraction of the atomic hydrogen is located in regions of density $n_{\rm H} \approx 0.5\,{\rm cm}^{-3}$ and temperature $T \approx 5000$ K – the so-called warm neutral medium (WNM). What fractional ionization is expected for such regions?

If there is no local source of x rays, the hydrogen ionization will be primarily due to cosmic rays, just as for the CNM. Just as in the CNM, photoionization of elements with ionization potentials $I < I_{\rm H}$ will contribute electrons, with $x_M \approx 1.7 \times 10^{-4}$. Figure 16.2 shows the resulting fractional ionization. For the cosmic ray ionization rate $\zeta_{\rm CR} \approx 1 \times 10^{-16}\,{\rm s}^{-1}$ inferred from observations of diffuse molecular clouds, about 2% of the hydrogen will be ionized in the WNM. This conclusion is not appreciably altered when cosmic ray ionization of He is included.

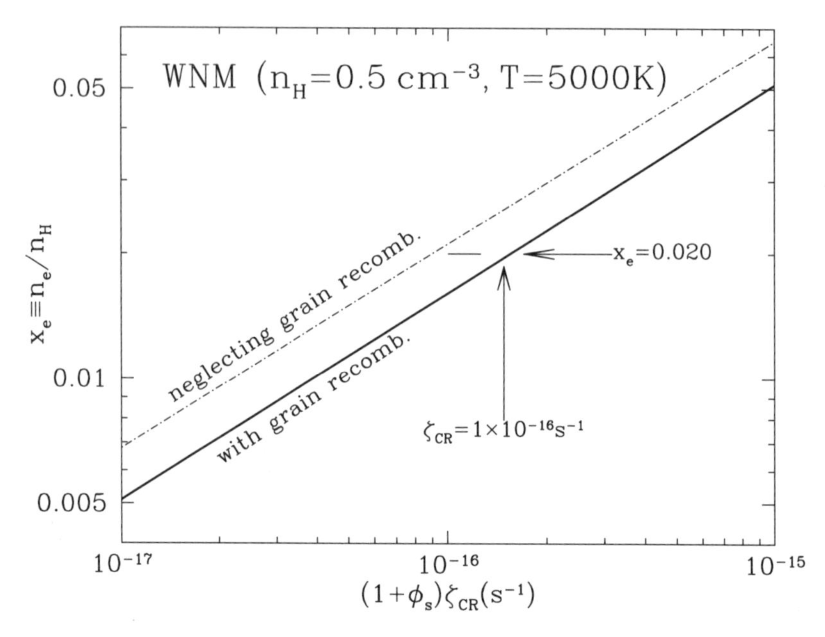

Figure 16.2 Fractional ionization of the WNM as a function of cosmic ray ionization rate $\zeta_{\rm CR}$. For $\zeta_{\rm CR} \approx 1 \times 10^{-16}\,{\rm s}^{-1}$ (indicated by observations of diffuse molecular clouds), the WNM electron density $n_e \approx 0.010\,{\rm cm}^{-3}$.

16.4 Diffuse Molecular Gas

In regions where the hydrogen is molecular, most ionizations produced by cosmic rays or x rays (including secondary electrons) detach an electron from H_2 to create ${H_2}^+$. If it encounters a free electron, the ${H_2}^+$ will dissociatively recombine, ${H_2}^+ + e^- \rightarrow {\rm H} + {\rm H}$, with a rate coefficient $k \approx 2 \times 10^{-8}\,{\rm cm}^3\,{\rm s}^{-1}$ (see Table 14.8). Because the electron fraction in molecular gas tends to be low, most of the ${H_2}^+$ will undergo the fast exothermic ion–molecule reaction:

$$ {\rm H_2}^+ + {\rm H_2} \rightarrow {\rm H}_3^+ + {\rm H} \quad , \quad k_{16.8} = 2.08 \times 10^{-9}\,{\rm cm}^3\,{\rm s}^{-1} \quad , \tag{16.8} $$

followed by dissociative recombination

$$ {\rm H}_3^+ + e^- \rightarrow {\rm H}_2 + {\rm H} \quad , \quad k_{16.9} = 5.0 \times 10^{-8} T_2^{-0.48}\,{\rm cm}^3\,{\rm s}^{-1} \quad , \tag{16.9} $$

$$ {\rm H}_3^+ + e^- \rightarrow {\rm H} + {\rm H} + {\rm H} \quad , \quad k_{16.10} = 8.9 \times 10^{-8} T_2^{-0.48}\,{\rm cm}^3\,{\rm s}^{-1} \quad , \tag{16.10} $$

or an exchange reaction such as

$$H_3^+ + O \rightarrow OH^+ + H_2 \quad , \quad k_{16.11} = 8.40 \times 10^{-10}\,\mathrm{cm^3\,s^{-1}} \; , \tag{16.11}$$

$$H_3^+ + O \rightarrow H_2O^+ + H \quad , \quad k_{16.12} = 3.60 \times 10^{-10}\,\mathrm{cm^3\,s^{-1}} \; , \tag{16.12}$$

followed by, for example,

$$OH^+ + e^- \rightarrow O + H \quad , \quad k_{16.13} = 6.5 \times 10^{-8} T_2^{-0.50}\,\mathrm{cm^3\,s^{-1}} \; . \tag{16.13}$$

Rate coefficients $k_{16.8}$, $k_{16.11}$, $k_{16.12}$, and $k_{16.13}$ are from the UMIST 2006 Database for Astrochemistry (Woodall et al. 2007), where the original sources are given, and $k_{16.9}$ and $k_{16.10}$ are from McCall et al. (2004).

In a diffuse region with fractional ionization $x_e \gtrsim x_M \approx 1.9 \times 10^{-4}$, destruction of H_3^+ will be dominated by Eqs. (16.9 and 16.10), and the abundance of H_3^+ will be given by

$$\frac{n(H_3^+)}{n(H_2)} \approx \frac{2\zeta_{CR}(1+\phi_s)}{(k_{16.9}+k_{16.10})n_H x_e} \quad . \tag{16.14}$$

Thus the cosmic ray ionization rate can be obtained:

$$\zeta_{CR}(1+\phi_s) \approx (k_{16.9}+k_{16.10}) n_H x_e \frac{N(H_3^+)}{2N(H_2)} \quad . \tag{16.15}$$

Consider now the diffuse molecular cloud located on the sightline toward the bright star ζ Per. Taking the observed $N(H_3^+) = 8 \times 10^{13}\,\mathrm{cm^{-2}}$ (McCall et al. 2003) and $N(H_2) = 5.0 \times 10^{20}\,\mathrm{cm^{-2}}$ (Bohlin et al. 1978), and adopting a density $n_H \approx 100\,\mathrm{cm^{-3}}$ and temperature $T \approx 60\,\mathrm{K}$ (van Dishoeck & Black 1986; Le Petit et al. 2004) and fractional ionization $x_e \approx x_M \approx 1.1 \times 10^{-4}$ (see Eq. 16.3), we obtain $\zeta_{CR}(1+\phi_s) \approx 2.2 \times 10^{-16}\,\mathrm{s^{-1}}$ in the diffuse molecular gas toward ζ Per. With $\phi_s \approx 0.67$ [see Eq. (13.12)], the primary ionization rate is estimated to be $\zeta_{CR} \approx 1.3 \times 10^{-16}\,\mathrm{s^{-1}}$.

This is strong evidence that in at least some interstellar clouds, the cosmic ray ionization rate $\zeta_{CR} \approx 1 \times 10^{-16}\,\mathrm{s^{-1}}$. H_3^+ detections on 14 diffuse cloud sightlines lead to ζ_{CR} estimates in the range $(0.5$ to $3) \times 10^{-16}\,\mathrm{s^{-1}}$, with an average of $2 \times 10^{-16}\,\mathrm{s^{-1}}$ (Indriolo et al. 2007). The free electrons in diffuse molecular clouds come almost entirely from photoionization of C, S, and other gas-phase atoms with $I < I_H$, with total abundance $x_M \approx 1.1 \times 10^{-4}$ – the molecular ions make only a very small contribution. For example, the H_3^+ ion has a fractional abundance $N(H_3^+)/2N(H_2) \approx 8 \times 10^{-8}$, and other molecular ions, such as OH^+ or HCO^+, have similarly low abundances.

16.5 Dense Molecular Gas: Dark Clouds

When the visual extinction exceeds $A_V \approx 3$ mag, ultraviolet starlight is sufficiently attenuated so that elements like C and S will be predominantly neutral in the gas phase, and most of the free electrons are the result of cosmic ray ionization. What fractional ionizations are present within these dark regions? Here we follow the treatment by McKee (1989).

Under these conditions, cosmic ray ionization will produce H_3^+ ions via

$$H_2 + CR \rightarrow H_2{}^+ + e^- \quad , \tag{16.16}$$

$$H_2{}^+ + H_2 \rightarrow H_3^+ + H \quad . \tag{16.17}$$

For fractional ionizations $x_e \lesssim 10^{-5}$, dissociative recombination of the H_3^+ ions will be of secondary importance, and most of the H_3^+ ions will react with atoms or molecules M (e.g., CO) to form

$$H_3^+ + M \rightarrow MH^+ + H_2 \; : \quad k_{16.18} \approx 2 \times 10^{-9}\,\mathrm{cm^3\,s^{-1}} \quad . \tag{16.18}$$

The generic molecular ion MH^+ (e.g., HCO^+) can recombine dissociatively:

$$MH^+ + e^- \rightarrow M + H \; : \quad k_{16.19} \approx 1 \times 10^{-7} T_2^{-0.5}\,\mathrm{cm^3\,s^{-1}} \quad , \tag{16.19}$$

or it can capture an electron from a grain:

$$MH^+ + \mathrm{grain}^- \rightarrow MH + \mathrm{grain} \; : \quad k_{16.20} \quad , \tag{16.20}$$

or it can exchange charge with a neutral metal atom such as S:

$$MH^+ + S \rightarrow MH + S^+ \; : \quad k_{16.21} \approx 1 \times 10^{-9}\,\mathrm{cm^3\,s^{-1}} \quad . \tag{16.21}$$

The S^+ will finally be neutralized by capturing an electron from a grain:

$$\left.\begin{array}{lcl} S^+ + \mathrm{grain}^- & \rightarrow & S + \mathrm{grain} \\ S^+ + \mathrm{grain} & \rightarrow & S + \mathrm{grain}^+ \end{array}\right\} \; : \; k_{16.22} \, , \tag{16.22}$$

where polycyclic aromatic hydrocarbon (PAH) particles are included in the grain population. If every H_2^+ that is created leads to formation of MH^+, then the formation rate of H_2^+ should equal the rate of destruction of MH^+:

$$n_H \zeta_{CR} (1 + \phi_s) = n(MH^+) \left(k_{16.19} n_e + k_{16.20} n_H + k_{16.21} n_S \right) \quad . \tag{16.23}$$

If the net charge contributed by the grains is small, then

$$n_e \approx n(MH^+) + n(S^+) \approx n(MH^+) \left(1 + \frac{k_{16.21} n_S}{k_{16.22} n_H} \right) \quad . \tag{16.24}$$

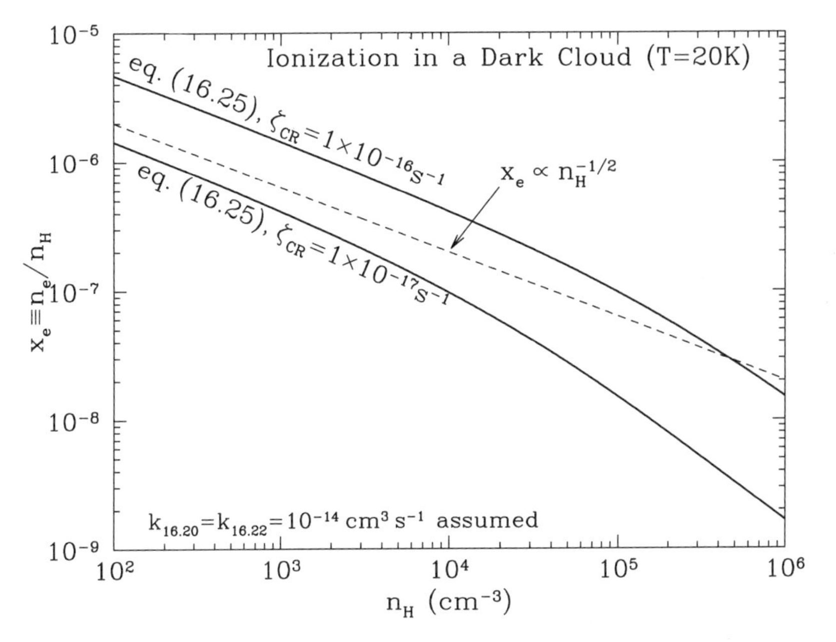

Figure 16.3 Fractional ionization in a dark cloud, estimated using Eq. (16.25), with the grain recombination rate coefficients set to $k_{16.20} = k_{16.22} = 10^{-14}\,\text{cm}^3\,\text{s}^{-1}$ (see Fig. 14.6). The dashed line is a simple power-law approximation $x_e \approx 2 \times 10^{-5}(n_\text{H}/\,\text{cm}^{-3})^{-1/2}$.

Substituting (16.24) into (16.23), we obtain a quadratic equation for $n(M\text{H}^+)/n_\text{H}$, from which we obtain

$$\frac{n_e}{n_\text{H}} = \frac{\left[B^2 + 4A\,\zeta_\text{CR}(1+\phi_s)/n_\text{H}\right]^{1/2} - B}{2k_{16.19}} \quad , \tag{16.25}$$

$$A = k_{16.19}\left(1 + \frac{k_{16.21}}{k_{16.22}}\frac{n_\text{S}}{n_\text{H}}\right) \quad , \tag{16.26}$$

$$B = k_{16.20} + k_{16.21}\frac{n_\text{S}}{n_\text{H}} \quad . \tag{16.27}$$

The fractional ionization estimated using Eq. (16.25) is shown in Figure 16.3, where it is seen to vary approximately as $1/\sqrt{n_\text{H}}$, reaching fractional ionizations of $\sim 10^{-7}$ for $n_\text{H} \approx 10^5\,\text{cm}^{-3}$ for an assumed cosmic ray ionization rate $\zeta_\text{CR} \approx 1 \times 10^{-16}\,\text{s}^{-1}$.

From this, we see that at points deep enough within a dark molecular cloud, where starlight cannot maintain a high ionization fraction for elements with $I_M < I_\text{H}$, cosmic rays can maintain only a very low fractional ionization. This has implications for the magnetohydrodynamics of the magnetized gas.

Chapter Seventeen

Collisional Excitation

Collisional excitation is important in the ISM for two reasons:

1. It puts ions, atoms, and molecules into excited states from which they may decay radiatively; these radiative losses result in cooling of the gas.
2. It puts species into excited states that can serve as diagnostics of the physical conditions in the gas. If the level populations can be determined observationally, from either emission lines or absorption lines, we may be able to infer the density, temperature, or radiation field in the region where the diagnostic species is located.

Throughout this chapter and the rest of the book, we will be making use of rate coefficients and transition rates. We will use the notation $k_{if} \equiv \langle \sigma v \rangle_{i \to f}$ for collisional rate coefficients, and $A_{if} \equiv A_{i \to f}$ to denote radiative transition probabilities, where the first subscript in k_{if} or A_{if} denotes the initial state, and the second the final state. For energy-level differences, we will set $E_{u\ell} \equiv E_u - E_\ell$.

17.1 Two-Level Atom

In some cases, it is sufficient to consider only the ground state and the first excited state – when attention is limited to these two states, we speak of the "two-level atom." Consider first the case where there is no background radiation present, and the only processes acting are collisional excitation, collisional deexcitation, and radiative decay. Let the ground state be level 0, and the excited state be level 1. Let n_j be the number density of the species in level j. For collisional excitation by some species (e.g., electrons) with density n_c, the population of the excited state must satisfy

$$\frac{dn_1}{dt} = n_c n_0 k_{01} - n_c n_1 k_{10} - n_1 A_{10} \quad . \tag{17.1}$$

The steady state solution ($dn_1/dt = 0$) is simply

$$\frac{n_1}{n_0} = \frac{n_c k_{01}}{n_c k_{10} + A_{10}} \quad . \tag{17.2}$$

The upward rate coefficient k_{01} is given in terms of the downward rate coefficient by

$$k_{01} = \frac{g_1}{g_0} k_{10} \, e^{-E_{10}/kT_{\rm gas}} \quad , \tag{17.3}$$

where g_0, g_1 are the level degeneracies, and $T_{\rm gas}$ is the kinetic temperature of the gas. In the limit $n_c \to \infty$, it is easy to see that $n_1/n_0 \to (g_1/g_0)\exp(-E_{10}/kT_{\rm gas})$.

Now suppose that radiation is present. Let u_ν be the specific energy density at frequencies near $\nu = E_{10}/h$. It is convenient to use instead the dimensionless (angle- and polarization-averaged) photon occupation number:

$$\bar{n}_\gamma \equiv \frac{c^3}{8\pi h\nu^3} u_\nu \quad . \tag{17.4}$$

Then,

$$\frac{dn_1}{dt} = n_0 \left[n_c k_{01} + \bar{n}_\gamma \frac{g_1}{g_0} A_{10} \right] - n_1 \left[n_c k_{10} + (1 + \bar{n}_\gamma) A_{10} \right] \quad . \tag{17.5}$$

The rate of photoabsorption is $\bar{n}_\gamma (g_1/g_0) A_{10} n_0$, and the rate of stimulated emission is $\bar{n}_\gamma A_{10} n_1$ (see Chapter 6). The **steady-state solution with radiation present** is

$$\frac{n_1}{n_0} = \frac{n_c k_{01} + \bar{n}_\gamma (g_1/g_0) A_{10}}{n_c k_{10} + (1 + \bar{n}_\gamma) A_{10}} \quad . \tag{17.6}$$

This is the fully general result for a two-level system.

It is instructive to examine Eq. (17.6) in various limits:

- If $\bar{n}_\gamma \to 0$, then we recover our previous result (17.2).
- If $n_c \to 0$, then $n_1/n_0 \to (g_1/g_0)\bar{n}_\gamma/(1 + \bar{n}_\gamma)$. If we have a blackbody radiation field with temperature $T_{\rm rad}$ [i.e., $n_\gamma = 1/(e^{E_{10}/kT_{\rm rad}} - 1)$], then $n_1/n_0 = (g_1/g_0) e^{-E_{10}/kT_{\rm rad}}$.
- If we have a blackbody radiation field with temperature $T_{\rm rad} = T_{\rm gas}$, then $n_1/n_0 = (g_1/g_0) e^{-E_{10}/kT_{\rm rad}}$ *independent* of the gas density n_c! The photons alone are sufficient to bring the two level system into LTE, and additional (thermal) collisions have no further effect on the level populations.

17.2 Critical Density $n_{{\rm crit},u}$

For a collision partner c, we define the **critical density** $n_{{\rm crit},u}(c)$ for an excited state u to be the density for which collisional deexcitation equals radiative deexcitation,

Table 17.1 Critical Densities for Fine-Structure Excitation in H I Regions

						$n_{\rm crit,\it u}$(H)		$n_{\rm crit,\it u}(e^-)$	
			E_ℓ/k	E_u/k	$\lambda_{u\ell}$	T=100 K	T=5000 K	T=100 K	T=5000 K
Ion	ℓ	u	(K)	(K)	(μm)	(cm^{-3})	(cm^{-3})	(cm^{-3})	(cm^{-3})
C II	^{2}P$^o_{1/2}$	^{2}P$^o_{3/2}$	0	91.21	157.74	2.7×10^3	1.5×10^3	6.8	40.
C I	^{3}P$_0$	^{3}P$_1$	0	23.60	609.7	620	170	76.	6.4
	^{3}P$_1$	^{3}P$_2$	23.60	62.44	370.37	720	150	75.	6.3
O I	^{3}P$_2$	^{3}P$_1$	0	227.71	63.185	2.5×10^5	4.9×10^4	1.8×10^5	4.8×10^4
	^{3}P$_1$	^{3}P$_0$	227.71	326.57	145.53	2.4×10^4	8.6×10^3	2.3×10^4	5.8×10^3
Si II	^{2}P$^o_{1/2}$	^{2}P$^o_{3/2}$	0	413.28	34.814	2.5×10^5	1.2×10^5	140.	1.5×10^3
Si I	^{3}P$_0$	^{3}P$_1$	0	110.95	129.68	4.8×10^4	2.8×10^4	2.9×10^4	830.
	^{3}P$_1$	^{3}P$_2$	110.95	321.07	68.473	9.9×10^4	3.6×10^4	4.4×10^4	1.9×10^3

including stimulated emission[1]:

$$n_{\rm crit,\it u}(c) \equiv \frac{\sum_{\ell<u}\left[1+(n_\gamma)_{u\ell}\right]A_{u\ell}}{\sum_{\ell<u}k_{u\ell}(c)} \quad . \tag{17.7}$$

Note that the definition (17.7) applies to multilevel systems, but each excited level u may have a different critical density. The definition (17.7) is appropriate when the gas is optically thin, so that the radiated photons can escape. When the emitting gas is itself optically thick at the emission frequency, we have "radiative-trapping," and the criterion for the critical density must be modified (see Chapter 19).

Note that this definition of $n_{\rm crit,\it u}$ depends on the intensity of ambient radiation at frequencies where level u can radiate. For many transitions of interest, we have $(\bar{n}_\gamma)_{u\ell} \ll 1$, and this correction is unimportant, but for radio frequency transitions – e.g., the 21-cm line of atomic hydrogen – it is important to include this correction for stimulated emission.

Critical densities $n_{\rm crit}$ for the fine structure levels of C I, C II, O I, Si I, and Si II are given in Table 17.1.

17.3 Example: H I Spin Temperature

Consider the ground state of the hydrogen atom (electron in the $1s$ orbital, electron spin antiparallel to nuclear spin, $g_0 = 1$), and the hyperfine excited state ($1s$ orbital, electron spin and nuclear spin parallel, $g_1 = 3$). The energy level structure is illustrated in Fig. 17.1.

The energy difference between the excited state (nuclear and electron spins parallel, $g_1 = 3$) and the ground state (nuclear and electron spins antiparallel) is only $E_{10} = 5.87\,\mu$eV, corresponding to a photon wavelength $\lambda = 21.11$ cm. The spontaneous decay rate is $A_{10} = 2.884 \times 10^{-15}\,{\rm s}^{-1}$, corresponding to a lifetime of $\sim 10^7$ yr.

[1] The definition of critical density is not completely standard. Some authors include collisional excitation channels in the denominator of Eq. (17.7).

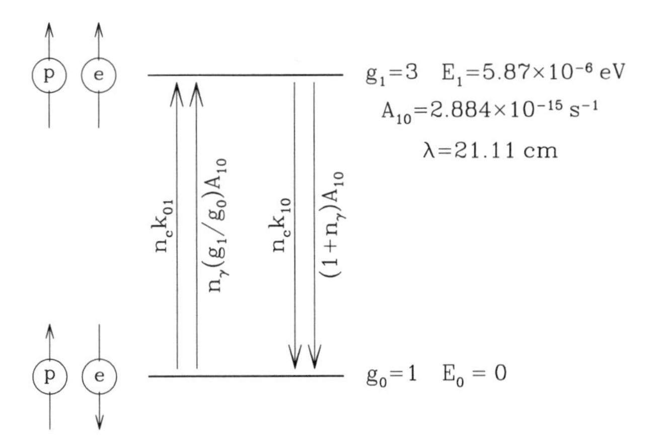

Figure 17.1 Hyperfine splitting of the $1s$ level of H.

The rate coefficient for collisional dexcitation of the hyperfine excited state due to collisions with other H atoms (Allison & Dalgarno 1969; Zygelman 2005) can be approximated by

$$k_{10} \approx \begin{cases} 1.19\times10^{-10} T_2^{0.74-0.20\ln T_2}\,\mathrm{cm^3\,s^{-1}} & (20\,\mathrm{K} < T < 300\,\mathrm{K}) \\ 2.24\times10^{-10} T_2^{0.207} \mathrm{e}^{-0.876/T_2}\,\mathrm{cm^3\,s^{-1}} & (300\,\mathrm{K} < T < 10^3\,\mathrm{K}) \end{cases} . \tag{17.8}$$

We obtain k_{01} from the principle of detailed balance (3.21):

$$k_{01} = 3k_{10}\,\mathrm{e}^{-0.0682\,\mathrm{K}/T} \quad . \tag{17.9}$$

What is the value of the photon occupation number $\bar{n}_\gamma$ in the diffuse ISM? The radiation field near 21 cm is dominated by the cosmic microwave background (CMB) plus Galactic synchrotron emission. Including the contribution[2] from synchrotron radiation, the angle-averaged background near 21 cm corresponds to an antenna temperature $T_A \approx T_{\mathrm{CMB}} + 1.04\,\mathrm{K}$ (see Chapter 12), where $T_{\mathrm{CMB}} = 2.73\,\mathrm{K}$. Thus

$$\bar{n}_\gamma \equiv \frac{1}{\exp(h\nu/kT_B) - 1} \tag{17.10}$$

$$\equiv \frac{kT_A}{h\nu} \approx \frac{3.77\,\mathrm{K}}{0.0682\,\mathrm{K}} \approx 55 \quad . \tag{17.11}$$

For the present case of a two-level system,

$$n_{\mathrm{crit}}(\mathrm{H}) \equiv \frac{(1+\bar{n}_\gamma)A_{10}}{k_{10}} \tag{17.12}$$

$$= 1.7\times10^{-3}(T/100\,\mathrm{K})^{-0.66}\,\mathrm{cm^{-3}} \quad (50\,\mathrm{K} \lesssim T \lesssim 200\,\mathrm{K}). \tag{17.13}$$

[2]Near 21 cm, even the CMB is in the Rayleigh-Jeans limit, so the antenna temperature T_A and brightness temperature T_B are approximately equal.

where we have taken $\bar{n}_\gamma = 55$, appropriate for H I in the diffuse interstellar medium. In the absence of collisions, the CMB plus Galactic synchrotron radiation would give optically thin H I an excitation temperature $T_{\rm exc} \approx 3.77\,{\rm K}$. If we consider only collisions with atomic H, we can solve Eq. (17.6) for various gas densities $n_{\rm H}$ and temperatures T. Figure 17.2 shows the resulting "spin temperature" $T_{\rm spin} \equiv 0.0682\,{\rm K}/\ln(n_0 g_1/n_1 g_0)$ as a function of density $n_{\rm H}$.

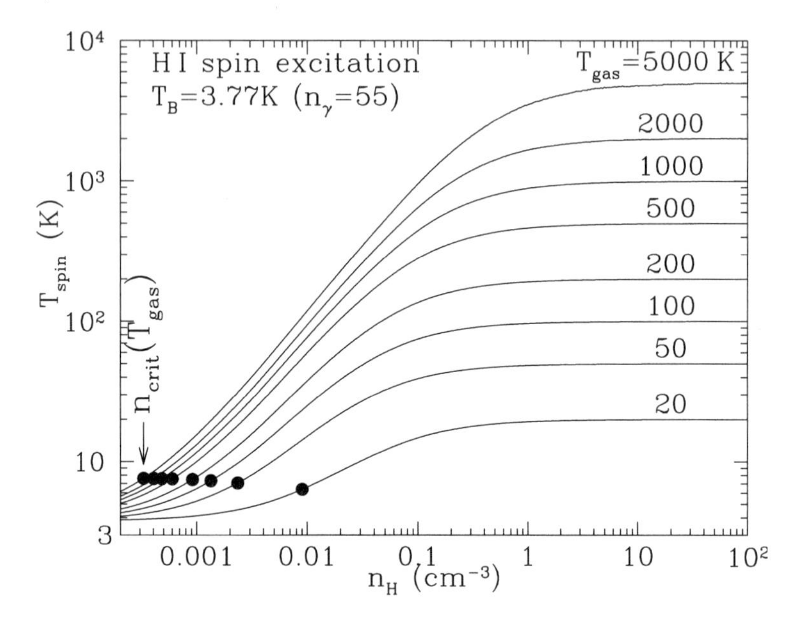

Figure 17.2 H I spin temperature as a function of density $n_{\rm H}$, including only 21 cm continuum radiation (with brightness temperature $T_B = 3.77\,{\rm K}$, i.e., $n_\gamma = 55$) and collisions with H atoms. Lyman α scattering is not included. Filled circles show $n_{\rm crit}$ for each value of $T_{\rm gas}$.

For densities $n \gg n_{\rm crit}$, we expect $T_{\rm spin} \approx T_{\rm gas}$, while for densities $n \ll n_{\rm crit}$, we expect $T_{\rm spin} \approx T_B(\nu)$, or 3.77K for the radiation field assumed here. The results in Fig. 17.2 are consistent with the expected asymptotic behavior, but it is important to note that one requires $n \gg n_{\rm crit}$ in order to have $T_{\rm spin}$ within, say, 10% of $T_{\rm gas}$, particularly at high temperatures. The points in Figure 17.2 show $n_{\rm crit}(T_{\rm gas})$ for each $T_{\rm gas}$; it is apparent that high values of $T_{\rm spin}$ are achieved only for $n \gg n_{\rm crit}$.

Collisions with protons and electrons can also be important for hyperfine excitation of H I; the rate coefficient for deexcitation by electrons is (Furlanetto & Furlanetto 2007)

$$k_{10}(e^-) \approx 2.26 \times 10^{-9} (T/100\,{\rm K})^{0.5}\,{\rm cm}^3\,{\rm s}^{-1} \quad (1 \lesssim T \lesssim 500\,{\rm K}), \qquad (17.14)$$

a factor ~ 10 larger than $k_{10}({\rm H})$; electrons will, therefore, be of minor importance in regions of fractional ionization $x_e \lesssim 0.03$, such as the CNM or WNM (see Figs. 16.1 and 16.2).

Resonant scattering of Lyman-α photons can also change the populations of the hyperfine levels (Wouthuysen 1952; Field 1958).[3] Let P_{01} be the probability per unit time of a transition from the hyperfine ground state to $2p$, followed by spontaneous decay from $2p$ to the hyperfine excited state, and P_{10} the probability per time of a transition from the hyperfine excited state to $2p$, followed by decay to the hyperfine ground state. The Lyman α profile depends on the kinetic temperature of the hydrogen atoms that are emitting and scattering the Lyman α. If the hydrogen atoms have a Maxwellian velocity distribution, Field (1959) showed that $P_{01} \approx 3P_{10}e^{-0.0682\,\mathrm{K}/T_{\mathrm{H}}}$, where $T_{\mathrm{H}} \equiv m_{\mathrm{H}}\sigma_V^2/k$, and σ_V is the one-dimensional velocity dispersion of the H atoms that are scattering the Lyman α photons. Therefore, this process acts essentially like a collisional process, except that the temperature T_{H} characterizing the Lyman α line profile includes a contribution from turbulent motions, in addition to microscopic thermal motions. Liszt (2001) estimates the Lyman α intensities expected in the warm neutral medium (WNM), and concludes that collisions and Lyman α together are not fast enough to thermalize the H I hyperfine transition. As a result, we should expect $T_{\mathrm{spin}} < T_{\mathrm{gas}}$ in the WNM.

17.4 Example: C II Fine Structure Excitation

The ground electronic state $1s^2 2s^2 2p\ ^2\mathrm{P}^{\mathrm{o}}$ of C II contains two fine-structure levels (see Fig. 17.3), $^2\mathrm{P}^{\mathrm{o}}_{1/2}$ and $^2\mathrm{P}^{\mathrm{o}}_{3/2}$. Will the populations of these two levels be thermalized in the ISM? Radiative decay of the $^2\mathrm{P}^{\mathrm{o}}_{3/2}$ excited fine-structure state produces a photon with $\lambda = 158\,\mu\mathrm{m}$. At this wavelength, the continuum background in the interstellar medium has $n_\gamma \ll 1$. In fact, from Figure 12.1, we estimate $n_\gamma \approx 10^{-5}$ in the diffuse ISM. Hence, if we are considering regions that are optically thin in the 158 μm line, we can neglect stimulated emission.

The $^2\mathrm{P}^{\mathrm{o}}_{3/2}$ fine-structure level can be excited by collisions of $^2\mathrm{P}^{\mathrm{o}}_{1/2}$ with electrons, H, He, and (in a molecular cloud) H_2. For electrons, the collision strength is (Tayal 2008)

$$\Omega(^2\mathrm{P}^{\mathrm{o}}_{1/2}, {}^2\mathrm{P}^{\mathrm{o}}_{3/2}) \approx 1.6 \quad , \tag{17.15}$$

[3] The Wouthuysen-Field effect can be understood semiclassically. When a Lyman α photon is absorbed, the H atom enters a $2p$ state with its electronic angular momentum $\mathbf{L}$ in some direction that is related to the direction of propagation and polarization of the absorbed photon, but is unrelated to the orientation of the nuclear or electron spins. During the $\sim 10^{-9}$ s lifetime of the $^2\mathrm{P}^{\mathrm{o}}_{1/2}$ or $^2\mathrm{P}^{\mathrm{o}}_{3/2}$ excited state, spin-orbit coupling will cause both the electron spin $\mathbf{S}$ and orbital angular momentum $\mathbf{L}$ to precess around $\mathbf{L}+\mathbf{S}$. When the Lyman α photon is emitted, the electron spin will be in a different direction, and thus its orientation relative to the nuclear spin will change some fraction of the time. Note that while the spin-orbit coupling in H is weak, it is not zero: the $0.366\,\mathrm{cm}^{-1}$ fine-structure splitting between $^2\mathrm{P}^{\mathrm{o}}_{3/2}$ and $^2\mathrm{P}^{\mathrm{o}}_{1/2}$ corresponds to an electron precession frequency $\sim 8\times10^{10}$ Hz – the ~ 1.6 ns lifetime of the excited state corresponds to $\sim 10^2$ precession periods.

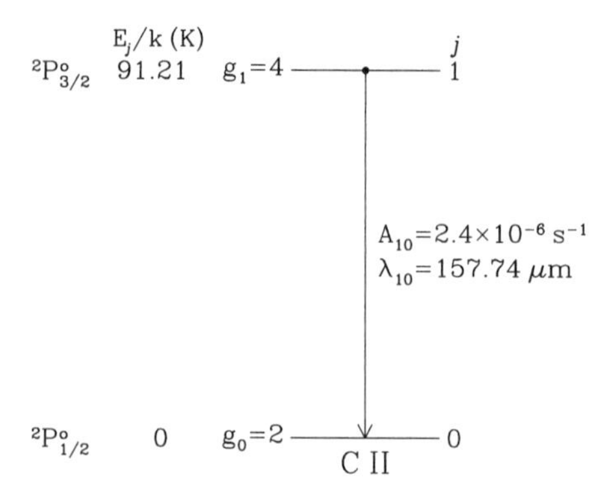

Figure 17.3 Fine-structure levels of C^+.

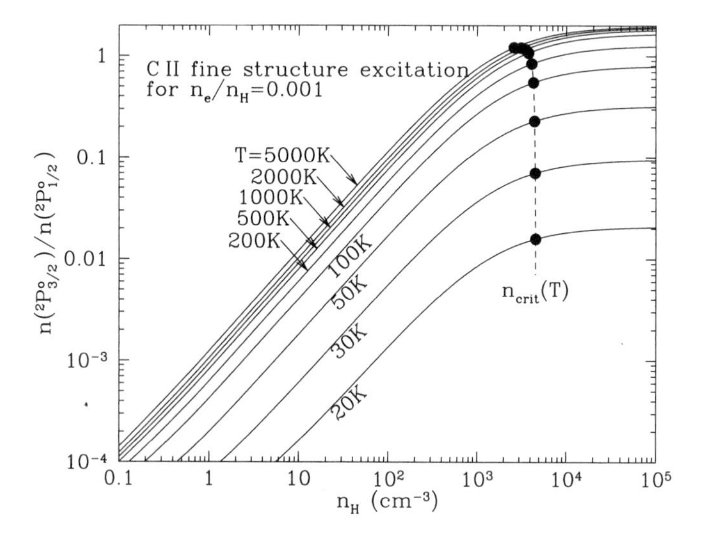

Figure 17.4 Excitation of the $^2P^o_{3/2}$ excited fine-structure level of C^+. The background radiation is assumed to have $n_\gamma \approx 10^{-5}$ at 158 μm.

so that

$$k_{10}(e^-) \approx 3.45 \times 10^{-8} T_4^{-1/2} \, \mathrm{cm^3\,s^{-1}} \quad , \tag{17.16}$$

while for H atoms (Barinovs et al. 2005):

$$k_{10}(\mathrm{H}) \approx 7.58 \times 10^{-10} T_2^{0.1281+0.0087 \ln T_2} \, \mathrm{cm^3\,s^{-1}} \quad . \tag{17.17}$$

Thus the critical densities are

$$n_{\rm crit}({\rm H}) \approx 3.2 \times 10^3 T_2^{-0.1281-0.0087\ln T_2}\,{\rm cm}^{-3} \quad , \tag{17.18}$$

$$n_{\rm crit}(e^-) \approx 70\,T_4^{1/2}\,{\rm cm}^{-3} \quad . \tag{17.19}$$

Therefore, we see that for both CNM and WNM conditions, the densities are well below critical, and the C II fine-structure levels will be subthermally excited. It follows that collisional excitations of C II $^2{\rm P}^{\rm o}_{3/2}$ will usually be followed by radiative decays, removing energy from the gas. The [C II] 158 μm transition is the principal cooling transition for the diffuse gas in star-forming galaxies. Plate 3c is an all-sky map of [C II] 158 μm emission from the Galaxy, made by the Far InfraRed Absolute Spectrophotometer (FIRAS) on the COsmic Background Explorer (COBE) satellite (Fixsen et al. 1999).

The preceding discussion neglects radiative excitation of C II $^2{\rm P}^{\rm o}_{3/2}$, appropriate for clouds that are optically thin in the [C II] 158 μm line. When the clouds become optically thick, the [C II] 158 μm line intensity can increase to the point where self-absorption makes an important contribution to the excitation of C II $^2{\rm P}^{\rm o}_{3/2}$. We will return to the question of excitation under such conditions in Chapter 19.

17.5★ Three-Level Atom

If we consider the ground state and two excited states, we refer to this as a "three-level atom." The equations for the evolution of the level populations are

$$\frac{dn_2}{dt} = R_{02}n_0 + R_{12}n_1 - (R_{20}+R_{21})\,n_2 \quad , \tag{17.20}$$

$$\frac{dn_1}{dt} = R_{01}n_0 + R_{21}n_2 - (R_{10}+R_{12})\,n_1 \quad , \tag{17.21}$$

where the rates R_{if} are:

$$R_{10} = C_{10} + (1+n_{\gamma,10})A_{10} \quad , \tag{17.22}$$

$$R_{20} = C_{20} + (1+n_{\gamma,20})A_{20} \quad , \tag{17.23}$$

$$R_{21} = C_{21} + (1+n_{\gamma,21})A_{21} \quad , \tag{17.24}$$

$$R_{01} = (g_1/g_0)\left[C_{10}{\rm e}^{-E_{10}/kT} + n_{\gamma,10}A_{10}\right] \quad , \tag{17.25}$$

$$R_{02} = (g_2/g_0)\left[C_{20}{\rm e}^{-E_{20}/kT} + n_{\gamma,20}A_{20}\right] \quad , \tag{17.26}$$

$$R_{12} = (g_2/g_1)\left[C_{21}{\rm e}^{-E_{21}/kT} + n_{\gamma,21}A_{21}\right] \quad . \tag{17.27}$$

The rates $C_{u\ell}$ for collisional deexcitation are summed over all collision partners c:

$$C_{u\ell} \equiv \sum_c n_c k_{u\ell}(c) \quad , \tag{17.28}$$

and we assume each colliding species to have a thermal velocity distribution corresponding to temperature T. For the three-level system, the solution is tractable:

$$\frac{n_1}{n_0} = \frac{R_{01}R_{20} + R_{01}R_{21} + R_{21}R_{02}}{R_{10}R_{20} + R_{10}R_{21} + R_{12}R_{20}} \quad , \tag{17.29}$$

$$\frac{n_2}{n_0} = \frac{R_{02}R_{10} + R_{02}R_{12} + R_{12}R_{01}}{R_{10}R_{20} + R_{10}R_{21} + R_{12}R_{20}} \quad . \tag{17.30}$$

For systems with more than three energy levels, the steady state level populations are usually found using numerical methods to solve the system of linear equations.

17.6★ Example: Fine Structure Excitation of C I and O I

C I ($1s^22s^22p^2$) and O I ($1s^22s^22p^4$) are two important examples of atoms with triplet ($S = 1$) ground states (see Fig. 17.5). Using collisional rate coefficients from Appendix F, we can solve for excitation of these levels. Results for n_1/n_0 for C I are shown in Fig. 17.6, and n_1/n_0 for O I are shown in Fig. 17.7. For both cases, we have assumed a fractional ionization $n_e/n_{\rm H} = 10^{-3}$ characteristic of H I clouds or photodissociation regions.

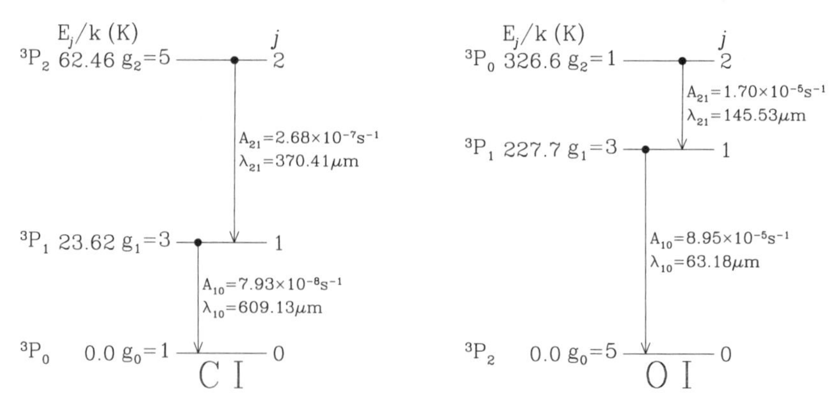

Figure 17.5 Fine-structure levels of C I and O I.

17.7★ Measurement of Density and Pressure Using C I

The fine-structure excited states of C I, with energies $E_1/k = 23.6\,{\rm K}$ and $E_2/k = 62.5\,{\rm K}$, can be collisionally populated even at low temperatures, and the level populations can be measured using C I's rich spectrum of ultraviolet absorption lines

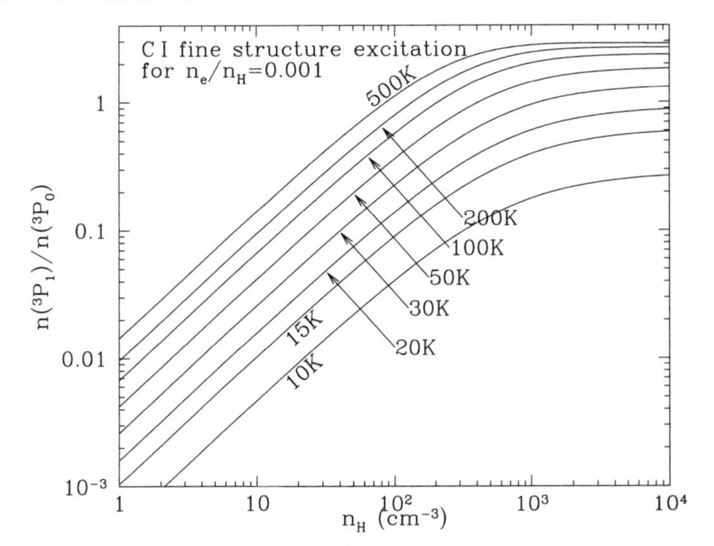

Figure 17.6 Excitation of C I $^3P_1$, source of 609.1 μm emission.

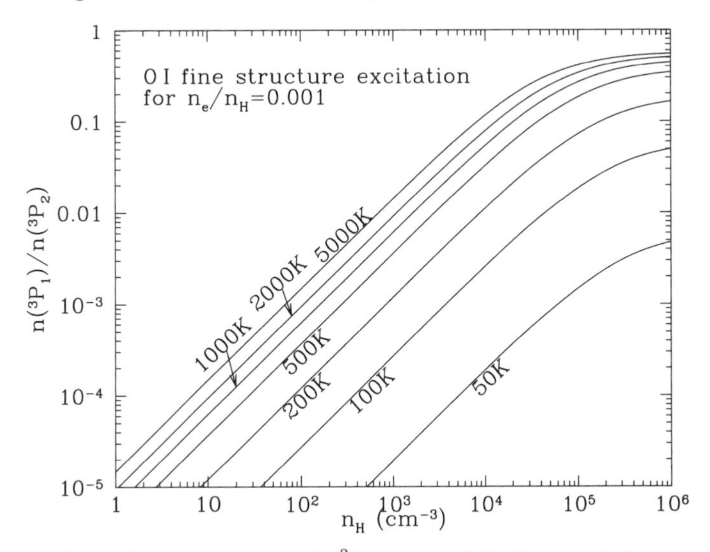

Figure 17.7 Excitation of O I $^3P_1$, source of 63.18 μm emission.

(see Appendix E). Because the critical density $n_{\rm crit,\it u}(\rm H)$ (see Table 17.1) is higher than the densities in typical diffuse clouds, the population of the C I fine structure levels can be used to constrain the density and temperature (Jenkins & Shaya 1979). A recent study by Jenkins & Tripp (2011) used high-quality spectra of UV absorption lines of C I on 89 sightlines to characterize the distribution of thermal pressures in diffuse clouds.

For a given gas composition (fractional ionization and H_2 fraction), temperature T, and density, the fractions $f_1 \equiv N(^3\rm P_1)/N(\rm C\,I)$ and $f_2 \equiv N(^3\rm P_2)/N(\rm C\,I)$ of the C I that are in the first and second excited states $^3P_1$ and $^3P_2$ of the ground electronic state ^3P can be calculated theoretically (e.g., Fig. 17.6 shows f_1/f_0). For a given T, varying the thermal pressure p will generate a track in the f_1–f_2 plane. Theoretical tracks for $T = 30\,\text{K}$, 80 K, and 300 K are shown in Figure 17.8, with each track generated by varying the pressure from $p/k = 10^2\,\text{cm}^{-3}\,\text{K}$ to $10^7\,\text{cm}^{-3}\,\text{K}$.

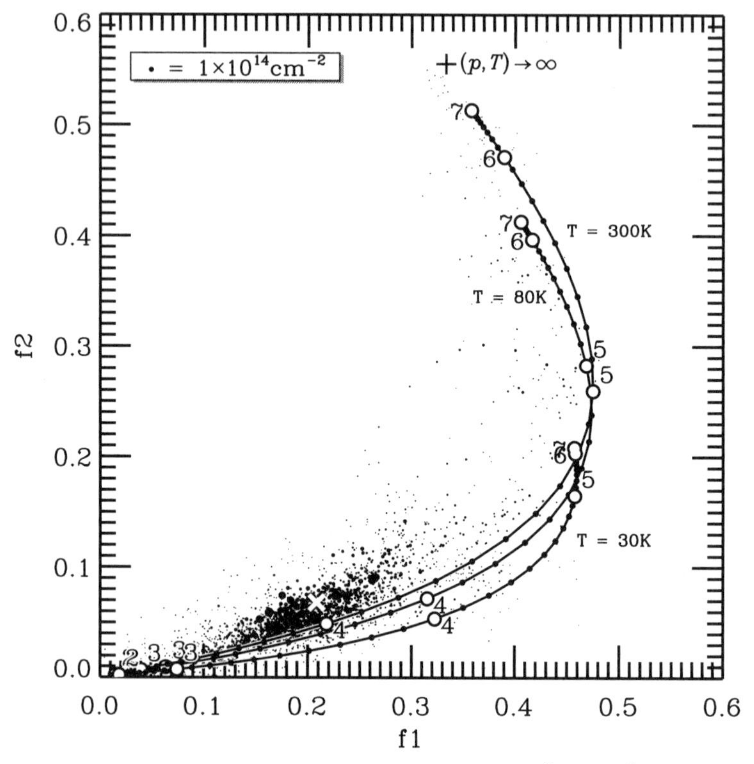

Figure 17.8 $f1$ and $f2$ are the fractions of C I that are in the $^3P_1$ and $^3P_2$ excited fine structure states. Solid lines are theoretical tracks for three different temperatures (30 K, 80 K, 300 K), as the pressure is varied from $p/k = 10^2\,\text{cm}^{-3}\,\text{K}$ to $10^7\,\text{cm}^{-3}\,\text{K}$, with numbers indicating the value of $\log_{10}[p/k(\text{cm}^{-3}\,\text{K})]$. Data points are measurements for different velocity components on 89 sightlines. The area of each dot is proportional to $N(\text{C I})$. The white $\times$ is the "center of mass" value $(f1, f2) = (0.21, 0.068)$. Taken from Jenkins & Tripp (2011).

Observed values of (f_1, f_2) are also plotted in Figure 17.8. Typically $f_1 \approx 0.20$ of the C I is found to be in the $^3P_1$ level, and $f_2 \approx 0.07$ is in the $^3P_2$ level, although on some sightlines f_2 can exceed 0.50. Note that the observed values of (f_1, f_2) usually fall somewhat *above and to the left* of the theoretical tracks. Jenkins & Tripp (2011) interpret this tendency as resulting from superposition of two components: a dominant component with moderate pressure p plus a small amount of high-pressure material.[4] UV pumping can also populate the excited fine-structure levels, and Jenkins & Tripp (2011) correct for this in their estimates for p.

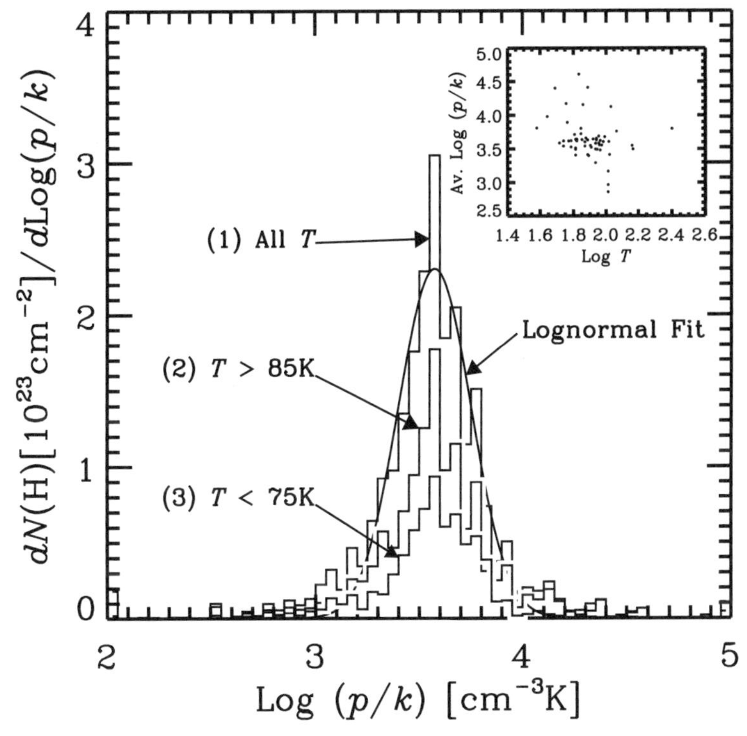

Figure 17.9 Distribution of thermal pressures measured using C I absorption lines (see text). Taken from Jenkins & Tripp (2011).

The distribution of pressures found by Jenkins & Tripp (2011) is shown in Fig. 17.9. The observed distribution can be approximated by a log-normal distribution with a peak at $p/k \approx 10^{3.575}\,\mathrm{cm}^{-3}\,\mathrm{K} \approx 3800\,\mathrm{cm}^{-3}\,\mathrm{K}$.

In many cases it was possible to determine the gas temperature using the H_2 $J = 1 - 0$ rotation temperature $T_{\rm rot}$. The inset in Fig. 17.9 shows the distribution of p and $T_{\rm rot}$; there appears to be no correlation between $T_{\rm rot}$ and p. For the "typical" $p/k \approx 3800\,\mathrm{cm}^{-3}\,\mathrm{K}$, the H_2 rotation temperatures range from $\sim 50\,\mathrm{K}$ to $\sim 250\,\mathrm{K}$.

Jenkins & Tripp (2011) conclude that interstellar clouds routinely contain a small amount of gas that is at the same bulk velocity but at a pressure that is much higher than the average pressure in the cloud – this is the only way that they can explain the tendency of the data points in Fig. 17.8 to fall above and to the left of the theoretical tracks. This is a very surprising result, as there is no obvious explanation for

[4]Jenkins & Tripp (2011) assume the high-pressure material to have $(f_1, f_2) = (0.38, 0.50)$.

why a small fraction of the cloud material should be overpressured without being at a significantly different velocity. It may be conjectured that the overpressured regions are the result of highly localized intermittent heating, perhaps due to turbulent dissipation, but the situation remains unclear. It is at least conceivable that the problem could be with the collisional rate coefficients – if, for example, the current theoretical rates have too small a value of C_{20}/C_{10}, then the true tracks at the low-pressure end of Fig. 17.8 would have a larger slope, perhaps passing through the cloud of points near (0.2,0.06) in Fig. 17.8, and removing the need to invoke an admixture of high pressure material.

Chapter Eighteen

Nebular Diagnostics

The populations of excited states of atoms and ions depend on the local density and temperature. Therefore, if we can determine the level populations from observations, we can use atoms and ions as probes of interstellar space. In this chapter, we focus on ions that allow us to probe the density and temperature of photoionized gas ("emission nebulae") in the temperature range $3000 \lesssim T \lesssim 3 \times 10^4$ K.

To be a useful probe, an atom or ion must be sufficiently abundant to observe, must have energy levels that are at suitable energies, and must have radiative transitions that allow us to probe these levels, either through emission lines or absorption lines.

There are two principal types of nebular diagnostics. The first (discussed in §18.1) uses ions with two excited levels that are both "energetically accessible" at the temperatures of interest, but with an energy difference between them that is comparable to kT, so that the populations of these levels are sensitive to the gas temperature. The level populations are normally observed by their line emission.

The second type of diagnostic (discussed in §§ 18.2 and 18.3) uses ions with two or more "energetically accessible" energy levels that are at nearly the same energy, so that the relative rates for populating these levels by collisions are nearly independent of temperature. The ratio of the level populations will have one value in the low-density limit, where every collisional excitation is followed by spontaneous radiative decay, and another value in the high-density limit, where the levels are populated in proportion to their degeneracies. If the relative level populations in these two limits differ (as, in general, they will), then the relative level populations (determined from observed emission line ratios) can be used to determine the density in the emitting region.

The temperature of ionized gas can also be measured using the "Balmer jump" in the emission spectrum of the recombining hydrogen (§18.4.1), or by observing emission lines that follow dielectronic recombination (§18.4.2).

Once the temperature of the gas has been determined, abundances of emitting species can be estimated using the strengths of collisionally excited emission lines relative to the emission from the ionized hydrogen (§18.5). When fine-structure emission lines are used, the inferred abundances are quite insensitive to uncertainties in the temperature determination.

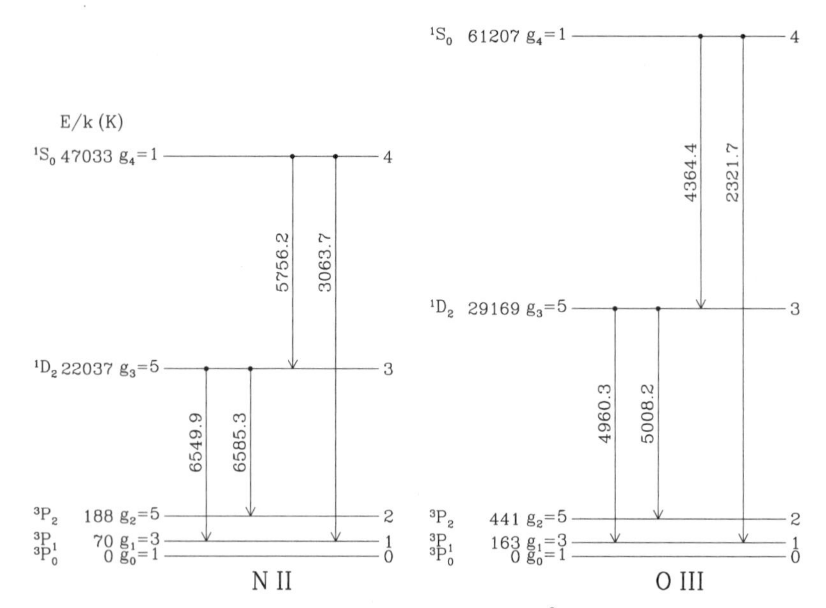

Figure 18.1 Energy levels for temperature diagnostic $2p^2$ ions N II and O III. Transitions are labeled by vacuum wavelength (Å).

18.1 Temperature Diagnostics: Collisionally Excited Optical/UV Lines

18.1.1 np^2 and np^4 Ions

Atoms or ions with six electrons have $2p^2$ as their lowest configuration: the ground state term is ^{3}P, and the first two excited terms are ^{1}D and ^{1}S. If the ^{1}S term is at a low enough energy ($E/k \lesssim 70,000\,$K), so that the rate for collisional excitation in gas with $T \approx 10^4\,$K is not prohibitively slow, and the abundance of the ion itself is not too low, then the ion can produce observable line emission from both the ^{1}D and ^{1}S levels. Because these levels are at very different energies, the relative strengths of the emission lines will be very sensitive to the temperature; the measured intensity ratio can be used to determine the temperature in the nebula.

Candidate $2p^2$ ions are C I, N II, O III, F IV, Ne V, and so on. C I is easily photoionized, and will have very low abundance in an H II region. The ionization potentials of F IV, Ne V, and so on exceed 54.4 eV, and we do not expect such high ionization stages to be abundant in H II regions excited by main-sequence stars with effective temperatures $kT_{\rm eff} \lesssim 5\,$eV. This leaves N II and O III as the only $2p^2$ ions that will be available in normal H II regions.[1]

Systems with eight electrons will have $2p^4$ configurations that will also have ^{1}D and ^{1}S as the first two excited terms. For O I, F II, and Ne III, the ^{1}S term is at $E/k < 70,000\,$K.

[1] F IV and Ne V may be present in planetary nebulae and active galactic nuclei, where the ionizing radiation is harder (with significant power above 54.4 eV).

Similar considerations for systems with 14 or 16 electrons in H II regions photoionized by main-sequence stars leave P II and S III as the only $3p^2$ ions, and Cl II, Ar III, and K IV as the only $3p^4$ ions, that can be used for temperature determination by comparison of emission lines from the ^{1}D and ^{1}S levels.

Figure 18.1 shows the first two electronic excited terms of the $2p^2$ ions N II and O III. It is easy to calculate what happens in the limit of very low densities, in which case essentially all of the N II and O III ions will be in the ground state $^3\mathrm{P}_0$. Let C_{03} and C_{04} be the rates for collisional excitation from the ground state to the $^1\mathrm{D}_2$ and $^1\mathrm{S}_0$ excited states. At low densities, every collisional excitation will be followed by radiative decays returning the ion to the ground state, with branching ratios that are determined by the Einstein coefficients $A_{u\ell}$. For example, after excitation of level 4, the probability of a $4 \to 3$ radiative transition is $A_{43}/(A_{41} + A_{43})$. Thus the power radiated per unit volume in the $4 \to 3$ and $3 \to 2$ transitions is

$$P(4 \to 3) = E_{43} \left[n_0 C_{04}\right] \frac{A_{43}}{A_{43} + A_{41}} \quad , \tag{18.1}$$

$$P(3 \to 2) = E_{32} \left[n_0 C_{03} + n_0 C_{04} \frac{A_{43}}{A_{43} + A_{41}}\right] \frac{A_{32}}{A_{32} + A_{31}} \quad , \tag{18.2}$$

where

$$C_{\ell u} = 8.629 \times 10^{-8} T_4^{-1/2} \frac{\Omega_{\ell u}}{g_\ell} \, \mathrm{e}^{-E_{u\ell}/kT} \, n_e \, \mathrm{cm}^3 \, \mathrm{s}^{-1} \quad . \tag{18.3}$$

Thus, in the limit $n_e \to 0$, the emissivity ratio

$$\frac{j(4 \to 3)}{j(3 \to 2)} = \frac{A_{43} E_{43}}{A_{32} E_{32}} \frac{(A_{32} + A_{31}) \Omega_{04} \, \mathrm{e}^{-E_{43}/kT}}{\left[(A_{43} + A_{41}) \Omega_{03} + A_{43} \Omega_{04} \, \mathrm{e}^{-E_{43}/kT}\right]} \quad . \tag{18.4}$$

Therefore, in the low-density limit, the emissivity ratio depends only on the atomic physics ($A_{u\ell}$, $E_{u\ell}$, $\Omega_{u\ell}$) and the gas temperature T. If the atomic physics is known, the observed emissivity ratio can be used to determine T. The low-density limit applies when the density is below the critical density for both $^1\mathrm{D}_2$ and $^1\mathrm{S}_0$. The critical densities for N II and O III are given in Table 18.1

The steady-state level populations have been calculated as a function of T for N II and O III, and the ratios of emission lines from the $^1\mathrm{S}_0$ and $^1\mathrm{D}_2$ levels are shown in Figure 18.2. We see that if $n_e \ll n_{\mathrm{crit}}$ for the $^1\mathrm{D}_2$ level ($n_{\mathrm{crit}} = 8 \times 10^4 \, \mathrm{cm}^{-3}$ for N II, and $6 \times 10^5 \, \mathrm{cm}^{-3}$ for O III), the line ratio is independent of the density, and depends only on the temperature. Fortunately, these values of n_{crit} are high enough so that these temperature diagnostics are useful in many ionized nebulae (e.g., the Orion Nebula, with $n_e \approx 3000 \, \mathrm{cm}^{-3}$ – see Chapter 28).

Note that the fine-structure excited states of the ground ^{3}P term have values of n_{crit} that are considerably lower than n_{crit} for $^1\mathrm{D}_2$ and $^1\mathrm{S}_0$ levels, because the fine-structure levels of the ground term have radiative lifetimes that are much longer than the excited terms.

Table 18.1 Critical Electron Density $n_{\rm crit}(e^-)$ ($\rm cm^{-3}$) for Selected np^2 and np^4 Ions

Configuration	Ion	$n_{\rm crit}(e)$ at $T=10^4$ K				
		$^3\rm P_0$	$^3\rm P_1$	$^3\rm P_2$	$^1\rm D_2$	$^1\rm S_0$
$1s^22s^22p^2$	C I	–	7.37×10^0	1.21×10^1		
	N II	–	1.67×10^2	2.96×10^2	7.68×10^4	1.23×10^7
	O III	–	1.74×10^3	3.79×10^3	6.40×10^5	2.78×10^7
	Ne V	–	3.19×10^5	3.48×10^5	1.44×10^8	9.58×10^8
$1s^22s^22p^4$	O I	3.11×10^3	2.87×10^4	–	1.62×10^6	4.04×10^8
	Ne III	3.02×10^4	2.76×10^6	–	9.47×10^6	1.37×10^8
	Mg V	4.36×10^6	4.75×10^7	–	1.07×10^9	8.07×10^9
$1s^22s^22p^63s^23p^2$	Si I	–	7.72×10^2	1.92×10^3		
	S III	–	4.22×10^3	1.31×10^4	7.33×10^5	1.52×10^7
	Ar V	–	1.09×10^7	1.16×10^7	3.65×10^8	2.49×10^8
$1s^22s^22p^23s^23p^4$	S I	1.04×10^5	1.55×10^5	–	4.12×10^7	1.38×10^9
	Ar III	2.49×10^5	2.67×10^6	–	1.26×10^7	4.54×10^8

Table 18.2 Critical Electron Density $n_{\rm crit}(e^-)$ ($\rm cm^{-3}$) for Selected np^3 Ions, for $T=10^4$ K

Configuration	Ion	$n_{\rm crit}(e)$ at $T=10^4$ K			
		$^2\rm D^o_{3/2}$	$^2\rm D^o_{5/2}$	$^2\rm P^o_{1/2}$	$^2\rm P^o_{3/2}$
$1s^22s^22p^3$	N I	2.18×10^4	1.19×10^4	7.11×10^7	3.15×10^7
	O II	4.49×10^3	3.31×10^3	5.30×10^6	1.03×10^7
	Ne IV	1.40×10^6	4.66×10^5	4.17×10^8	2.79×10^8
$1s^22s^22p^63s^23p^3$	S II	1.49×10^4	1.57×10^3	1.49×10^6	1.91×10^6
	Ar IV	1.35×10^6	1.55×10^4	1.06×10^7	1.81×10^7

However, when L-S coupling is a good approximation, quantum-mechanical calculations of the collision strengths $\Omega_{u\ell}$ for different fine-structure levels ℓ within a single term (e.g., $^3\rm P_{0,1,2}$) have $\Omega_{u\ell}/\Omega_{u\ell'} \approx g_\ell/g_{\ell'}$. When this is true, the collisional rate coefficients for excitation out of the different fine-structure levels will be nearly the same for the different fine-structure levels, so it does not matter whether these levels are populated thermally or whether the only level occupied is the ground state $^3\rm P_0$.

Ions with 14 electrons – S III is an example – have $1s^22s^22p^63s^23p^2$ configurations with the same term structure as $1s^22s^22p^2$, and therefore can be used for temperature determination in the same way. Figure 18.2 shows how the ratio [S III]6313.8/[S III]9533.7 serves as a temperature diagnostic.

A fundamental assumption is that the levels producing the observed lines are populated only by collisional excitation. The $^1\rm D_2$ level is at sufficiently high energy that as the temperature T is lowered below $\sim$5000 K, the rate of collisional excitation becomes very small. This means that the line becomes very weak and difficult to observe; it also means that if the next ionization state (N III, O IV, S IV) has an appreciable abundance, radiative recombination with electrons may make a significant contribution to population of the $^1\rm D_2$ level. As a result, the observed

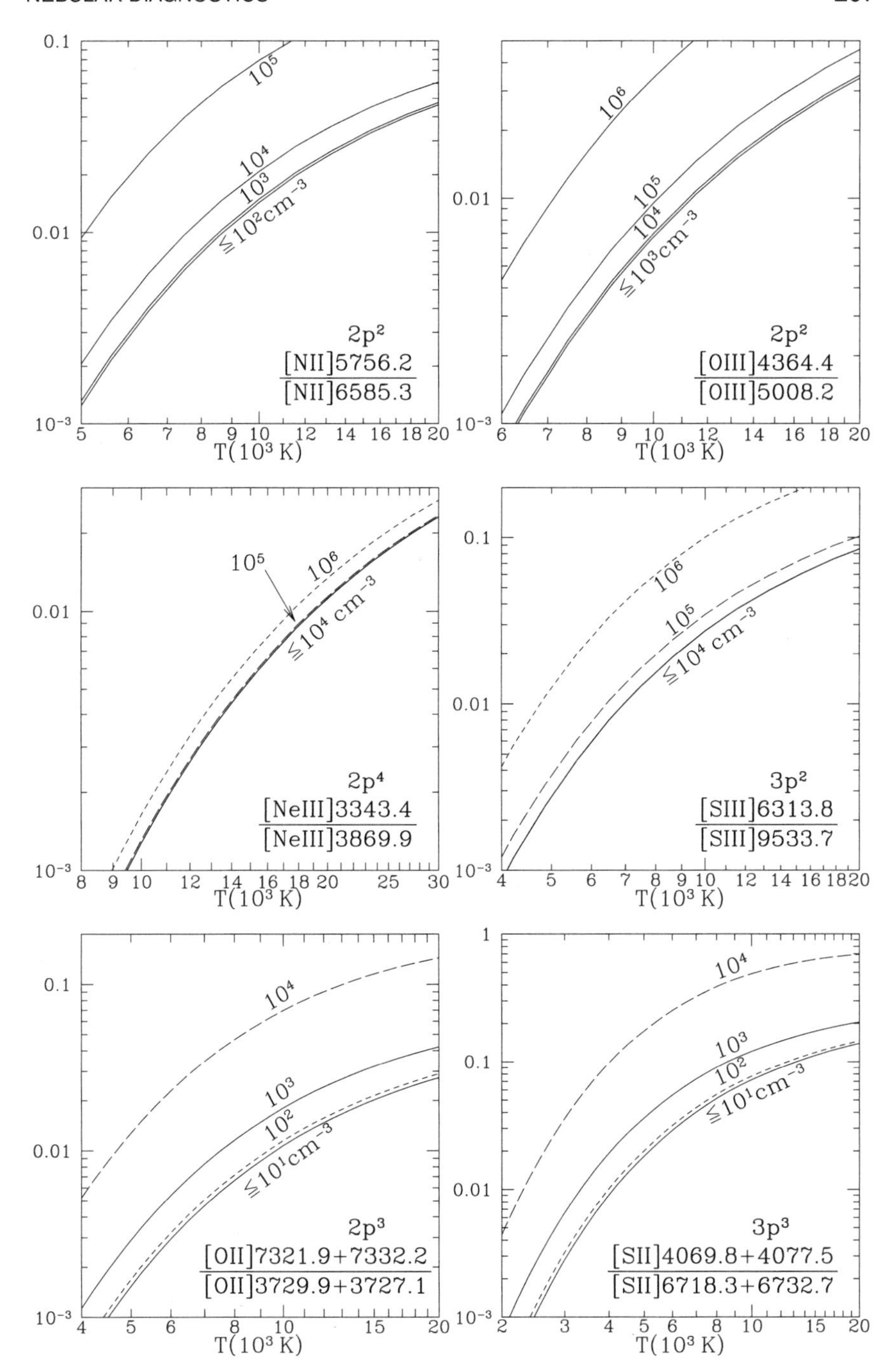

Figure 18.2 Line ratios that are useful as temperature diagnostics (see text). Curves are labeled by n_e (cm^{-3}). For each ion, the low density limit is shown, as well as results for higher densities, showing deviations from the low density behavior.

line ratios may not be suitable for temperature determination when the intensity $I(^1\mathrm{D}_2 \to {}^1\mathrm{S}_0) \lesssim 10^{-3} I(^1\mathrm{S}_0 \to {}^3\mathrm{P}_J)$.

18.1.2 np^3 Ions

Atoms or ions with seven electrons have $1s^2 2s^2 2p^3$ as their lowest configuration: the ground term is $^4\mathrm{S}^\circ_{3/2}$, and the first two excited terms are $^2\mathrm{D}^\circ_{3/2,5/2}$ and $^2\mathrm{P}^\circ_{1/2,3/2}$. Candidate ions are N I, O II, F III, Ne IV, and so on. N I will be photoionized in H II regions, leaving O II, F III, and Ne IV as the $2p^3$ ions suitable for observation in H II regions.

Atoms or ions with 15 electrons have $1s^2 2s^2 2p^6 3s^2 3p^3$ as their lowest configuration. Just as for $2p^3$, the ground term is $^4\mathrm{S}^\circ_{3/2}$, and the first two excited terms are $^2\mathrm{D}^\circ_{3/2,5/2}$ and $^2\mathrm{P}^\circ_{1/2,3/2}$. Candidate ions are P I, S II, Cl III, and Ar IV. P I is easily photoionized, leaving S II, Cl III, and Ar IV as the $3p^3$ ions that will be present in regions with $h\nu > 13.6$ eV radiation extending possibly up to 54.4 eV.

The ratio of the intensities of lines emitted by the $^2\mathrm{P}^\circ$ term to lines from the $^2\mathrm{D}^\circ$ term is temperature-sensitive. Figure 18.2 shows [O II](7322+7332)/(3730+3727) and [S II](6718+6733)/(4070+4077) as functions of temperature. For these two

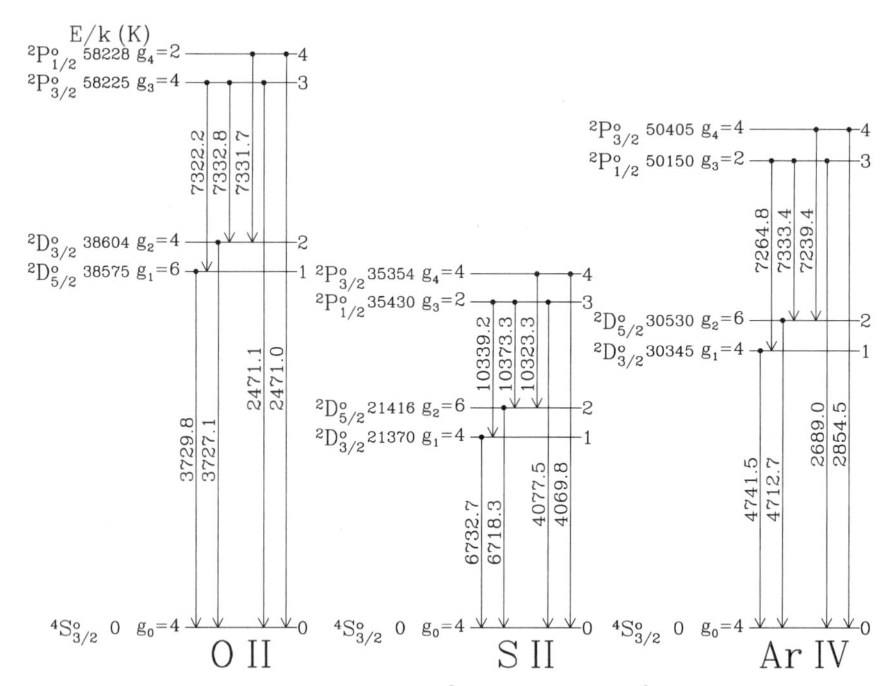

Figure 18.3 First five energy levels of the $2p^3$ ion O II, and the $3p^3$ ions S II and Ar IV. Transitions are labeled by wavelength in vacuo.

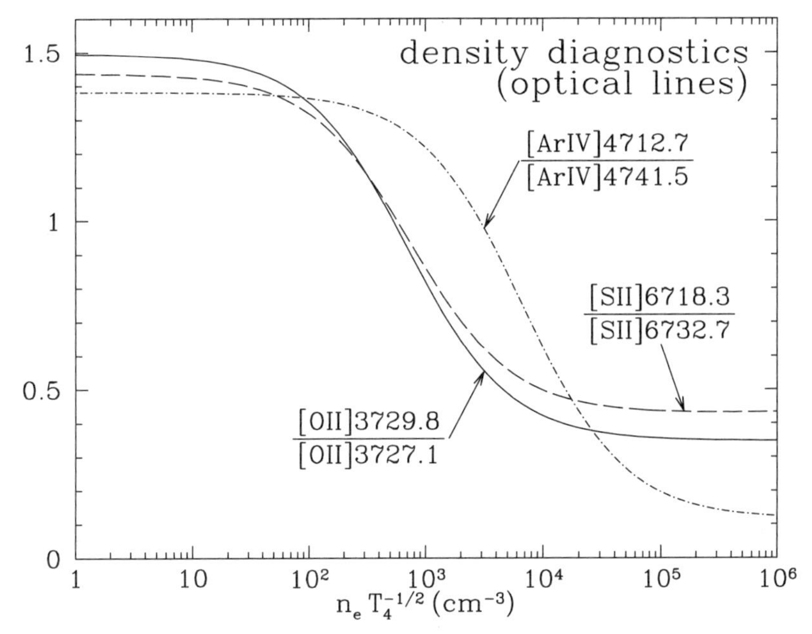

Figure 18.4 [O II], [S II], and [Ar IV] optical line intensity ratios useful for density determination. Wavelengths are in vacuo.

ions, the critical density for $^2\mathrm{D}^\circ$ is relatively low (see Table 18.2), so that these T-sensitive line ratios are also sensitive to n_e for $n_e \gtrsim 300\,\mathrm{cm}^{-3}$. Because of this sensitivity, the np^3 ions are only useful if n_e is known, or is known to be $\leq 10^2\,\mathrm{cm}^{-3}$.

18.2 Density Diagnostics: Collisionally Excited Optical/UV Lines

Ions with 7 or 15 electrons have $2s^22p^3$ and $3s^23p^3$ configurations, with energy-level structures that make them suitable for use as density diagnostics: the ground state is a singlet $^4\mathrm{S}^\circ_{3/2}$, and the first excited term is a doublet $^2\mathrm{D}^\circ_{3/2,5/2}$ (see Figure 18.3). At low densities, every collisional excitation of either the $^2\mathrm{D}^\circ_{3/2}$ or $^2\mathrm{D}^\circ_{5/2}$ level will be followed by radiative decay. Therefore, in the low-density limit, the power radiated in each of the two decay lines from the $^2\mathrm{D}^\circ$ term (the 3729.8 and 3727.1Å lines in the case of O II) is simply proportional to the collision rates. Because the fine-structure splitting is small ($E_{21} \ll E_{10}$, $E_{21} \ll kT$), the emissivity ratio

$$\frac{j(2\rightarrow 0)}{j(1\rightarrow 0)} = \frac{\Omega_{20}}{\Omega_{10}}\frac{E_{20}}{E_{10}}\mathrm{e}^{-E_{21}/kT} \approx \frac{\Omega_{20}}{\Omega_{10}} \quad . \tag{18.5}$$

At high densities, however, the levels become thermalized, and the emissivity ratio becomes

$$\frac{j(2 \to 0)}{j(1 \to 0)} = \frac{g_2}{g_1} \mathrm{e}^{-E_{21}/kT} \frac{E_{20} A_{20}}{E_{10} A_{10}} \approx \frac{g_2 A_{20}}{g_1 A_{10}} \quad . \tag{18.6}$$

Because the low-density and high-density limits (18.5 and 18.6) will in general differ, the observed intensity ratio provides information concerning the density: we can either determine the density, or establish an upper or lower limit. Figure 18.4 shows the density dependence of this emissivity ratio for O II, S II, and Ar IV.

The $^2\mathrm{D}^\mathrm{o}$ levels of O II have critical densities of $\sim 3300\,\mathrm{cm}^{-3}$ and $4500\,\mathrm{cm}^{-3}$, and we see from Figure 18.4 that the [O II] line ratio is sensitive to variations in the density over the range 10^2 to $4000\,\mathrm{cm}^{-3}$. In the case of S II, the $^2\mathrm{D}^\mathrm{o}_{5/2}$ level has $n_{\mathrm{crit}} = 1600\,\mathrm{cm}^{-3}$, so the line ratios begin to vary with density above $\sim 10^2\,\mathrm{cm}^{-3}$ – similar to O II – but the $^2\mathrm{D}^\mathrm{o}_{3/2}$ level has $n_{\mathrm{crit}} = 1.5 \times 10^4\,\mathrm{cm}^{-3}$, so the S II line ratio continues to be sensitive up to $\sim 10^4\,\mathrm{cm}^{-3}$. Because of larger radiative transition rates, the Ar IV line ratio is density-sensitive over 10^3 to $10^5\,\mathrm{cm}^{-3}$.

18.3 Density Diagnostics: Fine-Structure Emission Lines

Ions with triplet ground states – in particular, the $^3\mathrm{P}_{0,1,2}$ terms for np^2 and np^4 configurations – allow density determination from the ratios of mid-infrared and far-infrared fine-structure lines. Examples are the $2p^2$ ions N II,O III, and Ne V; the $2p^4$ ion Ne III; and the $3p^2$ ion S III. If these ions are approximated as three-level systems (i.e., neglecting population of the fine-structure levels by radiative decay from higher terms), then in the low-density limit, the emissivity intensity ratio is simply

$$\frac{j(2 \to 1)}{j(1 \to 0)} \approx \frac{\Omega_{20}\, \mathrm{e}^{-E_{21}/kT}}{\Omega_{10} + \Omega_{20}\, \mathrm{e}^{-E21/kT}} \frac{E_{21}}{E_{10}} \quad , \tag{18.7}$$

where we have assumed that $A_{20} \ll A_{21}$, which is the case because $\Delta J = 2$ radiative transitions are strongly suppressed. In the high-density limit, the levels are thermally populated, and

$$\frac{j(2 \to 1)}{j(1 \to 0)} = \frac{g_2 A_{21} E_{21}}{g_1 A_{10} E_{10}} \mathrm{e}^{-E_{21}/kT} \approx \frac{g_2 A_{21} E_{21}}{g_1 A_{10} E_{10}} \quad . \tag{18.8}$$

Figure 18.5 shows the fine-structure line ratios calculated for N II, O III, S III, Ne III, Ne V, Ar III, and Ar V (including transitions to and from the $^1\mathrm{D}$ and $^1\mathrm{S}$ terms). Because of their long wavelengths, these lines are relatively unaffected by dust extinction, and therefore n_e can be deduced from the observed line ratio, independent of uncertainties in the reddening.

Collisional deexcitation of these ions is dominated by electron–ion collisions, with rates that scale as $n_e \Omega_{ij}/\sqrt{T_e}$. If the collision strengths Ω_{ij} were independent of T, and the electron temperature is sufficiently high that $E_{21}/kT_e \ll 1$, each

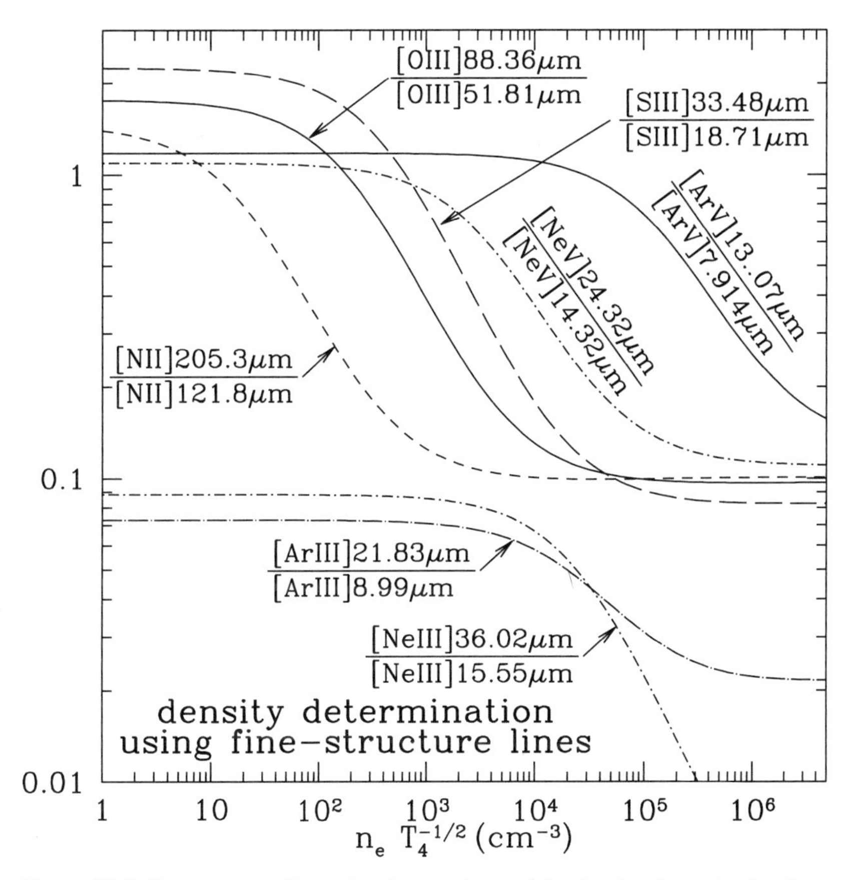

Figure 18.5 Fine-structure line ratios that can be used for density determination (see text).

line ratio would depend only on $n_e/\sqrt{T_e}$, and the measured line ratio allows one to infer $n_e/\sqrt{T_e}$. However, the collision strengths are not independent of T, and E_{21}/kT may also not be negligible (especially for ions with large fine-structure splitting, e.g., Ar V with $E_{21}/k = 1818\,\mathrm{K}$). Therefore the line ratio will depend on both $n_e/\sqrt{T}$ and T_e. In Figure 18.6 we show the [S III] line ratios as a function of $n_e/\sqrt{T_e}$ for several different values of the electron temperature T_e. Tommasin et al. (2008) measured [S III] fine-structure line ratios for a number of AGN. In a number of cases they obtained [S III]33.48/[S III]18.7 between 2.2 and 3.5, above the low-density limit for $T = 10^4\,\mathrm{K}$. Figure 18.6 shows that values as large as 3 can be produced if the emission is coming from regions with $T \lesssim 5000\,\mathrm{K}$. Because Auger emission allows a single x ray photon to ionize S I→S III (see Figure 13.3), it is reasonable to suppose that there may be "warm" X-ray-heated regions containing S III.

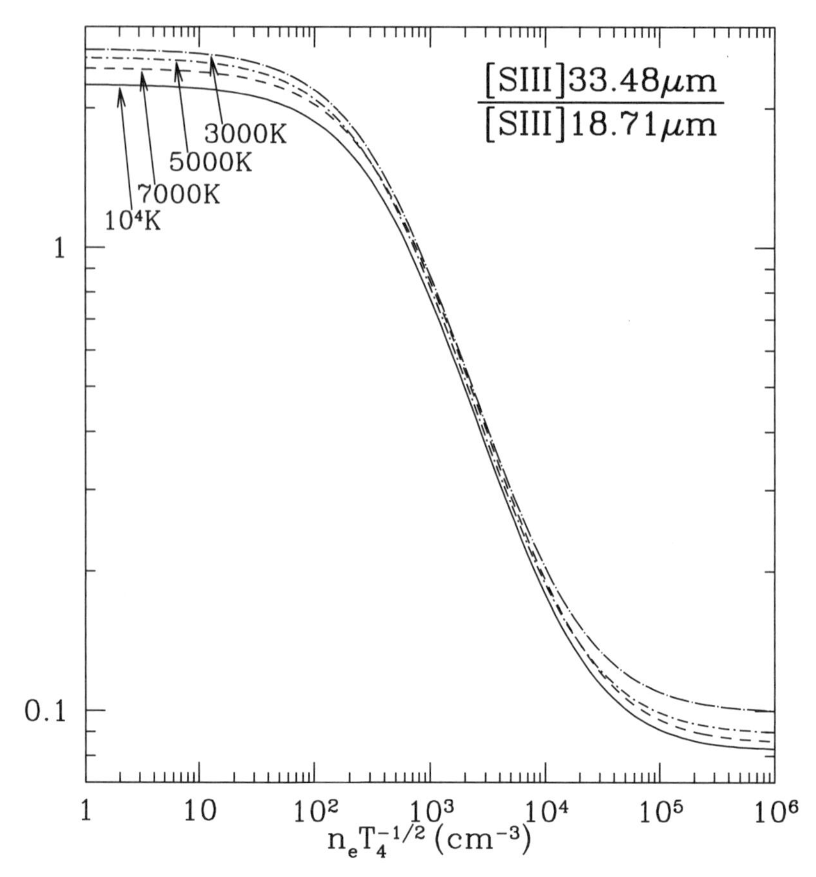

Figure 18.6 [S III]33.48 μm/[S III]18.71 μm as a function of $n_e/\sqrt{T_e}$, for four values of T_e. As the temperature is lowered, the line ratio rises both at low density and at high density.

18.4★ Other Diagnostic Methods

18.4.1★ Temperature from Recombination Continuum: The Balmer Jump

It is possible to determine the temperature from the strengths of the discontinuities in the recombination continuum relative to the strengths of recombination lines. The most commonly used discontinuity is the "Balmer jump" at $\lambda = 3647.0$ Å: $\mathrm{BJ} \equiv I_\lambda(\lambda_{\mathrm{BJ,blue}}) - I_\lambda(\lambda_{\mathrm{BJ,red}})$, where $\lambda_{\mathrm{BJ,blue}}$ is chosen to be just blueward of the jump, and $\lambda_{\mathrm{BJ,red}}$ is chosen to be slightly redward of the jump, and to be located between H recombination lines. For example, $\lambda_{\mathrm{BJ,red}} = 3682.1$ Å would fall midway between the H20 and H21 lines.

The "jump" discontinuity is produced by recombining electrons with zero kinetic energy, and is therefore proportional to the electron energy distribution at $E = 0$, and therefore BJ $\propto EM \times T^{-3/2}$, where EM is the emission measure.

The strength of a recombination line such as the H11 line ($n = 11 \rightarrow 2$ at $\lambda = 3769.7$ Å) is proportional to rates of radiative recombination to levels $n \geq 11$. The effective recombination rate coefficient for emitting H11 will vary approximately as $T^{-0.8}$ near the temperatures of interest, and the intensity of the recombination line $I(\mathrm{H11}) \propto EM \times T^{-0.8}$. Thus we expect $\mathrm{BJ}/I(\mathrm{H11}) \propto T^{-0.7}$: the dependence on T is strong enough that this is a useful diagnostic.

Allowance must be made for the contribution from helium: doubly ionized helium recombining to level $n = 4$ contributes to the observed Balmer jump, and He II $n = 22 \rightarrow 4$ coincides with H11.

This method has been used to determine the electron temperature in H II regions and planetary nebulae. In a sample of 23 planetary nebulae, temperatures $T_{e,\mathrm{BJ}}$ derived from the Balmer jump are generally lower than the temperature $T_{e,[\mathrm{O\,III}]}$ determined from collisional excitation of [O III] optical lines, with $T_{e,\mathrm{BJ}}/T_{e,\mathrm{CE}} \approx 0.75 \pm 0.25$ (Wesson et al. 2005). The reason for the discrepancy is unclear; Wesson et al. (2005) suggest that cool, dense, metal-rich knots may be present in planetary nebulae.

18.4.2★ Temperatures from Dielectronic Recombination

For some ions, it is possible to observe both collisionally excited lines and lines emitted following dielectronic recombination. For example, electrons colliding with C IV can produce collisionally excited levels of C IV, but can also produce excited levels of C III by dielectronic recombination. Because the rate coefficients for collisional excitation and for dielectronic recombination will have different temperature dependences, the ratio of dielectronic lines to collisionally excited lines will be temperature-sensitive, and therefore useful as a temperature diagnostic. Examples of useful line ratios are C III 2297/C IV 1549, O II 4705/O III] 1665, and C II 4267/C III] 1909 (see Fig. 5.4 of Osterbrock & Ferland 2006).

18.4.3★ Densities from the Balmer Decrement

The "Balmer decrement" refers to the sequence of line ratios $I(\mathrm{H}\alpha)/I(\mathrm{H}\beta)$, 1, $I(\mathrm{H}\gamma)/I(\mathrm{H}\beta)$, $I(\mathrm{H}\delta)/I(\mathrm{H}\beta)$, and so on. These line ratios are relatively insensitive to the electron temperature, and at low density are independent of density. Therefore, comparison of the observed line ratios to theoretical line ratios is usually used to determine the degree of reddening by dust. However, at high densities the line ratios are affected by collisional effects, with systematic enhancement of the high-n levels relative to Hα and Hβ, and these line ratios can therefore be used to constrain the electron density when $n_e > 10^4\,\mathrm{cm}^{-3}$.

18.5 Abundance Determination from Collisionally Excited Lines

The abundance of He relative to H is determined from comparison of the strengths of radiative recombination lines of H and He in regions ionized by stars that are sufficiently hot ($T_{\rm eff} \gtrsim 3.9 \times 10^4$ K – see Table 15.1) so that the He is ionized throughout the H II zone.

The abundances relative to H of elements heavier than He can be inferred by comparing the strengths of emission lines excited by collisions with electrons to emission resulting from recombination of electrons with H^+. We will consider oxygen as an example; similar considerations apply to other heavy elements.

The abundance of O^{++} relative to H can be obtained from the ratio of [O III]5008 to Hβ. In the low-density limit $n_e < 10^4\,{\rm cm}^{-3}$,

$$\frac{I([{\rm O\,III}]5008)}{I({\rm H}\beta)} = \frac{n_e n({\rm O\,III}) k_{03} E_{32} A_{32}/(A_{31}+A_{32})}{n_e n({\rm H}^+) \alpha_{{\rm eff,H}\beta} E_{{\rm H}\beta}} \quad , \tag{18.9}$$

where

$$k_{03} = 8.629 \times 10^{-8} T_4^{-1/2} \frac{\Omega_{03}}{g_0} \, {\rm e}^{-E_{30}/kT} \, {\rm cm}^3\,{\rm s}^{-1} \quad , \tag{18.10}$$

and $E_{30}/k = 29170$ K. Since $\alpha_{{\rm eff,H}\beta} \propto T_4^{-0.87}$ for $T_4 \approx 1$ (see Table 14.2), and $\Omega_{03} \approx T_4^{0.12}$ (see Appendix F), we have

$$\frac{n({\rm O\,III})}{n({\rm H}^+)} = C \, \frac{I([{\rm O\,III}]5008)}{I({\rm H}\beta)} \, T_4^{-0.49} \, {\rm e}^{2.917/T_4} \quad , \tag{18.11}$$

where C is a known constant. Therefore, if the temperature T is known, the abundance $n({\rm O\,III})/n({\rm H}^+)$ can be obtained from the measured line ratio. Unfortunately, the derived abundance is sensitive to the temperature: if $T_4 \approx 0.8$, the derived abundance in Eq. (18.11) varies as $\sim T_4^{3.3}$, so that a 10% uncertainty in T_4 translates into a 33% uncertainty in the derived abundance. This sensitivity to the uncertain temperature plagues abundance determinations based on collisionally excited optical lines.

Fine-structure lines can also be used to determine abundances of ions with fine-structure splitting of the ground state. [O III] is a good example: the $^3{\rm P}_1 \rightarrow {}^3{\rm P}_0$ and $^3{\rm P}_2 \rightarrow {}^3{\rm P}_1$ transitions at 88.36 μm and 51.81 μm can be used to determine the O III abundance. In the low-density limit $n_e \lesssim 10^2\,{\rm cm}^{-3}$ (see Fig. 18.5), the emissivity $j \propto n_e n({\rm O\,III}) T^{-0.5}$, and is therefore much less sensitive to uncertainties in the temperature; in fact, the $T^{-0.5}$ temperature dependence is similar to that of the H recombination lines. Unfortunately, these far-infrared lines cannot be observed from the ground. However, they have the advantage of being nearly unaffected by interstellar extinction; the fine-structure line emission can be compared to free–free emission observed at radio wavelengths, enabling abundance determinations that are not compromised by uncertain reddening corrections.

To determine the total abundance, one must sum over all the important ion stages. In an H II region ionized by a B0 star, most of the oxygen will be O II, because there will be few photons above the O II $\rightarrow$ O III ionization threshold of 35.1 eV. However, in H II regions ionized by hotter stars, or in planetary nebulae, much of the oxygen may be present as O III.

18.6★ Abundances from Optical Recombination Lines

Abundance determination from collisionally excited optical or ultraviolet lines requires knowledge of the temperature T, and we have seen that this is often uncertain. Optical recombination lines of an ion X^{+r} are the result of radiative recombination of X^{+r+1}, which will depend on T (and n_e) in a way very similar to radiative recombination of H^+, thus allowing straightforward comparison of the X^{+r+1}/H^+ ratio.

Care must be taken to select lines that will not be excited by optical pumping in the nebula; this can be done, for example, by using lines from levels with different total spin than the ground electronic state of X^{+r}. Another caution is that if the ground term of the recombining species X^{+r+1} has fine structure, the recombination spectrum will depend on the relative populations of the different fine-structure levels, and therefore on n_e. Alert to these concerns, observers have used what appear to be suitable recombination lines for abundance determinations. The results, however, are surprising. In H II regions, the abundances inferred from optical recombination lines (ORLs) are moderately higher than had been estimated from analysis of the collisionally excited lines (CELs). In planetary nebulae, however, the discrepancy between CEL- and ORL-based abundances estimates can be very large. The CEL-based estimates tend to be close to solar abundances, whereas the ORL-based abundances are often much higher, sometimes by factors as large as 10. The reason for this is not yet understood. Wesson et al. (2005) suggest that planetary nebulae may contain photoionized regions with elevated abundances of heavy elements; the enhanced cooling would keep them cool so that they do not contribute to the collisionally excited optical line emission, but they would be effective emitters of recombination radiation.

18.7★ Ionization/Excitation Diagnostics: The BPT Diagram

The optical line emission from star-forming galaxies is usually dominated by emission lines from H II regions. Some galaxies, however, have strong continuum and line emission from an active galactic nucleus (AGN). The line emission is thought to come from gas that is heated and ionized by x rays from the AGN. Even in a star-forming galaxy with line emission from H II regions, the emission lines from the AGN may dominate the overall spectrum of the galaxy.

The AGN spectrum normally includes strong emission lines from high-ionization species like C IV and Ne V, which are presumed to be ionized by x rays from the AGN; these are called **Seyfert galaxies**, after Seyfert (1943), who discovered that some galaxies had extremely luminous, point-like nuclei, with emission line widths in some cases exceeding $4000\,\mathrm{km\,s^{-1}}$.

In other cases the nucleus has strong emission lines but primarily from low-ionization species – these are termed **LINERS**, for "Low Ionization Nuclear Emission Region."

Baldwin, Phillips & Terlevich (1981) pointed out that one could distinguish star-forming galaxies from galaxies with spectra dominated by active galactic nuclei by plotting the ratio of [O III]λ5008/Hβ vs. [N II]λ6585/Hα – this is now referred to as the **BPT diagram**. Hα, Hβ, [N II]λ6585, and [O III]λ5008 have the advantage of being among the strongest optical emission lines from H II regions. Furthermore, the line ratios employ pairs of lines with similar wavelengths (5008 and 4863 Å; 6585 and 6565 Å) so that the line *ratios* are nearly unaffected by whatever dust extinction may be present.

From our understanding of H II regions (see Chapter 15), we can predict where H II regions should fall in the BPT diagram. As seen in Table 15.2, in H II regions where He is neutral (no photons above 24.6eV), N and O will be essentially 100% singly ionized throughout the zone where H is ionized. For O stars that are hot enough to have an appreciable zone where He is ionized, the stellar spectrum will extend to $54.4\,\mathrm{eV}$ (the ionization threshold for He II) and the N and O in this zone can be doubly ionized. Because N and O have similar second ionization potentials (29.6 and 35.1 eV, respectively), to a good approximation H II regions will have $\mathrm{N^+/N \approx O^+/O}$, and $\mathrm{N^{++}/N \approx O^{++}/O}$. Essentially all of the gas-phase O and N in the H II region will be either singly or doubly ionized.

If we assume the N abundance to be solar, and the O abundance to be 80% solar (the other 20% is presumed to be in silicate grains), then for an assumed electron temperature T we can produce a theoretical curve of [O III]5008/Hβ versus [N II]6585/Hα by varying the fraction $\xi \equiv \mathrm{N^{++}/(N^+ + N^{++}) = O^{++}/(O^+ + O^{++})}$ of the N and O that is doubly ionized. In the low-density limit,

$$\frac{[\mathrm{O\,III}]\lambda5008}{\mathrm{H}\beta} \approx 214\,\xi\, T_4^{0.494+0.089\ln T_4}\,\mathrm{e}^{-2.917/T_4}\left(\frac{n_{\mathrm{O}}/n_{\mathrm{H}}}{0.8\times5.37\times10^{-4}}\right) \qquad (18.12)$$

$$\frac{[\mathrm{N\,II}]\lambda6585}{\mathrm{H}\alpha} \approx 12.4(1-\xi)T_4^{0.495+0.040\ln T_4}\,\mathrm{e}^{-2.204/T_4}\left(\frac{n_{\mathrm{N}}/n_{\mathrm{H}}}{7.41\times10^{-5}}\right), \qquad (18.13)$$

where we have used Eqs. (14.8,14.9) for Hα and Hβ, electron collision strengths for N II and O III from Appendix F, and branching ratios $A_{32}/(A_{31}+A_{32}) = 0.745$ and 0.748 for O III and N II, respectively.

Figure 18.7a shows theoretical tracks calculated by varying ξ from 0 to 1 while holding T fixed. Curves are shown for $T = 7000$, 8000, and 9000 K. The curves in Fig. 18.7a were calculated for non-LTE excitation of the lowest 5 levels ($^3\mathrm{P}_{0,1,2}$, $^1\mathrm{D}_2$, $^1\mathrm{S}_0$) at $n_{\mathrm{H}} = 10^2\,\mathrm{cm^{-3}}$, but it can be verified that the tracks are close to Eqs. (18.12 and 18.13). The critical densities for [N II]λ6585 and [O III]λ5008 are $8\times10^4\,\mathrm{cm^{-3}}$ and $6\times10^5\,\mathrm{cm^{-3}}$ (see Table 18.1), hence the theoretical tracks should be valid for $n_{\mathrm{H}} \lesssim 10^4\,\mathrm{cm^{-3}}$.

H II regions with near-solar metallicity and densities $10^2 - 10^4\,\mathrm{cm^{-3}}$ are expected to have $T \approx 7000 - 8000\,\mathrm{K}$ (see Fig. 27.3), and this is confirmed by observations. Hence we expect star-forming galaxies with near-solar metallicity to fall between the 7000K and 8000K curves in Fig. 18.7.

Fig. 18.7b shows where real galaxies[2] fall in the BPT diagram. Many of the spectra are clustered in a well-defined locus lying within the zone

$$\log_{10}([\mathrm{O\,III}]5008/\mathrm{H}\beta) < 1.10 - \frac{0.60}{0.01 - \log_{10}([\mathrm{N\,II}]6585/\mathrm{H}\beta)} , \quad (18.14)$$

with line ratios characteristic of ordinary H II regions. Fully 76.6% (93820/122514) of the galaxies fall on the star-forming galaxy side of (18.14). Because for fixed T, the theoretical curves depend linearly on N/H and O/H, at first sight the compactness of the distribution seems remarkable, suggesting a narrow range in metallicities. However, there is a thermostatic effect operating: lowering the metal abundance results in increased T, because the metal lines have to radiate away the heat deposited by photoionization. The combined power in *all* the metal lines is essentially proportional to the H ionization rate, thus the combined power in the "metal lines" is proportional to the recombination lines Hα and Hβ. Thus if [N II]λ6585 and [O III]λ5008 carry an approximately constant fraction of the total metal line power, then scaling elemental abundances up or down will have essentially *no* effect on [N II]λ6585/Hα and [O III]λ5008/Hβ. Now the tightness of the distribution of the "star-forming" points in Fig. 18.7b seems less surprising.

The broken curve labeled K03 in Fig. 18.7b shows the empirical boundary to the star-forming region obtained by Kauffmann et al. (2003). Eq. (18.14) seems to provide a somewhat better boundary to the locus of star-forming galaxies.

Some galaxies have [O III]λ5008/Hβ and [N II]λ6584/Hα falling to the right and above Eq. (18.14). The "Seyfert" region is defined by [N II]λ6585/H$\alpha > 0.6$ and [O III]λ5008/H$\beta > 3$. Sources in the Seyfert region have both $T > 10^4$ K (required to have ([N II]λ6585/H$\alpha > 0.6$) *and* relatively high ionization (enhanced [O III]λ5008/Hβ), consistent with heating and ionization by x rays.

The "LINER" region is defined by [N II]λ6585/H$\alpha > 0.6$, [O III]λ5008/H$\beta <$ 3. LINERs have $T > 10^4$ K (required to have ([N II]λ6585/H$\alpha > 0.6$) but relatively low ionization ($[\mathrm{N\,II}]\lambda6585/\mathrm{H}\alpha > 0.6$ but $[\mathrm{O\,III}]5008/\mathrm{H}\beta < 3$). LINERs appear to be systems where an AGN is emitting hard X-rays that only partially ionize nearby gas (Veilleux & Osterbrock 1987). The hard X-rays, photoelectrons and secondary electrons partially ionize the gas, and the photoelectrons and electrons heat the partially ionized gas to $\sim 10^4$ K. While there is some O III, most of the oxygen is O I and O II, resulting in relatively weak [O III]λ5008. This interpretation is supported by the fact that LINERS have significant [O I]λ6302 emission, with [O I]λ6302/H$\alpha > 0.16$, whereas [O I]λ6302 emission is very weak in H II region spectra because there is very little neutral oxygen present. LINERS are nicely separated from H II galaxies in a diagram plotting [O III]λ5008/Hβ versus [O I]λ6302/Hα (Ho 2008).

[2]Data are from SDSS DR7 (Abazajian et al. 2009). Line fluxes were obtained from the MPA-JHU DR7 release http://www.mpa-garching.mpg.de/SDSS/DR7/raw_data.html. As recommended, formal line uncertainties were scaled by factors 1.882, 1.566, 2.473, and 2.039 for Hβ, [O III]5008, Hα, and [N II]6585. Of the 927552 galaxy spectra in DR7, 122514 have S/N> 5 for all 4 lines.

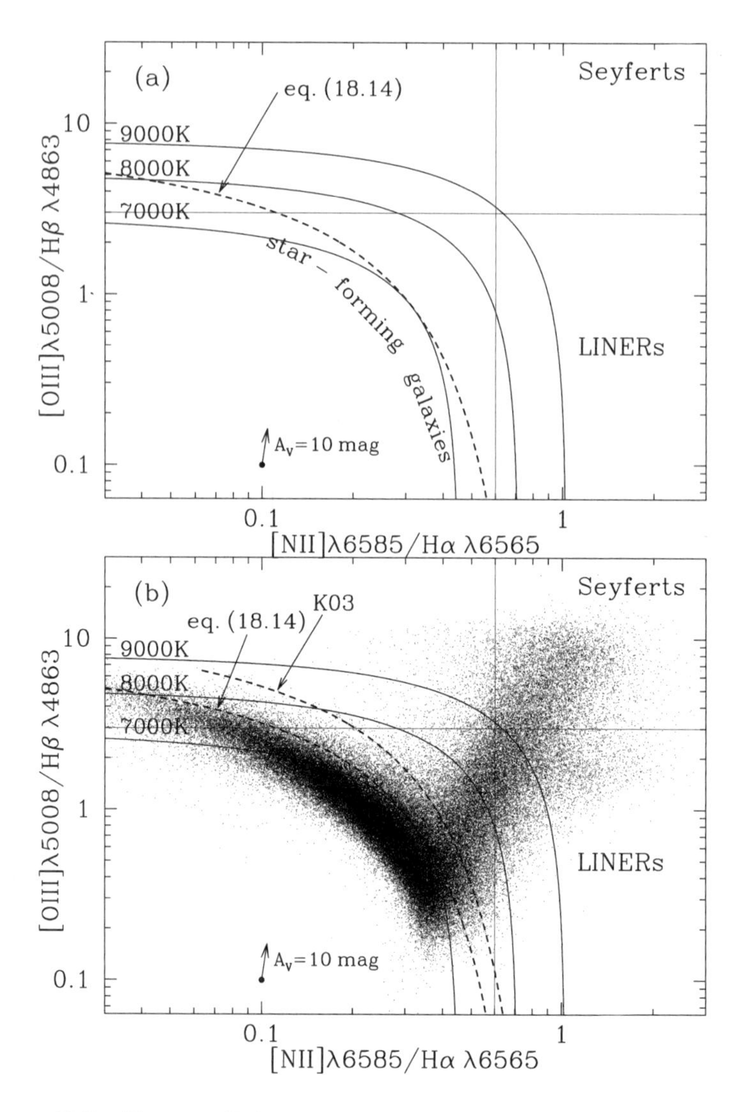

Figure 18.7 The so-called "BPT" diagram (Baldwin et al. 1981) showing [O III]5008/Hβ vs [N II]6585/Hα. The "reddening vector" shows the displacement on the plot due to reddening by Milky Way dust with $A_V = 10$ mag – it is clear that the BPT diagram is almost completely unaffected by reddening. (a) Solid curves show emission ratios calculated for gas with solar abundances, for three gas temperatures. Along each curve the oxygen and nitrogen vary from singly ionized at the bottom, to doubly ionized at the upper left. (b) Line ratios for 122514 galaxies in SDSS DR7 with S/N> 5. The curve labeled K03 is the boundary proposed by Kauffmann et al. (2003) to separate star-forming galaxies from AGN. Eq. (18.14) shows an improved boundary. 70.0% of the galaxies fall in the star-forming region defined by Eq. 18.14. 12.8% fall in the AGN region defined by [N II]$6585/\mathrm{H}\alpha > 0.6$.

Chapter Nineteen

Radiative Trapping

In many situations of astrophysical interest, there is sufficient gas present so that, for some species X, a photon emitted in a transition $X_u \rightarrow X_\ell$ will have a high probability of being absorbed by another X_ℓ somewhere nearby, and, therefore, a low probability of escaping from the emitting region. This phenomenon – referred to as **radiative trapping** – has two effects: (1) it reduces the emission in the $X_u \rightarrow X_\ell$ photons emerging from the region, and (2) it acts to increase the level of excitation of species X (relative to what it would be were the emitted photons to escape freely).

An exact treatment of the effects of radiative trapping is a complex problem of coupled radiative transfer and excitation – it is nonlocal, because photons emitted from one point in the cloud affect the level populations at other points. However, radiative trapping occurs frequently, and it is important to have an approximate treatment of its effects. The **escape probability approximation** is a simple way to take into account the effects of radiative trapping.

19.1 Escape Probability Approximation

Suppose that at some point $\mathbf{r}$ in the cloud, the optical depth $\tau_\nu(\hat{\mathbf{n}}, \mathbf{r})$ in direction $\hat{\mathbf{n}}$ and at frequency ν is known. We can define the "escape probability" β_ν for photons with frequency ν emitted from location $\mathbf{r}$:

$$\bar{\beta}_\nu(\mathbf{r}) \equiv \int \frac{d\Omega}{4\pi} \, e^{-\tau_\nu(\hat{\mathbf{n}},\mathbf{r})} \quad , \tag{19.1}$$

where the bar indicates averaging over direction $\hat{\mathbf{n}}$. Now let $\langle\beta(\mathbf{r})\rangle$ represent the direction-averaged escape probability $\bar{\beta}_\nu(\mathbf{r})$ averaged over the line profile:

$$\langle\beta(\mathbf{r})\rangle = \int \phi_\nu \bar{\beta}_\nu(\mathbf{r}) d\nu \quad , \tag{19.2}$$

where ϕ_ν is the normalized line profile ($\int \phi_\nu d\nu = 1$).

Now we make two approximations: First, we will approximate the excitation in the cloud as uniform. Second, we make the "on-the-spot" approximation: we assume that if a radiated photon is going to be absorbed, it will be absorbed so close to the point of emission that we can approximate it as being absorbed *at* the point of emission.[1] These approximations replace a difficult nonlocal excitation problem with a much simpler local one!

For simplicity, consider a two-level system, and a single collision partner c. (Generalization to multiple levels and collision partners is straightforward.) Recall that the rate of change of the level populations is given by

$$\frac{dn_u}{dt} = \left(n_c k_{\ell u} + n_\gamma \frac{g_u}{g_\ell} A_{u\ell} \right) n_\ell - \left(n_c k_{u\ell} + A_{u\ell} + n_\gamma A_{u\ell} \right) n_u \quad , \tag{19.3}$$

where n_γ is the photon occupation number.

Since we assume uniform excitation (and, therefore, uniform source function j_ν / κ_ν) within the cloud, the intensity at a point within the cloud is (see §7.4)

$$I_\nu = I_\nu(0) \mathrm{e}^{-\tau_\nu} + B_\nu(T_{\mathrm{exc}})(1 - \mathrm{e}^{-\tau_\nu}) \quad . \tag{19.4}$$

From the definitions of n_γ and T_{exc}, it follows that

$$n_\gamma(\nu) = n_\gamma^{(0)} \mathrm{e}^{-\tau_\nu} + \frac{1 - \mathrm{e}^{-\tau_\nu}}{(n_\ell g_u)/(n_u g_\ell) - 1} \quad , \quad n_\gamma^{(0)} \equiv \frac{c^2}{2h\nu^3} I_\nu(0) \quad . \tag{19.5}$$

Now $\mathrm{e}^{-\tau_\nu}$ is just the escape probability β_ν for a photon traveling in a particular direction. Replacing $\mathrm{e}^{-\tau_\nu} \rightarrow \beta_\nu$, averaging over direction, and averaging over the line profile, we obtain

$$\langle n_\gamma \rangle = \langle \bar{\beta} \rangle n_\gamma^{(0)} + \frac{1 - \langle \bar{\beta} \rangle}{(n_\ell g_u / n_u g_\ell) - 1} \quad , \tag{19.6}$$

where we have assumed the externally incident intensity $I_\nu(0)$ to be isotropic and constant across the line profile.

Substituting $\langle n_\gamma \rangle$ from Eq. (19.6) into (19.3), the equation for the rate of change of the level populations becomes

$$\frac{dn_u}{dt} = n_c k_{\ell u} n_\ell - n_c k_{u\ell} n_u - \langle \bar{\beta} \rangle A_{u\ell} n_u + n_\ell \frac{g_u}{g_\ell} \langle \bar{\beta} \rangle A_{u\ell} n_\gamma^{(0)} \left(1 - \frac{n_u g_\ell}{n_\ell g_u} \right) \quad . \tag{19.7}$$

This is called the **escape probability approximation**. It is a deceptively simple result! Recall that we started with an equation that included both photoexcitation and stimulated emission, including the effects of photons emitted by the cloud material itself – Eq. (19.7) therefore includes these physical processes. However, Eq. (19.7) is also the equation that we would write down *if*

1. There were no internally generated radiation field present

[1] This second assumption actually follows from the first: if the excitation is uniform, then it makes no difference how far an emitted photon travels before it is absorbed.

2. The cloud were transparent to the external radiation field $I_\nu(0)$

3. The actual Einstein A coefficient were replaced by an "effective" value $\langle\bar{\beta}\rangle A_{u\ell}$.

If the value of the escape probability $\langle\bar{\beta}\rangle$ is known, it allows us to write down the equation that the level populations must satisfy, taking into account all radiative processes (absorption, spontaneous emission, and stimulated emission), and photons both originating externally and emitted within the cloud.

Because Eq. (19.7) shows that the rate of change of the level populations is as though the rate of spontaneous decay is only $\langle\bar{\beta}\rangle A_{u\ell}$, the **critical density** for a level u (the density at which collisional deexcitation is equal to the effective rate of deexcitation by spontaneous decay) is, for a collision partner c,

$$n_{\mathrm{crit},u}(c) \equiv \frac{\sum_{\ell<u}\langle\bar{\beta}_{u\ell}\rangle A_{u\ell}}{\sum_{\ell<u} k_{u\ell}(c)} \quad . \tag{19.8}$$

19.2 Homogeneous Static Spherical Cloud

The angle-averaged escape probability $\bar{\beta}_\nu(\mathbf{r})$ defined in Eq. (19.1) depends on the geometry and velocity structure of the region. For the case of a finite cloud, $\bar{\beta}_\nu$ will depend on position – it will be highest at the cloud boundary, and smallest at the cloud center. We now define $\langle\beta\rangle_{\mathrm{cloud}}$ to be $\bar{\beta}_\nu$ averaged over the line profile *and* over the cloud volume. For a uniform density spherical cloud,

$$\langle\beta\rangle_{\mathrm{cloud}} = \frac{3}{4\pi R^3}\int_0^R \langle\beta(\mathbf{r})\rangle\, 4\pi r^2 dr \quad , \tag{19.9}$$

where $\langle\beta\rangle$ in the integrand is defined in Eq. (19.2). Figure 19.1 shows the mass-averaged escape probability $\langle\beta\rangle_{\mathrm{cloud}}$ calculated numerically for the case of a homogeneous spherical cloud, as a function of

$$\tau_0 \equiv \frac{g_u}{g_\ell}\frac{A_{u\ell}\lambda_{u\ell}^3}{4(2\pi)^{3/2}\sigma_V} n_\ell R\left(1-\frac{n_u g_\ell}{n_\ell g_u}\right) \quad , \tag{19.10}$$

the optical depth at line-center from the center of the cloud to the surface. The gas is assumed to have a Gaussian velocity distribution (6.31) with one-dimensional velocity dispersion σ_V.

As can be seen from Fig. 19.1, a satisfactory approximation is provided by the simple fitting function

$$\langle\beta\rangle_{\mathrm{cloud}} \approx \frac{1}{1+0.5\tau_0} \quad . \tag{19.11}$$

There is a simple interpretation of $\langle\beta\rangle_{\mathrm{cloud}}$: it is approximately the fraction of the

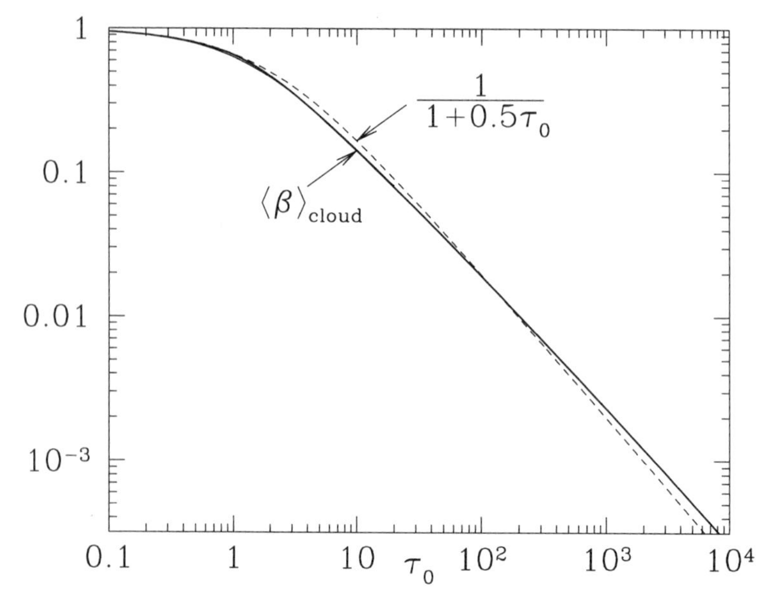

Figure 19.1 Average escape probability $\langle\beta\rangle_{\rm cloud}$ for a homogeneous spherical cloud. τ_0 is the line-center optical depth from the center of the cloud to the surface.

cloud mass that is within optical depth $\sim\frac{2}{3}$ of the cloud surface. Photons emitted near the surface have a high probability of escape, and photons emitted far below the surface have a negligible probability of escape; the average escape probability is, in effect, just the fraction of the emitted photons that are emitted from the surface layer.

Of course, before we can determine $\langle\beta\rangle_{\rm cloud}$, we need to know the level populations n_u and n_ℓ so that we can calculate the optical depth:

$$\tau_\nu = \frac{g_u}{g_\ell}\frac{A_{u\ell}}{8\pi}\lambda_{u\ell}^2\phi_\nu \int n_\ell\left(1-\frac{n_u}{n_\ell}\frac{g_\ell}{g_u}\right)ds \quad . \tag{19.12}$$

Therefore, in practice this becomes an iterative problem: guess a trial value of $\langle\beta\rangle$; then solve for the level populations n_u, n_ℓ; find the new value of $\langle\beta\rangle$; and iterate to find the self-consistent excitation n_u, n_ℓ.

19.3 Example: CO J =1–0

The $J=1\rightarrow 0$ rotational line of CO, at $\nu = 115\,{\rm GHz}$ ($\lambda = 0.260\,{\rm cm}$), is a frequently observed tracer of molecular gas. The $J=1$ level has $A_{10} = 7.16\times 10^{-8}\,{\rm s}^{-1}$ (see eq. 5.7). Consider a diffuse molecular cloud with $n_{\rm H} = 10^3 n_3\,{\rm cm}^{-3}$, radius $R = 10^{19}R_{19}\,{\rm cm}$, and CO abundance $n({\rm CO})/n_{\rm H}\approx 7\times10^{-5}$ (i.e., about

25% of the carbon is in CO).[2] If the CO has a Gaussian velocity distribution (6.31), then the attenuation coefficient

$$\kappa_\nu = n_\ell \left(1 - \frac{n_u}{n_\ell}\frac{g_\ell}{g_u}\right) \frac{\lambda^2}{8\pi}\frac{g_u}{g_\ell} A_{u\ell} \frac{1}{\sqrt{\pi}}\frac{\lambda}{b} e^{-(\Delta v/b)^2} \quad , \tag{19.13}$$

and the line-center optical depth, from cloud center to edge, is

$$\tau_0 = \kappa_{\nu_0} R = n_\ell R \left(1 - \frac{n_u}{n_\ell}\frac{g_\ell}{g_u}\right) \frac{\lambda^3}{8\pi^{3/2} b}\frac{g_u}{g_\ell} A_{u\ell} \tag{19.14}$$

$$= 297 n_3 R_{19} \left[\frac{n(\mathrm{CO})/n_\mathrm{H}}{7\times10^{-5}}\right]\left[\frac{n(J=0)}{n(\mathrm{CO})}\right]\left(\frac{2\,\mathrm{km\,s^{-1}}}{b}\right)\left(1 - \frac{n_u}{n_\ell}\frac{g_\ell}{g_u}\right). \tag{19.15}$$

Equation (19.15) requires an estimate of the fraction of the CO that is in the $J = 0$ and $J = 1$ levels.

19.3.1 CO Partition Function

Let us suppose that the CO rotational excitation is characterized by an excitation temperature (or "rotation temperature") T_exc. The fraction of CO in a given rotational level will then be

$$\frac{n(\mathrm{CO}, J)}{n(\mathrm{CO})} = \frac{(2J+1)e^{-B_0 J(J+1)/kT_\mathrm{exc}}}{\sum_J (2J+1)e^{-B_0 J(J+1)/kT_\mathrm{exc}}} \quad , \tag{19.16}$$

where B_0 is the "rotation constant" [see Eq. (5.3)]. For rotation temperatures of interest, we can approximate the partition function in the denominator of (19.16) by

$$Z \equiv \sum_J (2J+1)e^{-B_0 J(J+1)/kT_\mathrm{exc}} \approx \left[1 + (kT_\mathrm{exc}/B_0)^2\right]^{1/2} \quad . \tag{19.17}$$

The approximation is exact in the limits $kT_\mathrm{exc}/B_0 \to 0$ and $kT_\mathrm{exc}/B_0 \gg 1$, and is accurate to within ±6% for all T_exc. For $^{12}\mathrm{C}^{16}\mathrm{O}$, $B_0/k = 2.77\,\mathrm{K}$, so that

$$\tau_0 \approx 297\, n_3\, R_{19} \left[\frac{n(\mathrm{CO})/n_\mathrm{H}}{7\times10^{-5}}\right]\left\{\frac{\left(1 - e^{-5.53\,\mathrm{K}/T_\mathrm{exc}}\right)}{[1 + (T_\mathrm{exc}/2.77)^2]^{1/2}}\right\}\left(\frac{2\,\mathrm{km\,s^{-1}}}{b}\right). \tag{19.18}$$

For a typical CO rotation temperature $T_\mathrm{exc} \approx 8\,\mathrm{K}$, Eq. (19.18) becomes

$$\tau_0 \approx 50 n_3 R_{19} \left[\frac{n(\mathrm{CO})/n_\mathrm{H}}{7\times10^{-5}}\right]\left[\frac{2\,\mathrm{km\,s^{-1}}}{b}\right] \quad . \tag{19.19}$$

Thus the CO 1–0 transition is expected to be often quite optically thick.

---

[2]The mean visual extinction to the center of this cloud would be $A_V \approx 5 n_3 R_{19}$ mag. We will see in Chapter 29 that $A_V \approx 5$ is common.

19.3.2 Critical Density for CO

The rate coefficient for collisional deexcitation of $CO(J = 1)$ by collisions with H_2 is

$$k_{10} \approx 6 \times 10^{-11} T_2^{0.2} \,\mathrm{cm}^3\,\mathrm{s}^{-1} \tag{19.20}$$

for $10\,\mathrm{K} \lesssim T \lesssim 250\,\mathrm{K}$ (Flower & Launay 1985; Flower 2001). In an optically thin region, the critical density for thermalizing CO would be $A_{10}/k_{10} \approx 1100 T_2^{-0.2}\,\mathrm{cm}^{-3}$.

For a cloud with $n_3 R_{19} \approx 1$, $b \approx 2\,\mathrm{km\,s}^{-1}$, and $T_{\rm exc} = 8\,\mathrm{K}$, Eq. (19.19) gives $\tau_0 \approx 50$, corresponding to a cloud-averaged escape probabilty $\langle\beta_{10}\rangle_{\rm cloud} \approx 1/(1 + 0.5\tau_0) \approx 0.04$, and the effective critical density – after taking account of radiative trapping – would be

$$n_{\mathrm{crit,H_2}}(\mathrm{CO}, J{=}1) = \frac{\langle\beta_{10}\rangle A_{10}}{k_{10}} \approx 50\, T_2^{-0.2}\,\mathrm{cm}^{-3} \quad . \tag{19.21}$$

Therefore, we see that at least the $J{=}1$ level of CO is expected to be thermalized in molecular clouds with $n_{\rm H} \gtrsim 10^2\,\mathrm{cm}^{-3}$.

19.4★ LVG Approximation: Hubble Flow

Earlier, we considered a static cloud with a uniform Gaussian velocity distribution function. Let us now consider the case where a velocity gradient is present, with point-to-point velocity differences across the flow field that are large compared to the width of the velocity distribution at a given point. This is called the **large velocity gradient (LVG) approximation**. First introduced by Sobolev (1957), it is often referred to as the **Sobolev approximation**.

The simplest case to consider is the case of "Hubble flow": spherical expansion with velocity proportional to radius, $\mathbf{v} = (dv/dr)\mathbf{r}$, where dv/dr is assumed to be independent of r. The density is assumed to be uniform. We know that all points in this flow field are equivalent, so we need only consider the escape of photons emitted from $r = 0$. For a photon propagating radially, the local attenuation coefficient is

$$\kappa_\nu(r) = K\,\phi\left[\nu - \nu_0\left(1 + \frac{dv}{dr}\frac{r}{c}\right)\right] \quad , \tag{19.22}$$

where ν_0 is the resonant frequency for gas at rest, $\phi(x)$ is the local line profile, with normalization $\int \phi(\nu) d\nu = 1$, and

$$K \equiv \int \kappa_\nu d\nu = \frac{g_u A_{u\ell}}{8\pi} \lambda_{u\ell}^2 \left(\frac{n_\ell}{g_\ell} - \frac{n_u}{g_u}\right) \quad . \tag{19.23}$$

Consider a photon emitted with frequency ν from $r = 0$; as it travels, it appears increasingly redshifted relative to the local material through which it is traveling. The optical depth to infinity is

$$\tau(\nu) = \int_0^\infty drK\,\phi\left[\nu - \nu_0\left(1 + \frac{dv}{dr}\frac{r}{c}\right)\right] \quad . \tag{19.24}$$

Define

$$\xi \equiv \nu - \nu_0 - \nu_0 \frac{dv}{dr}\frac{r}{c}. \tag{19.25}$$

Then,

$$\tau(\nu) = \frac{K\lambda_{u\ell}}{dv/dr}\,y(\nu - \nu_0) \quad , \quad y(\xi) \equiv \int_{-\infty}^{\xi} \phi(\xi')d\xi' \quad . \tag{19.26}$$

The escape probability $e^{-\tau}$ averaged over the line profile is

$$\langle\beta\rangle = \int_{-\infty}^{\infty} \phi(\nu - \nu_0)\,e^{-\tau(\nu)}d\nu \quad . \tag{19.27}$$

The integral is easily evaluated by change of variable to $dy = \phi(\xi)d\xi$:

$$\langle\beta\rangle = \frac{1 - e^{-\tau_{\rm LVG}}}{\tau_{\rm LVG}} \quad , \quad \tau_{\rm LVG} \equiv \frac{K\lambda_{u\ell}}{|dv/dr|} \quad . \tag{19.28}$$

$\tau_{\rm LVG}$ has a simple interpretation: it is the total optical depth across the region for any frequency in $\nu_0\left[1 \pm (R/c)(dv/dr)\right]$.

Uniform spherical expansion is mathematically convenient but may not be the appropriate velocity field for other flows. For flows with well-defined monotonic velocity fields – stellar winds and shock waves would be two examples – the radiative transfer problem is purely local: if a photon is able to travel any appreciable distance, the Doppler shift carries it out of resonance, after which the photon is not affected by line absorption. Given a velocity field $\mathbf{v}(\mathbf{r})$, the optical depth $\tau_\nu(\hat{\mathbf{n}})$ can be evaluated, and the escape probability $\langle\beta\rangle$ calculated by averaging over direction $\hat{\mathbf{n}}$ and the local line profile ϕ_ν.

19.5 Escape Probability for Turbulent Clouds

The LVG approximation is sometimes applied to radiative trapping in turbulent molecular clouds, but in this case one faces the question of what value to use for the velocity gradient dv/dr in Eq. (19.28). Rather than use the LVG approximation, it is more appropriate to use the homogeneous spherical cloud result (19.11) but with the turbulent velocities included in the one-dimensional velocity dispersion σ_V.

A uniform density cloud of mass $M = 4.9 \times 10^3 n_3 R_{19}^3 \, M_\odot$ has gravitational self-energy $-3GM^2/5R$. If magnetic energy is not important, and the cloud is self-gravitating, the virial theorem yields a mass-weighted one-dimensional velocity dispersion $\sigma_V^2 = GM/5R$. If the velocities are mostly from fluid motions (i.e., turbulence) rather than thermal motions, then all species will have this same velocity dispersion, corresponding to a broadening parameter

$$b = 1.6 n_3^{1/2} R_{19} \, \mathrm{km\,s^{-1}} \quad . \tag{19.29}$$

The optical depth at line-center from center to edge of the cloud will be

$$\tau_0 = \frac{g_u}{g_\ell} \frac{A_{u\ell} \lambda_{u\ell}^3}{8\pi} \left(\frac{5}{2\pi G} \right)^{1/2} \frac{n_\ell R^{3/2}}{M^{1/2}} \left(1 - \frac{n_u}{n_\ell} \frac{g_\ell}{g_u} \right) \quad . \tag{19.30}$$

The luminosity of the cloud in a spectral line is

$$L_{u\ell} = \int dr 4\pi r^2 n_u A_{u\ell} h\nu_{u\ell} \langle \beta \rangle_{\mathrm{cloud}} \tag{19.31}$$

$$\approx \frac{4\pi}{3} R^3 n_u A_{u\ell} h\nu_{u\ell} \frac{1}{1 + 0.5\tau_0} \quad . \tag{19.32}$$

If we now assume $\tau_0 \gg 1$ (which we have seen was appropriate for the CO 1–0 transition) and use Eq. (19.30), then the line luminosity per unit mass is

$$\frac{L_{u\ell}}{M} \approx 32\pi^2 \left(\frac{2G}{15} \right)^{1/2} \frac{hc}{\lambda_{u\ell}^4} \frac{1}{\langle \rho \rangle^{1/2}} \frac{1}{\left(e^{h\nu/kT_{\mathrm{exc}}} - 1 \right)} \quad . \tag{19.33}$$

Note that for a given spectral line, the luminosity per unit mass depends on the cloud density ρ and on the excitation temperature T_{exc} characterizing the population ratio n_u/n_ℓ of the upper and lower levels in the transition. More importantly, the luminosity is *independent* of the actual abundance of the emitting species, provided that it is sufficiently large for the transition to have $\tau_0 \gg 1$.

If the cloud is larger than our antenna beam, we can relate the antenna temperature integrated over the $J = 1 \rightarrow 0$ line, $\int T_A(1-0) dv \equiv (\lambda^3/2k) \int I_\nu d\nu$, to the total H column density (averaged over the beam solid angle):

$$\frac{N_{\mathrm{H}}}{\int T_A(1-0) dv} = \frac{1}{4\pi} \frac{k\lambda}{hc} \left(\frac{15}{2.8 G m_{\mathrm{H}}} \right)^{1/2} (n_{\mathrm{H}})^{1/2} \left(e^{h\nu/kT_{\mathrm{exc}}} - 1 \right) \quad . \tag{19.34}$$

This theoretical estimate for the line luminosity of a cloud is about as good as can be done for this simple "one-zone" model of a cloud. Real molecular clouds are inhomogeneous, and the excitation of the emitting molecules will vary considerably between the lower-density outer layers of the cloud to the denser central regions.

19.6 CO 1–0 Emission as a Tracer of H_2 Mass: CO "X-Factor"

Equation (19.34) applies to any optically thick line radiated by a self-gravitating cloud. The CO 1–0 transition, with $\lambda = 0.260\,\mathrm{cm}$, is often used as a tracer of molecular cloud mass. If we assume $N(\mathrm{H_2}) = 0.5 N_\mathrm{H}$, Eq. (19.34) becomes

$$X_\mathrm{CO} \equiv \frac{N(\mathrm{H_2})}{\int T_A dv} = \frac{1}{8\pi} \frac{k\lambda}{hc} \left(\frac{15}{2.8 G m_\mathrm{H}}\right)^{1/2} (n_\mathrm{H})^{1/2} \left(e^{h\nu/kT_\mathrm{exc}} - 1\right) \quad (19.35)$$

$$= 1.58 \times 10^{20} n_3^{1/2} \left(e^{5.5\,\mathrm{K}/T_\mathrm{exc}} - 1\right) \frac{\mathrm{cm}^{-2}}{\mathrm{K\,km\,s^{-1}}} . \quad (19.36)$$

For $n_\mathrm{H} = 10^3\,\mathrm{cm}^{-3}$ and $T_\mathrm{exc} = 8\,\mathrm{K}$, this yields $N(\mathrm{H_2})/\int T_A dv = 1.56 \times 10^{20}\,\mathrm{cm}^{-2}/\,\mathrm{K\,km\,s^{-1}}$.

Recent observational determinations of $X_\mathrm{CO} \equiv N(\mathrm{H_2})/W_\mathrm{CO}$ (where $W_\mathrm{CO} \equiv \int T_A dv$ integrated over the CO line) find $X_\mathrm{CO} = (1.8 \pm 0.3) \times 10^{20}\,\mathrm{cm}^{-2}/\,\mathrm{K\,km\,s^{-1}}$ (Dame et al. 2001), where infrared emission from dust (Schlegel et al. 1998) was used as a mass tracer. This appears to be the most reliable determination for X_CO in the solar neighborhood. Earlier studies using diffuse galactic γ-ray emission as a mass tracer found similar values of X_CO: e.g., $X_\mathrm{CO} = (1.56 \pm 0.05) \times 10^{20}\,\mathrm{cm}^{-2}/\,\mathrm{K\,km\,s^{-1}}$ (Hunter et al. 1997). The most recent determination using γ rays finds $X_\mathrm{CO} = (1.76 \pm 0.04) \times 10^{20} \mathrm{H_2\,cm^{-2}}/(\,\mathrm{K\,km\,s^{-1}})$ for the Orion A GMC (Okumura et al. 2009).

Our theoretical value (19.35) for X_CO is sensitive to the values adopted for the cloud density n_H and the CO excitation temperature T_exc. The fact that the observed values of X_CO fall close to Eq. (19.36) suggests that $n_\mathrm{H} \approx 10^3\,\mathrm{cm}^{-3}$ and $T_\mathrm{exc} = 8\,\mathrm{K}$ may be representative of self-gravitating molecular clouds in the local ISM.

The study by Dame et al. (2001) included $\sim 98\%$ of the CO emission at $|b| < 32°$. Some of this emission presumably comes from clouds that are partially confined by the pressure of the ISM, in which case the observed value of X_CO would be expected to be smaller than the value of X_CO that is obtained in Eq. (19.36) if it is assumed that the observed velocity dispersion must be bound by self-gravity. It must, therefore, be regarded as somewhat fortuitous that the observed X_CO is numerically close to the value estimated in Eq. (19.36).

The theoretical X_CO factor depends explicitly on both n_H and T_exc. It would, therefore, not be at all surprising if the value of X_CO in other galaxies were to differ appreciably from the value found for the Milky Way, or if the value of X_CO showed cloud-to-cloud or regional variations within the Milky Way.

In addition, real molecular clouds have an outer layer where the hydrogen is molecular, but where the CO abundance is very low because it is not sufficiently shielded from dissociating radiation. This gas does not show up in H I 21-cm or CO 1–0 surveys – leading Wolfire et al. (2010) to refer to it as "the dark molecular gas"' – but since it does contain dust, it does contribute to the far-infrared emission. The thickness of this transition layer will depend on the dust abundance, which will in turn depend on the metallicity of the gas. Wolfire et al. (2010) estimate that

$\sim 30\%$ of the molecular mass in a typical Galactic molecular cloud may be in this envelope.

At constant n_3 and $T_{\rm exc}$, this low-CO layer will result in an increase in $X_{\rm CO}$. This is in agreement with recent observations indicating that $X_{\rm CO}$ in Local Group galaxies is of order $\sim 4 \times 10^{20}\,{\rm cm}^{-2}/\,{\rm K\,km\,s}^{-1}$ (Blitz et al. 2007), with large variations from one galaxy to another. An indirect estimate of the molecular gas mass based on modeling the infrared emission from a sample of nearby galaxies also favored $X_{\rm CO} \approx 4 \times 10^{20}\,{\rm cm}^{-2}/\,{\rm K\,km\,s}^{-1}$ (Draine et al. 2007).

Chapter Twenty

Optical Pumping

As seen in previous chapters, excited states of molecules, atoms, and ions can be populated by collisional excitation, and also by radiative recombination of ions to excited states. Collisional excitation is responsible for emission lines such as [N II]6550,6585Å, while radiative recombination is responsible for Hα6565Å.

A third process – **optical pumping** – can also be important for populating excited states. "Optical pumping" refers to the process of excitation from a lower level ℓ by absorption of a photon either directly to u, or to a level x that lies above u, but which can decay to u by spontaneous emission of one or more photons.

The basic idea is extremely simple: The species of interest has some lower level ℓ that can absorb radiation at wavelength $\lambda_{\ell x}$, with probability per unit time $\zeta_{\ell x}$, resulting in a transition to level x. Level x then has some probability $p_{x,u}$ of spontaneously emitting one or more photons to arrive in level u.

20.1 UV Pumping by Continuum

Under many circumstances, the radiation responsible for optical pumping forms a continuum – for example, the emission from an active galactic nucleus (AGN), the flash from a gamma-ray burst (GRB), starlight, or infrared continuum from dust grains. The photoexcitation rate is

$$\zeta_{\ell x} = \bar{n}_{\gamma,\ell x} \frac{g_x}{g_\ell} A_{x\ell} \quad , \tag{20.1}$$

where $\bar{n}_{\gamma,\ell x}$ is the angle- and polarization-averaged photon occupation number at wavelength $\lambda_{x\ell}$ [see eq. (6.12)]. Normally, $\bar{n}_{\gamma,\ell x} \ll 1$, and stimulated emission from level x can be neglected.

Considering spontaneous decays only, it is straightforward to calculate branching probabilities

$$p_{ij} = \frac{A_{ij}}{\sum_{k<i} A_{ik}} \quad , \tag{20.2}$$

where the notation assumes the levels to be indexed in order of increasing energy.

Then the probability per unit time that level x will be populated following photoexcitation out of level ℓ can be written

$$\beta_{\ell u} = \zeta_{\ell u} + \sum_{x>u} \zeta_{\ell x} \left[p_{xu} + \sum_{u<i<x} p_{xi} p_{iu} + \sum_{u<i<x} p_{xi} \sum_{\ell<j<i} p_{ij} p_{ju} + \ldots \right] ; \tag{20.3}$$

where the first term in the square brackets represents direct decays $x \to u$, the second term arrivals at u after emitting two photons, and so on. Often the "direct" term $\zeta_{\ell u}$ is completely negligible, either because $\ell \to u$ may be forbidden, or because the radiation field near wavelengths $\lambda_{\ell u}$ may be weak.

UV pumping by continuum radiation absorbed by atoms or ions can produce populations in excited states in excess of what would be produced by collisional excitation (or radiative recombination). One example of this would be excited fine-structure levels: in N II, for example (see Fig. 6.1), absorption of a 1084.6Å photon by the ground state $^3P_0$ excites the $^3D_1^o$ state; a fraction of the spontaneous decays then populate the excited fine-structure levels $^3P_1$ and $^3P_2$ of the ground state.

A more important example is H_2: optical pumping via UV lines is responsible

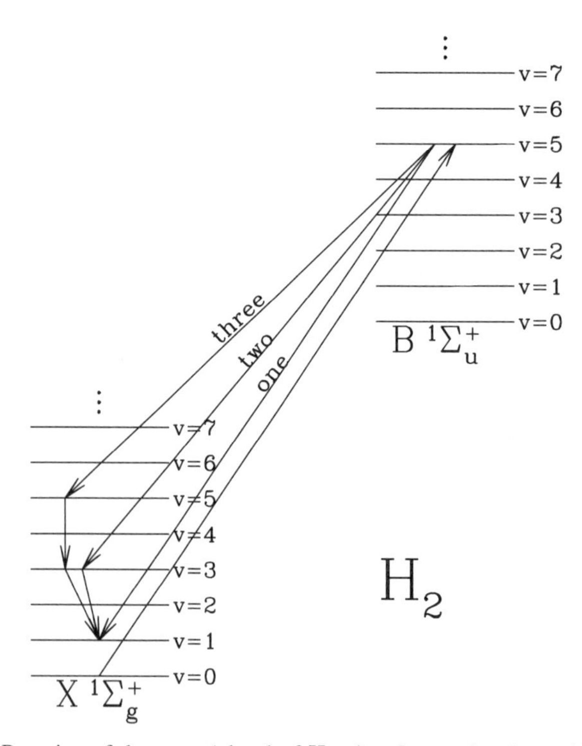

Figure 20.1 Pumping of the $v = 1$ level of H_2 via a Lyman-band transition to the $v = 5$ level of $B^1\Sigma_u^+$. The single-photon decay path (labeled "one") is shown, as well as sample decay paths involving two and three steps. The rotational substructure of the vibrational levels is not shown.

for much of the vibrationally excited H_2 in galaxies. The pumping is primarily via the strong Lyman-band and Werner-band transitions. Figure 20.1 shows some of the pathways that can lead to population of the $v = 1$ level of H_2. This excitation process will be discussed further in Chapter 31.

20.2★ Infrared Pumping: OH

Optical pumping can also be driven by infrared radiation. OH maser emission present in some regions is thought to be the result of infrared pumping. The maser emission is the result of population inversions among the Λ-doubling sublevels of the ground electronic state of OH. As discussed in §5.1.8, the ground electronic state of OH has two fine-structure levels, $^2\Pi_{3/2}$ (the ground state) and $^2\Pi_{1/2}$. Each of these fine structure states has two "Λ-doubling" levels, and each of these is split into two levels by interaction with the proton magnetic moment. In addition, the molecule can have rotational angular momentum, resulting in a "rotational ladder" of states with increasing rotational angular momentum.

Because OH has a large dipole moment (1.67 D), the rotational transitions are

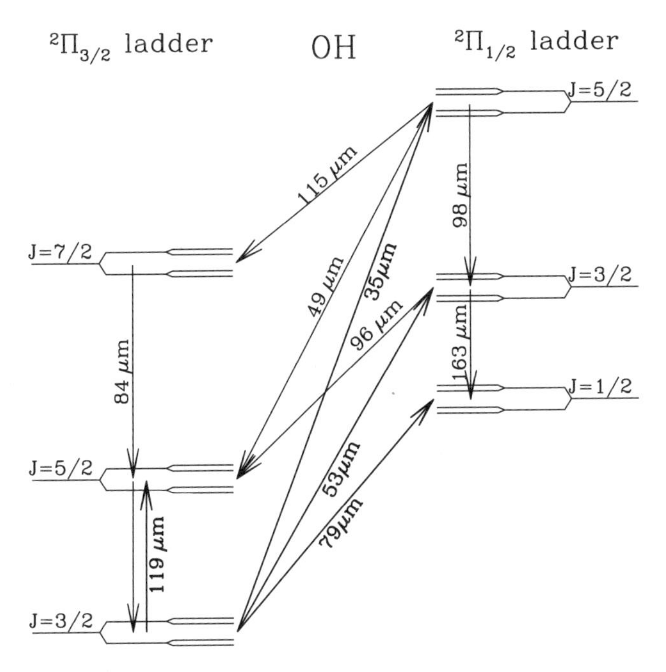

Figure 20.2 Pumping of the hyperfine levels of OH $^2\Pi_{3/2}(J = 3/2)$ via infrared pumping to $J = 1/2$, $J = 3/2$, and $J = 5/2$ levels of the $^2\Pi_{1/2}$ rotational ladder. Each of the "lines" shown here is actually a multiplet of transitions between hyperfine levels of the upper and lower rotational level.

strong. The OH ground state $^2\Pi_{3/2}(J = \frac{3}{2})$ can absorb infrared radiation in groups of lines near $119\,\mu$m, $79\,\mu$m, $53\,\mu$m, and $35\,\mu$m. These radiative excitations are followed by spontaneous decays, which end up repopulating the four sublevels of $^2\Pi_{3/2}(J = \frac{3}{2})$. Figure 20.2 outlines the pumping transitions and the radiative decay paths that follow the absorption.

OH masers come in a number of "flavors." Some OH masers are found in the photodissociation regions around "ultracompact" H II regions. These are usually strongest in the 1667 and 1665 MHz "mainline" transitions (see Fig. 5.3). OH masers present in the dusty outflows from some evolved stars ("OH-IR stars," see §38.2) are usually strongest in the 1612 MHz line. Last, very powerful "OH megamasers" are seen in a small fraction of galaxies, including both starburst galaxies and Seyfert 2 galaxies; OH megamasers are usually strongest in the 1667 MHz line. In some objects, OH maser emission has been seen in transitions involving rotationally excited levels.

The IR pumping of OH depends on the relative strength of the far-infrared emission at the four wavelengths $(119, 79, 53, 35\,\mu\text{m})$ where absorption from $^2\Pi_{3/2}$ can take place, and therefore the relative strengths of the different maser lines is expected to depend on the temperature of the dust emitting the far-infrared radiation exciting the OH. In addition, the small hyperfine splitting of the far-infrared lines allows line overlap to occur, depending on the velocity dispersion in the gas; when the absorption lines become optically thick, the relative importance of different pumping routes will depend on line overlap, and therefore on the velocity distribution of the OH. Thus object-to-object variations in the infrared spectrum and velocity dispersion can lead to variations in maser line ratio. In addition, of course, collisional processes can also affect the level populations.

20.3★ UV Pumping by Line Coincidence: Bowen Fluorescence

Because certain ultraviolet lines – such as Lyman α – can be very intense in and near ionized regions, these lines can also contribute to optical pumping. We have in fact already considered optical pumping when we discussed the fate of Lyman β photons in H II regions – a Lyman β photon will almost always be absorbed by an H atom, "pumping" that H atom to the $3p$ level. The $3p$ state has a finite probability of decaying to $2s$, which is in effect "ultraviolet pumping" of H $2s$. Thus Case B recombination is a situation where the $2s$ level is pumped by Lyman β, γ, and so on.

However, H and He resonance-line photons can also pump *other* species when there is an accidental coincidence with a resonance line, resulting in emission in specific longer-wavelength lines as part of the radiative decay cascade from the UV-pumped level. This mechanism was identified by I. Bowen (1934), who noted that He II emission could pump O III; another line coincidence allows O III emission to then pump N III (Bowen 1935). Lyman β radiation can pump O I and Mg II (Bowen 1947). The general process is referred to as **Bowen fluorescence**. In principle, there can be Bowen-type fluorescent excitation using photons that are emitted

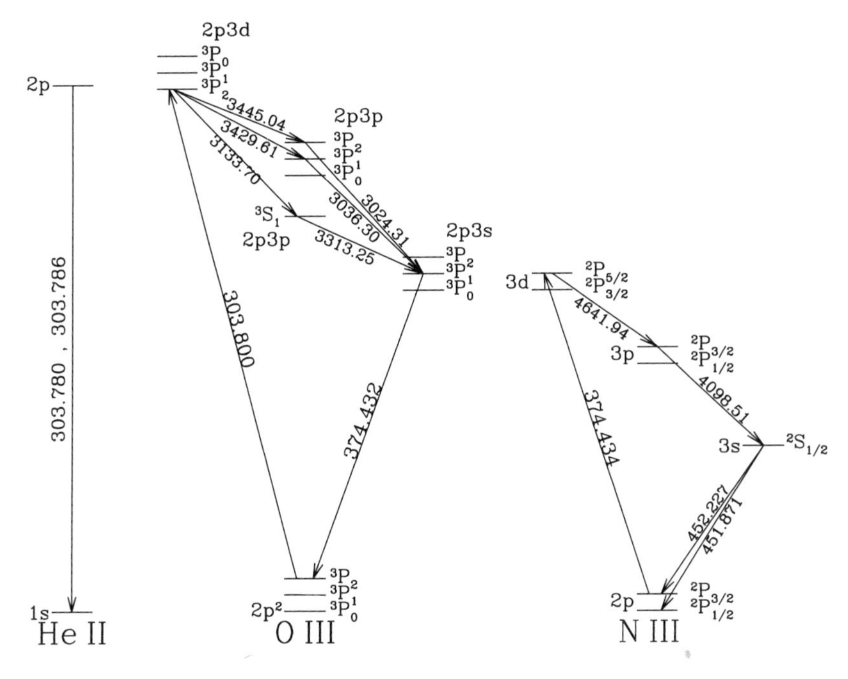

Figure 20.3 Pathway for pumping of O III by He II Lyα, and for pumping N III by O IIIλ374.432 Å. Pumping of O III by He II Lyα results in observable Bowen fluorescence of, e.g., O III$\lambda\lambda$3133.70, 3445.04, 3429.61 Å. Pumping of N III by O IIIλ374.432 Å results in observable Bowen fluorescence of N IIIλ4641.94Å and N IIIλ4098.51Å. Energy level diagram is not to scale, and some other possible transitions have been omitted for clarity. Wavelengths are all in vacuo, and in Å.

by any species, but, in practice, species other than H and He are simply not abundant enough for their radiation to affect the excitation of other species appreciably. Hence Bowen fluorescence in astrophysics is limited to excitation by emission lines from H I, He I, and He II.

20.3.1★ Bowen Fluorescent Excitation of O III and N III

Bowen fluorescence explains a number of observed emission lines of O III. The O III $2p^2\,{}^3\mathrm{P}_2 - 2p3d\,{}^3\mathrm{P}_2^{\mathrm{o}}$ $\lambda = 303.80$ Å transition happens to coincide with He II Lyman α at $\lambda = 303.78$ Å. Absorption of an He II Lyman α photon excites O III to the $2p3d\,{}^3\mathrm{P}_2^{\mathrm{o}}$ level. This level has probabilities $p_1 = 0.738$ of reemitting a 303.80 Å photon (which can again be resonantly scattered by He II) and $p_2 = 0.246$ of emitting a 303.62 Å photon in a transition to the ${}^3\mathrm{P}_1$ fine-structure level of the ground state, which can be reabsorbed by another O III ${}^3\mathrm{P}_1$, repopulating the ${}^3\mathrm{P}_2^{\mathrm{o}}$ level. For each excitation of $2p3d$ ${}^3\mathrm{P}_2^{\mathrm{o}}$, there is a probability $p_3 = 0.0105$ of radiating a 3133.77 Å photon, which then emerges as part of the observable spectrum of the H II region.[1] Summing over the full radiative cascade is tedious but straightforward, and provides an explanation for various observed emission lines of O III (e.g., $\lambda\lambda 2837.14, 3133.70, 3429.61, 3445.09$ Å) that cannot be explained by collisional excitation.

A fraction of the excitations of O III by He Lyα end with emission of a 374.432 Å photon that can be resonantly absorbed by N III $^2P_{3/2}$, resulting in N III Bowen fluorescence, with emission lines at 4641.94 Å and 4098.51 Å that would otherwise not be excited.

Bowen fluorescence of O III and then N III is driven by He II Lyα, produced by radiative recombination of He^{++}. Bowen fluorescence of O III and N III is not important in H II regions excited by O or B stars, as these have very little He^{++}. However, the central stars of planetary nebulae can be very hot, producing substantial amounts of He^{++}, and Bowen fluorescence of O III and N III is observed in some planetary nebulae.

20.3.2⋆ Bowen Fluorescent Excitation of O I

There is also a wavelength coincidence between H I Lyman β 1025.72 Å and O I $2p^4\,{}^3P_2 - 2p^3 3d\,{}^3D^o_{1,2,3}$ 1025.76 Å. About 78% of the excitations will be to O I $2p^3 3d\,{}^3D^o_3$; this level can of course decay by reemitting a 1025.76 Å photon (or 1027.43, 1028.16 Å photons in transitions to the fine-structure excited states $^3P_1$ and $^3P_0$), but 29% of the excitations of O I $2p^3 3d\,{}^3D^o_3$ decay instead by emission of a 1.1290 μm photon, followed by an 8448.68 Å photon, followed by emission of a photon in the 1302.17, 1304.86, 1306.03 Å triplet. Because there is appreciable O I present in the partially ionized gas at the edge of an H II region, the O I 1.1290 μm and 8448.68 Å lines are observed in H II region spectra.[2] The predicted intensity ratio is $I(8448.68\,\text{Å})/I(1.1290\,\mu\text{m}) = (1.1290/.8448) = 1.336$. An observed intensity ratio smaller than this value would be an indication of differential attenuation by dust.

20.3.3⋆ Bowen Fluorescent Excitation of Other Species

Other Bowen-type processes can also occur. Shull (1978) noted a near-coincidence in wavelength between H I Lyman α 1215.67 Å and the H_2 Lyman-band 1–2 P(5) 1216.07 Å absorption from $H_2(v=2, J=5)$ to $B^1\Sigma_u^+(v=1, J=4)$, which can potentially modify the H_2 level populations in regions where Lyman α is present.

Neufeld (1990) has discussed the transfer of H Lyman α radiation in astrophysical media, including the effects of resonant absorption and reemission by H, scattering and absorption by dust, and possibly resonant absorption by $H_2(v=2, J=5)$.

---

[1] Lines at 3445.05 Å and 3429.61 Å are also emitted, but are weaker than 3133.77 Å.

[2] The 1302.17, 1304.86, 1306.03 Å lines are also present, but are in the vacuum ultraviolet and hence more difficult to observe.

Chapter Twenty-one

Interstellar Dust: Observed Properties

Dust plays an important role in astrophysics, and the need to characterize and understand dust is increasingly appreciated. Historically, interstellar dust was first recognized for its obscuring effects, and the need to correct observed intensities for attenuation by dust continues today. But with the increasing sensitivity of IR, FIR, and sub-mm telescopes, dust is increasingly important as a diagnostic, with its emission spectrum providing an indicator of physical conditions, and its radiated power bearing witness to populations of obscured stars of which we might otherwise be unaware.

More fundamentally, dust is now understood to play many critical roles in galactic evolution. By sequestering selected elements in the solid grains, and by catalyzing formation of the H_2 molecule, dust grains are central to the chemistry of interstellar gas. Photoelectrons from dust grains can dominate the heating of gas in regions where ultraviolet starlight is present, and in dense regions the infrared emission from dust can be an important cooling mechanism. Last, dust grains can be important in interstellar gas dynamics, communicating radiation pressure from starlight to the gas, and coupling the magnetic field to the gas in regions of low fractional ionization.

We begin with a brief review of some of the observational evidence that informs our study of interstellar dust. Unfortunately, it is not yet possible to bring representative samples of interstellar dust into the laboratory, and we must rely on remote observations. Our strongest constraints on interstellar dust come from observations of its interaction with electromagnetic radiation:

- Wavelength-dependent attenuation ("extinction") of starlight by absorption and scattering, now observable at wavelengths as long as $20\,\mu$m ("mid-infrared"), and as short as $0.1\,\mu$m ("vacuum ultraviolet"). The extinction includes a number of spectral features that provide clues to grain composition.
- Polarization-dependent attenuation of starlight, resulting in wavelength-dependent polarization of light reaching us from reddened stars.
- Scattered light in reflection nebulae.
- Thermal emission from dust, at wavelengths ranging from the sub-mm to $2\,\mu$m.
- Small-angle scattering of x rays, resulting in "scattered halos" around X-ray point sources.

- Microwave emission from dust, probably from rapidly spinning ultrasmall grains.
- Luminescence when dust is illuminated by starlight – the so-called extended red emission.

In addition to these electromagnetic studies, our knowledge of dust is also informed by other, less direct, evidence:

- Presolar grains preserved in meteorites – a selective but not well-understood sampling of the interstellar grains that were present in the solar nebula 4.5 Gyr ago.
- "Depletion" of certain elements from the interstellar gas, with the missing atoms presumed to be contained in dust grains.
- The observed abundance of H_2 in the ISM, which can only be understood if catalysis on dust grains is the dominant formation avenue.
- The temperature of interstellar diffuse H I and H_2, in part a result of heating by photoelectrons ejected from interstellar grains.

21.1 Interstellar Extinction

Barnard (1907, 1910) was apparently the first to realize that some stars were dimmed by an "absorbing medium." This was confirmed by Trumpler (1930), who showed that the stars in distant open clusters were dimmed by something in addition to the inverse square law, and concluded that interstellar space in the galactic plane contained "fine cosmic dust particles of various sizes . . . producing the observed selective absorption." Over the succeeding eight decades, we have built on these pioneering studies, but many aspects of interstellar dust – including its chemical composition! – remain uncertain. Let us, therefore, begin by reviewing the different ways in which nature permits us to study interstellar dust.

Trumpler analyzed the interaction of light with interstellar dust, and this remains our most direct way to study interstellar dust. Using stars as "standard candles," we study the "selective extinction" – or "reddening" – of starlight by the dust. It is assumed that we know what the spectrum of the star is before reddening by dust takes place; this is usually accomplished by observation of another star with similar spectral features in its atmosphere but with negligible obscuration between us and the star. (This is known as the "pair method".)

With the assumption that the extinction ($\equiv$ absorption + scattering) goes to zero at wavelengths $\lambda \to \infty$, and including observations of the star at sufficiently long wavelength where extinction is negligible, one can determine the attenuation of the starlight by dust as a function of wavelength. Because atomic hydrogen absorbs strongly for $h\nu > 13.6\,\text{eV}$, it is possible to measure the contribution of dust to the attenuation of light only at $h\nu < 13.6\,\text{eV}$, or $\lambda > 912\,\text{Å}$. Astronomers customarily

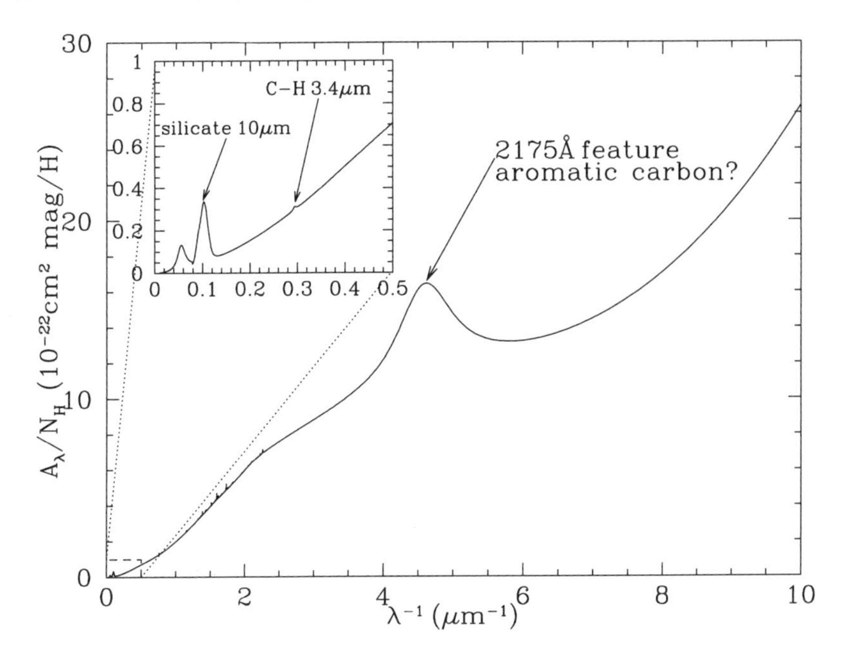

Figure 21.1 Extinction versus inverse wavelength λ^{-1} on a typical sightline in the local diffuse ISM. The inset shows the extinction at $\lambda > 2\,\mu$m.

characterize the attenuating effects of dust by the "extinction" A_λ at wavelength λ. The extinction A_λ – measured in "magnitudes" – is defined by

$$\frac{A_\lambda}{\rm mag} = 2.5 \log_{10} \left[F_\lambda^0/F_\lambda\right] \quad , \tag{21.1}$$

where F_λ is the observed flux from the star, and F_λ^0 is the flux that would have been observed had the only attenuation been from the inverse square law. The extinction measured in magnitudes is proportional to the optical depth:

$$\frac{A_\lambda}{\rm mag} = 2.5 \log_{10} \left[{\rm e}^{\tau_\lambda}\right] = 1.086\, \tau_\lambda \quad . \tag{21.2}$$

21.1.1 The Reddening Law

A typical "extinction curve" – the extinction A_λ as a function of wavelength or frequency — is shown in Figure 21.2, showing the rapid rise in extinction in the vacuum ultraviolet. Because the extinction increases from red to blue, the light reaching us from stars will be "reddened" owing to greater attenuation of the blue

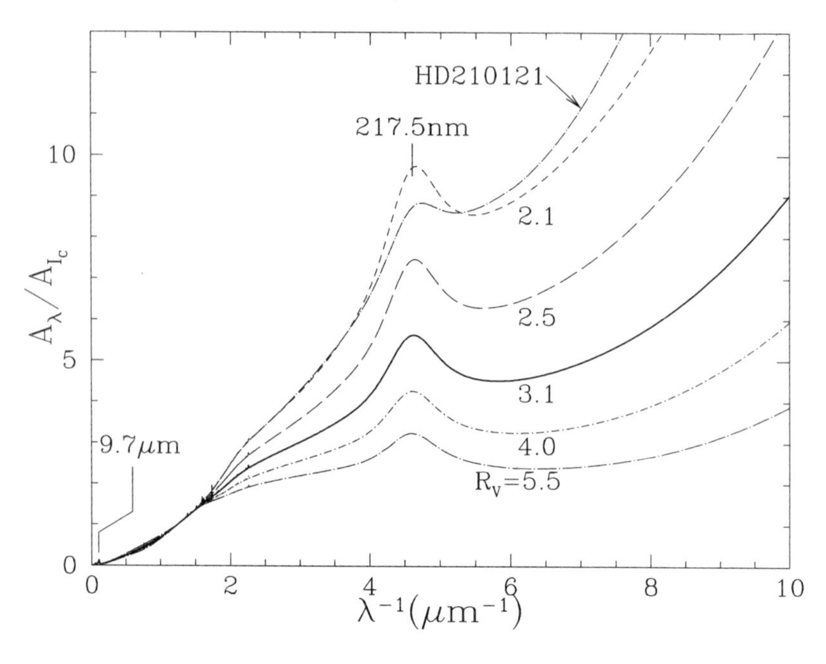

Figure 21.2 Extinction at wavelength λ, relative to the extinction in the Cousins I band ($I_C = 8020$ Å), as a function of inverse wavelength λ^{-1}, for Milky Way regions characterized by different values of $R_V \equiv A_V/(A_B - A_V) \equiv A_V/E(B-V)$, where A_B is the extinction at $B = 0.44\,\mu$m, A_V is the extinction at $V = 0.55\,\mu$m, and the "reddening" $E(B-V) \equiv A_B - A_V$. The curves shown are from the one-parameter family of curves $f_1^{\rm CCM}(\lambda)$ parameterized by R_V (see §21.2). Also shown is the extinction curve toward the star HD210121 (with $R_V = 2.1$), showing that it differs from the CCM extinction curve $f_1^{\rm CCM}$ for $R_V = 2.1$. Note the rapid rise in extinction in the vacuum ultraviolet ($\lambda \lesssim 0.15\,\mu$m) for regions with $R_V \lesssim 4$. The normalization per H nucleon is approximately $A_{I_C}/N_{\rm H} \approx 2.9 \times 10^{-22}$mag cm^2/H. The silicate absorption feature (see §23.3.2) at 9.7 μm and the diffuse interstellar bands (see §23.3.4) are barely visible.

light. The detailed wavelength dependence of the extinction – the "reddening law" – is sensitive to the composition and size distribution of the dust particles.

Observed extinction curves vary in shape from one line of sight to another. The slope of the extinction at visible wavelengths is characterized by the dimensionless ratio

$$R_V \equiv \frac{A_V}{A_B - A_V} \equiv \frac{A_V}{E(B-V)} \quad , \tag{21.3}$$

where A_B and A_V are the extinctions measured in the B (4405 Å) and V (5470 Å) photometric bands, and $E(B-V) \equiv A_B - A_V$ is the "reddening."

Sightlines through diffuse gas in the Milky Way have $R_V \approx 3.1$ as an average value. The extinction A_λ, relative to A_V, is given in Table 21.1 for a number of

Table 21.1 Extinction for Standard Photometric Bands for $R_V = 3.1$

Band	$\lambda(\mu\text{m})$	A_λ/A_{I_C}	Band	$\lambda(\mu\text{m})$	A_λ/A_{I_C}
M	4.75	0.0573	i	0.7480	1.125
L'	3.80	0.0842	R_C	0.6492	1.419
L	3.45	0.101	R_J	0.6415	1.442
K	2.19	0.212	r	0.6165	1.531
H	1.65	0.315	V	0.5470	1.805
J	1.22	0.489	g	0.4685	2.238
z	0.893	0.830	B	0.4405	2.396
I_J	0.8655	0.879	U	0.3635	2.813
I_C	0.8020	1.000	u	0.3550	2.867

standard photometric bands for sightlines characterized by $R_V \approx 3.1$. The smallest well-determined value is $R_V = 2.1$ toward the star HD 210121 (Welty & Fowler 1992); the extinction toward HD 201021 is shown in Fig. 21.2. Sightlines through dense regions tend to have larger values of R_V; the sightline toward HD 36982 has $R_V \approx 5.7$ (Cardelli et al. 1989; Fitzpatrick 1999).

21.2 Parametric Fits to the Extinction Curve

A very useful parametrization of the extinction curve within the Milky Way was provided by Cardelli et al. (1989), who showed that the extinction relative to some reference wavelength λ_{ref} can be well-described as a function of λ by a fitting function

$$A_\lambda/A_{\lambda_{\text{ref}}} \approx f_7^{\text{CCM}}(\lambda) \quad , \tag{21.4}$$

where f_7^{CCM} has seven adjustable parameters. At wavelengths $3.5\,\mu\text{m} > \lambda > 3030\,\text{Å}$, the function $f_7^{\text{CCM}}(\lambda)$ depends only on λ and the single parameter R_V.

Six parameters are required to describe the UV extinction. Three parameters specify the strength, central wavelength, and width of the 2175 Å "bump" (relative to A_V), and three specify the slope and curvature of the continuous extinction underlying the bump and extending to shorter wavelengths. So-called CCM extinction curves are obtained using the function $f_7^{\text{CCM}}(\lambda)$ with suitable choices for the seven fit parameters.

Cardelli et al. (1989) showed that if the single quantity R_V is known, it is possible to estimate the values of the other six parameters so that the optical-UV extinction can be approximated by a one-parameter family of curves:

$$A_\lambda/A_{\lambda_{\text{ref}}} \approx f_1^{\text{CCM}}(\lambda; R_V) \quad . \tag{21.5}$$

Fitzpatrick (1999) recommends a slightly modified function $f_1^{\text{CCM}}(\lambda, R_V)$, which has been used to generate the synthetic extinction curves in Fig. 21.2 for $R_V = 2.1$, 2.5, 3.1, 4.0, and 5.5. The extinction was extended into the infrared following Draine (1989*a*).

We will discuss dust grain optics in Chapter 22, but it is clear that if the dust grains were large compared to the wavelength, we would be in the "geometric optics" limit, and the extinction cross section would be independent of wavelength, with $R_V = \infty$. The tendency for the extinction to rise with decreasing λ, even at the shortest ultraviolet wavelengths where we can measure it, tells us that grains smaller than the wavelength must be making an appreciable contribution to the extinction at all observed wavelengths, down to $\lambda = 0.1\,\mu\mathrm{m}$. As we will see in the following, "small" means (approximately) that $2\pi a/\lambda \lesssim 1$. Thus interstellar dust must include a large population of grains with $a \lesssim .015\,\mu\mathrm{m}$.

The dust responsible for interstellar extinction appears to be relatively well-mixed with the gas; the gas and dust go together, with

$$\frac{N_{\mathrm{H}}}{E(B-V)} = 5.8 \times 10^{21}\mathrm{H\,cm^{-2}mag^{-1}} \tag{21.6}$$

(Bohlin et al. 1978; Rachford et al. 2009). For sightlines with $R_V \equiv A_V/E(B-V) \approx 3.1$, this implies that

$$\frac{A_V}{N_{\mathrm{H}}} = \frac{3.1}{5.8 \times 10^{21}\mathrm{H\,cm^{-2}mag^{-1}}} = 5.3 \times 10^{-22}\mathrm{mag\,cm^2H^{-1}} \quad . \tag{21.7}$$

21.3 Polarization by Interstellar Dust

The polarization of starlight was discovered serendipitously in 1949 (Hall 1949; Hall & Mikesell 1949; Hiltner 1949*a*,*b*). When it was realized that the degree of polarization tended to be larger for stars with greater reddening, and that stars in a given region of the sky tended to have similar polarization directions, it became obvious that the polarization is produced by the ISM: initially upolarized light propagating through the ISM becomes linearly polarized as a result of preferential extinction of one linear polarization mode relative to the other. Figure 21.3 shows the direction of polarization and the strength of polarization for 5453 stars with galactic latitudes b between $-80°$ and $+80°$. The large-scale organization of the polarization vectors can be understood if dust grains are somehow aligned by the interstellar magnetic field.

The polarization percentage typically peaks near the V band (5500 Å), and can be empirically described by the "Serkowski law" (Serkowski 1973):

$$p(\lambda) \approx p_{\max} \exp[-K \ln^2(\lambda/\lambda_{\max})] \quad , \tag{21.8}$$

with $\lambda_{\max} \approx 5500$ Å and $K \approx 1.15$. The peak polarization $p_{\max}$ is found to fall

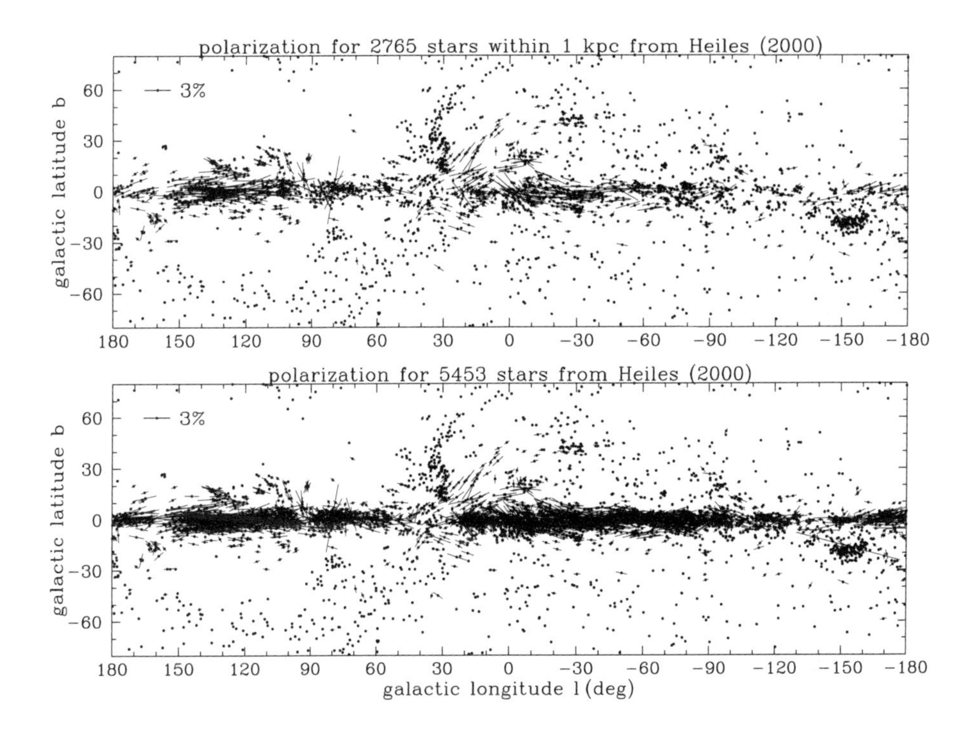

Figure 21.3 Linear polarization of starlight plotted in galactic coordinates, for stars within 1 kpc, and for all stars in the catalog of Heiles (2000). The length of each line segment is proportional to the degree of polarization.

within an envelope

$$0 < p_{\rm max} \leq 0.09 \left[\frac{E(B-V)}{\rm mag}\right] \approx 0.03 \left[\frac{A_V}{\rm mag}\right] \quad , \tag{21.9}$$

or $0 < p_V \lesssim 0.03\tau_V$.

The polarization is produced by dust grains that are somehow partially aligned by the interstellar magnetic field. It appears that the grains are aligned with their *shortest* axes tending to be parallel to the magnetic field direction. The largest values of $p_{\rm max}/E(B-V)$ are presumed to arise on sightlines where the magnetic field is uniform and perpendicular to the line of sight. While the Serkowski law was originally put forward as an empirical fit to the observed polarization at $0.3\,\mu{\rm m} \lesssim \lambda \lesssim 1\,\mu{\rm m}$, it turns out to give a surprisingly good approximation to the measured linear polarization in the vacuum ultraviolet (Clayton et al. 1992; Wolff et al. 1997), although there are some sightlines where the Serkowski law underpredicts the UV polarization, and one sightline where the 2175Å feature appears to be weakly polarized.

The mechanism responsible for the grain alignment remains a fascinating puzzle (see Chapter 26). Independent of the grain alignment mechanism, however, we can infer the sizes of the interstellar grains responsible for this polarization by noting that the extinction rises rapidly into the UV, whereas the polarization declines (**?**). This can be understood if the grains responsible for the polarization have diameters $2a$ such that $a \approx (\lambda_{\rm max}/2\pi) \approx 0.1\,\mu{\rm m}$: as one proceeds into the UV, one moves toward the "geometric optics" limit where both polarization modes suffer the same extinction, so the polarization goes to zero. Thus we conclude that:

- The extinction at $\lambda \approx 0.55\,\mu{\rm m}$ has an appreciable contribution from grains with sizes $a \approx 0.1\,\mu{\rm m}$. These grains are nonspherical and substantially aligned.
- The grains with $a \lesssim 0.05\,\mu{\rm m}$, which dominate the extinction at $\lambda \lesssim 0.3\,\mu{\rm m}$, are either spherical (which seems unlikely) or minimally aligned.

21.4 Scattering of Starlight by Interstellar Dust

When an interstellar cloud happens to be unusually near one or more bright stars, we have a **reflection nebula**, where we see starlight photons that have been scattered by the dust in the cloud. (The blue nebulosity to the North of the Trifid Nebula

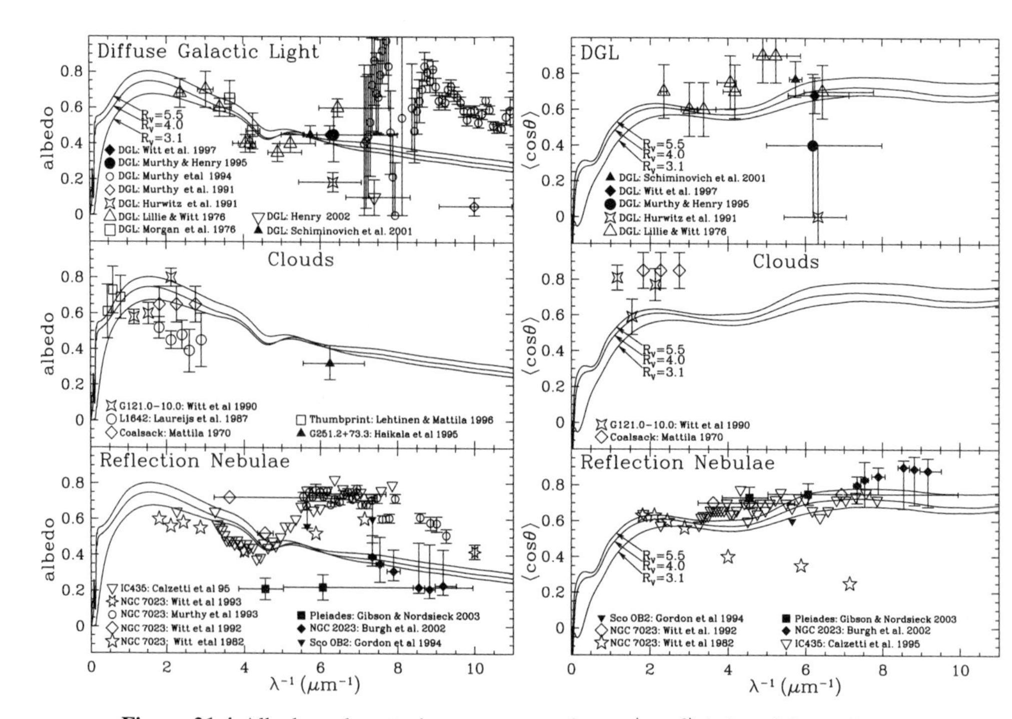

Figure 21.4 Albedo and scattering asymmetry factor $\langle\cos\theta\rangle$ inferred from observations of the diffuse galactic light, reflection nebulae, and dark clouds.

in Plate 11 is primarily scattered light.) The spectrum of the light coming from the cloud surface shows the stellar absorption lines, thus demonstrating that scattering rather than some emission process is responsible. By comparing the observed scattered intensity with the estimated intensity of the starlight incident on the cloud, it is possible to infer the **albedo** ω of the dust – the ratio of scattering cross section to extinction cross section. It is also possible to infer $\langle\cos\theta\rangle$ for the dust, where θ is the scattering angle.

Figure 21.4 shows ω and $\langle\cos\theta\rangle$ for (1) the dust in the general diffuse ISM producing the "diffuse galactic light." (2) dust in individual clouds illuminated by the general starlight, and (3) dust in clouds that are illuminated by a nearby bright star. In the optical, the interstellar dust mixture has an albedo $\omega \approx 0.5$ – scattering is about as important as absorption – and the grains are somewhat forward-scattering, with $\langle\cos\theta\rangle \approx 0.5$. Rayleigh scattering by particles small compared to the wavelength has $\langle\cos\theta\rangle \approx 0$, so this tells us that

- The particles dominating the scattering at $\lambda \approx 0.6\,\mu\text{m}$ have $a \gtrsim \lambda/2\pi \approx 0.1\,\mu\text{m}$.

21.5 Size Distribution of Interstellar Dust

Based on observations of ultraviolet extinction, scattering of visible light, and polarization of starlight, it is clear that the interstellar grain population must have a broad size distribution, extending from sizes as small as $a \approx 0.01\,\mu\text{m}$ (or even smaller) to sizes $a \approx 0.2\,\mu\text{m}$ (or even larger). In fact, we will see that observations of 3 to 12 μm infrared emission require that the size distribution extend down to grains containing as few as ~ 50 atoms, corresponding to volume-equivalent radii $a \approx 3.5\,\text{Å}$.

We will discuss the grain size distribution in more detail later; here it is sufficient to remark that the size distribution has most of the mass in the larger grains, and most of the surface area in the smaller grains.

21.6★ Purcell Limit: Lower Limit on Dust Volume

Purcell (1969) pointed out that the Kramers-Kronig relations can provide useful constraints in dust modeling. The Kramers-Kronig relations are general relations that apply to a "linear response function" (e.g., a dielectric function) which specifies the *response* (e.g., the electric polarization) to an applied *stress* (e.g., the applied electric field). The *only* assumptions are that (1) the response is *linear*, and (2) the system is *causal* – the response can depend on the stress applied in the past, but cannot depend on the future. The Kramers-Kronig relations can be derived from these very general assumptions [see Landau et al. (1993) for a derivation]. Applied to the complex dielectric function $\epsilon(\omega) = \epsilon_1 + i\epsilon_2$, the Kramers-Kronig relations are

$$\epsilon_1(\omega) = 1 + \frac{2}{\pi} P\int_0^\infty dx \, \frac{x\epsilon_2(x)}{x^2 - \omega^2} \quad , \tag{21.10}$$

$$\epsilon_2(\omega) = \frac{2}{\pi}\omega \, P\int_0^\infty dx \, \frac{[\epsilon_1(x) - 1]}{\omega^2 - x^2} \quad , \tag{21.11}$$

where P indicates that the principal value of the integral is to be taken.

Thus the real and imaginary parts of $\epsilon(\omega)$ are not independent – if one is specified at all frequencies, the other is fully determined.

21.6.1 Grain Volume per Hydrogen Atom

Purcell applied the Kramers-Kronig relations directly to the ISM itself. Let $\epsilon_{\rm ISM}(\omega)$ be the dielectric function of the ISM. Electromagnetic plane waves $E(x,t) \propto e^{ikx - i\omega t}$ propagate through the ISM, undergoing attenuation by scattering and absorption. If we consider only the response to electric fields, Maxwell's equations require that $k^2 = \epsilon_{\rm ISM}\omega^2/c^2$. The attenuation coefficient for power is

$$n_{\rm gr} C_{\rm ext}(\omega) = 2\,{\rm Im}(k) = 2(\omega/c)\,{\rm Im}\left(\sqrt{\epsilon_{\rm ISM}}\,\right) \approx (\omega/c)\,{\rm Im}(\epsilon_{\rm ISM}) \quad , \tag{21.12}$$

where $n_{\rm gr}$ is the number density of dust grains, and $C_{\rm ext}(\omega)$ is the extinction cross section of a dust grain. Purcell then used (21.10) with $\omega = 0$ to determine the contribution of dust to the static polarizability $\epsilon_{\rm ISM,1}(\omega = 0)$ of the ISM.[1] If $n_{\rm gr}$ is the number density of grains with volume $V_{\rm gr}$ and extinction cross section $C_{\rm ext}(\lambda)$, then

$$\frac{n_{\rm gr} V_{\rm gr}}{n_{\rm H}} = \frac{1}{3\pi^2 F({\rm shape}, \epsilon_0)} \int_0^\infty d\lambda \frac{n_{\rm gr}}{n_{\rm H}} C_{\rm ext}(\lambda) \quad , \tag{21.13}$$

where the dimensionless function $F({\rm shape}, \epsilon_0)$ is the ratio of the orientationally averaged static polarizability of grains of specified shape and composed of material with dielectric function ϵ_0 divided by the polarizability of an equal-volume conducting sphere. For spheroids,

$$F(a/b, \epsilon_0) = \frac{\epsilon_0 - 1}{3}\left[\frac{1}{(\epsilon_0 - 1)3L_a + 3} + \frac{2}{(\epsilon_0 - 1)3(1 - L_a)/2 + 3}\right] , \tag{21.14}$$

where L_a is a "shape factor," with $L_a = 1$ for $a/b \to 0$ (pancake), $L_a = \frac{1}{3}$ for a sphere, and $L_a \to 0$ for $a/b \to \infty$ (needle). Insulating materials have finite ϵ_0, and $\epsilon_0 = \infty$ for conductors.

The function F is shown for selected spheroidal shapes in Fig. 21.5. F depends on shape, but relatively weakly, except for conducting materials and extreme shapes.

[1] Considering only the response of the dust – it is assumed that no free charge is present.

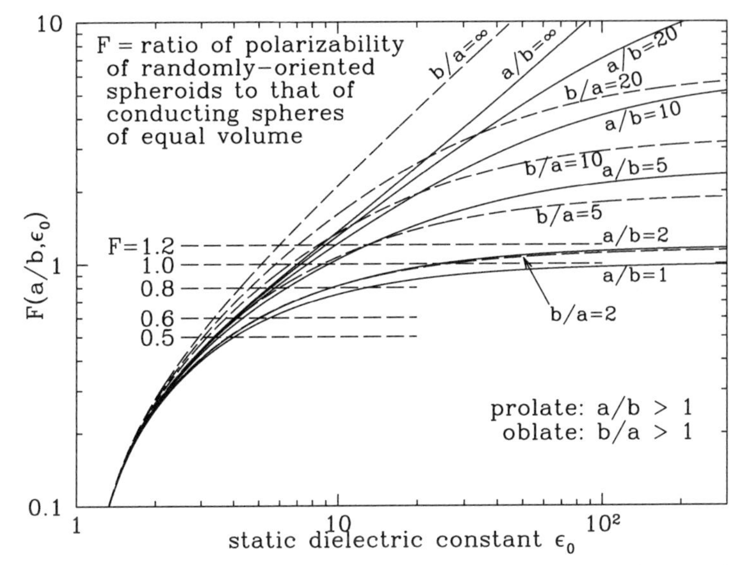

Figure 21.5 Function $F(a/b, \epsilon_0)$ for spheroids, for selected values of the axial ratio a/b. After Purcell (1969). For $0.5 \lesssim a/b \lesssim 2$ and $\epsilon_0 > 4$, we have $0.5 < F < 1.2$.

The extinction produced by dust is difficult to measure at very long wavelengths because it is small, and at wavelengths $\lambda < 912$ Å because of the strong absorption by atomic hydrogen, but we do have empirical knowledge of the extinction curve between $0.1\,\mu\text{m}$ and, say, $30\,\mu\text{m}$:

$$\int_{0.1\,\mu\text{m}}^{30\,\mu\text{m}} \frac{\tau_{\text{ext}}}{N_{\text{H}}} d\lambda \approx 1.1 \times 10^{-25}\,\text{cm}^3/\text{H} \quad . \tag{21.15}$$

Approximately half of this integral is contributed by $0.1 < \lambda < 1\,\mu\text{m}$, and half by $1 < \lambda < 30\,\mu\text{m}$. This gives us a lower bound on the volume of grain material per H nucleon:

$$\frac{n_{\text{gr}} V_{\text{gr}}}{n_{\text{H}}} \gtrsim 3.7 \times 10^{-27} F^{-1}\,\text{cm}^3/\text{H} \quad , \tag{21.16}$$

or, if the grain material has solid density ρ_{gr}, a lower bound on the mass of grain material relative to H mass:

$$\frac{M_{\text{gr}}}{M_{\text{H}}} \gtrsim 0.0056 \left(\frac{1.2}{F}\right) \left(\frac{\rho_{\text{gr}}}{3\,\text{g}\,\text{cm}^{-3}}\right) \quad , \tag{21.17}$$

where the reference density $\rho_{\text{gr}} = 3\,\text{g}\,\text{cm}^{-3}$ is intermediate between the density of graphite ($\rho = 2.24\,\text{g}\,\text{cm}^{-3}$) and olivine MgFeSiO_4 ($\rho \approx 3.8\,\text{g}\,\text{cm}^{-3}$).

This lower bound is model-independent, except through the dependence on the unknown shape factor F. However, we see from Figure 21.5 that for materials with $\epsilon_0 > 4$, grains with moderate shapes $0.5 < a/b < 2$ have $0.5 \lesssim F \lesssim 1.2$. The only way to have $F > 1.2$ is to have an extreme shape ($a/b > 10$ or $a/b < 0.1$) *and* a dielectric function $\epsilon_0 \gtrsim 10$. Note also that we have entirely ignored extinction at $\lambda < 0.1\,\mu\text{m}$ or $\lambda > 30\,\mu\text{m}$; these additional contributions to the integral in Eq. (21.13) are what make Eq. (21.17) only a lower limit. Thus, if we assume that $F \lesssim 1.2$, and if interstellar grains have solid densities $\rho_{\text{gr}} \approx 3\,\text{g}\,\text{cm}^{-3}$, the Kramers-Kronig integral implies a grain mass $M_{\text{gr}}/M_{\text{H}} \gtrsim 0.0056 \times (1.2/F)$. A reasonable estimate for F might be $F \approx 0.8$ (see Fig. 21.5), in which case the Kramers-Kronig argument gives

$$\frac{n_{\text{gr}} V_{\text{gr}}}{n_{\text{H}}} \gtrsim 4.6 \times 10^{-27}\,\text{cm}^3/\text{H} \quad , \tag{21.18}$$

$$\frac{M_{\text{gr}}}{M_{\text{H}}} \gtrsim 0.0083 \left(\frac{\rho_{\text{gr}}}{3\,\text{g}\,\text{cm}^{-3}} \right) \quad . \tag{21.19}$$

This *lower bound* on the grain mass places a strong constraint on grain models.

21.6.2 Asymptotic Behavior at Long Wavelengths

Suppose that $C_{\text{ext}} \propto \lambda^{-\beta}$ as $\lambda \to \infty$. The Kramers-Kronig integral $\int d\lambda\, C_{\text{ext}}$ in Eq. (21.13) would obviously be divergent unless $\beta > 1$, and if β is only slightly larger than 1, then the lower bound we obtained by considering only the $\lambda < 30\,\mu\text{m}$ extinction might seriously underestimate the total volume of grain material. It seems much more likely that $\beta \approx 2$, in which case wavelengths $\lambda > 30\,\mu\text{m}$ make only a modest contribution to the integral. We will see in §22.4 that $\beta \approx 2$ is expected for simple models of both insulating and conducting materials.

21.7 Infrared Emission

Dust grains are heated by starlight, and cool by radiating in the infrared. Plates 6, 7, and 8 show the $8\mu\text{m}$ emission from PAHs in the Andromeda galaxy, M31, the starburst galaxy M81, and the Whirlpool galaxy, M51.

An all-sky map of the dust emission at $100\,\mu\text{m}$ is shown in Plate 2. The emission from dust at high galactic latitudes has been studied by a number of satellites. Figure 21.6 shows the emission spectrum from $800\,\mu\text{m}$ to $3\,\mu\text{m}$. The 3 to $12\,\mu\text{m}$ spectrum is estimated from observations of the Galactic plane near $l \approx 45°$, if we assume that the ratio of 3 to $12\,\mu\text{m}$ emission to the $100\,\mu\text{m}$ emission is unchanged in going from observations of the Galactic plane to high galactic latitudes. The correlation of the IR emission with H I 21-cm emission at high latitudes (see Fig. 29.4) is used to estimate the power radiated per H nucleon: $5.0 \times 10^{-24}\,\text{erg}\,\text{s}^{-1}/\text{H}$.

Interstellar dust is heated primarily by starlight (as will be discussed in Chapter 24), and the total power radiated requires, therefore, that the absorption cross section of interstellar dust be such that the power absorbed per H (for the estimated spectrum of the starlight heating the dust) match the observed emission, $5 \times 10^{-24}\,\text{erg}\,\text{s}^{-1}\text{H}^{-1}$. The infrared spectrum provides very strong constraints

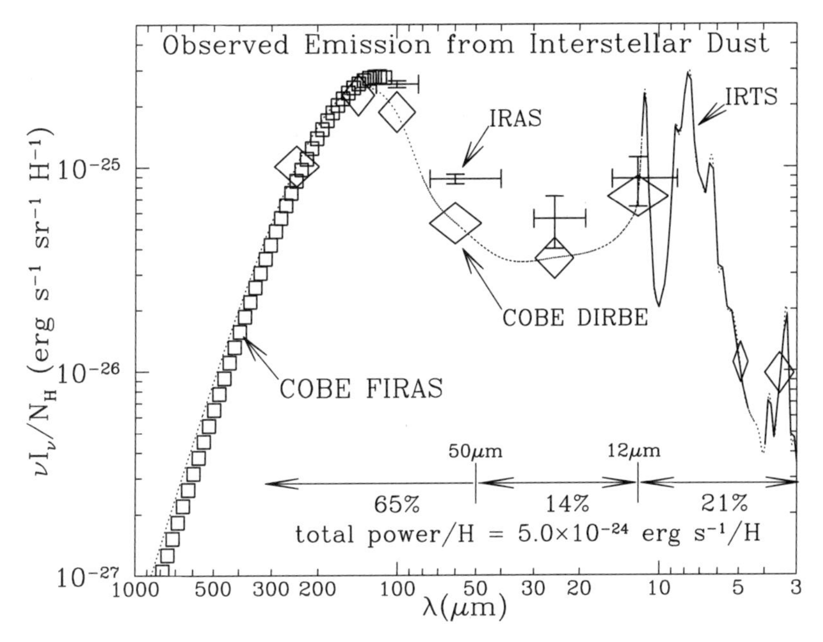

Figure 21.6 Observed infrared emission per H nucleon from dust heated by the average starlight background in the local Milky Way. Crosses: IRAS (Boulanger & Perault 1988); squares: COBE-FIRAS (Wright et al. 1991); diamonds: COBE-DIRBE (Arendt et al. 1998); heavy curve: IRTS (Onaka et al. 1996; Tanaka et al. 1996). The interpolated dotted line is used to estimate the total power.

on grain models, as the dust must include a component that can account for the fact that $\sim 35\%$ of the radiated power is shortward of $50\,\mu\text{m}$, including the strong emission features at $\sim 12\,\mu\text{m}$ and $6-8\,\mu\text{m}$. Models for interstellar dust that can reproduce this emission spectrum will be discussed in §24.

21.8★ Luminescence

The energy absorbed by dust grains is primarily reradiated in the mid- and far-IR, but there is evidence that dust grains also emit light at optical and near-IR wavelengths. Studies of reflection nebulae indicate that there is more light emerging at wavelengths $6000-8000\,\text{Å}$ than can be accounted for by scattering alone, and this excess is ascribed to **luminescence** from dust grains following absorption of shorter-wavelength photons [see the review by Witt & Vijh (2004)]. Luminescence at 6000 to 8000 Å is also termed "extended red emission," or ERE. Luminescence in the blue has also been reported (Vijh et al. 2005). Candidate materials to explain this luminescence must of course reproduce the observed luminescence spectrum. The luminescing materials have not yet been conclusively identified. The blue luminescence may be produced by neutral PAHs (Vijh et al. 2005), and PAH di-cations (PAH^{++}) may be responsible for the ERE (Witt et al. 2006).

Chapter Twenty-two

Scattering and Absorption by Small Particles

22.1 Cross Sections and Efficiency Factors

A number of different quantities are used to characterize the absorption, scattering, and emission of electromagnetic radiation by a (nonrotating) dust grain:

- The **absorption cross section** at wavelength λ, $C_{\rm abs}(\lambda)$.
- The **scattering cross section** $C_{\rm sca}(\lambda)$.
- The **extinction cross section** $C_{\rm ext}(\lambda) \equiv C_{\rm abs} + C_{\rm sca}$.
- The **albedo**

$$\omega \equiv \frac{C_{\rm sca}}{C_{\rm abs} + C_{\rm sca}} = \frac{C_{\rm sca}}{C_{\rm ext}} \quad . \tag{22.1}$$

- The **differential scattering cross section**

$$\frac{dC_{\rm sca}(\theta)}{d\Omega} \tag{22.2}$$

for incident unpolarized light to be scattered by an angle θ. This is related to the dimensionless **Muller matrix element** S_{11} by

$$\frac{dC_{\rm sca}}{d\Omega} \equiv \frac{S_{11}(\theta)}{k^2} \quad , \tag{22.3}$$

where $k \equiv 2\pi/\lambda$.

- The **mean value of $\cos\theta$** for scattered light

$$\langle \cos\theta \rangle = \frac{1}{C_{\rm sca}} \int_0^{\pi} \cos\theta \, \frac{dC_{\rm sca}}{d\Omega} \, 2\pi \sin\theta d\theta \quad . \tag{22.4}$$

- The **radiation pressure cross section**

$$C_{\rm pr}(\lambda) \equiv C_{\rm abs}(\lambda) + (1 - \langle\cos\theta\rangle)C_{\rm sca}(\lambda) \quad . \tag{22.5}$$

- The **degree of polarization** $P(\theta)$ for light scattered through an angle θ (for incident unpolarized light).

For a given direction of incidence relative to a fixed grain, we would obviously need two angles (θ, ϕ) to fully specify the scattering direction. However, for spherical grains, or for an ensemble of randomly oriented grains, the scattering properties can be described as a function of a single scattering angle θ.

In some cases, one wants to consider scattering of polarized light. For this case, it is usual to use the four-element **Stokes vector** to specify the intensity and state of polarization of radiation propagating in a particular direction. The ability of a grain to scatter radiation with incident Stokes vector $\mathbf{V}_{\rm in}$ to outgoing Stokes vector $\mathbf{V}_{\rm sca}$ is conveniently specified by a 4×4 dimensionless scattering matrix S_{ij}, known as the Muller matrix. [See Bohren & Huffman (1983) or Mishchenko et al. (2000) for discussions of scattering concepts and terminology.]

It is convenient to normalize the absorption and scattering cross sections $C_{\rm abs}$ and $C_{\rm sca}$ to some area characterizing the grain. In the case of a spherical grain, it is natural to use the grain geometric cross section πa^2.

For nonspherical grains, some authors choose to normalize using the geometric cross section as seen from the direction of the incident radiation; other authors choose to normalize using the average geometric cross section for random orientations.

Here, we will instead normalize to the **geometric cross section of an equal-solid-volume sphere**. For a target with solid volume V (V does not include the volume of any voids, if present), we define efficiency factors $Q_{\rm sca}$, $Q_{\rm abs}$ and $Q_{\rm ext} \equiv Q_{\rm abs} + Q_{\rm sca}$ by

$$Q_{\rm sca} \equiv \frac{C_{\rm sca}}{\pi a_{\rm eff}^2} \quad , \quad Q_{\rm abs} \equiv \frac{C_{\rm abs}}{\pi a_{\rm eff}^2} \quad , \quad a_{\rm eff} \equiv \left(\frac{3V}{4\pi}\right)^{1/3} \quad . \tag{22.6}$$

Here, $a_{\rm eff}$ is the radius of an equal-volume sphere. This is a natural choice, because it relates the scattering and absorption cross sections directly to the actual volume of grain material.

22.2 Dielectric Function and Refractive Index

In order to calculate scattering and absorption of electromagnetic waves by targets, we need to characterize the response of the target material to the local oscillating electric and magnetic fields. At submillimeter frequencies and above, real materials have only a negligible response to an applied magnetic field – this is because the magnetization of materials is the result of aligned electron spins and electron orbital currents, and an electron spin (or orbit) can change direction only on time scales longer than the period for the electron spin (or orbit) to precess in the local (microscopic) magnetic fields within atoms and solids. These fields are at most $B_i \lesssim 10\,\mathrm{kG}$, and the precession frequencies are $\omega_p \approx \mu_\mathrm{B} B_i/\hbar \lesssim 10^{10}\,\mathrm{s}^{-1}$, where μ_B is the Bohr magneton. When a weak applied field oscillates at frequencies $\omega \gg 10^{10}\,\mathrm{s}^{-1}$, the magnetization of the material cannot respond. As a result, for frequencies $\nu \gtrsim 10\,\mathrm{GHz}$ we normally set the magnetic permeability $\mu = 1$, and consider only the material's response to the oscillating electric field.

The response of material to an applied oscillating electric field $E = E_0 \mathrm{e}^{-i\omega t}$ is characterized by a complex **dielectric function**

$$\epsilon(\omega) = \epsilon_1 + i\epsilon_2 \quad . \tag{22.7}$$

The electrical conductivity σ, if any, can be absorbed within the imaginary part of the dielectric function, with the replacement

$$\epsilon \to \epsilon + \frac{4\pi i\sigma}{\omega} \quad . \tag{22.8}$$

The complex **refractive index** $m(\omega)$ is related to the complex dielectric function by $m = \sqrt{\epsilon}$.

There are two sign conventions for the imaginary part of the dielectric function or refractive index. If we choose to write oscillating quantities $\propto \mathrm{e}^{i\mathbf{k}\cdot\mathbf{r}-i\omega t}$, then $\mathrm{Im}(\epsilon) > 0$ and $\mathrm{Im}(m) > 0$ for absorbing, dissipative materials, where a propagating wave is attenuated. This is the convention that we will use.[1] In terms of the refractive index, the wave vector

$$k = m(\omega)\,\frac{\omega}{c} \tag{22.9}$$

and, therefore, the electric field

$$E \propto \mathrm{e}^{ikx-i\omega t} \propto \mathrm{e}^{-\mathrm{Im}(m)\omega x/c} \tag{22.10}$$

and the power in the wave ($\propto |E|^2$) decays as $\exp\left[-2\,\mathrm{Im}(m)\omega x/c\right]$. Therefore, the attenuation coefficient κ and attenuation length $L_\mathrm{abs} \equiv 1/\kappa$ for the wave are simply

$$\kappa(\omega) = 2\,\mathrm{Im}(m)\frac{\omega}{c} \quad , \quad L_\mathrm{abs}(\omega) = \frac{c}{2\omega\,\mathrm{Im}(m)} = \frac{\lambda}{4\pi\mathrm{Im}(m)} \quad . \tag{22.11}$$

where $\lambda = 2\pi c/\omega$ is the wavelength in vacuo.

[1] Alternatively, if one chooses to write quantities $\propto \mathrm{e}^{i\omega t-i\mathbf{k}\cdot\mathbf{r}}$, then $\mathrm{Im}(\epsilon) < 0$ and $\mathrm{Im}(m) < 0$ for absorbing materials. This sign convention is used, e.g., by van de Hulst (1957).

22.3 Electric Dipole Limit: Size $\ll \lambda$

We are often interested in situations where the grain is much smaller than the wavelength of the incident electromagnetic wave. In this situation, the small grain is subject to an incident applied electric field that is nearly uniform in space. The electric field inside the grain will be proportional to the applied external electric field $\mathrm{Re}\left(\mathbf{E}_0 e^{-i\omega t}\right)$. Averaged over one cycle, the rate per volume at which energy is absorbed within the grain is proportional to $\omega\epsilon_2 E_0^2$.

The absorption and scattering cross sections can be written

$$C_{\rm abs} = \frac{4\pi\omega}{c}\mathrm{Im}(\alpha) \quad , \tag{22.12}$$

$$C_{\rm sca} = \frac{8\pi}{3}\left(\frac{\omega}{c}\right)^4 |\alpha|^2 \quad , \tag{22.13}$$

where α is the **electric polarizability** of the grain: the electric dipole moment of the grain $\mathbf{P} = \alpha\mathbf{E}$, where $\mathbf{E}$ is the instantaneous applied electric field. Calculating the polarizability in the limit $\omega a/c \rightarrow 0$ becomes a problem in electrostatics.

22.3.1 Ellipsoids

Analytic solutions are known for ellipsoids (with spheres or spheroids as special cases). If the electric field is oriented parallel to one of the principal axes j of the ellipsoid, the polarizability is

$$\alpha_{jj} = \frac{V}{4\pi}\left[\frac{\epsilon - 1}{(\epsilon - 1)L_j + 1}\right] \quad , \tag{22.14}$$

where L_j is called the "shape factor" for $\mathbf{E}$ along axis j. For a spheroid with length $2a$ along the symmetry axis, and diameter $2b$ perpendicular to the symmetry axis, the shape factors are given by

$$L_a = \frac{1-e^2}{e^2}\left[\frac{1}{2e}\ln\left(\frac{1+e}{1-e}\right) - 1\right] \quad , \text{ for } a > b \text{ (prolate spheroid)}, \tag{22.15}$$

$$= \frac{1+e^2}{e^2}\left[1 - \frac{1}{e}\arctan(e)\right] \quad , \text{ for } a < b \text{ (oblate spheroid)}, \tag{22.16}$$

$$L_b = \frac{1}{2}(1 - L_a) \quad , \tag{22.17}$$

$$e^2 \equiv \left|1 - (b/a)^2\right| \quad . \tag{22.18}$$

Needles ($a/b \rightarrow \infty$) have $L_a \rightarrow 0$, and pancakes ($a/b \rightarrow 0$) have $L_a \rightarrow 1$. For a sphere, $L_a = L_b = 1/3$, and

$$C_{\rm abs} = 18\pi \frac{\epsilon_2}{(\epsilon_1+2)^2+\epsilon_2^2} \frac{V}{\lambda} \quad , \tag{22.19}$$

$$C_{\rm sca} = 24\pi^3 \left|\frac{\epsilon-1}{\epsilon+2}\right|^2 \frac{V^2}{\lambda^4} \quad . \tag{22.20}$$

From Eqs. (22.12 and 22.13), we see that $C_{\rm abs} \propto V$ and $C_{\rm sca} \propto V^2$. Therefore, in the limit $V \to 0$, absorption dominates: $C_{\rm abs} \gg C_{\rm sca}$ provided only that the material itself is absorptive (i.e., $\epsilon_2 > 0$). At wavelengths that are long compared to the particle size, the opacity is simply proportional to the total volume of grain material, independent of the sizes of the individual particles.

22.4 Limiting Behavior at Long Wavelengths

22.4.1 Insulators

At sufficiently long wavelengths (at frequencies well below the lowest frequency resonance in the solid), insulators tend to have

$$\epsilon_1 \to \epsilon_0 = \text{const.} \quad , \tag{22.21}$$

$$\epsilon_2 \to A\omega \quad , \tag{22.22}$$

where $A = \text{const.}$ (with dimensions of time). In this limit, it is easy to see from Eqs. (22.19) and (22.20) that, for a sphere of volume V,

$$C_{\rm abs} \to 36\pi^2 \frac{Ac}{(\epsilon_0+2)^2} \frac{V}{\lambda^2} \quad , \tag{22.23}$$

$$C_{\rm sca} \to 24\pi^3 \frac{(\epsilon_0-1)^2}{(\epsilon_0+2)^2} \frac{V^2}{\lambda^4} \quad . \tag{22.24}$$

Therefore, for insulators, we expect $C_{\rm abs} \propto \lambda^{-2}$ at long wavelengths. We also see that if $A \neq 0$, absorption will dominate over scattering as $\lambda \to \infty$.

22.4.2 Conductors

For conductors with conductivity σ_0 at zero frequency, the low-frequency limiting behavior of the dielectric function is[2]

$$\epsilon_1 \rightarrow \epsilon_0 = \text{const.} \quad , \tag{22.25}$$

$$\epsilon_2 \rightarrow A\omega + \frac{4\pi\sigma_0}{\omega} \quad . \tag{22.26}$$

In the long wavelength limit, we then have

$$C_{\rm abs} \rightarrow \frac{9\pi c}{\sigma_0} \frac{V}{\lambda^2} \quad , \tag{22.27}$$

$$C_{\rm sca} \rightarrow 24\pi^3 \frac{V^2}{\lambda^4} \quad . \tag{22.28}$$

Note that at long wavelengths, $C_{\rm abs} \propto 1/\sigma_0$: materials with high conductivities are weak absorbers. This is because for highly conducting materials, the electric field is screened from the interior by surface charge.

Note also that, just as in the case of insulators, conducting materials at long wavelengths have $C_{\rm abs} \propto \lambda^{-2}$. Therefore, it is rather natural to expect absorption by interstellar grains to vary as λ^{-2} at very long wavelengths.

22.5 Sizes Comparable to Wavelength: Mie Theory

At optical and ultraviolet wavelengths, the dust particles are not necessarily small compared to the wavelength, and the electric dipole approximation is no longer applicable. We must find the solution to Maxwell's equations with an incident plane wave, for an object of specified size and shape, composed of material with a specified dielectric function ϵ or refractive index m.

For the special case of a sphere, an elegant analytic solution was found by Mie (1908) and Debye (1909), and is known as **Mie theory**. In brief, the electromagnetic field inside and outside the sphere can be decomposed into spherical harmonics with appropriate radial functions, with coefficients determined by the need to

---

[2] A simple model for a conductor has

$$\epsilon(\omega) = 1 + \Delta\epsilon^{\rm bound}(\omega) + \frac{i\omega_p^2\tau}{\omega(1 - i\omega\tau)} \quad ,$$

$$\epsilon_1(\omega) = 1 + \Delta\epsilon_1^{\rm bound}(\omega) - \frac{(\omega_p\tau)^2}{1 + (\omega\tau)^2} \quad ,$$

$$\epsilon_2(\omega) = \Delta\epsilon_2^{\rm bound}(\omega) + \frac{\omega_p^2\tau}{\omega\left[1 + (\omega\tau)^2\right]} \quad ,$$

where $\Delta\epsilon^{\rm bound}$ is the contribution from the bound electrons, and the free electrons are characterized by a plasma frequency ω_p and damping time τ.

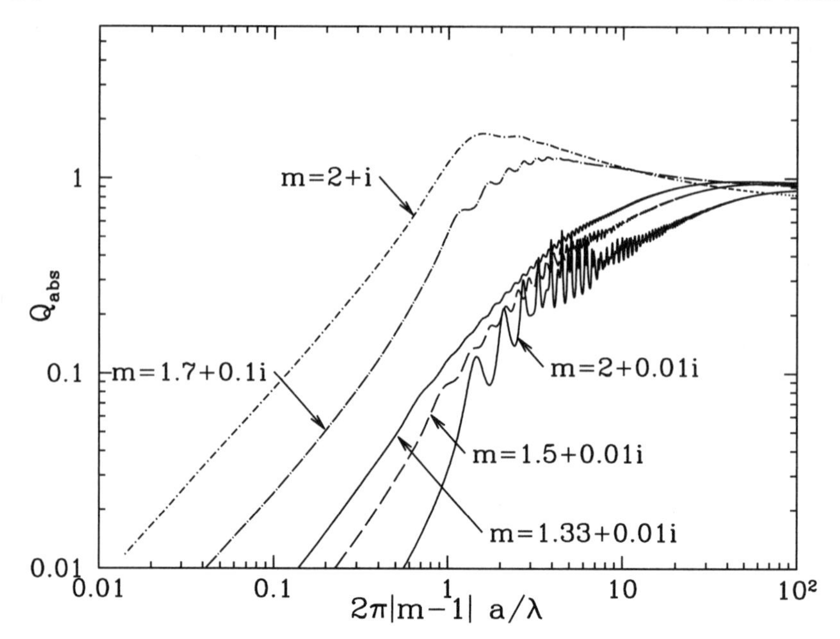

Figure 22.1 Absorption efficiency factors $Q_{\rm abs}$ for spheres with various refractive indices m.

give an incident plane wave at infinity and to satisfy the continuity conditions at the surface of the sphere.[3] Computer programs to evaluate the Mie theory solution are widely available.[4]

The character of the electromagnetic scattering will depend on the dimensionless ratio a/λ and on the dimensionless refractive index $m(\omega)$. One relevant parameter will be the phase shift of a wave traveling a distance equal to the grain radius within the grain, expressed in radians. For nonabsorptive material, this would be just $2\pi a|m-1|/\lambda$. Figures 22.1 to 22.3 show five examples, where we plot the absorption, scattering, and extinction efficiency factors against this phase shift.

The details depend on the refractive index m, but the general trend is for $Q_{\rm ext}$ to rise to a value $Q_{\rm ext} \approx 3-5$ near $|m-1|2\pi a/\lambda \approx 2$. For dielectric functions with small imaginary components [i.e., weakly absorbing material, $\mathrm{Im}(m) \ll 1$] $Q_{\rm ext}$ as a function of a/λ shows oscillatory behavior due to interference effects, but the oscillations are minimal for strongly absorbing materials [$\mathrm{Im}(m) \gtrsim 1$].

For $(a/\lambda) \to \infty$, all of these examples have $Q_{\rm ext} \to 2$. This is a general result, sometimes referred to as "the extinction paradox":

- For $x \equiv 2\pi a/\lambda \to \infty$ and $|m-1|x \to \infty$, the extinction cross section is equal to **exactly** twice the geometric cross section.

[3]The Mie theory solution is effectively a series expansion in powers of $x = 2\pi a/\lambda$. The series is convergent, but the number of terms that must be retained is $\sim O(x)$.

[4]For example, the program bhmie.f available at http://www.astro.princeton.edu/~draine/scattering.html .

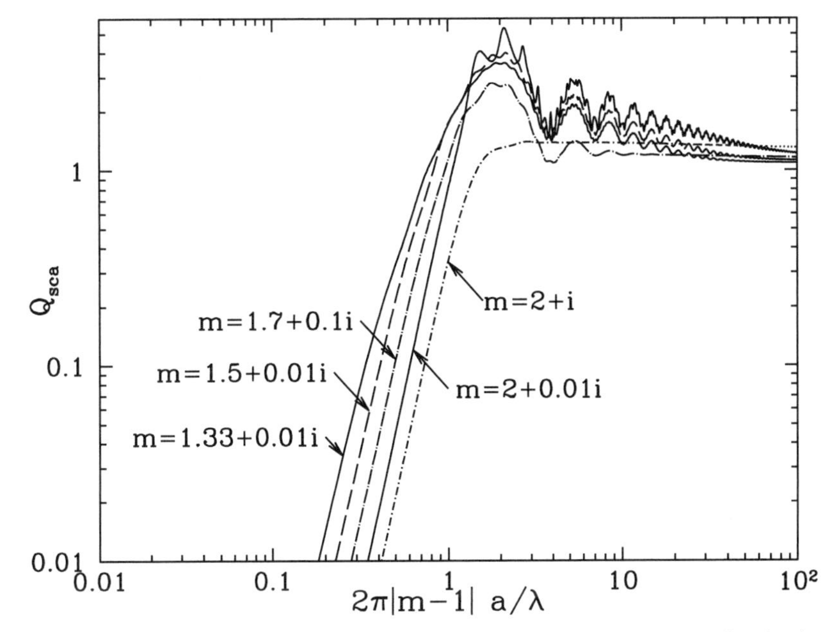

Figure 22.2 Scattering efficiency factors $Q_{\rm sca}$ for spheres with various refractive indices m.

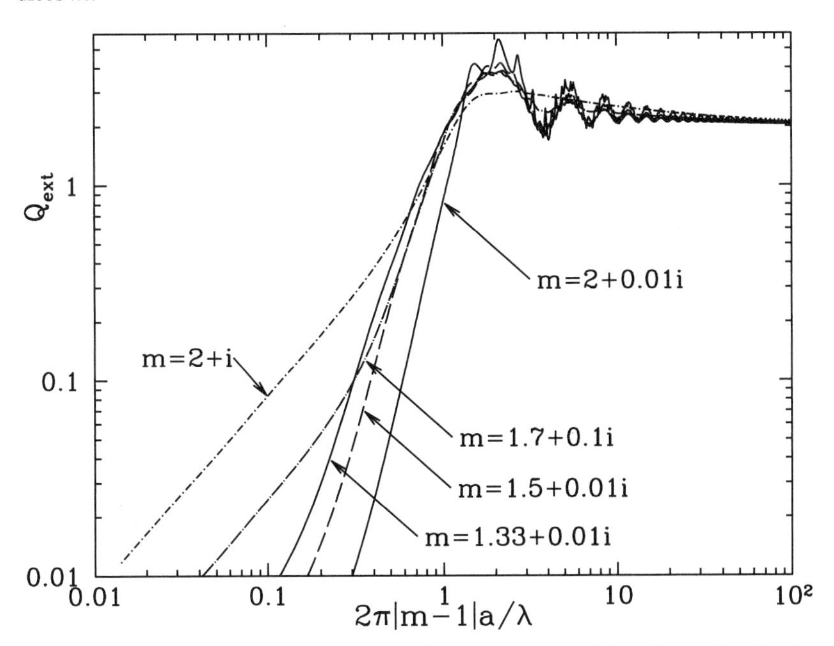

Figure 22.3 Extinction efficiency factors $Q_{\rm ext}$ for spheres with various refractive indices m. Note that $Q_{\rm ext} \to 2$ for $|m-1|a/\lambda \to \infty$.

Ray-tracing arguments would lead us to expect the extinction cross section to be equal to the geometric cross section, but diffraction around the target leads to additional small-angle scattering, with the total extinction cross section equal to *twice* the geometric cross section.

22.6★ Nonspherical Particles

Mie theory is a powerful and robust computational tool with which one can efficiently calculate scattering and absorption by spheres with a wide range of dielectric constants, for $x \equiv 2\pi a/\lambda \lesssim 10^4$. For $x > 10^4$, cancellation in the alternating series leads to roundoff errors on machines with 64-bit arithmetic, but for the size distributions that are present in the ISM, scattering by the dust mixture is usually dominated by particles with $x \approx 1$, and particles with $x \gg 1$ can generally be ignored except at X-ray energies.

However, one thing we know for certain about interstellar grains: the observed polarization of starlight implies that they are *not* spherical. If the grains are not spherical, how are we to calculate scattering and absorption cross sections? Elegant analytic treatments do exist for spheroids or infinite cylinders, but for more general shapes it is necessary to resort to brute force treatments. One approach that has proven useful is to approximate the actual target (with its particular geometry and dielectric function) by an array of "point dipoles." For a target illuminated by an incident monochromatic electromagnetic wave, each of these dipoles is assigned a complex polarizability $\alpha(\omega)$. Each dipole has an instantaneous dipole moment $\mathbf{P}_j = \alpha_j \mathbf{E}_j$, where α_j is the polarizability tensor for dipole j, and $\mathbf{E}_j$ is the electric field at location j due to the incident wave plus all of the other dipoles. This method, pioneered by Purcell & Pennypacker (1973), is known as the **discrete dipole approximation (DDA)** or **coupled dipole approximation**.[5]

DDA calculations are CPU-intensive, but many problems of practical interest can be handled by a desktop computer.[6] For example, the DDA has been used to study absorption and scattering by graphite particles (Draine & Malhotra 1993) and by random agglomerates (Shen et al. 2008).

22.6.1 X-Ray Regime

Figure 22.4 shows the real and imaginary components of the dielectric function for $MgFeSiO_4$ In the optical and ultraviolet, normal solids have refractive indices $|m-1| \gtrsim 0.3$. At X-ray energies, however, $|m-1| \ll 1$, and the character of the scattering changes considerably. The wavelength $\lambda = 0.00124(\mathrm{keV}/h\nu)\,\mu\mathrm{m}$ is small compared to the sizes $a \approx 0.2\,\mu\mathrm{m}$ of the particles containing most of the grain mass. The result is that the X-ray scattering is very strongly peaked in the

[5]For more details on the DDA, see the review by Draine & Flatau (1994).

[6]Public-domain DDA codes are available, e.g., DDSCAT (http://www.ddscat.org).

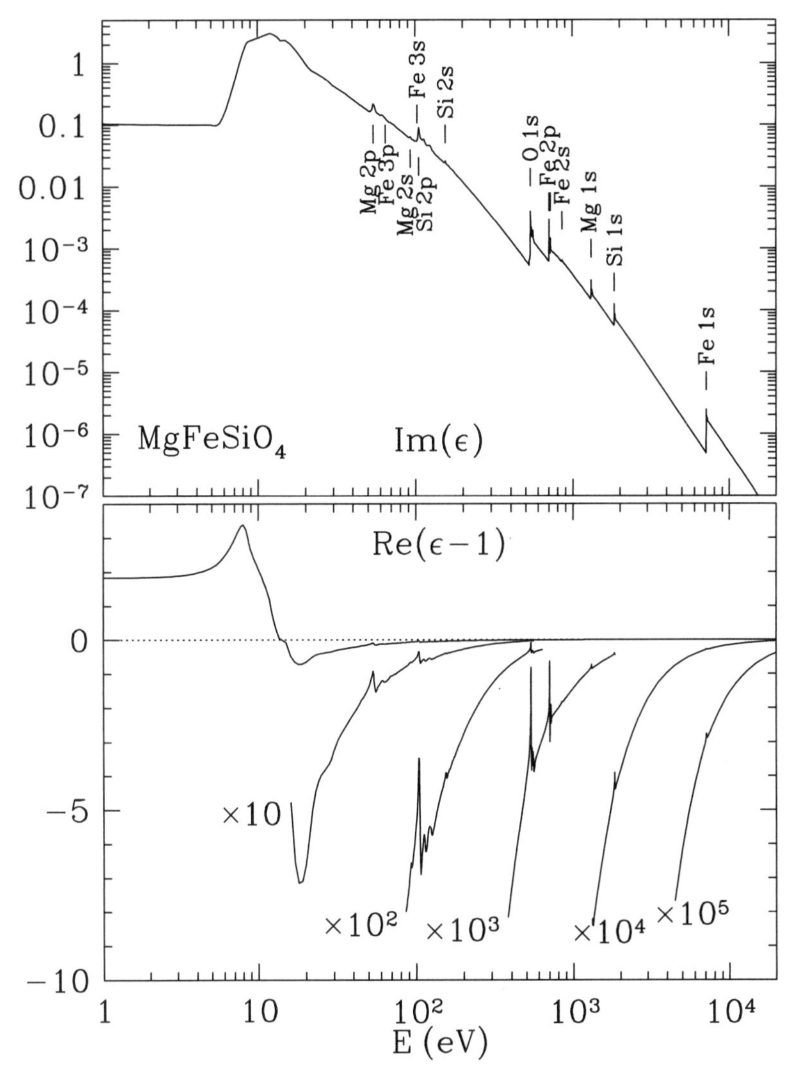

Figure 22.4 Dielectric function ϵ for $MgFeSiO_4$ material. Various absorption edges are labelled in the plot of $\mathrm{Im}(\epsilon)$. This dielectric function, and its continuation at lower energies, will be referred to as "astrosilicate". From Draine (2003*b*), reproduced by permission of the AAS.

forward direction, with a characteristic scattering angle

$$\theta \approx \frac{\lambda}{\pi a} \approx 800'' \left(\frac{\mathrm{keV}}{h\nu}\right)\left(\frac{0.1\,\mu\mathrm{m}}{a}\right) \quad . \tag{22.29}$$

For spherical grains, the scattering can be calculated using Mie theory. which, consists of a series expansion where $\sim 2\pi a/\lambda$ terms must be retained in the sums.

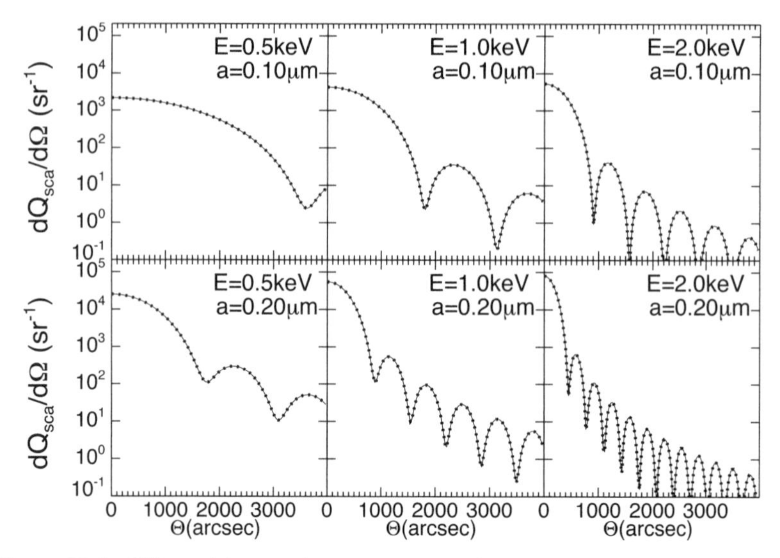

Figure 22.5 Differential scattering cross section for $a = 0.1$ and $0.2\,\mu$m silicate spheres at $E = 0.5$, 1.0, and 2.0 keV. Solid curve is Mie theory, dots show results calculated using anomalous diffraction theory (ADT). The results are indistinguishable, showing that ADT provides an excellent approximation. From Draine & Allaf-Akbari (2006), reproduced by permission of the AAS.

When $a/\lambda \gtrsim 10^3$, roundoff errors may prevent accurate evaluation of the necessary sums, limiting the practical applicability of Mie theory. However, when Mie theory becomes impractical, X-ray scattering and absorption by grains can be calculated using an approximation called "anomalous diffraction theory" (ADT), originally introduced by van de Hulst (1957) and recently applied by Draine & Allaf-Akbari (2006). The validity criteria for ADT are very simple:

$$\frac{2\pi a}{\lambda} \gg 1 \quad , \tag{22.30}$$

$$|m-1| \ll 1 \quad . \tag{22.31}$$

The first condition ensures that the ray-optics approximation is valid, and the second condition ensures that refraction (and reflection) are unimportant when the ray crosses the grain surface. As can be seen from Fig. 22.5, anomalous diffraction theory provides excellent accuracy within its domain of applicability. One advantage of ADT is that it can be applied to nonspherical grains.

22.7 Interstellar Grains

Above we have discussed calculational methods for various regimes. We can now calculate scattering and absorption cross sections for micron- or submicron-sized

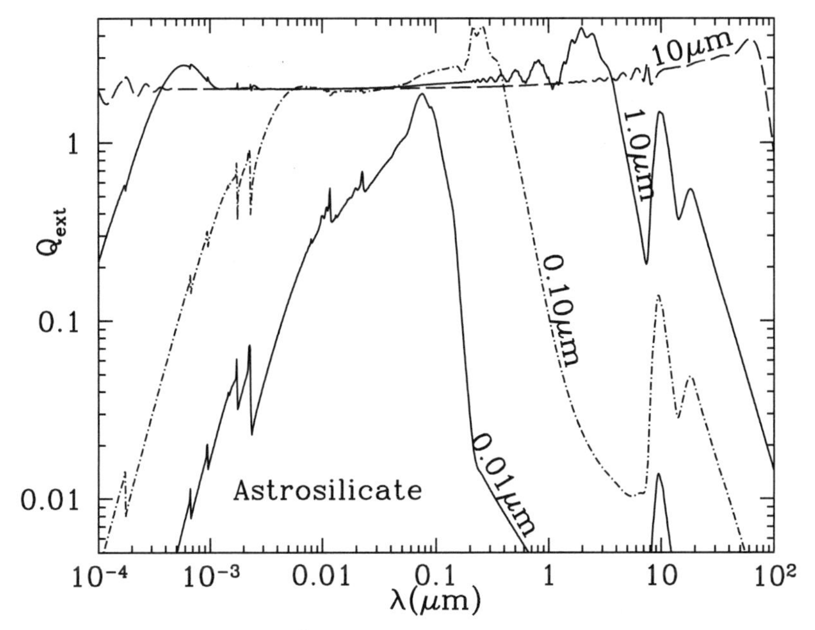

Figure 22.6 $Q_{\rm ext} \equiv C_{\rm ext}/\pi a^2$ for $a = 0.01$, 0.1, 1, and $10\,\mu$m amorphous silicate spheres, for wavelengths ranging from $\lambda = 10^{-4}\,\mu$m $= 1\,$Å ($h\nu = 12.4\,$keV) to $\lambda = 10^3\,\mu$m $= 1\,$mm. At short wavelengths, the $a = 0.01$ and $0.10\,\mu$m grains show discontinuities in $Q_{\rm ext}$ at X-ray absorption edges. In the IR, the $a = 0.01, 0.1, 1\,\mu$m grains show prominent silicate absorption features at 9.7 and $18\,\mu$m, but these features are suppressed when $a = 10\,\mu$m.

grains from X-ray to sub-mm wavelengths. Figure 22.6 shows the extinction efficiency $Q_{\rm ext}$ calculated for grains of amorphous silicate ("astrosilicate") from the X-ray to the submm, for four different sizes. There are several noteworthy features:

1. $Q_{\rm ext}$ shows sharp discontinuities at X-ray absorption edges. The amorphous silicate material is assumed to have composition $MgFeSiO_4$. Two conspicuous edges are the Fe K edge at 1.75 Å (7.1 keV) and the O K edge at 23 Å (528 eV). Note that the appearance of these edges depends on grain size. As the grains become larger, scattering makes an appreciable contribution to $Q_{\rm ext}$, and the long-wavelength side of the O K edge is "filled in" by scattering.

2. $a = 1\,\mu$m grains are, in effect, optically thick (with $Q_{\rm ext} \approx 2$) for $0.001 \lesssim \lambda \lesssim 2\,\mu$m; for $\lambda < 10^{-3}\,\mu$m ($h\nu > 1.24\,$keV), the absorption length exceeds the grain diameter, and for $\lambda > 2\,\mu$m, the grain is smaller than the wavelength. Similarly, the $a = 10\,\mu$m grain is optically thick for $10^{-4}\,\mu$m $\lesssim \lambda \lesssim 10\,\mu$m.

3. The silicate absorption features at 9.7 and $18\,\mu$m are prominent absorption

features for the $a = 0.01, 0.1, 1\,\mu\text{m}$ cases shown, but are suppressed in the $a = 10\,\mu\text{m}$ example, because the grain is, in effect, optically thick at wavelengths on either side of the silicate features.

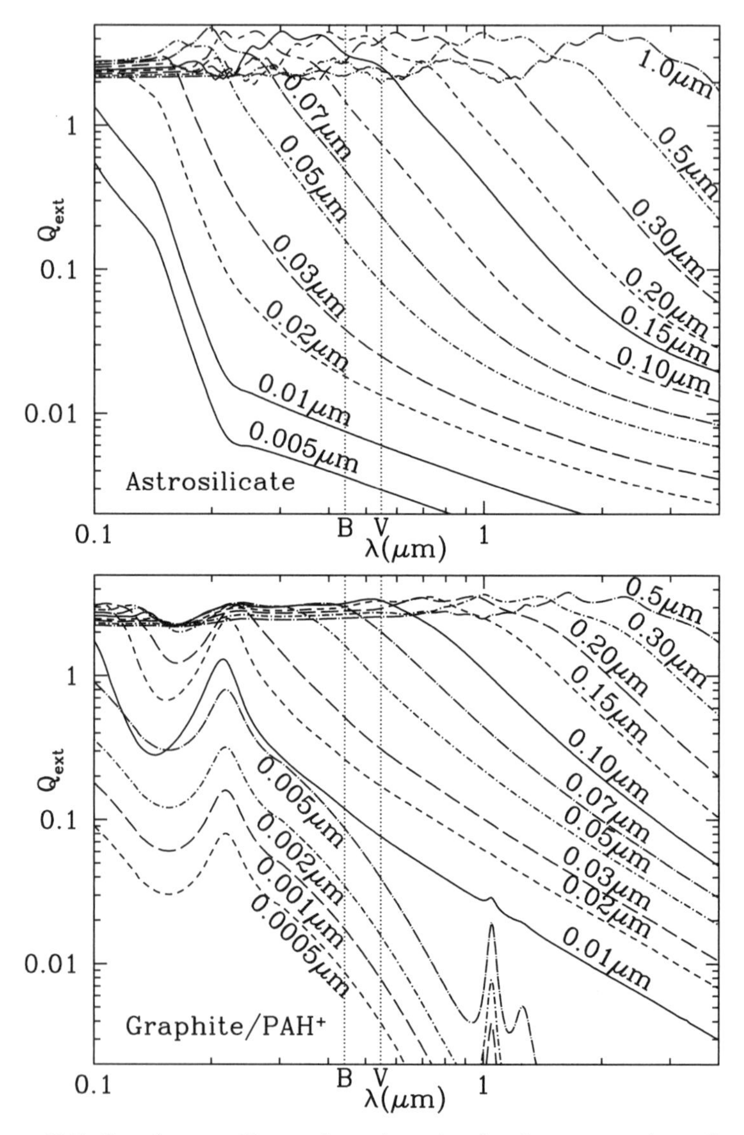

Figure 22.7 $Q_{\rm ext}$ for astrosilicate spheres (upper) and carbonaceous spheres (lower) for wavelengths ranging from $\lambda = 0.1\,\mu\text{m}$ to $\lambda = 4\,\mu\text{m}$. The locations of the B (4405 Å) and V (5470 Å) bands are shown. Curves are labeled by radius a.

Figure 22.7 shows the behavior of $Q_{\rm ext}$ for wavelengths running from the vacuum ultraviolet into the infrared. The upper panel shows the extinction efficiency factors $Q_{\rm ext}$ for spheres with the "astrosilicate" dielectric function. For the wavelength range shown here, there are no spectral features, although small particles do show a rise in extinction for $\lambda \lesssim 0.2\,\mu\text{m}$ due to the onset of ultraviolet absorption in silicates. Scattering becomes important for $\lambda \lesssim 2\pi a$ (i.e., $x = 2\pi a/\lambda \gtrsim 1$), and $Q_{\text{ext}} \gtrsim 2$ for $\lambda \lesssim 4a$.

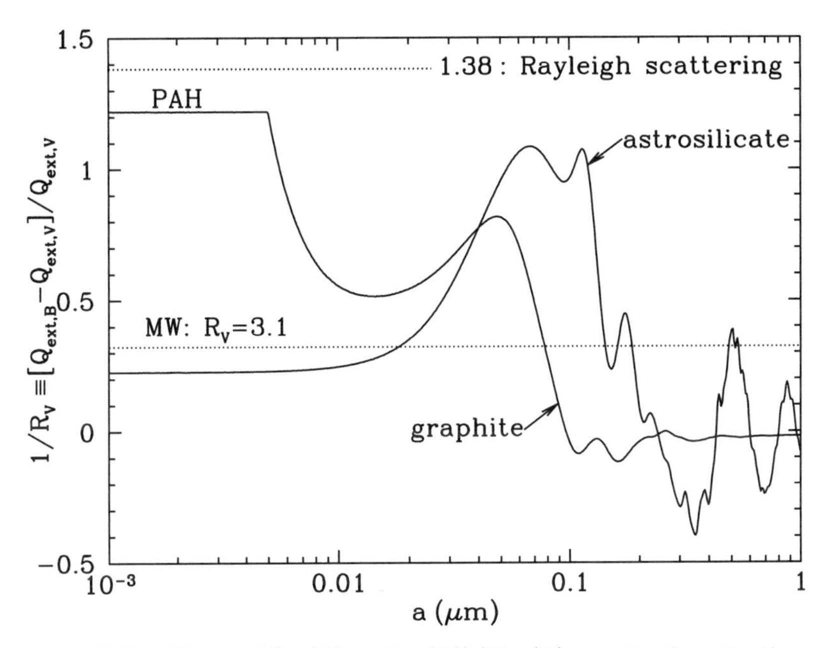

Figure 22.8 $1/R_V \equiv (C_{\text{ext}}(B) - C_{\text{ext}}(V))/C_{\text{ext}}(V)$ as a function of radius a for astrosilicate and carbonaceous spheres ($B = 0.44\,\mu\text{m}$, $V = 0.55\,\mu\text{m}$). The carbonaceous spheres are assumed to be graphitic for $a > 0.01\,\mu\text{m}$, and PAHs for $a < 0.005\,\mu\text{m}$, with a continuous transition between 0.005 and $0.01\,\mu\text{m}$. For $a \lesssim 0.02\,\mu\text{m}$, scattering is unimportant, and R_V is determined by the absorptive properties of the grain material. For $a \gtrsim 0.12\,\mu\text{m}$, scattering resonances move through the wavelength range between B and V, and $1/R_V$ has oscillatory behavior. The dust in the diffuse ISM is observed to have $R_V \approx 3.1$, shown by the dotted line. $R_V \approx 3.1$ for $a \approx 0.08\,\mu\text{m}$ or $0.14\,\mu\text{m}$ for graphitic and astrosilicate grains, respectively.

For the adopted optical constants (Draine & Li 2007), the small $a \lesssim 0.02\,\mu\text{m}$) carbonaceous particles show a strong absorption feature near 2175 Å, closely matching the observed interstellar feature near this wavelength. However, the theoretically-calculated feature broadens as the grain size increases to $0.03\,\mu\text{m}$, and disappears for larger grains because the grain becomes optically thick not only at the wavelength of the resonance, but also at wavelengths above and below the resonance.

As discussed earlier, interstellar extinction curves are often characterized by $R_V \equiv A_V/(A_B - A_V)$, and it is of interest to see what value of R_V would apply to the extinction produced by grains of a single size and composition. Because R_V is singular when $A_B = A_V$, it is preferable to instead consider $1/R_V \equiv$

$(A_B - A_V)/A_V$, which is proportional to the slope of the extinction curve between V and B. Figure 22.8 shows $1/R_V$ versus grain radius for carbonaceous grains and astrosilicate grains. For very small grains, scattering is negligible compared to absorption, and the value of R_V in the limit $a \to 0$ depends on the wavelength dependence of the optical constants – hence the very different limiting values for PAHs and astrosilicates. As the grain radius is increased, scattering begins to contribute significantly to the extinction, but we see that neither the silicate nor carbonaceous particles ever reach the value of $1/R_V = 1/0.726$ appropriate to Rayleigh scattering by particles with a polarizability that is wavelength independent. This is because, for our assumed dielectric functions, when the particles are small enough to be in the Rayleigh limit, absorption makes an important contribution to the extinction.

$R_V \approx 3.1$ is attained by graphitic grains for $a \approx 0.08\,\mu\text{m}$, and by astrosilicate[7] grains for $a \approx 0.15\,\mu\text{m}$. Although a broad size distribution is required to match the full extinction curve, grain models that reproduce the observed extinction should have the extinction in the visible dominated by grains with $a \approx 0.1\,\mu\text{m}$.

[7] Astrosilicate grains also have $R_V \approx 3.1$ for $a \approx 0.02\,\mu\text{m}$, but such small grains produce much more extinction in the ultraviolet than in the visible (see Fig. 22.7), implying that they contribute only a small fraction of the extinction in the visible.

Chapter Twenty-three

Composition of Interstellar Dust

There is ample evidence for the presence of substantial amounts of submicron-sized dust particles in interstellar space. What is this dust made of? This question has been difficult to answer.

The preferred approach would be spectroscopy: ideally, we would observe spectroscopic features that would uniquely identify the materials, and, furthermore, allow us to measure the amounts of each material present. This is the approach that is followed for atoms, ions, and small molecules, but unfortunately it is difficult to apply to solid materials because: (1) the optical and UV absorption is largely a continuum; and (2) the spectral features that do exist are broad, making them difficult to identify conclusively.

An alternative approach is to ask: What materials could plausibly be present in the interstellar medium in quantities sufficient to account for the observed extinction? We have seen in §21.6.1 that a Kramers-Kronig integral over the observed extinction indicates that the total grain mass relative to total hydrogen mass $M_{\rm dust}/M_{\rm H} \gtrsim 0.0083$.

In §9.11 we discussed that fact that certain elements appear to be underabundant, or "depleted," in the gas phase. In this chapter, we first consider what observed **depletions** tell us about the major elemental composition of interstellar dust. Following this, we review candidate materials, including spectroscopic evidence if any.

23.1 Abundance Constraints

The available evidence indicates that the overall abundances in the ISM are close to the values in the solar photosphere. Because there is no way to have hydrogen contribute appreciably to the grain mass [even polyethylene $(CH_2)_n$ is 86% carbon by mass], and He and Ne are chemically inert, the only way to have a dust/H mass ratio of 0.0056 or higher is to build the grains out of the most abundant condensible elements: C, O, Mg, Si, S, and Fe.

As discussed in §9.11, absorption-line spectroscopy of C, Mg, Si, and Fe in the gas phase shows that these elements are in fact underabundant in the gas ("depleted"), with about $\frac{2}{3}$ of the C and 90% or more of Mg, Si, and Fe presumed to be incorporated in dust grains in the typical diffuse interstellar cloud. Perhaps the best-studied sightline in the ISM is toward the star ζ Ophiuchi, a bright O9.5V

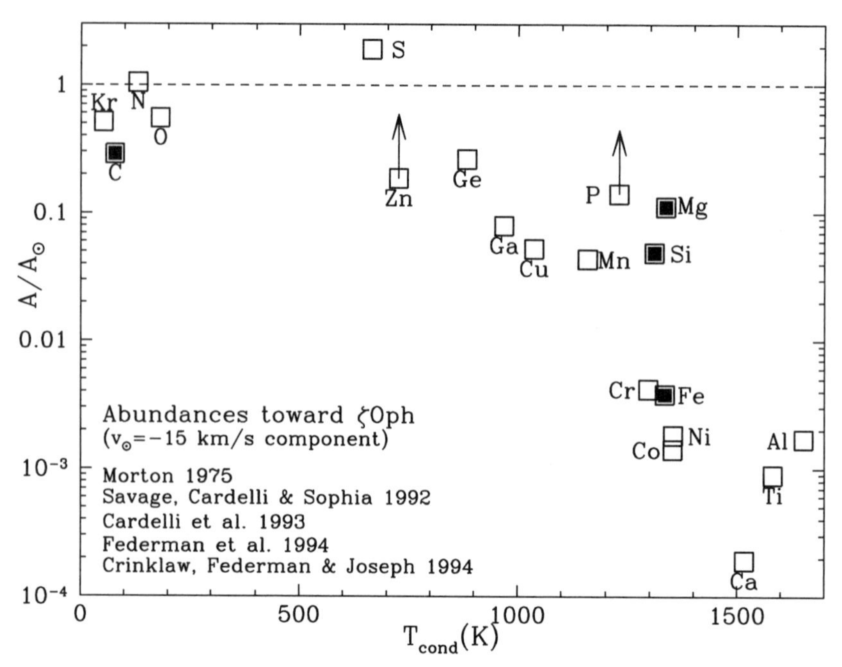

Figure 23.1 Gas-phase abundances (relative to solar) in the diffuse cloud toward ζ Oph, plotted versus "condensation temperature" $T_{\rm cond}$ (see text). Data from Morton (1975), Savage et al. (1992), Cardelli et al. (1993), Federman et al. (1993), Crinklaw et al. (1994). Solid symbols: major grain constituents C, Mg, Si, Fe. The C abundance has been calculated assuming $f(\text{C II}]2325\,\text{Å}) = 1.0 \times 10^{-7}$ (see text). The apparent overabundance of S may be due to observational error, but may also arise because of S II absorption in the H II region around ζ Oph.

star only 138 pc away. Absorption-line spectroscopy has allowed the *gas-phase* abundances of many of the elements to be measured. These abundances, relative to solar, are shown for 20 elements in Fig. 23.1.

Table 23.1 gives the gas-phase abundances for 10 major elements. Determination of the column density of C II (accounting for most of the gas-phase carbon) relies on knowledge of the oscillator strength of the weak intersystem line C II]2325 Å. Table 23.1 gives the gas-phase carbon abundances estimated using $f(\text{C II}]2325\,\text{Å}) = 4.78 \times 10^{-8}$ from Morton (2003), but also for $f(\text{C II}]2325\,\text{Å}) = 1.0 \times 10^{-7}$, as appears to be required to reconcile abundances estimated using the 2325 Å line (Sofia et al. 2004) with C II abundances estimated using strong lines (Sofia & Parvathi 2010).

If we assume that the total abundance of each element is equal to the current best-estimate of the solar abundance, then the difference between solar abundance and the observed gas-phase abundance will tell us what contribution that element makes toward the dust mass in the cloud toward ζ Oph. This inventory is carried out in Table 23.1.

Table 23.1 Inferred Elemental Composition of Dust toward ζOph

X	$(N_X/N_{\rm H})_\odot$ [a] (ppm)	$N_{X,\rm gas}/N_{\rm H}$ [b] (ppm)	$N_{X,\rm dust}/N_{\rm H}$ (ppm)	$10^3 M_{X,\rm dust}/M_{\rm H}$
C	295 ± 36	135 ± 33 [d,e]	160 ± 49	1.92 ± 0.59 [e]
		85 ± 20 [d,f]	210 ± 41	2.52 ± 0.49 [f]
N	74.1 ± 9.0	78 ± 13 [g]	-14 ± 16	0
O	537 ± 62	295 ± 36 [d]	242 ± 72	3.87 ± 1.15
		[383] [c]	154 ± 8 [c]	2.46 ± 0.13 [c]
Mg	43.7 ± 4.2	4.9 ± 0.5 [g]	39 ± 4	0.94 ± 0.10
Al	2.8 ± 0.2	0.005 ± 0.001 [h]	2.8 ± 0.2	0.08 ± 0.01
Si	35.5 ± 3.0	1.7 ± 0.5 [i]	34 ± 3	0.95 ± 0.08
S	14.5 ± 1.0	28 ± 16 [j]	-14 ± 16	0
Ca	2.3 ± 0.2	0.0004 ± 0.0001 [k]	2.2 ± 0.2	0.09 ± 0.008
Fe	34.7 ± 3.3	0.13 ± 0.01 [g]	35 ± 3	1.96 ± 0.17
Ni	1.7 ± 0.2	0.0030 ± 0.0002 [j]	1.7 ± 0.2	0.10 ± 0.01
Total if $f(\text{C II}]2325) = 4.78 \times 10^{-8}$ (see text)				9.9 ± 1.3 [e]
Total if $f(\text{C II}]2325) = 1.0 \times 10^{-7}$ (see text)				10.5 ± 1.3 [f]
Total if $f(\text{C II}]2325) = 1.0 \times 10^{-7}$, $N_{\rm O,dust}/N_{\rm H} = 154$ ppm (see text)				9.1 ± 0.6 [c]

[a] Asplund et al. (2009).
[b] Assuming $N({\rm H}) + 2N({\rm H_2}) = 10^{21.13\pm0.03}\,{\rm cm^{-2}}$.
[c] Assuming $N_{\rm O,dust}/N_{\rm H} = 154$ ppm.
[d] Cardelli et al. (1993).
[e] If $f(\text{C II}]2325\,\text{Å}) = 4.78 \times 10^{-8}$ (Morton 2003).
[f] If $f(\text{C II}]2325\,\text{Å}) = 1.00 \times 10^{-7}$ (see text).
[g] Savage et al. (1992).
[h] Morton (1975).
[i] Cardelli et al. (1994).
[j] Federman et al. (1993).
[k] Crinklaw et al. (1994).

The reported depletion of oxygen from the gas toward ζ Oph is difficult to understand, because it is not clear what compounds can account for it (Jenkins 2009; Whittet 2010). As will be discussed in the following, on a sightline like that toward ζ Oph, negligible amounts of H_2O are present, and gas-phase species such as CO and OH contain only a small fraction of the O. If we consider silicates with olivine-like composition $\text{Mg}_x\text{Fe}_{2-x}\text{SiO}_4$, we can account for only $\sim 4 \times (34 \pm 3) = 136 \pm 12$ ppm of O in silicates; adding other metal oxides,[1] the solid-phase oxygen can be raised to ~ 154 ppm, yet observations seem to indicate that 242 ± 72 ppm are missing from the gas on the sightline toward ζ Oph – some of the oxygen seems to have gone missing! On the other hand, according to Table 9.5, the typical CNM cloud appears to have a gas-phase O abundance of ~ 457 ppm; if the total O abundance is the current solar value of 537 ppm, this corresponds to only ~ 80 ppm of O in grains – only $\sim 60\%$ of the amount that we estimate to be present in the dust. Thus on these sightlines we seem to have somewhat *more* oxygen in the gas than we expect.

At this time, it is simply not clear where all of the oxygen resides. One should keep in mind the possibility that there may be some error in our estimation of the oxygen budget in Table 23.1.

If we use the carbon abundances estimated using the larger oscillator strength $f(\text{C II}]2325\,\text{Å}) = 1.0 \times 10^{-7}$ (Sofia & Parvathi 2010), we arrive at $M_{\rm dust}/M_{\rm H} \approx$

[1] E.g., if the composition is $34\text{Mg}_x\text{Fe}_{2-x}\text{SiO}_4 + 3\text{Fe}_2\text{O}_3 + 1.4\text{Al}_2\text{O}_3 + 2.2\text{CaO} + 0.85\text{Ni}_2\text{O}_3$, with $x = 39/34 = 1.15$.

0.0091 (see Table 23.1), with $\sim 28\%$ of the dust mass contributed by carbon, and 72% by compounds containing Mg, Al, Si, Ca, Fe, Ni and O, presumably mainly in silicates. If the carbonaceous material has a density $\rho \approx 2.2\,\mathrm{g\,cm^{-3}}$, and the silicate density is $\rho \approx 3.8\,\mathrm{g\,cm^{-3}}$, then the total grain volume per H is $V_{\rm tot} \approx 4.8 \times 10^{-27}\,\mathrm{cm^3/H}$, with silicates accounting for $\sim 60\%$ of $V_{\rm tot}$.

The grain volume $V_{\rm tot} \approx 4.8 \times 10^{-27}\,\mathrm{cm^3/H}$ and grain mass $M_{\rm dust}/M_{\rm H} \approx 0.0091$ are consistent with Purcell's Kramers-Kronig argument (§21.6.1), which obtained [see Eqs. 21.18 and 21.19] $V_{\rm tot} \gtrsim 4.6 \times 10^{-27}\,\mathrm{cm^3/H}$ and $M_{\rm dust}/M_{\rm H} \gtrsim 0.0083$. We must remember, however, that the Kramer-Kronig analysis is only a lower bound, as it neglected the extinction at $\lambda < 0.1\,\mu\mathrm{m}$ and $\lambda > 30\,\mu\mathrm{m}$.

Figure 23.1 shows gas-phase abundances, relative to solar abundances, plotted against the **condensation temperature** $T_{\rm cond}$, the temperature at which 50% of the element in question would be incorporated into solid material in a gas of solar abundances, at LTE at a pressure $p = 10^2\,\mathrm{dyn\,cm^{-2}}$ (Lodders 2003). The condensation temperature indicates whether an element is able to form stable solid compounds in gas of solar composition. We see that there is a strong tendency for elements with high $T_{\rm cond}$ to be underabundant in the gas phase, presumably because most of the atoms are instead in solid grains.

With the elements providing the bulk of the grain volume identified, we can limit consideration to the following possible materials:

- Silicates, e.g., **pyroxene** composition $\mathrm{Mg}_x\mathrm{Fe}_{1-x}\mathrm{SiO}_3$, or **olivine** composition $\mathrm{Mg}_{2x}\mathrm{Fe}_{2-2x}\mathrm{SiO}_4$ $(0 \leq x \leq 1)$
- Oxides of silicon, magnesium, and iron (e.g., $\mathrm{SiO_2}$, MgO, $\mathrm{Fe_3O_4}$)
- Carbon solids (graphite, amorphous carbon, and diamond)
- Hydrocarbons (e.g., polycyclic aromatic hydrocarbons)
- Carbides, particularly silicon carbide (SiC)
- Metallic Fe

Other elements (e.g., Ti, Cr) are also present in interstellar grains, but, because of their low abundances, they contribute only a minor fraction of the grain mass.

23.2 Presolar Grains in Meteorites

Certain meteorites have been found to contain grains whose formation predated the solar system. For the most part, these grains have been identified by virtue of anomalous isotopic composition, indicating not only that these grains are presolar, but also that many of them formed in outflows from stars with anomalous isotopic composition. The presolar grain abundances vary from meteorite to meteorite, but are highest in the meteorites that appear to be most primitive (i.e., have undergone the least amount of heating) – the meteorite class referred to as **carbonaceous chondrites**. Table 23.2 lists the major types of presolar materials found in

Table 23.2 Types and Properties of Major Presolar Materials[a,b] Identified in Meteorites and IDPs.

Material	Source	Grain Size (μm)	Abundance[c] (ppm)†
Amorphous silicates	Circumstellar	0.2–0.5	20–3600
Forsterite (Mg_2SiO_4) Enstatite ($MgSiO_3$)	Circumstellar	0.2–0.5	10–1800
Diamond		$\sim$0.002	$\sim$1400
P3 fraction	Not known		
HL fraction	Circumstellar		
Silicon carbide	Circumstellar	0.1–20	13–14
Graphite	Circumstellar	0.1–10	7–10
Spinel ($MgAl_2O_4$)	Circumstellar	0.1–3	1.2
Corundum (Al_2O_3)	Circumstellar	0.5–3	0.01
Hibonite ($CaAl_{12}O_{19}$)	Circumstellar	1–2	0.02

[a] Other presolar materials include TiC, MoC, ZrC, RuC, FeC, Si_3N_4, TiO_2, and Fe-Ni metal.
[b] See Huss & Draine (2007) for details and references therein.
[c] Abundance in fine-grained fraction (= matrix in primitive chondrites).

meteorites. Surprisingly, the principal carbonaceous material by mass consists of extremely small ($\sim$20 Å) particles of diamond. These "nanodiamonds" make up fully 0.14% of the mass of the fine-grained matrix material in primitive carbonaceous chondrites.

While some of the nanodiamonds are definitely of presolar origin, it is possible that the bulk of the nanodiamond material might have been produced in the solar system – the provenance of the nanodiamond material is not yet known.

23.3 Observed Spectral Features of Dust

23.3.1 The 2175 Å Feature

The extinction curves in Fig. 21.2 show a conspicuous extinction feature at $\lambda^{-1} = 4.6\,\mu\mathrm{m}^{-1}$, or $\lambda = 2175$ Å. The feature is well-described by a Drude profile (Eq. G.9). The central wavelength is nearly identical on all sightlines, but the width varies significantly from one region to another (Fitzpatrick & Massa 1986).

The strength of this feature implies that the responsible material must be abundant (Draine 1989*b*): it must be made from H, C, N, O, Mg, Si, S, or Fe. Small graphite grains would have a strong absorption peak at about this frequency, due to $\pi \to \pi^*$ electronic excitations in the sp^2-bonded carbon sheets (Stecher & Donn 1965; Draine 1989*b*). Because the carbon skeleton of polycyclic aromatic hydrocarbon (PAH) molecules resembles a portion of a graphite sheet, such molecules also tend to have strong electronic transitions at about this frequency. Although alternatives have been suggested [e.g., OH^- on small silicate grains (Steel & Duley 1987; Bradley et al. 2005)], it seems most likely that the 2175 Å feature is due to some form of sp^2-bonded carbon material.

23.3.2 Silicate Features at 9.7 μm and 18 μm

There is a conspicuous infrared absorption feature at 9.7 μm, shown in Fig. 23.2 (and later in Fig. 23.6). Silicate minerals generally have strong absorption resonances due to the Si-O stretching mode near 10 μm, and it seems virtually certain that the interstellar 9.7 μm feature is due to silicates. This conclusion is strengthened by the fact that the 10 μm emission feature is seen in the outflows from oxygen-rich stars (which would be expected to condense silicate dust) but not in the outflows from carbon-rich stars. The interstellar 9.7 μm feature is seen both in emission [e.g., in the Trapezium region in Orion (Gillett et al. 1975*a*)] and in extinction in the interstellar medium (Roche & Aitken 1984). Sightlines within a few kpc of the Sun have $A_V/\Delta\tau_{9.7} \approx 18.5 \pm 2$ (see Table 1 of Draine 2003*a*), but sightlines to sources near the Galactic Center have $A_V/\Delta\tau_{9.7} = 9 \pm 1$ (Roche & Aitken 1985).

Near 18 μm, interstellar dust shows another feature, attributable to the Si-O-Si bending mode in amorphous silicates.

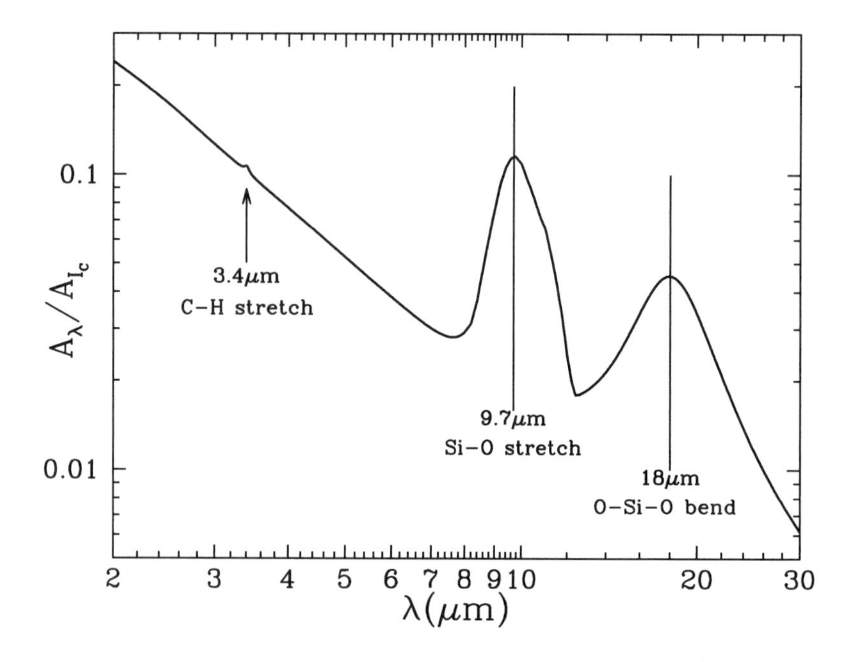

Figure 23.2 Infrared extinction curve. The 8 to 13 μm silicate profile is as observed toward the Galactic Center by Kemper et al. (2004), but with $A_V/\Delta\tau_{9.7\,\mu\mathrm{m}} = 18.5$, as appropriate for sightlines through diffuse gas within a few kpc of the Sun (see Table 1 of Draine 2003*a*). The 3.4 μm C–H stretching feature is indicated.

23.3.3 The 3.4 μm Feature

There is a broad absorption feature at 3.4 μm that is almost certainly due to the C–H stretching mode in hydrocarbons. Pendleton & Allamandola (2002) concluded that hydrocarbons with a mixed aromatic (ring) and aliphatic (chain) character provided a good fit to the observed interstellar absorption, including the $3.35 - 3.53\,\mu$m region. This included hydrocarbon films deposited following laser-ablation of amorphous carbon in Ar, followed by exposure to atomic H (Mennella et al. 1999) or from a weakly ionized plasma produced by laser-ablation of graphite in hydrogen (Scott & Duley 1996*a*; Duley et al. 1998). Pendleton & Allamandola (2002) concluded that the carbonaceous material was $\sim 85\%$ aromatic and $\sim 15\%$ aliphatic. However, a separate study by Dartois et al. (2004) concluded that *at most* 15% of the carbon is aromatic. The aromatic/aliphatic ratio remains uncertain.

Somewhat surprisingly, the 3.4 μm C–H feature is found to be weaker (relative to the overall extinction) in dark clouds than in diffuse clouds (Shenoy et al. 2003), which has been interpreted as evidence that the C–H bonds responsible for the 3.4 μm feature are destroyed in molecular clouds, perhaps as the result of cosmic ray irradiation, and regenerated when carbonaceous grains are exposed to atomic hydrogen in diffuse clouds (Mennella et al. 2003).

23.3.4 Diffuse Interstellar Bands

The three features at 2175 Å, 9.7 μm, and 18 μm are by far the strongest features seen in diffuse interstellar dust. There are, in addition, numerous weaker features in the optical known as the **diffuse interstellar bands** or **DIBs**. These are features that are too broad (FWHM $\gtrsim 1$ Å) to be absorption lines of atoms, ions, or small molecules. The first DIBs were discovered 88 years ago (Heger 1922), and their interstellar nature was established 76 years ago (Merrill 1934).

DIBs are present in Fig. 21.2, but appear much more clearly in the expanded plot in Fig. 23.3, showing the extinction for $1.5\,\mu\mathrm{m}^{-1} < \lambda^{-1} < 1.75\,\mu\mathrm{m}^{-1}$, with several conspicuous DIBs present, most notably the DIB at 5780 Å. The strongest DIB falls at 4430 Å (not shown in Fig. 23.3). Hobbs et al. (2009) report a total of 414 DIBs between 3900 and 8100 Å!

It is embarassing that Nature has provided astrophysicists with this wealth of spectroscopic clues, yet as of this writing not a single one of the DIBs has been convincingly identified! It seems likely that some of the DIBs may be due to free-flying large molecules (i.e., ultrasmall dust grains); this hypothesis has received support from high resolution spectra of the 5797 Å feature (see Figure 23.4) showing intrinsic ultrafine structure (Sarre et al. 1995; Kerr et al. 1998). Similar fine structure is also seen in some other bands (Galazutdinov et al. 2003, 2005), and it now seems likely that at least a substantial fraction of the DIBs are due to free-flying molecules, possibly ionized. However, one would expect that a given molecule would have multiple absorption lines to different vibrational states of the electronic excited state. In a careful correlation study, McCall et al. (2010) found what appears to be a nearly perfect correlation between the strengths of DIBs at 6196.0 and

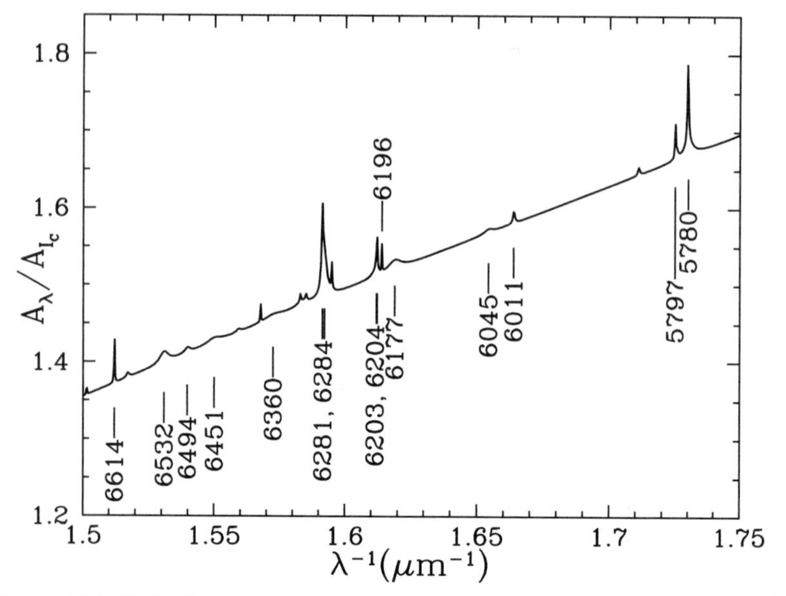

Figure 23.3 Extinction at wavelength λ (relative to the extinction at $I_C = 8020$ Å) for 6667 Å $> \lambda >$ 5714 Å, showing some of the diffuse interstellar bands, based on the compilation by Jenniskens & Desert (1994).

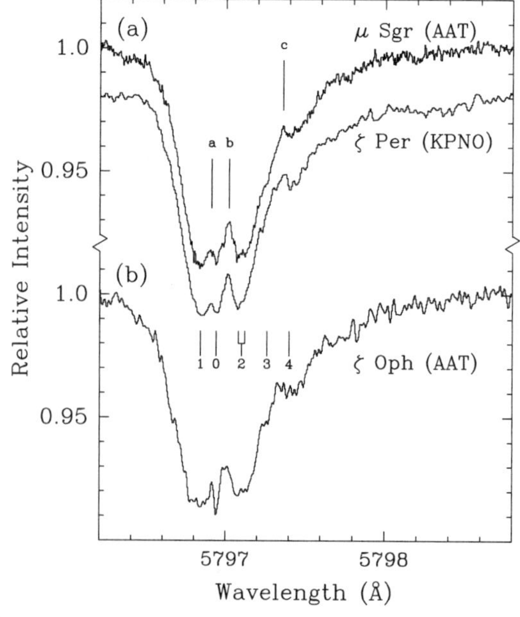

Figure 23.4 Fine structure, possibly due to molecular rotation, in the $\lambda\lambda$5797 Å DIB, on 3 different sightlines. This and similar structure seen in other bands strongly suggests that at least some DIBs arise in large free-flying molecules = ultrasmall dust grains. From Kerr et al. (1998), reproduced by permission of the AAS.

6613.6Å (air wavelengths), suggesting that these may be two absorption features produced by a single absorber.

23.3.5 Ice Features in Diffuse and Dark Regions

In dark molecular clouds, a number of additional absorption features appear, most notably a strong band at $3.1\,\mu\mathrm{m}$ which is produced by the O-H stretching mode in H_2O ice. The spectrum of the Becklin-Neugebauer (BN) object in Figure 23.5 shows a very strong H_2O absorption feature at $3.1\,\mu$m, as does the spectrum of Sgr A* in Figure 23.6.

However, the $3.1\,\mu\mathrm{m}$ feature is *not* seen on sightlines that pass only through diffuse interstellar clouds, even when the total extinction is large – the sightline to the B5 hypergiant star Cyg OB2-12 has $\Delta\tau(3.05\,\mu\mathrm{m})/\Delta\tau(9.7\,\mu\mathrm{m}) < 0.037$, implying that less than 0.4% of the O atoms on this sightline are in H_2O (Gillett et al. 1975*b*; Knacke et al. 1985; Whittet et al. 1997). We have seen [eq. (21.19)] that $M_{\rm gr}/M_{\rm H} > 0.0083$; it follows that if any H_2O is present on this sightline, it contributes $< 0.5\%$ of the grain mass. In the diffuse ISM, ices, if present at all, are not a significant part of the dust mixture.

Although dust in the diffuse ISM appears to be ice-free, H_2O *can* contribute a significant fraction of the dust mass in dark clouds. Whittet et al. (1988) found that in the Taurus dark cloud complex, the strength of the $3.1\mu\mathrm{m}$ feature is approximately given by

$$\Delta\tau_{3.1} \approx \begin{cases} 0 & \text{for } A_V \lesssim 3.3\text{ mag} , \\ 0.093(A_V - 3.3\text{mag}) & \text{for } A_V \gtrsim 3.3\text{ mag} , \end{cases} \tag{23.1}$$

which suggests that ice is present only in regions that are shielded from the diffuse starlight background by $A_V \gtrsim 1.65$ mag. The dust shielding is probably needed to suppress H_2O removal by photodesorption.

When a strong $3.1\,\mu\mathrm{m}$ feature appears in absorption, a number of other absorption features are also seen, including features due to CO ($4.67\,\mu\mathrm{m}$), CH_3OH ($3.53\,\mu\mathrm{m}$), and CO_2 ($15.2\,\mu\mathrm{m}$). The shape of the $3.1\,\mu\mathrm{m}$ H_2O feature is indicative of the type of ice and the impurities present in it. The relative strengths of the various features indicate that H_2O is the dominant "ice" species, with NH_3, CO, CH_3OH, and CO_2 as secondary constituents.

23.4 Silicates

Geologists are familiar with a great variety of silicate minerals, found in the crust of the Earth, the Moon, and Mars, and in meteorites. In all cases, crystalline silicate minerals exhibit strong absorption bands near $10\,\mu$m (stretching of the Si-O bond) and near $20\,\mu$m (bending of the O-Si-O structure).

There is unequivocal evidence for substantial amounts of silicate material in the interstellar medium. On sightlines in the Milky Way with sufficient column den-

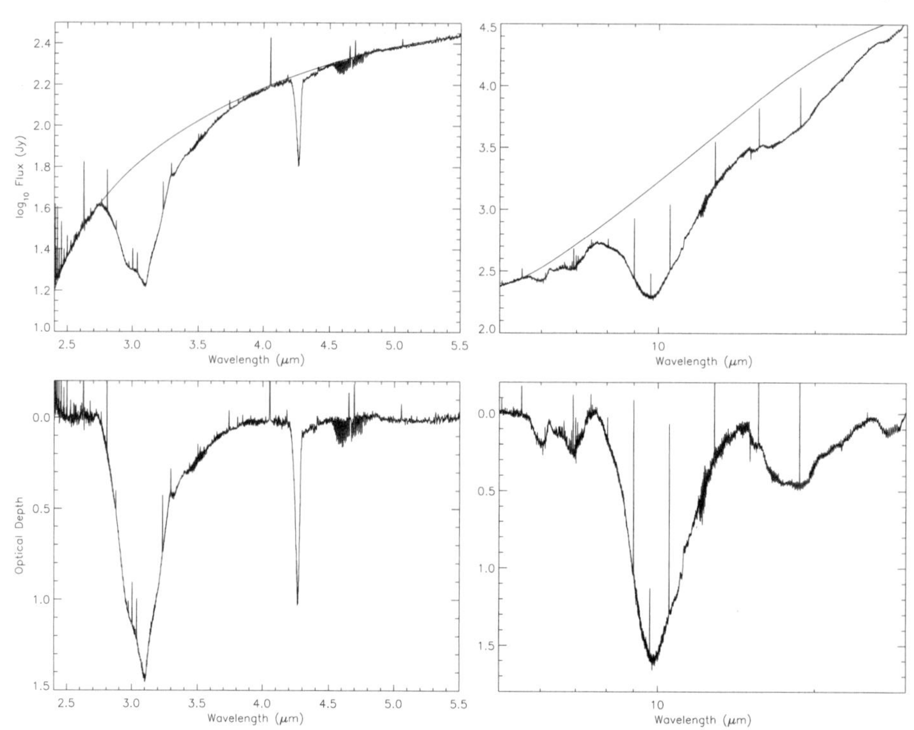

Figure 23.5 Upper panels: spectrum of the BN Object in the Orion Molecular Cloud OMC-1. Lowe panels: estimated absorption optical depth τ. Strong absorption features are seen at 3.1 μm (O-H stretching mode in H_2O), 4.27 μm (C-O stretch in CO_2), 6.02 μm (H-O-H bend in H_2O), 9.7 μm (Si-O stretch in amorphous silicate), 18 μm (O-Si-O bend in amorphous silicate), with additional weaker absorption features. Narrow lines from gas-phase H_2 and CO appear in emission. From Gibb et al. (2004), reproduced by permission of the AAS.

sities, we observe strong absorption with a broad profile, peaking at $\lambda = 9.7\,\mu$m, with FWHM $\approx 2.32\,\mu$m (Kemper et al. 2004). Figure 23.5 shows the 2.5–30 μm spectrum of the Becklin-Neugebauer object (a bright infrared source in the OMC-1 molecular cloud). The spectrum shows strong absorption features due to amorphous silicate material at 9.7 μm and 18 μm (as well as additional absorption features due to ices). Figure 23.6 shows spectrophotometry for three sources in the Galactic Center region; in each case, a deep absorption feature with a minimum at 9.7 μm is present, together with a weaker feature with a minimum near 18 μm.

Interstellar dust grains heated by intense radiation fields – for example, the dust near the Trapezium stars in the Orion Nebula (Gillett et al. 1975*a*) – exhibit strong emission near 10 μm, with a profile that resembles the absorption profile seen toward the Galactic Center. Hot dust present in outflows from stars with oxygen-rich atmospheres (with O/C $>$ I) also shows a very similar emission feature near 10 μm, whereas this emission feature is *not* seen in outflows from carbon stars (with

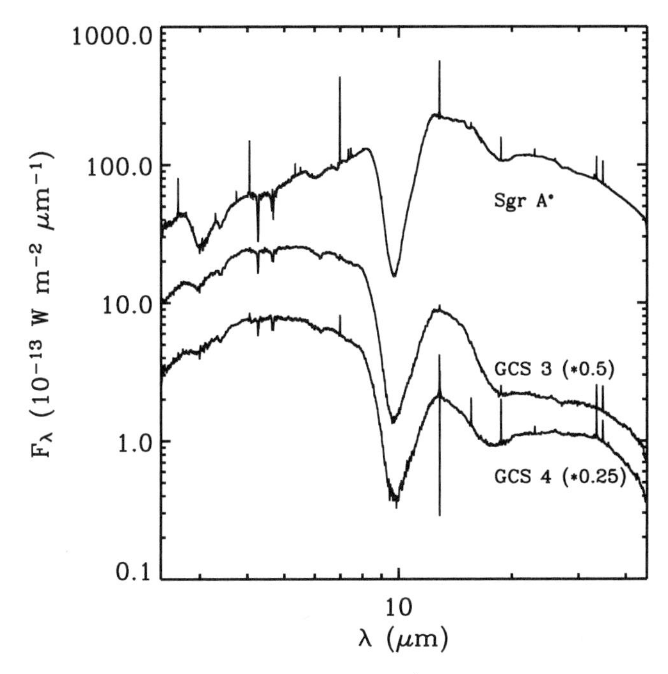

Figure 23.6 Spectra of the Galactic Center (Sgr A*), and two infrared sources GCS3 and GCS4 located near the Galactic Center. In all cases there is strong absorption in the 9.7 μm silicate feature, with associated weaker absorption in the 18 μm feature. There is also absorption in the 3.1 μm feature of H_2O ice toward Sgr A*, with weaker ice absorption seen toward GCS 3. From Kemper et al. (2004), reproduced by permission of the AAS.

C/O > 1), where the gas-phase chemistry is not expected to allow formation and growth of silicates.

Identification of silicate material as a major component of interstellar grains seems incontrovertible, but the specific chemical composition has been difficult to determine. The observed interstellar absorption is broad and smooth, quite unlike the highly structured absorption profiles measured for crystalline silicate minerals in the laboratory. It appears that the interstellar material is **amorphous** rather than crystalline. Amorphous silicates with absorption profiles that closely resemble the observed interstellar profiles can be produced in the laboratory by ion bombardment of initially crystalline material (Kraetschmer & Huffman 1979), formation in smokes (Day 1979), rapid quenching of a melt (Jaeger et al. 1994), or deposition following evaporation (Koike & Tsuchiyama 1992; Stephens et al. 1995; Scott & Duley 1996*b*).

Based on nondetection of sharp features that would be produced by crystalline silicates, upper limits can be placed on the fraction of interstellar silicates that are crystalline. Li & Draine (2001) found that not more than 5% of interstellar Si atoms could be in crystalline silicates, and more recent work has lowered the upper limit on the crystalline fraction to $\lesssim 2.2\%$ (Kemper et al. 2005).

However, the infrared spectra of some AGB stars (de Vries et al. 2010), as well as some comets (e.g., Comet Hale-Bopp: Wooden et al. 1999) and disks around T Tauri stars (Olofsson et al. 2009), do show fine structure characteristic of crystalline silicates. The fine structure indicates that the crystalline material present is predominantly of an olivine ($Mg_{2x}Fe_{2-2x}SiO_4$) structure, with a magnesium fraction $x \approx 0.8$.

In crystalline silicates, the Mg/Fe ratio can be diagnosed by well-defined shifts in spectral features, but determining the Mg/Fe ratio in amorphous silicates is much more challenging. From the observed interstellar extinction, Kemper et al. (2004) infer that Mg/(Mg+Fe) $\approx$ 0.5; Min et al. (2007), on the other hand, conclude that Mg/(Mg+Fe) $\approx$ 0.9 .

The overall strength of the silicate absorption feature requires that a substantial fraction of interstellar silicon atoms reside in amorphous silicate grains. See Henning (2010) for a recent review of silicates in the ISM and around stars.

23.5 Polycyclic Aromatic Hydrocarbons

The infrared emission spectra of spiral galaxies show conspicuous emission features at 3.3, 6.2, 7.7, 8.6, 11.3, and 12.7 μm that are attributable to vibrational transitions in polycyclic aromatic hydrocarbon (PAH) molecules. PAH molecules are planar structures consisting of carbon atoms organized into hexagonal rings, with hydrogen atoms attached at the boundary. Figure 23.7 shows the 5 to 15 μm

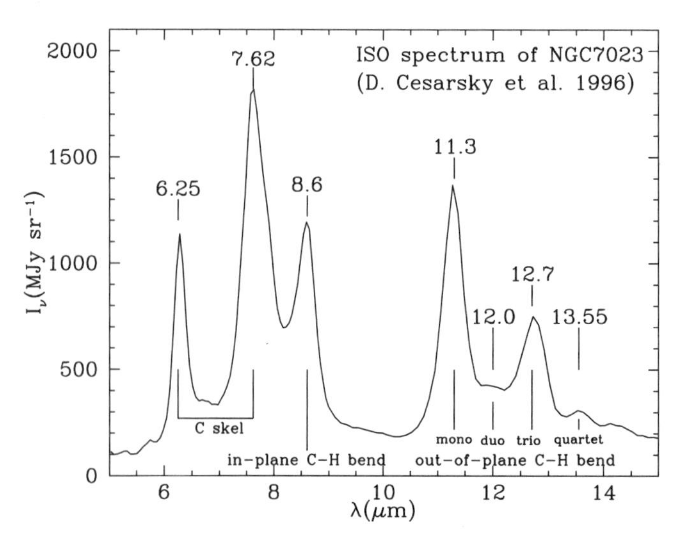

Figure 23.7 The 5 to 15 μm spectrum of the reflection nebula NGC 7023 (Cesarsky et al. 1996).

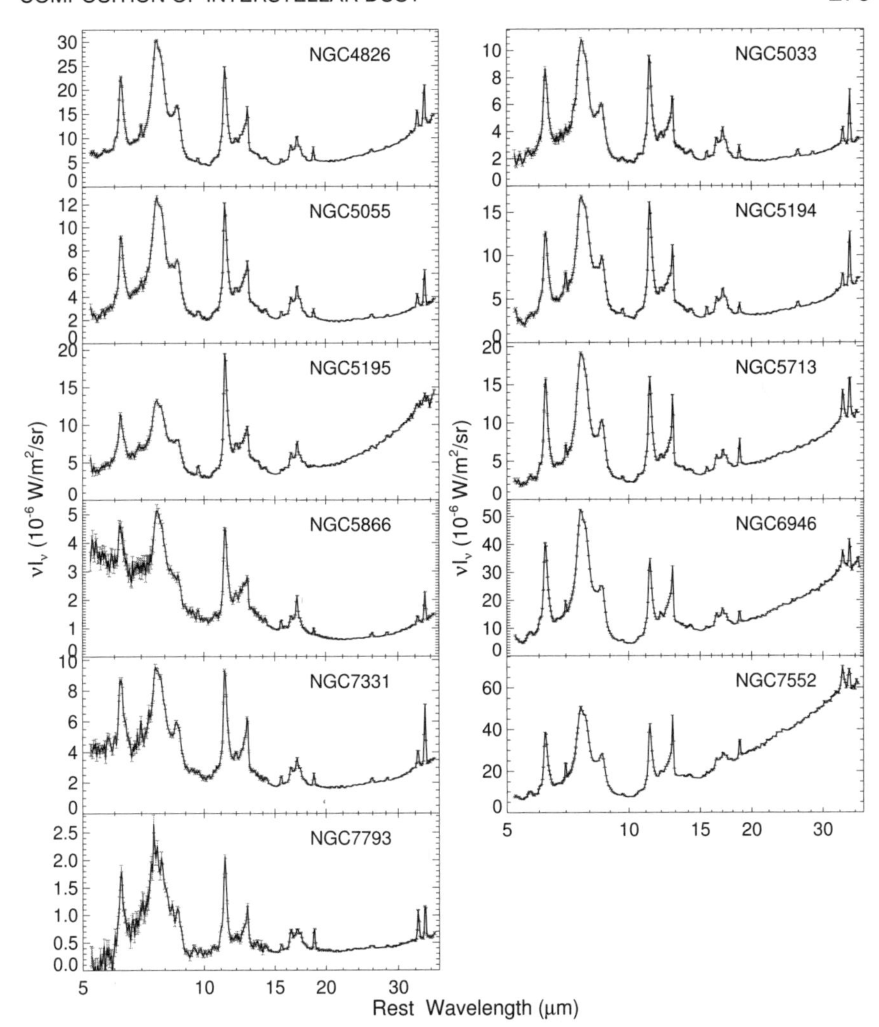

Figure 23.8 5.5 to 36.5 μm spectra of the central regions of various galaxies. Various emission lines (e.g., [Ne II]12.81 μm, [Ne III]15.55 μm, $H_2S(1)17.04\,\mu$m, [S III]18.71 μm, [Fe II]25.99 μm, [S III]33.48 μm, and [Si II]34.82 μm) are visible, but the spectra are dominated by strong PAH emission features peaking at 6.2, 7.7, 8.6, 11.3, 12.7, 16.4, and 17 μm. From Smith et al. (2007), reproduced by permission of the AAS.

spectrum of the bright reflection nebula NGC 7023, showing the features at 6.2, 7.7, 8.6, 11.3, and 12.7 μm. Weaker features are also present at 12.0 and 13.55 μm. The integrated emission from dusty spiral galaxies (see Figure 23.8) shows the same PAH emission features as seen in NGC 7023 – PAH emission features can account for as much as 20% of the total infrared luminosity of a star-forming galaxy. A complex of emission features is present near 17 μm; this complex correlates with the 6.2, 7.7, and 11.3 μm features, and is presumed to also be emitted by PAHs.[2]

The 3.3 μm feature (not shown in Fig. 23.7 or Fig. 23.8) is produced by the C–H stretching mode in PAHs. The features at 6.2 and 7.7 μm are produced by vibrational modes of the carbon skeleton. The feature at 8.6 μm is associated with in-plane C–H bending modes, and the features at 11.3, 12.0, 12.7, and 13.55 μm are due to out-of-plane C–H bending modes of H atoms at "mono," "duo," "trio," or "quartet" sites, defined by the number of adjacent H atoms. Figure 23.9 shows four examples of PAHs, with examples of mono, duo, trio, or quartet sites indicated.

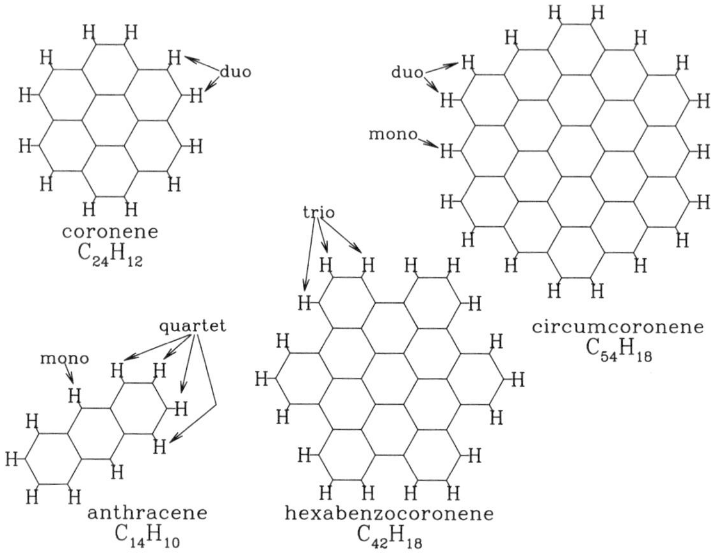

Figure 23.9 Structure of four PAHs. Examples of singlet, doublet, trio, and quartet H atoms are indicated.

Interstellar PAHs may not be as perfect as the examples in Fig. 23.9 – for example, one or more of the peripheral H atoms may be missing, perhaps replaced by radicals such as OH or CN, or one of the carbons may be replaced by a nitrogen (Hudgins et al. 2005).

A neutral PAH can be photoionized by the $h\nu < 13.6\,\mathrm{eV}$ starlight in diffuse clouds, creating a PAH$^+$ **cation**, and large PAHs can be multiply ionized. Collision of a neutral PAH with a free electron can create a PAH$^-$ **anion**. The fundamental vibrational modes – C–H stretching and bending, and vibrational modes of the carbon skeleton – remain at nearly the same frequency, although the electric dipole moment of the different modes can be sensitive to the ionization state. For example, the "solo" C–H out-of-plane bending mode at 11.3 μm is much stronger for neutral PAHs than for PAH ions, while the 7.7 μm vibrational mode of the carbon skeleton has a much larger electric dipole moment in PAH ions than in neutrals (see Draine & Li 2007, and references therein).

[2]Table 1 of Draine & Li (2007) has a list of PAH features found in galaxy spectra.

The fraction of interstellar carbon that is incorporated into PAH material is uncertain. Based on the observed strength of the PAH emission features, it appears that ~ 10–15% of the interstellar carbon resides in PAHs containing fewer than ~ 500 C atoms. For example, the dust model of Draine & Li (2007) has 20 ppm C in PAHs with 30 to 100 C atoms, and another 20 ppm C in PAHs with 100 to 500 C atoms; together, these account for $\sim 14\%$ of the total C abundance in Table 1.4.

The PAH emission features are excited only in PAHs that are sufficiently small so that absorption of a single optical or UV photon can heat the grain to $T \gtrsim 250\,\mathrm{K}$. Additional PAH material may be incorporated into larger grains. In principle, this material could be detected in absorption. The most-easily detected absorption feature may be the C-C bending mode at $6.2\,\mu\mathrm{m}$. Chiar & Tielens (2001) report an upper limit on the strength of an interstellar absorption feature at this wavelength that is slightly below the absorption strength predicted by the PAH model of Draine & Li (2007).

More information about interstellar PAHs can be found in the review by Tielens (2008).

23.6⋆ Graphite

Graphite is the most stable form of carbon (at low pressures), consisting of infinite parallel sheets of sp^2-bonded carbon. A single (infinite) sheet of carbon hexagons is known as **graphene**; each carbon atom in graphene has three nearest neighbors, with a nearest-neighbor distance of 1.421 Å. Crystalline graphite consists of regularly stacked graphene sheets, with an interlayer separation of 3.354 Å and a density $\rho = 2.26\,\mathrm{g\,cm^{-3}}$. The sheets are weakly bound to one another by van der Waals forces. Carbon in which the graphene sheets are parallel, but not regularly stacked, is known as **turbostratic carbon**; densities range from $2.21 - 2.26\,\mathrm{g\,cm^{-3}}$.

Graphite is a semimetal, with nonzero electrical conductivity even at low temperatures. It is a strongly anisotropic material; the response to applied electric fields depends on the orientation of the electric field relative to the "basal plane" (i.e., graphene plane). Stecher & Donn (1965) noted that small, randomly oriented graphite spheres would be expected to produce strong UV absorption with a profile very similar to the observed "extinction bump" near 2175 Å. This absorption is due to $\pi \rightarrow \pi^\star$ transitions in the graphite. A C atom in the "interior" of a large PAH molecule is bonded to 3 nearest-neighbor C atoms, just as in graphite. The electron orbitals of the C atoms in the interior of a large PAH molecule are therefore very similar to the electron orbitals in graphite, and PAHs therefore also have strong absorption near 2175 Å due to $\pi \rightarrow \pi^*$ electronic transitions.

Given the abundance of PAHs required to account for the observed IR emission features, it now seems possible that the observed interstellar 2175 Å extinction feature may be produced primarily by absorption in PAH molecules, or clusters of PAHs, rather than particles of graphite.

23.7★ Diamond

Diamond consists of sp^3-bonded carbon atoms, with each carbon bonded to four equidistant nearest neighbors. As mentioned earlier (see Table 23.2), diamond nanoparticles are relatively abundant in primitive meteorites. Based on isotopic anomalies associated with them, we know that some fraction of the nanodiamond was of presolar origin, and thus was present in the interstellar medium prior to the formation of the Sun; therefore, *some* nanodiamond is presumably present in the ISM today, but its abundance is not known (see Jones & D'Hendecourt 2004).

23.8★ Amorphous Carbons, Including Hydrogenated Amorphous Carbon

Graphite and diamond are ideal crystals, but carbonaceous solids are often disordered mixtures of both sp^2- and sp^3-bonded carbon, often with hydrogen also present (see Robertson 2003). **Amorphous carbon** is a mixture of sp^2- and sp^3-bonded carbon – one can think of it as a jumble of microcrystallites with more-or-less random orientations, with the microcrystallites connected haphazardly by interstitial carbon atoms. Amorphous carbon is not a well-defined material, and its properties depend on the method of preparation. Densities are typically in the range 1.8 to $2.1\,\mathrm{g\,cm^{-3}}$.

Hydrogenated amorphous carbon (HAC) is a class of materials obtained when sufficient hydrogen is present, with H:C ratios ranging from 0.2:1 to 1.6:1 (Angus & Hayman 1988). As with amorphous carbon, the properties of HAC depend on the method of preparation. HAC is a mixture of sp^2- and sp^3-bonded carbon, giving it a diamondlike character; the properties of HAC depend on the $sp^2 : sp^3$ ratio. HAC is a semiconductor, with a bandgap. Jones et al. (1990) discuss the properties of HAC as a candidate interstellar grain material.

Glassy or **vitreous carbon** (Cowlard & Lewis 1967), consisting primarily of sp^2-bonded carbon, but without long-range order, is another form of solid carbon that is generally considered to be distinct from amorphous carbon. The density is $\sim 1.5\,\mathrm{g\,cm^{-3}}$. Vitreous carbon is electrically conducting, with a conductivity similar to the conductivity of graphite for conduction in the basal plane. By contrast, "amorphous" carbon (with a significant sp^3 fraction) is an insulator.

23.9★ Fullerenes

Fullerenes are cage-like carbon molecules, including C_{60}, C_{70}, C_{76}, and C_{84}, where the carbon is sp^2-bonded with 3 near-coplanar nearest neighbors, but where a few of the hexagons are replaced by pentagons to allow the surface to close upon itself. C_{60}, also known as buckminsterfullerene, is the most stable fullerene. Fullerenes have been proposed as likely to be present in the ISM (Kroto & Jura 1992).

Foing & Ehrenfreund (1994) found diffuse interstellar bands at 9577 Å and 9632 Å that were consistent, within uncertainties, with lab measurements of absorption by matrix-isolated C_{60}^+, but the identification remains tentative because of uncertain "matrix shifts" in the lab measurements, and failure to detect associated features expected near 9366 Å and 9419 Å (Jenniskens et al. 1997). Foing & Ehrenfreund (1994) estimated that 0.3–0.9% of interstellar carbon in C_{60}^+ would be required to account for the observed DIBs at 9577 Åand 9632 Å.

Sellgren et al. (2010) reported observation of three infrared emission features, at 7.04, 17.4, and 18.9 μm, that appear to confirm the presence of neutral C_{60} in the reflection nebula NGC 7023. They estimate that 0.1–0.6% of interstellar carbon in C_{60} is required to account for the strength of the emission bands. Cami et al. (2010) report detection of IR bands of C_{60} and C_{70} in a young, carbon-rich, planetary nebula. In this source they estimate that at least 1.5% of the available carbon is present in each of the species, although the estimate is quite uncertain.

The detection of C_{60} in the reflection nebula NGC 7023 appears to confirm the presence of fullerenes in the general interstellar medium. The abundances are uncertain, but current evidence suggests that the fullerene family might contain as much as 1% of the interstellar carbon. While significant, this would be an order of magnitude below the estimated abundance of carbon in the PAH population. Fullerenes altogether probably account for less than 1% of the total dust mass.

23.10 Models for Interstellar Dust

A model for interstellar dust must specify the composition of the dust as well as the geometry (shape and size) of the dust particles. If the model is to reproduce the polarization of starlight, at least some of the grains should be nonspherical and aligned.

From the observational data available to us, it is not yet possible to arrive at a unique grain model. A class of models that has met with some success assumes the dust to consist of two materials: (1) amorphous silicate, and (2) carbonaceous material. Mathis et al. (1977) showed that models using silicate and graphite spheres with power-law size distributions $dn/da \propto a^{-3.5}$ for $a_{\rm min} < a < a_{\rm max}$ [frequently referred to as the Mathis-Rumpl-Nordsieck, or "MRN," size distribution] could reproduce the observed extinction from the near-infrared to the ultraviolet. Draine & Lee (1984) presented self-consistent dielectric functions for graphite and silicate, and showed that the graphite-silicate model appeared to be consistent with what was known about dust opacities in the far-infrared.

With the recognition that PAH particles are an important component, it is natural to add them to the graphite-silicate model, either as a third component, or as the small-particle extension of the graphitic material. (This is appropriate because PAHs are, essentially, graphene fragments with hydrogen attached at the periphery.) The carbonaceous material is assumed to be PAH-like when the particles are small ($a \lesssim 0.005\,\mu$m), but when the particles are large ($a \gtrsim 0.02\,\mu$m), the carbonaceous material is approximated by graphite. Grain models based on amorphous silicate, graphite, and PAHs have been put forward by a number of authors (Desert et al. 1990; Weingartner & Draine 2001*a*; Zubko et al. 2004; Gupta et al. 2005; Draine & Li 2007; Draine & Fraisse 2009). The size distributions should reproduce the observed extinction curve (see Fig. 21.1), using amounts of grain material that are consistent with the abundance limits in Table 23.1.

If a suitable shape is assumed (e.g., spheroids with some specified axial ratio), such models are capable of reproducing both the wavelength-dependent extinction and the wavelength-dependent polarization of starlight, provided the grains have a suitable size distribution, and the degree of alignment is allowed to vary with grain size (?Draine & Allaf-Akbari 2006; Draine & Fraisse 2009; Das et al. 2010).

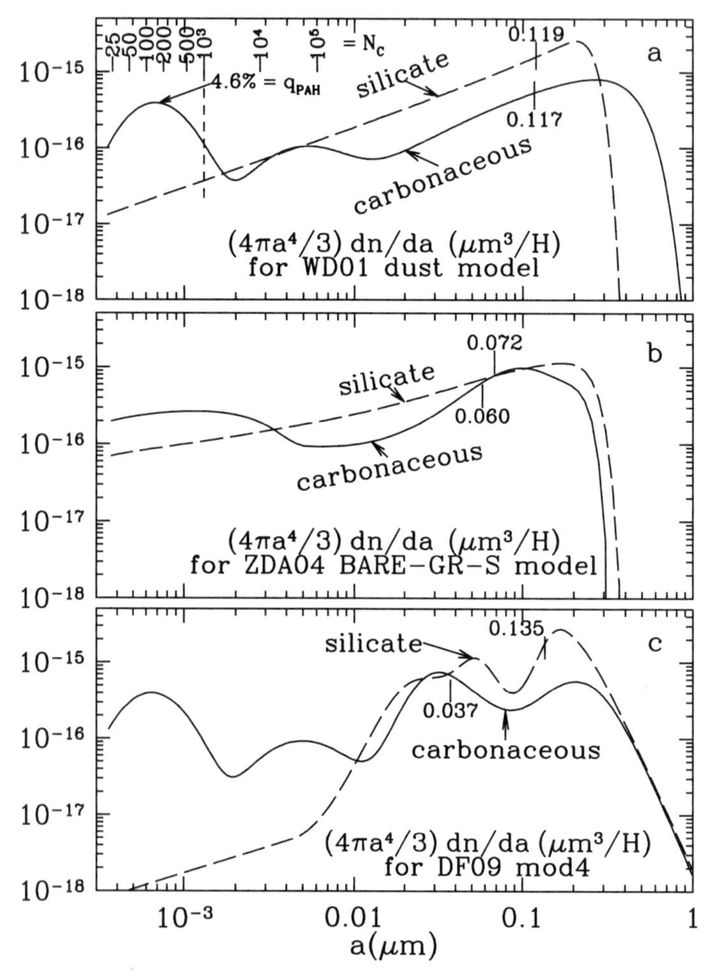

Figure 23.10 Size distributions for silicate and carbonaceous grains for dust models from (a) Weingartner & Draine (2001*a*), (b) Zubko et al. (2004), and (c) Draine & Fraisse (2009). The quantity plotted, $(4\pi a^3/3)dn/d\ln a$ is the grain volume per H per logarithmic interval in a. In each case, tick-marks indicate the "half-mass" radii for the silicate grains and carbonaceous grains.

Figure 23.10 shows size distributions found by three independent studies, all based on silicate, graphite, and PAH material. Weingartner & Draine (2001*a*, hereafter WD01) found size distributions that reproduce the observed extinction from the infrared ($4\,\mu$m) to the vacuum ultraviolet ($0.1\,\mu$m). The half-mass grain radius (50% of the mass in grains with $a > a_{0.5}$) is $a_{0.5} \approx 0.12\,\mu$m for both silicate and carbonaceous grains. The model does a good job of reproducing the observed extinction (see Figure 23.11), but the assumed mass in dust exceeds estimates based on elemental abundances and observed depletions. Table 23.3 shows the amounts of the different elements that the WD01 model consumes. The WD01 model appears to require $(231-186)/186=24\%$ more C in dust, and $(48-27)/27=78\%$ more Si in dust, than is indicated by observations of the CNM, although the discrepancy is less severe if we use gas phase abundances measured toward ζ Oph: the WD01 model requires only $(231–210)/210=10\%$ more C, and $(48–34)/34=41\%$ more Si, than is indicated by observations toward ζ Oph.

As previously discussed in §23.1, the oxygen abundance is very problematic (Jenkins 2009; Whittet 2010). The observed variations in the gas phase abundance of oxygen in diffuse clouds imply that the abundance of oxygen in the solid phase seemingly varyies from 80 ppm in the CNM to 242 ppm toward ζ Oph. The problem is that H_2O ice is *not* present on these diffuse sightlines (see §23.3.5), and the abundances of Mg, Fe, Si, etc. do not allow silicates or metal oxides to account for the large amount of oxygen that is missing from the gas. One possibility is that the oxygen is somehow associated with the hydrocarbon material. The large variations in gas-phase oxygen abundance, in regions where ices are not present, are, at this time, not understood, and the dust models do not pretend to account for them.

The size distribution of the "BARE-GR-S" model of Zubko et al. (2004, herafter ZDA04), composed of bare graphite grains, bare silicate grains, and PAHs, differs significantly from the WD01 size distribution. There is much less mass in grains with radii $a \gtrsim 0.2\,\mu$m – the half-mass radius is only $0.06\,\mu$m for carbonaceous grains, and $0.07\,\mu$m for silicate grains. Both numbers are significantly smaller than the WD01 values. The reduced abundance of the larger grains in the ZDA04 model appears to be the result of different values adopted for the "observed" extinction. This can be seen in Figure 23.11, showing the "observed" extinction adopted by

Table 23.3 Abundances of Major Elements in Grains

Model	V_{car} (cm^3/H)	V_{sil} (cm^3/H)	C/H[d] (ppm)	O/H[e] (ppm)	Mg/H[e] (ppm)	Si/H[g] (ppm)	Fe/H[f] (ppm)
WD01[a]	2.09×10^{-27}	3.64×10^{-27}	231	193	48	48	48
ZDA04[b]	2.21×10^{-27}	2.71×10^{-27}	244	144	36	36	36
DF09[c]	1.89×10^{-27}	3.20×10^{-27}	209	170	42	42	42
Observed for typical CNM. [f]			186	80	35	27	34
Observed toward ζ Oph. [g]			210	242	39	34	35

[a] Weingartner & Draine (2001*a*).
[b] Zubko et al. (2004).
[c] Draine & Fraisse (2009).
[d] Assumed carbon density $2.2\,\mathrm{g\,cm^{-3}}$.
[e] Nominal composition $MgFeSiO_4$, density $3.8\,\mathrm{g\,cm^{-3}}$.
[f] For CNM, from Table 9.5.
[g] Toward ζ Oph, from Table 23.1.

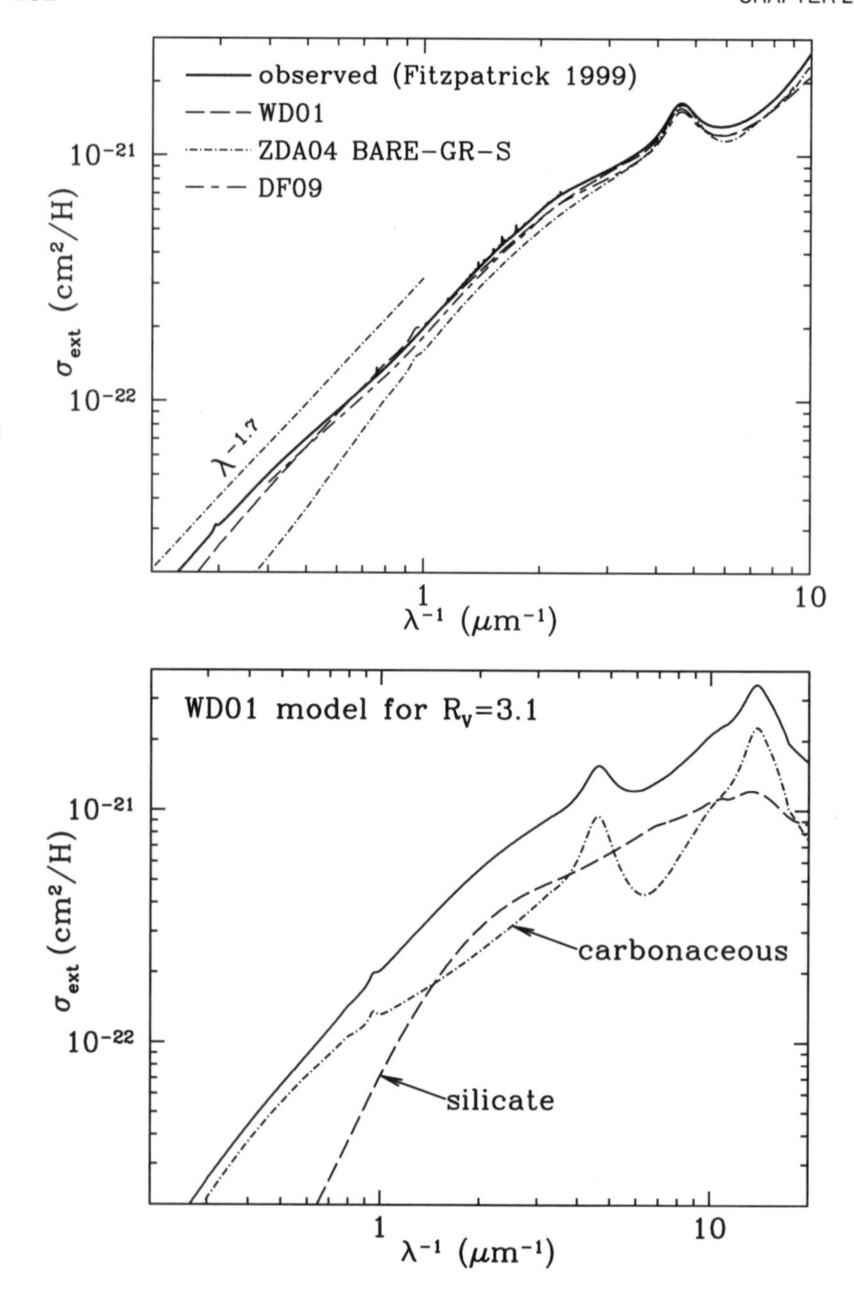

Figure 23.11 Upper: Average observed extinction for $R_V = 3.1$ (Fitzpatrick 1999) and extinction curves calculated for the WD01 silicate-carbonaceous model (Weingartner & Draine 2001*a*) and for the ZDA04 BARE-GR-S silicate-carbonaceous model (Zubko et al. 2004). The WD01 model provides considerably more extinction in the infrared (1 to 4 μm) than the ZDA04 model (see text). Lower: Separate contributions of silicate and carbonaceous grains.

WD01, as well as the extinction calculated for the WD01 and ZDA04 models. The WD01 and ZDA04 models have similar extinction for $\lambda < 1\,\mu\mathrm{m}$, but the ZDA04 model does not reproduce the "observed" extinction in the infrared, where the WD01 model provides substantially more extinction than the ZDA04 model. The extra extinction in the infrared in the WD01 model is provided by additional mass in ($a \gtrsim 0.2\,\mu\mathrm{m}$) grains, which appears to be required in order to reproduce the adopted "observed" extinction in the infrared.

It is sometimes suggested that nonspherical grains might be able to account for the observed extinction using less mass. Draine & Fraisse (2009, hereafter DF09) used grain models with spheroidal graphite and silicate grains to reproduce both the observed extinction and polarization. The resulting size distributions for one of their models is shown in Figure 23.10, and the elemental abundances consumed are given in Table 23.3. The DF09 model uses $(209-186)/186 = 12\%$ more C than is estimated to be available, well within the uncertainties, but overconsumes Si by 55%, which is a large enough discrepancy to be worrisome. It may indicate that there is a problem with the grain model, or it may indicate that the actual extinction/H in the infrared may not be as large as the "observed" value that was adopted to constrain the WD01 and DF09 dust models. Measuring both IR extinction *and* the total hydrogen column density is difficult, and further study of the absolute extinction in the infrared would be very valuable. It is also possible that the interstellar Si abundance may be larger than the current estimate for the Solar abundance of Si.

23.10.1 Radiation Pressure Cross Sections

The radiation pressure cross section per H nucleon is

$$\sigma_{\rm rad.pr.}(\nu) = \sigma_{\rm abs}(\nu) + (1 - \langle\cos\theta\rangle)\sigma_{\rm sca}(\nu) \quad , \tag{23.2}$$

where $\sigma_{\rm abs}$ and $\sigma_{\rm sca}$ are absorption and scattering cross sections per H, and θ is the scattering angle. For a spectrum L_ν, the spectrum-averaged radiation pressure cross section per H is

$$\langle\sigma_{\rm rad.pr.}\rangle \equiv \frac{\int d\nu L_\nu \sigma_{\rm rad.pr.}(\nu)}{\int d\nu L_\nu} \quad . \tag{23.3}$$

Figure 23.12 shows the spectrum-averaged radiation-pressure cross section calculated for the WD01 and ZDA04 dust models, for blackbody spectra L_ν, as a function of the blackbody temperature $T_{\rm rad}$. Results are shown both for the WD01 model and for the ZDA04 model; note that the results are very similar.

For low values of the blackbody temperature, the radiation pressure cross section is small, but it rises to a peak value $\sim 2\times10^{-21}\,\mathrm{cm}^2/\mathrm{H}$ for blackbody temperatures $\sim 5000\,\mathrm{K}$. As we move to higher temperatures, $\langle\sigma_{\rm rad.pr.}\rangle$ declines for two reasons: the scattering is increasingly in the forward direction ($\langle\cos\theta\rangle \rightarrow 1$), and the absorption cross section begins to decline for high photon energies.

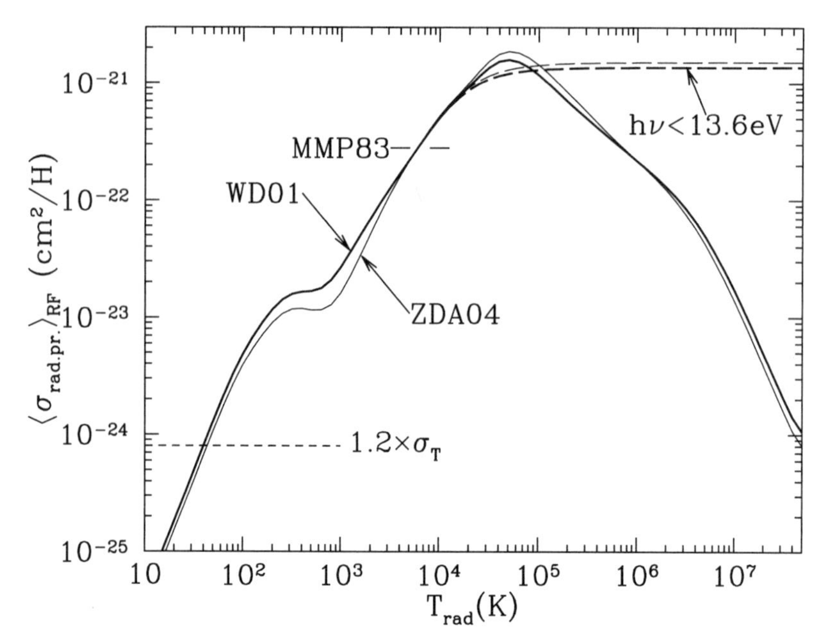

Figure 23.12 Planck-averaged radiation pressure cross section per H $\langle\sigma_{\rm rad.pr.}\rangle$ for the WD01 dust model and the ZDA04 BARE-GR-S dust model, as a function of the blackbody temperature $T_{\rm rad}$. Dashed lines show $\langle\sigma_{\rm rad.pr.}\rangle$ averaged over only the $h\nu < 13.6\,{\rm eV}$ portion of the spectrum, as would be appropriate in neutral gas. Also indicated is the value $\langle\sigma_{\rm rad.pr.}\rangle = 2.80 \times 10^{-22}\,{\rm cm}^2/{\rm H}$ calculated for the ISRF of Mathis et al. (1983). For comparison, $1.2\sigma_T$ is shown, where σ_T is the Thompson cross section.

The interstellar radiation field is contributed by stars of many different temperatures, with the stellar radiation filtered through differing amounts of interstellar dust. The radiation pressure cross section appropriate to the spectrum of the interstellar radiation field in the solar neighborhood (taken from Mathis et al. 1983) is found to be

$$\langle\sigma_{\rm rad.pr.}\rangle_{\rm ISRF} = 2.80 \times 10^{-22}\,{\rm cm}^2\,{\rm H}^{-1} \quad , \tag{23.4}$$

with nearly the same value found for the WD01 and ZDA04 size distributions. This value is the same as for a $\sim 6000\,{\rm K}$ blackbody.

The radiation pressure cross section appropriate to fully ionized gas with He/H = 0.1 is also shown in Figure 23.12. It is apparent that for the dust abundance of the local Milky Way, radiation pressure on dust dominates over free-electron scattering for a broad range of radiation fields.

Chapter Twenty-four

Temperatures of Interstellar Grains

The "temperature" of a dust grain is a measure of the internal energy $E_{\rm int}$ present in vibrational modes and possibly also in low-lying electronic excitations. If the dust grain has internal energy E, the temperature T can be taken to be the thermodynamic temperature for which the expectation value $\langle E_{\rm int}\rangle_T = E$. For very small internal energies $E_{\rm int}$, the concept of "temperature" becomes problematic, but this is only an issue when $E_{\rm int}$ becomes comparable to the energy of the first excited state of the grain.[1] Such low degrees of excitation are rarely of interest. Therefore, we will find it convenient to specify the energy content of a grain by its temperature T.

Energy can be added or removed from the grain by absorption or emission of photons, or by inelastic collisions with atoms or molecules from the gas.[2] In diffuse regions, where ample starlight is present, grain heating is dominated by absorption of starlight photons; however, in dense dark clouds, grain heating can be dominated by inelastic collisions.

The dust grains responsible for the bulk of the observed extinction at optical wavelengths – grains with radii $a \gtrsim 0.03\,\mu{\rm m}$ – can be considered "classical." These grains are macroscopic – absorption or emission of single quanta do not appreciably change the total energy in vibrational or eleconic excitations – and their temperatures are discussed in §24.1.

The grain population also includes ultrasmall particles, ranging down to large molecules, where quantum effects are important (this includes the "spinning" dust grains responsible for microwave emission). Heating and cooling of these ultrasmall grains is the subject of §24.2.

24.1 Heating and Cooling of "Classical" Dust Grains

24.1.1 Radiative Heating

When an optical or ultraviolet photon is absorbed by a grain, an electron is raised into an excited electronic state. If the electron is sufficiently energetic, it may be able to escape from the solid as a "photoelectron." In rare cases, the grain will

[1]The assignment of a nominal "temperature" to grains in very low states of excitation is discussed by Draine & Li (2001).

[2]Grain–grain collisions will also heat grains, but are too infrequent to contribute significantly to the infrared emission.

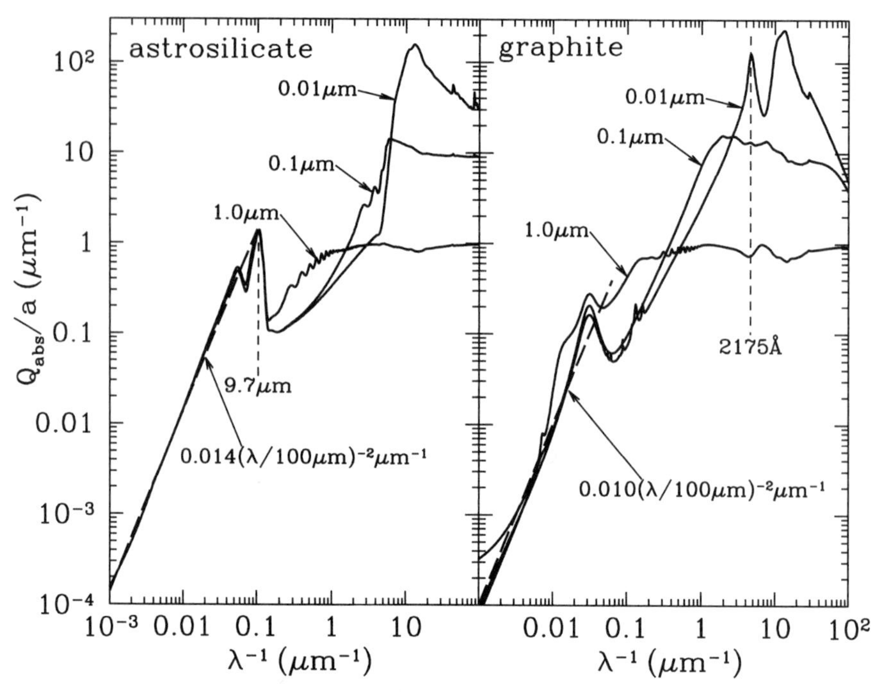

Figure 24.1 Absorption efficiency $Q_{\rm abs}(\lambda)$ divided by grain radius a for spheres of amorphous silicate (left) and graphite (right). Also shown are power-laws that provide a reasonable approximation to the opacity for $\lambda \gtrsim 20\,\mu$m.

"luminesce": the excited state will decay radiatively, emitting a photon of energy less than or equal to the energy of the absorbed photon.[3] In most solids or large molecules, however, the electronically excited state will deexcite nonradiatively, with the energy going into many vibrational modes – i.e., heat.

Ignoring the small fraction of energy appearing as luminescence or photoelectrons, the rate of heating of the grain by absorption of radiation can be written

$$\left(\frac{dE}{dt}\right)_{\rm abs} = \int \frac{u_\nu d\nu}{h\nu} \times c \times h\nu \times Q_{\rm abs}(\nu)\pi a^2 \ . \tag{24.1}$$

Here, $u_\nu d\nu / h\nu$ is the number density of photons with frequencies in $[\nu, \nu + d\nu]$; the photons move at the speed of light c and carry energy $h\nu$; and the grain has absorption cross section $Q_{\rm abs}(\nu)\pi a^2$.

Figure 24.1 shows $Q_{\rm abs}(\lambda)/a$ (which is proportional to the absorption cross section per unit volume) for graphite and silicate spheres with radii $a = 0.01, 0.1$, and

[3]Luminescence is referred to as "fluorescence" when it occurs promptly, and "phosphorescence" when it involves slow decay from a metastable level.

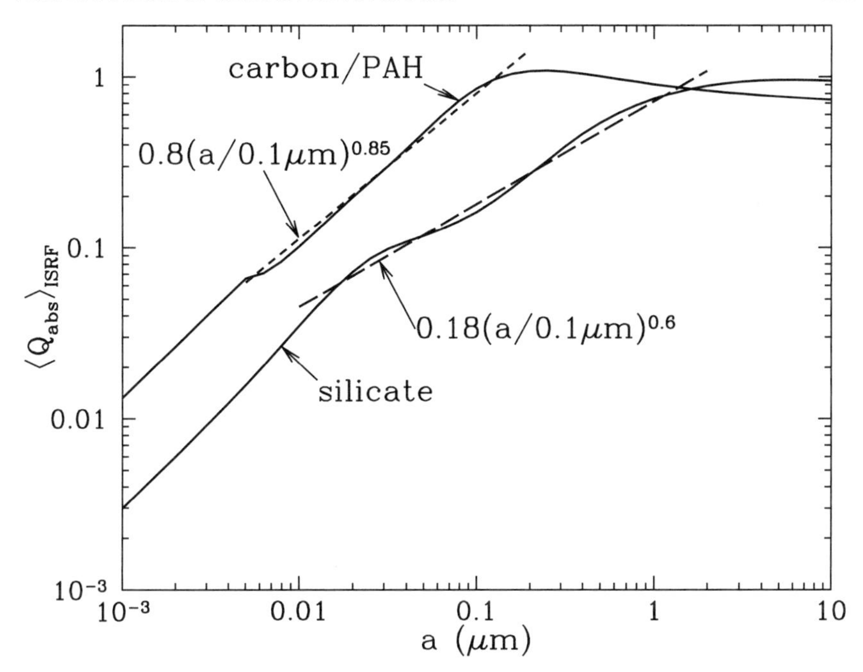

Figure 24.2 Absorption efficiency $\langle Q_{\rm abs}\rangle_{\rm ISRF}$ averaged over the ISRF spectrum of Mathis et al. (1983) for spheres of amorphous silicate and aromatic carbonaceous material (graphite/PAH), as a function of radius a.

$1\,\mu{\rm m}$. It is convenient to define a **spectrum-averaged absorption cross section**:

$$\langle Q_{\rm abs}\rangle_\star \equiv \frac{\int d\nu\, u_{\star\nu} Q_{\rm abs}(\nu)}{u_\star} \quad , \quad u_\star \equiv \int d\nu\, u_{\star\nu} \quad , \tag{24.2}$$

so that the radiative heating rate is simply

$$\left(\frac{dE}{dt}\right)_{\rm abs} = \langle Q_{\rm abs}\rangle_\star \pi a^2 u_\star c \quad . \tag{24.3}$$

We use the subscript $\star$ because starlight is often the dominant source of radiation heating the dust. Figure 24.2 shows $\langle Q_{\rm abs}\rangle_\star$ as a function of radius a for graphite and silicate grains, and the spectrum of the interstellar radiation field (ISRF) from Mathis et al. (1983) (see Chapter 12). The numerical results in Fig. 24.2 can be approximated by

$$\langle Q\rangle_{\rm ISRF} \approx 0.18(a/0.1\,\mu{\rm m})^{0.6} \quad , \quad \text{for silicate}, 0.01 \lesssim a \lesssim 1\,\mu{\rm m}\,, \tag{24.4}$$

$$\approx 0.8(a/0.1\,\mu{\rm m})^{0.85} \quad , \quad \text{for graphite}, 0.005 \lesssim a \lesssim 0.15\,\mu{\rm m}\,. \tag{24.5}$$

The starlight energy density $u_\star = 1.05\times10^{-12} U\,{\rm erg\,cm^{-3}}$, where $U = 1$ for the ISRF estimated by Mathis et al. (1983) (see Table 12.1).

24.1.2 Collisional Heating

Consider a neutral, spherical grain of radius a, at rest in a gas with temperature $T_{\rm gas}$. The net rate of collisional heating by the gas can be written

$$\left(\frac{dE}{dt}\right)_{\rm gas} = \sum_i n_i \left(\frac{8kT_{\rm gas}}{\pi m_i}\right)^{1/2} \pi a^2 \times \alpha_i \times 2k(T_{\rm gas} - T_{\rm dust}) \quad , \qquad (24.6)$$

where the sum is over different gas species i (H, He, H_2, e^-, H^+, ...).[4] The term $(8kT_{\rm gas}/\pi m_i)^{1/2}$ is the mean speed of species i. The mean kinetic energy of thermal particles colliding with a surface is $2kT_{\rm gas}$ – larger than the mean kinetic energy $\frac{3}{2}kT_{\rm gas}$ in the gas because the more energetic particles collide more frequently.

The net rate of energy transfer vanishes if $T_{\rm gas} = T_{\rm dust}$. The *accommodation coefficient* $0 \leq \alpha_i \leq 1$ measures the degree of inelasticity for collisions of particle i with the solid surface. Perfectly elastic collisions would have $\alpha_i = 0$, whereas if impinging particles "stick" to the surface for more than $\sim 10^{-12}$ s, the accommodation coefficient $\alpha_i \approx 1$. The value of α_i for H, He, and H_2 incident on interstellar grain materials depends on $T_{\rm gas}$ and on the (uncertain) composition and roughness of the grain surface (and on the grain temperature). At this time, our ignorance of the surface physics is considerable, and it is usual to assume simply that for $T_{\rm gas} \lesssim 10^4$ K, α_i will be less than 1 but of order unity – say, $\alpha_i \approx 0.5$.

In atomic H, the ratio of collisional heating to radiative heating is

$$\frac{(dE/dt)_{\rm gas}}{(dE/dt)_{\rm abs}} = \frac{n_{\rm H}(8kT/\pi m_{\rm H})^{1/2} 2\alpha_{\rm H} kT}{\langle Q_{\rm abs}\rangle_\star u_\star c} \times 1.05 \qquad (24.7)$$

$$= \frac{3.8 \times 10^{-6}}{U} \frac{\alpha_{\rm H}}{\langle Q_{\rm abs}\rangle_\star} \left(\frac{n_{\rm H}}{30\,{\rm cm}^{-3}}\right) \left(\frac{T_{\rm gas}}{10^2\,{\rm K}}\right)^{3/2} \quad , \qquad (24.8)$$

where the factor 1.05 allows for He with $n({\rm He})/n_{\rm H} = 0.1$.

We see, then, that for CNM conditions (Table 1.3), collisional heating is $\lesssim 10^{-5}$ of radiative heating, and can be ignored. Collisional heating can, however, be important within dark clouds, where the high gas density and very low intensity of optical or UV radiation make collisional heating competitive.

24.1.3 Radiative Cooling

Grains lose energy by infrared emission, at a rate

$$\left(\frac{dE}{dt}\right)_{\rm emiss.} = \int d\nu\, 4\pi B_\nu(T_d) C_{\rm abs}(\nu) = 4\pi a^2 \langle Q_{\rm abs}\rangle_{T_d} \sigma T_d^4 \ , \qquad (24.9)$$

[4] The contribution of ions or electrons to the heating is modified by the effects of electrostatic focusing or repulsion. In the case of a neutral grain, the collision rate and heating rate are modified by polarization effects. In the case of a charged grain, ions and electrons are subject to either Coulomb focusing or repulsion. See Draine & Sutin (1987) for further details.

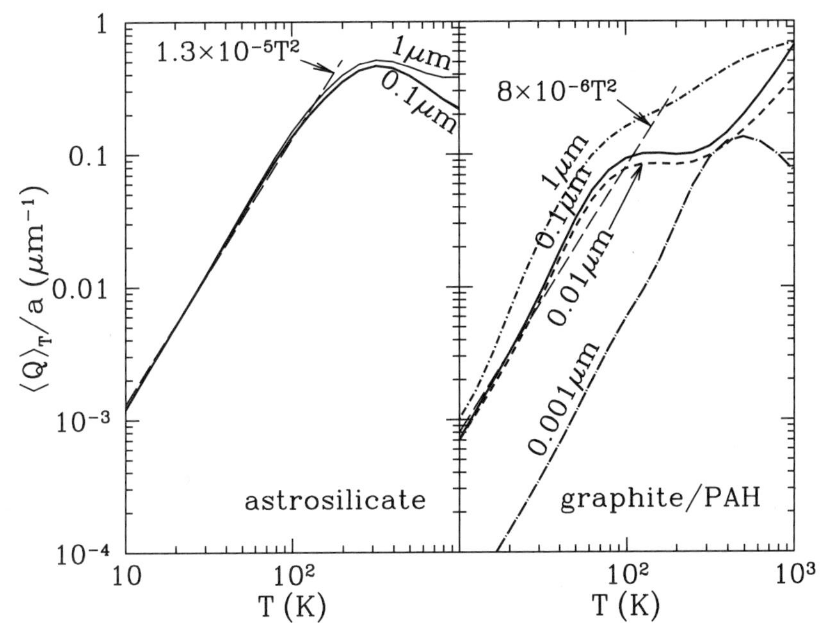

Figure 24.3 Planck-averaged absorption efficiency divided by grain radius as a function of grain temperature T for spheres of amorphous silicate (left) and graphite/PAH (right).

where σ is the Stefan-Boltzmann constant, and the **Planck-averaged emission efficiency** is defined by

$$\langle Q_{\rm abs}\rangle_T \equiv \frac{\int d\nu B_\nu(T) Q_{\rm abs}(\nu)}{\int d\nu B_\nu(T)} \quad . \tag{24.10}$$

If $Q_{\rm abs}(\nu)$ can be approximated as a power-law in frequency,

$$Q_{\rm abs}(\nu) = Q_0\,(\nu/\nu_0)^\beta = Q_0\,(\lambda/\lambda_0)^{-\beta} \quad , \tag{24.11}$$

then the Planck average can be obtained analytically:

$$\langle Q\rangle = \frac{15}{\pi^4}\Gamma(4+\beta)\zeta(4+\beta)Q_0\left(\frac{kT}{h\nu_0}\right)^\beta \quad , \tag{24.12}$$

where $\Gamma(x)$ and $\zeta(x)$ are the usual gamma function and Riemann ζ-function, respectively. From Fig. 24.1, we see that

$$Q_{\rm abs} \approx 1.4\times10^{-3}\left(\frac{a}{0.1\,\mu{\rm m}}\right)\left(\frac{\lambda}{100\,\mu{\rm m}}\right)^{-2}, \quad \text{silicate}, \lambda \gtrsim 20\,\mu{\rm m}, \tag{24.13}$$

$$\approx 1.0\times10^{-3}\left(\frac{a}{0.1\,\mu{\rm m}}\right)\left(\frac{\lambda}{100\,\mu{\rm m}}\right)^{-2}, \quad \text{graphite}, \lambda \gtrsim 30\,\mu{\rm m}. \tag{24.14}$$

The Planck averages are then[5]

$$\langle Q_{\rm abs}\rangle_T \approx 1.3\times10^{-6}(a/0.1\,\mu{\rm m})(T/\,{\rm K})^2 \quad ({\rm silicate}) \tag{24.15}$$

$$\approx 8\times10^{-7}(a/0.1\,\mu{\rm m})(T/\,{\rm K})^2 \quad ({\rm graphite}). \tag{24.16}$$

These are plotted in Fig. 24.3, and are seen to agree very well at $T \lesssim 10^2\,{\rm K}$ with the Planck averages calculated using the actual $Q_{\rm abs}(\nu)$.

24.1.4 Steady State Grain Temperature

We can now determine the steady state grain temperature $T_{\rm ss}$ by requiring that cooling balance heating:

$$4\pi a^2\langle Q_{\rm abs}\rangle_{T_{\rm ss}}\sigma T_{\rm ss}^4 = \pi a^2\langle Q_{\rm abs}\rangle_\star u_\star\, c \quad , \tag{24.17}$$

where $\langle Q_{\rm abs}\rangle_\star$ is the grain absorption cross section averaged over the spectrum of the radiation (usually starlight) heating the grain (see Eq. 24.2), with energy density $u_\star$. If $Q_{\rm abs} = Q_0(\lambda/\lambda_0)^{-\beta}$ in the infrared, the solution is

$$T_{\rm ss} = \left(\frac{h\nu_0}{k}\right)^{\beta/(4+\beta)}\left[\frac{\pi^4\langle Q_{\rm abs}\rangle_\star\, c}{60\Gamma(4+\beta)\zeta(4+\beta)Q_0\sigma}\right]^{1/(4+\beta)} u_\star^{1/(4+\beta)} \quad . \tag{24.18}$$

If we assume the spectrum of the ISRF, and the absorption properties of silicate and graphite, then

$$T_{\rm ss} \approx 16.4\,(a/0.1\,\mu{\rm m})^{-1/15}\,U^{1/6}\,{\rm K} \quad , \quad {\rm silicate}, 0.01 \lesssim a \lesssim 1\,\mu{\rm m} \tag{24.19}$$

$$\approx 22.3\,(a/0.1\,\mu{\rm m})^{-1/40}\,U^{1/6}\,{\rm K} \quad , \quad {\rm graphite}, 0.005 \lesssim a \lesssim 0.15\,\mu{\rm m} \tag{24.20}$$

for $U \lesssim 10^4$. Figure 24.4 shows that the above approximations agree very well with numerical results for the steady-state grain temperature.

24.2 Heating and Cooling of Ultrasmall Dust Grains: Temperature Spikes

Above we evaluated the steady-state temperature $T_{\rm ss}$ at which the time-averaged power radiated by the grain is equal to the time-averaged rate of heating the grain. When the grain is at this temperature, the vibrational energy content of the grain is

$$E_{\rm vib}(T_{\rm ss}) = \int_0^{T_{\rm ss}} C(T)dT \quad , \tag{24.21}$$

---

[5] $\Gamma(6) = 5! = 120$ and $\zeta(6) = 1.0173$

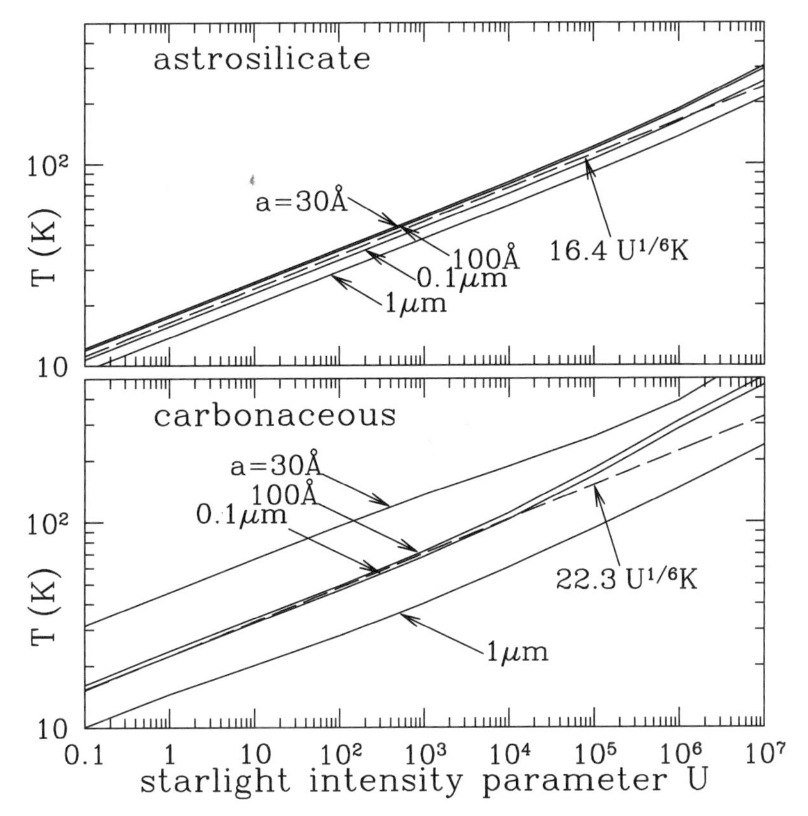

Figure 24.4 Equilibrium temperature for astrosilicate and carbonaceous grains heated by starlight with the spectrum of the local radiation field, and intensity U times the local intensity. Also shown are the power-laws $T = 16.4U^{1/6}$ K and $T = 22.3U^{1/6}$ for $a = 0.1\,\mu$m from Eqs. (24.19 and 24.20).

where $C(T)$ is the heat capacity of the grain at temperature T. If $E_{\rm vib}(T_{\rm ss}) \lesssim \langle h\nu\rangle_{\rm abs}$, where $\langle h\nu\rangle_{\rm abs}$ is the mean energy per absorbed photon, then individual photon absorptions will cause pronounced upward jumps in the grain temperature. It will also be the case that substantial radiative cooling of the grain will take place between photon absorptions. As the result, the grain temperature T will be a strongly fluctuating quantity, with large excursions above and below $T_{\rm ss}$.

Figure 24.5 shows the temperature histories of five graphitic grains over the span of 10^5 s (~ 1 day). For the grain sizes shown here, $Q_{\rm abs}(\lambda) \propto a$, so that $C_{\rm abs} \propto a^3$, and the starlight photon absorption rate $\propto a^3 U \approx 1 \times 10^{-6} U (a/10\,\text{Å})^3\,{\rm s}^{-1}$. Because the time to cool below ~ 5 K is only $\sim 10^4$ s – independent of grain size for $a \lesssim 200$ Å – a small grain can cool to a very low temperature between photon absorptions, as seen for $a \lesssim 50$ Å for $U = 1$. When a photon absorption does take place, the small heat capacity of the grain results in a high peak temperature. It is clear that one cannot speak of a representative grain temperature under these

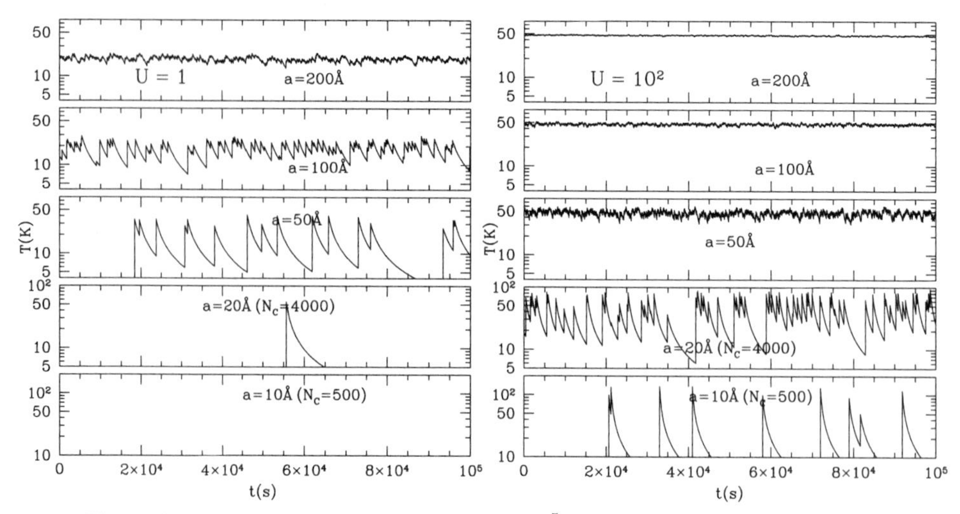

Figure 24.5 Temperature versus time during 10^5 s (~ 1 day) for five carbonaceous grains in two radiation fields: the local starlight intensity ($U = 1$; left panel) and 10^2 times the local starlight intensity ($U = 10^2$; right panel). The importance of quantized stochastic heating is evident for the smaller sizes.

conditions – one must instead use a temperature distribution function. As the grain size is increased, however, photon absorption events occur more frequently, the temperature rise at each event is reduced by the increased heat capacity, and the temperature varies over only a small range, as seen for the $a = 200$ Å grain in Figure 24.5.

To calculate the emission from small, stochastically heated grains, one requires the probability distribution function dP/dT, where $P(T)$ is the probability that the grain will have temperature less than or equal to T. The temperature distribution function dP/dT will depend on grain size a, composition, and the intensity (and spectrum) of the radiation illuminating the grains. While dP/dT can be obtained from a Monte Carlo simulation of $T(t)$ (Draine & Anderson 1985), it is far more efficient to solve directly for the discretized steady state distribution function (Guhathakurta & Draine 1989; Draine & Li 2001).

Figure 24.6 shows temperature distribution functions calculated for graphite/PAH dust grains of selected radii, exposed to the ISRF of Mathis et al. (1983). We see that dP/dT for a grain with $a = 10$ Å extends to $T = 400$ K – this is the temperature that this grain will be heated to when it absorbs a single photon with energy just below the Lyman limit cutoff at 13.6 eV. While the probability of finding the grain at $T > 100$ K is very small, nevertheless most of the energy radiated by the grain is radiated while it is at temperatures above 100 K – the typical absorbed photon raises the grain temperature to $T \gtrsim 200$ K, and the grain must radiate away most of this heat before the temperature drops to 100 K. Grains such as these radiate

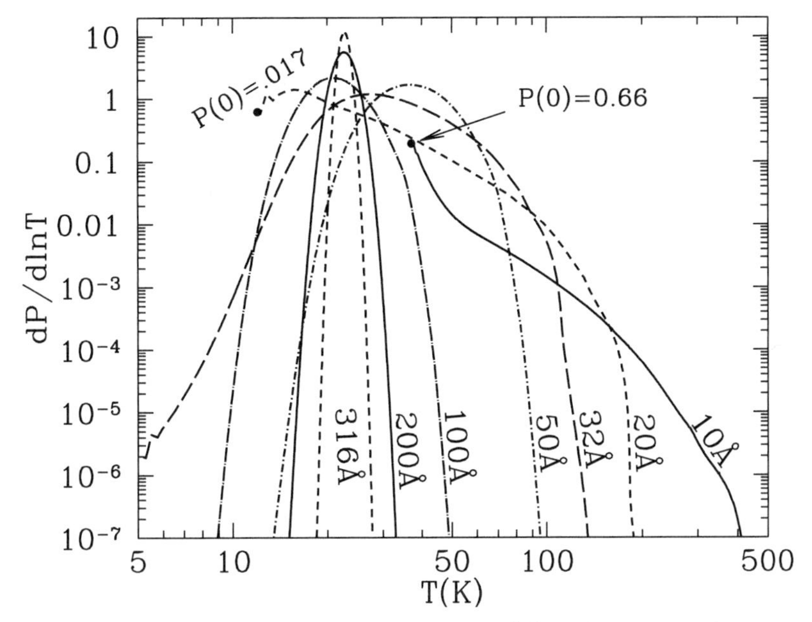

Figure 24.6 Temperature distribution function $dP/d\ln T$ for seven carbonaceous grains in ISRF with $U = 1$. Curves are labeled by grain radius a. For the $a = 10$ and 20 Å curves, the dot indicates the first excited state, and $P(0)$ is the fraction in the ground state.

strongly in the PAH features at 7.7, 8.6, and 11.3 μm.

It should also be noted that dP/dT shown for the $a = 10$ Å grain does not extend below ~ 35 K. This is because this "temperature" corresponds to the grain having a single vibrational quantum in the lowest vibrational mode.

For $U = 1$ and radii $a \gtrsim 50$ Å, dP/dT has a well-defined peak, and the distribution becomes narrower as a is increased. Note that for $a = 316$ Å, $dP/d\ln T$ peaks at 22 K, close to $T_{\rm ss} = 23$ K predicted by Eq. (24.20).

24.3 Infrared Emission from Grains

In a typical spiral galaxy, perhaps $\frac{1}{3}$ of the energy radiated by stars is absorbed by dust grains, and reemitted in the infrared. The spectrum of this emission is determined by the temperatures and composition of the dust grains.

Ultimately, infrared emission is a quantum process – a radiative transition between an upper and lower energy level of the grain. However, it has been shown (Draine & Li 2001) that a "thermal" approach provides an excellent approximation – the power per unit frequency radiated by a grain containing energy E_u is approximated as $P_\nu = 4\pi B_\nu(T) C_{\rm abs}(\nu)$, where T is the temperature for which

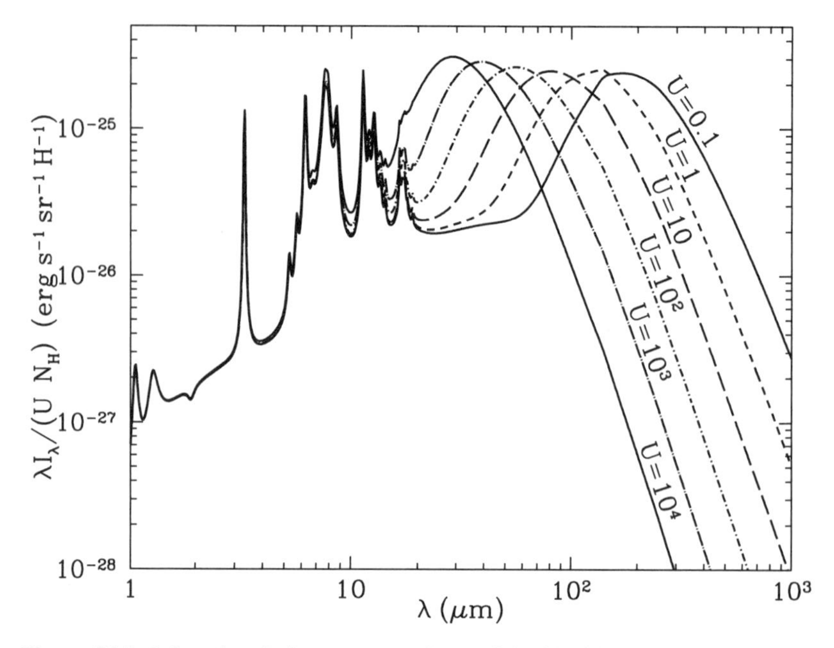

Figure 24.7 Infrared emission spectrum for model with silicate and graphite/PAH grains in ISRF intensity scale factor U from 0.1 to 10^4 ($U = 1$ is the local ISRF). Spectra are scaled to give power per H nucleon per unit U, calculated using the model of Draine & Li (2007).

$\langle E_{\rm int}\rangle_T = E_u$. Therefore, the emissivity of a population of grains can be written

$$j_\nu = \sum_i \int da \frac{dn_i}{da} \int dT \left(\frac{dP}{dT}\right)_{i,a} C_{\rm abs}(\nu; i, a) B_\nu(T) \quad , \tag{24.22}$$

where $(dn_i/da)da$ is the number density of grains of type i with radii in $[a, a+da]$.

Calculation of j_ν, therefore, requires a grain model to provide the size distributions dn_i/da for each composition i, the absorption cross sections $C_{\rm abs}(\nu; i, a)$, and the temperature distribution functions $(dP/dT)_{i,a}$. For large grains, the temperature distribution function is sufficiently narrow that it may be approximated by a delta function $dP/dT \rightarrow \delta(T - T_{\rm ss})$, where $T_{\rm ss}$ is the steady-state temperature for which the time-averaged cooling equals the time-averaged heating, but for $a \lesssim 0.01\,\mu$m, one should use a realistic temperature distribution dP/dT rather than a delta-function.

Model infrared emission spectra have been calculated by Draine & Li (2007) for a grain model that consists of carbonaceous grains and amorphous silicate grains, with size distributions that reproduce the average interstellar extinction curve. Spectra for this grain model are shown in Fig. 24.7. The spectra are normalized by U. We see that the power per logarithmic interval has a peak in the far-infrared that corresponds to emission from "large" grains with more-or-less steady temperatures $T_{\rm ss} \approx 20U^{1/6}\,{\rm K}$. As expected, this peak shifts toward shorter wavelengths as U is increased and the grains become warmer. The long wavelength peak in λI_λ occurs at $\lambda_{\rm peak} \approx 140U^{-1/6}\,\mu{\rm m}$, consistent with the expected variation of $T_{\rm dust} \propto U^{1/6}$ from Eqs. (24.19 and 24.20).

There are additional peaks at shorter wavelengths that are due to vibrational modes of PAH grains (see §23.5); these features account for $\sim 25\%$ of the total power, but (when normalized by the total power) these features hardly change as the radiation intensity is changed. The PAH emission occurs following single-photon heating of very small grains. Increasing U causes the "temperature spikes" to occur more frequently, but even for $U = 10^4$, grains with $a \lesssim 10\,\text{Å}$ are able to cool thoroughly during the time between photon absorptions, losing all of the absorbed photon energy before the next absorption takes place. Therefore, the PAH features each contain a nearly constant fraction of the total energy radiated.

Figure 24.7 shows emission spectra for dust heated by different starlight intensities U. A star-forming galaxy will contain regions with a wide range of starlight intensities, and a weighted sum over U is appropriate, although at this time we lack an a-priori understanding of what distribution function over U is appropriate. Using simple distribution functions, model spectra can be obtained that appear to be consistent with the observed spectra for star-forming galaxies (Draine et al. 2007).

24.4 Collisionally Heated Dust

In an ionized gas, the ratio of collisional heating to radiative heating by starlight is

$$\frac{(dE/dt)_{\rm coll}}{(dE/dt)_\star} \approx \frac{2n_ekT}{u_\star} \times \frac{\gamma_e}{\langle Q_{\rm abs}\rangle_\star} \times \frac{(8kT/\pi m_e)^{1/2}}{c} \quad , \tag{24.23}$$

where the factor γ_e allows for Coulomb focusing/repulsion of the electrons, as well as for the possibility that only a fraction of the kinetic energy of the impacting electron may be converted to heat. Normally, the thermal pressure $2n_ekT$ is comparable to the starlight energy density $u_\star$, and collisional heating is negligible because the electron thermal speed $(8kT_e/\pi m_e)^{1/2}$ is small compared to c. However, in a hot dense plasma, collisional heating can dominate – an example is the shock-heated gas produced by the interaction between the ejecta from SN 87a and the preexisting equatorial ring. Dwek et al. (2010) show that the IR emission between days 6000 and 8000 was dominated by $\sim 180\,{\rm K}$ silicate dust, heated by a plasma with $n_e \approx 3 \times 10^4\,{\rm cm}^{-3}$ and $T \approx 5 \times 10^6\,{\rm K}$. Collisionally heated dust is also seen in the supernova remnant Cas A (Arendt et al. 1999; Rho et al. 2008), where $n_e \approx 400\,{\rm cm}^{-3}$ and $T \approx 4 \times 10^6\,{\rm K}$ in the shocked knots.

Stochastic heating can be important for small grains, because a single electron impact may deposit $\sim 2kT_e \approx 700(T_e/4 \times 10^6\,{\rm K})\,{\rm eV}$ of heat.

Chapter Twenty-five

Grain Physics: Charging and Sputtering

25.1 Collisional Charging

Consider a spherical dust grain of radius a and charge $Z_{\rm gr}e$. If we assume that an approaching "projectile" with charge $Z_{\rm proj}e$ does not perturb the charge distribution on the grain, then we can approximate the interaction potential as a monopole interaction:

$$V(r) = \frac{Z_{\rm proj} Z_{\rm gr} e^2}{r} \quad . \tag{25.1}$$

Let the projectile have initial kinetic energy E, and impact parameter b, and suppose the distance of closest approach is $r_{\rm min}$. Angular momentum is conserved, so that the speed at closest approach is $(2E/m_{\rm proj})^{1/2} b/r_{\rm min}$; energy conservation then requires

$$E = \left(\frac{b}{r_{\rm min}}\right)^2 E + \frac{Z_{\rm gr} Z_{\rm proj} e^2}{r_{\rm min}} \quad . \tag{25.2}$$

Consider projectiles with energy E: they will collide with the grain surface if, and only if, $r_{\rm min} \leq a$. By setting $r_{\rm min} = a$, we can solve Eq. (25.2) for the maximum impact parameter $b_{\rm max}(E)$, and obtain the collision cross section

$$\sigma(E) = \pi b_{\rm max}^2(E) = \pi a^2 \left[1 - \frac{Z_{\rm gr} Z_{\rm proj} e^2}{aE}\right] \quad . \tag{25.3}$$

The rate at which projectiles collide with the grain in a thermal gas is obtained by integrating over the Maxwellian distribution of energies, [see Eq. (2.5)]:

$$\left(\frac{dN}{dt}\right)_{\rm proj} = n_{\rm proj} \int_{E_{\rm min}}^{\infty} \sigma(E)\, v f_E \, dE \tag{25.4}$$

$$= \pi a^2 n_{\rm proj} \left(\frac{8kT}{\pi m_{\rm proj}}\right)^{1/2} F(Z_{\rm proj}\phi) \quad , \tag{25.5}$$

$$\phi \equiv \frac{Z_{\rm gr} e^2}{akT} \quad , \quad F(x) \equiv \begin{cases} (1-x) & \text{if } x < 0 \quad , \\ {\rm e}^{-x} & \text{if } x > 0 \quad . \end{cases} \tag{25.6}$$

The function F defined in Eq. (25.6) is the amount by which Coulomb focusing changes the collision rate relative to the rate for an uncharged grain. Consider now a plasma. Let n_e be the electron density, and n_i be the density of ions with charge $Z_i e$. An initially uncharged grain will collide more frequently with electrons than with ions, and will acquire negative charge until the rate of collisions with ions balances the rate of collisions with electrons. The condition for this is

$$Z_i n_i s_i m_i^{-1/2}(1 - Z_i\phi) = n_e s_e m_e^{-1/2} e^{\phi} \quad , \tag{25.7}$$

where s_i and s_e are the "sticking" coefficients for ions and electrons. For a hydrogen plasma, if we assume $s_e = s_i$, this becomes

$$(1 - \phi)e^{-\phi} = \sqrt{1836.1} \quad . \tag{25.8}$$

This transcendental equation has solution $\phi = -2.504$: the grain will charge to a potential $U = -2.504kT/e = -2.16T_4$V. The grain charge is

$$Z_{\rm gr} = \frac{Ua}{e} = -2.504\frac{akT}{e^2} = -150\left(\frac{a}{0.1\,\mu{\rm m}}\right)T_4 \quad . \tag{25.9}$$

Charge quantization will be important when $aT \lesssim 10^2\ \mu{\rm m\,K}$.

When the grain is neutral, an approaching electron or ion will polarize the grain, attracting the projectile to the grain. This can significantly enhance the charging rate for neutral grains (Draine & Sutin 1987) and is especially important for very small grains, which may have an appreciable probability of being neutral.

25.2 Photoelectric Emission

When an energetic photon is absorbed in the grain, it may excite an electron to a sufficiently high energy so that the electron may escape from the grain. Such an ejected electron is referred to as a "photoelectron." The "photoelectric yield" $Y_{\rm pe}(h\nu, a, U)$ is just the probability that absorption of a photon of energy $h\nu$ will result in a photoelectron; Y will depend on the photon energy, the composition of the grain, the grain size a, and the potential U to which the grain is already charged. The rate at which photoelectrons are ejected is

$$\left(\frac{dN}{dt}\right)_{\rm pe} = \int d\nu \frac{u_\nu c}{h\nu}\pi a^2 Q_{\rm abs} Y_{\rm pe} \quad . \tag{25.10}$$

Photoelectric emission from solids can be measured in the laboratory, but is difficult to calculate theoretically because it involves calculating the probability that an excited electron will travel from the point of photon absorption and across the grain surface without losing too much of its initial kinetic energy along the way. In the case of photoelectric emission from small grains, there are three additional complications:

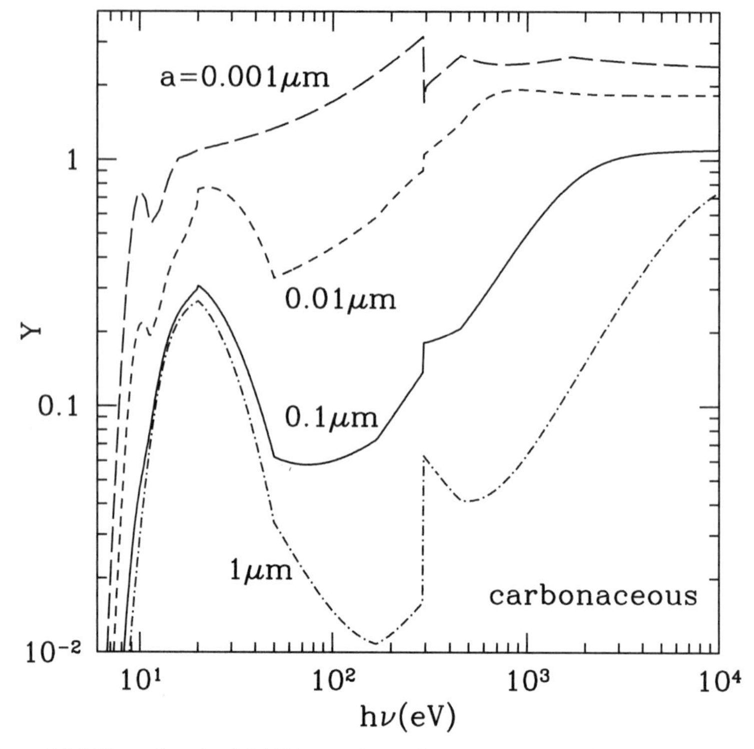

Figure 25.1 Photoelectric yield Y for uncharged carbonaceous grains for selected radii, as a function of photon energy. For sufficiently small grains, Y can exceed unity because of secondary electron emission for $h\nu \gtrsim 14\,\mathrm{eV}$, and Auger electron emission for $h\nu > 291\,\mathrm{eV}$. From Weingartner et al. (2006), reproduced by permission of the AAS.

1. Even if the grain is initially neutral, an escaping photoelectron leaves behind a positively charged grain; after traversing the grain surface, an electron must overcome a Coulomb potential in order to escape to infinity. This causes the "ionization potential" for very small grains to be larger than the "work function" for the bulk solid.

2. An electron outside a solid polarizes the nearby solid material, inducing a positive charge density that produces an attractive potential. In the case of a small grain, this "image charge" effect is modified relative to that for bulk material. This alters the potential that an electron has to overcome in order to escape to infinity.

3. When the grain is small compared to the photon absorption length $L_{\rm abs}$ [see Eq. (22.11)], photon absorption events will, in general, be closer to the surface than in a bulk sample. Therefore, the probability of being able to travel to the surface without energy loss is increased, resulting in an increase in the

photoelectric yield $Y_{\rm pe}$.

The bottom line is that we have only approximate estimates for $Y_{\rm pe}(h\nu, a, U)$ for small particles of likely interstellar grain materials. Weingartner et al. (2006) have recently estimated photoelectric yields for silicate grains and carbonaceous grains.

25.3 Grain Charging in the Diffuse ISM

In the diffuse ISM, the density of UV photons is comparable to the electron density, but photons move at the speed of light, much faster than the electrons. If $Q_{\rm abs}$ is of order unity, the photon absorption rate will be large compared to the rate at which electrons collide with the grain. Therefore, if $Y_{\rm pe}$ is not small, photoelectric charging can drive the grain to a positive potential. As the grain becomes positively charged, $Y_{\rm pe}$ declines, because some electrons that would have been able to escape from a neutral grain do not have enough energy to overcome the potential from a positively charged grain. Furthermore, Coulomb focusing increases the rate at which electrons collide with the grain. The resulting grain potential U will satisfy the equation

$$n_i s_i \left(\frac{8kT}{\pi m_i}\right)^{1/2} F(eU/kT) + \int d\nu \left(\frac{u_\nu c}{h\nu}\right) Q_{\rm abs} Y_{\rm pe}(h\nu, a, U) = n_e s_e \left(\frac{8kT}{\pi m_e}\right)^{1/2} F(-eU/kT) \,. \quad (25.11)$$

where F is defined in Eq. (25.6). This equation can be solved for U, and the grain charge $Z_{\rm gr} = Ua$ can be calculated. In Eq. (25.11), the grain charge has been treated as continuous. If the resulting $|Z_{\rm gr}| \gg 1$, this is a good approximation, but if $|Z_{\rm gr}| \lesssim 10$, then charge quantization will be important, and it is necessary to solve for the charge distribution function $f(Z_{\rm gr}; a)$. This can then be used to calculate charge-dependent processes, such as the Lorentz force on a moving grain.

Figure 25.2 shows the charge distribution function $f(Z_{\rm gr})$ for selected grain sizes for conditions corresponding to the CNM. Charge quantization is important for $a \lesssim 200$ Å.

The time-averaged electrostatic potential $\langle U \rangle$ as a function of grain size is shown in Fig. 25.3 for grains exposed to the average starlight background in CNM, WNM, and WIM conditions. For our current best estimates of grain photoelectric yields, it appears that the ultraviolet radiation in the average starlight background is able to drive grains with radii $a \gtrsim 0.01\,\mu$m to positive potentials.

25.4★ Secondary Electron Emission

Because infrared emission is so effective at cooling grains, they can remain cool even in very hot (e.g., $T \gtrsim 10^6$ K) plasma, for the plasma densities that are found

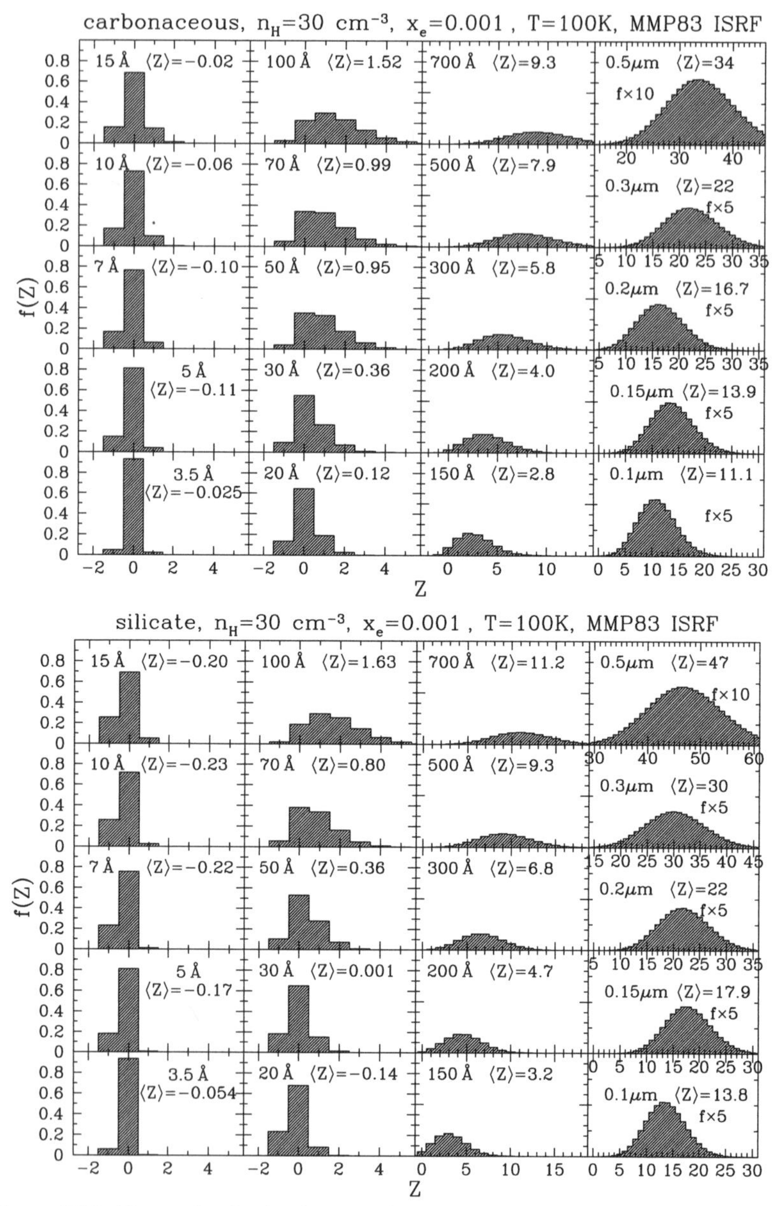

Figure 25.2 Charge distribution function for carbonaceous grains and silicate grains of selected radii in the diffuse ISM with the MMP83 interstellar radiation field (Mathis et al. 1983). The grain radius and mean charge $\langle Z \rangle$ is given in each panel. After Draine (2004).

in the interstellar or intergalactic medium. However, just as energetic electrons can collisionally ionize an atom, they can also eject bound electrons from a solid grain. This process is called **secondary electron emission**. Secondary emission yields have been estimated by Draine & Salpeter (1979). For selected energies and

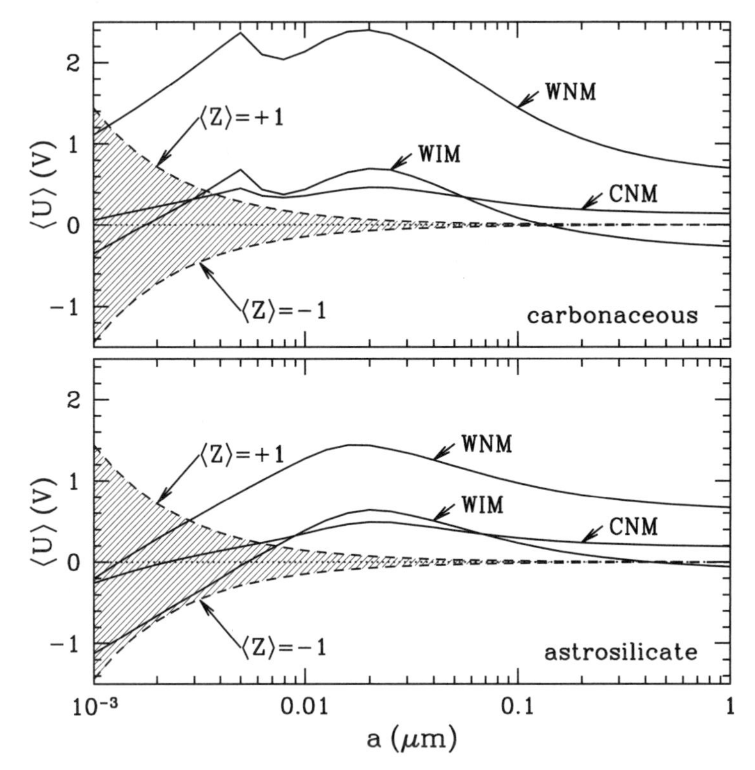

Figure 25.3 Time-averaged potential U as a function of grain size for silicate and carbonaceous grains for 3 different environments: CNM, WNM, and WIM. Also shown are potentials for $Z = \pm 1$; away from the (shaded) region bounded by these two curves, charge quantization is of secondary importance.

grain sizes, the secondary electron yield can exceed the sticking coefficient for the incident electron, so that the net effect of electron collisions is to cause the grain to become positive, even in the absence of photoelectric emission!

25.5★ Electron Field Emission

When bombarded by energetic electrons, the grain potential may be driven to large negative values, resulting in a large electric field at the grain surface. If the electric field exceeds $\sim 10^7\,\mathrm{V\,cm^{-1}}$, electrons begin to be ejected from the grain via a quantum-mechanical tunneling process – this is known as **electron field emission**.

25.6★ Ion Field Emission and Coulomb Explosions

If collisions with hot plasma or X-ray irradiation drive the grain to sufficiently large positive potential, the electric field at the grain surface can become large enough so that one of two things will happen. If the grain is structurally weak, then the Coulomb repulsion between the positive charge on different parts of the grain may result in a **Coulomb explosion**, where the grain fragments into two or more parts.

If, however, the mechanical strength of the grain is comparable to the strengths of ideal materials, the situation is different: before the electric field is strong enough to fracture the grain, individual ions will begin to be emitted from the surface, carrying away positive charge. This process is known as **ion field emission**. The electric field has to approach $3 \times 10^8\,\mathrm{V\,cm^{-1}}$ for ion field emission to occur.

Grains near gamma-ray bursts are exposed to very high X-ray fluxes, which can lead to either ion field emission (Waxman & Draine 2000) or Coulomb explosions (Fruchter et al. 2001). Current uncertainties concerning the structure of interstellar grains leave both possibilities open.

25.7 Sputtering in Hot Gas

If the gas temperature is sufficiently high, then impinging atoms or ions can erode the grain, one atom at a time, through a process known as **sputtering**. The **sputtering yield** $Y_{\rm sput}(E)$, depends on the impact energy E of the projectile ion, on the composition of the target, and also on the mass and charge of the impacting ion. It is not yet possible to calculate $Y_{\rm sput}(E)$ reliably from first principles, but experiments have been carried out for various materials, and formulae have been found that approximately reproduce the measured yields and that allow us to estimate yields for projectile-target combinations that have not yet been studied – see Draine (1995) for a review.

In high-temperature gas, sputtering rates can be rapid. For a stationary grain, the thermal sputtering rate is just

$$\frac{da}{dt} = -\frac{m_x}{4\rho}\sum_i n_i \int_{E_{\rm min}}^{\infty} dE f_E \left(\frac{2E}{m_i}\right)^{1/2} \left(1 - \frac{Z_i eU}{E}\right) Y_{\rm sput}(E - Z_i eU), \tag{25.12}$$

where f_E is the energy distribution function [Eq. (2.5)], the sum is over projectile species i (e.g., $\mathrm{H^+}, \mathrm{He^{++}}$), the $(1 - Z_i eU/E)$ term is the Coulomb focusing factor, the sputtering yield is evaluated using the kinetic energy at impact, and $E_{\rm min} = \max(0, Z_i eU)$. If the grain is moving relative to the gas, as in shocked gas, the sputtering rate is calculated using the velocity distribution as seen by the grain (Draine & Salpeter 1979).

Figure 25.4 shows thermal sputtering rates estimated for graphite and silicate grains. For $10^5 \lesssim T \lesssim 10^9$ K the calculated sputtering rates for graphite, silicate,

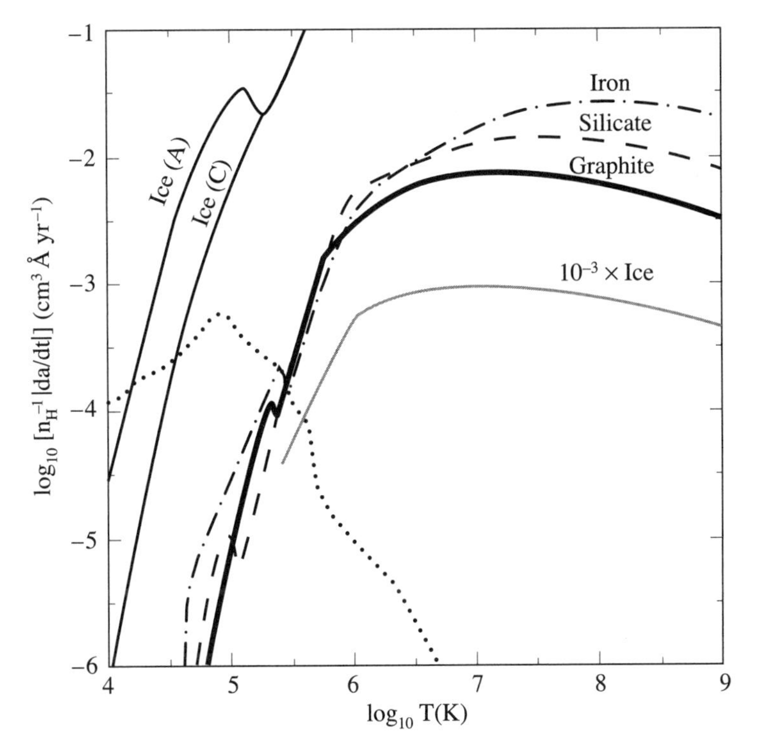

Figure 25.4 Thermal sputtering rates (Draine & Salpeter 1979). For $10^6 \lesssim T \lesssim 10^9$ K all three refractory materials (iron, silicate, graphite) have $(1/n_{\rm H})|da/dt| \approx 10^{-2}\,{\rm cm}^3\,{\rm Å}\,{\rm yr}^{-1}$

or iron grains can be approximated by

$$\frac{da}{dt} \approx -\frac{1 \times 10^{-6}}{1 + T_6^{-3}} \left(\frac{n_{\rm H}}{{\rm cm}^{-3}}\right) \mu{\rm m}\,{\rm yr}^{-1} \quad , \tag{25.13}$$

corresponding to a grain lifetime

$$\tau = \frac{a}{|da/dt|} \approx 1 \times 10^5 \left[1 + T_6^{-3}\right] \frac{(a/0.1\,\mu{\rm m})}{(n_{\rm H}/{\rm cm}^{-3})}\,{\rm yr} \quad . \tag{25.14}$$

Thus, in a supernova remnant with $n_{\rm H} \approx 1\,{\rm cm}^{-3}$ and $T_6 \gtrsim 1$, a grain with initial radius $a = 0.1\,\mu{\rm m}$ could survive for $\sim 10^5$ yr. In the X-ray emitting gas of the Coma cluster, with $T \approx 9.6 \times 10^7$ K ($kT \approx 8.25$ keV) and $n_{\rm H} = 2.9 \times 10^{-3}\,{\rm cm}^{-3}$ (Hughes et al. 1988; Herbig et al. 1995), a $0.1\mu{\rm m}$ dust grain would have a lifetime $\tau \approx 3 \times 10^7$ yr, short compared to the cluster age.

Chapter Twenty-six

Grain Dynamics

The motion and rotation of interstellar grains is determined by the forces and torques that act upon them. Grain motion is important because it can result in separation of dust from gas, because rapid motions can lead to grain destruction, and because the moving grains may help couple the neutral gas to the magnetic field in regions where the fractional ionization is low.

26.1 Translational Motion

26.1.1 Gas Drag

A grain moving relative to gas or plasma is subject to drag forces. If the grain is spherical, and if the colliding atoms or ions are assumed to either stick or undergo specular reflection, the drag force is given by (Spitzer 1962; Baines et al. 1965; Draine & Salpeter 1979)

$$F_{\rm drag} = 2\pi a^2 kT \left\{ \sum_i n_i \left[G_0(s_i) + z_i^2 \phi^2 \ln(\Lambda/z_i) G_2(s_i) \right] \right\} , \tag{26.1}$$

$$G_0(s) \equiv \left(s^2 + 1 - \frac{1}{4s^2} \right) {\rm erf}(s) + \frac{1}{\sqrt{\pi}} \left(s + \frac{1}{2s} \right) {\rm e}^{-s^2} , \tag{26.2}$$

$$\approx \frac{8s}{3\sqrt{\pi}} \left(1 + \frac{9\pi}{64} s^2 \right)^{1/2} , \tag{26.3}$$

$$G_2(s) \equiv \frac{{\rm erf}(s)}{s^2} - \frac{2}{s\sqrt{\pi}} {\rm e}^{-s^2} , \tag{26.4}$$

$$\approx \frac{s}{(3\sqrt{\pi}/4 + s^3)} , \tag{26.5}$$

$$\phi \equiv eU/kT , \tag{26.6}$$

$$s_i \equiv \left(m_i v^2 / 2kT \right)^{1/2} , \tag{26.7}$$

$$\Lambda \equiv \frac{3}{2ae|\phi|} \left(\frac{kT}{\pi n_e} \right)^{1/2} . \tag{26.8}$$

The approximation (26.3) is accurate to within 1%, and (26.5) to within 10%; both are exact in the limits $s \to 0$ and $s \to \infty$. The sums over i run over the atomic and ionic species in the gas, with masses m_i and charges $z_i e$. $U = z_{\rm grain} e/a$ is the electrostatic potential at the grain surface. The electron contribution to $F_{\rm drag}$ is generally negligible. Note that the "Coulomb drag" term $\propto \phi^2 \ln \Lambda\ G_2$ can be large if $s \lesssim 1$ and $\phi^2 \ln \Lambda \gg 1$.

Let τ_M be the time for a stationary grain to collide with its own mass $M_{\rm gr}$ of gas. In gas with $n({\rm H})/n_{\rm H} = 1$, $n({\rm He})/n_{\rm H} = 0.1$:

$$\tau_M = \frac{(4\pi/3)\rho a^3}{\pi a^2 n_{\rm H} (8 m_{\rm H} kT/\pi)^{1/2} \times 1.2} \tag{26.9}$$

$$= 1.45 \times 10^5 \left(\frac{\rho}{3\,{\rm g\,cm^{-3}}}\right) a_{-5} \left(\frac{30\,{\rm cm^{-3}}}{n_{\rm H}}\right) T_2^{-1/2}\,{\rm yr}\ , \tag{26.10}$$

where

$$a_{-5} \equiv \frac{a}{10^{-5}\,{\rm cm}} \equiv \frac{a}{0.1\,\mu{\rm m}}\ . \tag{26.11}$$

In the absence of other forces, the slowing-down time, or **drag time**, is

$$\tau_{\rm drag} \equiv \frac{M_{\rm gr} v}{F_{\rm drag}}\ . \tag{26.12}$$

In neutral gas, for subsonic motion ($s_{\rm H} \ll 1$):

$$\tau_{\rm drag} = \frac{3}{4} \tau_M \tag{26.13}$$

$$= 1.1 \times 10^5 \left(\frac{\rho}{3\,{\rm g\,cm^{-3}}}\right) a_{-5} \left(\frac{30\,{\rm cm^{-3}}}{n_{\rm H}}\right) T_2^{-1/2}\,{\rm yr}\ . \tag{26.14}$$

In diffuse clouds, gas drag is able to decelerate grains on relatively short time scales of only $\sim 10^5$ yr.

26.1.2 Lorentz Force

Grains with a velocity component transverse to the local magnetic field **B** will experience a Lorentz force if they have a nonzero charge, as is often the case. We write

$$F_B = M_{\rm gr} \vec{v} \times \vec{\omega}_B \qquad \vec{\omega}_B \equiv \frac{Q}{M_{\rm gr} c} \vec{B}\ . \tag{26.15}$$

A grain charged to a potential $U = Q/a$ will orbit around the magnetic field with a period

$$\tau_B = \frac{2\pi}{\omega_B} = 2\pi \frac{M_{\rm gr} c}{|Q| B} = 450 \left(\frac{\rho}{3\,{\rm g\,cm^{-3}}}\right) a_{-5}^2 \left(\frac{\rm Volt}{|U|}\right) \left(\frac{5\,\mu{\rm G}}{B}\right) {\rm yr.} \tag{26.16}$$

26.1.3 Radiation Pressure and Recoil Forces

The interstellar radiation field is, in general, quite anisotropic. The radiation pressure force on a grain is

$$F_{\rm rad} = \langle Q_{\rm pr} \rangle \pi a^2 \Delta u_{\rm rad} \ , \tag{26.17}$$

where $\Delta u_{\rm rad} = F_{\rm net}/c$, $F_{\rm net}$ is the net (vector) radiative flux, and $\langle Q_{\rm pr} \rangle$ is $Q_{\rm pr}(\lambda)$ averaged over the spectrum of $F_{\rm net}$. Radiation pressure can drive grains through the gas with appreciable velocities.

In addition to the direct force due to absorption or redirection of photon momenta, anisotropic radiation fields can result in large forces on dust grains if there is preferential ejection of either atoms or electrons from the "illuminated" side of the grain. The recoil from ejected electrons or atoms can be large, resulting in a large time-averaged force (Weingartner & Draine 2001*c*).

26.1.4 Poynting-Robertson Effect

Consider a particle in a circular orbit with radius R and velocity $v_{\rm orb}$ around a star with luminosity $L_\star$. Because of aberration of starlight, in the instantaneous rest frame of the orbiting particle, the radiative flux from the star has a component $\beta L/4\pi R^2$ in the direction antiparallel to the motion of the grain, where $\beta \equiv v_{\rm orb}/c$. This radiation therefore acts to reduce the orbital angular momentum J of the particle. This is called the **Poynting-Robertson effect**; the torque

$$\left(\frac{dJ}{dt}\right)_{\rm PR} = -\beta \frac{L_\star}{4\pi R^2 c} \langle Q_{\rm pr} \rangle_\star \pi a^2 R \ , \tag{26.18}$$

leads to orbital decay on a time scale that depends on the grain size:

$$\tau_{\rm PR} = \frac{J}{-(dJ/dt)_{\rm PR}} = \frac{16\pi}{3} \frac{\rho a c^2 R^2}{\langle Q_{\rm pr} \rangle_\star L_\star} \tag{26.19}$$

$$= 8.3 \times 10^7 \,{\rm yr} \left(\frac{\rho}{3\,{\rm g\,cm^{-3}}}\right) \left(\frac{a}{\rm cm}\right) \left(\frac{R}{\rm AU}\right)^2 \frac{1}{\langle Q_{\rm pr} \rangle_\star} \frac{L_\odot}{L_\star} \ , \tag{26.20}$$

where $\langle Q_{\rm pr} \rangle_\star$ is the radiation pressure efficiency factor averaged over the stellar spectrum. The mean photon energy for a blackbody of temperature $T_\star$ is $\sim 3kT_\star$, and we expect $Q_{\rm pr} \approx 1$ for $x \equiv 2\pi a/\lambda \gtrsim 1$. Thus $\langle Q_{\rm pr} \rangle_\star \approx 1$ for $a \gtrsim (1/2\pi) hc/3kT_\star \approx 0.15\,\mu{\rm m} \times (5000\,{\rm K}/T_\star)$. Poynting-Robertson drag can lead to orbital decay on $\lesssim$ Gyr time scales for particles up to ~ 10 cm in size. Micron-sized particles have very short orbital lifetimes near stars.

26.1.5 Radiation-Pressure-Driven Drift

If an anisotropic radiation field is present, it will cause grain drift. If a magnetic field is present, we must distinguish between motions parallel and perpendicular to

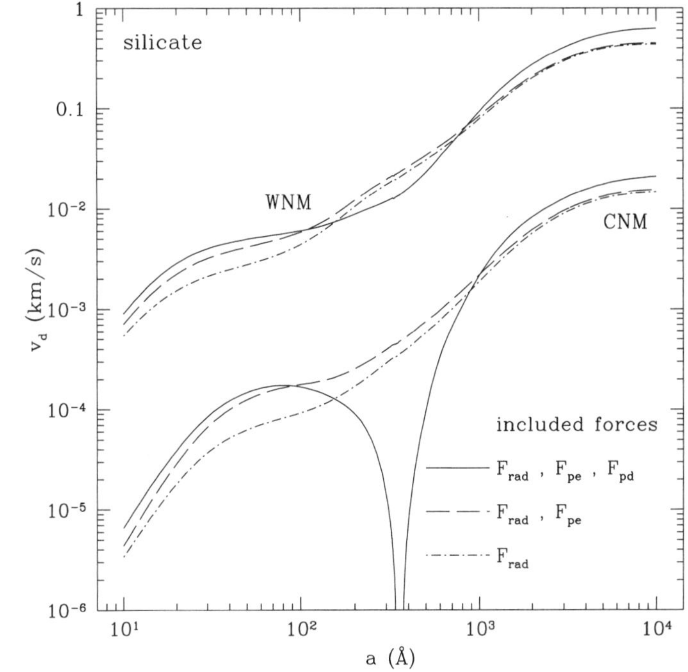

Figure 26.1 Drift velocity for silicate grains, as a function of grain size, in two environments, cold neutral medium (CNM) and warm neutral medium (WNM), showing drift velocity calculated for radiation pressure only (dotted-dashed curve), and including force from photoelectric emission and photodesorption (solid curves). The minimum in v_d at $a \approx 350$ Å occurs because for this size photodesorption occurs primarily on the "far" side of the grain, so that the photodesorption recoil force acts opposite to the radiation pressure force. From Weingartner & Draine (2001*c*), reproduced by permission of the AAS.

B: drift perpendicular to **B** is strongly suppressed if $\omega_B \tau_{\rm drag} \gg 1$. Figure 26.1 shows estimated drift speeds for silicate grains, as a function of grain size, for the case where the magnetic field is parallel to the direction of starlight anisotropy. The starlight intensity and anisotropy are appropriate for the solar neighborhood.

26.2 Rotational Motion

A spherical dust grain with rotational kinetic energy $E_{\rm rot} = \frac{3}{2}kT_{\rm rot}$ will have rotation frequency

$$\frac{\omega}{2\pi} = \frac{1}{2\pi}\left(\frac{45}{8\pi}\frac{kT_{\rm rot}}{\rho a^5}\right)^{1/2} \tag{26.21}$$

$$= 4.6 \times 10^4\,{\rm Hz}\left(\frac{T_{\rm rot}}{100\,{\rm K}}\right)^{1/2}\left(\frac{0.1\,\mu{\rm m}}{a}\right)^{5/2} \quad . \tag{26.22}$$

For grain sizes $\sim 0.1\,\mu\mathrm{m}$, this rotation rate is not extreme. A point on the equator has a speed $\omega a \approx 2.9\,\mathrm{cm\,s^{-1}} a_{-5}^{-3/2}$.

If, however, we consider very small grains, the rotation rates can be extreme:

$$\frac{\omega}{2\pi} = 4.6\,\mathrm{GHz} \left(\frac{T_{\mathrm{rot}}}{100\,\mathrm{K}}\right)^{1/2} \left(\frac{0.001\,\mu\mathrm{m}}{a}\right)^{5/2} . \tag{26.23}$$

The observed infrared emission from interstellar grains requires a large population of PAH particles with very small heat capacities, so that single-photon heating can raise them to high vibrational temperatures. (See the size distribution shown in Fig. 23.10.) If these particles have rotational temperatures $T_{\mathrm{rot}} \approx 10^2\,\mathrm{K}$, and if they have electric dipole moments, they will generate detectable levels of electric dipole radiation.

26.2.1 Suprathermal Rotation of Large Grains

The rotational dynamics of dust grains is rich in physics. For many years it was assumed that the rotational excitation of "classical" grains would be dominated by random collisions with gas atoms, and that this would result in "Brownian rotation" with $T_{\mathrm{rot}} \approx T_{\mathrm{gas}}$. However, Purcell (1979) pointed out that because the ISM is not in LTE, interstellar grains will act as "heat engines," and can achieve "suprathermal" rotation rates with $T_{\mathrm{rot}} \gg T_{\mathrm{gas}}$. There are at least four processes that can drive rapid rotation:

1. Formation of H_2 on the grain surface, followed immediately by impulsive ejection of the newly formed H_2 (Purcell 1979).

2. Emission of photoelectrons from grains exposed to UV radiation (Purcell 1979).

3. Irregular grain surfaces, together with $T_{\mathrm{gas}} \neq T_{\mathrm{grain}}$ (Purcell 1979).

4. Radiative torques due to absorption and scattering of (possibly anisotropic) starlight by irregular grains (Draine & Weingartner 1996).

Each of these processes results in a **systematic** torque on the grain. The first three torques are fixed in "body-coordinates," and the fourth would also be fixed in body coordinates if the starlight were isotropic. Such torques, because they are systematic, can drive suprathermal rotation. However, the effects of the first three of these torques can be suppressed by the phenomenon of "thermal flipping" (see below).

When achieved, suprathermal rotation is not rapid enough to disrupt the grain (unless the grain is extremely fragile), but can have very important consequences for our understanding of the process of grain alignment with the interstellar magnetic field.

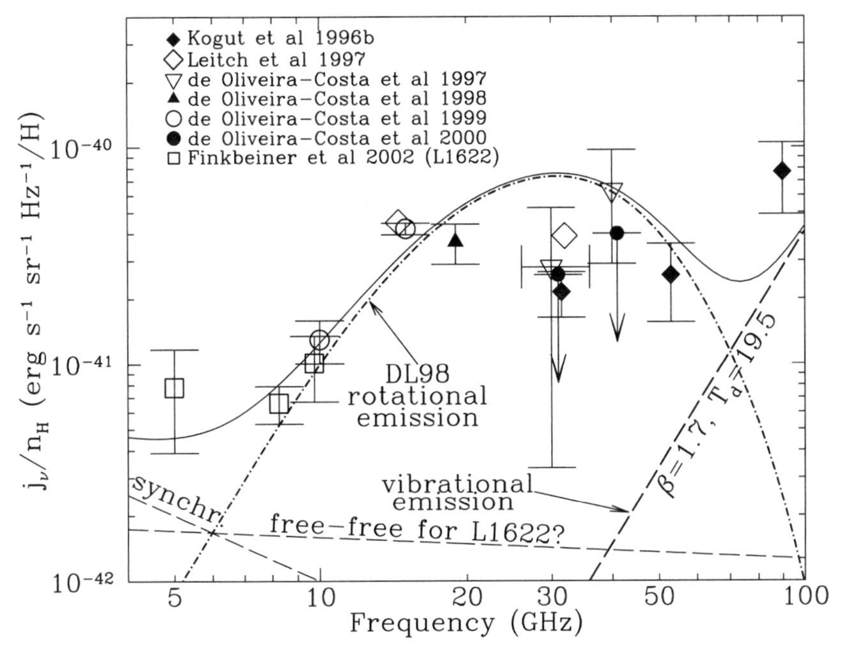

Figure 26.2 Dust-correlated microwave emission, expressed as microwave intensity per unit column density of hydrogen. The curve labeled DL98 is a theoretical estimate of rotational emission (Draine & Lazarian 1998*a*,*b*). Also shown are the low-frequency tail of the thermal emission from larger grains, and possible contributions of synchrotron and free–free emission toward L1622. The solid curve is the sum of all components. Observational data from Kogut et al. (1996), Leitch et al. (1997), de Oliveira-Costa et al. (1997, 1998, 1999, 2000), and Finkbeiner et al. (2002).

26.2.2 Rotation of Small Grains: Microwave Emission

The systematic torques identified above are *not* sufficient to drive grains with $a \lesssim 0.05\,\mu\text{m}$ to suprathermal rotation rates, because of thermal fluctuations within the grain that cause the grain to change its orientation at constant angular momentum – a phenomenon referred to as "thermal flipping" – resulting in what has been termed "thermal trapping" (Lazarian & Draine 1999*b*).

The rotational dynamics of the smallest grains $a \lesssim 0.001\,\mu\text{m}$, is complex. Processes that have important effects on the angular momentum of a very small grain include

1. Random collisions with neutral gas atoms.
2. Collisions of neutral or negatively charged grains with positive ions, where the colliding ion can hit the grain with a substantial angular momentum.
3. Fluctuating electric fields due to passing ions acting on the permanent electric dipole moment of the grain.

4. Angular momentum deposition by absorbed starlight photons.
5. Angular momentum loss to radiated infrared photons.
6. Angular momentum loss in electric dipole radiation from the spinning grain.

The balance among these processes depends on the grain size and upon the local environmental conditions (Draine & Lazarian 1998*b*). The net result is that the smallest grains generally tend to have rotational distribution functions that are actually somewhat *subthermal*, with $T_{\rm rot} < T_{\rm gas}$. Nevertheless, the rotation rates are large enough so that the electric dipole radiation from the smallest grains in the interstellar grain mixture appears to be able to account for what has been referred to as the "anomalous" microwave emission associated with interstellar dust. Figure 26.2 shows the emission spectrum predicted by Draine & Lazarian (1998*a*,*b*), together with some of the early observational determinations. A number of subsequent papers have confirmed the existence of dust-correlated microwave emission from dust in our Galaxy with intensity and spectrum consistent with the spinning dust model (see, e.g., Finkbeiner 2004; Finkbeiner et al. 2004; Watson et al. 2005; Fernández-Cerezo et al. 2006; Davies et al. 2006; Bonaldi et al. 2007; Miville-Deschênes et al. 2008; Casassus et al. 2008; Dickinson et al. 2009; Dobler et al. 2009; Scaife et al. 2009). Microwave emission from spinning dust in the nearby galaxy NGC 6946 has also been reported (Murphy et al. 2010). Recent studies (Ali-Haïmoud et al. 2009; Hoang et al. 2010) have improved the theoretical treatment of the rotational excitation process.

26.3★ Alignment of Interstellar Dust

Interstellar grains are observed to be systematically aligned with their *short* axis (i.e., the axis of largest moment of inertia) tending to be parallel to the local magnetic field. How this comes about has not yet been worked out in detail, but we *think* that we have identified the important elements of the dynamical story.

Interstellar grain alignment actually involves two separate alignment problems: (1) alignment of the grain angular momentum $\mathbf{J}$ with the magnetic field $\mathbf{B}_0$, and (2) alignment of the nonspherical body of the grain with the grain angular momentum.

A theory of grain alignment must account for the observed alignment of grains as a function of grain size: grains with sizes $a \gtrsim 0.1\,\mu$m are observed to be substantially aligned, $a \lesssim 0.05\,\mu$m grains are not appreciably aligned.

26.3.1 Precession of $\vec{J}$ around $\vec{B}$

The shortest time scale in the rotational dynamics of the grain is just the rotation period itself – a time that can range from milliseconds to less than a nanosecond. In astrophysical problems, it is always reasonable to average over the grain rotation.

A spinning grain will in general have a magnetic moment parallel (or antiparallel) to the instantaneous angular velocity. This can arise from two independent effects. If the grain has a net charge $Q \neq 0$ (with the net charge tending to reside on the grain surface), grain rotation implies an electric current that will generate a magnetic dipole field (Martin 1971). For a spherical grain, with uniform surface charge density $Q/4\pi a^2 = U/4\pi a$, rotation will result in a magnetic moment

$$\mu = \frac{Qa^2\omega}{3c} \quad . \tag{26.24}$$

In the presence of a static magnetic field, this will cause the angular momentum $\vec{J}$ to precess around B with a precession frequency

$$\Omega_L = \frac{5UB}{8\pi\rho a^2 c} = 3.7\times 10^{-10}\left(\frac{3\,\mathrm{g\,cm^{-3}}}{\rho}\right)\left(\frac{U}{\mathrm{Volt}}\right)\left(\frac{B}{5\,\mu\mathrm{G}}\right)\left(\frac{0.1\,\mu\mathrm{m}}{a}\right)^2 \mathrm{s}^{-1}. \tag{26.25}$$

The precession period, $2\pi/\Omega_L \approx 500\,\mathrm{yr}$, is relatively short.

In fact, however, a spinning grain is expected to have a much larger magnetic moment arising from the **Barnett effect** – a spinning body will spontaneously magnetize with a magnetization that is proportional to the rotation rate. This occurs because the unpaired electrons – which in a nonrotating body would be evenly divided between spin-up and spin-down – preferentially occupy the spin state with spin angular momentum parallel to the angular velocity, resulting in a net magnetization that is antiparallel to the angular velocity. The role of the Barnett effect in grain dynamics was pointed out by Dolginov & Mytrophanov (1976).

The rotationally induced magnetic moment μ is proportional to ω and to the zero-frequency magnetic susceptibility $\chi(0)$ of the grain material:

$$\mu = -\chi(0) V \frac{\hbar\omega}{g\mu_{\mathrm{B}}} \quad , \tag{26.26}$$

where μ_{B} is the Bohr magneton, and $g \approx 2$ is the gyromagnetic ratio. Normal paramagnetism in interstellar grains is expected to result in

$$\chi(0) \approx 10^{-4}\left(\frac{20\,\mathrm{K}}{T_{\mathrm{gr}}}\right) . \tag{26.27}$$

The precession period of a spherical grain in a static magnetic field will then be

$$\tau_{\mathrm{Barnett}} = \frac{2\pi}{\Omega_{\mathrm{B}}} = \frac{2\pi I\omega}{\mu B_0} = \frac{4\pi g\mu_{\mathrm{B}}\rho a^2}{5\hbar\chi(0)B_0} \tag{26.28}$$

$$\approx 0.8 a_{-5}^2 \left(\frac{\rho}{3\,\mathrm{g\,cm^{-3}}}\right)\left[\frac{10^{-4}}{\chi(0)}\right]\left(\frac{5\,\mu\mathrm{G}}{B_0}\right)\ \mathrm{yr} \quad . \tag{26.29}$$

This is a *very short* precession period – whatever population of partially aligned grains is present, the properties should be averaged over the precession cone for each grain.

The rapid precession resulting from the Barnett effect has another important dynamical consequence: if the direction of the magnetic field $\vec{B}_0$ changes slowly (on time scales longer than the precession period $\tau_{\rm Barnett}$), the projection of the angular momentum along the magnetic field will be an adiabatic invariant, and the precessing grains will maintain a constant "cone angle" θ_{BJ}, where θ_{BJ} is the instantaneous angle between the grain angular momentum $\vec{J}$ and the ambient magnetic field $\vec{B}$.

Hydromagnetic waves will usually change the field direction only on time scales longer than $\tau_{\rm Barnett}$; even C-type shock waves (to be discussed in Chapter 36) may change the magnetic field direction only on time scales longer than $\tau_{\rm Barnett}$. Aligned grains will, therefore, remain aligned with the *local* magnetic field in the presence of MHD waves and weak shock waves.

26.3.2 Alignment of the Grain Body with $\vec{J}$

The moment of inertia tensor of the grain has three eigenvalues, $I_1 \geq I_2 \geq I_3$. Let $\hat{a}_1$, $\hat{a}_2$, and $\hat{a}_3$ be unit vectors along the three principal axes of the grain. If the grain has fixed angular momentum $\mathbf{J}$, the rotational kinetic energy is minimized when $\hat{a}_1 \parallel \mathbf{J}$, with value $E_{\rm rot} = J^2/2I_1$. For the simple case of an oblate spheroid, with $I_1 > I_2 = I_3$, the rotational kinetic energy is

$$E_{\rm rot} = \frac{J^2}{2I_1} + \frac{J^2(I_1 - I_2)}{2I_1 I_2} \sin^2\theta \ , \tag{26.30}$$

where θ is the angle between $\vec{J}$ and $\hat{a}_1$. If $\vec{J}$ is not aligned with one of the principal axes of the grain, then the grain will "tumble," and several different mechanisms will lead to dissipation of the rotational kinetic energy, causing $\theta \to 0$; the lost rotational kinetic energy appears as heat, i.e., an increase in the vibrational energy of the grain. The fluctuation–dissipation theorem, however, requires that the rotational kinetic energy can also be increased by exchange of energy from the vibrational modes to the rotational modes. Thus the grain alignment will not be perfect. However, if the rotation is suprathermal ($J^2/I_1 \gg kT$), then

$$\langle \sin^2\theta \rangle \rightarrow\sim \frac{I_1 kT}{J^2} \frac{I_2}{I_1 - I_2} \ll 1 \ , \tag{26.31}$$

assuming that the grain is appreciably nonspherical (I_1 is appreciably larger than I_2). This is one important consequence of suprathermal rotation: excellent alignment between $\vec{J}$ and $\hat{a}_1$ is rapidly achieved by internal dissipation within the grain. Dissipation processes include viscoelastic damping, damping associated with the Barnett effect acting on the electrons, and damping associated with the Barnett effect acting on the nuclear spin system (Lazarian & Draine 1999*a*).

26.3.3 Alignment of $\vec{J}$ with $\vec{B}$

The angle θ_{BJ} measures the alignment of the angular momentum with B, and the major challenge of grain alignment is to explain why this $\langle \cos^2\theta_{BJ} \rangle \ll 1/3$. Davis

& Greenstein (1951) showed that ordinary paramagnetic dissipation within a grain spinning in a static magnetic field produces a torque acting to reduce the component of the angular momentum perpendicular to $\vec{B}$, leaving the component parallel to $\vec{B}$ unchanged. In grain "body coordinates," the grain material sees a magnetic field with a rotating component, and the material tries to become magnetized in response to this field. Because the magnetization is varying, there will be dissipation, with rotational kinetic energy converted to heat. The rate of dissipation is proportional to $\mathrm{Im}[\chi(\omega)]$. In the absence of other torques acting to change θ_{BJ}, one can show that the angle θ_{BJ} decays, with

$$\tan\theta_{BJ}(t) = \tan\theta_{BJ}(0)\, \mathrm{e}^{-t/\tau_{\rm DG}} \tag{26.32}$$

$$\tau_{\rm DG} = \frac{2\rho a^2}{5KB_0^2} \quad , \quad K(\omega) \equiv \frac{\mathrm{Im}[\chi(\omega)]}{\omega} \tag{26.33}$$

$$\tau_{\rm DG} = 1.5\times 10^6\, a_{-5}^2 \left(\frac{\rho}{3\,{\rm g\,cm^{-3}}}\right) \left[\frac{10^{-13}\,{\rm s}}{K(\omega)}\right] \left(\frac{5\,\mu{\rm G}}{B_0}\right)^2 \,{\rm yr} \quad . \tag{26.34}$$

For normal paramagnetic materials,

$$K \approx 10^{-13} \left(\frac{18\,{\rm K}}{T_{\rm gr}}\right) \,{\rm s}, \tag{26.35}$$

(Jones & Spitzer 1967; Draine 1996); $K \approx 10^{-13}$ s therefore appears to be a reasonable estimate that would apply to carbonaceous materials as well as amorphous silicate material. We see that for these materials, in the absence of other torques that could change θ_{BJ}, the Davis-Greenstein alignment process would bring about grain alignment on a time scale $\tau_{\rm DG} \approx 1.5 \times 10^6 a_{-5}^2$ yr. This time is short compared to the lifetimes of interstellar dust or interstellar clouds, but unfortunately random collisions with gas atoms will also act to change θ_{BJ}. If the angular momentum J is purely thermal [i.e., $J^2 \approx \frac{3}{2}IkT$], then random collisions with gas atoms will act to change the direction of $\vec{J}$ on the time scale τ_M for the grain to collide with its own mass of gas [see Eq. (26.9)]. If grains were indeed rotating thermally, and the Davis-Greenstein alignment mechanism operated, then we would expect the degree of grain alignment to be an *increasing* function of the ratio $\tau_M/\tau_{\rm DG} \propto a^{-1}$; we would not expect alignment of $a \gtrsim 0.1\,\mu$m grains because these have $\tau_M/\tau_{\rm DG} \lesssim 0.1$, but we *would* expect alignment of $\vec{J}$ with $\vec{B}$ for $a \lesssim 0.01\,\mu$m grains. However, the opposite is observed: large grains tend to be aligned, while the small grains are not! Obviously, there must be more to the story.

Purcell (1979) pointed out the importance of suprathermal rotation (see §26.2.1). If systematic torques that are *fixed in body coordinates* act to give the grain $J^2 \gg \frac{3}{2}IkT$, then the time scale for random collisions with gas atoms to change the direction of $\vec{J}$ is increased by a factor $\sim J^2/IkT$. Note that the time scale $\tau_{\rm DG}$ does not depend on the grain rotation rate if $K = constant$ (as is expected to be a good approximation), and therefore suprathermal rotation supresses the disalignment from random collisions without affecting the Davis-Greenstein alignment rate, and we therefore expect grains that rotate suprathermally to become aligned on time scales of a few Myr.

The difficulty here is that the systematic torques identified by Purcell – H_2 formation on the grain surface, photoelectric emission, and variations in accomodation coefficient over the grain surface – are all sensitive to the surface properties of the grain, and these appear likely to change on time scales $\ll \tau_{DG}$, as the result of accretion or desorption of species from the grain, or sticking of a small grain to the surface of a large grain. If the grain surface properties change, then the systematic torques will change in a random way. This may lead to "spin down" of the grain to a state with low angular momentum. During the time that the grain has very low angular momentum state – known as a "crossover event" – the direction of $\vec{J}$ can be very easily changed by random collisions with gas atoms. If the only systematic torques are the three identified by Purcell, then it seems likely that crossovers would be frequent, and only minimal grain alignment would take place. Furthermore, the process of **thermal flipping** (Lazarian & Draine 1999*b*) appears able to prevent $a \lesssim 1\,\mu\text{m}$ grains from ever achieving suprathermal rotation: at constant $\vec{J}$, the grain flips rapidly from one "flip state" to the other. Since the three torques identified by Purcell are essentially fixed in body coordinates, if the grain spends 50% of the time in each flip state, then in inertial coordinates the time-average torque is zero. For systematic torques that are fixed in grain coordinates, thermal flipping appears to be rapid enough to suppress suprathermal rotation for grain radii $a \lesssim 1\,\mu\text{m}$ (Lazarian & Draine 1999*b*).

Radiative torques appear to save the day: if the starlight background is anisotropic (as is generally the case) then the starlight torque does not average to zero if the grain spends 50% of the time in each of the two flip states. The radiative torque alone can drive the grain to suprathermal rotation – provided that the radiative torque is sufficiently strong. In the "electric dipole" limit, the radiative torque goes to zero: for the radiative torque to be appreciable, the grain must be large enough that its response to an incident electric field cannot be simply approximated as an induced electric dipole. This requires $a \gtrsim \lambda/2\pi$. Because the starlight background has a characteristic wavelength $\lambda \approx 0.5\,\mu\text{m}$ (determined by the typical temperatures of the stars that dominate the diffuse starlight background), the radiative torque is only strong for grains that have $a \gtrsim \lambda/2\pi \approx 0.1\,\mu\text{m}$. This is wonderful – we appear to have identified the reason why large grains are aligned, but small grains are not: Large grains are driven to suprathermal rotation by radiative torques, but small grains do not achieve suprathermal rotation, and are subject to rapid disalignment by collisions with gas atoms.

This would already be sufficient to establish radiative torques as a central part of the grain alignment story, but it turns out they have an even more direct effect: when an anisotropic radiation field is present, radiative torques can change the *direction* of $\vec{J}$ as well as its magnitude. The dynamics are complex, but it appears that radiative torques can often bring an irregular grain into alignment with $\vec{B}$ on the gas-drag time scale $\tau_M \ll \tau_{DG}$ (Draine & Weingartner 1997; Weingartner & Draine 2003; Hoang & Lazarian 2009).

Chapter Twenty-seven

Heating and Cooling of H II Regions

Hot stars photoionize the gas around them; the photoionized gas is referred to as an H II region, because the hydrogen is predominantly ionized. We discussed the ionization conditions in H II regions in Chapter 15, but in that discussion we presupposed that the temperature was $T \approx 10^4$ K. In Chapter 18, we discussed nebular diagnostics that allow us to determine the gas temperature from observations.

Here, we discuss the physical processes that actually regulate the temperature in H II regions. The dominant heating process is photoionization: ionizing photons have energies larger than the ionization threshold, and the resulting photoelectrons will have nonzero kinetic energy, adding to the thermal energy of the gas. At the same time, recombination processes (primarily radiative recombination) are removing electrons from the plasma, along with the kinetic energy that they possessed just before the recombination event, and thermal energy is also lost when electron collisions excite ions from lower to higher energy levels, followed by emission of photons. The temperature of the gas is determined by a balance between the heating and cooling processes.

27.1 Heating by Photoionization

Consider photoelectric absorption creating a photoelectron with kinetic energy KE:

$$X^{+r} + h\nu \rightarrow X^{+r+1} + e^- + KE \quad . \tag{27.1}$$

If $\sigma_{\rm pe}(\nu)$ is the photoionization cross section for X^{+r} in its ground electronic state, the probability per unit time for photoionization is [see Eq. (13.5)]

$$\zeta(X^{+r}) = \int_{\nu_0}^{\infty} \sigma_{\rm pe}(\nu) c \left[\frac{u_\nu}{h\nu}\right] d\nu \quad , \tag{27.2}$$

where $h\nu_0$ is the threshold energy for photoionization from the ground state. Each photoionization event injects a photoelectron with kinetic energy $(h\nu - h\nu_0)$ into the plasma; the heating rate per unit volume from this process is

$$\Gamma_{\rm pe} = n(X^{+r}) \int_{\nu_0}^{\infty} \sigma_{\rm pe}(\nu) c \left[\frac{u_\nu}{h\nu}\right] (h\nu - h\nu_0)\, d\nu \quad . \tag{27.3}$$

The mean photoelectron energy is

$$E_{\rm pe}(X^{+r}) = \frac{\Gamma_{\rm pe}}{n(X^{+r})\zeta(X^{+r})} \ , \tag{27.4}$$

which will depend on the spectrum u_ν of the photons responsible for the photoionization.

To gain an understanding of what we expect for $E_{\rm pe}$ in the photoionized nebula around a hot star, let us suppose first of all that the radiation leaving the star can be approximated by a blackbody spectrum with color temperature T_c, and define the dimensionless ratio

$$\psi \equiv \frac{E_{\rm pe}}{kT_c} \ . \tag{27.5}$$

We anticipate that ψ will be of order unity, but we need to calculate it.

Near the star, the effects of absorption on the radiation field will be negligible; the stellar radiation will be attenuated by the inverse-square law, but will have the same shape: $u_\nu \propto B_\nu(T_c)$. Here, we will have $\psi = \psi_0$, where

$$\psi_0 kT_c \equiv \frac{\int_{\nu_0}^\infty \left[B_\nu(T_c)/h\nu\right]\sigma_{\rm pe}(\nu)(h\nu - h\nu_0)d\nu}{\int_{\nu_0}^\infty \left[B_\nu(T_c)/h\nu\right]\sigma_{\rm pe}(\nu)d\nu} \ . \tag{27.6}$$

$\psi_0 kT_c$ is the photoelectron energy weighted by $(B_\nu/h\nu)\sigma_{\rm pe}(\nu)$. For the important case of photoionization of H, we know from the discussion of Strömgren spheres that essentially all of the photons with $h\nu > h\nu_0 = I_{\rm H}$ will produce a photoionization somewhere in the H II region. We can obtain the average photoelectron energy

$$\langle\psi\rangle kT_c = \frac{\int_{\nu_0}^\infty \left[B_\nu(T_c)/h\nu\right](h\nu - h\nu_0)d\nu}{\int_{\nu_0}^\infty \left[B_\nu(T_c)/h\nu\right]d\nu} \ . \tag{27.7}$$

Values of ψ_0 and $\langle\psi\rangle$ are given in Table 27.1 for selected values of T_c. The important thing to note is that both ψ_0 and $\langle\psi\rangle$ are of order unity over a broad range of T_c. This is because the values of T_c in the table are all small compared to $I_{\rm H}/k = 157,800\,{\rm K}$, so the dominant behavior in the integrands is the rapid falloff in B_ν with increasing ν.

The heating rate $\Gamma_{\rm pe}$ in Eq. (27.3) depends on the abundance $n(X^{+r})$ of the species that is being photoionized. If the H II region is in photoionization equilibrium, then the condition

$$\zeta(X^{+r})n(X^{+r}) = \alpha\, n_e\, n(X^{+r+1}) \ , \tag{27.8}$$

(where α is the rate coefficient for recombination $X^{+r+1} + e^- \rightarrow X^{+r}$) can be used to obtain the local heating rate:

$$\Gamma_{\rm pe} = \alpha\, n_e\, n(X^{+r+1})\, \psi kT_c \ . \tag{27.9}$$

Equation (27.9) simply states that $\Gamma_{\rm pe}$, the rate per unit volume of injection of photoelectron energy, is equal to the recombination rate per unit volume times the mean photoelectron energy $E_{\rm pe} = \psi k T_c$. The advantage of writing it this way is that in an H II region, the dominant element, H, will be nearly fully ionized, $n({\rm H}^+) \approx n_{\rm H}$, so that the rate of heating due to photoionization of H is

$$\Gamma_{\rm pe}({\rm H} \rightarrow {\rm H}^+) \approx \alpha_B \, n_{\rm H} \, n_e \, \psi k T_c \quad . \tag{27.10}$$

It should be kept in mind that real stellar spectra are not blackbodies, but it is reasonable to use the stellar effective temperature in place of T_c. A precision calculation of $E_{\rm pe}$ using a model atmosphere will give numerical values of $E_{\rm pe}$ that differ slightly from $\psi k T_c$, even for a pure hydrogen nebula. Real nebulae of course have He, and if the star is hot enough to photoionize He, many of the photons with $h\nu > 24.6\,{\rm eV}$ will be absorbed by He rather than H. Furthermore, the calculation of $E_{\rm pe}$ for hydrogen should include the photoionizations of H by He recombination radiation. Last, real nebulae are dusty, and the dust grains will absorb some of the ionizing photons. However, for normal H II regions these complications result in only minor numerical corrections to the rate of photoelectric heating.

27.2 Other Heating Processes

In "normal" H II regions [with $\tau_{S0} \gtrsim 10^2$ – see Eq. (15.12)], heating of the gas is dominated by photoionization of H and He. However, other heating processes can sometimes make important contributions to the heating, and we briefly review them here.

27.2.1 Photoelectric Emission from Dust

Photoelectric absorption by dust grains can produce energetic photoelectrons. These can be important under two circumstances: (1) when the radiation field in the H II region has lots of UV photons with $10 \lesssim h\nu < 13.6\,{\rm eV}$, unable to ionize H, but able to ionize a dust grain; and (2) when the parameter $\tau_{d0} \gtrsim 1$, so that an appreciable fraction of the $h\nu > 13.6\,{\rm eV}$ photons are absorbed by dust rather than H.

Table 27.1 Mean Photoelectron Energy $E_{\rm pe}/kT_c$

T_c(K)	8000	16000	32000	64000
ψ_0	0.959	0.922	0.864	0.775
$\langle\psi\rangle$	1.101	1.199	1.380	1.655

From Spitzer (1978).

Calculating the photoelectric heating by the dust requires determination of the dust charging, because the photoelectric yield is suppressed if a dust grain is driven to a large positive potential by the photoelectric emission process itself, and those photoelectrons that do escape to infinity lose part of their initial kinetic energy in overcoming the attractive Coulomb potential.

The contribution of photoelectric emission from dust grains to the heating of photoionized gas has been examined in several recent papers. It can be important in planetary nebulae (Dopita & Sutherland 2000; Stasińska & Szczerba 2001), in the diffuse "warm ionized medium" (WIM) (Weingartner & Draine 2001*d*), and perhaps even in intergalactic Lyman α clouds (Inoue & Kamaya 2010).

27.2.2★ Cosmic Rays

Cosmic rays can heat the gas by two processes: (1) interaction with bound electrons resulting in ejection of an energetic secondary electron; and (2) transfer of kinetic energy to free electrons by elastic scattering. An H atom exposed to the cosmic ray background will have a probability per unit time $\zeta_{\rm CR}$ of being ionized by a passing cosmic ray. These cosmic rays will also transfer energy to the electron plasma at a rate that can be written (Goldsmith et al. 1969)

$$\Gamma_{\rm CR} = A\, n_e\, \zeta_{\rm CR} \ , \tag{27.11}$$

where the coefficient A depends weakly on the energy spectrum of the cosmic rays, with $A = 5.6 \times 10^{-10}$ erg for 2 MeV protons, or $A = 3.8 \times 10^{-10}$ erg for 2 GeV protons. Note that for fixed cosmic ray flux (i.e., fixed $\zeta_{\rm CR}$), the heating rate $\Gamma_{\rm CR} \propto n_e$, whereas $\Gamma_{\rm pe} \propto n_e^2$. Plasma drag may therefore make a significant contribution to the overall heating in low-density H II regions, but is likely to be negligible in high-density regions.

27.2.3★ Damping of MHD Waves

In inhomogeneous (clumpy) H II regions, pressure gradients will drive gas flows, which may excite MHD waves, in some cases even shock waves. Shock waves will directly heat the gas, and other MHD waves will in general be damped, with conversion of the wave energy into heat. Suppose that the plasma contains waves with energy density $\Delta u_{\rm wave}$, that the wave propagates at speed $v_{\rm wave}$, and that $L_{\rm damp}$ is the damping length for the wave. The heating rate due to wave dissipation would then be

$$\Gamma_{\rm wave} \approx (\Delta u_{\rm wave}) v_{\rm wave} / L_{\rm damp} \tag{27.12}$$

$$\approx 8.9 \times 10^{-25} n_{\rm H} T_4 \left(\frac{\Delta u_{\rm wave}}{2 n_{\rm H} kT}\right) \left(\frac{v_{\rm wave}}{10\,{\rm km\,s}^{-1}}\right) \left(\frac{\rm pc}{L_{\rm damp}}\right) {\rm erg\,s}^{-1} . \tag{27.13}$$

The factor $(\Delta u_{\rm wave})/2 n_{\rm H} kT$ is the ratio of the wave pressure to the gas pressure, and is expected to be smaller than unity. The wave speed $v_{\rm wave}$ will not significantly exceed the thermal sound speed $\sim 10\,{\rm km\,s^{-1}}$ for $T \approx 10^4\,{\rm K}$.

27.3 Cooling Processes

Earlier, we considered various heating mechanisms – processes that add to the translational kinetic energy of the particles in the gas. Now, we consider processes that remove thermal kinetic energy from the gas.

In a time-dependent flow, calculating the thermal evolution of the gas requires careful accounting for changing numbers of particles if molecules are forming or being dissociated, or if electrons are recombining with ions, or being liberated by ionization processes. If, however, the gas is in a steady state, with the number of free particles constant, discussing the heating and cooling is much more straightforward. Here we will consider the balance between heating and cooling processes in a plasma that is in ionization equilibrium, where every ionization event is balanced by a recombination event.

27.3.1 Recombination Radiation

Every time an electron recombines with an ion, the plasma loses the kinetic energy of the recombining electron: the rate per unit volume at which thermal energy is lost is

$$\Lambda_{\rm rr} = \alpha_B \, n_e \, n({\rm H}^+) \, \langle E_{\rm rr} \rangle \quad , \tag{27.14}$$

where α_B is the rate coefficient for Case B radiative recombination (see Chapter 14), and $\langle E_{\rm rr} \rangle$ is the mean kinetic energy of the recombining electrons.

Suppose that we approximate the cross section for radiative recombination by a power-law:

$$\sigma_{\rm rr}(E) = \sigma_0 (E/E_0)^\gamma \quad . \tag{27.15}$$

The rate coefficient is then

$$\langle \sigma v \rangle = \left(\frac{m_e}{2\pi kT}\right)^{3/2} \int_0^\infty 4\pi v^2 dv \, {\rm e}^{-E/kT} \sigma v \tag{27.16}$$

$$= \left(\frac{8kT}{\pi m_e}\right)^{1/2} \sigma_0 \left(\frac{kT}{E_0}\right)^\gamma \Gamma(2+\gamma) \tag{27.17}$$

$$\propto T^{\gamma+1/2} \quad ,$$

where Γ is the usual gamma function.

Consider now recombination of the most important ion, H^+. The rate coefficients α_A and α_B (see Fig. 14.1) are not actually power-laws in T, and therefore the recombination cross sections $\sigma_{\rm rr}(E)$ are not strictly power-laws in E. Nevertheless, we can obtain an effective power-law index by taking the logarithmic derivative of the rate coefficient:

$$\gamma + \frac{1}{2} = \frac{d\ln\langle\sigma v\rangle}{d\ln T} \quad . \tag{27.18}$$

The fits (14.5 and 14.6) to α_A and α_B give

$$\gamma_A = -1.2130 - 0.0230\ln(T_4/Z^2) \ , \tag{27.19}$$

$$\gamma_B = -1.3163 - 0.0416\ln(T_4/Z^2) \ , \tag{27.20}$$

for the effective values of the exponent γ for Case A and Case B recombination.

For the power-law dependence from Eq. (27.15), the mean kinetic energy per recombining electron is

$$\langle E_{\rm rr}\rangle = \frac{\int v^2 dv\, e^{-E/kT}\sigma v E}{\int v^2 dv\, e^{-E/kT}\sigma v} = \frac{\Gamma(3+\gamma)}{\Gamma(2+\gamma)}kT = (2+\gamma)kT \quad . \tag{27.21}$$

Thus, the mean energy per recombining electron is

$$\langle E_{\rm rr}\rangle_A = \left[0.787 - 0.0230\ln(T_4/Z^2)\right] kT \ , \tag{27.22}$$

$$\langle E_{\rm rr}\rangle_B = \left[0.684 - 0.0416\ln(T_4/Z^2)\right] kT \ . \tag{27.23}$$

for Case A and Case B, respectively. The recombining electrons are, on average, *less* energetic than the mean electron kinetic energy of $1.5kT$, for the simple reason that the cross section for radiative recombination is larger for low energy electrons.

27.3.2 Free–Free Emission

Free electrons scattering off free ions produce free–free emission with power per unit volume given by Eq. (10.12). In a pure H plasma near 10^4 K the free–free emission energy per recombining electron is (using Eqs. 10.10 and 14.6)

$$\frac{\Lambda_{\rm ff}}{n_e n({\rm H}^+)\alpha_B} = 0.54\,T_4^{0.37}\,kT \quad . \tag{27.24}$$

Thus, for $T_4 \approx 1$, radiative recombination and free–free emission together cause the plasma to lose $(0.68+0.54)kT = 1.22kT$ of kinetic energy per Case-B recombination.

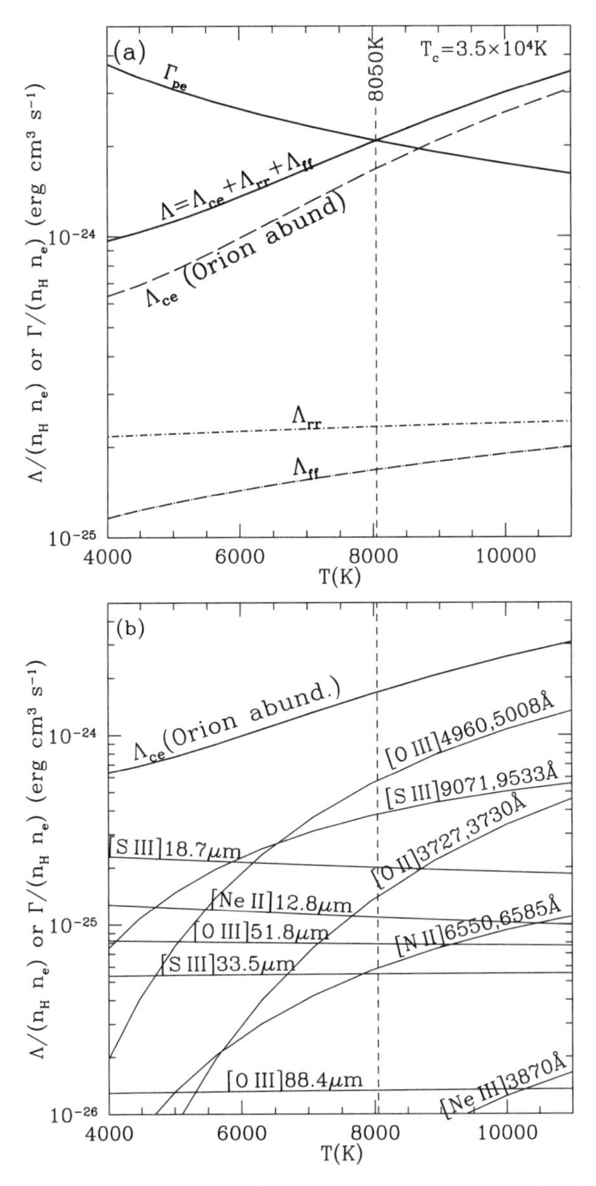

Figure 27.1 (a) Photoelectric heating function Γ_{pe} and radiative cooling function Λ as functions of gas temperature T in an H II region with Orion-like abundances and density $n_{\rm H} = 4000\,{\rm cm}^{-3}$. Heating and cooling balance at $T \approx 8050\,{\rm K}$. (b) Contributions of individual lines to Λ_{ce}.

27.3.3 Collisionally Excited Line Radiation

In an H II region, most of the hydrogen will be ionized. Even if some He or He$^+$ is present, the energy of the first excited state is so far above the ground state that the rate for collisional excitation is negligible. However, if heavy elements such as oxygen are present, there may be ions – such as O II or O III – with energy levels that can be collisionally excited by electrons with kinetic energies of just a few eV. If the collisional excitation is followed by a collisional deexcitation, the kinetic energy of the gas will be unchanged. If the only excitation process is collisional (i.e., optical pumping and recombination to excited states are both negligible), the rate of energy loss by the gas by collisional excitation is just

$$\Lambda_{ce} = \sum_X \sum_i n(X,i) \sum_{j<i} A_{ij}(E_i - E_j) \ , \qquad (27.25)$$

where the sum is over species X and excited states i. Collisionally excited emission lines have been discussed in Chapter 18. In Eq. (27.25), we sum over all line emission that results from collisional excitation.

27.4 Thermal Equilibrium

The gas will stabilize at a temperature where heating balances the total cooling: $\Gamma_{\rm pe} = \Lambda = \Lambda_{ce} + \Lambda_{\rm rr} + \Lambda_{\rm ff}$. Figure 27.1 shows the T-dependence of $\Gamma_{\rm pe}$ and $\Lambda(T)$ for gas with composition ($\sim$ solar), density ($n_{\rm H} = 4000\,{\rm cm}^{-3}$), and ionization balance similar to that in the Orion Nebula H II region, assuming heating by a $T = 3.5 \times 10^4\,{\rm K}$ blackbody. If we have identified all the heating and cooling processes and evaluated them accurately, it appears that the gas temperature should be $T \approx 8050\,{\rm K}$.

Table 27.2 Principal Collisionally Excited Cooling Lines in H II Regions

ion	lines
N II	6585 Å, 6550 Å
O II	3730 Å, 3727 Å
O III	88.36 μm, 51.81 μm, 5008 Å, 4960 Å
Ne II	12.81 μm
S II	6733 Å, 6718 Å
S III	33.48 μm, 18.71 μm, 9071 Å, 9533 Å

For abundances that are similar to solar abundances, and ionizations similar to those in the Orion Nebula, Table 27.2 lists some of the stronger collisionally excited lines. Some of these are fine-structure transitions, appearing in the mid- and far-infrared, and some are transitions from electronic excited states, with lines in the near-infrared through ultraviolet.

The equilibrium temperature is obviously sensitive to the abundance of the coolant species – lowering the abundances will cause the steady state temperature to increase. As seen in Figure 27.2, if we reduce the abundances of elements beyond helium by a factor of 10, as might be appropriate in a low-metallicity galaxy, the thermal equilibrium temperature rises to $\sim 15,600\,{\rm K}$. Conversely, if we *raise* the

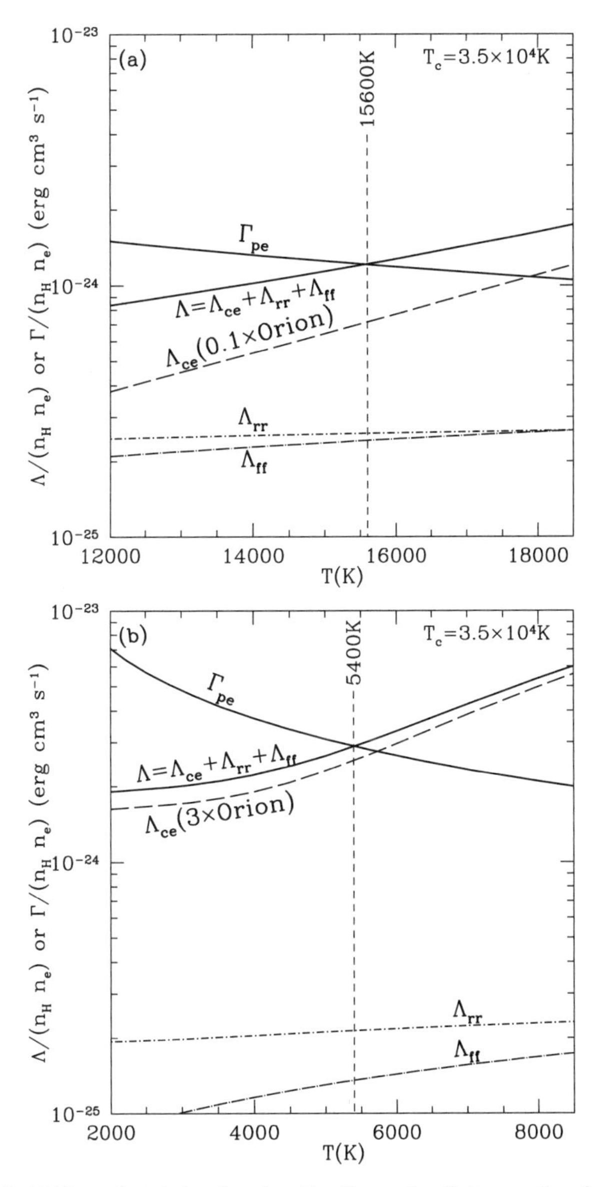

Figure 27.2 (a) Photoelectric heating function Γ_{pe} and radiative cooling function Λ as functions of temperature T in an H II region with abundances that are (a) only 10%, or (b) enhanced by a factor of 3 relative to the Orion Nebula. A density $n_{\rm H} = 4000\,{\rm cm}^{-3}$ is assumed. Thermal equilibrium occurs for $T \approx 15600\,{\rm K}$ and $\sim 5400\,{\rm K}$ for the two cases.

heavy-element abundances by a factor of 3, as might be appropriate in the central regions of a mature spiral galaxy, the H II region temperature drops to just $\sim 5400\,{\rm K}$.

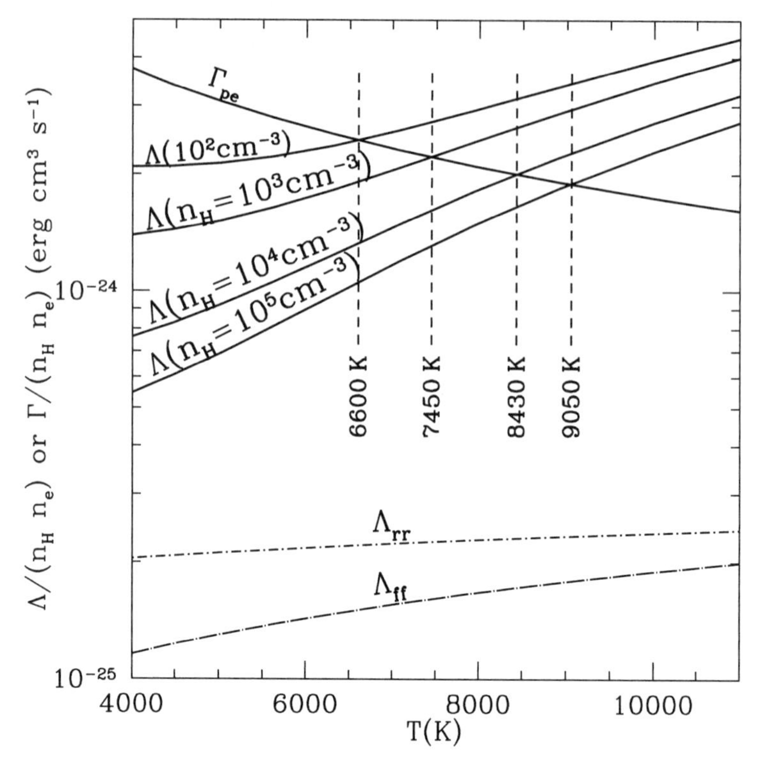

Figure 27.3 Cooling function $\Lambda(T)$ for different densities $n_{\rm H}$. The gas is assumed to have Orion-like abundances and ionization conditions. As the gas density is varied from $10^2\,{\rm cm}^{-3}$ to $10^5\,{\rm cm}^{-3}$, the equilibrium temperature varies from 6600 K to 9050 K, because of collisional deexcitation of excited states.

For a given ionizing star and gaseous abundances, the H II region temperature will also be sensitive to the gas density. When the density exceeds the critical density of some of the cooling levels, the cooling will be suppressed, and the equilibrium temperature will rise. Figure 27.3 shows how, at fixed Orion Nebula-like abundances and ionization balance, the cooling function $\Lambda(T)/n_{\rm H}n_e$ responds to changes in the density. As the density is increased from $n_{\rm H} = 10^2\,{\rm cm}^{-3}$ to $n_{\rm H} = 10^5\,{\rm cm}^{-3}$, the thermal equilibrium temperature shifts from 6600 K to 9050 K.

27.5 Emission Spectrum of an H II Region

When the gas is near thermal equilibrium, the principal cooling lines are shown in Fig. 27.1(b), and listed in Table 27.2.

The optical spectrum of an H II region is dominated by the major hydrogen recombination lines (H α 6565 Å, H β 4863 Å, H γ 4342 Å), and collisionally excited lines of [S II]6733,6718 Å, [N II]6585,6550 Å, and [O II]3727,3730 Å. If there is an appreciable He^+ ionization zone, then there will also be He recombination lines (He I 5877 Å, He I 4473 Å, and He I 3890 Å are the strongest) as well as strong [O III]4960,5008 Å.

In the near-infrared, the [S III]9071,9533 Å doublet is usually strong. If there is an He^+ ionization zone, the He I 1.083 μm triplet will be very strong. The Paschen α $1.876\,\mu\mathrm{m}$ recombination line is also strong, but does not fall in an atmospheric window.

The fine-structure lines are major coolants, principally [Ne II]12.81 μm, [S III]18.71 μm, [O III]51.81 μm, and [S III]33.48 μm. [Ne II]12.81 μm can be observed from the ground through an atmospheric window, but the other fine structure lines require observations from above the atmosphere.

The [O III] and [S III] lines will be suppressed in H II regions around cooler stars, where the oxygen and sulfur will be mainly singly ionized.

27.6 Observed Temperatures in H II Regions

As discussed in Chapter 18, observed emission line ratios can be used to measure the actual temperatures in H II regions. The most useful species for this purpose are [O III], [N II], and [S III] (see Fig. 18.2). The gas temperature can also be estimated from the spectrum of the recombination continuum (see §18.4.1).

Because the ionization structure and the radiation field in an H II region are not uniform, the temperature in the H II region is expected to also vary radially. Different ions will measure the temperature in different regions. Therefore, we do not expect temperatures estimated from different diagnostics to be in perfect agreement.

Osterbrock & Ferland (2006) compare temperature determinations in H II regions and planetary nebulae. When more than one estimator is used for a single nebula, differences are typically at the $\pm 10\%$ level, which seems satisfactory. The estimated temperatures at different locations in the Orion Nebula are in the 8000 to 10,000 K range (Baldwin et al. 1991), only slightly higher than $T = 8050\,\mathrm{K}$ determined from Figure 27.1. At the densities of the Orion Nebula, it seems unlikely that heating by cosmic rays or wave dissipation could be significant, but dust photoelectric heating might not be negligible, helping to bring the theoretical temperature closer to the observed values.

In summary, both theory and observation tell us that the photoionized gas in H II regions and planetary nebulae should be heated to temperatures $T \approx 6000$ to $15,000\,\mathrm{K}$, depending on the metallicity, the density of the gas, and the spectrum of the radiation from the star.

Chapter Twenty-eight

The Orion H II Region

The Orion Nebula (= M 42 = NGC 1976) is the brightest H II region in our sky – not because it is an especially luminous H II region, but simply because it happens to be the *nearest* dense H II region, at a distance of only $\sim 414 \pm 7\,\mathrm{pc}$ (Menten et al. 2007). Its proximity has allowed studies at relatively high spatial resolution, so that the Orion Nebula now informs much of our understanding of H II regions. O'Dell (2001) has a nice review of M 42. Plates 9 and 10 show the Orion Nebula at optical wavelengths.

28.1 Trapezium Stars

The Orion H II region contains a cluster of stars referred to as the **Orion Nebula Cluster**, or **ONC**. With a core radius of $\sim 0.2\,\mathrm{pc}$ and a central stellar density $\sim 2 \times 10^4\,\mathrm{stars\,pc^{-3}}$, the ONC contains ~ 2300 stars within a radius $2.06\,\mathrm{pc}$, with a total stellar mass $\sim 1800\,M_\odot$, formed within the past $\sim 2 \times 10^6\,\mathrm{yr}$ (Hillenbrand & Hartmann 1998). The median stellar mass in the ONC is probably $\sim 0.3\,M_\odot$.

The four brightest members at the center of the ONC form the θ^1 Ori "Trapezium" system – see Table 28.1. The most massive star, θ^1 Ori C, of spectral type O7V (Stahl et al. 2008), produces $\sim 80\%$ of the H-ionizing photons. θ^1 Ori C is $\sim 10''$ ($0.02\,\mathrm{pc}$) S of the center of the ONC. The second most massive star, θ^1 Ori D, of spectral type O9.5V, contributes $\sim 15\%$ of the ionization. The radiative properties of the O stars are from Table 15.1. Those for the B stars are from Smith et al. (2002).

The O9V star θ^2 Ori is located $135''$ SE of θ^1 Ori C. Although not located in the

Table 28.1 Trapezium Stars and θ^2 Ori

Star	Spectral Type	T_{eff} (K)	$L(10^5\,L_\odot)$	$Q_0(10^{48}\,\mathrm{s^{-1}})$
θ^1 Ori C	O7V	36,900	1.4	5.6
θ^1 Ori D	O9.5V	31,880	0.48	0.76
θ^1 Ori A	B0.5V	28,100	0.4	0.10
θ^1 Ori B	B3V	17,900	0.017	—
θ^2 Ori	O9V	32,830	0.59	1.15
Total			2.9	7.6

brightest part of M 42, θ^2 Ori contributes to ionization of the lower density portion of the H II region.

28.2 Distribution of Ionized Gas

The Trapezium stars ionize the hydrogen around them, maintaining an H II region. The H II region was presumably initially surrounded (and contained) by molecular gas. The combination of ionizing radiation and the overpressure of the ionized gas resulted in expansion of the H II region, and development of a "blister" geometry (Israel 1978). The high-pressure ionized gas was eventually able to break out of the blister. Once it has broken out of the confining molecular gas, the ionized gas is able to exhaust into the lower density interstellar medium, in what is termed a "champagne flow" (Tenorio-Tagle 1979). As a result, the H II region does not have the simple geometry of an idealized Strömgren sphere. The actual geometry is best visualized by maps of the free–free continuum emission at radio frequencies, unaffected by foreground dust. Figure 28.1 shows an image of the free–free emission from the central region of M 42.

The geometry is obviously not a simple uniform-density sphere, but the principal ionizing source, θ^1 Ori C, is near the center of the emission, and most of the emission is coming from a region with an angular diameter of about $4'$, or a diameter $\sim 0.5\,\mathrm{pc}$. The peak emission measure, averaged over a $43''$ beam, is $EM \approx 5\times10^6\,\mathrm{cm}^{-6}\,\mathrm{pc}$ (Felli et al. 1993). If the line-of-sight path through the ionized gas is $\sim 0.5\,\mathrm{pc}$, the rms electron density along this path is $\langle n_e^2\rangle^{1/2} \approx 3200\,\mathrm{cm}^{-3}$.

The electron density and temperature in the H II region can be determined using [S II]6718/6733 and other line ratios. Esteban et al. (2004) estimate $n_e \approx 8900\pm200\,\mathrm{cm}^{-3}$ for the region with the highest surface brightness ($\sim 20''$ SW of θ^1 Ori C). The temperature is found to be $T_e \approx 10000\pm400\,\mathrm{K}$ from emission lines of N II, S II, and O II, and $T_e \approx 8320\pm40\,\mathrm{K}$ from emission lines of O III, S III, and Ar III. Radio observations of the H64α recombination line give an average temperature $T_e = (8300\pm200)\,\mathrm{K}$ (Wilson et al. 1997).

The total H recombination rate can be estimated from the total free–free radiation, or from extinction-corrected observations of H recombination lines. Observations of Hα integrated over a $16'\times16'$ area have been used to estimate a total H recombination rate of $8.2\times10^{48}\,\mathrm{s}^{-1}$ (Wen & O'Dell 1995) (corrected for a distance $D = 414\,\mathrm{pc}$). This is slightly *above* the estimated total rate $Q_0 \approx 7.6\times10^{48}\,\mathrm{s}^{-1}$ of emission of $h\nu > 13.6\,\mathrm{eV}$ photons by the Trapezium stars plus θ^2 Ori (see Table 28.1), but well within the uncertainties of the estimate for Q_0.

Given that the estimate for Q_0 does not exceed the recombination rate derived from the observed Hα, it appears that the fraction of the $h\nu > 13.6\,\mathrm{eV}$ photons emitted by θ_1 Ori C, θ_1 Ori D, and θ_2 Ori that produce photoionizations is $f_{\rm ion} \approx 1$.

What is expected? θ_1 Ori has $Q_{0,49} = 0.64$, and the H II region has $n_{\rm rms} \approx 3200\,\mathrm{cm}^{-3}$. From Eq. (15.30), a static H II region with $Q_{0,49}n_{\rm rms} \approx 2000\,\mathrm{cm}^{-3}$ would have $\tau_{d0} \approx 2.6\sigma_{d,-21}$. If $\sigma_{d,-21} \approx 1$, then from Fig. 15.2, we estimate

$f_{\rm ion} \approx 0.45$, whereas observations indicate that $f_{\rm ion} \approx 1$. This suggests that perhaps the stellar ionizing output Q_0 may have been underestimated, or perhaps σ_d in the ionized gas is significantly smaller than $1 \times 10^{-21}\,{\rm cm^2 H^{-1}}$.

28.3 Orion Bar

A conspicuous feature in optical images of the Orion H II region (see Plate 10) is the bright bar-like feature running NE to SW, passing $\sim 110''$ SE of the Trapezium. This feature shows up as a local maximum just interior to a conspicuous discontinuity in the free–free radio emission (see Fig. 28.1). For a point on the Orion Bar, Pogge et al. (1992) found $n_e \approx 3000\,{\rm cm^{-3}}$ from [S II]6718/6733, significantly lower than $n_e \approx 9000\,{\rm cm^{-3}}$ in the bright area $30''$ SW of θ^1 Ori C.

The Orion Bar is a photoionization/photodissociation front viewed nearly edge-on. Infrared spectroscopy of the Orion Bar shows strong emission in vibration–rotation transitions of H_2 just outside the ionized gas (van der Werf et al. 1996; Allers et al. 2005). The zone with strong H_2 line emission is $\sim 5''$ thick, corresponding to a projected thickness $\Delta r \approx 3 \times 10^{16}$ cm. Outside the region with strong H_2 emision, there is an extended region of molecular gas seen in CO $J = 1 \rightarrow 0$.

While the Orion Bar stands out, it is important to realize that a photodissociation region (PDR – see §31.7) is present at all points where the ionized gas abuts the molecular material that partially surrounds the H II region – the PDR extends to the E, NE, and N of the Trapezium and, in fact, also *behind* the central portion of the H II region. The Orion Bar is conspicuous as a result of the geometric accident that our line of sight happens to be approximately tangential to the PDR, leading to enhanced observed intensities.

Wen & O'Dell (1995) have constructed a three-dimensional model of the Orion Nebula. In their model, the distance from θ^1 Ori C to the Orion Bar ionization front is $\sim 7.8 \times 10^{18}$ cm. The region of highest surface brightness, $\sim 30''$ SW of θ^1 Ori C, corresponds to a point where the ionization front along the "back" of the H II region is located only 3.6×10^{17} cm from θ^1 Ori C: the nebula is brighter there because the flux of ionizing photons is $\sim (7.8/3.6)^2 \approx 4.7$ times higher than at the Bar.

28.4 Gas Kinematics

The gas in the brightest part of M 42 has a thermal pressure $p/k \approx 1.6 \times 10^8\,{\rm cm^{-3}\,K}$; this is $\sim 10^{4.5}$ times greater than the typical pressures in the diffuse ISM. The H II region is able to "vent" into the diffuse ISM as though it were a near-vacuum, with flow speeds approaching or exceeding the isothermal sound speed $(kT/\mu)^{1/2} = 11.4 T_4^{1/2}\,{\rm km\,s^{-1}}$ in the ionized gas.

The dense molecular gas adjacent to M 42 has a heliocentric radial velocity of $(v_{\rm OMC})_r = +28\,{\rm km\,s^{-1}}$. H64$\alpha$ observations (Wilson et al. 1997) of the ionized

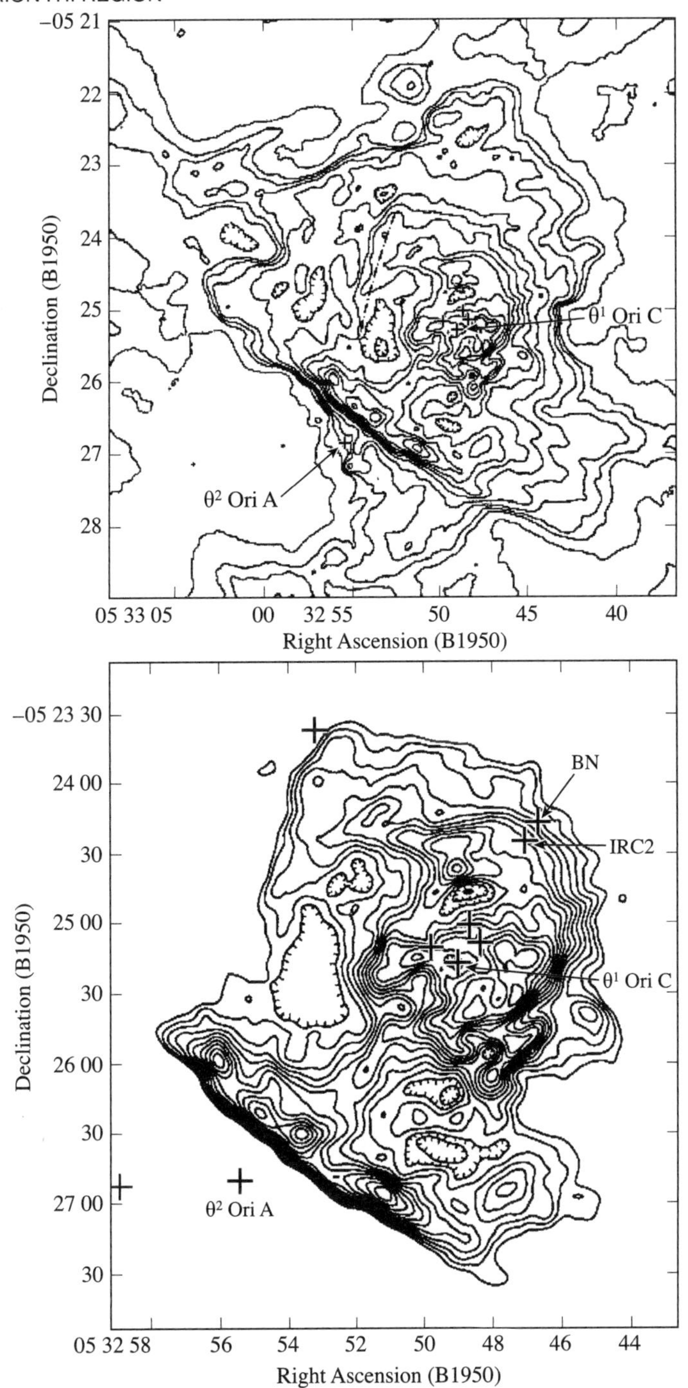

Figure 28.1 Maps of M 42 at $\lambda = 20\,\mathrm{cm}$, with HPBW $\approx 6.2''$, from Felli et al. (1993). The Orion Bar ionization front is $\sim 110''$ SE of θ^1Ori C, with a projected separation of $0.22\,\mathrm{pc} = 6.8 \times 10^{17}\,\mathrm{cm}$. The lower map is an expanded view of the brightest region, with only the brightest contours shown.

gas show a radial velocity $v_r = 18.4\,\mathrm{km\,s^{-1}}$ at the radio brightness peak. Thus, relative to the molecular cloud, the ionized gas is streaming *toward* us, with $(v - v_{\rm OMC})_r = -10\,\mathrm{km\,s^{-1}}$. At a location 80″ SW of θ^1 Ori C, the ionized gas has $(v - v_{\rm OMC})_r = -14\,\mathrm{km\,s^{-1}}$. It seems likely that the gas flow has a component in the plane of the sky, so that $|\mathbf{v} - \mathbf{v}_{\rm OMC}| \approx 20\,\mathrm{km\,s^{-1}}$, of order 1.5–2 times the isothermal sound speed $\sim 11\,\mathrm{km\,s^{-1}}$ in the ionized gas. This Mach ~ 2 flow is approximately what is expected for an H II region that is "venting" into a low-pressure surrounding medium.

28.5 PIGS, Proplyds, and Shadows

High-resolution imaging of the Orion Nebula reveals a number of curious small-scale structures:

- **Herbig-Haro objects:** These are small emission regions located near the end of high velocity jet-like outflows from young stellar objects. The H-H object is thought to be shocked gas where the material in the jet is stopped by the ambient medium.

- **Proplyds: Proplyd** (from "Protoplanetary disk") refers to a young stellar object with a dusty gaseous disk. A number of these are found exposed to the ultraviolet radiation in the H II region. They can be seen in silhouette against the bright background from the H II region, but some show evidence of luminosity from a star at the center of the disk. Such objects were actually first detected in radio continuum maps made with the Very Large Array (VLA) interferometer. These maps revealed about 25 ultracompact sources of free–free emission within 30″ of θ^1 Ori C, with apparent sizes $\sim 10^{15}$ cm ($\sim 0.15''$), which Garay et al. (1987) interpreted as partially ionized globules (**PIGs**). The PIGs appear to be protoplanetary disks undergoing photoionization and "photoevaporation" resulting from being illuminated by the ionizing radiation from θ^1 Ori C. This interpretation was confirmed by optical imaging with Hubble Space Telescope (O'Dell et al. 1993), as can be seen in Plate 11.

- **Radiation Shadows:** These are shadows cast by the proplyds. Because the bulk of the ionizing radiation is coming from a single star, θ^1 Ori C, the proplyd casts a shadow. The shadowed region will still be exposed to a diffuse ionizing background, produced by recombination to the H ground state, but this radiation is much "softer" than the emission from hot star, and not able to ionize O II $\rightarrow$ O III, which requires $h\nu > 35.1\,\mathrm{eV}$. As a result, the gas in the shadow does not radiate in [OIII]5008. These shadows can be seen as linear dark features in a map of [OIII]5008/Hα (O'Dell et al. 2009).

Chapter Twenty-nine

H I Clouds: Observations

Approximately 60% of the gas in the Milky Way is in H I regions – regions where the hydrogen is predominantly atomic.[1] The H I can be surveyed using the 21-cm line (in emission or absorption), by measuring absorption lines in the spectra of stars, and by observing infrared emission from dust that is mixed with the H I. The magnetic field in the clouds can be visualized through the starlight polarization produced by aligned dust grains, and in some cases the line-of-sight component of the magnetic field can be measured using the Zeeman effect in H I.

29.1 21-cm Line Observations

Except for directions and radial velocities where the 21-cm line becomes optically thick, observations of 21-cm emission directly measure the total column density of H I. Plate 3a is an all-sky map of the 21-cm line emission. If a background radio source is available, observations of the 21-cm line in absorption can be used to determine the H I spin temperature $T_{\rm spin}$, which is normally close to the kinetic temperature (see §8.3 and §17.3). The spin temperature is a function of position, and therefore of the radial velocity of the gas. Some observers report radial velocities in a heliocentric coordinate frame where the Sun is at rest, but sometimes the radial velocity is measured relative to a hypothetical **Local Standard of Rest** (LSR).[2]

Figure 29.1 shows the emission and absorption profiles measured toward the quasar 3C48, and the derived velocity distribution dN/dv. The most recent emission–absorption surveys (Heiles & Troland 2003) support the idea that, in the solar neighborhood (i.e., within $\sim 500\,{\rm pc}$ of the Sun), interstellar H I is found primarily in two distinct phases: the cold neutral medium (CNM) and the warm neutral medium (WNM). About 40% of the H I (by mass) is in the CNM, with a median spin temperature $T \approx 70\,{\rm K}$. The remaining 60% of the H I is in the WNM phase, which appears to have a volume filling factor $f_{\rm WNM} \approx 50\%$ near the disk midplane.

---

[1] Approximately 23% of the hydrogen is ionized, and $\sim 17\%$ is in molecular clouds (see Table 1.2).

[2] The LSR is defined to be the velocity of a closed orbit around the Galactic Center passing through the location of the Sun (Binney & Merrifield 1998). Unfortunately, the motion of the LSR is not yet well-determined, and authors are not always clear on what they assume. Some authors (e.g., Allen 1973) take the LSR to be such that velocity of the Sun relative to the LSR is $(\mathbf{v}_\odot)_{\rm LSR} = 19.7\,{\rm km\,s^{-1}}$ toward $(\ell, b) = (57°, 22°)$. A more recent estimate (Binney & Merrifield 1998) has $(\mathbf{v}_\odot)_{\rm LSR} = 13.4\,{\rm km\,s^{-1}}$ toward $(\ell, b) = (27.5°, 32.6°)$.

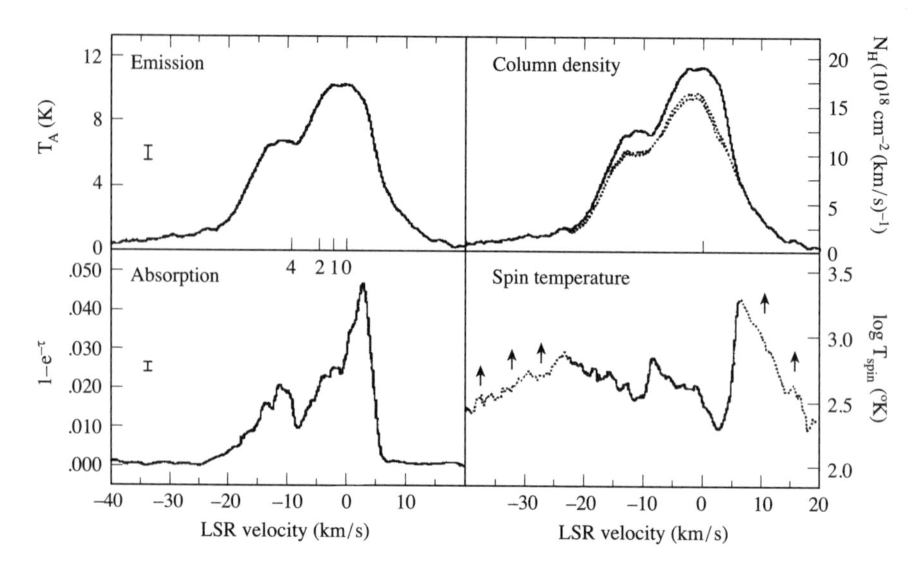

Figure 29.1 Left panels: Observed H I emission (off the quasar 3C48) and absorption (toward 3C48, at $\ell = 134^\circ$, $b = -28.7^\circ$). Lower right: spin temperature $T_{\rm spin}(v)$ as a function of LSR velocity. Tick marks labeled 0, 1, 2, and 4 on abscissa of left panels show the LSR velocity expected for gas at a distance of 0,1,2,4 kpc (for an assumed Galactic rotation curve). Upper right: $dN(\text{H I})/dv$ for different assumptions regarding the relative (foreground/background) locations of cold absorbing gas and warm gas seen only in emission. From Dickey et al. (1978).

The Sun is, fortuitously, very near the midplane of the disk. If the disk were plane-parallel, then toward galactic latitude b we would expect the H I column density to vary as $N(\text{H I}, b) = N(\text{H I}, b = 90^\circ)/\sin|b| = N_0 \csc|b|$. Figure 29.2 shows that there are appreciable deviations from the plane-parallel ideal, but a reasonable estimate for the half-thickness of the H I disk at the solar circle is $N_0 \approx 3 \times 10^{20}\,\text{cm}^{-2}$.

Because warm H I absorbs very weakly, for some of the WNM material it is only possible to determine a lower bound on $T_{\rm spin}$. Figure 29.2 shows the distribution of spin temperatures found by Dickey et al. (1978). Heiles & Troland (2003) conclude that $> 48\%$ of the WNM has $500 < T_{\rm spin} < 5000\,\text{K}$ – as we will see in §30.4, at these temperatures the gas is expected to be thermally unstable.

29.2 Distribution of the H I

Differential rotation of gas in the Galactic disk means that – except for the directions $\ell = 0$ or $\ell = 180^\circ$ – regions at different distances from the Sun will have different radial velocities (see, e.g., Binney & Merrifield 1998). Therefore, for an

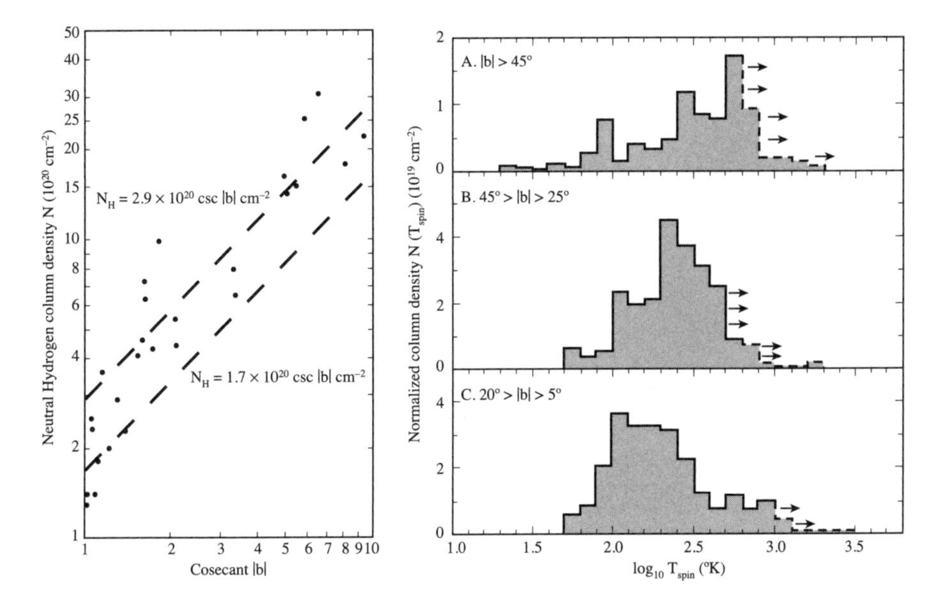

Figure 29.2 Left: $N(\mathrm{H\,I})$ versus $\csc|b| \equiv 1/\sin|b|$. The dashed line $2.9 \times 10^{20} \csc|b|\,\mathrm{cm}^{-2}$ shows the average relation from a southern survey (Radhakrishnan et al. 1972). Right: Distribution of T_{spin} for gas at high (10 directions), intermediate (8 directions), and low (9 directions) galactic latitudes. Gas with only a lower bound on T_{spin} constitutes 54% of the emission at $|b| > 45°$, and 26% at $25° < |b| < 45°$. After Dickey et al. (1978).

assumed Galactic rotation curve, the measured 21-cm intensity vs. radial velocity can be used to map out the 3-dimensional distribution of H I in the Galaxy.

Because the gas in fact has noncircular motions, the inferred distance is uncertain; in addition, for the inner galaxy ($90° < \ell < -90°$ and $R < 8.5\,\mathrm{kpc}$) there are distance ambiguities, where two different distances correspond to the same radial velocity. Nevertheless, maps of the distribution of H I gas within the Milky Way have been produced (see, e.g., Fig. 9 of Nakanishi & Sofue 2003). The radial distribution of H I derived by Nakanishi & Sofue (2003) is shown in Figure 29.3.

29.3 Zeeman Effect

For the weak ($\sim 5\mu$G) B fields in the diffuse ISM, the Zeeman effect (see §4.7) results in a very small frequency shift between the left- and right-circularly polarized 21-cm emission. The frequency shift is proportional to $B_{\|}$, the magnetic field component along the emission direction. The frequency shift is small compared to the line widths, even for narrow 21-cm lines, but the systematic shift can be obtained from the *difference* of the left- and right-circularly polarized spectra. In regions

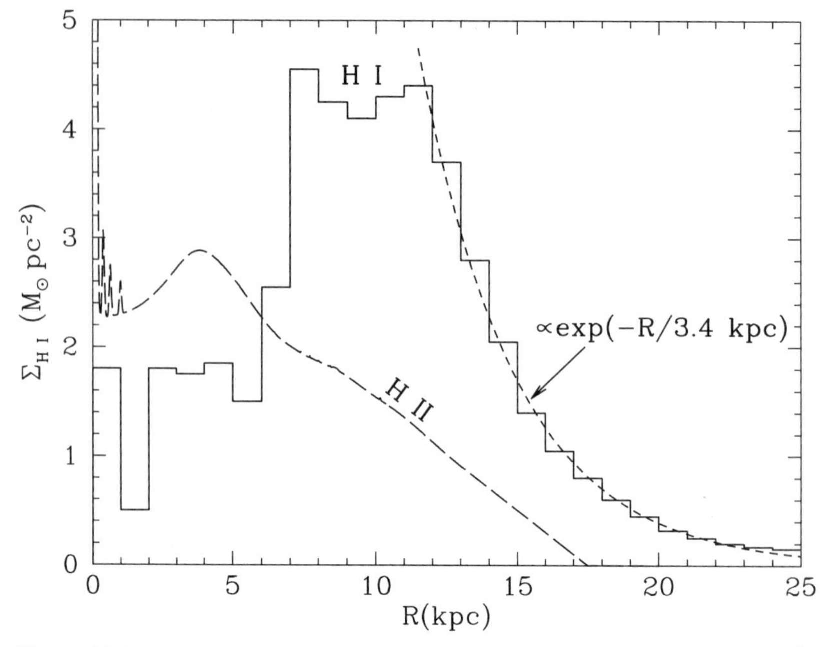

Figure 29.3 Radial distribution of H I from Nakanishi & Sofue (2003). At $R \gtrsim 11\,\mathrm{kpc}$, the H I surface density declines exponentially. Also shown is the radial distribution of H II from Figure 11.4

with simple line profiles, it is possible to measure this frequency shift and thereby determine $B_{\parallel}$.

The number of Zeeman measurements using H I is very limited, because the observations are difficult. Some selected results are given here:

- Diffuse H I clouds studied in absorption against background extragalactic radio sources have a median $B \approx 6.0 \pm 1.8\,\mu\mathrm{G}$ (Heiles & Crutcher 2005). The implied magnetic pressure is large: $B^2/8\pi k \approx 1.0 \times 10^4\,\mathrm{cm}^{-3}\,\mathrm{K}$ – several times larger than the gas pressure $nT \approx 3000\,\mathrm{cm}^{-3}\,\mathrm{K}$.

- The H I gas surrounding the Orion A GMC (see Chapter 32) has $B_{\parallel} = +10\,\mu\mathrm{G}$ (+ means **B** points away from the observer). (Heiles & Troland 1982)

- The magnetic fields in five shell-like structures [including the North Celestial Pole Loop, the North Polar Spur loop (= "Loop I"), and the shell around the Eridanus superbubble] were found to be strong, with $\langle |B_{\parallel}| \rangle \approx 6.4\,\mu\mathrm{G}$ (Heiles 1989). The total fields are expected to be larger, on average, by a factor 2, so that we estimate $|B| \approx 13\,\mu\mathrm{G}$ in these shells. Each H I shell is thought to be compressed gas behind expanding shock waves, with the cavity inside the shell (largely devoid of H I) pressurized by multiple supernova explosions. The compression of the H I in the shells appears to be limited by the magnetic pressure.

29.4 Optical and UV Absorption Line Studies

The H I gas can also be studied using absorption lines. Lyman α observations measure $N(\mathrm{H\,I})$ directly. O I is a reliable tracer of the H I because (1) most of the oxygen remains in the gas phase, and (2) both O I and H I are the dominant ionization states (and in partially ionized gas, the O I/O II ratio is nearly the same as the H I/H II ratio – see Fig. 14.5). Other species, such as Na I or Ca II are useful for studying the kinematics of the H I, but do not provide accurate information on the total amount of H I.

Measurement of the fine-structure excitation of species such as C I and C II can be used to constrain the density and temperature in the H I. An extensive study of C I excitation in the CNM (see §17.7) found $nT \approx 3800\,\mathrm{cm}^{-3}\,\mathrm{K}$ (Jenkins & Tripp 2011). For the median $T_{\mathrm{spin}} \approx 70\,\mathrm{K}$ found above, this implies $n_{\mathrm{H}} \approx 50\,\mathrm{cm}^{-3}$.

29.5 Infrared Emission

Because the H I gas is dusty, dust can be used a tracer of the ISM. The dust can be detected via reddening and polarization of starlight. It can also be detected through infrared emission. At high galactic latitudes, the dust is heated by diffuse starlight,

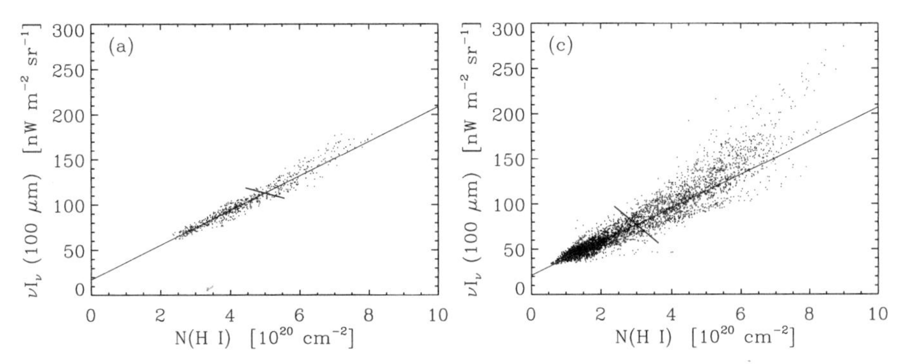

Figure 29.4 100 μm intensity measured by DIRBE, after subtraction of zodiacal emission, plotted against $N(\mathrm{H\,I})$ obtained from 21-cm observations. Left: The north ecliptic pole region ($\ell \approx 124^\circ$, $b \approx 27^\circ$). Right: The region with $|b| > 25^\circ$ and ecliptic latitude $|\beta| > 25^\circ$ covered by the Bell Laboratories H I survey. The best-fit linear relationship is shown in each case; data above and to the right of the short lines intersecting the fitted lines were not used in the fit. From Arendt et al. (1998), reproduced by permission of the AAS.

and the dust temperatures are expected to be similar from one diffuse region to another. If the dust to gas ratio is constant, and the dust temperature is uniform, then the infrared emission will be proportional to the 21-cm emission.

The Diffuse Infrared Background Experiment (DIRBE) on the Cosmic Background Explorer (COBE) satellite made an all-sky map of the 100-μm sky brightness. Figure 29.4 shows the 100-μm intensity, after subtraction of zodiacal emission, versus 21-cm column density, toward high-latitude regions. It is clear that the 100-μm emission and 21-cm emission are highly correlated. The good correlation tells us that (1) dust and gas are well-mixed, and (2) the starlight heating the dust must be fairly uniform. In the right-hand panel there are points with substantial ($\sim$ 30%) excess 100-μm emission relative to H I – these are likely to be diffuse regions where a significant amount of H_2 is present, with the dust in the H_2 contributing to the 100-μm emission.

Chapter Thirty

H I Clouds: Heating and Cooling

Most of the interstellar gas in the Milky Way is neutral, and $\sim 78\%$ of the neutral hydrogen is atomic, or H I. What temperatures do we expect H I gas to be at in a galaxy like the Milky Way? Here, we discuss the heating and cooling processes, and the equilibrium temperature where heating and cooling can balance. We will find that, under some circumstances, more than one equilibrium solution is possible. This leads to a model where the H I in the interstellar medium can be thought of as two distinct "phases," in pressure equilibrium.

30.1 Heating: Starlight, Cosmic Rays, X Rays, and MHD Waves

Possible mechanisms for heating H I regions include

- Ionization by cosmic rays
- Photoionization of H and He by x rays
- Photoionization of dust grains by starlight UV
- Photoionization of C, Mg, Si, Fe, etc. by starlight UV
- Heating by shock waves and other MHD phenomena

The observed spectrum of cosmic rays was discussed in §13.5, where we saw that the primary ionization rate for an H atom is $\zeta_{\rm CR} \gtrsim 7 \times 10^{-18}\,{\rm s}^{-1}$, with a substantially larger rate being allowed by uncertainties regarding the flux of cosmic rays below 1 GeV/nucleon. In §16.4, we saw that observations of H_3^+ appear to indicate a cosmic ray ionization rate $\zeta_{\rm CR} \approx (0.5 - 3) \times 10^{-16}\,{\rm s}^{-1}$ in diffuse molecular gas.

As discussed in §13.5, each "primary" ionization by a cosmic ray creates a "secondary" electron with mean kinetic energy $\sim 35\,{\rm eV}$. Some of this kinetic energy will go into secondary ionization and excitation of bound states of H, H_2, and He that will then deexcite radiatively, but a fraction of the secondary electron energy will ultimately end up as thermal kinetic energy. The heating efficiency depends upon the fractional ionization. If the ionization is high, then the primary electron has a high probability of losing its energy by long-range Coulomb scattering off free electrons, and $\sim 100\%$ of the initial kinetic energy will be converted to heat. However, when the gas is neutral, a fraction of the primary electron energy goes into secondary ionizations or excitation of bound states. In the limit where the ionization fraction $x \to 0$, the fraction of the energy going into heat is ≈ 0.2. The heat per primary ionization in partially ionized atomic gas has been calculated by Dalgarno & McCray (1972). Their numerical results can be approximated by

$$E_h \approx 6.5\,\mathrm{eV} + 26.4\,\mathrm{eV}\left(\frac{x_e}{x_e+0.07}\right)^{1/2} \quad , \quad x_e \equiv \frac{n_e}{n_\mathrm{H}} \quad . \tag{30.1}$$

The heating rate due to cosmic ray ionization is then

$$\Gamma_{\mathrm{CR},n} \approx \left[n(\mathrm{H}^0) + n(\mathrm{He}^0)\right] \zeta_\mathrm{CR} E_h \tag{30.2}$$

$$\approx 1.03 \times 10^{-27} n_\mathrm{H} \frac{\mathrm{erg}}{\mathrm{s}} \left(\frac{\zeta_\mathrm{CR}}{10^{-16}\,\mathrm{s}^{-1}}\right)\left[1 + 4.06\left(\frac{x_e}{x_e+0.07}\right)^{1/2}\right] , \tag{30.3}$$

where we have taken the cosmic ray ionization cross section for He to be similar to H. Cosmic rays also interact with free electrons. The heating rate is

$$\Gamma_{\mathrm{CR},e} \approx A\,\zeta_\mathrm{CR}\,n_e \quad , \tag{30.4}$$

where the coefficient A depends weakly on the CR energy (Goldsmith et al. 1969) with $A \approx 4.6 \times 10^{-10}\,\mathrm{erg}$ if the ionization is dominated by $\sim 50\,\mathrm{MeV}$ protons, as appears likely if the primary ionization rate is as large as $\sim 10^{-16}\,\mathrm{s}^{-1}$ (see Fig. 13.5).

X rays emitted by compact objects or hot interstellar plasma can impinge on neutral regions. Photoelectrons produced by X-ray ionization of H will have energies $E = h\nu - 13.6\,\mathrm{eV}$, and, per primary ionization, secondary ionizations and heating are greater than for cosmic ray ionization. The photoabsorption cross section per H nucleon for a 0.4 keV x ray is $\sim 4 \times 10^{-22}\,\mathrm{cm}^2$, with H, He, and the combined heavy elements (C, O, Ne, Mg, Si) each contributing about $\frac{1}{3}$ of the total. Thus, a 0.4 keV x ray can penetrate a column $N_\mathrm{H} \approx 2.5 \times 10^{21}\,\mathrm{cm}^{-2}$. Higher energies are more penetrating, but lower energy x rays will only be able to heat the surface layer of a neutral cloud.

From Fig. 12.1, we see that the local X-ray background can be approximated by $\nu u_\nu \approx 1 \times 10^{-18} (h\nu/400\,\mathrm{eV})^2\,\mathrm{erg\,cm}^{-3}$ for $(0.4 - 1)\,\mathrm{keV}$. The $h\nu > 0.4\,\mathrm{keV}$ X-ray background contributes some ionization and heating, but at a level well below the effects of cosmic ray ionization. X rays will be an important source of heating only in clouds that happen to be close to strong sources of $\lesssim 200\,\mathrm{eV}$ X-rays.

30.2 Photoelectric Heating by Dust

Photoelectrons emitted by dust grains dominate the heating of diffuse H I in the Milky Way. The work function for graphite is $4.50 \pm 0.05\,\mathrm{eV}$ (e.g., Moos et al. 2001), and, therefore, large neutral carbonaceous grains can in principle be photoionized by photons with energies down to $\sim 4.5\,\mathrm{eV}$. Similarly, the work function for lunar surface material has been measured to be $5.0\,\mathrm{eV}$ (Feuerbacher et al. 1972). However, near-threshold yields are small, and we expect photoelectric heating by dust to be dominated by photons with $h\nu \gtrsim 8\,\mathrm{eV}$.

For an order-of-magnitude estimate of the photoelectric heating rate due to dust, let $n(8-13.6\,\mathrm{eV})$ be the number density of $8 < h\nu < 13.6\,\mathrm{eV}$ photons; let $\langle\sigma_{\rm abs}\rangle$ be the total dust photoabsorption cross section per H nucleon, averaged over the 8 to 13.6 eV spectrum; let $\langle Y\rangle$ be the photoelectric yield averaged over the spectrum of 8 to 13.6 eV photons absorbed by the interstellar grain mixture; let $\langle E_{\rm pe}\rangle$ be the mean kinetic energy of escaping photoelectrons; and let $\langle E_c\rangle$ be the mean kinetic energy of electrons captured from the plasma by grains. Then,

$$\frac{\Gamma_{\rm pe}}{n_{\rm H}} \approx 1.4\times10^{-26}\,\frac{\rm erg}{\rm s}\left[\frac{n(8-13.6\,{\rm eV})}{3\times10^{-3}\,{\rm cm}^{-3}}\right]\frac{\langle\sigma_{\rm abs}\rangle}{10^{-21}\,{\rm cm}^2}\,\frac{\langle Y\rangle}{0.1}\,\frac{(\langle E_{\rm pe}\rangle-\langle E_c\rangle)}{1\,{\rm eV}} . \quad (30.5)$$

For the nominal values of $\langle Y\rangle$ and $(\langle E_{\rm pe}\rangle - \langle E_c\rangle)$ appearing in Eq. (30.5), comparison with Eq. (30.3) shows that photoelectric heating from dust may be *an order of magnitude larger* than the cosmic ray heating rate, Eq. (30.3), even for cosmic ray ionization rates $\zeta_{\rm CR} \approx 10^{-16}\,\mathrm{s}^{-1}$. Photoelectrons from dust grains appear to be the dominant heating mechanism in the diffuse neutral ISM.

Grain charging was discussed in Chapter 25. To model the charging of grains in the ISM or IGM, it is necessary to estimate the photoelectric yield and distribution of photoelectron kinetic energies as a function of grain size, grain charge, and photon energy, for photon energies extending from the ultraviolet to x rays (Weingartner et al. 2006). With this in hand, one must next determine the probability distribution $f(Z)$ for the grain charge, again as a function of grain size and composition. Then, one can evaluate $\Gamma_{\rm pe}/n_{\rm H}$ as a function of n_e, T and the spectrum of the incident radiation field. In diffuse neutral gas, only $h\nu < 13.6\,\mathrm{eV}$ photons need be considered.

Because the photoelectric heating power per volume is proportional to the dust density, the starlight intensity, and a function that depends on the grain charge, we may write

$$\Gamma_{\rm pe} = n_{\rm H}\, G_0 \times g\,(G_0/n_e, T_e) \ , \quad (30.6)$$

where G_0 is a measure of the starlight intensity. The function $g(G_0/n_e, T_e)$ depends on G_0/n_e and T_e because the grain charge depends on these two quantities; g also depends on the composition and size distribution of the grains.

Because small grains account for most of the UV absorption to begin with, and because photoelectric yields $Y(a, h\nu)$ are enhanced for small grains, $\Gamma_{\rm pe}$ is dominated by photoelectrons from very small grains, including the PAHs.

30.3 Cooling: [C II] 158 μm, [O I] 63 μm, and Other Lines

An example of the "cooling function" Λ for predominantly neutral gas, as a function of temperature, is shown in Figure 30.1 for abundances appropriate to diffuse

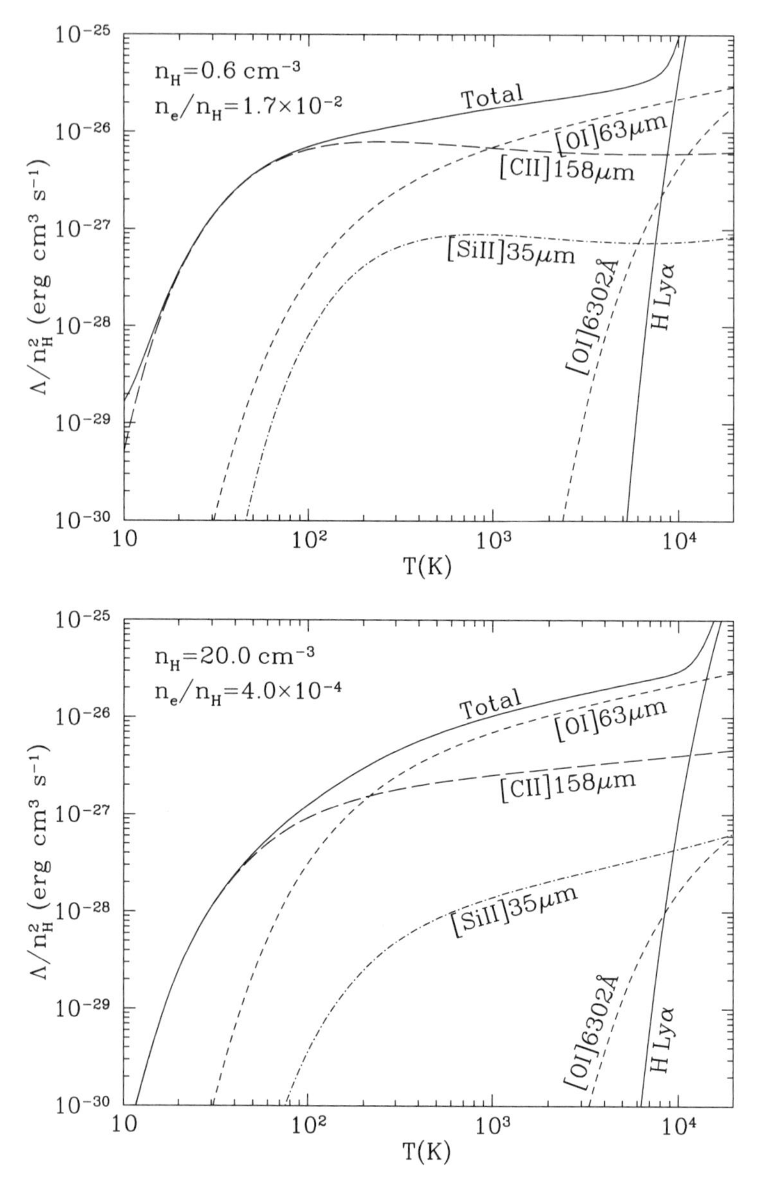

Figure 30.1 Cooling rate for neutral H I gas at temperatures $10 \lesssim T \lesssim 2 \times 10^4$ K for two fractional ionizations. For $T < 10^4$ K, the cooling is dominated by two fine structure lines: [C II]158 μm and [O I]63 μm.

H I in the Milky Way, and for two different fractional ionizations: $x_e = 0.017$ (WNM conditions) and $x_e = 4 \times 10^{-4}$ (CNM conditions). For $10 \lesssim T \lesssim 10^4$ K, the [C II]158 μm fine structure line is a major coolant. The [O I]63 μm fine structure line is important for $T \gtrsim 100$ K. Lyman α cooling dominates only at $T \gtrsim 1 \times 10^4$ K.

The critical densities for [C II]158 μm and [O I]63 μm are $\sim 4 \times 10^3$ cm^{-3} and $\sim 10^5$ cm^{-3}, respectively (see Table 17.1), implying that collisional deexcitation of these levels is unimportant in the diffuse ISM of the Milky Way. Thus, for fixed composition (and ionization fraction x_e), the cooling power per volume $\Lambda \propto {n_{\rm H}}^2 \times \lambda(T)$, where the cooling rate coefficient λ depends only on T.

30.4 Two "Phases" for H I in the ISM

A thermal equilibrium must have heating and cooling balanced, i.e., $\Gamma = \Lambda(T)$. If we vary the H nucleon density $n_{\rm H}$, we can find the thermal equilibrium as a function of $n_{\rm H}$. We include cosmic ray ionization of the H, and we take the oxygen ionization to be coupled to the hydrogen ionization by charge exchange, as discussed in §14.7.1.

The resulting steady state temperatures $T_{\rm eq}$ are plotted versus $n_{\rm H}$ in Figure 30.2(a), with $T_{\rm eq}(n_{\rm H})$ seen to be a monotonically decreasing function of $n_{\rm H}$.

Now we ask the question: if we fix the pressure $p = nkT$, what will be the temperature T where heating and cooling balance? Figure 30.2(b) shows T versus pressure p. At low pressures, heating balances cooling at $T \approx 6000$ K – these are warm neutral medium (WNM) conditions. At high pressures, heating and cooling balance for $T \approx 100$ K – these are cold neutral medium (CNM) conditions. However, there is an intermediate pressure range where, for a given pressure, there are *three* possible solutions. The upper and lower solutions are stable – if the gas temperature is perturbed away from the equilibrium, it will return to it. However, the intermediate solution is *thermally unstable* – if T is perturbed upward, the gas will warm up to the stable WNM solution, and if T is perturbed downward, it will cool to the stable CNM branch. Thus, for our current best estimates of cosmic ray ionization, photoelectric heating, and cooling processes in the diffuse ISM, we conclude that an ISM that is in thermal equilibrium and dynamic equilibrium (uniform pressure) would have diffuse atomic gas in two distinct phases, provided the pressure p falls in the range $3175 \lesssim p/k \lesssim 4425$ cm^{-3} K.

This "two-phase" model of the ISM was first developed by Field et al. (1969). Because their model did not include photoelectric heating by dust, they required a cosmic ray ionization rate $\zeta_{\rm CR} \approx 4 \times 10^{-16}$ s^{-1} to sustain a two-phase medium at pressure $p/k \lesssim 1800$ cm^{-3} K. After the importance of photoelectric heating by dust was recognized, the two-phase model was revisited a number of times (e.g., Draine 1978; Wolfire et al. 1995, 2003) with differing assumptions regarding grain photoelectric heating, grain-assisted recombination, inelastic collision cross sections for cooling processes, and abundances of coolants (particularly C$^+$). The steady-state temperaure and ionization equilibria shown in Figure 30.2 were calcu-

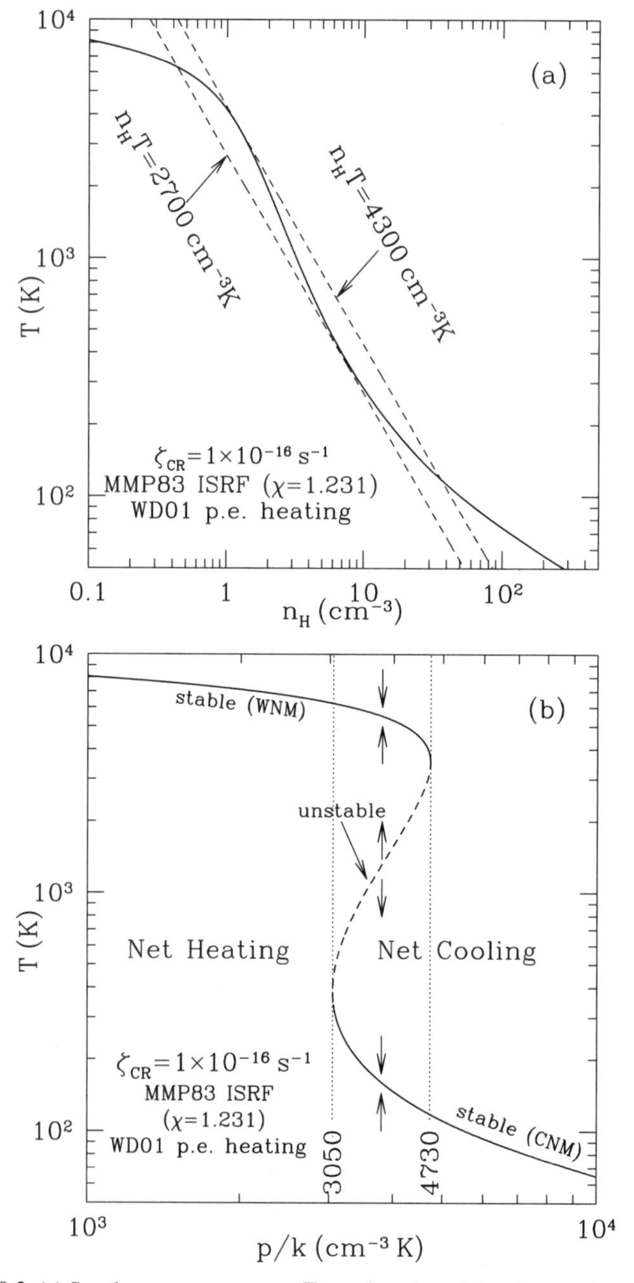

Figure 30.2 (a) Steady state temperature T as a function of density $n_{\rm H}$, for gas heated by cosmic rays and photoelectric heating by dust grains. Two lines of constant $n_{\rm H}T$ are shown. (b) Steady state temperature T as a function of thermal pressure p. For $3200 \lesssim p/k \lesssim 4400\,{\rm cm}^{-3}\,{\rm K}$ there are three possible equilibria – a high-T WNM solution, a low-T CNM solution, and an intermediate temperature equilibrium that is thermally unstable.

lated using the dust photoelectric heating of Weingartner & Draine (2001*d*), with PAHs containing $\sim$25% of the interstellar carbon (60 ppm of C per H, out of a total $\sim$250 ppm); with grain-assisted recombination of H^+ following Weingartner & Draine (2001*b*); with a gas-phase carbon abundance $C/H = 1.0 \times 10^{-4}$ (Sofia & Parvathi 2010); and for cosmic ray ionization rate $\zeta_{CR} = 1 \times 10^{-16}\,s^{-1}$, as indicated by observations of H_3^+ in diffuse molecular clouds (see §16.4).

30.5 Emission Spectrum of an H I Cloud

According to the preceding discussion, we may expect that much of the H I gas in the Milky Way has density and temperature characteristic of the WNM or CNM solutions for $p/k \approx 3800\,cm^{-3}\,K$. In Table 30.1 we list the emission per H nucleon from each phase. Note that the 100 μm emission from dust is the same for the two

Table 30.1 Conditions at Stable Thermal Equilibria for $p/k = 3800\,cm^{-3}\,K$

		CNM	WNM
T(K)		160.	5512
n_H(cm^{-3})		21.5	0.626
n_e(cm^{-3})		0.00925	0.0116
n_e/n_H		0.00043	0.0185
$n(H^+)/n_H$		0.000272	0.0167
$4\pi\nu j_\nu(\mathrm{dust}, 100\,\mu m)/n_H$ ($10^{-26}\,erg\,s^{-1}H^{-1}$)		240.	240.
$4\pi j/n_H$ ($10^{-26}\,erg\,s^{-1}H^{-1}$):	[C II]158 μm	2.85	0.385
	[O I]63.2 μm	2.00	1.05
	[O I]145 μm	0.119	0.0875
	[O I]6302 Å	—	0.0317
	[Si II]34.8 μm	0.0341	0.0474
	[S II]6733 Å	—	0.100
	[S II]6718 Å	—	0.148
	[Fe II]5.34 μm	—	0.0216
	[Fe II]26.0 μm	0.00101	0.00904

phases, as the dust is heated by the same radiation field. The line emission varies considerably between the two phases. For the CNM solution at $T = 160\,K$, the strongest coolant is [C II]158 μm, but for the WNM solution at $T = 5512\,K$, the strongest coolant is [O I]63.2 μm. Note that the $T = 160\,K$ temperature of the CNM solution is above the median spin temperature $T \approx 70\,K$ found in 21-cm studies, but the distribution of spin temperatures in Figure 29.2 shows that it is not uncommon to have temperatures in the 100 to 200 K range. At $T = 160\,K$, [O I]63.2 μm emission is providing 40% of the total cooling. It is also of interest to note that the total line cooling power per H is lower for the WNM solution than for the CNM solution – this is because positive charging of dust grains in the WNM is lowering the dust photoelectric heating rate per H.

Chapter Thirty-one

Molecular Hydrogen

31.1 Gas-Phase Formation of H_2

When two free H atoms, both in the ground electronic state, approach one another, by symmetry there is no electric dipole moment. As a result, there is no electric dipole radiation that could remove energy from the system and leave the two H atoms in a bound state. Electric quadrupole transitions are possible, but the rates are very low. Thus the rate coefficient for $\mathrm{H+H} \rightarrow \mathrm{H_2}+h\nu$ is so small that this reaction can be ignored in astrochemistry. The three-body reaction $\mathrm{3H} \rightarrow \mathrm{H_2} + \mathrm{H} + \mathrm{KE}$ can occur, with the third body carrying off the energy released when H_2 is formed, but the rate for this three-body reaction is negligible at interstellar or intergalactic densities.[1]

The dominant channel for H_2 formation in the gas phase begins by formation of H^- by **radiative association**:

$$\mathrm{H} + e^- \rightarrow \mathrm{H}^- + h\nu \quad , \quad k_{31.1} \approx 1.9 \times 10^{-16} T_2^{0.67} \, \mathrm{cm^3\,s^{-1}} , \tag{31.1}$$

followed by formation of H_2 by **associative detachment**:

$$\mathrm{H}^- + \mathrm{H} \rightarrow \mathrm{H_2}(v, J) + e^- + \mathrm{KE} \quad , \quad k_{31.2} \approx 1.3 \times 10^{-9} \, \mathrm{cm^3\,s^{-1}} \tag{31.2}$$

(Le Teuff et al. 2000); this is an exothermic ion–molecule reaction.

The rate for formation of H_2 by associative detachment is proportional to the density of H^-, which tends to be very low in diffuse regions because formation of H^- by radiative association is slow and destruction of H^- is rapid. In addition to the H_2-forming reaction (31.2), H^- can also be destroyed by reaction with protons: (Moseley et al. 1970)

$$\mathrm{H}^- + \mathrm{H}^+ \rightarrow \mathrm{H} + \mathrm{H} \quad , \quad k_{31.3} \approx 6.9 \times 10^{-7} T_2^{-1/2} \, \mathrm{cm^3\,s^{-1}} \tag{31.3}$$

or other positive ions (Dalgarno & McCray 1972):

$$\mathrm{H}^- + M^+ \rightarrow \mathrm{H} + M \quad , \quad k_{31.4} \approx 4 \times 10^{-7} T_2^{-1/2} \, \mathrm{cm^3\,s^{-1}} \quad , \tag{31.4}$$

[1] At the high densities of a protostar or protoplanetary disk, $\mathrm{3H} \rightarrow \mathrm{H_2} + \mathrm{H}$ is able to convert H to H_2, but at interstellar or intergalactic densities, three-body reactions are extremely slow.

or by photodetachment:

$$\mathrm{H}^- + h\nu \rightarrow \mathrm{H} + e^- \quad , \quad \zeta_{31.5} \approx 2.4 \times 10^{-7} G_0\,\mathrm{s}^{-1} \quad , \tag{31.5}$$

in the interstellar radiation field (Le Teuff et al. 2000). In the diffuse ISM, where $n(\mathrm{H}^+) \approx 0.01\,\mathrm{cm}^{-3}$, most of the H^- that is formed by (31.1) is destroyed by photodetachment, resulting in a very low formation rate for H_2.

In the absence of dust (e.g., in the early universe), $\mathrm{H}^- + \mathrm{H} \rightarrow \mathrm{H}_2 + e^-$ is the dominant channel for forming H_2. Glover et al. (2006) discuss uncertainties in the rate coefficients for destruction of H^-.

31.2 Grain Catalysis of H_2

The dominant process for H_2 formation in the Milky Way and other galaxies is via **grain catalysis**, a process first discussed by Gould & Salpeter (1963) and Hollenbach & Salpeter (1971). The idea is that a first H atom arrives at a grain and becomes bound to the grain surface. Initially, the binding may be weak enough that the H atom is able to diffuse (i.e., random-walk) some distance on the grain surface, until it happens to arrive at a site where it is bound strongly enough that it becomes "trapped" – thermal fluctuations at the low temperature ($T_{\mathrm{gr}} \approx 20\,\mathrm{K}$) of the grain are unable to free it for further exploration of the grain surface. Subsequent H atoms arrive at random locations on the grain surface and undergo their own random walks until they also become trapped, but eventually one of the newly arrived H atoms encounters a previously bound H atom before itself becoming trapped. When the two H atoms encounter one another, they react to form H_2. The energy released when two free H atoms react to form H_2 in the ground state is $\Delta E = 4.5\,\mathrm{eV}$. This energy is large enough to overcome the forces that were binding the two H atoms to the grain, and the H_2 molecule is ejected from the grain surface.

Let

$$\Sigma_{\mathrm{gr}} \equiv \Sigma_{-21} \times 10^{-21}\,\mathrm{cm}^2\mathrm{H}^{-1} \equiv \frac{1}{n_{\mathrm{H}}} \int da \frac{dn_{\mathrm{gr}}}{da} \pi a^2 \tag{31.6}$$

be the total grain geometric cross section per H nucleon. The observed UV extinction reaches values $\tau(\lambda = 0.1\,\mu\mathrm{m}) \approx 2 \times 10^{-21}\,\mathrm{cm}^2/\mathrm{H}$. This suggests that the grain population has

$$\Sigma_{-21} \gtrsim 1 \quad . \tag{31.7}$$

This is a lower bound: Σ_{-21} could conceivably be much larger than the lower bound (31.7) if a large population of very small $a \lesssim 50\,\text{Å}$ grains is present. All grain models that reproduce the interstellar extinction have $\Sigma_{-21} > 1$; models that include PAHs to also explain the observed IR emission have $\Sigma_{-21} \gtrsim 5$, with $\Delta\Sigma_{-21} \approx 0.5$ coming from the $a \gtrsim 0.01\,\mu\mathrm{m}$ grains.

Suppose that a fraction $\epsilon_{\rm gr}$ of the H atoms that collide with a grain in the ISM depart from the grain as H_2 – this fraction may be a function of both grain size a and composition. The rate for H_2 formation via grain catalysis would then be

$$\left(\frac{dn({\rm H}_2)}{dt}\right)_{\rm gr} = R_{\rm gr} n_{\rm H} n({\rm H}) \quad , \tag{31.8}$$

where the effective "rate coefficient" $R_{\rm gr}$ is given by

$$R_{\rm gr} = \frac{1}{2}\left(\frac{8kT}{\pi m_{\rm H}}\right)^{1/2} \langle\epsilon_{\rm gr}\rangle \Sigma_{\rm gr} \quad ; \tag{31.9}$$

the leading factor of $\frac{1}{2}$ is because two H atoms are required to form H_2, and

$$\langle\epsilon_{\rm gr}\rangle \equiv \frac{1}{\Sigma_{\rm gr}} \int da \frac{dn_{\rm gr}}{da} \pi a^2 \epsilon_{\rm gr}(a) \tag{31.10}$$

is the formation efficiency averaged over the grain surface area. Numerically,

$$R_{\rm gr} = 7.3 \times 10^{-17}\,{\rm cm}^3\,{\rm s}^{-1} \left(\frac{T}{100\,{\rm K}}\right)^{1/2} \langle\epsilon_{\rm gr}\rangle \Sigma_{-21} \quad . \tag{31.11}$$

Jura (1975) used ultraviolet spectroscopy of diffuse clouds to determine that $R_{\rm gr} \approx 3 \times 10^{-17}\,{\rm cm}^3\,{\rm s}^{-1}$ in gas with $T \approx 70\,{\rm K}$, which is consistent with Eq. (31.11) if $\langle\epsilon_{\rm gr}\rangle\Sigma_{-21} \approx 0.5$. The silicate-graphite-PAH grain model of Weingartner & Draine (2001*a*) has $\Sigma_{-21} \approx 6.0$; thus for this grain model, the observed $R_{\rm gr}$ would appear to indicate $\langle\epsilon_{\rm gr}\rangle \approx 0.08$. This average could be the result of a very low value of $\epsilon_{\rm gr}$ for the PAHs, which dominate the surface area, and $\epsilon_{\rm gr} \gtrsim 0.5$ for the $a \gtrsim 0.01\,\mu{\rm m}$ "classical" silicate and carbonaceous grains.

31.3 Photodissociation of H_2

Photodissociation ($H_2 + h\nu \rightarrow {\rm H} + {\rm H} + KE$) is the principal process destroying interstellar H_2 in galaxies. The first step in H_2 photodissociation is absorption of a resonance line photon, raising the H_2 from an initial level ${\rm X}(v, J)$ of the ground electronic state ${\rm X}\,^1\Sigma_g^+$ to a level $B(v, J)$ or $C(v, J)$ of the first and second electronic excited states, ${\rm B}\,^1\Sigma_u^+$ and ${\rm C}\,^1\Pi_u$. The original photoexcitation is via a permitted absorption line, and therefore the newly excited level $B(v', J')$ or $C(v', J')$ is guaranteed to have electric dipole-allowed decay channels. In general, the excited level $B(v', J')$ or $C(v', J')$ is most likely to decay to vibrationally excited bound levels $X(v'', J'')$ of the ground electronic state, and such decays occur $\sim 85\%$ of the time. Sometimes, however, spontaneous decay of the excited level $B(v', J')$ will be to the **vibrational continuum** of the ground electronic state: the H_2 molecule will fly apart in $\sim 10^{-14}$ s, separating into two free H atoms. Each

Table 31.1 Photoexcitation and Photodissociation Rates[a] for Unshielded H_2

level ℓ (v,J)	$\zeta_{\rm photoexc,\ell}/\chi$ [b] $(10^{-10}\,{\rm s}^{-1})$	$\zeta_{\rm diss,\ell}/\chi$ $(10^{-11}\,{\rm s}^{-1})$	$\langle p_{\rm diss}\rangle_\ell$
(0,0)	3.08	4.13	0.134
(0,1)	3.09	4.20	0.136
(0,2)	3.13	4.23	0.135
(0,3)	3.15	4.57	0.145
(0,4)	3.21	4.94	0.154
(0,5)	3.26	5.05	0.155

[a] From Draine & Bertoldi (1996).
[b] $\chi \equiv (\nu u_\nu)_{1000\text{Å}}/(4\times 10^{-14}\,{\rm erg\,cm^{-3}})$

electronically excited level u has some probability $p_{{\rm diss},u}$ of spontaneous decay to the vibrational continuum.

The probability per unit time of photoexcitation of H_2 from lower level ℓ to upper level u is given by

$$\zeta_{\ell\to u} = \frac{\pi e^2}{m_e c^2 h} f_{\ell u}\lambda_{\ell u}^3 (u_\lambda)_{\ell u} \quad . \tag{31.12}$$

There are many transitions out of a given lower vibration-rotation level ℓ. The total rate of photoexcitation out of ℓ is

$$\zeta_{{\rm photoexc},\ell} = \sum_u \zeta_{\ell\to u} \quad , \tag{31.13}$$

and the photodissociation rate is obtained by summing over all of the photoexcitation channels, each multiplied by the probability $p_{{\rm diss},u}$ that the upper level will decay to the vibrational continuum:

$$\zeta_{{\rm diss},\ell} = \sum_u \zeta_{\ell\to u} p_{{\rm diss},u} \quad . \tag{31.14}$$

The dissociation probability averaged over the photoexcitation channels is just

$$\langle p_{\rm diss}\rangle_\ell \equiv \frac{\zeta_{{\rm diss},\ell}}{\zeta_{{\rm photoexc},\ell}} \quad . \tag{31.15}$$

In the absence of shielding from the interstellar radiation field, the rates for photoexcitation and photodissociation of H_2 in various (v,J) levels are given in Table 31.1. Note that the unshielded rates $\zeta_{{\rm photoexc},\ell}$ and $\zeta_{{\rm diss},\ell}$ are nearly independent of the level ℓ. The ultraviolet radiation field was taken to be

$$\nu u_\nu = \chi \times \left(4\times 10^{-14}\,{\rm erg\,cm^{-3}}\right)\left(\frac{\lambda}{1000\,\text{Å}}\right) \tag{31.16}$$

over the 1100–912Å wavelength range where H_2 absorbs UV in a neutral region.

Near 1000Å the spectrum (31.16) has a color temperature of 29000K, corresponding to a B0 star. Because H_2 only absorbs strongly over a limited range of wavelengths, the photodissociation rates depend mainly on the intensity near 1000Å measured by

$$\chi \equiv \frac{(\nu u_\nu)_{1000\,\text{Å}}}{4 \times 10^{-14}\,\mathrm{erg\,cm^{-3}}} \tag{31.17}$$

and are insensitive to modest variations in the spectral slope. The scaling factor $\chi = 1$ for the Habing (1968) radiation field, $\chi = 1.71$ for the ISRF of Draine (1978), and $\chi = 1.23$ for the ISRF of Mathis et al. (1983). Thus in the local diffuse neutral ISM the H_2 photodissociation rate $\zeta_{\rm diss} \equiv \langle p_{\rm diss} \rangle \zeta_{\rm photoexc} \approx 4 \times 10^{-11} \chi\,\mathrm{s^{-1}} \approx 5 \times 10^{-11}\,\mathrm{s^{-1}}$.

The steady state abundance of H_2 will be determined by a balance between formation on grains and photodissociation, resulting in a very low steady state abundance:

$$\zeta_{\rm diss} n(\mathrm{H_2}) = R_{\rm gr} n_{\rm H} n(\mathrm{H}) , \tag{31.18}$$

$$\frac{n(\mathrm{H_2})}{n_{\rm H}} = \frac{R_{\rm gr} n(\mathrm{H})}{\zeta_{\rm diss}} \tag{31.19}$$

$$\approx 1.8 \times 10^{-5} \left(\frac{n(\mathrm{H})}{30\,\mathrm{cm^{-3}}} \right) \left(\frac{R_{\rm gr}}{3 \times 10^{-17}\,\mathrm{cm^3\,s^{-1}}} \right) \left(\frac{5 \times 10^{-11}\,\mathrm{s^{-1}}}{\zeta_{\rm diss}} \right) . \tag{31.20}$$

In the absence of self-shielding, diffuse H I clouds will contain only trace amounts of H_2.

31.4 Self-Shielding

Self-shielding refers to the phenomenon where the photoexcitation transitions become optically thick, so that the molecule in question is "shielded" from starlight by other molecules. The H_2 molecule is the most important example of self-shielding. Suppose that the ultraviolet radiation is coming from a single direction, and that the gas between the point of interest and the illuminating stars has column density $N[\mathrm{H_2}(v, J)]$ in the different rotation-vibration levels of H_2. If we ignore the possibility that two different lines may accidentally overlap, then the rate of photoexcitation from level $\ell = \mathrm{X}(v, J)$ to level $u = \mathrm{B}(v', J')$ or $\mathrm{C}(v', J')$ is

$$\zeta_{{\rm photoexc},\ell\rightarrow u} = \left(\frac{c u_\lambda}{hc/\lambda} \right)_{\ell u} \frac{d(W_\lambda)_{\ell u}}{dN_\ell} \tag{31.21}$$

$$= \left(\frac{\lambda c u_\lambda}{hc/\lambda} \right)_{\ell u} \frac{dW_{\ell u}}{dN_\ell} , \tag{31.22}$$

where $W_{\ell u}$ [see Eq. (9.3)] is the dimensionless equivalent width in the $\ell \rightarrow u$

line. In the optically thin limit [see Eq. (9.13)], $dW_{\ell u}/dN_\ell \to (\pi e^2/m_e c^2) f_{\ell u}\lambda_{\ell u}$. Therefore, relative to the optically thin value, the photoexcitation rate for a specific transition $\ell \to u$ is reduced by the self-shielding factor:

$$f_{\text{shield},\ell u} \equiv \frac{dW_{\ell u}/dN_\ell}{(\pi e^2/m_e c^2) f_{\ell u}\lambda_{\ell u}} < 1 \quad , \tag{31.23}$$

where u_λ is the radiation energy density per unit wavelength in the absence of H_2 line absorption. Self-shielding occurs on a line-by-line basis, with stronger self-shielding (i.e., smaller $f_{\text{shield},\ell u}$) for the stronger lines (large oscillator strengths $f_{\ell u}$) from levels with large populations N_ℓ.

The photodissociation rate for H_2 in level ℓ is reduced by self-shielding:

$$\zeta_{\text{diss},\ell} = \frac{\pi e^2}{m_e c^2 h} \sum_u f_{\ell u}\lambda_{\ell u}^3 u_\lambda f_{\text{shield},\ell u}\, p_{\text{diss},u} \quad . \tag{31.24}$$

The photodissociation rate per H_2 is obtained by averaging (31.24) over the populated levels,

$$\zeta_{\text{diss}} = \frac{\sum_\ell n_\ell \zeta_{\text{diss},\ell}}{\sum_\ell n_\ell} \quad .$$

A reasonably accurate approximation is given by (Draine & Bertoldi 1996):

$$\zeta_{\text{diss}} \approx \zeta_{\text{diss},0} f_{\text{shield,diss}} e^{-\tau_{d,1000}} \quad , \tag{31.25}$$

$$f_{\text{shield,diss}} \approx \frac{0.965}{(1+x/b_5)^2} + \frac{0.035}{(1+x)^{0.5}} \exp\left[-8.5\times 10^{-4}(1+x)^{0.5}\right] , \tag{31.26}$$

$$x \equiv \frac{N(\mathrm{H_2})}{5\times 10^{14}\,\mathrm{cm^{-2}}} \quad , \quad b_5 \equiv \frac{b}{\mathrm{km\,s^{-1}}} \quad , \tag{31.27}$$

where $\zeta_{\text{diss},0}$ is the photodissociation rate in the absence of dust extinction or self-shielding, and $\tau_{d,1000}$ is the optical depth for attenuation of the radiation field by dust at 1000 Å.

31.5★ Excitation of Vibration and Rotation by UV Pumping

Photoexcitation to some level $\mathrm{B}(v', J')$ or $\mathrm{C}(v', J')$ will be followed, within a few nanoseconds, by spontaneous decay to some level $\mathrm{X}(v'', J'')$ of the ground electronic state. Every excited level $\mathrm{B}(v', J')$ can decay into a number of different rotation-vibration levels. The $\mathrm{B}(v=2, J=1)$ state, for example, has a probability $p = 0.959$ of decaying to a vibrationally excited level of the ground electronic state. As a result, UV pumping of H_2 populates the vibrationally excited levels of the ground electronic state. At low densities, these vibrationally excited levels

Table 31.2 Einstein A Coefficients and Critical Densities for $H_2(v=0,J)$ at $T=70\,K$

J	$A_{J\rightarrow J-2}$ [a] (s^{-1})	$n_{crit,H}$ [b] (cm^{-3})	n_{crit,H_2} [c] (cm^{-3})
2	2.94×10^{-11}	1.5×10^{3}	4.1×10^{1}
3	4.76×10^{-10}	1.2×10^{4}	9.2×10^{2}
4	2.76×10^{-9}	6.8×10^{4}	2.0×10^{4}
5	9.83×10^{-9}	1.1×10^{6}	3.4×10^{5}

[a] Wolniewicz et al. (1998)
[b] Forrey et al. (1997)
[c] Le Bourlot et al. (1999)

will spontaneously decay to lower vibrational levels via electric quadrupole transitions, usually with $\Delta v=-1$. This "radiative cascade" populates many lower levels, finally reaching the ground vibrational level.

In the low density limit, the radiative cascade process is completely determined by the Einstein A coefficients, and an electronically excited level u has some probability $q(u;v',J')$ of populating level $X(v',J')$ in the course of the radiative cascade. In the absence of collisions, the population of level $X(v,J)$ can, therefore, be calculated from

$$N(v,J)A_{tot}(v,J) = \sum_{\ell} N_{\ell} \sum_{u} \zeta_{pump,\ell\rightarrow u} q(u;v,J) \quad , \tag{31.28}$$

where $A_{tot}(v,J)$ is the total spontaneous decay rate from level (v,J).

31.6★ Rotational Level Populations

The vibrationally excited levels have radiative lifetimes of only $\sim 10^6$ s, and collisional deexcitation by collisions with H, H_2, or He is unlikely at densities $n_H \lesssim 10^4\,cm^{-3}$. In the ground vibrational state, however, the lifetimes of the lowest rotational levels are long enough that collisional effects can play a role in depopulating the lowest J levels. The Einstein A coefficients and critical densities for a few of the rotationally excited levels of H_2 are given in Table 31.2. The populations of the lowest J levels are, therefore, sensitive to the density n_H and temperature T of the gas.

The pumping rate is of course affected by self-shielding, so we cannot discuss the rotational excitation of the H_2 without specifying (1) the ultraviolet intensity in the absence of self-shielding and (2) the amount of H_2 between the point of interest and the ultraviolet source. We must specify not only the total column $N(H_2)$ providing the self-shielding, but also the amount in each rotational level. For illustration, the H_2 providing the shielding is assumed to have a thermal rotational distribution,

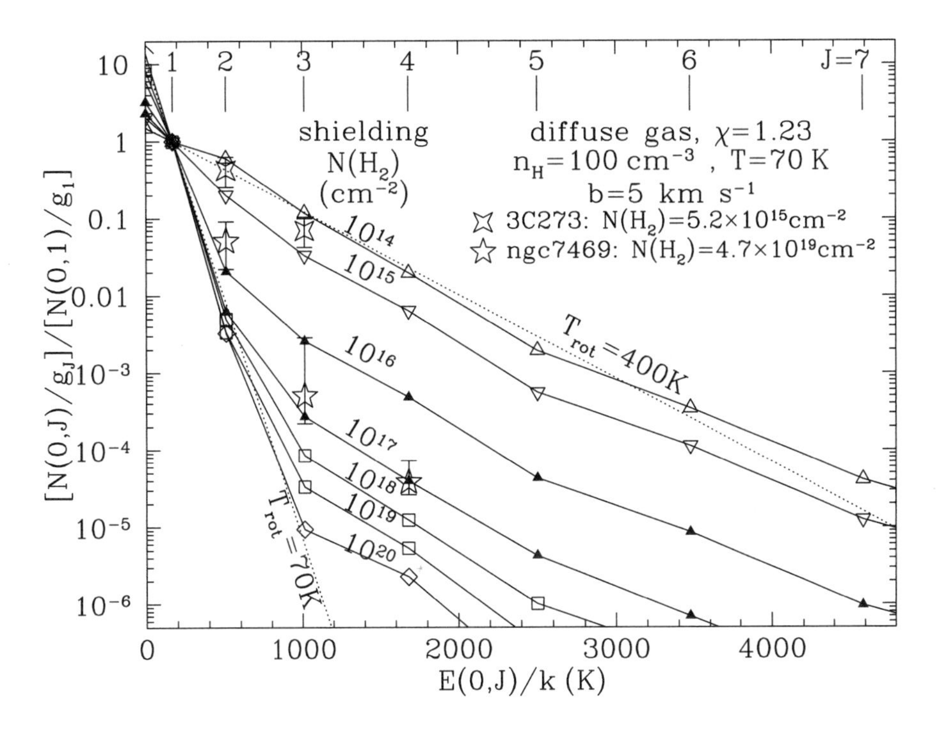

Figure 31.1 Rotational excitation of H_2 in diffuse clouds, for various $N(H_2)$. Also shown is the rotational excitation of H_2 in diffuse clouds falling on sightlines to the AGNs 3C273 and NGC 7469 (Gillmon et al. 2006).

with $T_{rot} = 100\,K$. Figure 31.1 shows the relative rotational level populations for H_2 exposed to the MRN radiation field but shielded by various column densities $N(H_2)$.

This plot shows several interesting features:

- For low levels of self-shielding $[N(H_2) \lesssim 10^{15}\,cm^{-2}]$ the rotational distribution function for $J > 2$ is relatively insensitive to the gas temperature – the rotational excitation for $J \geq 2$ is the result of UV pumping.

- The rotational levels $J \geq 3$ have relative populations that can be approximately characterized by rotational temperature $T_{rot} \approx 400\,K$, but this has nothing to do with the actual kinetic temperature of the gas: it is entirely the result of the branching ratios in the vibration-rotation "cascade" that populates the high J levels.

- As the shielding column density $N(H_2)$ increases, the UV pumping rates decline, and the fraction of H_2 in levels $J > 3$ declines.

- For $N(\mathrm{H}_2) \gtrsim 10^{18}\,\mathrm{cm}^{-2}$, the UV pumping rates are small enough so as to not appreciably raise the abundance of $J = 2$, and the relative populations of levels $J = 0$ and $J = 2$ can be used as a thermometer to estimate the gas temperature:

$$T_{\mathrm{gas}} \approx \frac{510\,\mathrm{K}}{\ln\left[5N(0,0)/N(0,2)\right]} \quad . \tag{31.29}$$

- The ratio of $J=1$ to $J=0$ tends to be *larger* than the thermal equilibrium value. This is because the larger statistical weight of the $J=1$ level leads to $N(J=1) > N(J=0)$, resulting in more effective self-shielding and a reduced photodissociation rate for $J = 1$.

When a sufficiently bright ultraviolet source is located behind a gas cloud, the H_2 rotational-level populations can be determined by the usual techniques of absorption line spectroscopy. Figure 31.1 shows the degree of rotational excitation in two Milky Way gas clouds, using absorption lines in the spectra of background AGNs.

For the two examples shown, the rotational excitation is larger than what was expected for the actual column densities. In part, this is because the observations sum over all the H_2, some of which is closer to the cloud surface, less self-shielded, and therefore more strongly pumped. In the following, we will consider models where we take into account the variation in the pumping rate with distance from the cloud surface.

31.7★ Structure of a Photodissociation Region

Stars are formed out of molecular gas, and when a massive star forms, it may strongly irradiate the remaining molecular clouds with ultraviolet radiation, resulting in photodissociation and photoionization. The photoionized gas, heated to $\sim 10^4$ K, will be overpressured, which will drive a compressive wave (possibly a shock wave) in the molecular cloud, and will also cause the ionized gas to try to flow toward lower-pressure regions nearby.

The interface between the H II region and the dense molecular cloud is called a **photodissociation region**, or **PDR**. It will be bounded by an **ionization front** – the surface where the hydrogen is 50% ionized – and will contain a **photodissociation front** – the surface where the hydrogen is 50% atomic and 50% molecular (by mass). If we adopt a frame of reference in which the photodissociation front is at rest, then the molecular gas will flow toward the photodissociation front where it is dissociated, after which the atomic gas flows away from the photodissociation front toward the ionization front. The structure of a PDR at the interface between an H II region and a dense molecular cloud is illustrated in Fig. 31.2.

If the flow velocities are sufficiently small, the ionization, chemistry, and heating and cooling may all be considered to be in steady state balance. In particular, H_2

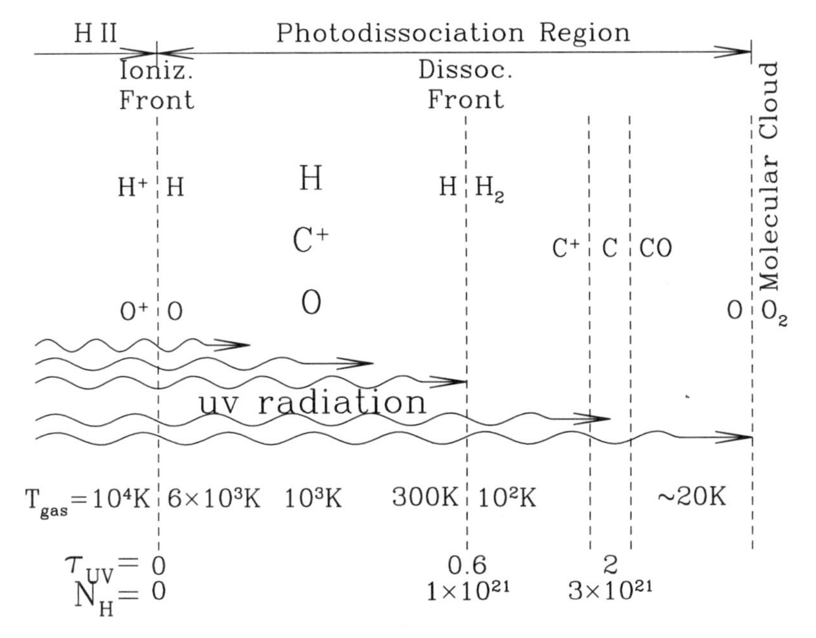

Figure 31.2 Structure of a PDR at the interface between an H II region and a dense molecular cloud.

formation and destruction must locally balance, as in Eq. (31.18). Diffuse molecular clouds have a qualtitatively similar structure, although they may lack both the hottest and the coolest regions shown in Fig. 31.2, depending on whether they are bounded by photoionized gas, on the one hand, and how thick they are, on the other.

Fig. 31.3 shows the profile of a model plane-parallel cloud with an H I/H_2 transition. For simplicity, the cloud is assumed to be illuminated from one side by unidirectional radiation with the energy density and spectrum of the interstellar radiation field. The gas is assumed to be at uniform pressure $p/k = 3000\,\mathrm{cm}^{-3}\,\mathrm{K}$, with a cosmic ray ionization rate $\zeta_{\mathrm{CR}} = 2\times10^{-16}\,\mathrm{s}^{-1}$ and standard dust properties for attenuation of starlight, photoelectric heating, and formation of H_2. The gas is further assumed to be in thermal and chemical equilibrium at each point, with heating = cooling, ionization = recombination, and H_2 destruction = formation. As the radiation field entering from the left is attenuated by dust, the gas makes a transition from the warm (WNM) phase to the cool (CNM) phase.

The H_2 abundance in the WNM is very low, $\lesssim 10^{-6}$. The steady state H_2 abundance rises as one enters the CNM phase, as a result of both the increased gas density (promoting H_2 formation), and growing self-shielding (lowering the photodissociation rate). The zone where the gas is more than 50% atomic has $N_{\mathrm{H}} = 3.9 \times 10^{20}\,\mathrm{cm}^{-2}$, and a dust column with $E(B-V) = 0.066\,\mathrm{mag}$, $A_V = 3.1E(B-V) = 0.2\,\mathrm{mag}$. The H_2 in the cloud is undergoing UV pumping, which

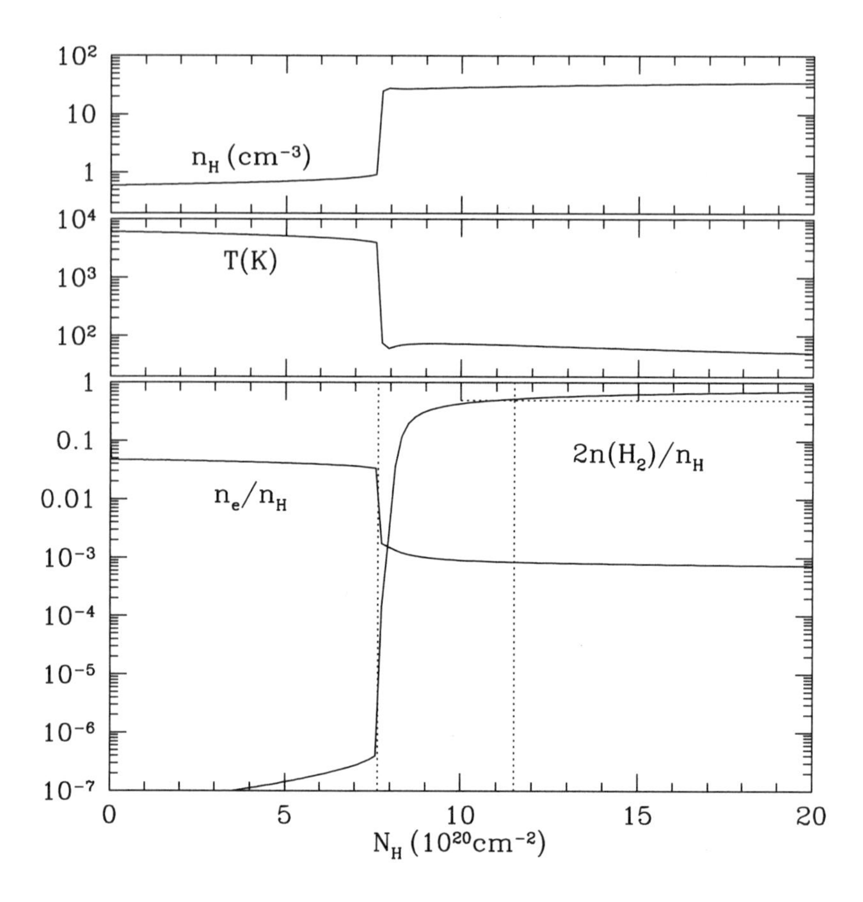

Figure 31.3 Profile of the H I/H_2 transition in a diffuse molecular cloud, for an assumed pressure $p/k = 3000\,\mathrm{cm}^{-3}\,\mathrm{K}$. The dotted lines delimit the surface layer of cool gas where more than 50% of the hydrogen is H I.

results in destruction of the H_2 ~ 15% of the time; the remaining ~ 85% of the UV excitations create a population of rotationally excited H_2 in the cloud.

Ultraviolet spectroscopy of extragalactic sources frequently shows absorption lines from H_2 in diffuse gas in the Galaxy. This diffuse H_2 will usually be excited by an ultraviolet radiation field resembling the local ISRF, and hence we expect rotationally excited H_2 to be present in these clouds due to the effects of UV pumping. Figure 31.4 shows the column densities in the $J = 2, 3, 4, 5$ rotational levels of the ground vibrational state, plotted against the total H_2 column density. The solid curves are adapted from the model cloud in Fig. 31.3, where we show the integrated columns of $(2/\cos\theta) \times N(\mathrm{H}_2, J)$ versus $(2/\cos\theta) \times N(\mathrm{H}_2, \mathrm{total})$ integrated from the WNM to a point within the cloud to simulate clouds of varying

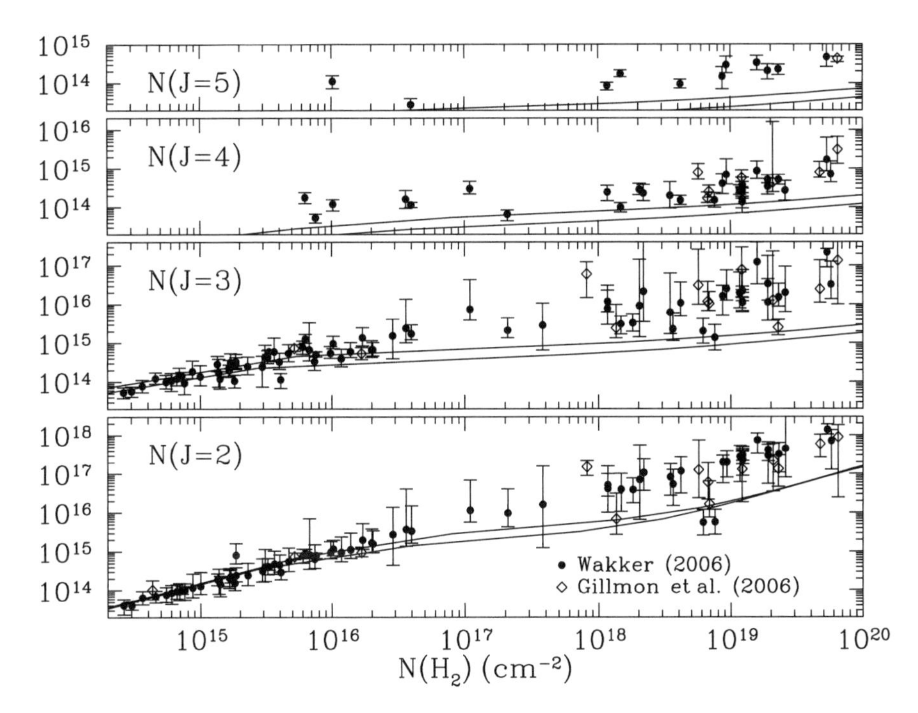

Figure 31.4 Solid curve: H_2 column densities $N(J)$ for rotational levels $J = 2, 3, 4, 5$ for diffuse clouds of varying total thickness, for the pressure and UV illumination assumed in Fig. 31.3. $N(H_2)$ is the total H_2 column density through the cloud. In each panel, the lower curve is for sightlines normal to the cloud, and the upper curve is for sightlines 60° away from the normal. Also shown are $N(J)$ in diffuse clouds observed toward various AGNs (Wakker 2006; Gillmon et al. 2006). The observed column densities of excited H_2 are often significantly higher than predicted by the UV pumping model.

total thicknesses. The factor of 2 allows for the fact that a cloud will have two surfaces; the factor of $1/\cos\theta$ allows for inclination of the plane-parallel cloud relative to the sightline to the AGN. The curves shown are for $\cos\theta = 0$ and 0.5.

Also shown are column densities measured for a number of AGN sightlines (Wakker 2006; Gillmon et al. 2006). For $N(H_2, \mathrm{total}) \lesssim 10^{17}\,\mathrm{cm}^{-2}$, the observed column densities of rotationally excited H_2 appear to be in agreement with the UV pumping model, whereas clouds with $N(H_2) \gtrsim 10^{18}\,\mathrm{cm}^{-2}$ almost always show more rotationally excited H_2 than predicted by the UV pumping model. In clouds with $N(H_2, \mathrm{total}) \gtrsim 10^{17}\,\mathrm{cm}^{-2}$, the amount of H_2 in the $J = 3$ level is often an order of magnitude larger than predicted by UV pumping alone.

The reason for this discrepancy is unknown; one possibility is that a large fraction of diffuse clouds contain small regions with high gas temperatures $T \gtrsim 500\,\mathrm{K}$ where the rotational levels $J = 3, 4, 5$ can be collisionally excited. This interpretation requires some mechanism to heat the gas locally; MHD shock waves or intermittent decay of strong turbulence are two possibilities.

31.8 Dense PDRs

In star-forming galaxies, an appreciable fraction ($\sim 10\%$) of the total luminosity of the galaxy is reprocessed through dense PDRs at the interface between molecular clouds and H II regions. Here, energy originally radiated by hot stars is absorbed by molecules and dust grains in the PDR, and reradiated at longer wavelengths as IR emission from dust and PAHs, and line emission from atoms and molecules in the gas. Part of the starlight energy goes into changing the physical state of the gas from cold and molecular to hot, photodissociated, and possibly photoionized if an ionization front is present.

The Orion Bar (see §28.3) is an example of a dense PDR. Moving outward from θ^1 Ori C, it includes a high pressure layer of photoionized gas, an ionization front, a photodissociation zone where the hydrogen is neutral but primarily atomic, and a photodissociation front. Tielens et al. (1993) provide a nice overview of the Orion Bar.

The overall physics and chemistry of PDRs is complex – see the review by Hollenbach & Tielens (1999). There are a number of outstanding issues, such as whether clumpiness of the gas is of major importance, how the dust population evolves as the photodissociation and photoionization fronts approach and pass by, and the processes, such as photoelectric heating, responsible for heating the gas in the PDR. For example, Allers et al. (2005) discuss the heating of the gas and evidence for evolution of the grain properties in the Orion Bar PDR.

Chapter Thirty-two

Molecular Clouds: Observations

32.1 Taxonomy and Astronomy

Molecular gas is abundant in star-forming galaxies like ours, and occurs over a very wide range of densities. Individual clouds are separated into categories based on their *optical appearance*: diffuse, translucent, or dark, depending on the visual extinction A_V through the cloud, as shown in Table 32.1.

Table 32.1 Cloud Categories

Category	A_V (mag)	Examples
Diffuse Molecular Cloud	$\lesssim 1$	ζ Oph cloud, $A_V = 0.84$ [a]
Translucent Cloud	1 to 5	HD 24534 cloud, $A_V = 1.56$ [b]
Dark Cloud	5 to 20	B68 [c], B335 [d]
Infrared Dark Cloud (IRDC)	20 to $\gtrsim 100$	IRDC G028.53-00.25 [e]

[a] van Dishoeck & Black (1986).
[b] Rachford et al. (2002).
[c] Lai et al. (2003).
[d] Doty et al. (2010).
[e] Rathborne et al. (2010).

Diffuse and **translucent** clouds have sufficient ultraviolet radiation to keep gas-phase carbon mainly photoionized throughout the cloud. Such clouds are usually pressure-confined, although self-gravity may be significant in some cases. The typical **dark cloud** has $A_V \approx 10\,\mathrm{mag}$, and is self-gravitating. Some dark clouds contain dense regions that are extremely opaque, with $A_V \gtrsim 20$ mag. In some cases, dark clouds with $A_V \gtrsim 10^2$ mag are observed; these **infrared dark clouds** (**IRDC**s) are opaque even at $8\,\mu\mathrm{m}$, and can be seen in silhouette against a background of diffuse $8\,\mu\mathrm{m}$ emission from PAHs in the ISM (see Plate 11).

The terminology in Table 32.1 describes the total surface density of the cloud, in terms of the visual extinction A_V through the cloud. Because molecular clouds do not form a one-parameter family, terminology has developed to describe other characteristics of the clouds. Unfortunately, the terminology has not been standardized, and different investigators may use the terms "clump" and "core" differently. We follow the usage outlined by Bergin & Tafalla (2007).

The **giant molecular cloud (GMC)** and **dark cloud** categories are distinguished mainly by total mass. Groups of distinct clouds are referred to as **cloud complexes**. Structures within a cloud (self-gravitating entities) are described as **clumps**.

Table 32.2 Terminology for Cloud Complexes and Their Components

Categories	Size (pc)	$n_{\rm H}$ (cm^{-3})	Mass ($M_\odot$)	Linewidth (km s^{-1})	A_V (mag)	Examples
GMC Complex	$25-200$	$50-300$	$10^5-10^{6.8}$	$4-17$	$3-10$	M17, W3, W51
Dark Cloud Complex	$4-25$	10^2-10^3	$10^3-10^{4.5}$	$1.5-5$	$4-12$	Taurus, Sco-Oph
GMC	$2-20$	10^3-10^4	$10^3-10^{5.3}$	$2-9$	$9-25$	Orion A, Orion B
Dark Cloud	$0.3-6$	10^2-10^4	$5-500$	$0.4-2$	$3-15$	B5, B227
Star-forming Clump	$0.2-2$	10^4-10^5	$10-10^3$	$0.5-3$	$4-90$	OMC-1, 2, 3, 4
Core	$0.02-0.4$	10^4-10^6	$0.3-10^2$	$0.3-2$	$30-200$	B335, L1535

Clumps may or may not be forming stars; in the former case they are termed **star-forming clumps**. **Cores** are density peaks within star-forming clumps that will form a single star or a binary star. Table 32.2 gives representative properties for the different categories.

Molecular clouds are sometimes found in isolation, but in many cases molecular clouds are grouped together into **complexes**. Since large clouds generally have substructure, the distinction between "cloud" and "cloud complex" is somewhat arbitrary. Delineation of structure in cloud complexes is guided by the intensities and radial velocities of molecular lines (e.g., CO $J = 1-0$) as well as maps of thermal emission from dust at submm wavelengths. Table 32.2 provides a guide to the terminology.

Much of the molecular mass is found in large clouds known as "giant molecular clouds" (GMCs), with masses ranging from $\sim 10^3\, M_\odot$ to $\sim 2\times10^5\, M_\odot$. These have reasonably well-defined boundaries, but the molecular gas within them has considerable substructure.

A **GMC complex** is a gravitationally bound group of GMCs (and smaller clouds) with a total mass $\gtrsim 10^{5.3}\, M_\odot$. The largest GMC complexes have masses $\sim 6\times 10^6\, M_\odot$.

The nearest example of a GMC complex is the Orion Molecular Cloud (OMC) complex, with a total mass $M \approx 3\times10^5\, M_\odot$, located $\sim 414\,\rm pc$ from the Sun. A map of the distribution of molecular gas in the OMC complex is shown in Figure 32.1. There are six GMCs shown on the map, three of which (Orion A, Orion B, and Northern Filament) form the Orion GMC complex; the other three GMCs on the map have different radial velocities and are thought to be background objects.

For the currently favored distance $d = 414\,\rm pc$, the Orion A, Orion B, and Northern Filament GMCs have virial masses $1.2{\times}10^5\, M_\odot$, $0.6{\times}10^5\, M_\odot$, and $0.8{\times}10^5\, M_\odot$ if magnetic fields are neglected. If magnetic fields are dynamically important, as appears to be the case, the virial mass estimates will increase by a factor of up to ~ 2. The GMCs are embedded within a lower density H I envelope, with a total H I mass $\sim 6\times10^4\, M_\odot$. The Orion A GMC, the most massive of the three GMCs in the Orion complex, hosts the famous Orion Nebula (M42 = NGC1976) H II region (see Chapter 28). Plates 13 and 14 show the Orion Nebula; the dust around it is made visible both by scattering light and by obscuring some parts of both M 42 and M 43.

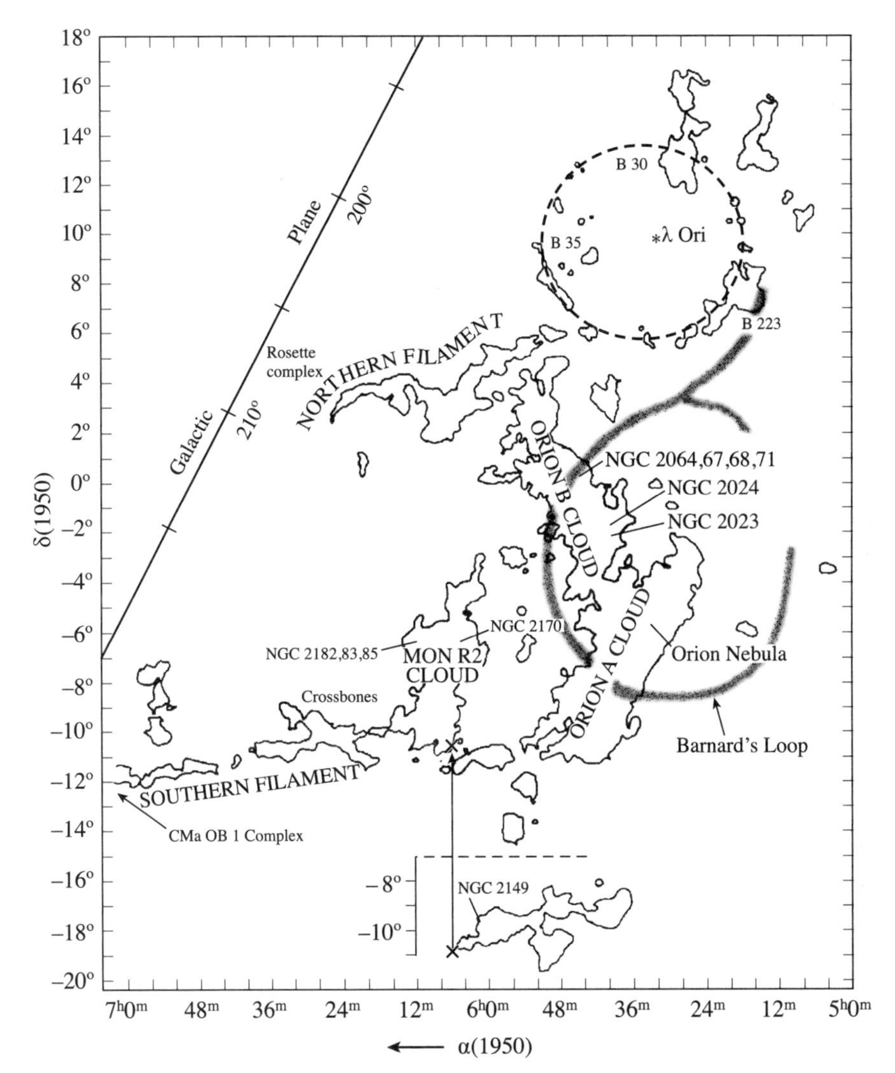

Figure 32.1 Schematic diagram showing the boundaries of molecular clouds in the Orion region. There are three GMCs forming the Orion GMC complex: Orion A, Orion B, and the Northern Filament, with virial masses $\sim 1.2\times10^5\,M_\odot$, $0.6\times10^5\,M_\odot$, and $0.8\times10^5\,M_\odot$. The Orion A and Orion B clouds have similar radial velocities and may be connected. The Orion Nebula is associated with the Orion A cloud. The NGC 2149 cloud, the Southern Filament, and Mon R2 are thought to be background features not associated with Orion A or Orion B. For clarity, the NGC 2149 cloud is shown 8° south of its actual location. The shaded arc is Barnard's Loop, seen in Hα (and visible in Plate 3b). After Maddalena et al. (1986).

Each of the GMCs in Orion contains a number of clumps. In projection, the Orion A cloud is $\sim 20\,\mathrm{pc}\times 75\,\mathrm{pc}$. About 50% of the total mass of Orion A is in the Orion A Ridge, a $\sim 3\,\mathrm{pc}\times 32\,\mathrm{pc}$ filament of enhanced density running along

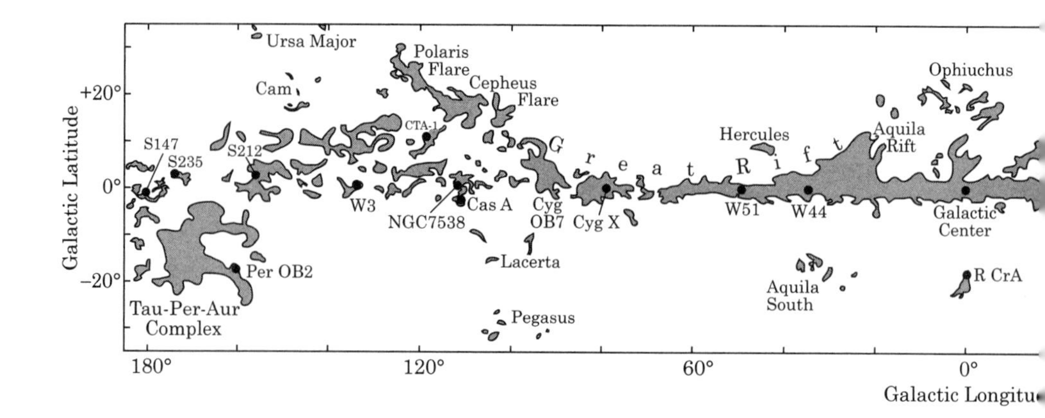

Figure 32.2 Locations of prominent molecular clouds along the Milky Way. From Dame et al. (2001), reproduced by permission of the AAS.

the long axis of the cloud. There are a number of density peaks, or clumps, along this filament. The most massive is OMC-1 ($\sim 10^3\, M_\odot$) centered behind the Orion H II region. OMC-2 and OMC-3, with masses $\sim 10^2\, M_\odot$, are located ~ 1.7 pc and 2.8 pc N of OMC-1, and OMC-4 is located ~ 1.4 pc S of OMC-1.

OMC-1 appears to be the site of the most vigorous current star formation in the Orion A molecular cloud, containing within it a cluster of young stars with total luminosity $L \approx 10^5\, L_\odot$. The most luminous sources in OMC-1 are the Becklin-Neugebauer object, a B3-B4 star (8–12 $M_\odot$, $L \approx 2500 - 10^4\, L_\odot$), and Source I, a heavily obscured star or protostar with $L \approx 5\times10^4\, L_\odot$. Source I appears to be responsible for a spectacular high velocity outflow in OMC-1, expanding radially outward, and visible in line emission from vibrationally excited H_2, rotationally excited OH, and high-J CO. Genzel & Stutzki (1989) give an excellent review of the molecular gas and star-formation in the Orion GMC complex.

Figure 32.2 shows the location of prominent molecular clouds projected onto the sky, and Fig. 32.3 shows the distribution of molecular clouds within 1 kpc of the Sun, projected onto the disk. The nearest molecular clouds are the Taurus Molecular Cloud complex, at a distance of ~ 140 pc, and the R Cor A, ρ Oph, and Lupus clouds, at $D \approx 150$, 165, and 170 pc, respectively. The Taurus, Lupus, and ρ Oph clouds each have $M \approx 3\times10^4\, M_\odot$; the R Cor A cloud is considerably less massive, with $M \approx 3\times10^3\, M_\odot$.

CO line surveys can detect GMCs at large distances, allowing the total number in the Galaxy to be estimated. Excluding the molecular material within a few hundred pc of the Galactic Center, the overall mass distribution of GMCs in the Milky Way

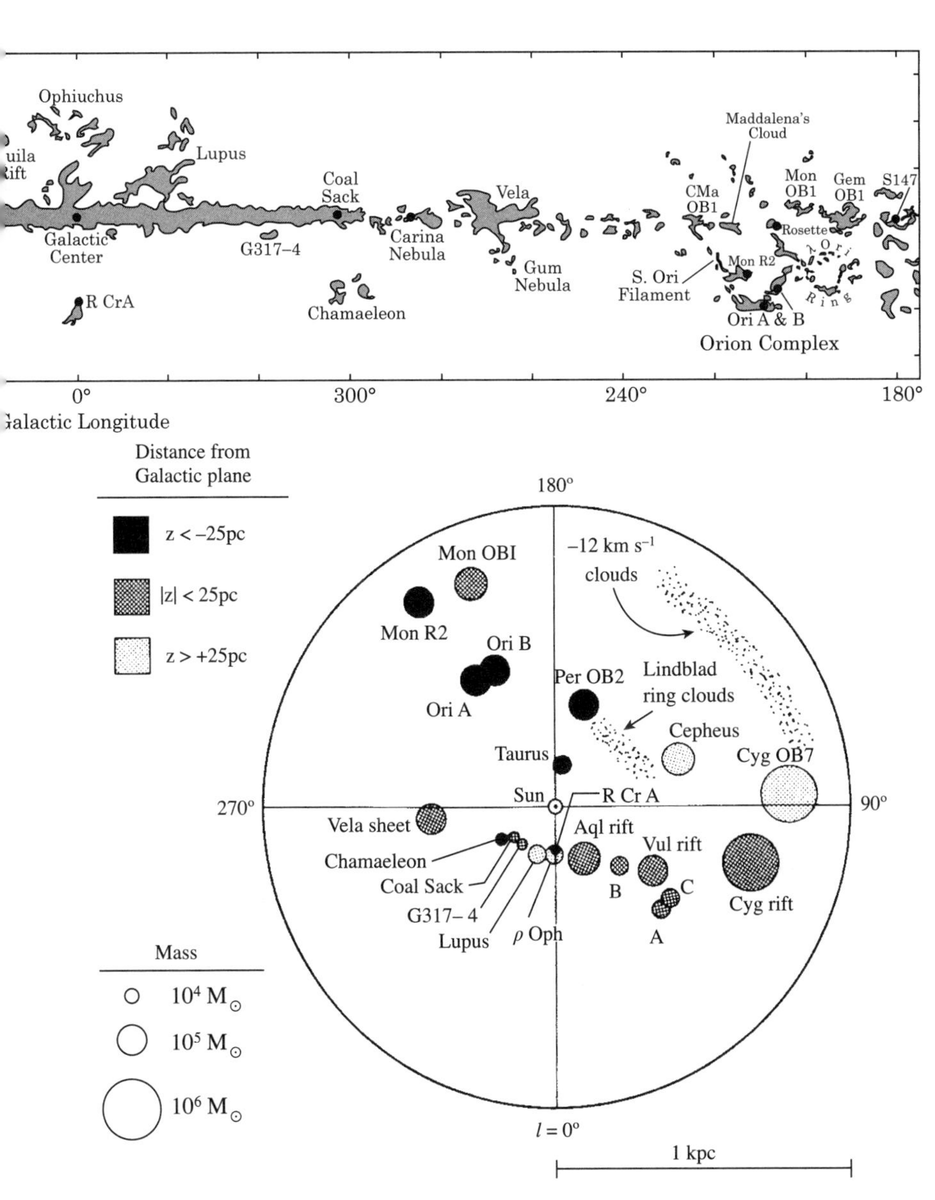

Figure 32.3 Molecular clouds within 1 kpc of the Sun. From Dame et al. (1987), reproduced by permission of the AAS.

can be approximated by a power-law:

$$\frac{dN_{\rm GMC}}{d\ln M_{\rm GMC}} \approx N_u \left(\frac{M_{\rm GMC}}{M_u}\right)^{-\alpha} \quad \text{for } 10^3\,M_\odot \lesssim M_{\rm GMC} < M_u \ , \qquad (32.1)$$

with $M_u \approx 6 \times 10^6\, M_\odot$, $N_u \approx 63$, and $\alpha \approx 0.6$ (Williams & McKee 1997). For the distribution (32.1), most of the mass is in the most massive GMCs: $\sim$80% of the molecular mass is in GMCs with $M > 10^5\, M_\odot$.

Plate 8 shows the distribution of GMCs and GMC complexes in the face-on spiral galaxy M51, with numerous GMC complexes with masses $M > 10^7\, M_\odot$ (Koda et al. 2009).

32.2 Star Counts

Molecular clouds were originally discovered by star counts: Herschel (1785) noticed that there were patches along the Milky Way where very few stars were seen. Herschel incorrectly attributed this to a real absence of stars; we now understand that the apparent deficiency of stars is the result of obscuration by dusty clouds. Star counts using background stars continue to be a good way to study the structure of these regions. Because the visual obscuration can be very large, studies of dark clouds using star counts are now usually done in the J, H, or K bands (e.g., the study of the Pipe Nebula by Lombardi et al. 2006).

32.3 Molecular Radio Lines

The most common way to study molecular gas is through molecular line emission, and the primary line used is the $J = 1 \rightarrow 0$ transition of CO. This transition is often optically thick, but, as discussed in §19.6, the CO 1–0 luminosity of a cloud is approximately proportional to the total mass.

Velocity-resolved mapping of CO 1–0 together with an assumed rotation curve and an adopted value of the "CO to H_2 conversion factor" X_{CO} have been used to infer the surface density of H_2 over the Milky Way disk (Nakanishi & Sofue 2006). The H_2 surface density shown in Figure 32.4 was obtained for the value of $X_{CO} = 1.8 \times 10^{20} H_2\,\mathrm{cm}^{-2}/\,\mathrm{K\,km\,s}^{-1}$ recommended by Dame et al. (2001). It is important to keep in mind that the value of X_{CO} is quite uncertain (see Chapter 19).

Table 32.3 gives the total masses of H I and H_2 in the Milky Way from the analysis of Nakanishi & Sofue (2003, 2006). Molecular gas accounts for $\sim$22% of the

Table 32.3 Mass of H I and H_2 in the Milky Way

Phase	$M(10^9\, M_\odot)$
Total H II (not including He)	1.1
Total H I (not including He)	2.9
Total H_2 (not including He)	0.84
Total H II, H I and H_2 (not including He)	4.8

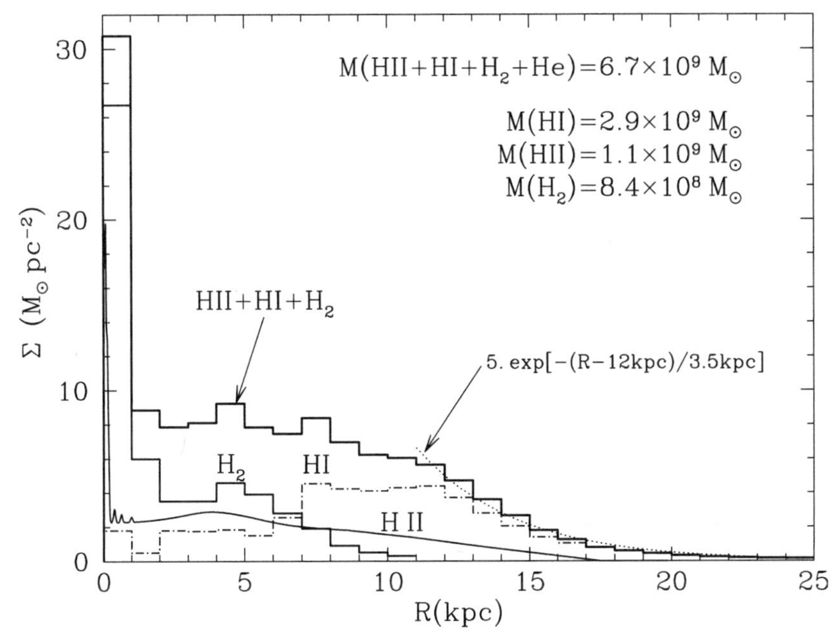

Figure 32.4 Gas surface densities Σ as a function of galactocentric radius R. The Sun is assumed to be at $R = 8.5\,\mathrm{kpc}$. **H_2:** Surface density of H_2 estimated from CO 1–0 observations (Nakanishi & Sofue 2006). **H II:** Surface density of H II derived from pulsar dispersion measures (Cordes & Lazio 2003) (see Fig. 11.4). **H I:** Surface density of H I from 21-cm studies (Nakanishi & Sofue 2003). **H II + H I + H_2:** Total gas surface density. Beyond ~11 kpc, the total gas surface density (dominated by H I) declines approximately exponentially, with a ~3.5 kpc scale length.

mass of the ISM in the Milky Way, contributing a mass $M(\mathrm{H_2}) \approx 8.4 \times 10^8\,M_\odot$. For comparison, the total molecular gas mass of M31 is $\sim 3.6 \times 10^8\,M_\odot$ (Nieten et al. 2006, using $X_{\mathrm{CO}} = 1.9 \times 10^{20}\mathrm{H_2\,cm^{-2}/\,K\,km\,s^{-1}}$) – about 40% of the molecular mass in the Milky Way.

32.4 FIR Emission from Dust

The CO 1–0 line is the classic tracer of molecular gas, but the observed line intensity is usually limited by radiative trapping effects, and estimation of the total molecular mass requires adoption of a value for the X_{CO} factor relating CO $J = 1 \rightarrow 0$ luminosity to H_2 mass. As we have seen previously (§19.6), the actual value of X_{CO} should, in principle, depend both on cloud density and on the excitation temperature of the CO. Cloud mass estimates based on the CO 1–0 luminosity must therefore be treated with caution.

We would like an independent way to estimate masses of molecular clouds. One way is through the far-infrared and submillimeter continuum emission from dust grains. Except in dark clouds where grains acquire ice mantles, measured depletions of elements like C, Mg, Si, and Fe from the gas show that H I clouds and H_2 clouds have very similar dust/gas mass ratios. At wavelengths $\lambda \gtrsim 300\,\mu\text{m}$, this emission is generally optically thin, so that radiative transfer corrections are unnecessary, and the *dust* mass $M_{\rm dust} = F_\nu D^2/\kappa_\nu B_\nu(T_{\rm dust})$, where κ_ν is the dust **opacity** (absorption cross section per unit mass of dust) at frequency ν, and $B_\nu(T)$ is the blackbody function. From the shape of the emission spectrum, or our general expectations regarding dust temperatures, we can usually estimate $T_{\rm dust}$ to within a factor 1.3 (say). If we are in the Rayleigh-Jeans limit, $B_\nu(T) \propto T$, then the main uncertainty is the value of the dust opacity κ_ν. If the dust were identical to that in H I clouds, then we could use κ_ν determined from observations of the H I "cirrus," which has been reasonably well-determined from comparisons of sub-mm emission and H I 21-cm emission. However, the dust in molecular clouds could differ in composition from dust in diffuse clouds, and, in principle, κ_ν might also depend on the temperature of the dust. Nevertheless, in practice, sub-mm observations are a good way to estimate the total mass of dust present in a molecular cloud.

32.5 γ rays

There are four principal channels for production of γ rays by the ISM:

$$\begin{aligned} {\rm CR}p + p &\rightarrow {\rm CR}p + p + \pi^0 \,, \\ &\qquad\qquad \pi^0 \rightarrow 2\gamma \quad \text{(pion decay)}, \qquad (32.2) \\ {\rm CR}e + p &\rightarrow {\rm CR}e + p + \gamma \qquad \text{(bremsstrahlung)}, \qquad (32.3) \end{aligned}$$

$${\rm CR}e + \left\{ \begin{array}{c} \text{CMB} \\ \text{FIR from dust} \\ \text{starlight} \end{array} \right\} \rightarrow {\rm CR}e + \gamma \qquad \text{(inverse Compton)}, \qquad (32.4)$$

$$e^+ + e^- \rightarrow 2\gamma \;\text{ or }\; 3\gamma \qquad (e^+e^- \text{ annihilation}), \qquad (32.5)$$

where CRp is a cosmic ray proton (or He nucleus), and CRe is a cosmic ray electron. The first two channels arise from cosmic rays colliding with interstellar gas. If the density of cosmic ray nuclei and electrons is assumed to be spatially uniform, then the γ-ray intensity from a given direction will have a component proportional to the mass surface density $\int \rho dr$. By calibrating on H I clouds where we can determine $\int \rho dr$ from 21-cm observations, we use the observed γ-ray intensity to determine the mass surface density $\int \rho dr$ of a molecular cloud. This relies on the assumption that the cosmic ray density inside the molecular cloud is the same as in diffuse regions.

Comparing cloud surface densities inferred from $E > 70\,\text{MeV}$ γ-ray images with H I 21-cm and CO 1–0 images, Bloemen et al. (1986) estimated $X_{\rm CO} =$

$2.8 \times 10^{20} \mathrm{H_2\,cm^{-2}/(K\,km\,s^{-1})}$. Subsequent work using 1-30 MeV data from COMPTEL gave $X_{CO} = (1.26 \pm 0.3) \times 10^{20} \mathrm{H_2\,cm^{-2}/(K\,km\,s^{-1})}$ (Strong et al. 1994), but the result from the higher-energy data may be more reliable as the cosmic rays required to produce it have higher penetrating power. Observations with the Fermi Gamma-Ray Space Telescope of $E > 100\,\mathrm{MeV}$ γ rays from the Orion molecular cloud have recently been used to estimate $X_{CO} = (1.76 \pm 0.04) \times 10^{20} \mathrm{H_2\,cm^{-2}/(K\,km\,s^{-1})}$ for the Orion A GMC, and $X_{CO} = (1.27 \pm 0.06) \times 10^{20} \mathrm{H_2\,cm^{-2}/(K\,km\,s^{-1})}$ for the Orion B GMC (Okumura et al. 2009). The value of X_{CO} for Orion A agrees well with $X_{CO} = (1.8 \pm 0.3) \times 10^{20} \mathrm{H_2\,cm^{-2}/(K\,km\,s^{-1})}$ determined using infrared emission from dust as the mass tracer (Dame et al. 2001). The discrepancy between the values of X_{CO} for Orion A and Orion B is very puzzling.

32.6★ Compact, Ultracompact, and Hypercompact H II Regions

Most of the gas and dust in a GMC is relatively cold. However, most GMCs have already had some star formation prior to the time when we observe them. The most conspicuous sites of recent star formation will be those where one or more massive ($M \gtrsim 30 M_\odot$) O-type or B-type stars have recently formed. The ionizing photons from an O-type or early B-type star will create an H II region, which will initially be very small because the gas is dense and dusty. In order of decreasing density, H II regions in dense clouds are termed **hypercompact** ($n_e \gtrsim 10^6\,\mathrm{cm^{-3}}$), **ultracompact** ($10^5 - 10^6\,\mathrm{cm^{-3}}$), or **compact** ($10^4 - 10^5\,\mathrm{cm^{-3}}$).

These objects were first discovered as compact sources of free–free radio emission, but are very bright in the far-infrared and mid infrared, and are now easily detected as bright infrared sources.

From the theory of dusty H II regions (see §15.4), it is clear that if the dust to gas ratio is anything like "normal," then an ultracompact or hypercompact H II region is expected to be strongly affected by radiation pressure, and should exhibit a shell-like morphology if it is static. In dense H II regions, we expect the dust to absorb a significant fraction of the ionizing radiation, as well as a substantial fraction of the recombination radiation, particularly Lyman α [see Eq. (15.53)]. Because dense H II regions are small, the radiation intensities are high and the dust can be quite warm. Thus, these regions stand out as sources that are bright at $24\,\mu\mathrm{m}$ or even $10\,\mu\mathrm{m}$.

If the stellar source of ionizing radiation is stationary relative to the gas, then the time scale for expansion of a hypercompact or ultracompact H II region is very short – of order the radius of the dusty Strömgren sphere divided by the $\sim 15\,\mathrm{km\,s^{-1}}$ sound speed – and we would expect these objects to be relatively rare. The observed number of these sources is, however, larger than expected, and it is thought that this is likely to be due to motion of the star relative to the gas: in the direction of motion of the star, the H II region ceases to expand when the expansion velocity of the gas is equal to the velocity of the star relative to the gas. In this scenario, the ionized gas should have a "cometary" appearance: flattened on the leading edge of the H II region, with a "tail" trailing behind the star. This morphology is sometimes seen. Alternatively, some of the ultracompact H II regions appear to be cases where a disk or other dense structure near the star is gradually being ablated by photoionization, providing a reservoir of gas to replace the gas removed by the expanding ionized outflow. If a disk is involved, the outflows may be bipolar. A nice review of ultracompact H II regions is provided by Hoare et al. (2007).

32.7★ IR Point Sources

Low-mass stars are much more numerous than the massive stars that power H II regions. Because of the dust that is present, stellar sources will produce IR nebulae with characteristic sizes

$$R \approx (n_{\rm H}\sigma_{\rm dust})^{-1} \approx 2\times 10^{18}\frac{10^3\,{\rm cm}^{-3}}{n_{\rm H}}\,{\rm cm}, \tag{32.6}$$

for a dust attenuation cross section (at optical/UV wavelengths) $\sigma_{\rm dust} \approx 5\times 10^{-22}\,{\rm cm}^2/{\rm H}$. In high density regions ($n_{\rm H} \gtrsim 10^4\,{\rm cm}^{-3}$), the resulting IR nebulae will be small, and may appear point-like depending on the distance and the angular resolution of the telescope.

32.8★ Masers

OH and H_2O masers are frequently found near the boundaries of compact and ultracompact H II regions. In a survey of ultracompact H II regions, 67% had H_2O masers and 70% had NH_3 masers located outside the ionized region. H_2CO and CH_3OH masers are also frequently seen.

The physical conditions in the masing regions remain poorly understood. In some cases, the pumping can result from the infrared emission from dust; in other cases, the population inversions are thought to result from collisional excitation in warm gas. In all cases, the maser emission should peak along maximum-gain paths, where projected velocity gradients are minimum.

32.9 Size–Linewidth Relation in Molecular Clouds

Larson (1981) noted that observations of molecular clouds in spectral lines of CO, H_2CO, NH_3, OH, and other species, were broadly consistent with a **size–linewidth** relation, where a density peak of characteristic size L tends to have a *three*-dimensional velocity dispersion given by

$$\sigma_v \approx 1.10\, L_{\rm pc}^{\gamma}\,{\rm km\,s}^{-1} \quad , \quad \gamma \approx 0.38 \quad {\rm for}\ 0.1 \lesssim L_{\rm pc} \lesssim 10^2 \quad , \tag{32.7}$$

where $L = L_{\rm pc}\,{\rm pc}$ is the maximum projected dimension of the density peak. Larson noted that the power-law index $\gamma \approx 0.38$ was curiously close to the index $\frac{1}{3}$ found by Kolmogorov ($\sigma_v \propto L^{1/3}$) for a turbulent cascade in an incompressible fluid. It therefore is tempting to refer to the observed fluid motions as "turbulence," although in reality the motions are some combination of thermal motions, rotation, MHD waves, and turbulence.

The power-law index $\gamma \approx 0.38$ found by Larson has been questioned. A study of 273 molecular clouds (Solomon et al. 1987) found $\sigma_v \approx (1.0\pm0.1) L_{\rm pc}^{0.5\pm0.05}\,{\rm km\,s^{-1}}$, somewhat steeper than Larson's result. A recent study by Heyer & Brunt (2004) found $\sigma_v \approx (0.96 \pm 0.17) L_{\rm pc}^{0.59\pm0.07}\,{\rm km\,s^{-1}}$, again somewhat steeper than Larson's original result. The following discussion will leave γ as a variable, but, for illustration, will evaluate expressions assuming Larson's result $\gamma \approx 0.38$. The reader should keep in mind that the power-law approach is only an approximate representation of complicated data. Note that the various studies do agree that $\sigma_v \approx 1\,{\rm km\,s^{-1}}$ when $L = 1\,{\rm pc}$.

For scales $L \gtrsim 0.02\,{\rm pc}$, σ_v from Eq. (32.7) exceeds the isothermal sound speed $(kT/\mu)^{1/2} \approx 0.23(T/15\,{\rm K})^{1/2}\,{\rm km\,s^{-1}}$ in the cold gas – the fluid motions are **supersonic**. Extending the studies to scales as small as 0.01 pc, the linewidth σ_v appears to go to a constant $\sim 0.2\,{\rm km\,s^{-1}}$ for very small clumps, $L \lesssim 0.02\,{\rm pc}$: the linewidths are nearly thermal, with only a small contribution from rotation, waves, or turbulence (e.g., Goodman et al. 1998).

The density peaks are generally self-gravitating. If we assume them to be in approximate virial equilibrium, and consider only the kinetic energy associated

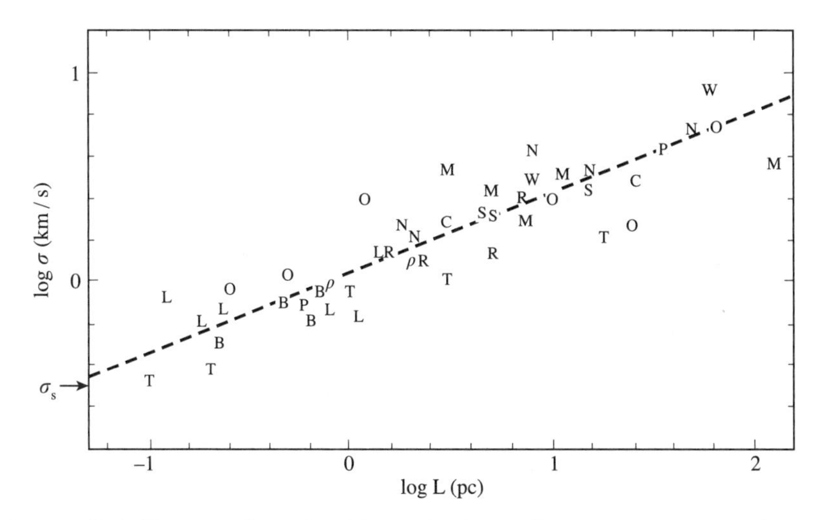

Figure 32.5 The three-dimensional internal velocity dispersion σ versus maximum linear dimension L of the density peak. The dashed line is given by Eq. (32.7). From Larson (1981).

with fluid motions (i.e., we neglect magnetic fields and external pressure) we can estimate the clump mass using the virial theorem (see §35.5). For a uniform density sphere with diameter L, virial equilibrium requires $\langle\sigma_v^2\rangle = 6GM/5L$, and we can therefore estimate the clump mass

$$M \approx \frac{5\sigma_v^2 L}{6G} \approx 230 L_{\rm pc}^{2\gamma+1}\, M_\odot \quad \rightarrow 230 L_{\rm pc}^{1.76}\, M_\odot \quad , \tag{32.8}$$

the characteristic density

$$n_{\rm H} \approx 1.3 \times 10^4 L_{\rm pc}^{2\gamma-2}\,{\rm cm}^{-3} \quad \rightarrow 1.3 \times 10^4 L_{\rm pc}^{-1.24}\,{\rm cm}^{-3} \quad , \tag{32.9}$$

and the characteristic column density

$$N_{\rm H} = n_{\rm H} L = 4.0 \times 10^{22} L_{\rm pc}^{2\gamma-1}\,{\rm cm}^{-2} \quad \rightarrow 4.0 \times 10^{22} L_{\rm pc}^{-0.24}\,{\rm cm}^{-2} \quad . \tag{32.10}$$

If $\gamma < 0.5$, smaller clouds tend to be darker, whereas for $\gamma = 0.5$, small clouds and large clouds would all have the same $N_{\rm H}$. We recall that for the dust in diffuse clouds, $A_V/N_{\rm H} = 5.3 \times 10^{-22}{\rm mag\,cm}^2$, and we would have

$$A_V \approx 21 L_{\rm pc}^{2\gamma-1}\ {\rm mag} \rightarrow 21 L_{\rm pc}^{-0.24}\ {\rm mag}. \tag{32.11}$$

The dust in dense clouds differs from that in the diffuse ISM, and $A_V/N_{\rm H}$ could be either larger or smaller than in the diffuse ISM, but this gives a reasonable estimate of the visual extinction through the cloud.

According to Eqs. (32.9 and 32.11), if $\gamma \approx 0.38$, then a GMC complex with $L \approx 50\,{\rm pc}$ would have $M \approx 2 \times 10^5\, M_\odot$, mean density $\langle n_{\rm H}\rangle \approx 100\,{\rm cm}^{-3}$ and $\langle A_V\rangle \approx 8$ mag, whereas a core with diameter $L = 0.1\,{\rm pc}$ would have $M \approx 4\, M_\odot$, $n_{\rm H} \approx 2 \times 10^5\,{\rm cm}^{-3}$, and $A_V \approx 40$ mag.

The scaling relation (32.7) for σ_v is only an approximate description of observed trends, and individual regions may deviate from it by factors of two, but the overall trend does generally describe the molecular structures in the nearby Milky Way.

Expressing these same relations with density as the independent variable, we obtain (again, using Larson's relation with $\gamma = 0.38$)

$$L_{\rm pc} = (n_3/13)^{1/(2\gamma-2)} \quad \rightarrow 7.94\, n_3^{-0.81} \quad , \tag{32.12}$$

$$\sigma_v = 1.1(n_3/13)^{\gamma/(2\gamma-2)}\,{\rm km\,s}^{-1} \quad \rightarrow 2.43\, n_3^{-0.31}\,{\rm km\,s}^{-1} \quad , \tag{32.13}$$

$$M = 230(n_3/13)^{(2\gamma+1)/(2\gamma-2)}\, M_\odot \quad \rightarrow 8940\, n_3^{-1.42}\, M_\odot \quad , \tag{32.14}$$

$$A_V = 21(n_3/13)^{(2\gamma-1)/(2\gamma-2)}\ {\rm mag} \quad \rightarrow 13\, n_3^{0.19}\ {\rm mag} \quad , \tag{32.15}$$

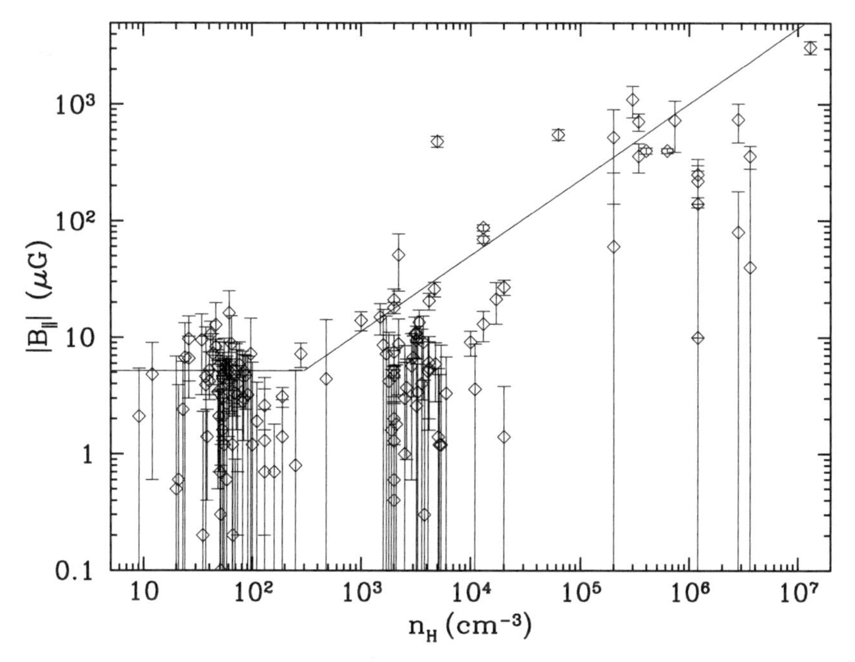

Figure 32.6 $|B_{\parallel}|$ versus $n_{\rm H}$ from 137 Zeeman measurements (69 HI, 54 OH, and 14 CN) with $1\,\sigma$ uncertainties, from Crutcher et al. (2010) (original data from Crutcher 1999; Heiles & Troland 2004; Troland & Crutcher 2008; Falgarone et al. 2008). The solid line [Eqs. (32.16), (32.17)] is the median total magnetic field strength $B_{0.5}$ as a function of $n_{\rm H}$ from the Bayesian analysis by Crutcher et al. (2010), with $B_{0.5} \propto {n_{\rm H}}^{0.65}$ for $n_{\rm H} > 300\,{\rm cm}^{-3}$.

where $n_3 \equiv n_{\rm H}/(10^3\,{\rm cm}^{-3})$, and the relations on the extreme right are for Larson's value $\gamma \approx 0.38$.

Note that A_V in Eq. (32.15) is only weakly dependent on n_3. As already noted above, if γ in Larson's relation is increased to $\sim$0.5 (as advocated by Solomon et al. (1987) and Heyer & Brunt (2004)), the virial analysis above gives $A_V \propto n_3^0$. Observationally, many self-gravitating molecular clouds – from individual dark clouds to GMC complexes – tend to have $A_V \approx 10$ mag. There is, however, an observed tendency for smaller structures to have higher A_V, consistent with exponent $\gamma < 0.5$. For this reason, we will retain Larson's value $\gamma = 0.38$ in the discussion below.

32.10 Magnetic Fields in Molecular Clouds

The virial estimate for the mass in Eq. (32.8) assumes that the cloud is supported by turbulence alone. However, there is strong evidence showing that magnetic fields are dynamically important in molecular clouds. In H I gas the line-of-sight component $B_{\|}$ of the magnetic field can be measured using the Zeeman effect on the 21-cm line. The Zeeman effect on the OH Λ-doubling lines $(1.665, 1.667, 1.720\,\mathrm{GHz})$ or on the CN $1-0$ rotational transition $(113\,\mathrm{GHz})$ can be used to measure the line-of-sight component $B_{\|}$ of the magnetic field in molecular clouds.

Crutcher et al. (2010) collected Zeeman measurements of $B_{\|}$ for 66 H I clouds and 72 molecular clouds, with density estimates for each case; the data are shown in Figure 32.6. Crutcher et al. (2010) show that near any given density $n_{\rm H}$, the distribution of the measured $|B_{\|}|$ requires that there be a distribution of magnetic field strengths. For clouds with density $n_{\rm H} = 10^4 n_4\,\mathrm{cm}^{-3}$, Crutcher et al. (2010) deduce a **median** field strength

$$B_{0.5} \approx 5\,\mu\mathrm{G} \qquad \text{for } n_4 < 0.03 \tag{32.16}$$

$$\approx 49 n_4^{0.65}\,\mu\mathrm{G} \qquad \text{for } 0.03 < n_4\,. \tag{32.17}$$

Thus, for $n_4 \gtrsim 0.03$, the median magnetic field strength estimated by Crutcher et al. (2010) implies a median Alfvén speed $v_A \equiv B/\sqrt{4\pi\rho}$ given by

$$(v_A)_{0.5} \approx 0.90\, n_4^{0.15}\,\mathrm{km\,s^{-1}} \qquad \text{for } n_4 \gtrsim 0.03; \tag{32.18}$$

with v_A exceeding this value $\sim 50\%$ of the time.

The ratio of the magnetic energy density $B^2/8\pi$ to the kinetic energy density $\frac{1}{2}\rho\sigma_v^2$ is simply $(v_A/\sigma_v)^2$, where σ_v is the 3-dimensional velocity dispersion. The dynamical importance of the magnetic field can, therefore, be seen by comparing v_A to σ_v from Eq. (32.13):

$$\frac{(v_A)_{0.5}}{\sigma_v} \approx 0.85 \left(\frac{n_4}{1.3}\right)^{0.15+\gamma/(2-2\gamma)} \rightarrow 0.75\, n_4^{0.46} \tag{32.19}$$

where the $n_4^{0.46}$ dependence is for $\gamma = 0.38$.[1]

If v_A is comparable to σ_v, it follows that the energy in the magnetic field is comparable to the turbulent kinetic energy, and the magnetic field is contributing significantly to supporting the cloud against self-gravity. Given the observed strength of the magnetic field for $n_{\rm H} \gtrsim 10^4\,\mathrm{cm}^{-3}$, what we have been referring to as "turbulence" should instead be thought of as MHD waves. If the magnetic field energy is comparable to the kinetic energy, it also follows that Eq. (32.8) *under*estimates the mass for $L_{\rm pc} \lesssim 5$, affecting the derived scaling relations Eqs. (32.9 – 32.15) for $L_{\rm pc} \lesssim 5$, or $n_4 \gtrsim 1.6$. However, if the magnetic and kinetic energy are comparable, then the cloud mass estimate is increased by only a factor $\sim \sqrt{2}$, which will not lead to any qualitative changes in the discussion.

An independent way to estimate the magnetic field strength is to use observations of aligned dust grains – either through maps of the polarization of background stars seen through the cloud, or maps of the polarization of the far-infrared or sub-mm

[1] The scaling exponent $\gamma \approx 0.5$ favored by some authors would give $(v_A)_{0.5}/\sigma_v \approx 0.72 n_4^{0.65}$.

emission from dust in the cloud. For example, Houde et al. (2004) used polarized $350\,\mu\mathrm{m}$ emission from dust to map the dust distribution and trace the magnetic field in the Orion A cloud.

If the magnetic field were weak, then the turbulence in the cloud would result in dispersion in directions of polarization over the map; if the dispersion is small, this indicates that the magnetic field is strong enough to resist substantial distortion by the turbulence. This is known as the **Chandrasekhar-Fermi** (CF) method to estimate the strength of the magnetic field. Crutcher (2004) applied the CF method to clouds with N_{H} ranging from $10^{21.4}\,\mathrm{cm}^{-2}$ to $10^{24}\,\mathrm{cm}^{-2}$, finding that the magnetic energy is comparable to the turbulent kinetic energy for $N_{\mathrm{H}} \gtrsim 10^{21.6}\,\mathrm{cm}^{-2}$, or $A_V \gtrsim 2$ mag. Novak et al. (2009) have applied the CF method to two GMCs, again finding that the magnetic energy must be at least as large as the turbulent kinetic energy.

The CF method is based on variations in the direction of linear polarization across the surface of the cloud. However, the relatively large (several %) linear polarization of the submillimeter thermal emission from dust (Novak et al. 2009, and references therein) requires the field to be coherent *along* the line of sight. The observed polarization, of course, requires that the grains must be fairly well-aligned by the local magnetic field at each point.[2] The large observed polarization implies that the transverse component of the magnetic field is relatively uniform in direction *along* the line of sight, since if the magnetic field were "tangled," the net polarization measured along a sightline would be small, even if the grains are well-aligned with the *local* magnetic field.

In summary, all of the evidence points to the presence of dynamically important magnetic fields in molecular cloud clumps with $n_{\mathrm{H}} \gtrsim 3000\,\mathrm{cm}^{-3}$.

32.11 Energy Dissipation in Molecular Clouds

As we have seen, the fluid motions in molecular clouds are supersonic for $L \gtrsim 0.02\,\mathrm{pc}$, and strongly supersonic for $L \gtrsim 1\,\mathrm{pc}$. If the magnetic field were not present, and the observed velocity dispersion were due to "random" motions of fluid elements, one would expect strong shocks to develop on a time scale

$$\tau_{\mathrm{cross}} \approx L/\sigma_v \approx 9 \times 10^5 L_{\mathrm{pc}}^{0.62}\,\mathrm{yr} \quad . \qquad (32.20)$$

For shock speeds of a few $\mathrm{km\,s}^{-1}$ and the densities of molecular clouds, the shocked gas would cool quickly, and much of the original kinetic energy would be radiated away. Unless there is an additional energy source to inject energy into the turbulent motions, clumps and cores would be expected to dissipate their kinetic energy and to collapse on time scales given by Eq. (32.20). This would result in what would appear to be unacceptably short lifetimes for molecular clouds.

[2]Lacking a quantitative understanding of the grain alignment mechanism, we are not yet able to determine the field strength from the local alignment.

One possibility is that once the first protostars form in a cloud, outflows from them inject enough kinetic energy to sustain the "turbulence." Another possibility is that the time scale for dissipation of the "turbulent" kinetic energy is in fact much longer than given by Eq. (32.20). The strong magnetic field implies that the fluid motions are subalfvénic, and the time scale for the waves to damp could be considerably longer than suggested by Eq. (32.20). While this conjecture seems in many ways attractive, numerical simulations appear to find rapid dissipation of MHD turbulence (see Stone et al. 1998, and references therein).

The actual rate of dissipation of kinetic energy in molecular clouds remains unclear. Perhaps the rapid dissipation in MHD simulations is an artifact of assumptions regarding the turbulent spectrum or magnetic field geometry. Alternatively, Elmegreen (2007) argues that dense molecular clouds *do* collapse rapidly, once they are assembled – this can be seen from the fact that giant molecular clouds in M51 appear to form stars very rapidly after they are assembled in spiral arms. In addition, the small spread of ages of stars in dense clusters suggests that most of the star formation is over in $\sim 3\,\mathrm{Myr}$ (although this could be in part because star formation near O stars is suppressed as soon as they begin to shine).

We recall the hierarchical structure of molecular clouds (see Table 32.1): GMCs have sizes $3 \lesssim L_{\rm pc} \lesssim 20$. A GMC consists of a low-density molecular envelope and denser "clumps". The clumps have sizes $0.3 \lesssim L_{\rm pc} \lesssim 3$. Upon entering a spiral arm density wave, the GMCs grow by accretion and the clumps appear to begin forming stars on approximately the turbulent "crossing time" given by Eq. (32.20). OB associations form, and disperse some, but not all, of the GMC envelope. Some fraction of the molecular gas in the GMC envelope survives and travels through the interarm region, eventually reaching the next spiral arm.

The nearby face-on spiral galaxy M51 (see Plates 10-12) has been mapped at many wavelengths. Sensitive interferometric observations in CO 1–0 using the CARMA array, together with single-dish observations using the Nobeyama 45-m telescope, reveal the detailed distribution of CO 1–0 emission in M51, allowing individual GMCs to be identified (Koda et al. 2009) – see Plate 8. With the usual assumption of a constant CO "X-factor" (see §19.6) the CO 1–0 map shows the distribution of molecular mass.

Combining the CO map with a high-resolution 21-cm map, Koda et al. (2009) show that the molecular *fraction* does not appear to change as the gas moves from spiral arm to interarm region: the majority of the molecular gas remains molecular from arm entry, star formation in the arm, and travel through the interarm region to the next spiral arm. Evidently, most of the molecular mass survives the star formation processes that are concentrated in the spiral arms.

Chapter Thirty-three

Molecular Clouds: Chemistry and Ionization

In the Milky Way, about 22% of the interstellar gas is in molecular clouds, where the bulk of the hydrogen is in H_2 molecules. As discussed in Chapter 31, in the Milky Way, H_2 is formed primarily by dust grain catalysis. Destruction of H_2 is primarily due to photodissociation, but self-shielding results in very low photodissociation rates in the central regions of molecular clouds.

Once the H_2 has been formed, other chemistry can follow. Most of the gas will be neutral, but, because of the presence of cosmic rays, there will always be some ions present in the gas. In the outer layers of molecular clouds, there may also be sufficient ultraviolet radiation to photoionize species with ionization potentials $I < 13.6\,\mathrm{eV}$.

There are five types of reactions that can be important:

1. **Photoionization:** $AB + h\nu \rightarrow AB^+ + e^-$.
   Many molecules (but not H_2) can be photoionized by photons with $h\nu < 13.6\,\mathrm{eV}$ present in the ISRF in H I or diffuse H_2 clouds. The ionization threshold for H_2 is 15.43 eV, hence H_2 will not be photoionized even in H I regions. For molecules with ionization threshold $< 13\,\mathrm{eV}$, photoionization rates in the ISRF are typically in the range 10^{-11}–$10^{-9}\,\mathrm{s}^{-1}$.

2. **Photodissociation:** $AB + h\nu \rightarrow AB^* \rightarrow A + B$.
   For some molecules (including the important species H_2 and CO), photoexcitation leading to dissociation occurs via *lines* rather than continuum. This allows such species either to self-shield (H_2 being the prime example), or to be partially shielded if there is accidental overlap between important absorption lines with strong lines of H_2, which will be self-shielded. This is the case with CO.

3. **Neutral–neutral exchange reactions:** $AB + C \rightarrow AC + B$.
   Most species in molecular clouds are neutral, and neutral–neutral collisions are, therefore, frequent. However, even when a neutral–neutral reaction is exothermic, there will often (although not always) be an energy barrier that must be overcome for the reaction to proceed: even though the reaction is exothermic, the ABC "complex" must pass through intermediate states that may have higher energy than the initial $AB{+}C$. For example, the exothermic reaction

$$\mathrm{OH}+\mathrm{H} \rightarrow \mathrm{O}+\mathrm{H}_2\,,\ k_{33.1} = 1.5\times10^{-12} T_2^{2.8} \mathrm{e}^{-19.50/T_2}\,\mathrm{cm}^3\,\mathrm{s}^{-1}, \qquad (33.1)$$

(for $300 < T/\mathrm{K} < 2500$; Woodall et al. 2007) has an energy barrier $\Delta E/k = 1950\,\mathrm{K}$, which causes the reaction to be negligibly slow at $T \lesssim 10^2\,\mathrm{K}$. On the other hand, the neutral–neutral reaction responsible for CO formation,

$$\mathrm{C+OH \rightarrow CO+H} \ , \ \ k_{33.2} = 1\times10^{-10}\,\mathrm{cm^3\,s^{-1}} \quad , \tag{33.2}$$

($10\,\mathrm{K} < T < 300\,\mathrm{K}$; Woodall et al. 2007) does not have any significant energy barrier, and can proceed at the low temperatures of molecular clouds.

4. **Ion–neutral exchange reactions:** $AB^+ + C \rightarrow AC^+ + B$.
   Ion–neutral reactions are very important in interstellar chemistry, for two reasons: (1) exothermic reactions generally lack energy barriers, allowing the reaction to proceed rapidly even at low temperature, and (2) the induced-dipole interaction results in ion–neutral rate coefficients that are relatively large, of order $\sim 2 \times 10^{-9}\,\mathrm{cm^3\,s^{-1}}$.

5. **Radiative association reactions:**

$$A + B \underset{k_d}{\overset{k_f}{\rightleftarrows}} (AB)^* \overset{k_r}{\rightarrow} AB + h\nu \quad . \tag{33.3}$$

The reaction first creates an excited complex AB^*. The complex will fly apart in one vibrational period, $\sim 10^{-14}\,\mathrm{s}$, unless a photon is emitted first; it can be thought of as having a probability per unit time $k_d \approx 10^{14}\,\mathrm{s^{-1}}$ of spontaneously dissociating. If AB has an apreciable electric dipole moment (i.e., is not homonuclear), the probability per unit time that AB^* will emit a photon in an electronic transition is $k_r \approx 10^6\,\mathrm{s^{-1}}$. Thus, a fraction $k_\mathrm{r}/(k_\mathrm{r} + k_\mathrm{d})$ of the AB^* complexes formed will result in formation of stable AB. The effective rate coefficient for radiative association will be

$$k_\mathrm{ra} = \frac{k_\mathrm{f} k_\mathrm{r}}{k_\mathrm{d} + k_\mathrm{r}} \quad . \tag{33.4}$$

For ion–neutral reactions, orbiting collisions (see §2.4) provide a rate coefficient for "complex formation" $k_\mathrm{f} \approx 10^{-9}\,\mathrm{cm^3\,s^{-1}}$, and hence the rate coefficient for radiative association will be $k_\mathrm{ra} \approx 10^{-17}\,\mathrm{cm^3\,s^{-1}}$. For example, formation of $\mathrm{CH^+}$ by radiative association

$$\mathrm{C^+ + H \rightarrow CH^+ + h\nu} \ , \ \ k_{33.5} \approx 4.46 \times 10^{-17} T_2^{-1/2} \mathrm{e}^{-0.229 T_2^{-2/3}}\,\mathrm{cm^3\,s^{-1}} \tag{33.5}$$

(Barinovs & van Hemert 2006, for $5\,\mathrm{K} < T < 10^3\,\mathrm{K}$), with $k_{33.5} = 3.5 \times 10^{-17}\,\mathrm{cm^3\,s^{-1}}$ at $T = 100\,\mathrm{K}$.

Table 33.1 Photodissociation and Photoionization Rates for Selected Species

Reaction	Rates (s^{-1})[a]				p_M[f]	Reference
	$A_V=0$[b]	$A_V=0.5$[c]	$A_V=1$[d]	$A_V=3$[e]		
$H_2 + h\nu \rightarrow 2H$	4.2(-11)	g	g	g	–	DB96
$CH + h\nu \rightarrow C + H$	8.6(-10)	2.2(-10)	3.9(-11)	1.2(-13)	730	R91
$CH^+ + h\nu \rightarrow C + H^+$	2.5(-10)	2.2(-11)	1.6(-12)	2.2(-18)	180	R91
$C_2 + h\nu \rightarrow 2C$	1.5(-10)				240	vD88
$CN + h\nu \rightarrow C + N$	1.1(-9)	9.7(-11)	7.0(-12)	7.5(-18)	11000	R91
$CO + h\nu \rightarrow C + O$	2.6(-10)	h	h	h	–	V09
$OH + h\nu \rightarrow O + H$	3.5(-10)	6.3(-11)	8.9(-12)	6.1(-15)	510	R91
$H_2O + h\nu \rightarrow H + OH$	5.9(-10)	1.1(-10)	1.7(-11)	1.2(-14)	970	R91
$HCN + h\nu \rightarrow H + CN$	1.3(-9)	1.8(-10)	2.1(-11)	2.1(-15)	3100	R91
$HCO + h\nu \rightarrow CO + H$	1.1(-9)				420	vD88
$H_2CO + h\nu \rightarrow H_2 + CO$	1.4(-9)	3.0(-10)	4.4(-11)	2.7(-14)	2700	R91
$O_2 + h\nu \rightarrow O + O$	6.9(-10)	1.4(-10)	2.1(-11)	9.4(-15)	750	R91
$CH + h\nu \rightarrow CH^+ + e^-$	7.6(-10)				0	vD88
$OH + h\nu \rightarrow OH^+ + e^-$	1.6(-12)				0	W07
$H_2O + h\nu \rightarrow H_2O^+ + e^-$	3.3(-11)				0	vD88
$C_2 + h\nu \rightarrow C_2^+ + e^-$	1.5(-10)	2.2(-11)	2.6(-12)	4.1(-16)	0	R91
$CN + h\nu \rightarrow CN^+ + e^-$	0[i]				0	
$CO + h\nu \rightarrow CO^+ + e^-$	0[j]				0	
$OH + h\nu \rightarrow OH^+ + e^-$	1.6(-12)				0	W07
$H_2O + h\nu \rightarrow H_2O^+ + e^-$	3.3(-11)				0	vD88
$HCN + h\nu \rightarrow HCN^+ + e^-$	0[k]				0	
$HCO + h\nu \rightarrow HCO^+ + e^-$	5.6(-10)	7.7(-11)	8.4(-12)	5.3(-16)	1170	R91
$H_2CO + h\nu \rightarrow H_2CO^+ + e^-$	4.7(-10)				–	vD88
$O_2 + h\nu \rightarrow O_2^+ + e^-$	5.6(-11)	4.6(-12)	3.1(-13)	2.2(-19)	120	R91

[a] $X \times 10^Y$ is written $X(Y)$
[b] For ISRF of Draine (1978).
[c] At center of slab with $A_V = 1$ mag.
[d] 10% of way through slab with $A_V = 10$ mag.
[e] 30% of way through slab with $A_V = 10$ mag.
[f] $\Delta\zeta_M = p_M \zeta_{CR}$ due to cosmic rays (see text).
[g] Self-shielding is important.
[h] Line shielding by H_2 and self-shielding are important.
[i] $I = 14.1$ eV
[j] $I = 14.01$ eV.
[k] $I = 13.65$ eV.
DB96=Draine & Bertoldi (1996).
R91=Roberge et al. (1991).
vD88=van Dishoeck (1988).
V09=Visser et al. (2009).
W07=Woodall et al. (2007).

33.1 Photoionization and Photodissociation of Molecules

The general ultraviolet background provided by starlight is lethal to small molecules, with either photodissociation or photoionization occurring rapidly. Rates for photodissociation and photoionization of selected species are given in Table 33.1. In the diffuse interstellar radiation field, small molecules have photodissociation rates that range from $\sim 4 \times 10^{-11}\,s^{-1}$ (e.g., H_2) to $\sim 10^{-9}\,s^{-1}$ (e.g., CN and H_2CO). In clouds, the ultraviolet radiation is attenuated by dust, and the photodissociation rates are reduced. The rate depends on the overall column density through the cloud, and the location within the cloud. For some of the species in Table 33.1, Roberge et al. (1991) have carried out radiative transfer calculations using realistic dust properties for clouds with plane-parallel geometry, with each surface illuminated by the average ISRF (over 2π sr). Rates in Table 33.1 are given in the unattenuated ISRF, at the center of a diffuse molecular cloud with total $A_V \approx 1$, and at points in a cloud of total thickness $A_V = 10$ (typical of GMCs, as seen in §32.9). In GMCs, the photodissociation rate at a "depth" $A_V \approx 3$ can be reduced by factors of $\sim 10^3 - 10^5$, rendering unimportant photodissociation by photons originating *outside* the cloud.

This does not, however, mean that photodissociation and photoionization are unimportant within dark clouds. Cosmic rays penetrating the cloud not only ionize H_2 and He, they also cause electronic excitation of H_2 by two processes. First, the electric field of passing cosmic rays can excite electrons to bound states (e.g., the $^1\Sigma_u^+$ and $^1\Pi_u$ states of H_2) followed by spontaneous emission of an ultraviolet photon. Second, the secondary electrons produced by cosmic ray ionization events can themselves excite electronic states of H_2. Together, these two processes result in generation of H_2 Lyman- and Werner-band UV photons with a rate that is proportional to the CR ionization rate (Prasad & Tarafdar 1983). These photons will either be absorbed by dust or by a permitted transition of an atom or molecule present in the medium. Let the total cosmic ray ionization rate per volume be $\zeta_{\rm CR} n_{\rm H}$. For a given molecule M, the increase to the photoionization or photodissociation rate contributed by cosmic-ray-generated Lyman- and Werner-band photons can be written $\Delta\zeta_M = p_M \zeta_{\rm CR}$. The coefficients p_M given in Table 33.1 are based on Woodall et al. (2007).[1]

The cosmic ray ionization rate was discussed in §13.5, with observations pointing to a cosmic ray primary ionization rate $\zeta_{\rm CR} \approx (0.5-3) \times 10^{-16}\,{\rm s}^{-1}$ in diffuse molecular clouds. The interiors of dark clouds may be partially shielded from low-energy cosmic rays, but it now appears reasonable to consider that $\zeta_{\rm CR} \approx 10^{-16}\,{\rm s}^{-1}$ may prevail in at least some dark clouds. The photodissociation rate for HCN, for example, would then be $\sim 3 \times 10^{-13}\,{\rm s}^{-1}$ for $A_V \gtrsim 2$.

33.2★ Ion–Molecule Chemistry in Cold Gas

Molecular gas is usually (although not always) quite cold, with $T < 100\,{\rm K}$. If H_2 is already present, and ultraviolet radiation is present, then the chemical network is dominated by ion–molecule reactions because they are fast ($k \approx 10^{-9}\,{\rm cm}^3\,{\rm s}^{-1}$), even at low temperatures.

33.2.1★ Formation of CH and CO

As an example, let us look at the part of the chemical network that is important for formation of the abundant and important CO molecule. In diffuse molecular clouds, most of the gas-phase carbon is in the form of C^+, and most of the H is in the form of H_2. CO is formed primarily by the sequence of reactions: (rates from

[1] The p_M values from Woodall et al. (2007) have been multiplied by a factor 2, because our $\zeta_{\rm CR}$ is the CR ionization rate per H, while their ζ is the CR ionization rate per H_2.

Woodall et al. (2007))

$$C^+ + H_2 \rightarrow CH_2^+ + h\nu \quad , \quad k_{33.6} = 5.0 \times 10^{-16} T_2^{-0.2}\,\mathrm{cm^3\,s^{-1}} \ , \tag{33.6}$$

$$CH_2^+ + e^- \rightarrow \left\{ \begin{array}{ll} CH + H & (25\%) \\ C + H_2 & (12\%) \\ C{+}H{+}H & (63\%) \end{array} \right\} , k_{33.7} = 1.24 \times 10^{-6} T_2^{-0.60} \frac{\mathrm{cm^3}}{\mathrm{s}} , \tag{33.7}$$

$$CH + O \rightarrow CO + H \quad k_{33.8} = 6.6 \times 10^{-11}\,\mathrm{cm^3\,s^{-1}} \ , \tag{33.8}$$

$$CO + h\nu \rightarrow C + O \quad k_{33.9} = 2.3 \times 10^{-10}\,\mathrm{s^{-1}} \times f_{\rm shield}(CO) \ , \tag{33.9}$$

$$C + h\nu \rightarrow C^+ + e^- \quad k_{33.10} = 2.6 \times 10^{-10}\,\mathrm{s^{-1}} \ . \tag{33.10}$$

The first step, radiative association, is slow but steadily produces the radical CH_2^+, which then reacts rapidly with electrons, producing CH about 25% of the time. The CH_2^+ produced in (33.6) can also be removed by photodissociation or reaction with H_2 (rates from Woodall et al. 2007):

$$CH_2^+ + h\nu \rightarrow \left\{ \begin{array}{ll} CH + H^+ & (1/3) \\ CH^+ + H & (1/3) \\ C^+ + H_2 & (1/3) \end{array} \right\} , \ k_{33.11} = 1.38 \times 10^{-10}\,\mathrm{s^{-1}} , \tag{33.11}$$

$$CH_2^+ + H_2 \rightarrow CH_3^+ + H \ , \ k_{33.12} = 1.60 \times 10^{-9}\,\mathrm{cm^3\,s^{-1}} \ . \tag{33.12}$$

For densities typical of diffuse molecular gas ($n_e \approx 0.01\,\mathrm{cm^{-3}}$, $n(H_2) \approx 30\,\mathrm{cm^{-3}}$), photodestruction of CH_2^+ is of secondary importance compared to dissociative recombination (33.7) or reaction with H_2 (33.12), the two channels that dominate the destruction of CH_2^+.

The CH produced by dissociative recombination of CH_2^+ can then react with O to produce CO via reaction (33.8), but it is also susceptible to photoionization and photodissociation (see Table 33.1):

$$CH + h\nu \rightarrow \left\{ \begin{array}{ll} CH^+ + e^- & (47\%) \\ C + H & (53\%) \end{array} \right\} k_{33.13} = 1.62 \times 10^{-9}\,\mathrm{s^{-1}} \ . \tag{33.13}$$

These are the chemical pathways that appear to account for production of CO in diffuse clouds where the carbon is primarily in the form of C^+. Aside from the initial production of H_2 via grain catalysis, it is assumed that all other reactions resulting in formation of CO take place in the gas phase. Whether this is actually the case is uncertain. For example, one could imagine that C and O atoms might stick to silicate grains and react to form CO, with the CO molecules returned to the gas phase either by the energy released in formation of CO, or by photodesorption.

33.2.2★ Formation of OH

Another important species is OH. In a diffuse H_2 cloud, OH is formed primarily by the reaction sequence: (rates from Woodall et al. 2007)

$$\mathrm{O + H_3^+ \rightarrow OH^+ + H_2} \ , \ k_{33.14} = 8.40 \times 10^{-10}\,\mathrm{cm^3\,s^{-1}}, \tag{33.14}$$

$$\mathrm{OH^+ + H_2 \rightarrow H_2O^+ + H} \ , \ k_{33.15} = 1.01 \times 10^{-9}\,\mathrm{cm^3\,s^{-1}}, \tag{33.15}$$

$$\mathrm{H_2O^+} + e^- \rightarrow \left\{ \begin{array}{lr} \mathrm{OH + H} & (20\%) \\ \mathrm{O + H_2} & (9\%) \\ \mathrm{O + H + H} & (71\%) \end{array} \right\} k_{33.16} = 4.30 \times 10^{-7}\,\mathrm{cm^3\,s^{-1}}, \tag{33.16}$$

$$\mathrm{H_2O^+ + H_2 \rightarrow H_3O^+ + H} \ , \ k_{33.17} = 6.40 \times 10^{-10}\,\mathrm{cm^3\,s^{-1}}, \tag{33.17}$$

$$\mathrm{H_3O^+} + e^- \rightarrow \left\{ \begin{array}{lr} \mathrm{O + H_2 + H} & (1\%) \\ \mathrm{OH + H_2} & (14\%) \\ \mathrm{OH + H + H} & (60\%) \\ \mathrm{H_2O + H} & (25\%) \end{array} \right\} k_{33.18} = 7.48 \times 10^{-7} T_2^{-0.5} \frac{\mathrm{cm^3}}{\mathrm{s}}. \tag{33.18}$$

The OH formed by reactions (33.16) and (33.18) is destroyed primarily by photodissociation (Table 33.1):

$$\mathrm{OH} + h\nu \rightarrow \mathrm{O + H} \ , \ k_{33.19} = 3.50 \times 10^{-10}\,\mathrm{s^{-1}} \ . \tag{33.19}$$

The reaction chain is initiated by H_3^+, which is generated by cosmic ray ionization of $\mathrm{H_2 + CR \rightarrow H_2{}^+} + e^-$ followed rapidly by $\mathrm{H_2{}^+ + H_2 \rightarrow H_3^+ + H}$; as discussed in §16.4. Diffuse molecular clouds appear to have $n(\mathrm{H_3^+})/n_\mathrm{H} \approx 5 \times 10^{-8}$ (Indriolo et al. 2007).

33.2.3★ Formation of H_2O

In a dense molecular cloud, shielded from external ultraviolet radiation, the ionization is produced by cosmic rays. Gas-phase production of H_2O is mainly by the reaction sequence (33.14 to 33.18), where $\sim$25% of the dissociative recombinations of H_3O^+ produce H_2O. For an assumed $n(\mathrm{H_3^+}) \approx 1 \times 10^{-5}\,\mathrm{cm^{-3}}$, this sequence of gas-phase reactions would convert the gas-phase O to H_2O on time scales $(1 \times 10^{-5} \times 8.4 \times 10^{-10})^{-1} \approx 4 \times 10^6\,\mathrm{yr}$, yielding $n(\mathrm{H_2O})/n_\mathrm{H} \approx 10^{-4}$.

The problem is that observations indicate very low gas-phase abundances of H_2O, $n(\mathrm{H_2O})/n_\mathrm{H} \lesssim 3 \times 10^{-8}$ (Snell et al. 2000), far below the predicted abundance of H_2O. The solution appears to be that the gas-phase H_2O is removed by freeze-out on grains, producing H_2O ice mantles which can contain a substantial fraction of the total oxygen. Hollenbach et al. (2009) model this in detail, including the role of CO, and desorption of frozen CO and H_2O by ultraviolet radiation and cosmic rays.

33.3★ The CH^+ Problem

The first interstellar molecules to be identified were CH (Swings & Rosenfeld 1937), CN (McKellar 1940), and CH^+ (Douglas & Herzberg 1941) because these three species have resonance lines at optical wavelengths. As modern models of chemistry in diffuse molecular clouds were developed, it was found that the observed abundances of CH and CN were in general agreement with what was expected from steady state chemistry in diffuse molecular clouds. However, this is emphatically *not* the case for CH^+. In molecular gas with $T \lesssim 10^2$ K, the principal channels for producing CH^+ would be radiative association $C^+ + H \rightarrow CH^+ + h\nu$ (Eq. 33.5) or photoionization of CH (Eq. 33.13).

Once formed, CH^+ is rapidly destroyed by

$$CH^+ + h\nu \rightarrow C + H^+ \qquad \text{(see Table 33.1)}\,, \qquad (33.20)$$

$$CH^+ + e^- \rightarrow C + H \qquad k_{33.21} = 2.38 \times 10^{-7} T_2^{-0.42}\,\mathrm{cm^3\,s^{-1}}\,, \qquad (33.21)$$

$$CH^+ + H \rightarrow C^+ + H_2 \qquad k_{33.22} = 7.50 \times 10^{-10}\,\mathrm{cm^3\,s^{-1}}\,, \qquad (33.22)$$

$$CH^+ + H_2 \rightarrow CH_2^+ + H \qquad k_{33.23} = 1.20 \times 10^{-9}\,\mathrm{cm^3\,s^{-1}}\,. \qquad (33.23)$$

In diffuse H I or H_2, the CH^+ lifetime is short, $\lesssim 10^9$ s. If radiative association were the dominant source of CH^+, the abundance of CH^+ would be very low.

The abundance of CH^+ varies from sightline to sightline, but is typically two or more orders of magnitude larger than the predictions of steady-state chemistry in cool, quiescent molecular clouds (Dalgarno 1976).

The most likely explanation is that CH^+ is formed in diffuse molecular regions that have been temporarily heated to high enough temperatures so that the endothermic reaction

$$C^+ + H_2 + 0.40\,\mathrm{eV} \rightarrow CH^+ + H \quad . \qquad (33.24)$$

can take place. For this reaction to proceed rapidly, the gas must have $T \gtrsim 10^3$ K to overcome the 0.40 eV energy barrier. The mechanism responsible for heating the regions where CH^+ formation takes place remains uncertain. The hot zone could be the postshock gas behind a shock wave that is propagating through the molecular cloud. Single-fluid shock models (Elitzur & Watson 1978) succeeded in producing CH^+ but produced OH in excess of observations, but models based on two-fluid MHD shocks (see §36.6) with shock speeds in the range $7\,\mathrm{km\,s^{-1}} \lesssim v_s \lesssim 12\,\mathrm{km\,s^{-1}}$ may be able to reproduce the observed CH^+ column densities ($10^{12}\,\mathrm{cm^{-2}} \lesssim N(CH^+) \lesssim 10^{13.5}\,\mathrm{cm^{-2}}$) without violating other observational constraints (Draine & Katz 1986*a*,*b*; Draine 1986; Pineau des Forets et al. 1986).

If the observed CH^+ is produced by MHD shocks, the frequency with which CH^+ is detected requires that a sightline traversing $E(B-V) = 1$ mag of reddening would be expected to intersect one shock front with $v_s \gtrsim 8\,\mathrm{km\,s^{-1}}$ (Draine & Katz 1986*b*). This shock frequency is, in fact, approximately what is expected based on the observed motions of interstellar gas, thus supporting the hypothesis that interstellar CH^+ is produced mainly in shocks.

However, there are clear kinematic expectations for the shock models: they require that sightlines with CH^+ should have two distinct velocity components – one representing the preshock gas, and one the postshock gas. The CH^+ should be at a velocity close to that of the postshock gas, but slightly displaced. Whether the observed kinematics are consistent with the shock models is not yet clear, and alternative explanations for CH^+ have been sought.

It is known that dissipation of turbulence in a fluid does not take place uniformly: "intermittency" refers to the phenomenon of the dissipation at any instant being concentrated in a small fraction of the volume of the fluid. It has been suggested (Godard et al. 2009) that intermittency in turbulent interstellar clouds creates small regions where turbulent dissipation is able to raise the gas temperature to $\gtrsim 10^3$ K, allowing the endothermic reaction (33.24) to generate CH^+.

Dissipation of turbulence in low-fractional-ionization molecular gas is complicated by the decoupling of the ion and neutral fluids on small scales (i.e., ambipolar diffusion), and the resulting ion–neutral "friction" may cause the turbulent dissipation to be spread out over a large enough volume that it cannot raise the gas temperature to the $T \gtrsim 10^3$ K required to drive CH^+ production. As of yet, there have not been three-dimensional numerical simulations of turbulent dissipation in magnetized molecular gas that would be capable of revealing the proposed levels of intermittency – this question remains unresolved.

Chapter Thirty-four

Physical Processes in Hot Gas

Within the Galaxy, hot ($T \gtrsim 10^6$ K) gas is produced by fast stellar winds, and by blastwaves from novae and supernovae. Beyond the Galaxy, much of the intergalactic medium is thought to be at $\sim 10^6$ K temperatures, and enormous masses of very hot ($10^7 - 10^8$ K) plasma are present in galaxy clusters. In some cases, the gas is detected via absorption lines of highly ionized species, such as O VI; in some cases, X-ray emission is observable.

In Chapter 25, we discussed the physics of dust grains immersed in hot gas. Here, we discuss gas-phase processes that affect the temperature of the gas: radiative cooling, thermal conduction, and electron–ion energy equipartition.

34.1 Radiative Cooling

Collisional excitation of ions in low-density plasma results in radiative cooling. The emitted power depends on the ionization state, and the plasma is often assumed to be in **collisional ionization equilibrium**, or **CIE**. CIE assumes that the plasma is in a steady state, and that collisional ionization, charge exchange, radiative recombination, and dielectronic recombination are the only processes altering the ionization balance, in which case the ionization fractions for each element depend only on the gas temperature, with no dependence on the gas density.

At temperatures $T > 10^4$ K, ionization of hydrogen provides enough free electrons so that collisional excitation of atoms or ions is dominated by electron collisions. At low densities, every collisional excitation is followed by a radiative decay, and the rate of removal of thermal energy per unit volume can therefore be written

$$\Lambda = n_e n_{\rm H} \times f_{\rm cool}(T) \ , \tag{34.1}$$

where the **radiative cooling function** $f_{\rm cool}(T) \equiv \Lambda / n_{\rm H} n_e$ is a function of temperature and of the elemental abundances relative to hydrogen.

At high densities, radiative cooling can be suppressed by collisional deexcitation, and $f_{\rm cool}$ will then depend on density $n_{\rm H}$, in addition to T and elemental abundances. Finally, if ionizing radiation is present, the ionization balance may depart from CIE, and the radiative cooling function $\Lambda / n_e n_{\rm H}$ will also depend on the spectrum and intensity of the ionizing radiation.

The cooling function $\Lambda / n_e n_{\rm H}$ for $T > 10^4$ K has been calculated by a number of groups over the years (e.g., Boehringer & Hensler 1989; Schmutzler & Tscharnuter

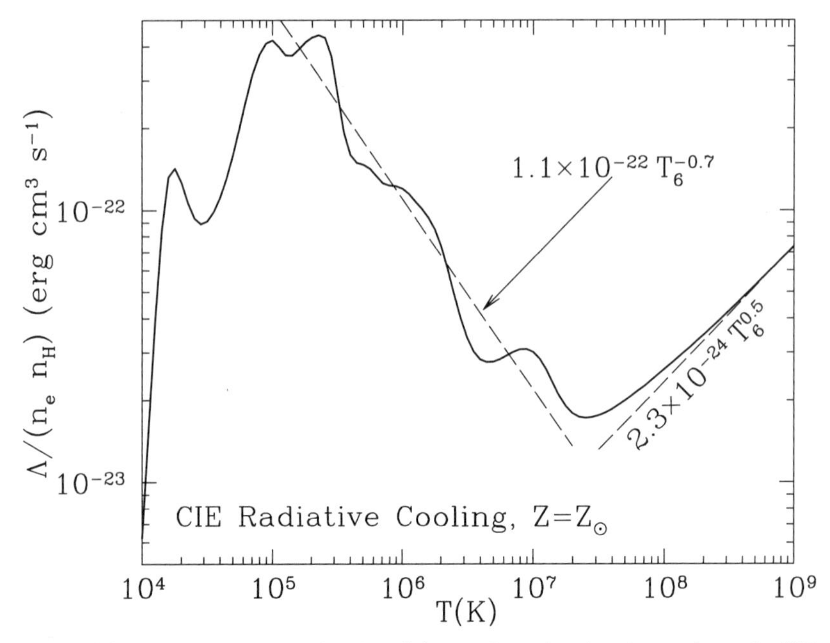

Figure 34.1 Radiative cooling function $\Lambda/n_e n_{\rm H}$ for solar-abundance plasma in CIE, computed with the CHIANTI code (Dere et al. 2009). Dashed lines show simple power-law approximations for $10^5 < T < 10^{7.3}$ K and for $T > 10^{7.5}$ K.

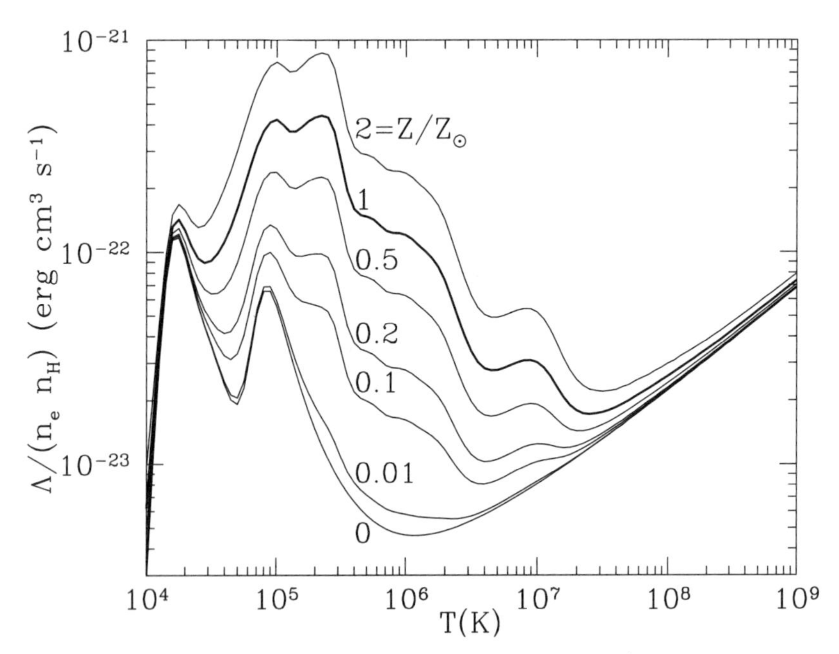

Figure 34.2 Same as Fig. 34.1, but for different abundances. $Z/Z_\odot$ is the abundance of elements heavier than He relative to solar abundances.

1993; Sutherland & Dopita 1993; Landi & Landini 1999; Smith et al. 2008; Dere et al. 2009; Schure et al. 2009). While the calculated cooling rates are in approximate agreement with one another, with some differences attributable to different adopted elemental abundances, there appear to also be significant differences in the adopted atomic physics. For example, near 10^5 K the cooling function of Schure et al. (2009) is a factor ~ 2.5 higher than the cooling function of Dere et al. (2009).

Figure 34.1 shows the radiative cooling function $\Lambda/n_{\rm H}n_e$ for plasma with solar abundances, based on calculations with the CHIANTI atomic database (Dere et al. 2009), kindly calculated for $n_{\rm H} = 1\,{\rm cm}^{-3}$ by K. Dere (2009, private communication). At temperatures $T < 10^7$ K, the cooling is dominated by collisional excitation of bound electrons. At high temperatures, the ions are fully stripped of electrons, and bremsstrahlung (i.e., free–free) cooling dominates, with $\Lambda/n_e n_{\rm H} \propto T^{0.5}$. The cooling function for solar abundances can be approximated by

$$\Lambda/n_e n_{\rm H} \approx 1.1 \times 10^{-22} T_6^{-0.7}\,{\rm erg\,cm^3\,s^{-1}}\ , \quad {\rm for}\ 10^5 < T < 10^{7.3}\,{\rm K}\ , \tag{34.2}$$

$$\Lambda/n_e n_{\rm H} \approx 2.3 \times 10^{-24} T_6^{0.5}\,{\rm erg\,cm^3\,s^{-1}}\ , \quad {\rm for}\ T > 10^{7.3}\,{\rm K}\ . \tag{34.3}$$

These two power-law fits are shown in Fig. 34.1.

In some cases (e.g., supernova ejecta), the plasma may have unusual abundance patterns, but in most applications the abundances of elements beyond He can be assumed to scale up and down together. Figure 34.2 shows radiative cooling functions for metallicities ranging from zero (H and He only) to twice solar.

Before using a rate coefficient for radiative cooling such as that shown in Figure 34.1, it is important to recognize its limitations:

- CIE requires that photoionization be unimportant.
- CIE requires that the plasma had time to attain collisional ionization equilibrium. If the gas has been suddenly shock-heated, time is required for collisional ionization to raise the ionization level to CIE. If the gas is cooling, the cooling rate should be slow enough that recombination processes are able to keep the ionization from lagging too far behind the ionization state corresponding to CIE.
- Figure 34.3 shows the contribution to the cooling from each of the 10 most important elements: H, He, C, N, O, Ne, Mg, Si, S, and Fe. For $10^{5.8} < T < 10^{7.2}$ K, radiative cooling is dominated by Mg, Si, and Fe – elements that in cold gas are normally depleted by factors of 5 or more. If these elements are underabundant in the plasma, the radiative cooling function will be suppressed. An accurate treatment of radiative cooling requires following the changing abundances in the gas phase as dust grains undergo sputtering in $T \gtrsim 10^6$ K plasma (see Fig. 25.4).
- In real problems, heating and cooling may be sufficiently rapid so that the ionization state of the gas lags behind CIE. This is particularly likely if the

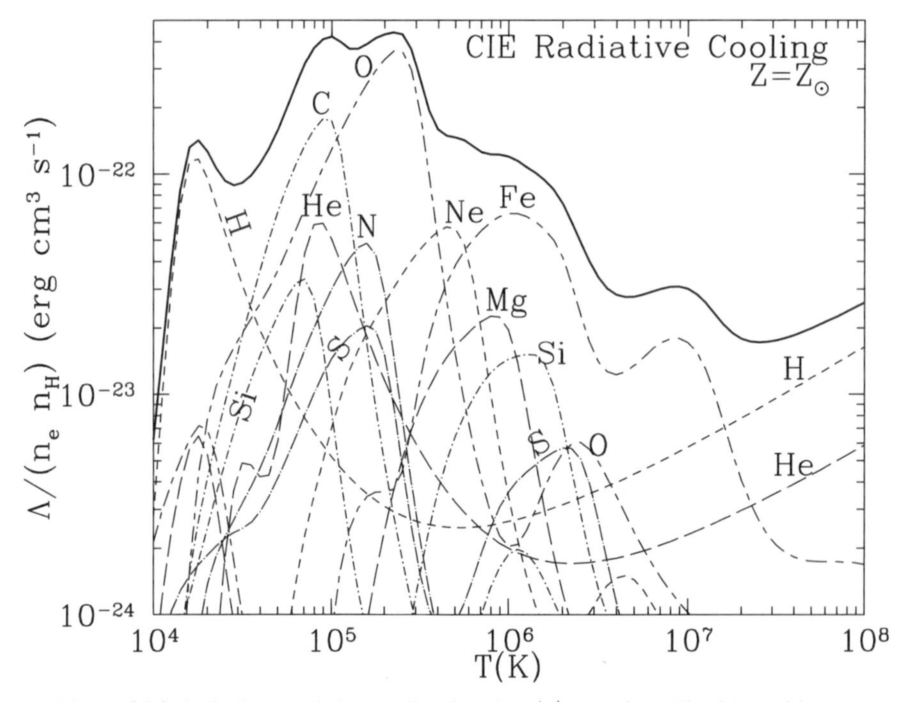

Figure 34.3 Solid line: radiative cooling function $\Lambda/n_e n_{\rm H}$ from Fig. 34.1, with contributions from selected elements shown.

gas has just been shock-heated, resulting in a sudden increase in the kinetic temperature of the gas, but can also be true when the gas is cooling rapidly, e.g., at $10^{4.9} < T < 10^{5.4}$ K, where the radiative cooling function peaks.

In these cases, the actual radiative cooling rate can be *slower* than CIE (when the gas is cooling faster than it can recombine, so that heavy elements have fewer bound electrons than they would in CIE), or *faster* than CIE (when the gas temperature has been suddenly increased, so that heavy elements are under-ionized). Underionization of elements such as Mg, Si, or Fe can also be an issue when atoms are being sputtered off of dust grains, since they are expected to enter the hot gas as neutral atoms.

34.2 Radiative Cooling Time

If the plasma cools at constant volume, the cooling time is given by Eq. (35.34). The cooling time is shown in Fig. 34.4 for metallicities ranging from 0 to 3 times

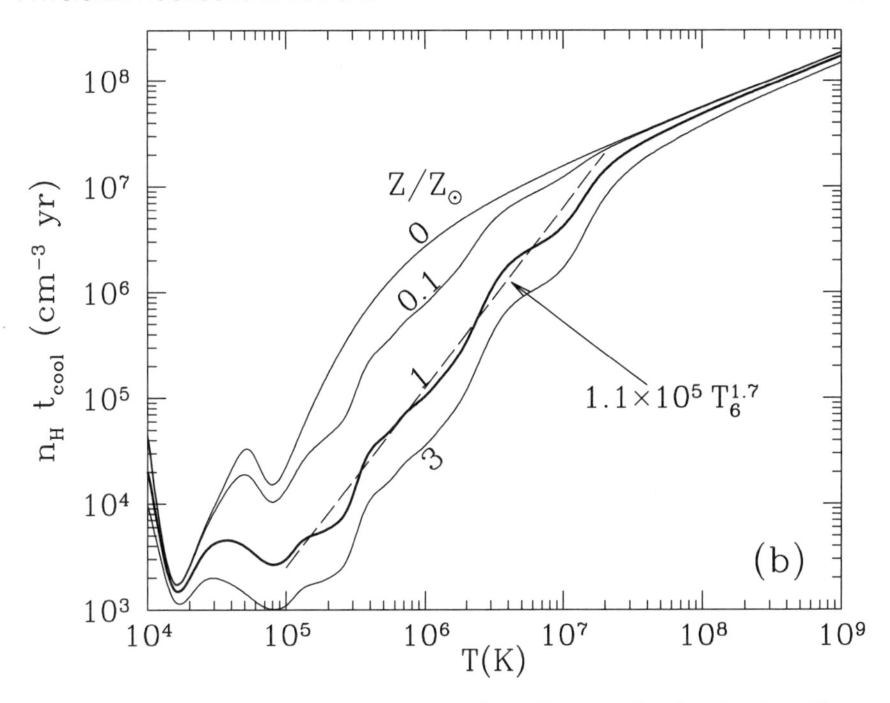

Figure 34.4 $n_{\rm H}t_{\rm cool,isochoric}$ for isochoric cooling with the cooling function from Fig. 34.1, for different metal abundances relative to solar. The dashed line is the approximation (34.4) for $Z \approx Z_\odot$ and $10^5\,{\rm K} < T < 10^{7.3}\,{\rm K}$.

solar. The cooling time for solar abundance plasma can be approximated by

$$\tau_{\rm cool,isochoric} \approx \frac{1.1\times10^5 T_6^{1.7}\,{\rm yr}}{n_{\rm H}/\,{\rm cm}^{-3}} \quad \text{for } 10^5\,{\rm K} \lesssim T \lesssim 10^{7.3}\,{\rm K} \quad . \tag{34.4}$$

34.3★ Thermal Conduction

In the absence of magnetic fields, a fully ionized H-He plasma has a "classical" thermal conductivity (Spitzer 1962):

$$\kappa_{\rm class}(T) \approx 0.87 \frac{k^{7/2} T_e^{5/2}}{m_e^{1/2} e^4 \ln\Lambda} \quad , \tag{34.5}$$

where $\ln\Lambda \approx 30$ is the usual Coulomb logarithm, given by Eq. (2.17), and the numerical factor 0.87 includes the effects of the electric field that normally accompanies a temperature gradient in a uniform-pressure plasma (Spitzer 1962).

If a magnetic field is present, as is normally the case in the ISM, the thermal conductivity becomes a tensor. In a coordinate frame where $\hat{\mathbf{x}} \parallel \mathbf{B}$, the thermal conductivity tensor is diagonal, with diagonal elements $(\kappa_\parallel, \kappa_\perp, \kappa_\perp)$, where $\kappa_\parallel = \kappa_{\rm class}$. The ratio $\kappa_\perp/\kappa_\parallel \approx 1/(\omega_B t_{\rm coll})^2$, where $\omega_B = eB/m_e c$ is the electron gyrofrequency, and $t_{\rm coll}$ is the mean collision time. For normal interstellar conditions, $\omega_B t_{\rm coll} \gg 1$, hence $\kappa_\perp \ll \kappa_\parallel$, and we need consider only the heat flow resulting from the component of ∇T along the local magnetic field direction.

When the temperature gradient becomes very large, the heat conduction "saturates," with the heat flux approaching a value $q_{\rm sat} \approx 5\rho(kT/\mu)^{3/2}$ (Cowie & McKee 1977). For finite ∇T, the heat flux can be written

$$\mathbf{q} \approx -\frac{\kappa_{\rm class} \nabla T \cdot \hat{\mathbf{b}}\hat{\mathbf{b}}}{1+\sigma_T} \quad , \tag{34.6}$$

where $\hat{\mathbf{b}} \equiv \mathbf{B}/|\mathbf{B}|$ is a unit vector parallel to the magnetic field, and the "saturation parameter"

$$\sigma_T \equiv \frac{\kappa_{\rm class} |\nabla T \cdot \hat{\mathbf{b}}|}{q_{\rm sat}} \tag{34.7}$$

$$= 2.53 \frac{k^2 T |\nabla T \cdot \hat{\mathbf{b}}|}{n_{\rm H} e^4 \ln \Lambda} \tag{34.8}$$

$$= 0.39 \left(\frac{T}{10^7\,{\rm K}}\right) \left(\frac{|\nabla T \cdot \hat{\mathbf{b}}|}{10^7\,{\rm K\,pc^{-1}}}\right) \left(\frac{\rm cm^{-3}}{n_{\rm H}}\right) \left(\frac{30}{\ln \Lambda}\right) . \tag{34.9}$$

34.4★ Cloud Evaporation in Hot Gas

Consider now a cold spherical cloud, with temperature T_c and radius R_c, surrounded by very hot gas, with density $n_{{\rm H},h}$ and temperature $T_h \gg T_c$. Suppose that $T_h \gtrsim 10^6$ K, so that the rate coefficient for radiative cooling (see Fig. 34.1) is relatively low. What is the temperature profile near the cloud?

Let us first neglect radiative losses. Further assume that there is negligible heat flow into the cold cloud itself. Then, at each radius, the inward thermal conduction must be balanced by outward advective transport of heat. For a steady constant-pressure outward flow, mass conservation and energy conservation give

$$\frac{5\dot{M}kT}{2\mu} = -4\pi r^2 \kappa \frac{dT}{dr} \quad , \tag{34.10}$$

where $\dot{M}$ is the rate of mass flow out of the cloud. If we assume a thermal conductivity $\kappa = \kappa_h (T/T_h)^{5/2}$, then one finds

$$T = T_h \left(1 - \frac{R_c}{r}\right)^{2/5} , \tag{34.11}$$

$$\dot{M} = \frac{16\pi\mu R_c \kappa_h}{25k} \propto R_c T_h^{5/2} . \tag{34.12}$$

What about cooling? The temperature just outside the cloud surface must pass through $T \approx 10^5$ K where the cooling function peaks (see Fig. 34.1). There are two limiting cases: (1) the small temperature gradient regime, where thermal conduction is small and cannot balance radiative cooling from the region where $T \approx 10^5$ K; and (2) the evaporative regime where radiative cooling is negligible, and there is an "evaporative flow" away from the cloud.

Cowie & McKee (1977) defined a "global saturation parameter":

$$\sigma_0 \equiv \frac{(2/25)\kappa_h T_h}{\rho_h (kT_h/\mu)^{3/2} R_c} = 0.4 \left(\frac{T_h}{10^7\,\mathrm{K}}\right)^2 \left(\frac{\mathrm{cm}^{-3}}{n_{\mathrm{H},h}}\right) \left(\frac{\mathrm{pc}}{R_c}\right) \tag{34.13}$$

and found three regimes:

$$\sigma_0 \lesssim 0.027 \quad : \text{cooling flow onto the cloud,} \tag{34.14}$$

$$0.027 \lesssim \sigma_0 \lesssim 1 \quad : \text{classical evaporation flow,} \tag{34.15}$$

$$1 \lesssim \sigma_0 \quad : \text{saturated evaporation flow.} \tag{34.16}$$

If we are in the regime $0.03 \lesssim \sigma_0 \lesssim 1$ where the classical evaporation mass loss rate (34.12) applies, the cloud lifetime against evaporation is

$$t_{\mathrm{evap}} = \frac{3M}{2\dot{M}} = \frac{25 \times 2.3\, (n_{\mathrm{H}})_c\, R_c^2\, m_e^{1/2}\, e^4 \ln\Lambda}{8 \times 0.87\, (kT_h)^{2.5}} \tag{34.17}$$

$$= 5.1 \times 10^4\,\mathrm{yr} \left(\frac{(n_{\mathrm{H}})_c}{30\,\mathrm{cm}^{-3}}\right) \left(\frac{R_c}{\mathrm{pc}}\right)^2 \left(\frac{T_h}{10^7\,\mathrm{K}}\right)^{-2.5} \left(\frac{\ln\Lambda}{30}\right) . \tag{34.18}$$

Therefore, if the classical thermal conductivity applies, small ($R_c \lesssim 1$ pc) clouds will be thermally evaporated by hot ($T_h \gtrsim 10^7$ K) gas in a few $\times 10^4$ yr.

34.5★ Conduction Fronts

Earlier, we discussed a spherical cloud in an infinite medium. Other geometries yield different behavior. Consider a planar cold cloud brought suddenly into contact with an infinite region of hot gas. At $t = 0$, the temperature profile is a step function. The conduction front then begins to heat some of the cold gas, bringing it to intermediate temperatures. As long as the front is thin, radiative losses are negligible compared to the heat flux into the intermediate temperature zone, and the continuing heat flux causes this intermediate temperature zone to grow. As the conduction front thickens and temperature gradients drop, the radiative losses increase, the conductive flux decreases, and the front approaches an asymptotic solution where evaporation ceases and thermal conduction is balanced by radiative losses.

If $T_h \gtrsim 10^6$ K, then the conduction front will contain ions such as C IV, Si IV, N V, S VI, and O VI, which have strong ultraviolet absorption lines that can be used as diagnostics. Borkowski et al. (1990) carried out simulations of planar conduction fronts including magnetic fields. The O VI abundance, for example, rises to $N(\mathrm{O\,VI}) \approx 10^{13}\,\mathrm{cm}^{-2}$ on a time scale of $\sim 10^5$ yr if $\mathbf{B} = 0$ or if $\mathbf{B}$ is perpendicular to the interface. If a magnetic field is present and inclined to the normal, the conduction front becomes thinner, and the ionic column densities are reduced.

Chapter Thirty-five

Fluid Dynamics

The interstellar medium is a fluid, and we need to understand how this fluid moves in response to pressure gradients within it, gravitational forces, and electromagnetic stresses. This chapter will develop the basic equations of motion for magnetized, conducting fluids.

As a conceptual aid, it is helpful to consider the dynamics of a *fluid element* – a small region of the fluid bounded by a closed surface S. The region is assumed to be large compared to the mean free path of the particles making up the fluid, but small compared to the length scales over which the fluid properties vary. The surface S is assumed to move with the fluid: the velocity of the surface at each point on the surface is equal to the local fluid velocity, so that there is no motion of fluid *across* the surface. Therefore, the mass within the fluid element is conserved. The fluid is decribed by

$$\rho(\mathbf{r},t) = \text{mass density}, \tag{35.1}$$

$$\mathbf{v}(\mathbf{r},t) = \text{velocity}, \tag{35.2}$$

$$p(\mathbf{r},t) = \text{pressure}. \tag{35.3}$$

We will want to evaluate the rate of change of variables associated with the **moving fluid element**. This is provided by the **convective derivative**

$$\frac{D}{Dt} \equiv \frac{\partial}{\partial t} + \mathbf{v}\cdot\nabla \quad . \tag{35.4}$$

$Df/Dt \equiv \partial f/\partial t + (v\cdot\nabla)f$ is the rate of change of a local variable f (e.g., ρ or T) that would be measured by an observer moving with the fluid.

35.1 Mass Conservation

Let $\Omega(t)$ be the **comoving volume** of our fluid element. The mass of fluid in the comoving volume is constant: $\rho\Omega = constant$, so that

$$\frac{D}{Dt}(\rho\Omega) = \rho\frac{D\Omega}{Dt} + \Omega\frac{D\rho}{Dt} = 0 \quad , \tag{35.5}$$

but

$$\frac{D\Omega}{Dt} = \int \mathbf{v}\cdot d\mathbf{S} = \int (\nabla\cdot\mathbf{v})d\Omega \;\rightarrow\; \Omega\nabla\cdot\mathbf{v} \quad . \tag{35.6}$$

Substitution into (35.5) gives $D\rho/Dt = -\rho\nabla\cdot\mathbf{v}$, or

$$\frac{\partial\rho}{\partial t} + \nabla\cdot(\rho\mathbf{v}) = 0 \ . \tag{35.7}$$

35.2 Conservation of Momentum: MHD Navier-Stokes Equation

Let $\Omega(t)$ be the volume of the moving fluid element, and let $\mathbf{F}$ be the total force applied to the matter in this fluid element by the rest of the universe. Then Newton's law $m\mathbf{a} = \mathbf{F}$ can be written

$$(\rho\Omega)\frac{D}{Dt}\mathbf{v} = \mathbf{F}_{\text{pressure}} + \mathbf{F}_{\text{EM}} + \mathbf{F}_{\text{grav}} + \mathbf{F}_{\text{viscosity}} \ , \tag{35.8}$$

where the terms on the right-hand side are, respectively, the force exerted on the fluid element by (1) the pressure of the surrounding fluid, (2) the electromagnetic field, (3) the gravitational field, and (4) viscous stresses. We will now examine each of these in turn.

Let the surface element $d\mathbf{S}$ be a vector pointing in the outward direction from the closed surface S. The pressure of the external fluid pushes *inward* at each point on the surface, and the net pressure force on the fluid element is

$$\mathbf{F}_{\text{pressure}} = \int(-p\,d\mathbf{S}) \tag{35.9}$$

$$= \int -\nabla p\,d\Omega \quad \text{(by Gauss's theorem)} \ \rightarrow \ -\Omega\,\nabla p \ . \tag{35.10}$$

We will assume the fluid element to have zero net charge, in which case there is no net force exerted by whatever $\mathbf{E}$ field may be present. However, there may be an electric current density $\mathbf{J}$ in the fluid element, in which case there is a force per volume $(1/c)\mathbf{J}\times\mathbf{B}$:

$$\mathbf{F}_{\text{EM}} = \int d\Omega\frac{\mathbf{J}}{c}\times\mathbf{B} \rightarrow \Omega\frac{\mathbf{J}}{c}\times\mathbf{B} \ . \tag{35.11}$$

Writing the force in this way requires that $\mathbf{J}$ be known. It is convenient to eliminate $\mathbf{J}$ using Maxwell's equation

$$\nabla\times\mathbf{B} = \frac{4\pi}{c}\mathbf{J} + \frac{1}{c}\frac{\partial\mathbf{D}}{\partial t} \ . \tag{35.12}$$

For the moment, let us assume the fluid to have infinite electrical conductivity. Then *in the fluid frame*, finite $\mathbf{J}$ implies $\mathbf{E} \to 0$ and $\mathbf{D} \to 0$, and

$$\mathbf{J} = \frac{c}{4\pi} \nabla \times \mathbf{B} \quad , \tag{35.13}$$

$$\mathbf{F}_{\text{EM}} = \frac{\Omega}{4\pi} (\nabla \times \mathbf{B}) \times \mathbf{B} \tag{35.14}$$

$$= \frac{\Omega}{4\pi} \left[-\frac{1}{2} \nabla B^2 + (\mathbf{B} \cdot \nabla) \mathbf{B} \right] \quad , \tag{35.15}$$

where we have used the vector identity $(\nabla \times \mathbf{B}) \times \mathbf{B} \equiv \left[(\mathbf{B} \cdot \nabla)\mathbf{B} - \frac{1}{2}\nabla B^2 \right]$.

For future reference, let us reverse Gauss's theorem and write the force $\mathbf{F}_{\text{EM}}$ as an integral over the surface S bounding the volume: Using $\nabla \cdot \mathbf{B} = 0$, one can show that[1]

$$F_{\text{EM},i} = -\int dS_i \frac{B_j B_j}{8\pi} + \int dS_j \frac{B_i B_j}{4\pi} \quad , \tag{35.16}$$

where summation over repeated indices is implied in (35.16). Thus, the magnetic field can be thought of as exerting forces on the boundaries of a fluid element.

Let $\Phi_{\text{grav}}(\mathbf{r}, t)$ be the gravitational potential. The gravitational force is just

$$\mathbf{F}_{\text{grav}} = (\rho\Omega)(-\nabla\Phi_{\text{grav}}) \quad . \tag{35.17}$$

Last, we turn to viscous stresses. These are, in general, described by the **viscous stress tensor** $\sigma_{ik}(\mathbf{r}, t)$, defined so that the viscous force on the fluid element is

$$(F_{\text{viscous}})_i \equiv \int \sigma_{ij} dS_j \quad , \tag{35.18}$$

$$= \int \frac{\partial}{\partial x_j} \sigma_{ij} d\Omega \quad \text{(by Gauss's theorem),} \tag{35.19}$$

$$\to \Omega \frac{\partial}{\partial x_j} \sigma_{ij} \quad , \tag{35.20}$$

with summation over the repeated index j in Eqs. (35.18 to 35.20) again implied. This result is for a general stress tensor σ_{ik}. The viscous stress tensor depends on gradients of the velocity field. If the "bulk" viscosity coefficient is assumed to be zero, the viscous stress tensor $\sigma_{ij} = -\eta[\partial v_i/\partial x_j + \partial v_j/\partial x_i - (2/3)\delta_{ij}\nabla \cdot \mathbf{v}]$, where η is the viscosity coefficient. We will usually neglect viscous stresses, but they become very important when fluid variables (e.g., velocity) change appreciably over length scales comparable to the mean-free-path of the fluid particles, as is often the case in shock waves.

We now note that each of the force expressions is proportional to Ω, so we can divide by Ω to obtain the general equation of momentum conservation[2]:

$$\rho \frac{D\mathbf{v}}{Dt} = -\nabla \left(p + \frac{B^2}{8\pi} \right) + \frac{1}{4\pi} (\mathbf{B} \cdot \nabla) \mathbf{B} - \rho \nabla \Phi_{\text{grav}} + \hat{\mathbf{x}}_i \frac{\partial}{\partial x_j} \sigma_{ij} \quad . \tag{35.21}$$

---

[1]Making use of $[(B \cdot \nabla)B - (1/2)\nabla B^2]_i = (B_j \partial/\partial x_j)B_i - (1/2)(\partial/\partial x_i)B_j B_j = (\partial/\partial x_j)B_i B_j - (1/2)(\partial/\partial x_i)B_j B_j$, where we have used $(\partial/\partial x_j)B_j = 0$.

[2]If $\mathbf{B} = 0$ and $\nabla\Phi_{\text{grav}} = 0$, this is called the **Navier-Stokes equation**.

We will usually neglect the viscous term on the right-hand side, but we include it here to show how viscous effects are treated.

35.3 Heating and Cooling

The preceding equations involve unknown fields $\mathbf{v}$, ρ, and p. Thus far, we have equations for the rate-of-change of $\mathbf{v}$ and ρ; we now need one for p. Let $U(t)$ be the "internal energy" of the fluid element Ω. This includes the kinetic energy in random motions of the particles, the energy in internal degrees of freedom of the molecules (rotation, and vibration), and "chemical energy." The pressure itself arises only from the kinetic energy in random translational motions, and we will need to know the relationship between U and the gas pressure.

Recall that we use the convective derivative D/Dt for the rate of change when moving with the fluid element. The first law of thermodynamics is simply

$$DU = -p\,D\Omega + DQ \quad . \tag{35.22}$$

$p\,D\Omega$ is the work done on the rest of the fluid when our fluid element expands by an amount $D\Omega$. DQ is the energy added to the fluid element by the rest of the universe by processes other than compressive work:

$$\frac{DQ}{Dt} = +\Omega\Gamma - \Omega\Lambda + \Omega\nabla\cdot(\kappa\nabla T) \quad . \tag{35.23}$$

Here, Γ is the rate/volume of deposition of thermal energy by processes such as photoionization, photoelectron emission from dust grains, etc., plus heating by viscous dissipation; Λ is the rate/volume of energy removal by radiation; and $-\kappa\nabla T$ is the conductive heat flux.

Let $N = n\Omega$ be the number of free particles in Ω, and let f be the number of effective degrees of freedom per free particle. A monatomic gas has $f = 3$ (only translational degrees of freedom); a diatomic gas at temperatures high enough to excite rotation but not vibration has $f = 3 + 2 = 5$. The internal thermal energy is $U_{\rm thermal} = NfkT/2$. If the chemical state of the gas does not change ($N = const$, $f = const$), then

$$DU = D\left(\frac{f}{2}NkT\right) \quad , \tag{35.24}$$

$$\frac{D}{Dt}\left(\frac{f}{2}n\Omega kT\right) = -nkT\frac{D\Omega}{Dt} + (\Gamma-\Lambda)\Omega + \Omega\nabla\cdot(\kappa\nabla T) \quad , \tag{35.25}$$

$$\frac{D}{Dt}\left(\frac{f+2}{2}n\Omega kT\right) - \Omega\frac{D}{Dt}(nkT) = (\Gamma-\Lambda)\,\Omega + \Omega\nabla\cdot(\kappa\nabla T) \quad , \tag{35.26}$$

$$\frac{DT}{Dt} = \frac{(\Gamma-\Lambda) + \nabla\cdot(\kappa\nabla T) + (D/Dt)(nkT)}{(f+2)nk/2} \quad . \tag{35.27}$$

Note that Eq. (35.27) for the time evolution of the temperature depends not only on the cooling but also on the time-dependence of the pressure.

Isobaric cooling: If the pressure remains constant, Eq. (35.27) becomes

$$\frac{DT}{Dt} = \frac{(\Gamma - \Lambda) + \nabla \cdot (\kappa \nabla T)}{(f+2)nk/2} \tag{35.28}$$

$$\rightarrow \frac{(\Gamma - \Lambda) + \nabla \cdot (\kappa \nabla T)}{(5/2)nk} \quad \text{for a monatomic gas.} \tag{35.29}$$

The numerator is the rate/volume of removal of heat, and the denominator $(f + 2)nk/2$ in Eq. (35.28) is just the heat capacity per unit volume for heating or cooling at *constant pressure*.

Isochoric cooling: If the density n remains constant, then Eq. (35.27) becomes

$$\frac{DT}{Dt} = \frac{(\Gamma - \Lambda) + \nabla \cdot (\kappa \nabla T)}{(f/2)nk} \tag{35.30}$$

$$\rightarrow \frac{(\Gamma - \Lambda) + \nabla \cdot (\kappa \nabla T)}{(3/2)nk} \quad \text{for a monatomic gas.} \tag{35.31}$$

The **cooling time** $\tau_{\rm cool}$ for a fluid element is defined to be the characteristic time for the temperature to decrease if $\Gamma = 0$ and there is no thermal conduction:

$$\tau_{\rm cool} \equiv \frac{T}{|DT/Dt|_{\Gamma=0,\kappa\nabla T=0}} \tag{35.32}$$

$$\rightarrow \frac{(f+2)nkT}{2\Lambda} \rightarrow \frac{5nkT}{2\Lambda} \quad \text{for isobaric cooling } (nkT = \text{const}), \tag{35.33}$$

$$\rightarrow \frac{fnkT}{2\Lambda} \rightarrow \frac{3nkT}{2\Lambda} \quad \text{for isochoric cooling } (n = \text{const}), \tag{35.34}$$

where n is the number density of free particles (molecules + atoms + ions + electrons), and the number of degrees of freedom per particle $f = 3$ if no molecules are present.

The radiative cooling time depends on the density, temperature, and composition of the gas. $\tau_{\rm cool}$ is evaluated for $10^4 < T < 10^9$ K in Fig. 34.4, for metal abundances Z ranging from $Z = 0$ to $Z = 3Z_\odot$.

35.4 Electrodynamics in a Conducting Fluid: Flux-Freezing

The fluid state is characterized by two vector fields, $\mathbf{v}(\mathbf{r}, t)$ and $\mathbf{B}(\mathbf{r}, t)$, and two scalar fields, $\rho(\mathbf{r}, t)$ and $T(\mathbf{r}, t)$. Thus far, we have obtained expressions for the time derivatives of ρ (35.7), $\mathbf{v}$ (35.21), and T (35.27). We now consider evolution of the magnetic field $\mathbf{B}$.

Any astrophysical fluid will have some nonzero electrical conductivity σ, defined so that the electric current

$$\mathbf{J} = \sigma \mathbf{E}' \quad . \tag{35.35}$$

In general, magnetic fields will be present and σ will be a tensor, but for purposes of this discussion we will treat σ as a scalar. The electric field $\mathbf{E}'$ in Eq. (35.35) is the field that would be measured by an observer *moving with the fluid*:

$$\mathbf{E}' = \mathbf{E} + \frac{1}{c}\mathbf{v} \times \mathbf{B} \quad , \tag{35.36}$$

where $\mathbf{v}$ is the fluid velocity. Thus

$$\mathbf{E} = \frac{\mathbf{J}}{\sigma} - \frac{1}{c}\mathbf{v} \times \mathbf{B} \quad . \tag{35.37}$$

We now use one of Maxwell's equations:

$$\frac{1}{c}\frac{\partial \mathbf{B}}{\partial t} = -\nabla \times \mathbf{E} = -\nabla \times \left(\frac{\mathbf{J}}{\sigma} - \frac{1}{c}\mathbf{v} \times \mathbf{B} \right) \quad , \tag{35.38}$$

and use another of Maxwell's equations to replace $\mathbf{J}$ by $(c/4\pi)\nabla\times\mathbf{B} - (1/4\pi)\partial\mathbf{D}/\partial t$ to obtain

$$\frac{\partial \mathbf{B}}{\partial t} = \nabla \times (\mathbf{v} \times \mathbf{B}) - \frac{c^2}{4\pi\sigma}\nabla \times (\nabla \times \mathbf{B}) + \frac{c}{4\pi\sigma}\nabla \times \frac{\partial \mathbf{D}}{\partial t} \quad . \tag{35.39}$$

Now, $\nabla \times (\nabla \times \mathbf{B}) \equiv \nabla (\nabla \cdot \mathbf{B}) - \nabla^2 \mathbf{B}$. With $\nabla \cdot \mathbf{B} = 0$, we obtain

$$\frac{\partial \mathbf{B}}{\partial t} \approx \nabla \times (\mathbf{v} \times \mathbf{B}) + \frac{c^2}{4\pi\sigma}\nabla^2 \mathbf{B} \quad , \tag{35.40}$$

where we have dropped the term $(c/4\pi\sigma)\nabla \times \partial\mathbf{D}/\partial t$.[3]

Suppose now that we draw a closed loop L in the fluid, and assume that the loop L moves *with the fluid*. Let $\Phi(t)$ be the magnetic flux through the loop:

$$\Phi(t) \equiv \int d\mathbf{S} \cdot \mathbf{B} \quad , \tag{35.41}$$

where the integral is over a surface S bounded by the loop L. The rate of change of Φ comes from (1) changes in $\mathbf{B}$, and (2) motion of the loop:

$$\frac{d\Phi}{dt} = \int d\mathbf{S} \cdot \frac{\partial \mathbf{B}}{\partial t} + \oint \mathbf{B} \cdot (\mathbf{v} \times d\mathbf{L}) \tag{35.42}$$

$$= \int d\mathbf{S} \cdot \frac{\partial \mathbf{B}}{\partial t} + \oint d\mathbf{L} \cdot (\mathbf{B} \times \mathbf{v}) \quad , \tag{35.43}$$

[3] $(c/4\pi\sigma)\nabla \times \partial\mathbf{D}/\partial t$ is smaller than $(c^2/4\pi\sigma)\nabla^2\mathbf{B}$ by a factor $(v/c)^2$.

where we have used the vector identity $\mathbf{B} \cdot (\mathbf{v} \times d\mathbf{L}) \equiv d\mathbf{L} \cdot (\mathbf{B} \times \mathbf{v})$. Now use Stokes's theorem to convert the integral over the loop to a surface integral:

$$\frac{d\Phi}{dt} = \int d\mathbf{S} \cdot \frac{\partial \mathbf{B}}{\partial t} + \int d\mathbf{S} \cdot \nabla \times (\mathbf{B} \times \mathbf{v}) \tag{35.44}$$

$$= \int d\mathbf{S} \cdot \left[\frac{\partial \mathbf{B}}{\partial t} + \nabla \times (\mathbf{B} \times \mathbf{v}) \right] \tag{35.45}$$

$$\approx \int d\mathbf{S} \cdot \frac{c^2}{4\pi\sigma} \nabla^2 \mathbf{B} \quad , \tag{35.46}$$

where we have used Eq. (35.40) to go from (35.45) to (35.46). Thus, we see that in the limit of infinite conductivity $\sigma \to \infty$, we have $d\Phi/dt = 0$: *the flux Φ through the loop is a conserved quantity.* By assumption, the loop moves with the fluid; the constancy of the flux is exactly what we would have if every magnetic field line were "frozen" into the fluid. Equation (35.40) with $\sigma \to \infty$ therefore describes **"flux-freezing."**

The conductivity is, of course, finite. The time scale for magnetic field decay is

$$\tau_{\rm decay} \approx \frac{B}{(\partial B/\partial t)_{\rm decay}} \approx \frac{B}{(c^2/4\pi\sigma)(B/L^2)} \approx \frac{4\pi\sigma L^2}{c^2} \quad . \tag{35.47}$$

For a fully ionized hydrogen plasma, the electrical conductivity is

$$\sigma \approx 0.59 \frac{(kT)^{3/2}}{e^2 m_e^{1/2} \ln\Lambda} = 4.6 \times 10^9\,\mathrm{s}^{-1} \left(\frac{T}{100\,\mathrm{K}}\right)^{3/2} \left(\frac{30}{\ln\Lambda}\right) \quad , \tag{35.48}$$

where the "Coulomb logarithm" $\ln\Lambda$ is given by Eq. (2.17). The time scale for magnetic field decay is

$$\tau_{\rm decay} = \frac{4\pi\sigma L^2}{c^2} \approx 5 \times 10^8\,\mathrm{yr} \left(\frac{T}{100\,\mathrm{K}}\right)^{3/2} \left(\frac{30}{\ln\Lambda}\right) \left(\frac{L}{\mathrm{AU}}\right)^2 \quad . \tag{35.49}$$

In ionized gas, such magnetic field decay due to finite conductivity (referred to as "Ohmic" decay) might possibly be of importance on very short length scales ($L \lesssim$ AU), but for $L \gtrsim 10^2$ AU $\approx 10^{-3}$ pc, $\tau_{\rm decay}$ exceeds the age of the Universe!

35.5★ Virial Theorem

In this section, we present a global dynamical condition that must be satisfied for an inviscid fluid. The *virial theorem* is a powerful mathematical result, based on the equation of momentum conservation:

$$\rho \left(\mathbf{v} \cdot \nabla + \frac{\partial}{\partial t} \right) \mathbf{v} = -\nabla \left(p + \frac{B^2}{8\pi} \right) + \frac{1}{4\pi} (\mathbf{B} \cdot \nabla) \mathbf{B} - \rho \nabla \Phi_{\rm grav} \quad . \tag{35.50}$$

We consider a system bounded by a closed surface S. The surface S is assumed to move with the local fluid velocity $\mathbf{v}$, so that there is no mass flow across the surface: the mass within S is conserved. The pressure p and magnetic field $\mathbf{B}$ are not required to vanish at S, but we will assume that the material exterior to S contributes negligibly to the gravitational acceleration $\nabla\Phi_{\rm grav}$. Assume that viscous stresses can be neglected. Define the quantity

$$I \equiv \int dV r^2 \rho \quad ; \tag{35.51}$$

I has the same dimensions as a moment of inertia. The virial theorem (see Appendix J for the derivation) states that

$$\begin{aligned} \frac{1}{2}\ddot{I} &= 2E_{\rm KE} + 3\Pi + E_{\rm mag} + E_{\rm grav} \\ &\qquad - \oint d\mathbf{S}\cdot\mathbf{r}\left(p + \frac{B^2}{8\pi}\right) + \frac{1}{4\pi}\oint d\mathbf{S}\cdot\mathbf{B}(\mathbf{r}\cdot\mathbf{B}) \quad , \end{aligned} \tag{35.52}$$

$$E_{\rm KE} \equiv \int dV \rho\frac{v^2}{2} \,, \qquad \Pi \equiv \int dV p \,, \qquad E_{\rm mag} \equiv \int dV \frac{B^2}{8\pi} \,, \tag{35.53}$$

$$E_{\rm grav} \equiv -\frac{1}{2}\sum_i \sum_{j\neq i} \frac{G m_i m_j}{|\mathbf{r}_i - \mathbf{r}_j|} = -\frac{G}{2}\int dV_1 \int dV_2 \frac{\rho(\mathbf{r}_1)\rho(\mathbf{r}_2)}{|\mathbf{r}_1 - \mathbf{r}_2|} \,, \tag{35.54}$$

where the factor of $\frac{1}{2}$ in Eq. (35.54) allows for the fact that each pair appears twice in the double sum. If the pressure and magnetic field have uniform values p_0 and $\mathbf{B}_0$ at the bounding surface S, then the surface integrals can be evaluated, and the virial theorem becomes

$$\frac{1}{2}\ddot{I} = 2E_{\rm KE} + 3(\Pi - \Pi_0) + (E_{\rm mag} - E_{\rm mag,0}) + E_{\rm grav} \quad , \tag{35.55}$$

$$\Pi_0 \equiv p_0 V \;, \qquad E_{\rm mag,0} \equiv \frac{B_0^2}{8\pi} V \;, \tag{35.56}$$

where V is the enclosed volume. The usual application of the virial theorem is to systems that are in a steady state, with time-average $\langle \ddot{I} \rangle = 0$. Then,

$$0 = 2\langle E_{\rm KE}\rangle + 3\langle \Pi - \Pi_0\rangle + \langle E_{\rm mag} - E_{\rm mag,0}\rangle + \langle E_{\rm grav}\rangle \quad . \tag{35.57}$$

Chapter Thirty-six

Shock Waves

A **shock wave** is defined here to be a pressure-driven disturbance propagating faster than the signal speed for compressive waves, resulting in an irreversible change (i.e., an increase in entropy). The literature on shock waves is vast (e.g., Landau & Lifshitz 2006; Zeldovich & Raizer 1968; Whitham 1974) with a review of interstellar shock waves by Draine & McKee (1993). This chapter will focus on the basic equations applicable to shocks, and will mention the different types of shocks that arise in different astrophysical settings – adiabatic shocks; radiative shocks; collisionless shocks; C-, C*-, and J-type two-fluid shocks.

We will speak of **shock fronts** – the shock front is the transition zone where the fluid properties (e.g., density) change from preshock values to postshock values.

36.1 Sources of Interstellar Shocks

Shocks are common in the ISM, occurring in a number of different circumstances, for example:

- Stellar explosions – novae and supernovae – drive shocks into the ISM.
- The gas in fast stellar winds is eventually decelerated when it "runs into" the ISM; the deceleration involves passage of the wind material through a shock front.
- Expanding H II regions can drive shock waves into surrounding neutral gas.
- In spiral galaxies, the gas in the interstellar medium is accelerated by gravity as it enters a spiral arm, and may pass through a shock transition involving deceleration (relative to the spiral arm) and compression.
- The ISM in a star-forming galaxy is "stirred up" by injection of energy from stars (H II regions, stellar winds, supernovae) and clouds moving supersonically will sometimes collide, driving shock waves into both of the colliding clouds.

36.2 Jump Conditions: Rankine-Hugoniot Relations

The sharp change in fluid properties (density, velocity, pressure) in a shock transition is called a **jump**. Our goal in this section is to obtain a set of **jump conditions** that relate the preshock and postshock state of the fluid. The jump conditions are often referred to as **Rankine-Hugoniot relations**, after 19th century physicists Rankine and Hugoniot.

36.2.1 Conservation of Mass

Shock waves are intrinsically transient phenomena, but it is often useful to approximate real time-dependent shocks by steady (time-independent) shocks. When this is the case, it is advantageous to adopt a frame of reference that is *moving with the shock*: we call this the **shock frame**.

In the shock frame, the shock structure appears to be stationary, although the fluid itself is moving; in the shock frame, a steady shock has $\partial/\partial t = 0$ for all variables. It is further advantageous if we can approximate the shock front as **plane-parallel**; then, if $\hat{\mathbf{x}}$ is the direction normal to the shock, we have $\partial/\partial y = \partial/\partial z = 0$. With these conditions, the equation of mass conservation $\partial\rho/\partial t = -\nabla\cdot(\rho\mathbf{v})$ reduces to

$$0 = -\frac{\partial}{\partial x}(\rho v_x) \quad , \tag{36.1}$$

which is readily integrated from x_1 to x_2 to obtain

$$(\rho v_x)_1 = (\rho v_x)_2 \quad . \tag{36.2}$$

36.2.2 Conservation of Momentum

The equation for momentum conservation, *including* viscous stresses, is a vector equation, Eq. (35.21). For a steady, plane-parallel shock, the x component of the equation reduces to

$$\rho v_x \frac{\partial v_x}{\partial x} = -\frac{\partial}{\partial x}\left(p + \frac{B^2}{8\pi}\right) + \frac{1}{4\pi}B_x\frac{\partial}{\partial x}B_x - \rho\frac{\partial}{\partial x}\Phi_{\rm grav} + \frac{\partial}{\partial x}\sigma_{xx} \quad , \tag{36.3}$$

which can be written (making use of $\rho v_x = const$) in terms of the divergence of a conserved momentum flux

$$\frac{\partial}{\partial x}\left(\rho v_x^2 + p + \frac{(B_y^2 + B_z^2)}{8\pi} - \sigma_{xx}\right) = -\rho\frac{\partial\Phi_{\rm grav}}{\partial x} \quad , \tag{36.4}$$

with a "source term" on the right-hand side. We now integrate across the shock transition. The viscous stress tensor σ_{xx} is large only in a narrow transition layer; we choose a preshock location x_1 and postshock location x_2 just outside this shock transition layer, so that $(\sigma_{xx})_1 \approx 0$, $(\sigma_{xx})_2 \approx 0$. We further assume that x_1 and x_2 are sufficiently close together so that we can neglect the integral over $\rho \partial \Phi_{\text{grav}}/\partial x$. The equation of momentum conservation then becomes

$$\left(\rho v_x^2 + p + \frac{B_y^2 + B_z^2}{8\pi} \right)_1 = \left(\rho v_x^2 + p + \frac{B_y^2 + B_z^2}{8\pi} \right)_2 . \tag{36.5}$$

36.2.3 Conservation of Energy

We now consider energy conservation. To calculate *all* of the mechanical work done on the fluid element (including viscous stresses), it is convenient to calculate what is happening at the surface, since the pressure force acts across the surface, and we have seen in Eq. (35.16) that the electromagnetic forces can also be evaluated as acting at the surface. The work per unit time done by the force acting at each point is obtained by taking the dot product of the force vector with the velocity *at that point*. The rate at which work is being done on the fluid element by external forces is therefore

$$\left(\frac{dE}{dt} \right)_{\text{mech}} = \int dS_i v_i \left(-p - \frac{B^2}{8\pi} \right) + \int dS_j v_i \left(\frac{B_i B_j}{4\pi} + \sigma_{ij} \right) - \int dV \rho v_i \frac{\partial}{\partial x_i} \Phi_{\text{grav}} . \tag{36.6}$$

In addition to the mechanical power, there is a net heating rate per volume $(\Gamma - \Lambda)$ plus heat conduction across the surface. These all add up to change the total energy in the comoving fluid element. Using Gauss's theorem to convert the surface integrals to volume integrals, we can now evaluate the rate of change of the sum of bulk kinetic energy, internal energy, and magnetic energy in the fluid element:

$$\frac{D}{Dt} \left[\frac{1}{2} \rho \Omega v^2 + \Omega U + \Omega \frac{B^2}{8\pi} \right] = \Omega \frac{\partial}{\partial x_i} \left[-p v_i - \frac{B^2}{8\pi} v_i + v_j \frac{B_i B_j}{4\pi} + v_j \sigma_{ji} \right] - \Omega \rho v_i \frac{\partial}{\partial x_i} \Phi_{\text{grav}} + \Omega(\Gamma - \Lambda) - \Omega \nabla \cdot (-\kappa \nabla T) . \tag{36.7}$$

Equation (36.7) is fully general. If we now set $\partial/\partial y = \partial/\partial z = \partial/\partial t = 0$, it simplifies to

$$\frac{\partial}{\partial x} \left[\frac{1}{2} \rho v_x v^2 + U v_x + p v_x + \frac{(B_y^2 + B_z^2)}{4\pi} v_x - \frac{B_x B_y v_y}{4\pi} - \frac{B_x B_z v_z}{4\pi} - v_j \sigma_{jx} - \kappa \frac{dT}{dx} + \rho v_x \Phi_{\text{grav}} \right] = \Gamma - \Lambda . \tag{36.8}$$

We now integrate this from preshock point 1 to postshock point 2, where we again assume that 1 and 2 are sufficiently far from the shock transition that $(\sigma_{jx})_1 \approx (\sigma_{jx})_2 \approx 0$. We also assume that 1 and 2 are sufficiently close together that $\int_1^2(\Gamma - \Lambda)dx \approx 0$, and $(\Phi_{\rm grav,2} - \Phi_{\rm grav,1}) \approx 0$. We further assume that the internal energy per unit volume is proportional to the thermal pressure:

$$U = \frac{p}{(\gamma - 1)} \quad , \tag{36.9}$$

where $\gamma \equiv c_p/c_v = (f + 2)/f$ is the ratio of specific heat at constant pressure to specific heat at constant volume. For a monatomic gas or plasma, $\gamma = 5/3$.

The energy jump condition is then

$$\left\{\left[\frac{\rho v^2}{2} + \frac{\gamma p}{(\gamma - 1)}\right]v_x + \frac{(B_y^2 + B_z^2)}{4\pi}v_x - \frac{(B_x B_y v_y + B_x B_z v_z)}{4\pi} - \kappa\frac{dT}{dx}\right\}_1 =$$

$$\left\{\left[\frac{\rho v^2}{2} + \frac{\gamma p}{(\gamma - 1)}\right]v_x + \frac{(B_y^2 + B_z^2)}{4\pi}v_x - \frac{(B_x B_y v_y + B_x B_z v_z)}{4\pi} - \kappa\frac{dT}{dx}\right\}_2 . \tag{36.10}$$

36.2.4 Conservation of Magnetic Flux

In addition to mass, momentum, and energy conservation, we have one more constraint from Maxwell's equations and the assumption of infinite electrical conductivity: Eq. (35.40) gives

$$\frac{\partial \mathbf{B}}{\partial t} = \nabla \times (\mathbf{v} \times \mathbf{B}) \quad . \tag{36.11}$$

With $\partial/\partial t = \partial/\partial y = \partial/\partial z = 0$ the y and z components of this equation give

$$0 = -\frac{\partial}{\partial x}(v_x B_y - v_y B_x) \quad , \tag{36.12}$$

$$0 = -\frac{\partial}{\partial x}(v_z B_x - v_x B_z) \quad . \tag{36.13}$$

36.2.5 Jump Conditions

The preceding equations are general for plane-parallel steady shocks. We now specialize to the case $B_x = 0$ (purely transverse magnetic field) and $\kappa\nabla T = 0$

(negligible thermal conductivity), and we choose a frame where $v_y = v_z = 0$ in the preshock gas (if $B_x = 0$, the postshock gas will then also have $v_y = v_z = 0$), and $B_z = 0$. The conservation laws for mass, momentum, energy, and magnetic flux can then be rewritten

$$\rho_1 u_1 = \rho_2 u_2 \quad , \tag{36.14}$$

$$\rho_1 u_1^2 + p_1 + \frac{B_1^2}{8\pi} = \rho_2 u_2^2 + p_2 + \frac{B_2^2}{8\pi} \quad , \tag{36.15}$$

$$\frac{\rho_1 u_1^3}{2} + \frac{\gamma}{\gamma-1} u_1 p_1 + \frac{u_1 B_1^2}{4\pi} = \frac{\rho_2 u_2^3}{2} + \frac{\gamma}{\gamma-1} u_2 p_2 + \frac{u_2 B_2^2}{4\pi} \quad , \tag{36.16}$$

$$u_1 B_1 = u_2 B_2 \quad , \tag{36.17}$$

where $u \equiv v_x$, and $B = B_y$. We assume that ρ_1, u_1, p_1 and B_1 are given. We have four equations in four unknowns: ρ_2, u_2, p_2, and B_2. The trivial solution ($\rho_2 = \rho_1, u_2 = u_1, p_2 = p_1$, and $B_2 = B_1$) always exists, but what is the condition for a second physical solution to exist? Define $v_s \equiv u_1$, and let $x \equiv \rho_2/\rho_1$ be the compression ratio. From mass conservation, we have $u_2 = v_s/x$, and flux conservation gives $B_2 = xB_1$. The momentum and energy equations become

$$\rho_1 v_s^2 + p_1 + \frac{B_1^2}{8\pi} = \frac{\rho_1 v_s^2}{x} + p_2 + \frac{B_1^2}{8\pi} x^2 \quad , \tag{36.18}$$

$$\frac{1}{2}\rho_1 v_s^3 + \frac{\gamma}{\gamma-1} p_1 v_s + \frac{B_1^2}{4\pi} v_s = \frac{1}{2}\frac{\rho_1 v_s^3}{x^2} + \frac{\gamma}{\gamma-1}\frac{p_2 v_s}{x} + \frac{B_1^2}{4\pi} v_s x \quad . \tag{36.19}$$

Now we have two equations in two unknowns: compression x and postshock pressure p_2. If we solve the first for p_2, and substitute into the second, we obtain a cubic equation for x. Remembering that $x = 1$ must be a solution, we can factor out $(x-1)$ to reduce this to a quadratic. The solution for the compression ratio x is

$$x = \frac{2(\gamma+1)}{D + \sqrt{D^2 + 4(\gamma+1)(2-\gamma)M_A^{-2}}} \quad , \tag{36.20}$$

$$D \equiv (\gamma - 1) + 2M^{-2} + \gamma M_A^{-2} \quad ,$$

$$M \equiv \frac{v_s}{\sqrt{\gamma p_1/\rho_1}} \quad ,$$

$$M_A \equiv \frac{v_s}{B_1/\sqrt{4\pi\rho_1}} \quad .$$

The condition for a compressive (i.e., $x > 1$) solution to exist is that

$$v_s > V_{\rm ms} \equiv \sqrt{\frac{\gamma p_1}{\rho_1} + \frac{B_1^2}{4\pi\rho_1}} \quad , \tag{36.21}$$

where the **magnetosonic speed** $V_{\rm ms}$ is the speed of small-amplitude compressive waves propagating perpendicular to $\mathbf{B}$. For a shock wave to exist, it must propagate *faster* than small-amplitude compressive waves in the preshock medium.

With the compression factor x from Eq. (36.20), one can immediately obtain the postshock velocity $v_2 = v_s/x$ and density $\rho_2 = x\rho_1$. The pressure p_2 can be obtained from Eq. (36.18), and the temperature $T_2 = p_2\mu/\rho_2 k$, where μ is the mass per particle.

When $\mathcal{M} \equiv v_s/V_{\rm ms} \gg 1$, the shock is termed a **strong shock**; the **strong shock jump conditions** are

$$x \equiv \frac{\rho_2}{\rho_1} \rightarrow \frac{\gamma+1}{\gamma-1} = 4 \text{ for } \gamma = 5/3 \quad , \tag{36.22}$$

$$u_2 \rightarrow \frac{\gamma-1}{\gamma+1} v_s = \frac{1}{4} v_s \text{ for } \gamma = 5/3 \quad , \tag{36.23}$$

$$T_2 \rightarrow \frac{2(\gamma-1)}{(\gamma+1)^2}\frac{\mu v_s^2}{k} = \frac{3}{16}\frac{\mu v_s^2}{k} \text{ for } \gamma = 5/3 \quad . \tag{36.24}$$

For a strong shock, the postshock temperature T_2 depends on both the shock speed and the molecular weight. Including helium (with $n_{\rm He}/n_{\rm H} = 0.1$):

$$\mu = \frac{1.4 m_{\rm H}}{1.1} = 1.273\, m_{\rm H} \quad \text{for neutral H I,} \tag{36.25}$$

$$\mu = \frac{1.4 m_{\rm H}}{2.3} = 0.609\, m_{\rm H} \quad \text{for fully ionized gas.} \tag{36.26}$$

For a strong shock, the postshock temperature is

$$T_2 \approx \frac{3}{16}\frac{\mu v_s^2}{k} = 2890\,{\rm K}\left(\frac{\mu}{1.273 m_{\rm H}}\right)\left(\frac{v_s}{10\,{\rm km\,s^{-1}}}\right)^2 \tag{36.27}$$

$$= 1.38 \times 10^7\,{\rm K}\left(\frac{\mu}{0.609 m_{\rm H}}\right)\left(\frac{v_s}{1000\,{\rm km\,s^{-1}}}\right)^2 \quad . \tag{36.28}$$

36.2.6 Postshock Cooling and "Isothermal" Shocks

We have used the conservation laws to obtain the immediate postshock compression and temperature. Suppose now that the hot gas cools by emitting radiation, cooling until it reaches a temperature $T_3 = T_1$. A shock where the postshock gas cools to the initial temperature T_1 is sometimes referred to as an **isothermal shock**. This is misleading – the gas temperature does *not* remain constant.

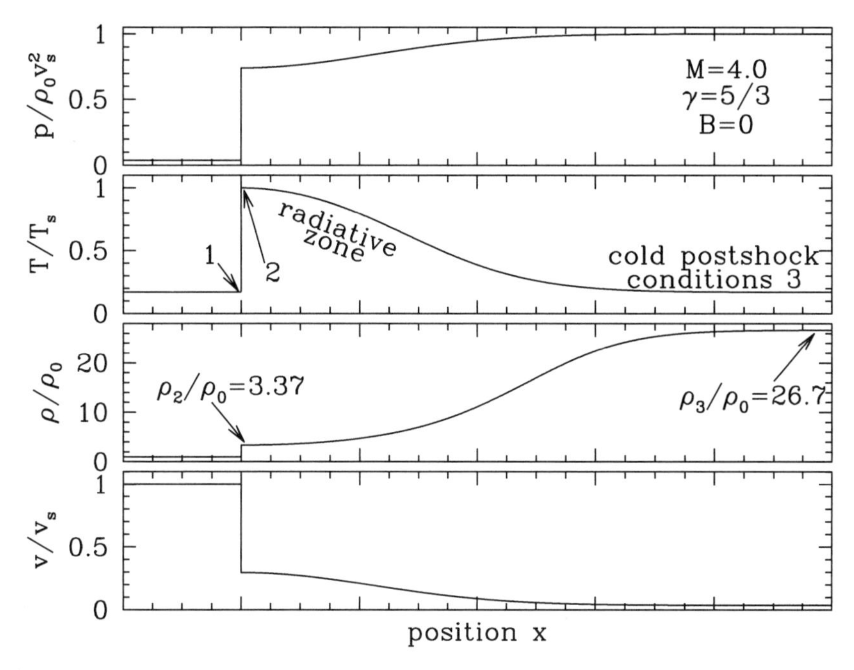

Figure 36.1 Structure of a nonmagnetic radiative shock. The fluid is assumed to have $\gamma = 5/3$, and the shock has Mach number $M = 4$.

Energy has been removed, so energy conservation equations do not apply, but mass, momentum, and magnetic flux remain conserved. Thus set $T_3 = T_1$, $\rho_3 = x_3\rho_1$, $p_3 = x_3 p_1$; momentum conservation gives

$$\rho_1 v_s^2 + p_1 + \frac{B_1^2}{8\pi} = \frac{\rho_1 v_s^2}{x_3} + x_3 p_1 + \frac{B_1^2}{8\pi} x_3^2 \quad . \tag{36.29}$$

The nonmagnetic case ($B_1 = 0$) gives a quadratic equation for x_3, with one trivial root ($x_3 = 1$) and the nontrivial root

$$x_3 = M_{\rm iso}^2 \equiv \frac{\rho_1 v_s^2}{p_1} \quad , \tag{36.30}$$

where the "isothermal Mach number" $M_{\rm iso}$ is the ratio of the shock speed v_s to the isothermal sound speed $\sqrt{p_1/\rho_1}$.

Now consider the more general case where $B_1 \neq 0$. Equation (36.29) now gives a cubic equation for x_3; one root is $x_3 = 1$, the other physical root is

$$x_3 = \sqrt{2} M_A \left[1 + \frac{1}{8M_A^2} \left(1 + 2\frac{M_A^2}{M_{\rm iso}^2} \right)^2 \right]^{1/2} - \frac{1}{2} - \frac{M_A^2}{M_{\rm iso}^2} \quad (36.31)$$

$$\rightarrow \begin{cases} \sqrt{2} M_A & \text{for } M_{\rm iso}^2 \gg M_A^2 > 1 \\ M_{\rm iso}^2 & \text{for } M_A \gg M_{\rm iso}^2 > 1 \end{cases} . \quad (36.32)$$

The structure of a single-fluid nonmagnetic shock is shown in Figure 36.1.

36.3 Cooling Time and Cooling Length

For a steady shock, the peak temperature occurs immediately behind the shock front. If the shock has been propagating long enough, the shocked gas is able to cool radiatively, resulting in further compression of the gas. The cooling behind the shock is approximately isobaric, so the cooling time for this gas is given by Eq. (35.33). For a strong shock in ionized gas, with $\rho_2 \approx 4\rho_1$, and temperature given by Eq. (36.28), the time $t_{\rm cool}$ for a just-shocked fluid element scales as $1/n_{\rm H,0}$ where $n_{\rm H,0}$ is the preshock density. Figure 36.2a shows $n_{\rm H,0}t_{\rm cool}$ as a function of shock speed. We see that shocks with $v_s \lesssim 100\,{\rm km\,s^{-1}}$ cool very rapidly, but the gas behind a high-velocity shock takes very long to cool.

For $80 \lesssim v_s \lesssim 1200\,{\rm km\,s^{-1}}$, the cooling time can be approximated by (see Fig. 36.2a)

$$t_{\rm cool} \approx 7000 \left(\frac{\rm cm^{-3}}{n_{\rm H,0}} \right) \left(\frac{v_s}{100\,{\rm km\,s^{-1}}} \right)^{3.4} {\rm yr} \quad . \quad (36.33)$$

The cooling length – the spatial extent of the radiative zone in Fig. 36.1 – is $L_{\rm cool} = (v_s/4)t_{\rm cool} \propto 1/n_{\rm H,0}$. The column density of material in the radiative zone $N_{\rm H,cool} = (4n_{\rm H,0})L_{\rm cool} = n_{\rm H,0}v_s t_{\rm cool}$ depends on v_s but not the preshock density $n_{\rm H,0}$. The cooling column density $N_{\rm H,cool}$ is shown as a function of shock speed in Fig. 36.2b.

36.4★ Collisionless Shocks

In the discussion thus far, we have assumed that viscous dissipation is capable of effecting the shock transition, dissipating much of the initial kinetic energy and converting it to heat. In **collisional** shocks this is accomplished by elastic scattering of molecules, atoms, or ions, and the shock transition has a thickness that is of order the mean free path against 90° scattering. From Chapter 2, the mean free path (mfp) in ionized gas at the immediate postshock temperature $T_s = (3/16)\mu v_s^2/k$

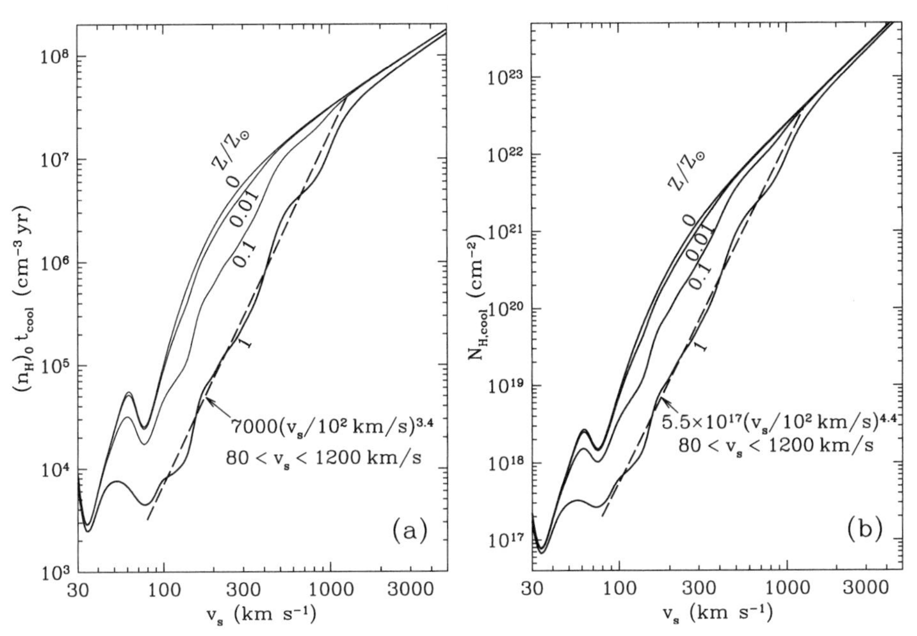

Figure 36.2 (a) $n_{\mathrm{H},0}t_{\mathrm{cool}}$ as a function of shock speed v_s for shocks propagating into H II, for the cooling function shown in Fig. 34.1. Strong shock jump conditions (for $\gamma = 5/3$) are assumed. $n_{\mathrm{H},0}$ is the preshock density, and t_{cool} is the isobaric cooling time (Eq. 35.33) for the immediate postshock gas. The cooling function $\Lambda(T)$ is taken from Fig. 34.1. Curves are labeled by $Z/Z_\odot$, the metallicity relative to solar. (b) The cooling column density $N_{\mathrm{H,cool}} = n_{\mathrm{H},0}v_s t_{\mathrm{cool}}$.

and immediate postshock density $n = 4n_1$ is

$$\mathrm{mfp} = v_1 t_{\mathrm{defl}} = \frac{3^4\mu^2 v_s^4}{2^{13}\pi n_1 e^4 \ln\Lambda} \tag{36.34}$$

$$= 2.4\times 10^{17}\,\mathrm{cm}\left(\frac{v_s}{10^3\,\mathrm{km\,s^{-1}}}\right)^4\left(\frac{\mathrm{cm}^{-3}}{n_e}\right) , \tag{36.35}$$

where we have evaluated the mfp for a proton with $E = (3/2)kT_s$. For shock speeds $v_s \gtrsim 10^3\,\mathrm{km\,s^{-1}}$, the mean free path against collisional scattering is very long, yet we see evidence of relatively well-defined shock transitions in young supernova remnants, and in interplanetary space.

The shock transition requires that the particle velocities be isotropized (in the postshock frame), but this can be done by large-scale electric and magnetic fields resulting from collective motions in the plasma. Shocks where the deflection of ions is accomplished by collective effects are referred to as **collisionless shocks**. The collective motions that generate the large-scale electric and magnetic fields are excited by instabilities driven by the streaming of the undeflected ions and electrons through previously decelerated plasma. If the preshock plasma is already magnetized, then the already-present magnetic field is available to deflect streaming ions and electrons. Instabilities in collisionless shocks can substantially amplify the magnetic field, beyond the factor of ~ 4 compression that would result in a plane-parallel strong shock with flux-freezing (Bell 2004). Even if no magnetic field is initially present, the Weibel instability (Weibel 1959) is able to generate magnetic fields in strong collisionless shocks.

36.5★ Electron Temperature

The existence of collisionless shocks is established by observations, showing that the bulk kinetic energy of the preshock flow is substantially converted to kinetic energy in random motions. Elastic scattering processes will relatively quickly isotropize the velocity distribution functions. Scattering of ions by ions, and electrons by electrons, will cause the ions and electrons to each relax to Maxwellian velocity distributions.

Almost all of the kinetic energy is originally in the ions. Isotropization of the velocity distribution function would result in initial ion and electron temperatures with $T_i/T_e \approx (m_p/m_e) = 1836$. Coulomb scattering will cause the electrons to be heated on an equilibration time scale

$$\tau_{\rm equil} \equiv \frac{T_i - T_e}{d(T_e - T_i)/dt} \approx \frac{m_p}{m_e} \times t_{\rm defl}({\rm e\ by\ p}) \tag{36.36}$$

$$\approx 1.4 \times 10^{10}\,{\rm s} \left(\frac{T_e}{10^6\,{\rm K}}\right)^{3/2} \left(\frac{{\rm cm}^{-3}}{n_e}\right) \left(\frac{25}{\ln \Lambda}\right) . \tag{36.37}$$

The long time scale for electron–ion temperature equilibration prevents the electrons from equilibrating with the ions just behind fast shock waves in young supernova remnants, or following shock-heating of the intergalactic medium in galaxy clusters.

When the electrons are much colder than the ions, there may be collective instabilities that will raise the electron temperature up to some fraction (e.g., $\sim 1/10$) of the ion temperature, but the final electron-ion temperature equilibration appears to take place via elastic scattering.

36.6★ Two-Fluid MHD Shocks in Low Fractional Ionization Gas

Earlier we discussed the jump conditions in ordinary shock waves, where some process – either collisional viscosity or some collective instability – decelerates and heats the inflowing preshock gas. This applies to fast shocks, or to shocks in ionized gas.

Much of the ISM has an appreciable magnetic pressure but low fractional ionization (see Chapter 16). In H I clouds (the "CNM"), the fractional ionization

$x_e \approx 3 \times 10^{-4}$ (see Fig. 16.1). In diffuse molecular clouds, ultraviolet radiation maintains ionization of species such as C^+, and the fractional ionization $x_e \approx 1.5 \times 10^{-4}$. In dark clouds, the fractional ionization is much lower, with $x_e \approx 10^{-7}$ in a region with density $n_H \approx 10^4 \,\mathrm{cm}^{-3}$ (see Fig. 16.3).

The gas can be thought of as two distinct fluids coexisting in space: a neutral fluid consisting of atoms and molecules, and a plasma, consisting of the ions, electrons, and magnetic field. When the fractional ionization is sufficiently low, a neutral particle will undergo many scatterings by other neutrals before it is scattered by an ion, so the neutrals act like a fluid with a Maxwellian velocity distribution. The long range of the Coulomb interaction means that the ions or electrons may undergo many Coulomb scatterings off other ions or electrons before interacting with a neutral – so the ions and electrons act like a separate fluid, although the mass mismatch between ions and electrons means that energy exchange between ions and electrons is slow, so that the electron and ion velocity distributions can be characterized by separate temperatures, T_i and T_e. Hence, it makes sense to speak of the neutrals and ions as distinct fluids, weakly coupled to one another via ion–neutral collisions.

Because the fractional ionization is low, even a modest magnetic field strength can imply a large value for the **magnetosonic speed** for compressive waves in the plasma:

$$V_{\rm ims} \approx \frac{B_0}{\sqrt{4\pi\rho_i}} \approx 112 \,\mathrm{km\,s^{-1}} \left(\frac{B_0}{5\,\mu\mathrm{G}}\right) \left(\frac{10^2\,\mathrm{cm}^{-3}}{n_{\rm H}} \frac{10^{-4}}{x_i}\right)^{1/2} , \tag{36.38}$$

where we have assumed a mass per ion $m_i \approx 12 m_{\rm H}$. If the shock speed $v_s < V_{\rm ims}$, then compressive waves in the plasma can propagate upstream ahead of the shock, producing compression in the plasma before it arrives at the shock transition in the neutral gas.

The neutral particles are not directly affected by the magnetic field, but the charged particles experience Lorentz forces when they move perpendicularly to the magnetic field. This means that when the magnetic field is strong enough to be dynamically important, it can only affect the neutral gas via ion–neutral scattering. But when the fractional ionization is low, the collisional coupling between neutrals and ions is weak. This leads to an interesting new type of shock structure – two-fluid MHD shocks.

To understand the structure of two-fluid MHD shocks, it is helpful to think about a sequence of two-fluid shocks, starting from a shock where the magnetic field is very weak, and considering the effect of gradually increasing the magnetic field strength – see Figure 36.3. In the upper panel, the magnetic field is assumed to be dynamically weak, but to be large enough that $V_{\rm ims} > v_s$. The compressive waves propagating upstream in the plasma ahead of the shock are damped by ion–neutral friction since the neutral gas is not participating in the compression. This results in a **magnetic precursor** of finite extent. We can estimate the extent L of the magnetic precursor by simply looking at the equation of motion for the ions:

$$\rho_i \frac{D\mathbf{v}_i}{Dt} = -\nabla p_i - \nabla \frac{B^2}{8\pi} + \frac{(\mathbf{B}\cdot\nabla)\mathbf{B}}{4\pi} + n_n n_i \langle \sigma v \rangle_{in} \mu (\mathbf{v}_n - \mathbf{v}_i) , \tag{36.39}$$

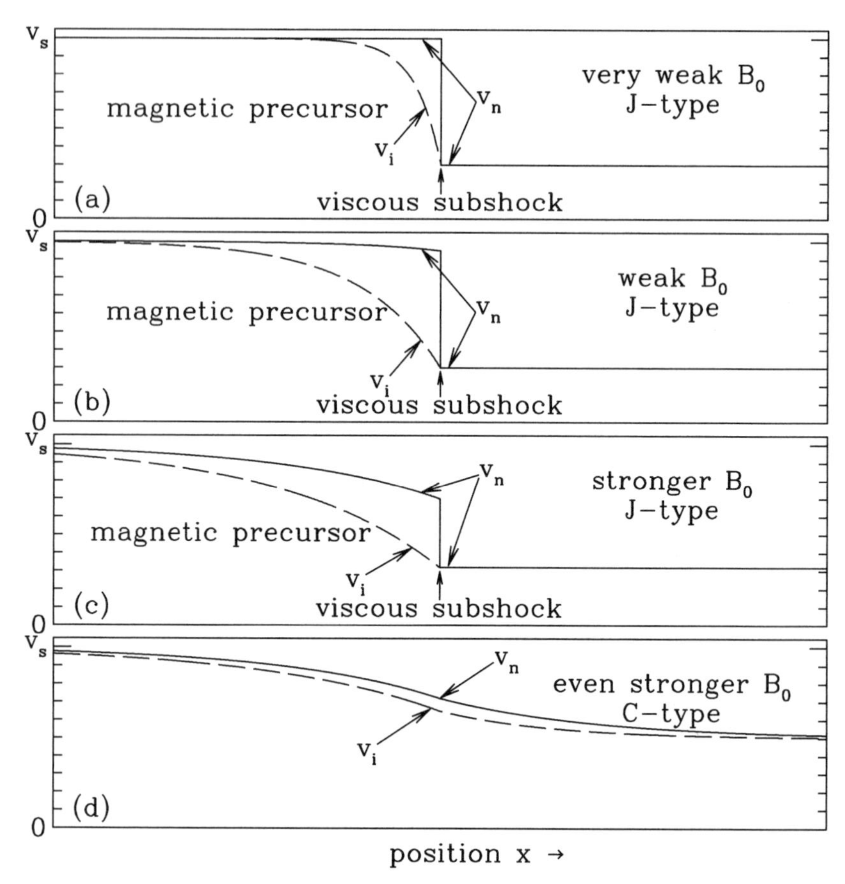

Figure 36.3 Sequence of two-fluid shocks, with different values of preshock magnetic field strength B_0. As the magnetic field strength B_0 is increased, the magnetic precursor extends further ahead of the viscous subshock (the discontinuity in the neutral velocity), and produces more deceleration of the neutrals. For large enough B_0, the neutrals are decelerated entirely by the magnetic precursor, and the shock becomes C-type.

where we now have a **source term** on the right-hand side, representing the rate per volume at which the ion momentum is changed as a result of collisions with neutrals. For plane-parallel, steady flows this becomes (using $\nabla \cdot \mathbf{B} = 0$)

$$\rho_i v_{ix} \frac{dv_{ix}}{dx} = -\frac{d}{dx} p_i - \frac{d}{dx} \frac{B^2}{8\pi} + n_n n_i \langle \sigma v \rangle_{in} \mu_{in} (v_{nx} - v_{ix}) \quad , \qquad (36.40)$$

where $\mu_{in} \equiv m_n m_i / (m_n + m_i)$ is the reduced mass in collisions between neutrals of mass m_n and ions of mass m_i. Because ρ_i and p_i are small, Eq. (36.40) requires approximate balance between the two dominant terms, and we can solve for the length scale L:

$$\frac{d}{dx}\left(\frac{B^2}{8\pi}\right) \approx n_n n_i \langle\sigma v\rangle_{in} \mu_{in} (v_{nx} - v_{ix}) \quad , \tag{36.41}$$

$$\frac{(2B_0)^2}{8\pi L} \approx n_{\rm H}(2x_i n_{\rm H})\langle\sigma v\rangle_{in} \frac{m_n m_i}{(m_n + m_i)} \frac{v_s}{2} \, , \tag{36.42}$$

$$L \approx 2\frac{v_{A0}^2}{v_s} \frac{(m_n + m_i)}{m_i} \frac{1}{n_{i0}\langle\sigma v\rangle_{in}} \tag{36.43}$$

$$\approx 1 \times 10^{15}\,{\rm cm} \left(\frac{v_{A0}}{{\rm km\,s^{-1}}}\right)^2 \left(\frac{10\,{\rm km\,s^{-1}}}{v_s}\right) \left(\frac{0.01\,{\rm cm^{-3}}}{n_{i0}}\right) , \tag{36.44}$$

where

$$v_{A0} \equiv \frac{B_0}{\sqrt{4\pi\rho_0}} = 1.84 \frac{B_0}{\mu{\rm G}} \left(\frac{{\rm cm^{-3}}}{(n_{\rm H})_0}\right)^{1/2} {\rm km\,s^{-1}} \tag{36.45}$$

is the Alfvén speed in the preshock medium, and n_{i0} is the number density of ions (= number density of electrons) in the preshock medium.

If the magnetic field is weak (v_{A0}^2 is small), the magnetic precursor does not extend very far upstream and the neutral gas, with its much greater inertia, is hardly affected by the ion–neutral collisions taking place in the very short precursor (see Fig. 36.3a). The deceleration and heating in the overall shock is almost entirely due to the viscous stresses arising in a transition layer that we term the **viscous subshock**, with thickness of order the mean free path for a neutral atom against elastic scattering by another neutral. Two-fluid shocks where a viscous "jump" is present are referred to as **J-type** shocks (Draine 1980).

If the preshock magnetic field strength is increased, the precursor extends farther (see Fig. 36.3b). The zone within which ion–neutral collisions are occurring is enlarged, and the amount of momentum removed from the streaming neutrals becomes appreciable, causing the flow velocity of the neutrals to decline noticeably *before* the neutrals reach the viscous shock transition. In addition to reducing the momentum (and bulk kinetic energy) of the preshock neutral flow, frictional heating by ion–neutral scattering also heats the neutral gas. Because of these two effects, the viscous subshock is weakened, with a reduced compression ratio, but the shock is still J-type.

As the magnetic field strength is further increased (Fig. 36.3c), the magnetic precursor extends further, the momentum transfer and frictional heating in the magnetic precursor region increases, and the strength of the viscous subshock further decreases. If the magnetic field is increased sufficiently (Fig. 36.3d), the viscous subshock entirely disappears, and we have an entirely new type of shock – referred to as a **C-type** shock (Draine 1980) – where all of the momentum transfer and dissipation take place as the result of ion–neutral streaming.

For a shock to be C-type, the neutral flow must remain supersonic throughout. If the magnetic field is strong enough (i.e., the preshock Alfvén number v_s/v_{A0} is low enough), this can be accomplished even without radiative cooling. However, even when the Alfvén number is large, the distributed heating by ion–neutral scattering may be countered by local radiative cooling, and it is often possible for radiative cooling to keep the neutral fluid cool enough that it remains supersonic.

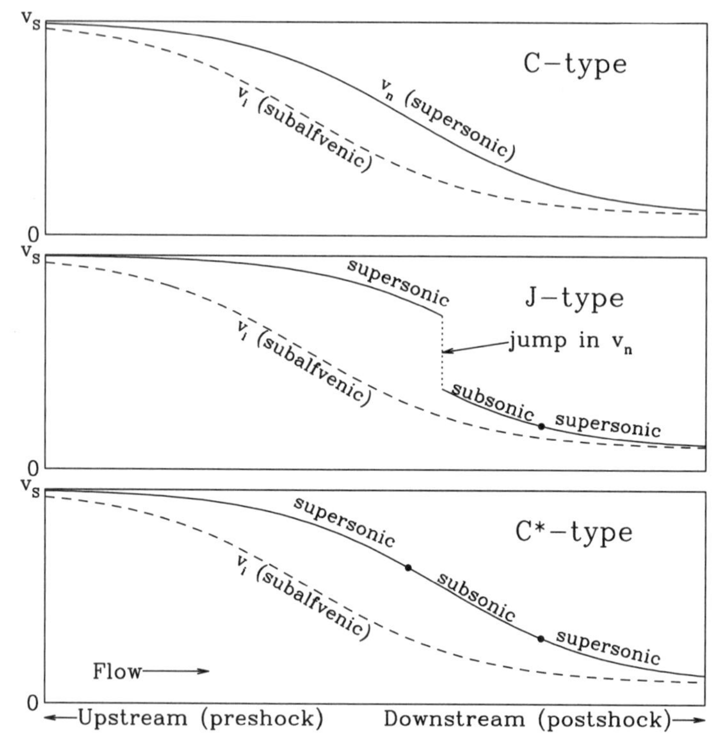

Figure 36.4 Structure of J-type, C*-type, and C-type 2-fluid MHD shocks (see text). In C-type shocks the neutral flow remains supersonic throughout. In J-type shocks the neutral flow makes a supersonic to subsonic transition via a viscous subshock, and as it cools it passes smoothly through a sonic point to become supersonic downstream. In C*-type shocks, the neutral flow passes smoothly through two sonic points as it first heats up from ion-neutral friction, and then cools down.

For fixed preshock conditions ($n_{\rm H}$, T, B_0, n_e), low-speed shocks will be C-type, but as the shock speed increases, heating of the neutral flow will make it subsonic. If we assume that radiative cooling in the postshock region will cause the downstream neutral flow to be supersonic, then we see that the neutral flow must have a supersonic$\rightarrow$subsonic transition at one point in the flow, followed by a subsonic$\rightarrow$supersonic transition further downstream (see Fig. 36.4). In J-type shocks, the supersonic$\rightarrow$subsonic transition is accomplished by a viscous subshock, but it is also possible for the transition to take place smoothly, without a viscous subshock. Such a shock is termed **C*-type** (Chernoff 1987; Roberge & Draine 1990).

Single-fluid shocks dissipate most of the ordered kinetic energy in a narrow viscous transition, with the heating taking place too rapidly for cooling to affect the peak temperature reached. In two-fluid shocks, on the other hand, much of the energy dissipation is continuous rather than impulsive, and radiative cooling is able to keep the gas cool even when the shock velocity is quite high.

The Orion Molecular Cloud shows strong infrared emission in vibration–rotation lines of H_2, with line profiles indicating that some of the emitting H_2 is moving at $> 30\,\mathrm{km\,s^{-1}}$. Models of single-fluid shocks propagating into H_2 gas showed that the H_2 will be fully dissociated if the shock speed exceeds $\sim 24\,\mathrm{km\,s^{-1}}$ (Kwan 1977), but two-fluid MHD shocks can have minimal H_2 dissociation and strong H_2 line emission for shock speeds as large as $\sim 40\,\mathrm{km\,s^{-1}}$ (Chernoff et al. 1982; Draine & Roberge 1982).

Because radiative cooling is effective, especially in molecular gas where the rotational lines of H_2 are strong coolants, shocks in interstellar H I clouds and molecular clouds are expected to be often C-type or C*-type. Shocks are expected to be common: Draine & Katz (1986*b*) estimated that a sightline traversing a path through one or more clouds with total $E(B-V) = 1$ mag would be expected to intersect one shock wave with $v_s > 8\,\mathrm{km\,s^{-1}}$, and argued that C-type shocks in interstellar clouds could account for the observed abundance of the enigmatic species CH^+ via the endothermic reaction $C^+ + H_2 \rightarrow CH^+ + H$ (see §33.3).

Chapter Thirty-seven

⋆ Ionization/Dissociation Fronts

Chapter 15 discussed the sizes of Strömgren spheres, assuming them to be static, with a stationary boundary separating the ionized from the neutral gas. This boundary is known as an **ionization front**, or **I-front**. I-fronts are common in the interstellar medium. The Orion Bar in Plate 10 is an example of an I-front, there are I-fronts bounding the bright Hα emission in the Trifid Nebula (Plate 11), and there are I-fronts at the surfaces of the many neutral cometary filaments in the Helix Nebula (Plate 13).

In general, an I-front is *not* stationary relative to the gas. This chapter discusses the gas dynamics that is associated with moving I-fronts.

Stars are formed by gravitational collapse in cold, dense molecular gas. Massive stars reach the main sequence quickly, and begin emitting hard ultraviolet radiation while still surrounded by dense molecular gas. The molecules in the nearby gas are photodissociated, and the gas is photoionized. The photodissociation/photoionization process increases the number of free particles ($H_2 \rightarrow 2H \rightarrow 2H^+ + 2e^-$) and raises the gas temperature to $\sim 10^4$ K; the thermal pressure therefore increases by a factor $\sim 10^3$, and this high-pressure gas will expand by pushing away the surrounding lower-pressure atomic and molecular gas. Under some circumstances, this high pressure can drive a shock wave into the surrounding medium, so that the neutral gas that the I-front encounters has already been shock-compressed – this is called a **D-type** I-front. Under other circumstances, the I-front is not preceded by a shock – this is referred to as an **R-type** I-front.

37.1⋆ Ionization Fronts: R-Type and D-Type

Consider a plane-parallel geometry, and adopt a coordinate system that is moving at the same speed as the I-front. Let J be the flux/area of ionizing photon arriving at the I-front. Let subscript 1 denote fluid variables in the neutral gas *ahead* of the I-front, and subscript 2 denotes fluid variables in the ionized gas *behind* the I-front. Mass conservation gives us

$$\rho_1 u_1 = \rho_2 u_2 = J\mu_i \quad , \tag{37.1}$$

where ρ and u are the density and velocity of the gas relative to the I-front, and μ_i is the mass *per ion* in the ionized gas. If He remains neutral, $\mu_i = 1.4 m_{\rm H}$; if He is singly ionized, $\mu_i = (1.4/1.1) m_{\rm H}$.

Equation (36.5) applies with only a slight modification:

$$\rho_1 u_1^2 + \rho_1 c_1^2 + \frac{B_1^2}{8\pi} = \rho_2 u_2^2 + \rho_2 c_2^2 + \frac{B_2^2}{8\pi} - \frac{Jh\nu}{c} \quad , \tag{37.2}$$

where c_1 and c_2 are the isothermal sound speeds in region 1 and region 2, and we have included the momentum flux $Jh\nu/c$ from the ionizing photons that are assumed to be absorbed at the I-front. We assume that ρ_1 and $u_1 = J\mu/\rho_1$ are known, and we seek to solve for the unknown u_2 and $\rho_2 = \rho_1 u_1/u_2$.

If all the terms in Eq. (37.2) are retained, one obtains a cubic equation, which is tractable but tedious to solve. Cases of astrophysical interest will normally have $u_2 \gg u_1$, $\rho_2 \ll \rho_1$, and, as a consequence, $B_2^2/8\pi$ (the magnetic pressure in the ionized gas) will be small compared to the other terms. In addition, it is easy to see that $Jh\nu/c = \rho_1 u_1 h\nu/\mu_i c \ll \rho_1(u_1^2+c_1^2)+B_1^2/8\pi$. If we drop $B_2^2/8\pi$ and $Jh\nu/c$ from Eq. (37.2), we obtain a simple quadratic equation for $x \equiv u_2/u_1 = \rho_1/\rho_2$:

$$u_1^2 x^2 - \left(u_1^2 + c_1^2 + \frac{v_{A1}^2}{2}\right) x + c_2^2 = 0 \quad , \tag{37.3}$$

with roots

$$x = \frac{1}{2u_1^2}\left\{\left(u_1^2 + c_1^2 + \frac{v_{A1}^2}{2}\right) \pm \left[\left(u_1^2 + c_1^2 + \frac{v_{A1}^2}{2}\right)^2 - 4u_1^2 c_2^2\right]^{1/2}\right\} . \tag{37.4}$$

The roots are real if and only if

$$u_1^2 + c_1^2 + \frac{v_{A1}^2}{2} > 2u_1 c_2 \quad , \tag{37.5}$$

which requires

$$u_1 > u_{\rm R} \equiv c_2 + \left[c_2^2 - c_1^2 - \frac{v_{A1}^2}{2}\right]^{1/2} \quad , \tag{37.6}$$

or

$$u_1 < u_{\rm D} \equiv c_2 - \left[c_2^2 - c_1^2 - \frac{v_{A1}^2}{2}\right]^{1/2} \quad . \tag{37.7}$$

Normally, $c_2 \approx 10\,{\rm km\,s^{-1}}$, and $c_2^2 \gg v_{A1}^2$, $c_2^2 \gg c_1^2$, and

$$\begin{aligned} u_{\rm R} &\approx 2\,c_2 \quad , & x_{\rm R} &\approx \tfrac{1}{2} + \tfrac{2c_1^2+v_{A1}^2}{16c_2^2} \quad , \\ u_{\rm D} &\approx \tfrac{2c_1^2+v_{A1}^2}{4c_2} \quad , & x_{\rm D} &\approx \tfrac{4c_2^2}{2c_1^2+v_{A1}^2} \quad , \end{aligned} \tag{37.8}$$

where $x_{\rm R}$ and $x_{\rm D}$ are the density ratios ρ_1/ρ_2 for I-fronts with $u = u_{\rm R}$ and $u = u_{\rm D}$, respectively. Ionization fronts with $u = u_{\rm R}$ are called **R-critical**, and I-fronts with $u = u_{\rm D}$ are termed **D-critical**. The solutions with $u_1 \geq u_{\rm R}$ and the solutions with $u_1 \leq u_{\rm D}$ are quite different in nature.

37.1.1 R-Type I-Fronts

If

$$J = \frac{\rho_1 u_1}{\mu} > \frac{\rho_1}{\mu} u_{\rm R} \quad , \tag{37.9}$$

then $u_1 > u_{\rm R}$ and the I-front is called **R-type**. There are two mathematical solutions, depending on which root of (37.4) is taken. The solutions corresponding to the larger root are called **weak R-type**. Weak R-type solutions have only a small density change across the I-front, with the density change going to zero in the limit of very fast-moving I-front. Both upstream and downstream flows are supersonic, and waves cannot propagate upstream.

What about the other solution branch, obtained if the negative root in (37.4) is taken? These solutions, referred to as **strong R-type**, also satisfy conservation of mass and momentum, but for $u_1 \gg u_{\rm R}$, the ionized gas is much denser than the neutral gas ($\rho_1/\rho_2 \ll 1$ in Figure 37.1). This can only occur if there is a large pressure gradient across the I-front – the I-front is also a shock front. Because the downstream flow is subsonic, it is physically possible for the postshock region to be pressurized by some "piston" that is acting on it (e.g., a stellar wind), but such a combined ionization–shock front requires that the pressure be just right so that the compressive shock front advances at the same speed as the I-front, whose speed depends on J. Such a situation generally does not arise in astrophysical I-fronts. Hence, only weak R-type I-fronts are physically relevant.

37.1.2 D-Type I-Fronts

Conversely, if

$$J = \frac{\rho_1 u_1}{\mu} < \frac{\rho_1}{\mu} u_{\rm D} \quad , \tag{37.10}$$

then $u_1 < u_{\rm D}$ and the I-front is called **D-type**. The solution corresponding to the larger root in Eq. (37.4) is termed **strong D-type**. The ionized gas is at much lower density than the neutral gas, and it flows away from the I-front at a speed that is close to the sound speed c_2. In this case, it is the "rocket effect" that is responsible for the large pressure *increase* as one moves from the ionized gas into the neutral gas – it is this pressure gradient (which we do not resolve when we treat the I-front as a discontinuity) that accelerates the ionized gas away from the I-front.

The other solution, obtained by taking the smaller root in (37.4), is termed **weak D-type**. For $u_1 \ll u_{\rm D}$, this solution corresponds to very slow flow in the neutral

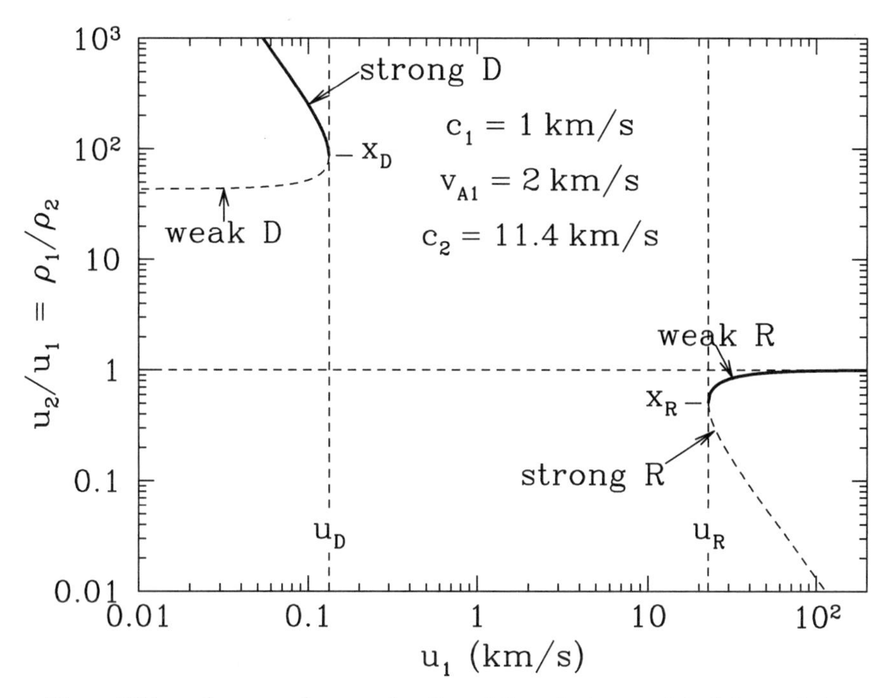

Figure 37.1 $u_2/u_1 = \rho_1/\rho_2$, as a function of the velocity u_1 of the I-front relative to the neutral gas just ahead of the I-front, for D-type and R-type ionization front solutions (see text) for an example with $c_1 = 1\,\mathrm{km\,s^{-1}}$, $v_{A1} = 2\,\mathrm{km\,s^{-1}}$, and $c_2 = 11.4\,\mathrm{km\,s^{-1}}$. The astrophysically relevant solutions are the strong D-type and weak R-type cases, shown as heavy curves. There are no solutions with u_1 between u_D and u_R.

medium, with approximate pressure equilibrium between the neutral gas and the ionized gas. This is the solution that would correspond to the late-stage evolution of an H II region in an initially uniform medium, where the ionized gas is essentially in pressure equilibrium with the neutral gas around it.

We have discussed solutions for $u_1 < u_\mathrm{D}$ and $u_1 > u_\mathrm{R}$. What happens if we have a neutral medium with density ρ_0, and we suddenly "turn on" an ionizing flux J such that $u_\mathrm{D} < J\mu/\rho_0 < u_\mathrm{R}$? Since J is imposed, what has to happen is that the density ρ_1 of the neutral gas just ahead of the I-front must be increased so that $J\mu/\rho_1 < u_\mathrm{D}$. This is accomplished by a pressure wave that travels forward into the neutral gas, compressing it from density ρ_0 to a density ρ_1 such that $J\mu/\rho_1 < u_\mathrm{D}$. Because compression requires that the pressure be significantly larger than the initial pressure, the pressure wave will be a shock wave.

In this case, there are *two* fronts to consider: (1) the shock front where the as-yet unperturbed medium with density ρ_0 is compressed to a density ρ_1, and accelerated in the process, and (2) the I-front, propagating into the compressed postshock gas. The shock front must, of course, move faster than the I-front so that it remains ahead of it.

The solution will be **D-critical**, with $u_1 \approx u_\mathrm{D}$. Substituting u_D into Eq. (37.4), one finds a density ratio

$$\left(\frac{\rho_1}{\rho_2}\right)_\mathrm{D} \equiv x_\mathrm{D} \approx \frac{1}{2} + \frac{4c_2^2}{(2c_1^2 + v_{A1}^2)} \quad . \tag{37.11}$$

The density in the neutral gas must be increased to a value $\rho_1 \approx J\mu/u_\mathrm{D}$. If the initial density is ρ_0, then the shock compression factor must be $\rho_1/\rho_0 = J\mu/(u_\mathrm{D}\rho_0)$. The compression in an isothermal shock is given by Eqs. (36.31 and 36.32).

37.2★ Expansion of an H II Region in a Uniform Medium

We can now discuss the expansion of an H II region in an initially uniform medium of density $(n_\mathrm{H})_0 \equiv n_0$. For simplicity, we will neglect helium. Suppose that a star suddenly turns on, emitting H-ionizing photons at a rate Q_0. At early times, the I-front will be weak R-type, propagating at a speed $u \gg u_R \approx 2c_2$, where $c_2 = 11.4 T_4^{1/2}\,\mathrm{km\,s^{-1}}$ is the isothermal sound speed if He remains neutral. Let $R_i(t)$ be the radius of the I-front. With the gas remaining at rest, the propagation of the I-front is determined by

$$n_0 4\pi R_i^2 dR_i = \left[Q_0 - \frac{4\pi}{3} R_i^3 \alpha_B n_0^2\right] dt \quad , \tag{37.12}$$

$$\frac{dR_i}{dt} = \frac{Q_0}{4\pi n_0 R_i^2} - \frac{n_0 \alpha_B R_i}{3} \quad , \tag{37.13}$$

which can be integrated to obtain $R_i(t)$ and the velocity $V_i(t)$:

$$R_i^3 = R_\mathrm{S0}^3 \left[1 - \mathrm{e}^{-t/\tau}\right] \;\; ; \;\; R_\mathrm{S0} \equiv \left(\frac{3Q_0}{4\pi\alpha_B n_0^2}\right)^{1/3} \quad , \;\; \tau \equiv (n_0\alpha_B)^{-1}, \tag{37.14}$$

$$V_i \equiv \frac{dR_i}{dt} = \left(\frac{Q_0 n_0 \alpha_B^2}{36\pi}\right)^{1/3} \frac{\mathrm{e}^{-t/\tau}}{\left(1 - \mathrm{e}^{-t/\tau}\right)^{2/3}} \tag{37.15}$$

$$= 842 \left(\frac{Q_0}{10^{48}\,\mathrm{s^{-1}}}\right)^{1/3} \left(\frac{n_0}{10^3\,\mathrm{cm^{-3}}}\right)^{1/3} \frac{\mathrm{e}^{-t/\tau}}{\left(1 - \mathrm{e}^{-t/\tau}\right)^{2/3}}\,\mathrm{km\,s^{-1}} \, . \tag{37.16}$$

The velocity V_i of the I-front decreases with increasing time, until it approaches $u_R \approx 2c_2 \approx 23\,\mathrm{km\,s^{-1}}$. From Eq. (37.16), we see that (for our nominal parameters $Q_0 = 10^{48}\,\mathrm{s^{-1}}$ and $n_0 = 10^3\,\mathrm{cm^{-3}}$), the I-front remains R-type until $\mathrm{e}^{-t/\tau} \approx 23/842$, or $t/\tau \approx \ln(842/23) \approx 3.6$, by which time the I-front has reached $R_i \approx (1 - 23/842)^{1/3} R_\mathrm{S0} = 0.99 R_\mathrm{S0}$.

Until this time, the neutral gas remained at rest as the I-front advanced into it, and the growth law Eq. (37.14) remains valid. However, when V_i falls below u_R, the character of the expansion will suddenly change: a shock wave will now move ahead of the I-front, and will compress and accelerate the neutral gas so that the I-front is moving into gas that has been both compressed *and* set into motion. The I-front will now be strong D-type, but it will be advancing into neutral gas that is already moving radially outward as the result of the shock. This geometry is shown in Figure 37.2.

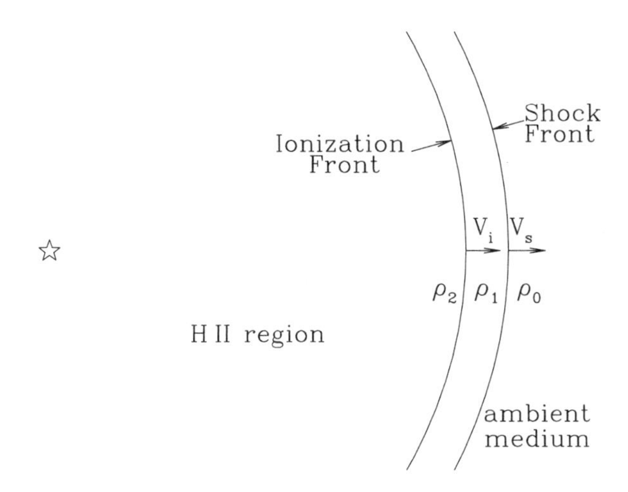

Figure 37.2 A D-type ionization front preceded by a shock wave.

Once the I-front becomes D-type, and a shock precedes it, analytic approximations to the expansion become less accurate even if the original medium is assumed to be uniform. Here, we follow the treatment developed by Spitzer (1978).

Let R_i be the radius of the I-front, moving with velocity (relative to the star) $V_i = dR_i/dt$. After the I-front becomes D-type, it is subsonic, and the ionized gas, in particular, can be assumed to be in pressure equilibrium, i.e., with a nearly uniform density interior to the I-front. Let $v_i(r_i)$ be the velocity (relative to the star), of ionized gas at radius $r_i < R_i$ interior to the I-front. Because the density is uniform, it follows that $\rho_i r^3$ is constant for a fluid element within the expanding H II region, hence

$$\frac{1}{\rho_i}\frac{d\rho_i}{dt} = -\frac{3}{r_i}\frac{dr_i}{dt} = -3\frac{v_i(r_i)}{r_i} \quad . \tag{37.17}$$

As the density of the ionized gas goes down, the stellar radiation is able to ionize more material, and the I-front advances into neutral gas. Neglecting absorption by dust, and assuming the recombination time $1/(n_e\alpha_B)$ is short compared to the expansion time R_i/V_i, radiative recombinations approximately balance photoionizations, hence $\rho_i^2 R_i^3 \approx const$, and

$$\frac{3}{R_i}\frac{dR_i}{dt} \approx -\frac{2}{\rho_i}\frac{d\rho_i}{dt} \quad , \tag{37.18}$$

$$V_i \equiv \frac{dR_i}{dt} \approx -\frac{2}{3}\frac{R_i}{\rho_i}\frac{d\rho_i}{dt} = 2v_i(R_i) \quad . \tag{37.19}$$

Since $V_i = v_i + u_{i2}$, the velocity of the ionized gas relative to the I-front is

$$u_{i2} = \frac{1}{2}V_i \quad . \tag{37.20}$$

The density of the neutral gas into which the I-front is advancing will be determined by compression behind the shock wave moving ahead of the I-front. The shock wave will be moving at a speed V_s that will be close to (just slightly larger than) the speed of the I-front:

$$V_s \approx V_i \quad . \tag{37.21}$$

After the I-front becomes D-type, the expanding H II region will be surrounded by a shell of shock-compressed gas. Let us assume that the ambient medium, with density ρ_0, is magnetized with a magnetic field B_0 that is transverse to the direction of propagation of the shock. Further suppose that the magnetic pressure is large compared to the thermal pressure in the ambient medium — $B_0^2/8\pi \gg \rho_0 c_{s0}^2$, where c_{s0} is the isothermal sound speed in the ambient medium. If the shock front and the I-front are propagating at almost the same speed, we can apply the conservation of momentum flux across the two fronts:

$$\rho_0 V_s^2 \approx \rho_2\left(u_{i2}^2 + c_{s2}^2\right) = \rho_2\left(\frac{V_i^2}{4} + c_{s2}^2\right) \quad . \tag{37.22}$$

If we now assume that $V_i \approx V_s$, we obtain

$$V_s^2 = \frac{\rho_2}{\rho_0}\frac{c_{s2}^2}{(1 - \rho_2/4\rho_0)} \quad . \tag{37.23}$$

Since $\rho_2 \leq \rho_0$, we may neglect $\rho_2/4\rho_0$ in the denominator, and obtain

$$V_i \approx V_s \approx \left(\frac{\rho_2}{\rho_0}\right)^{1/2} c_{s2} \quad . \tag{37.24}$$

The rate at which new ions must be created at the ionization front is $4\pi R_i^2 n u_{i2} = 2\pi R_i^2 n V_i$. If $Q_0 \gg 2\pi R_i^2 n V_i$, then the expanding H II region has $\rho_2^2 R_i^3 = const$, hence $\rho_2 = \rho_0(R_D/R_i)^{3/2}$, and

$$R_i^{3/4}\frac{dR_i}{dt} = R_D^{3/4} c_{s2} \quad , \tag{37.25}$$

$$R_i = R_D \left[1 + \frac{7}{4}\frac{c_{s2}(t - t_D)}{R_D}\right]^{4/7} \quad \text{for } t > t_D \quad , \tag{37.26}$$

where $R_D \approx R_{S0}$ and t_D are the radius and time when the I-front first becomes D-type. This result was first obtained by Spitzer (1978).

To illustrate the evolution of an H II region, Fig. 37.3 shows the radius R_i and velocity V_i of the ionization front as a function of time for the H II region created by a star that at $t = 0$ suddenly begins emitting ionizing photons at a rate $Q_0 = 10^{49}\,\mathrm{s}^{-1}$, in an initially uniform medium of density $n_0 = 10^3\,\mathrm{cm}^{-3}$. The R-type phase lasts only to $t \approx 540\,\mathrm{yr}$, at which time the I-front which has become R-critical, switching from R-type to D-type. Immediately after becoming D-type, the velocity V_i of the ionization front is $\sim 10\,\mathrm{km\,s}^{-1}$, but it gradually slows, dropping to $V_i < 2\,\mathrm{km\,s}^{-1}$ at $t = 10^6$ yr.

The evolution of real H II regions may depart substantially from Eq. (37.26), for two important reasons:

1. The ambient gas around a real H II region will, in general, not be uniform and at rest. Nonuniformities will allow the H II region to expand more rapidly in lower-density regions, with a "champagne flow" (Tenorio-Tagle 1979) resulting if the high-pressure H II region is able to "vent" into a low-density region.

2. The illuminating star will in general be moving relative to the ambient medium, resulting in a nonspherical geometry even if the ambient density is uniform.

37.3★ Photodissociation Fronts

Above we have discussed the expansion of an H II region surrounded by neutral gas. It is often the case that the ambient gas was originally molecular. Under some circumstances, there will be a **photodissociation front** (**PD-front**) preceding the I-front.

As discussed in §31.3, H_2 can be photodissociated by photons with $11.2\,\mathrm{eV} < h\nu < 13.6\,\mathrm{eV}$. Within the H II region, these photons suffer only extinction by dust. In neutral gas with a high fraction of H_2, a large fraction of the 11.2 to 13.6 eV photons will be absorbed by H_2, and $\sim 15\%$ of the photoabsorptions will be followed by a photodissociation. If F_i is the flux of $h\nu > 13.6\,\mathrm{eV}$ photons at the I-front, and F_d is the flux of $11.2\,\mathrm{eV} < h\nu < 13.6\,\mathrm{eV}$ photons, then the PD-front will be able to propagate ahead of the I-front if $2 \times 0.15 F_d > F_i$. For the expanding

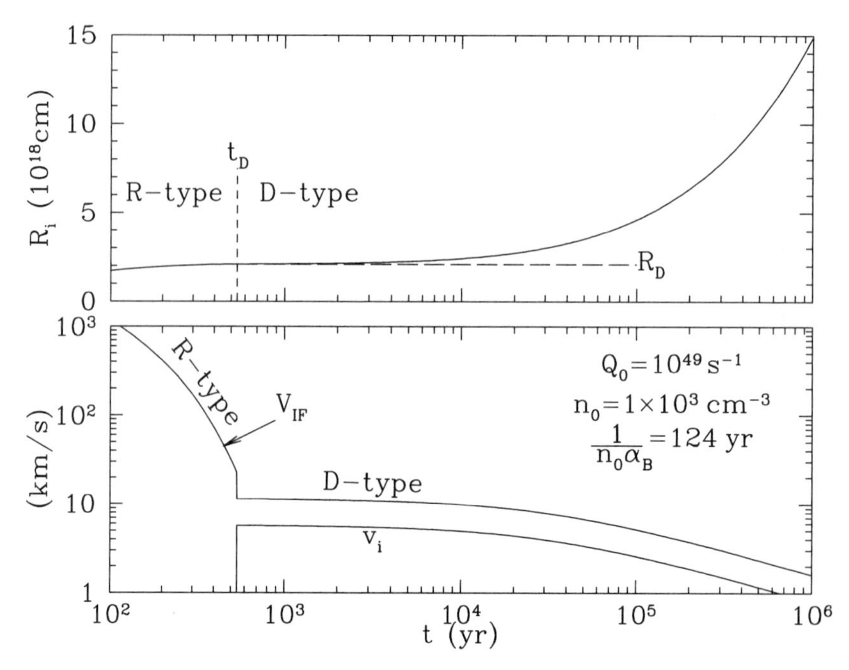

Figure 37.3 Upper panel: radius R_I of the ionization front versus time for expansion of an H II region into an initially uniform neutral cloud with $n_0 = 10^3\,\mathrm{cm}^{-3}$, for an ionizing source with $Q_0 = 10^{49}\,\mathrm{s}^{-1}$. Dust absorption is neglected. For this example, the ionization front makes the transition from R-type to D-type at time $t_\mathrm{D} \approx 540\,\mathrm{yr}$. Lower panel: Velocity V_IF of the ionization front, and velocity v_i (relative to the star) of the gas just interior to the ionization front.

H II region discussed here, using $V_i = 2v_i$ from Eq. (37.19),

$$F_i = (V_i - v_i)\frac{\rho_2}{\mu} = \frac{\rho_2 V_i}{2\mu} \tag{37.27}$$

$$= \frac{\rho_0 c_{s2}}{2\mu}\left(\frac{R_\mathrm{D}}{R_i}\right)^{9/4} \tag{37.28}$$

$$= \frac{\rho_0 c_{s2}}{2\mu}\left(1 + \frac{7}{4}\frac{c_{s2}(t - t_\mathrm{D})}{R_\mathrm{D}}\right)^{-9/7} , \tag{37.29}$$

while the flux of H_2-dissociating photons crossing the I-front is

$$F_d = \frac{\dot{N}(11.2 - 13.6\,\mathrm{eV})}{4\pi R_\mathrm{D}^2}\left(1 + \frac{7}{4}\frac{c_{s2}(t - t_\mathrm{D})}{R_\mathrm{D}}\right)^{-8/7} . \tag{37.30}$$

The PD-front will move ahead of the I-front if $F_d > F_i/0.30$, or

$$\frac{\dot{N}(11.2 - 13.6\,\mathrm{eV})}{4\pi R_\mathrm{D}^2} > \frac{1}{0.30}\frac{\rho_0 c_{s2}}{2\mu}\left(1 + \frac{7}{4}\frac{c_{s2}(t - t_\mathrm{D})}{R_\mathrm{D}}\right)^{-1/7} . \tag{37.31}$$

Using $Q_0 \approx (4\pi/3)\alpha_B n_0^2 R_{\rm D}^3$, it follows that the stellar spectrum must have

$$\frac{\dot{N}(11.2-13.6\,{\rm eV})}{\dot{N}(>13.6\,{\rm eV})} > 5\left(\frac{c_{s2}/R_{\rm D}}{n_0\alpha_B}\right)\left(1+\frac{7}{4}\frac{c_{s2}(t-t_{\rm D})}{R_{\rm D}}\right)^{-1/7} \qquad (37.32)$$

in order for a photodissociation front to propagate ahead of the IF. A typical H II region has $n_0\alpha_B \gg c_{s2}/R_{\rm S0}$ (recombination time $\ll$ sound-crossing time); therefore the condition (37.32) will generally be satisfied. Note, however, that this analysis has assumed a D-type I-front, and therefore is valid only for $t > t_{\rm D}$.

At very early times $t \ll 1/(n_0\alpha_B)$, the PD-front will propagate ahead of the I-front only if

$$\frac{\dot{N}(11.2-13.6\,{\rm eV})}{\dot{N}(>13.6\,{\rm eV})} > \frac{1}{0.30} = 3.3 \quad . \qquad (37.33)$$

For a blackbody, this condition is satisfied only for $T < 13800\,{\rm K}$, which is well below the $T \gtrsim 3\times 10^4\,{\rm K}$ temperature of a star powering a typical H II region. Therefore, in the very early stages of an R-type I-front, the PD-front will be merged with the I-front, rather than propagating ahead of it as it will do when the I-front becomes D-type.

Chapter Thirty-eight

⋆ Stellar Winds

Stellar winds inject mass, momentum, and energy into the ISM, and can carve out conspicuous structures in the ISM in the vicinity of the star. The most important stellar winds come from hot, young, massive O stars; from red giants and supergiants; and from the progenitors of planetary nebulae. We do not attempt to discuss either stellar evolution or the mechanisms driving the winds; our concern here is on the impact of the wind on the ISM.

38.1⋆ Winds from Hot Stars: Stellar Wind Bubbles

Stellar winds normally have terminal speeds V_w (the asymptotic speed of the wind after it has traveled many stellar radii from the star) that are of order the escape velocity from the stellar surface. For O-type massive stars, $1500 \lesssim V_w/\,\mathrm{km\,s^{-1}} \lesssim 2500$, with estimated mass loss rates $\dot{M}$ ranging from $10^{-6.5}$ to $10^{-5}\,M_\odot\,\mathrm{yr}^{-1}$ (Markova et al. 2004). B supergiants also have fast winds, with velocities ranging from $\sim 1500\,\mathrm{km\,s^{-1}}$ for B0.5 I to $\sim 300\,\mathrm{km\,s^{-1}}$ for B9 I, and estimated $\dot{M}$ in the range 10^{-7} to $10^{-6}\,M_\odot\,\mathrm{yr}^{-1}$ (Markova & Puls 2008). However, the mass loss rates for OB stars may actually be as much as an order of magnitude lower than the above estimates if, as suspected, clumping is important (Puls et al. 2008).

The structure of the stellar wind bubble around an O-type star is illustrated in Fig. 38.1. The ionizing radiation from the star will have already created an H II region into which the wind will blow – this is the ambient medium in Fig. 38.1, with density ρ_0. Assume that the gas in the H II region is at rest. Assume that the star is also at rest, so that the bubble is spherical.

Because V_w greatly exceeds the sound speed in the H II, the wind will initially drive a shock into the H II region – the outer shock in Fig. 38.1 – with radius R_s and velocity V_s. The shocked ambient medium accumulates in a shell behind the outer shock front. There will also be a second shock front where the stellar wind is decelerated; the shocked stellar wind material accumulates in a shell exterior to the shock front, and interior to the spherical contact discontinuity separating the shocked ambient medium from the shocked wind material. If there were no instabilities, and diffusion can be neglected, these two regions would not mix. The innermost region contains the unshocked stellar wind.

Suppose that the wind begins abruptly at $t = 0$ with constant mass loss rate $\dot{M} \equiv 10^{-6}\dot{M}_6\,M_\odot\,\mathrm{yr}^{-1}$ and wind velocity $V_w \equiv 10^3 V_{w,8}\,\mathrm{km\,s^{-1}}$. The mechanical

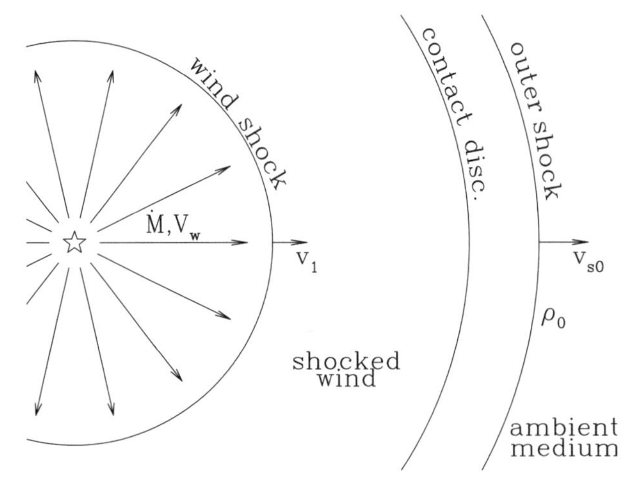

Figure 38.1 Structure of the bubble produced by a fast stellar wind (not to scale).

power in the wind is considerable: $\dot{E} = \dot{M}V_w^2/2 = 3.2 \times 10^{35}\dot{M}_{-6}V_{w,8}^2\,\mathrm{erg\,s^{-1}}$.

Let the ambient medium have density $n_{\mathrm{H}} = \rho_0/1.4m_{\mathrm{H}} = 10^3 n_3\,\mathrm{cm^{-3}}$. At *very* early times, the wind undergoes essentially free expansion, with $R_{\mathrm{s}} \approx V_w t$ until the wind mass and swept-up mass are comparable: $\dot{M}t \approx (4\pi/3)\rho_0(V_w t)^3$. The free expansion phase ends at a time

$$t_0 \approx \left(\frac{3\dot{M}}{4\pi\rho_0 V_w^3}\right)^{1/2} = 2.54\, n_3^{-1/2}\dot{M}_{-6}^{1/2}V_{w8}^{-3/2}\,\mathrm{yr}\,. \tag{38.1}$$

This is such a short time that the free expansion phase is unlikely to be observed.

At early times, radiative losses from the shocked wind region will be negligible – this is the **energy-conserving phase**. The stellar wind is the energy source, hence the total energy $E(t) = (1/2)\dot{M}V_w^2 t$. It is reasonable to estimate that this will be divided approximately equally between thermal energy and ordered kinetic energy.[1] Most of the mass will be swept-up gas, thus $M(t) \approx (4\pi/3)\rho_0 R_{\mathrm{s}}^3$. Let us assume power-law behavior: $R_{\mathrm{s}} = At^\eta$. Then $V_{\mathrm{s}} = \eta R_{\mathrm{s}}/t$. The gas just interior to the outer shock will be moving radially with velocity $\frac{3}{4}V_{\mathrm{s}}$. Taking this as the rmsvelocity of the shocked ISM, and assuming an equal amount of energy in thermal motions, we have

$$E(t) \approx 2 \times \frac{1}{2}\frac{4\pi}{3}R_{\mathrm{s}}^3\rho_0\left(\frac{3\eta R_{\mathrm{s}}}{4t}\right)^2 = \frac{1}{2}\dot{M}V_w^2 t\ , \tag{38.2}$$

$$R_{\mathrm{s}}^5 = A^5 t^{5\eta} \approx \frac{2}{3\pi\eta^2}\frac{\dot{M}V_w^2}{\rho_0}t^3\ , \tag{38.3}$$

[1]This is confirmed by detailed modeling (e.g., Koo & McKee 1992).

from which it is evident that $\eta = 3/5$ and

$$R_\mathrm{s} \approx \left(\frac{50\dot{M}V_w^2}{27\pi\rho_0}\right)^{1/5} t^{3/5} = 3.5\times10^{17}\frac{\dot{M}_{-6}^{1/5}V_{w8}^{2/5}}{n_3^{1/5}}\left(\frac{t}{10^3\,\mathrm{yr}}\right)^{3/5}\mathrm{cm}, \quad (38.4)$$

$$V_\mathrm{s} = \eta\frac{R_\mathrm{s}}{t} \approx 66\frac{\dot{M}_{-6}^{1/5}V_{w8}^{2/5}}{n_3^{1/5}}t_3^{-2/5}\,\mathrm{km\,s^{-1}}, \quad (38.5)$$

at time $t \equiv 10^3 t_3$ yr. The outer shock speed $V_\mathrm{s} \equiv 100V_{s7}\,\mathrm{km\,s^{-1}}$ is decreasing with t. Figure 36.2a shows that the cooling time becomes shorter as V_s decreases, with $t_\mathrm{cool} \approx 7V_{s7}^{3.4}n_3^{-1}$ yr for $0.8 < v_{s7} < 12$. The energy-conserving phase will end at a time $t = t_\mathrm{rad}$ such that $t = t_\mathrm{cool}(t)$, or

$$t_3 \approx 0.007n_3^{-1}V_{s7}^{3.4} \approx 0.007n_3^{-1}\left(0.66\frac{\dot{M}_{-6}^{1/5}V_{w8}^{2/5}}{n_3^{1/5}}t_3^{-2/5}\right)^{3.4}, \quad (38.6)$$

$$t_\mathrm{rad} \approx 67\left(\dot{M}_{-6}V_{w8}^2\right)^{0.29} n_3^{-0.71}\,\mathrm{yr}, \quad (38.7)$$

$$R_\mathrm{s,rad} \approx 6.9\times10^{16}\left(\dot{M}_{-6}V_{w8}^2\right)^{0.37} n_3^{-0.63}\,\mathrm{cm}, \quad (38.8)$$

$$V_{s,\mathrm{rad}} \approx 190\left(n_3\dot{M}_{-6}V_{w8}^2\right)^{0.08}\,\mathrm{km\,s^{-1}}. \quad (38.9)$$

The outer shock, therefore, becomes radiative almost immediately, and the gas cools and collapses into a thin shell. The ionizing radiation from the O star will keep the material in the shell photoionized and at $T \approx 10^4$ K. As the zone with shocked ISM cools and collapses into a thin shell, the expansion briefly slows, but then resumes as the pressure from the hot gas in the interior continues to act.

The shocked wind material is very hot (just beyond the wind shock the gas has $T \approx 1.4\times10^7 V_{w8}^2$ K), and the density is low. As a result, radiative cooling is very slow (again, see Fig. 36.2a). There will be a long phase during which the shocked wind material loses energy primarily by continuing to do work on the dense shell. Aside from a short time at the onset of cooling behind the outer shock (when the shell expansion will have a sudden drop, followed by a partial recovery), the subsequent shell expansion has the same power-law dependence on t as in the energy-conserving phase:

$$R_\mathrm{s} \approx 0.85\left(\frac{50\dot{M}V_w^2}{27\pi\rho_0}\right)^{1/5} t^{3/5} = 3\times10^{17}\left(\frac{\dot{M}_{-6}V_{w8}^2}{n_3}\right)^{1/5} t_3^{3/5}\,\mathrm{cm}, \quad (38.10)$$

$$V_\mathrm{s} = \eta\frac{R_\mathrm{s}}{t} = 60\left(\frac{\dot{M}_{-6}V_{w8}^2}{n_3}\right)^{1/5} t_3^{-2/5}\,\mathrm{km\,s^{-1}}, \quad (38.11)$$

where the prefactor 0.85 in Eq. (38.10) is introduced to reproduce the results of more detailed modeling (e.g., Koo & McKee 1992).

As the expansion velocity drops, the shock becomes weaker, with a decreasing compression ratio. When V_s falls to the ambient sound speed, $\sim 15\,\mathrm{km\,s^{-1}}$ in an H II region, the outer shock will disappear, with the pressure pulse propagating away as an acoustic wave. The inner bubble of fast stellar wind and hot shocked stellar wind will remain as an X-ray-emitting cavity in pressure equilibrium with the H II region. For our nominal parameters ($\dot{M}_{-6}V_{w8}^2/n_3 = 1$), the outer shock disappears at $t \approx 3 \times 10^4\,\mathrm{yr}$, leaving behind the inner bubble containing the stellar wind and the shell of X-ray-emitting plasma created when the fast stellar wind is shocked.

When will the gas in the hot interior of the bubble begin to cool by radiative losses? Unlike the outer shell, the hot gas in the interior does not make a sudden transition from predominantly hot to predominantly cold.[2] The wind material near the contact discontinuity cools rapidly to $T \approx 10^4\,\mathrm{K}$, resulting in a thin layer of cooled stellar wind material in contact with the cool shell of ISM material, even while much or most of the shocked stellar wind material is still hot.

An O7 Ia star ($Q_0 \approx 10^{49.41}\,\mathrm{s^{-1}}$; see Table 15.1) would create a Strömgren sphere with radius $3 \times 10^{18} n_3^{-2/3}\,\mathrm{cm}$; the stellar wind bubble will be easily contained within the H II region.

The preceding discussion has allowed for radiative cooling but not for thermal conduction. In the absence of magnetic fields, thermal conduction would be expected to be important at the inner edge of the cool shell, where heat could flow from the hot gas into the cool shell. The resulting conduction front (see §34.5) contains material at intermediate temperatures, and at intermediate ionization states. Weaver et al. (1977) estimated that the conduction front would have $N(\mathrm{O\,VI}) \approx 2 \times 10^{13}\,\mathrm{cm^{-2}}$, and proposed that such circumstellar shells could account for the O VI absorption lines frequently seen in the spectra of hot stars. However, surveys of O VI (e.g., Bowen et al. 2008) find the O VI to be predominantly of interstellar rather than circumstellar origin, with distant stars having larger $N(\mathrm{O\,VI})$ than more nearby ones.

The role of thermal conduction remains unclear. Conduction perpendicular to magnetic field lines is strongly suppressed. If the thermal conduction is magnetically suppressed, the conduction front becomes thinner, with smaller column densities of species like O VI.

[2]The reason is that even in the absence of radiative cooling, there is a temperature gradient in the shocked wind material, as a result of the shock speed itself increasing with time, and differing amounts of adiabatic cooling.

38.2★ Winds from Cool Stars

The winds from cool stars can also produce bubbles. These winds are molecular and dusty, and typically have low outflow velocities $V_w \approx 15 - 30\,\mathrm{km\,s^{-1}}$. Mass loss rates can vary from $\sim 10^{-7}\,M_\odot\,\mathrm{yr}^{-1}$ for M giant stars to $\sim 10^{-4}\,M_\odot\,\mathrm{yr}^{-1}$ for AGB stars (e.g., "OH/IR" stars). The free-expansion phase will end at

$$t_0 \approx \left(\frac{3\dot{M}}{4\pi\rho_0 V_w^3}\right)^{1/2} = 8.0 \times 10^4 \dot{M}_{-6} V_{w6}^{-3/2} n_0^{-1/2}\,\mathrm{yr} \ , \tag{38.12}$$

$$R_{s0} \approx 2.5 \times 10^{18} \dot{M}_{-6}^{1/2} n_0^{-1/2}\,\mathrm{cm} \ , \tag{38.13}$$

where $V_{w6} \equiv V_w/10\,\mathrm{km\,s^{-1}}$. For $t \gtrsim t_0$, the bubble expansion will decelerate. The gas is relatively dense, and molecules are effective coolants. As a result, these bubbles are generally strongly radiative, consisting of a supersonic stellar wind impacting a cold dense shell. The "push" on the inner edge of the cold shell now comes from the momentum deposited by the wind.

If the pressure of the ambient medium can be ignored (i.e., the cool shell is moving supersonically) and the stellar wind is continuing, then

$$\frac{4\pi}{3}\rho_0 R_s^3 v_s \approx \dot{M} V_w t \ . \tag{38.14}$$

If we assume $R_s \propto t^\eta$, we find $\eta = \frac{1}{2}$ and

$$R_s \approx \left(\frac{3\dot{M}V_w}{2\pi\rho_0}\right)^{1/4} t^{1/2} = 3.4 \times 10^{17} \left(\frac{\dot{M}_{-6} V_{w6}}{n_0}\right)^{1/2} t_3^{1/2}\,\mathrm{cm} \ , \tag{38.15}$$

$$V_s = 53 \left(\frac{\dot{M}_{-6} V_{w,6}}{n_0}\right)^{1/2} t_3^{-1/2}\,\mathrm{km\,s^{-1}} \ , \tag{38.16}$$

for $t > t_0$.

An AGB star might have $\dot{M} = 10^{-4}\,M_\odot\,\mathrm{yr}^{-1}$ and $V_w \approx 20\,\mathrm{km\,s^{-1}}$, losing $\sim 1\,M_\odot$ over $\sim 10^4$ yr. From Eq. (38.12), we see that in the general ISM ($n_0 \approx 1$), such an outflow will be in the free-expansion phase during the $\sim 10^4$ yr duration of the wind. The gas leaving the photosphere is molecular. If $C/O < 1$, the O in the stellar wind will be mainly in CO and H_2O. As it moves away from the star, the molecules become exposed to the general interstellar radiation field. An H_2O molecule will therefore travel a only distance

$$R_{\mathrm{OH}} \approx \frac{V_w}{\zeta_{d,\mathrm{H_2O}}} \approx 3 \times 10^{15} V_{w,6} \left(\frac{6 \times 10^{-10}\,\mathrm{s^{-1}}}{\zeta_{d,\mathrm{H_2O}}}\right)\,\mathrm{cm} \ , \tag{38.17}$$

where $\zeta_{d,\mathrm{H_2O}}$ is the photodissociation rate for H_2O. At the radius R_{OH}, the H_2O

is destroyed and OH is produced. In the ISRF, $\zeta_{d,\mathrm{H_2O}} \approx 6 \times 10^{-10}\,\mathrm{s^{-1}}$ (see Table 33.1). However, the dusty outflow will provide some attenuation of the ultraviolet from external starlight, and $R_{\mathrm{OH}} \approx 10^{16}\,\mathrm{cm}$. The OH is also subject to photodissociation $\mathrm{OH} + h\nu \rightarrow \mathrm{O} + \mathrm{H}$, with $\zeta_{d,\mathrm{OH}} \approx 3 \times 10^{-10}\,\mathrm{s^{-1}}$. As a result, the outflow will have a spherical shell where the OH abundance is high. The Λ-doubling levels of OH can be inverted by infrared radiation from warm dust in the outflow (see §20.2), resulting in OH maser emission. For effective maser amplification, the velocity gradient should be small along the direction of propagation. For the expanding shell geometry, this favors radial paths, and paths that are tangential to the OH shell. Thus, OH/IR stars can have maser emission in a ring around the star, and also in a spot coinciding with the position of the star.

38.3★ Stellar Wind Bow-Shock

Suppose now that a star is moving through the medium with velocity $V_\star$. If $V_\star$ exceeds the ISM sound speed c_{ISM}, a **bow shock** will develop ahead of the star. Stellar winds are usually supersonic, with wind speed $V_w > c_w$, where c_w is the sound speed in the wind from the star, in which case the stellar wind will be decelerated in a shock front, usually referred to as the **termination shock**. Thus there are *two* shock waves, with a contact discontinuity separating the shocked stellar wind from the shocked ISM. (The contact discontinuity is an idealization – in a real flow there will be some mixing of the two shocked fluids.) The geometry is shown schematically in Figure 38.2. The region within the bow shock is termed the **astrosphere**; in the case of the Sun, this is called the **heliosphere**.

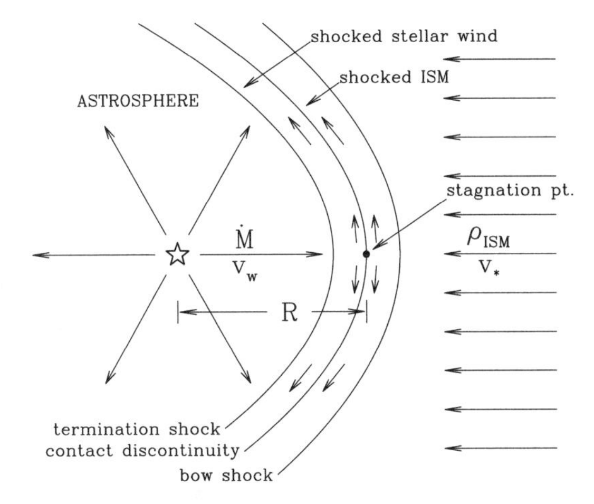

Figure 38.2 Stellar wind and interstellar wind in a reference frame where the star is stationary. R is the standoff distance of the bow shock.

How far ahead of the star does the bow shock extend? It is easiest to analyze the flow in the reference frame comoving with the star. In a steady flow situation, there will be a **stagnation point** on the contact discontinuity where the gas velocity is zero. Let R be the distance of the stagnation point from the star. The thickness of the two shocked layers will depend on the Mach number of each of the shocks, and also on how the cooling time compares with the characteristic flow times $\sim R/V_\star$ (in the case of the shocked ISM) or $\sim R/V_w$ (in the case of the shocked wind), but the pressure at the stagnation point (see Fig. 38.2) can be estimated from the momentum flux in the stellar wind evaluated at radius $\sim R$, which must be balanced by the momentum flux from the ISM:

$$\frac{\dot{M}V_w}{4\pi R^2} \approx \rho_{\rm ISM}\left(c_{\rm ISM}^2 + V_\star^2\right) \quad , \tag{38.18}$$

$$R = \left[\frac{\dot{M}V_w}{4\pi\rho_{\rm ISM}\left(c_{\rm ISM}^2 + V_\star^2\right)}\right]^{1/2} \tag{38.19}$$

$$\approx 1.5 \times 10^{18} \left(\frac{\dot{M}_{-6}\, V_{w,6}}{n_0}\right)^{1/2} \frac{1}{V_{*,6}}\ {\rm cm}\ . \tag{38.20}$$

The preceding estimates assume that both the ISM and the stellar wind can be treated as collisional fluids. In some cases – including the heliosphere of the Sun – the size of the astrosphere will be smaller than or comparable to the collisional mean free path for neutral particles. In this case, the neutral atoms in the interstellar "wind" will be able to traverse the bow shock and enter into the astrosphere on ballistic trajectories. In such a situation, the bow shock standoff distance estimate in Eq. (38.18) will be an underestimate unless the mass density $\rho_{\rm ISM}$ is limited to the mass in charged particles. If the fractional ionization of the ISM is low, the bow shock may be C-type, with deflection of the ions upstream from the shock.

Our own heliosphere, including the termination shock and the bow shock are the subject of intense study. The *Voyager* 1 and 2 spacecraft have crossed the termination shock, providing in-situ measurements in both the unshocked and shocked solar wind. Studies of backscattered resonance lines from ballistic H and He atoms in the interstellar wind allow an accurate determination of the velocity $V_\star = 26.2 \pm 0.5\,{\rm km\,s^{-1}}$ of the Sun relative to the local ISM (Möbius et al. 2004).

Other astrospheres have also been studied, both by absorption line spectroscopy (e.g., Wood et al. 2003) and by direct imaging. Spectacular examples include ultraviolet imaging of the bow shock and turbulent wake produced by Mira (o Ceti) moving through the ISM at $\sim 130\,{\rm km\,s^{-1}}$ (Martin et al. 2007), and the carbon star IRC +10216 = CW Leo, with a huge mass-loss rate $\dot{M}_w \approx 2\times10^{-5}\,M_\odot\,{\rm yr^{-1}}$, seen in both the ultraviolet (Sahai & Chronopoulos 2010) and the far-infrared (Ladjal et al. 2010).

Chapter Thirty-nine

Effects of Supernovae on the ISM

The dynamical state of the ISM in the Milky Way and other galaxies is strongly affected by supernova explosions. The light emitted by a supernova (SN) is spectacular, but it is the high-velocity ejecta that have the dominant effect on the ISM. Here, we discuss the blastwaves created by these explosions – referred to as **supernova remnants**, or **SNR**s. Plate 12 shows images of the Cas A SNR in radio synchrotron emission and in X-ray emission.

39.1 Evolution of a Supernova Remnant in a Uniform ISM

Consider first the simplest case of a spherically symmetric explosion of a star in a uniform medium with hydrogen density $n_{\rm H} \equiv n_0 \,{\rm cm}^{-3}$ and temperature T_0.

39.1.1 Free-Expansion Phase

The SN explosion ejects a mass $M_{\rm ej}$ with a kinetic energy $E_0 \equiv 10^{51} E_{51}\,{\rm erg}$. Although some unusual Type II supernovae (SNe) have kinetic energies as large as $\sim 10^{52}\,{\rm erg}$ (e.g., Rest et al. 2011), it is thought that the typical SN has $E_{51} \approx 1$.

Depending on the supernova type, the ejecta mass can range from $\sim 1.4 M_\odot$ (a Type Ia supernova, produced by explosion of a white dwarf near the Chandrasekhar limit) to perhaps $\sim 10 - 20 M_\odot$ for Type II supernovae following core collapse in massive stars. The ejecta will have a range of velocities, with the outermost material moving fastest. The rms velocity of the ejecta is

$$\langle v_{\rm ej}^2 \rangle^{1/2} = \left(\frac{2E_0}{M_{\rm ej}}\right)^{1/2} = 1.00 \times 10^4\,{\rm km\,s^{-1}}\, E_{51}^{1/2} \left(\frac{M_\odot}{M_{\rm ej}}\right)^{1/2} \quad . \tag{39.1}$$

This velocity is far greater than the sound speed in the surrounding material, and the expanding ejecta will therefore drive a fast shock into the circumstellar medium. We will refer to all of the matter interior to this shock surface as the **supernova remnant**, or **SNR**. In the first days after the explosion, the density of the ejecta far exceeds the density of the circumstellar medium, and the ejecta continue to expand ballistically at nearly constant velocity – this is referred to as the "free expansion phase." At these early times, there is only one shock of interest – the shock wave propagating outward into the ambient medium.

As the density of the expanding ejecta drops (as t^{-3}), the pressure of the shocked circumstellar medium soon exceeds the thermal pressure in the ejecta, and a **reverse shock** is driven into the ejecta. The remnant now contains two shock fronts: the original outward-propagating shock (the **blastwave**) expanding into the circumstellar/interstellar medium, and the reverse shock propagating inward, slowing and shock-heating the ejecta (which had previously been cooled by adiabatic expansion). The reverse shock becomes important when the expanding ejecta material has swept up a mass of circumstellar or interstellar matter comparable to the ejecta mass. The radius of the blastwave when this occurs is

$$R_1 = \left(\frac{3M_{\rm ej}}{4\pi\rho_0}\right)^{1/3} = 5.88\times10^{18}\,{\rm cm}\left(\frac{M_{\rm ej}}{M_\odot}\right)^{1/3} n_0^{-1/3} \quad , \tag{39.2}$$

and the time when it occurs is

$$t_1 \approx \frac{R_1}{\langle v_{\rm ej}^2\rangle^{1/2}} = 186\,{\rm yr}\left(\frac{M_{\rm ej}}{M_\odot}\right)^{5/6} E_{51}^{-1/2} n_0^{-1/3} \quad . \tag{39.3}$$

The free-expansion phase applies only for $t \lesssim t_1$.

In the case of Type II supernovae resulting from core collapse in massive stars, the supernova explosion is often preceded by a red supergiant phase, leaving a relatively dense circumstellar medium with a $\sim r^{-2}$ density profile rather than the uniform density ambient medium considered here. The Cas A SNR (see Plate 11), resulting from an explosion occurring in 1681 ± 19 (Fesen et al. 2006) has been modeled with $M_{\rm ej} \approx 4\,M_\odot$, $E_{51} \approx 2$, expanding into a circumstellar medium with $n_{\rm H} \approx 7(r/\,{\rm pc})^{-2}\,{\rm cm}^{-3}$ left by a red supergiant phase (van Veelen et al. 2009). The reverse shock is now located at $\sim 60\%$ of the outer shock radius – much of the Cas A ejecta is still in the free expansion phase.

39.1.2 Sedov-Taylor Phase

For $t \gtrsim t_1$, the reverse shock has reached the center of the remnant, all of the ejecta are now very hot, and the free-expansion phase is over. The pressure in the supernova remnant is far higher than the pressure in the surrounding medium. The hot gas has been emitting radiation, but if the densities are low, the radiative losses at early times are negligible.

The SNR now enters a phase that can be approximated by idealizing the problem as a "point explosion" injecting only energy E_0 into a uniform-density zero-temperature medium of density ρ_0: we neglect (1) the finite mass of the ejecta, (2) radiative losses; and (3) the pressure in the ambient medium.

We can use simple **dimensional analysis** to find out the form of the time evolution of the remnant. Let the radius of the spherical shock front be R_s, and let the explosion occur at $t = 0$. Suppose that

$$R_s = A\,E^\alpha\,\rho^\beta\,t^\eta \quad , \tag{39.4}$$

where A is dimensionless. We can determine α, β, and η by dimensional analysis, by equating the powers to which mass, length, and time appear in Eq. (39.4):

$$\text{Mass}: \quad 0 = \alpha + \beta \quad , \tag{39.5}$$

$$\text{Length}: \quad 1 = 2\alpha - 3\beta \quad , \tag{39.6}$$

$$\text{Time}: \quad 0 = -2\alpha + \eta \quad , \tag{39.7}$$

which are easily solved to obtain $\alpha = 1/5$, $\beta = -1/5$, and $\eta = 2/5$:

$$R_s = A\, E^{1/5}\, \rho_0^{-1/5}\, t^{2/5} \quad , \tag{39.8}$$

where the dimensionless coefficient $A = 1.15167$ is found from the exact solution (see below). Thus

$$R_s = 1.54 \times 10^{19}\,\mathrm{cm}\; E_{51}^{1/5} n_0^{-1/5} t_3^{2/5} \quad , \tag{39.9}$$

$$v_s = 1950\,\mathrm{km\,s^{-1}}\; E_{51}^{1/5} n_0^{-1/5} t_3^{-3/5} \quad , \tag{39.10}$$

$$T_s = 5.25 \times 10^{7}\,\mathrm{K}\; E_{51}^{2/5} n_0^{-2/5} t_3^{-6/5} \quad , \tag{39.11}$$

where $t_3 \equiv t/10^3\,\mathrm{yr}$.

We have been assuming that the internal structure of the remnant is given by a **similarity solution**: by this we mean that the hydrodynamical variables (density, velocity, pressure) can be written

$$\rho(r) = \rho_0\, f(x) \quad , \tag{39.12}$$

$$v(r) = \frac{R_s}{t}\, g(x) \quad , \tag{39.13}$$

$$p(r) = \frac{\rho_0 R_s^2}{t^2}\, h(x) \quad , \tag{39.14}$$

where $x \equiv r/R_s(t)$, and $f(x)$, $g(x)$, and $h(x)$ are dimensionless functions. If Eqs. (39.12 to 39.14) are inserted into the fluid equations, with the Rankine-Hugoniot relations (see §36.2) used for boundary conditions on $f(1)$, $g(1)$, and $p(1)$, one obtains a set of ordinary differential equations that can be solved numerically to obtain $f(x)$, $g(x)$, and $h(x)$ for $0 < x < 1$. This is known as the **Sedov-Taylor solution**, having been found independently by Taylor (1950) and Sedov (1959) in connection with the development of nuclear weapons. The Sedov-Taylor solution is shown in Fig. 39.1.

The initial conditions for a real blastwave will of course differ from the assumptions of the Sedov-Taylor solution. However, provided the ambient medium is uniform away from the immediate neighborhood of the explosion, the blastwave will

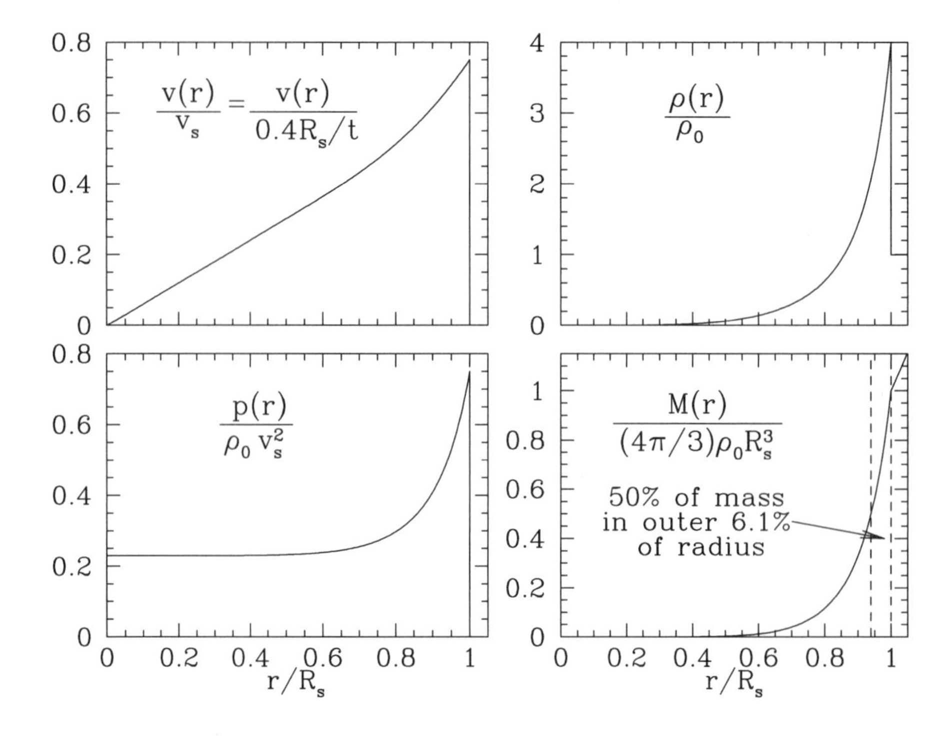

Figure 39.1 Sedov-Taylor solution for $\gamma = \frac{5}{3}$ gas. The temperature profile (not shown) can be obtained from the ratio of the pressure and density profiles. The density falls inward, and the temperature rises inward, with $\rho/\rho_0 \to 0$ and $T/T_s \to \infty$ as $x \to 0$.

evolve *toward* the Sedov-Taylor solution. Even in the absence of radiative cooling, the actual SNR will deviate from the Sedov-Taylor solution because the original explosion injected a mass $M_{\rm ej}$ (as opposed to the point injection of pure energy at $t = 0$ assumed for the Sedov-Taylor solution), leaving waves reverberating in the remnant even after the reverse shock has reached the center (Cioffi et al. 1988). Nevertheless, once the swept-up mass exceeds $M_{\rm ej}$, the Sedov-Taylor solution will be reasonably close to the actual density and temperature over most of the remnant. This is known as the **Sedov-Taylor phase** (or often just "Sedov phase") in the evolution of the blastwave.

The hot gas interior to the shock front is, of course, radiating energy. When the radiative losses become important the blastwave will enter a "radiative" phase, where the gas in the shell just interior to the shock front is now able to cool to temperatures much lower than the temperature $T_s = \frac{3}{16}\mu v_s^2/k$ at the shock front.

To estimate when the SNR will leave the Sedov-Taylor phase, we idealize the cooling function as

$$\Lambda \approx C\, T_6^{-0.7} n_{\rm H} n_e \quad , \quad C = 1.1 \times 10^{-22}\,{\rm erg\,cm^3\,s^{-1}} \quad , \tag{39.15}$$

As shown in Fig. 34.1, this is a fair approximation for solar-metallicity gas for temperatures $10^5 < T < 10^{7.3}$ K. Thus

$$\frac{dE}{dt} = -\int_0^{R_s} \Lambda\, 4\pi r^2 dr \tag{39.16}$$

$$= -1.2C(n_{\rm H})_0^2 \left(\frac{10^6\,{\rm K}}{T_s}\right)^{0.7} \frac{4\pi}{3} R_s^3 \left\langle \left(\frac{\rho}{\rho_0}\right)^2 \left(\frac{T_s}{T}\right)^{0.7} \right\rangle \quad , \tag{39.17}$$

where $\langle ... \rangle$ denotes a volume-weighted average over the blastwave. As long as the energy loss is small, we can use the density and temperature profile for the Sedov-Taylor solution from Fig. 39.1 to obtain $\langle (\rho/\rho_0)^2 (T_s/T)^{0.7} \rangle = 1.817$, and we can take $R_s(t)$ and $v_s(T)$ from the Sedov-Taylor solution to obtain

$$\Delta E(t) = -1.2C \frac{4\pi}{3} (n_{\rm H})_0^2 \times 1.817 \int_0^t dt\, R_s^3\, T_{s6}^{-0.7} \quad , \tag{39.18}$$

where $T_{s6} \equiv T_s/10^6$ K. Now use Eqs. (39.9 and 39.11) to obtain the fractional energy loss by time t:

$$\frac{\Delta E(t)}{E_0} \approx -2.38 \times 10^{-6} n_0^{1.68} E_{51}^{-0.68} t_3^{3.04} \quad . \tag{39.19}$$

If we now suppose that we leave the Sedov-Taylor phase and enter the "radiative" phase when $\Delta E(t_{\rm rad})/E_0 \approx -1/3$, we can solve for the cooling time $t_{\rm rad}$, and for the radius and shock speed at the end of the Sedov-Taylor phase:

$$t_{\rm rad} = 49.3 \times 10^3\,{\rm yr}\, E_{51}^{0.22} n_0^{-0.55} \quad , \tag{39.20}$$

$$R_{\rm rad} = 7.32 \times 10^{19}\,{\rm cm}\, E_{51}^{0.29} n_0^{-0.42} \quad , \tag{39.21}$$

$$v_s(t_{\rm rad}) = 188\,{\rm km\,s^{-1}}\, \left(E_{51} n_0^2\right)^{0.07} \quad , \tag{39.22}$$

$$T_s(t_{\rm rad}) = 4.86 \times 10^5\,{\rm K}\, \left(E_{51} n_0^2\right)^{0.13} \quad , \tag{39.23}$$

$$kT_s(t_{\rm rad}) = 41\,{\rm eV}\, \left(E_{51} n_0^2\right)^{0.13} \quad . \tag{39.24}$$

39.1.3 Snowplow Phase

When $t \approx t_{\rm rad}$, cooling causes the thermal pressure just behind the shock to drop suddenly, and the shock wave briefly stalls. However, the very hot gas in the interior of the SNR has not cooled, and its outward pressure forces the SNR to continue its expansion. The blastwave now leaves the Sedov-Taylor solution and enters what is called the **snowplow phase**, with a dense shell of cool gas enclosing a hot central volume where radiative cooling is unimportant. This is called the snowplow phase because the mass of the dense shell increases as it "sweeps up" the ambient gas. Let M_s be the mass of the shell; the gas in the shell has a radial velocity that is almost the same as the shock speed. The gas in the hot interior cools by adiabatic expansion, with $pV^\gamma = const$, or $p \propto V^{-\gamma} \propto R_s^{-3\gamma} = R_s^{-5}$, so that the pressure p_i in the interior evolves as

$$p_i = p_0(t_{\rm rad}) \left(\frac{R_{\rm rad}}{R_s}\right)^5 \quad . \tag{39.25}$$

The pressure exerted by the hot center causes the "radial momentum" of the shell to increase:

$$\frac{d}{dt}(M_s v_s) \approx p_i 4\pi R_s^2 = 4\pi p_0(t_{\rm rad}) R_{\rm rad}^5 R_s^{-3} \quad . \tag{39.26}$$

Suppose that there is a power-law solution, $R_s \propto t^\eta$. Equation (39.26) then requires $4\eta - 2 = -3\eta$, or $\eta = 2/7$. Thus we approximate the expansion by

$$R_s \approx R_s(t_{\rm rad})(t/t_{\rm rad})^{2/7} \quad , \tag{39.27}$$

$$v_s \approx \frac{2}{7}\frac{R_s}{t} = \frac{2}{7}\frac{R_s(t_{\rm rad})}{t_{\rm rad}}\left(\frac{t}{t_{\rm rad}}\right)^{-5/7} \quad , \tag{39.28}$$

for $t > t_{\rm rad}$. Because the effect of the internal pressure has been included, this solution is referred to as the **pressure-modified snowplow phase**. Note that with this construction, $R_s(t)$ is continuous from the Sedov-Taylor phase to the pressure-modified snowplow phase, but $v_s(t)$ undergoes a discontinuous drop by $\sim 30\%$ at $t = t_{\rm rad}$; this mimics the behavior seen in time-dependent fluid-dynamical simulations (e.g., Cioffi et al. 1988).

39.1.4 Fadeaway

For typical interstellar parameters, the shock speed at the beginning of the snowplow phase is $\sim 150\,{\rm km\,s^{-1}}$, which results in a very strong shock when propagating through interstellar gas with $T \lesssim 10^4\,{\rm K}$. However, the shock front gradually slows, and the shock compression declines. This proceeds until the shock speed approaches the effective sound speed in the gas through which the blastwave is propagating, at which point the compression ratio $\rightarrow 1$, and the shock wave turns into a sound wave. Suppose that c_s is the one-dimensional velocity dispersion in the preshock medium (including both thermal and turbulent motions) over scales of a few pc. It is reasonable to estimate that the snowplow blastwave simply "fades away" when its expansion velocity has slowed to c_s. This gives a "fadeaway time"

$$t_{\rm fade} \approx \left(\frac{(2/7)R_{\rm rad}/t_{\rm rad}}{c_s}\right)^{7/5} t_{\rm rad} \tag{39.29}$$

$$\approx 1.87 \times 10^6\,{\rm yr}\,E_{51}^{0.32} n_0^{-0.37} \left(\frac{c_s}{10\,{\rm km\,s^{-1}}}\right)^{-7/5} , \tag{39.30}$$

$$R_{\rm fade} \approx 2.07 \times 10^{20}\,{\rm cm}\,E_{51}^{0.32} n_0^{-0.37} \left(\frac{c_s}{10\,{\rm km\,s^{-1}}}\right)^{-2/5} . \tag{39.31}$$

39.2★ Overlapping of SNRs

We have been thinking about the evolution of a single SNR, and we have seen that it does not fade away until it reaches an age $t_{\rm fade}$, by which time it has expanded to fill a volume $(4\pi/3)[R_s(t_{\rm fade})]^3$. What is the probability that another SN will occur within this volume and affect the evolution of the original SNR *before* it has faded away?

Suppose that supernovae (SNe) occur at random in the disk of the Galaxy, with a SN rate per volume $S \equiv 10^{-13} S_{-13}\,{\rm pc^{-3}\,yr^{-1}}$. The SN rate in the Milky Way has been estimated from records of historical SNe and from observations of similar galaxies; the SN frequency in the Galaxy is estimated to be one event every 40 ± 10 yr (Tammann et al. 1994).[1] Suppose that we consider the "disk" to have a radius 15 kpc and a thickness 200 pc, and suppose that the SN rate within this volume is 1/60 yr. This gives a SN rate per volume $S \approx 1.2 \times 10^{-13}\,{\rm pc^{-3}\,yr^{-1}}$, or $S_{-13} \approx$ 1.2.

The expectation value for the number of additional SNe that will explode within the volume $V_{\rm fade}$ during the lifetime $t_{\rm fade}$ of the original SNR is

$$N_{\rm SN} = S\,\frac{4\pi}{3} R_{\rm fade}^3\, t_{\rm fade} \tag{39.32}$$

$$\approx 0.24\, S_{-13}\, E_{51}^{1.26}\, n_0^{-1.47} \left(\frac{c_s}{10\,{\rm km\,s^{-1}}}\right)^{-2.6} . \tag{39.33}$$

The two-phase model of the ISM postulated that most of the interstellar volume was filled by warm H I gas (the warm neutral medium, WNM) with density $n_0 \approx 1$

[1] We know of only 3 or 4 SNRs associated with SNe occurring in the past 500 years – Tycho's SN (1572), Kepler's SN (1604), Cas A (1681 ± 19; Fesen et al. 2006), and perhaps SNR G1.9+0.3 (1880 ± 30; Reynolds et al. 2008; Borkowski et al. 2010) – whereas if the SN rate for the Galaxy is one per 40 ± 10 yr the expected number over this period would be 10 to 17. If the estimate for the rate is correct, historical records and surveys for young SNRs must have missed ∼75% of the Galactic SNe occurring in the past 5 centuries. Obscuration by dust, and the fact that half of the events would be more than 10 kpc away, make this plausible.

and $c_s \approx 6\,\mathrm{km\,s^{-1}}$, corresponding to the one-dimensional velocity dispersion for $T = 6000\,\mathrm{K}$. For $E_{51} = 1$, $S_{-13} \approx 1.2$, $n_0 \approx 1$, and $c_s \approx 6\,\mathrm{km\,s^{-1}}$, we obtain $N_{\rm SN} \approx 1.1$. Therefore, we must conclude that if we were to start with the gas distributed as in the two-phase model, in a matter of $\sim$2 Myr, the initially near-uniform WNM will be destroyed by the effects of SNe: SNRs will overlap and ocupy a major fraction of the disk volume. This implies that, at least for the Milky Way, SNRs will create a multiphase ISM, consisting of low-density regions in the interior of the SNRs, and dense regions containing most of the gas mass. We will consider the global structure of the ISM further in §39.4.

39.3★ Supernova Remnants in an Inhomogeneous Medium

The preceding discussion of a spherical blastwave assumed that the SN explosion took place in a uniform medium, but we have just seen that SN explosions will disrupt an initially uniform ISM in only $\sim$2 Myr. How will the evolution of a SNR be altered by preexisting inhomogeneities in the ISM? The importance of inhomogeneities was first emphasized by Cox & Smith (1974), who idealized the ISM as a "hot network of tunnels." McKee & Ostriker (1977) instead envisaged an ISM where most of the volume was occupied by the hot phase, with discrete cool clouds dispersed through it; this is the geometry we will consider here.

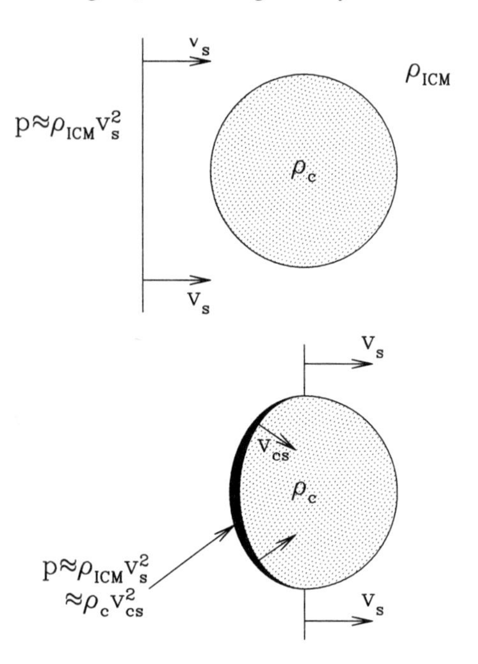

Figure 39.2 Cloud engulfed by supernova blastwave, shown before the blastwave arrival, and as the blastwave is passing around it. A shock is driven into the cloud with shock speed $v_{\rm cs} \approx (\rho_{\rm ICM}/\rho_{\rm c})^{1/2}\, v_s$.

The blastwave propagates more rapidly in a low density medium. The lowest density phase in the ISM is the hot ionized medium (HIM), with temperatures $T \approx 10^6$ K, a substantial volume filling factor, and a density $(n_{\rm H})_{\rm HIM} \approx 0.005\,{\rm cm}^{-3}$. A typical supernova blastwave will be expanding into such gas, with the blastwave passing *around* any high-density clouds that may be present. If the clouds were rigid, and had a small filling factor, they would have little effect on the propagation of the blastwave, but of course real clouds are compressible, and can be heated and "evaporated" by contact with hot gas. These effects will now be briefly considered.

Let $\rho_{\rm ICM}$ be the density of the intercloud medium (ICM). The blastwave is propagating through the ICM with shock speed v_s. When the blastwave comes in contact with a cloud, it applies a pressure $p \approx \rho_{\rm ICM} v_s^2$ to the cloud surface. This overpressure will drive a shock into the cloud. The speed v_{cs} of this "cloud shock" can be readily estimated by noting that the pressure $\sim \rho_{cs} v_{cs}^2$ in the shocked cloud material must be in approximate pressure equilibrium with the pressure $\rho_{\rm ICM} v_s^2$ in the shocked intercloud medium. This gives a very simple result:

$$v_{\rm cs} = \sqrt{\frac{\rho_{\rm ICM}}{\rho_c}}\, v_s \quad . \tag{39.34}$$

Consider an intercloud medium with $(n_{\rm H})_{\rm ICM} \approx 0.005\,{\rm cm}^{-3}$, and a cloud density $(n_{\rm H})_c \approx 30\,{\rm cm}^{-3}$. A blastwave propagating at $1000\,{\rm km\,s}^{-1}$ through the ICM will drive a shock into the cloud with a shock speed of only $v_{\rm cs} \approx 13\,{\rm km\,s}^{-1}$. The density contrast protects the material in the cloud from the direct effects of very fast shocks.

The shock passing through the cloud will set the cloud material into motion, but with velocity gradients in the shocked material. If the cloud is not magnetized and not self-gravitating, these velocity gradients can act to "shred" the cloud. However, the magnetic field that is almost certainly already present in the cloud may be able to oppose these shearing effects.

After the blastwave has passed over the cloud, the cloud finds itself engulfed in hot gas resulting from the shock-heating of the low density intercloud medium. As discussed in §34.3 and §34.4, thermal conduction will transport heat from the hot plasma into the cool cloud. If the cloud is sufficiently small, this thermal conduction can lead to "evaporation" of cloud material, resulting in mass loss from the cloud, and an increase in the mass density of the shocked intercloud medium. The increase in the mass density will be accompanied by a drop in the temperature, as the thermal energy is shared by more particles. The combined effects of increased density and lowered temperature act to promote radiative cooling.

39.4 Three-Phase Model of the ISM

In §39.2, we saw that an initially uniform ISM consisting of warm H I would be transformed by SNRs into a medium consisting of low-density hot gas and dense shells of cold gas. This transformation would take place in just a few Myr. McKee

& Ostriker (1977) developed a model of the ISM that took into account the effects of these blastwaves. They envisaged an ISM consisting of three distinct phases: cold gas – the cold neutral medium (CNM); warm gas – the warm neutral medium (WNM) and warm ionized medium (WIM); and hot gas – the hot ionized medium (HIM). A SNR blastwave expands into this composite medium, as illustrated in Fig. 39.3.

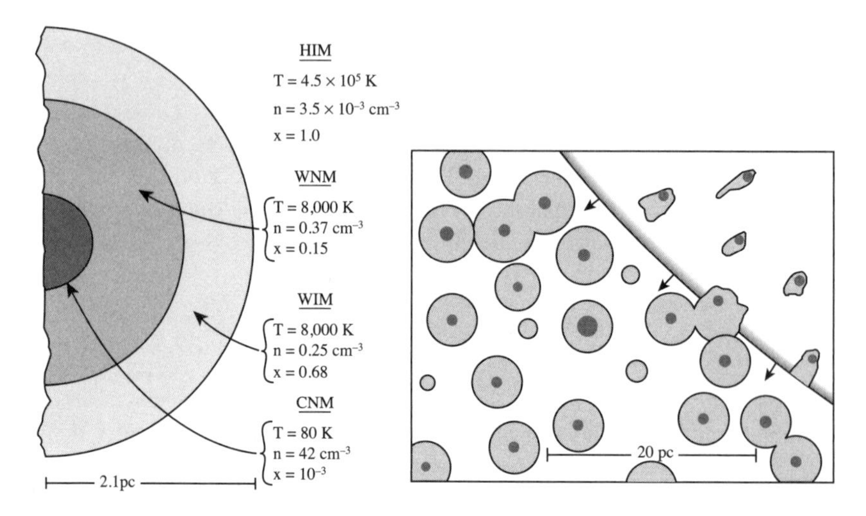

Figure 39.3 Left: Structure of a typical cold cloud in the three-phase model of McKee & Ostriker (1977). Right: Close-up of a supernova blastwave. From McKee & Ostriker (1977).

McKee & Ostriker (1977) argued that the pressure in the ISM was maintained by SNe – if initially the ISM had a low pressure, then SNRs would expand to large radii before "fading," with resulting overlap. The pressure in the ISM will rise until the SNRs tend to overlap just as they are fading, at which point the pressure in the ISM is the same as the pressure in the SNR. According to this argument, the condition $N_{\rm SN} \approx 1$ can be used to *predict* the pressure in the ISM.

We previously obtained an equation (39.32) for $N_{\rm SN}$ in terms of the supernova rate per volume S, E_{51}, n_0, and c_s. If we write $p = 1.4 n_{\rm H} m_{\rm H} c_s^2$, we can eliminate c_s in favor of the pressure p. The expectation value for overlap then becomes

$$N_{\rm SN} = 0.24\, S_{-13}\, E_{51}^{1.26}\, n_0^{-1.47}\, c_{s,6}^{-13/5} \tag{39.35}$$

$$= 0.48\, S_{-13}\, E_{51}^{1.26}\, n_0^{-0.17}\, p_4^{-1.30} \quad , \quad p_4 \equiv \frac{p/k}{10^4\,{\rm cm}^{-3}\,{\rm K}} \, . \tag{39.36}$$

Setting $N_{\rm SN} = 1$, we solve for the pressure:

$$\frac{p}{k} = S_{-13}^{0.77}\, E_{51}^{0.97}\, n_0^{-0.13} \times 5700\,{\rm cm}^{-3}\,{\rm K} \ . \tag{39.37}$$

The derived pressure depends weakly on n_0; if we set $n_0 \approx 1$ (the mean density in the ISM in the solar neighborhood) and $S_{-13} \approx 1.2$, we obtain $p/k \approx 6600\,\mathrm{cm}^{-3}\,\mathrm{K}$, comparable to the observed thermal pressure ($p/k \approx 3800\,\mathrm{cm}^{-3}\,\mathrm{K}$ – see §17.7) in the ISM today. This is a remarkable result: given (1) the *observed* SN rate/volume ($S_{-13} \approx 1.2$); (2) the *observed* kinetic energy per supernova ($E_{51} \approx 1$); and (3) the atomic physics of the cooling function (using observed abundances) – from these alone McKee & Ostriker (1977, hereafter MO77) were able to predict the interstellar pressure!

The MO77 model envisaged three phases of the ISM: cold neutral gas (CNM), warm neutral and ionized gas (WNM and WIM), and hot ionized gas (HIM). MO77 did not explicitly consider molecular gas, because it occupied a very small volume filling factor, and can be considered part of the CNM. Our current view of the ISM continues to identify these as major phases, and follows the central ideas of the MO77 model:

- Pressurization of the ISM by SNRs.
- Mass exchange between the phases: cold clouds "evaporated" and converted to diffuse gas, and diffuse gas swept up by SN blastwaves and compressed in the high-pressure shells of radiative SNRs.
- Injection of high-velocity clouds by fragmentation of the dense shell present in radiative SNRs.

The principal shortcoming of the MO77 model is the failure to predict the substantial amount of warm H I that is present in the ISM. The model parameters given by MO77 have only 4.3% of the H I mass in the warm phase (WNM and WIM). As seen in Chapter 29, 21-cm line observations indicate that more than 60% of the H I within 500 pc of the Sun is actually in the warm phase (Heiles & Troland 2003).

Chapter Forty

⋆ Cosmic Rays and Gamma Rays

40.1 Cosmic Ray Energy Spectrum and Composition

The ISM is pervaded by **cosmic rays** – a population of very energetic nuclei and electrons. The energy spectrum of low-energy cosmic rays was discussed in connection with ionization processes in the ISM (see Fig. 13.5). Most of the energy density in cosmic rays comes from transrelativistic particles with kinetic energies per nucleon $E \approx 1\,\mathrm{GeV}$. However, the cosmic ray energy spectrum extends to extraordinarily high energies, as shown in Fig. 40.1. Between $\sim 10\,\mathrm{GeV}$ and $\sim 10^7\,\mathrm{GeV}$, the observed particle flux $\Phi_{\rm CR}$ is well-described by a power law $d\Phi_{\rm CR}/dE \propto E^{-2.65}$. There is a slight steepening at $E \approx 10^{6.5}\,\mathrm{GeV}$, referred to as the "knee," with $d\Phi_{\rm CR}/dE$ changing from $\sim E^{-2.65}$ to $\sim E^{-3}$. There appear to also be further changes in slope at higher energies ("2nd knee" and "ankle" in Fig. 40.1).

At all energies (except possibly the very highest), the cosmic ray composition is dominated by protons. Relativistic electrons are also present, but with a flux that is small compared to the protons. Figure 40.2 shows the composition of cosmic ray nuclei at $1\,\mathrm{GeV/nucleon}$. The plot is normalized to the abundance of Si (= 100). Note two points:

1. As noted above, protons dominate, with He next in importance.

2. The elements Li, Be, and B ($Z = 3, 4, 5$) are greatly overabundant relative to solar abundances. A large fraction of the Li, Be, and B nuclei in cosmic rays are the result of **spallation**, where (for example) a cosmic ray oxygen nucleus, colliding with a proton at rest, can fragment into spallation products that include ^{6}Li, ^{7}Li, ^{9}Be, ^{10}B, and ^{11}B. The observed abundances of these isotopes in the cosmic rays, in primitive meteorities, and in stellar atmospheres are consistent with spallation by cosmic rays over the age of the Galaxy, with cosmic rays being generated continuously and traversing a column density of $\sim 6\,\mathrm{g\,cm^{-2}}$ of interstellar matter (Meneguzzi et al. 1971) before escaping from the Galaxy. Big-bang nucleosynthesis appears able to account for the observed ^{7}Li, but ^{6}Li, ^{9}Be, ^{10}B, and ^{11}B appear to be primarily spallation products.

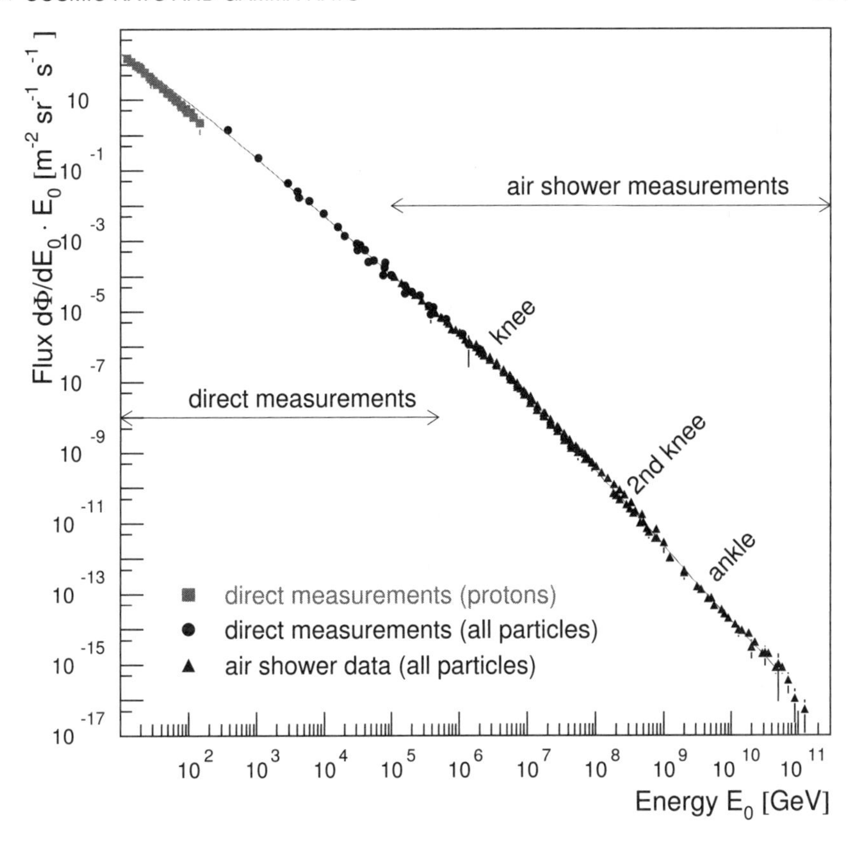

Figure 40.1 $Ed\Phi_{\rm CR}/dE$ for cosmic rays (summed over all species) with kinetic energy $E > 10\,{\rm GeV}$. From Blümer et al. (2009).

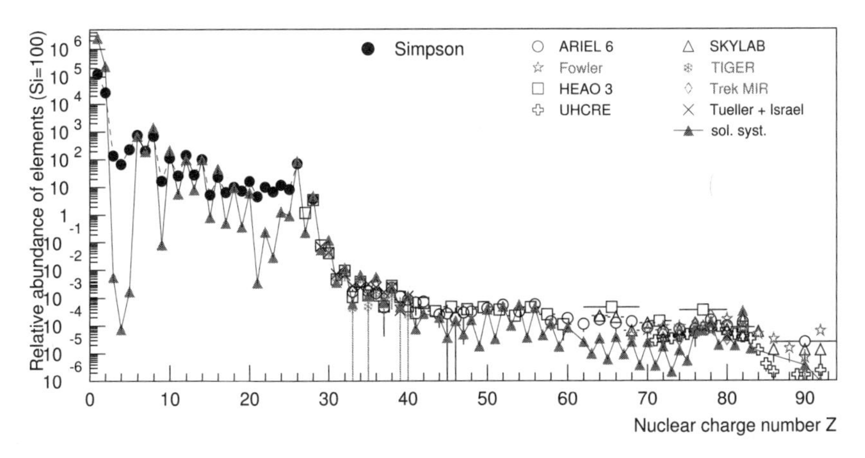

Figure 40.2 The composition of cosmic rays with $\sim 1\,{\rm GeV/nucleon}$, normalized to Si = 100. The solid triangles show solar abundances, for comparison. From Blümer et al. (2009).

40.2★ Theory of Diffusive Shock Acceleration

The origin of high-energy cosmic rays has been a long-standing problem. Fermi (1949) proposed that cosmic rays could be accelerated by scattering off the random motions of magnetized gas clouds, but this mechanism, now referred to as **second-order Fermi acceleration**, produced a random walk in energy, and was unable to explain the observed spectrum. In 1977–1978, several authors (Axford et al. 1977; Krymskii 1977; Bell 1978; Blandford & Ostriker 1978) showed that interstellar shock waves could systematically accelerate particles, and could naturally produce a spectrum close to that observed. This mechanism – particle acceleration in the converging flow of a shock wave – is referred to as **diffusive shock acceleration**. It is an example of **first-order Fermi acceleration**. Here, we outline the theory.

Consider a shock in a frame of reference where the shock front is stationary. The "upstream" (preshock) fluid is flowing toward the shock front with velocity v_s. Consider an energetic particle with kinetic energy $E_0 \gg kT_s$ in the postshock region, moving with speed $w \gg v_s/r$. If the velocity of this particle happened to be directed back toward the shock front, and the particle is sufficiently close to the shock, the particle could, in principle, travel back across the shock front and reenter the upstream preshock region. Electromagnetic fields in the preshock flow will eventually scatter the particle (elastically), and it will be carried back toward the shock and will enter the postshock region with an increased energy $E > E_0$. These scattering processes (due to electromagnetic fluctuations) also occur in the postshock region, so that the particle in question may return to the shock front and again cross into the preshock region. With each repetition, the particle energy is increased.

We can develop a simple theory to determine the form of the steady-state energy spectrum of the cosmic rays near the shock front. Let p be the (scalar) momentum of a particle. For a point at the shock front, let

$$f(p)dp \equiv \text{ the density of particles with momentum } \in \ [p, p + dp] \quad .$$

Because the cosmic rays are moving freely back and forth across the shock front, with gyroradii that are large compared to the thickness of the shock transition, the function $f(p)$ is continuous across the shock.

A fraction of the particles at the shock front will return to the preshock region and be scattered back to the shock, returning to the shock front and entering the postshock region with increased energy. It is as though the particle was trapped between two converging mirrors, one moving with the velocity of the preshock fluid, and the other moving with the velocity of the postshock fluid. Let

$$\Delta v = (1 - 1/r)v_s \tag{40.1}$$

denote the velocity difference between the preshock and postshock fluids, where r is the compression ratio. Consider a particle that has just crossed the shock from downstream to upstream, and is reflected back downstream. Averaging over the direction of motion of the particle, assumed to be at an angle θ relative to the shock normal both before and after reflection, the reflection will increase p by an amount Δp, with

$$\langle \Delta p \rangle = \frac{\int_0^{\pi/2} (w \cos\theta)(\frac{2\Delta v}{w} p \cos\theta) \sin\theta d\theta}{\int_0^{\pi/2} (w \cos\theta) \sin\theta d\theta} \tag{40.2}$$

$$= 2\frac{\Delta v}{w} p \frac{\int_0^{\pi/2} \cos^2\theta \sin\theta d\theta}{\int_0^{\pi/2} \cos\theta \sin\theta d\theta} = \frac{4\Delta v}{3w} p \quad . \tag{40.3}$$

The particle has some probability per unit time $t_{\rm refl}^{-1}$ of being reflected back toward the shock by turbulent fluctuations in the B field. The acceleration time $t_{\rm acc}$ is

$$t_{\rm acc}^{-1}(p) \equiv \langle (d/dt) \ln p \rangle \tag{40.4}$$

$$= \frac{1}{p} \langle \Delta p \rangle \, t_{\rm refl}^{-1} = \frac{4\Delta v}{3w} \, t_{\rm refl}^{-1} \quad . \tag{40.5}$$

The cosmic rays are scattered by fluctuations in the magnetic field, which is essentially a stochastic process as far as a single cosmic ray is concerned. Let $t_{\rm esc}^{-1}(p)$ be the probability per unit time that a particle with momentum p will get swept downstream by the postshock flow, not to cross the shock front again.

The continuity equation in phase space is

$$\frac{\partial}{\partial p}(f \dot{p}) + \frac{\partial f}{\partial t} = \left(\frac{\partial f}{\partial t}\right)_{\rm source-sink} \quad , \tag{40.6}$$

where the source-sink term represents the escape process:

$$\left(\frac{\partial f}{\partial t}\right)_{\rm source-sink} = -\frac{f}{t_{\rm esc}} \quad . \tag{40.7}$$

If we assume a steady state $\partial f / \partial t = 0$, the continuity equation becomes

$$\frac{d}{dp}\left(f \frac{p}{t_{\rm acc}}\right) = -\frac{f}{t_{\rm esc}} \quad . \tag{40.8}$$

If we now assume that $(d/dp)t_{\rm acc} = 0$, we obtain

$$\frac{p}{f}\frac{df}{dp} = -\left(1 + \frac{t_{\rm acc}}{t_{\rm esc}}\right) \quad , \tag{40.9}$$

with solution

$$f \propto p^{-\alpha} \quad , \quad \alpha = 1 + \frac{t_{\rm acc}}{t_{\rm esc}} \quad . \tag{40.10}$$

We will now see that the ratio $t_{\rm acc}/t_{\rm esc}$ can be related to the compression in the shock. Let $f_s(p)$ be the phase space density at the shock front. Because the relativistic particles cross the shock front with impunity, $f(p)$ is continuous across the shock front. The CR flux from upstream crossing the shock is $(w/4)f_s(p)$. The CR flux far downstream from the shock is $v_2 f_2$. Therefore the probability that a CR particle crossing the shock will "escape" downstream (i.e., fail to be reflected back across the shock) is

$$\text{escape probability} = \frac{v_2 f_2}{w f_s/4} \approx \frac{4v_2}{w} \quad . \tag{40.11}$$

Let $r = v_s/v_2$ be the compression ratio for the shock. Then,

$$\text{escape probability} = \frac{4v_s}{rw} \quad . \tag{40.12}$$

Therefore, the probability per unit time that a given CR will escape is

$$t_{\rm esc}^{-1} = \frac{4v_s}{rw} \times t_{\rm refl}^{-1} \quad , \tag{40.13}$$

and

$$\frac{t_{\rm acc}}{t_{\rm esc}} = \frac{(3w/4\Delta v)t_{\rm refl}}{(rw/4v_1)t_{\rm refl}} = \frac{3v_s}{r\Delta v} = \frac{3}{r-1} \quad . \tag{40.14}$$

Thus we expect the cosmic rays to have $f \propto p^{-\alpha}$, with

$$\alpha = 1 + \frac{t_{\rm acc}}{t_{\rm esc}} = 1 + \frac{3}{r-1} = \frac{r+2}{r-1} \quad . \tag{40.15}$$

A strong nonradiative shock has $r = 4$, giving $\alpha = 2$, which is not far from the observed power-law index $\alpha \approx 2.65$ for $10 \lesssim E/\,\text{GeV} \lesssim 10^7$, lending support to the notion that interstellar shock waves play an important role is acceleration of cosmic rays. Cosmic rays can be lost to diffusion out of the Galaxy; since this would be more rapid at increasing energy, this process will steepen the interstellar cosmic ray spectrum – α will be somewhat larger than given by Eq. (40.15).

In order for cosmic rays to be scattered by the magnetic field, the magnetic field must be "turbulent" on a length scale $\sim pc/eB$. It appears that these fluctuations can be generated by the cosmic rays themselves – cosmic rays can excite the MHD waves that are needed for cosmic ray acceleration.

40.3★ Injection Problem

The preceding section discussed acceleration of cosmic rays that are already present. The details of "injection" into the low-energy end of the spectrum are not yet understood. Because very low energy particles can be quickly thermalized by Coulomb

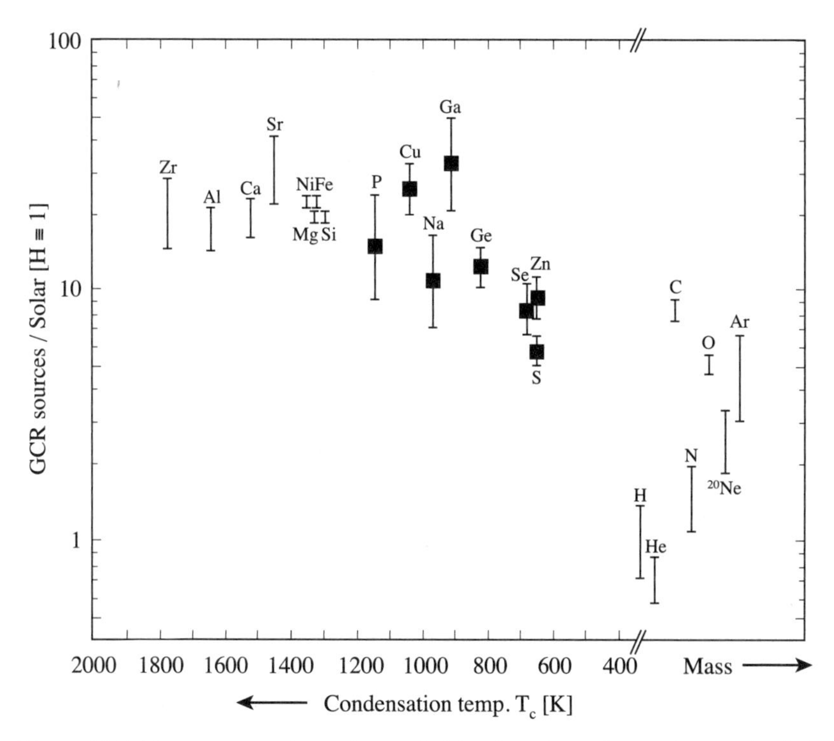

Figure 40.3 Elemental abundance in ~ 1 GeV/nucleon Galactic cosmic rays (GCRs) relative to solar abundance, versus "condensation temperature" (see Fig. 23.1 for a plot of gas-phase abundances versus condensation temperature). Elements with high condensation temperature appear to be enhanced in cosmic rays. After Meyer et al. (1997).

scattering, the injection problem consists of explaining how a small fraction of the ions can enter the cosmic ray energy spectrum with enough energy so that diffusive shock acceleration can overcome loss of energy to Coulomb scattering off the thermal plasma.

Figure 40.3 shows abundances in cosmic rays, relative to solar, vs. condensation temperature. Elements that are normally "depleted" into refractory grains appear to be overrepresented in cosmic rays. Epstein (1980) proposed that this may be connected to the injection process. The plasma passing through the shock transition is more-or-less thermalized, either collisionally or by the fluctuating electromagnetic fields in a collisionless shock transition. Dust grains, with a much lower ratio of charge to mass, travel across the shock transition almost undeflected. With their large "rigidity" pc/q, where q is the charge, dust grains in some ways behave like cosmic rays. Compression in the cooling postshock flow can lead to "betatron acceleration" of the dust grains (Spitzer 1976), and in some cases the dust grain trajectory may return it to the preshock medium, where it will undergo first-order Fermi acceleration and be returned to the postshock medium with an increased momentum (Slavin et al. 2004). As a result, dust grains may be moving through the postshock gas with velocities exceeding v_s. Sputtering will inject high-velocity atoms into the plasma, which will quickly be ionized; the ions will populate the extreme tail of the velocity distribution function, and should be candidates for subsequent acceleration. This may explain the apparent enhancement of elements like Mg, Si, and Fe in cosmic rays.

40.4★ Upper Limits on Cosmic Ray Energy

Interstellar shock waves are produced by discrete events, such as supernova explosions, and, therefore, have a finite size and duration. The gyroradius for a cosmic ray in a magnetic field $B_\perp$ is

$$R_{\rm gyro} = \frac{pc}{eB_\perp} = 1.11 \times 10^{12} \left(\frac{pc}{\rm GeV}\right) \left(\frac{3\,\mu{\rm G}}{B_\perp}\right) {\rm cm} \quad . \tag{40.16}$$

For any acceleration to occur, a cosmic ray must be scattered by the magnetic field in the postshock gas, which can only occur if the postshock region has a spatial extent $L \gtrsim R_{\rm gyro}$. If the magnetic field strength in the SNR blastwave is $B_{\rm SNR}$, then there is a critical energy $E_{\rm max}$ and momentum $p_{\rm max}$ above which acceleration should be ineffective:

$$E_{\rm max} = cp_{\rm max} = eB_{\rm SNR}L \quad . \tag{40.17}$$

The compressed shell in a SN blastwave with radius R has a typical thickness $\sim 0.05R$ (see Fig. 39.1). We might take $L \approx 0.05R_{\rm rad}$, where $R_{\rm rad}$ is the radius at which a SNR becomes radiative [see Eq. (39.21)]. Then, for typical parameters ($E_{51} \approx 1$, $n_0 \approx 1$), we find

$$E_{\rm max} \approx 10^{7.0}\,{\rm GeV} \left(\frac{R_{\rm rad}}{7 \times 10^{19}\,{\rm cm}}\right) \left(\frac{B_{\rm SNR}}{10\,\mu{\rm G}}\right) \quad . \tag{40.18}$$

At this energy, a cosmic ray would be scattered only once by a given expanding SNR, but acceleration could proceed by scattering off one expanding SNR after another. The observed cosmic ray energy spectrum in fact has a "knee" at $\sim 10^7$ GeV, with the spectrum steepening from $\sim p^{-2.65}$ to $\sim p^{-3}$. It is at least plausible that this steepening could be related to the maximum energy (40.18).

Supernova blastwaves appear to be very effective at particle acceleration, with cosmic rays making an appreciable contribution to the pressure in the shocked gas. In the SNR RCW 86, it has been argued that $\gtrsim 50\%$ of the postshock pressure is provided by cosmic rays rather than thermal motions (Helder et al. 2009).

Relativistic electrons are accelerated by the same processes that accelerate nuclei. However, very energetic electrons lose energy by synchrotron emission and inverse Compton scattering, which can suppress the population of such electrons.

The local spectrum of cosmic ray electrons can be directly measured. At $E = 1\,{\rm TeV}$ the electrons have $Ed\Phi_e/dE \approx 1.2 \times 10^{-4}\,{\rm m^{-2}\,s^{-1}\,sr^{-1}}$ (Ackermann et al.

2010),smaller than the 1 TeV hadronic cosmic ray flux in Fig. 40.1 by a factor ~ 250. The electron spectrum falls off more steeply with increasing energy than does the proton spectrum.

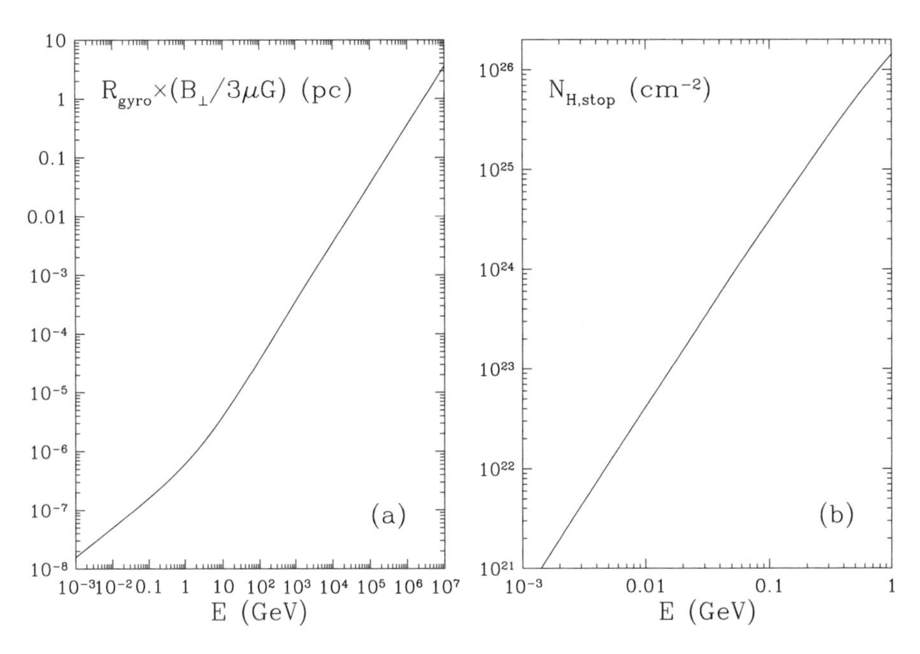

Figure 40.4 (a) The gyroradius of a cosmic ray proton with kinetic energy E in a magnetic field with $B_\perp = 3\,\mu$G. Even 10^6 GeV protons have gyroradii small compared to 1 pc. (b) The H column density to stop a proton of energy E due to ionization losses in neutral gas. Above ~0.3 GeV, energy loss due to pion production becomes important.

40.5★ Cosmic Ray Propagation

The gyroradius of a cosmic ray proton is shown in Fig. 40.4a, where we have assumed a $3\,\mu$G magnetic field, typical of the ISM within $\sim 150\,$pc of the midplane. For $E \lesssim 10^5$ GeV, cosmic rays are effectively trapped on whatever magnetic field line they are orbiting. Scattering (both in pitch angle and onto adjacent magnetic field lines) will be dominated by magnetic fluctuations (MHD waves) with wavelengths $\sim R_{\rm gyro}$ and occasional nuclear collisions. Protons with $E \lesssim 10^3\,$GeV have $R_{\rm gyro} < 10^{-4}\,$pc. Propagation of cosmic rays is normally approximated as a diffusive process (Strong et al. 2007). Because of their small gyroradii, the arrival directions of low-energy cosmic rays are essentially isotropic.

In the absence of acceleration, cosmic rays will gradually lose energy to ionization of neutral gas and Coulomb scattering by the thermal plasma. In neutral regions, the ionization losses dominate. Above proton kinetic energy $E = 0.28\,$GeV, pion production in p-p collisions becomes possible, and dominates the energy loss at higher energies (Mannheim & Schlickeiser 1994). Using the cross section in Eq. (13.15), and assuming an average energy loss per ionization of $13.6\,\mathrm{eV}$ due to ionization and $\sim 35\,\mathrm{eV}$ of kinetic energy to the electron, we can evaluate the column density $N_{\mathrm{H,stop}}$ required to stop a cosmic ray of initial kinetic energy E. The result is shown in Fig. 40.4b. A 100-MeV proton can penetrate even a dense cloud with $N_{\mathrm{H}} \approx 10^{23}\,\mathrm{cm}^{-3}$, whereas a 1-MeV proton would be stopped by even a diffuse H I cloud.

Compression of the magnetic field in the cloud can lead to "magnetic mirror" reflection of cosmic rays with large pitch angles. But the magnetic pinch does not exclude cosmic rays from the compressed region. If the magnetic field is static, and the incident cosmic ray flux is isotropic in pitch angle, then the small pitch angle particles will enter the compressed regions, resulting in a cosmic ray density within the cloud that is the same as outside the cloud. The only way to exclude cosmic rays from dense regions is to have magnetic field lines that do not connect to the diffuse ISM where the particles are presumed to be accelerated, or to have such large column densities that the lower energy cosmic rays are stopped by ionization losses.

40.6★ Synchrotron Emission and Supernova Remnants

Direct evidence for particle acceleration in shocks is provided by synchrotron emission produced by relativistic electrons in supernova remnants. In some objects, such as the Crab Nebula, the relativistic electrons may be coming from the rotating neutron star, but in many SNRs, such as Cas A (see Plate 12) the synchrotron emission appears to come from the shocked gas, and provides evidence for acceleration of electrons in supernova blastwaves.

The synchrotron emission from SNRs is easily observed at radio frequencies, but in some cases can be detected at optical (e.g., the Crab Nebula, although in this case the relativistic electrons appear to come from the pulsar wind) or X-ray energies (e.g., the Crab, Tycho's SNR, and Cas A (see Plate 12)).

40.7★ Gamma Ray Emission from Interstellar Clouds

High-energy cosmic rays ions (mainly protons) or electrons colliding with nuclei in the gas or dust can result in emission of gamma rays. There are three principal channels for gamma-ray production (see, e.g., Bloemen 1987):

$$CRp + p \rightarrow p + p + \mathrm{pions} \quad ; \quad \pi^0 \rightarrow 2\gamma \qquad (\pi^0 \ \mathrm{decay}), \tag{40.19}$$

$$CRe + p \rightarrow e + p + \gamma \qquad (\mathrm{bremsstrahlung}), \tag{40.20}$$

$$CRe + \gamma \rightarrow e + \gamma' \qquad (\mathrm{inverse\ Compton}). \tag{40.21}$$

γ rays in the 50 MeV to 3 GeV range are produced primarily by cosmic rays with energies of 1 to 10 GeV/nucleon. For the interstellar spectrum of cosmic ray protons and electrons, the $\pi^0 \rightarrow 2\gamma$ channel dominates production of $E > 150\,\mathrm{MeV}$ γ-rays, while bremstrahhlung dominates for $E \lesssim 150\,\mathrm{MeV}$.

If the $E \gtrsim 1\,\mathrm{GeV}$ cosmic ray flux is more-or-less uniform, then an interstellar cloud will emit γ rays with an intensity $I(\gamma) = C \times N_{\rm H}$. The constant C can be calibrated by observing H I clouds for which $N_{\rm H}$ can be determined from 21-cm emission; gamma ray observations of molecular clouds can then be used to measure their column densities. This is one of the methods used to determine empirically the ratio between CO J=1–0 luminosity and molecular gas mass.

40.8★ ^{26}Al in the ISM

The stable isotope of Al is ^{27}Al; the isotope ^{26}Al is unstable against

$$^{26}\mathrm{Al} \rightarrow {}^{26}\mathrm{Mg}^* + e^+ + \bar{\nu}_e(1.16\,\mathrm{MeV}) \tag{40.22}$$

$$^{26}\mathrm{Mg}^* \rightarrow {}^{26}\mathrm{Mg} + \gamma(1.81\,\mathrm{MeV}) \quad , \tag{40.23}$$

with a half-life $\tau_{1/2}(^{26}\mathrm{Al}) = 7.4 \times 10^5\,\mathrm{yr}$. Interest in interstellar ^{26}Al first arose when it was discovered that certain Ca-Al-rich inclusions (CAIs) in the Allende meteorite contained excess ^{26}Mg relative to ^{24}Mg, the dominant isotope of Mg. The implication was that the inclusions solidified with ^{26}Al present, which then decayed into ^{26}Mg. The observed ^{26}Mg/^{27}Al in some CAIs implies that they had $^{26}\mathrm{Al}/^{27}\mathrm{Al} \approx 5 \times 10^{-5}$ (MacPherson et al. 1995) when they solidified. This was surprising, because it required enrichment of the gas with ^{26}Al not more than a few Myr before solidification of the CAIs.

Recent mapping of the ^{26}Mg* 1.80865 MeV line by the INTEGRAL satellite shows that the Milky Way has $\sim 2.7 \pm 0.7\,M_\odot$ of ^{26}Al in the ISM. (Wang et al. 2009) – this corresponds to an average $^{26}\mathrm{Al}/^{27}\mathrm{Al} \approx 9 \times 10^{-6}$ in the ISM. This average value of $^{26}\mathrm{Al}/^{27}\mathrm{Al}$ is a factor of 6 below the highest values in the CAIs, but with the short lifetime of ^{26}Al it would not be suprising if the $^{26}\mathrm{Al}/^{27}\mathrm{Al}$ ratio in the ISM showed strong spatial variations.

The origin of the interstellar ^{26}Al is uncertain. Some is produced by spallation of, e.g., ^{28}Si and ^{40}Ar by cosmic ray protons, but the bulk is thought to be injected by winds from massive stars and core-collapse supernovae (Voss et al. 2008). If massive stars dominate the injection of ^{26}Al, we expect the ^{26}Al abundance to be higher in massive star-forming regions.

40.9★ Positrons and Positronium in the ISM

Decaying ^{26}Al injects positrons into the ISM at a global rate $\sim 4 \times 10^{42}\,\mathrm{s}^{-1}$. Interstellar positrons are also generated by decay of other radionuclides, by cosmic ray interactions with the ISM (CR$p + p \rightarrow$ CR$p + p + \pi^+ + \pi^-$, followed by, e.g. $\pi^+ \rightarrow \mu^+ + \bar{\nu}_\mu$, followed by $\mu^+ \rightarrow e^+ + \bar{\nu}_e + \nu_\mu$), and by pair production in the jets from some compact objects. What is the fate of these positrons?

A relativistic positron traveling through the ISM can lose energy to synchrotron radiation, to inverse Compton scattering, and to Coulomb scattering. There are several different annihilation channels:

1. Direct annihilation on free electrons: $e^+ + e^- \rightarrow 2\gamma$.
2. Direct annihilation on atomic H: $e^+ + \mathrm{H} \rightarrow \mathrm{H}^+ + 2\gamma$.
3. Formation of positronium, followed by annihilation.

Positronium (Ps) consists of an electron and a positron in a bound state. The energy spectrum of Ps is just like that of H, except that the energies are all a factor of 2 smaller because the reduced mass of positronium is $m_e/2$. Positronium can form either by radiative recombination

$$e^+ + e^- \rightarrow \mathrm{Ps} + h\nu \quad , \tag{40.24}$$

or by charge exchange

$$e^+ + \mathrm{H} \rightarrow \mathrm{Ps} + \mathrm{H}^+ \quad . \tag{40.25}$$

The charge exchange process has an energy barrier of $6.8\,\mathrm{eV}$, and therefore requires energetic positrons.

The positronium can form with the electron and positron spins parallel (the triplet state $^3\mathrm{S}_1$, with spin $S = 1$) or antiparallel (the singlet state $^1\mathrm{S}_0$, with spin $S = 0$); $\sim 75\%$ of radiative recombinations will form the triplet state, and $\sim 25\%$ will form the singlet state.

The distinction between the singlet and triplet states is important, because they have different annihilation channels:

$$\mathrm{Ps}\,^3\mathrm{S}_1 \rightarrow 3\gamma \quad , \quad \tau = 1.4 \times 10^{-7}\,\mathrm{s} \quad , \tag{40.26}$$

$$\mathrm{Ps}\,^1\mathrm{S}_0 \rightarrow 2\gamma \quad , \quad \tau = 1.25 \times 10^{-10}\,\mathrm{s} \quad . \tag{40.27}$$

The triplet state decays into a three γ-ray continuum, extending from 0 to $511\,\mathrm{keV}$. The singlet state decays into two $511\,\mathrm{keV}$ photons. The γ rays from positronium annihilation have been observed from the inner Galaxy, and exhibit the expected 3:1 ratio of triplet to singlet decay channels (for a review, see Diehl & Leising 2009).

Chapter Forty-one

Gravitational Collapse and Star Formation: Theory

Gravity is responsible for gathering gas into self-gravitating structures ranging in size from stars to giant molecular cloud complexes. Star formation involves extreme compression: part of a gas cloud collapses from a size $\sim 10^{18}$ cm down to a stellar size, $\sim 10^{11}$ cm, with an accompanying increase in density by a factor $\sim 10^{21}$.

Here, we consider the conditions necessary for gravitational collapse to occur. There are several barriers to gravitational collapse. Gravity must of course overcome the resistance of pressure, both gas pressure and magnetic pressure. If the collapse is to produce a huge increase in density (as is necessary to form a star), then nearly all of the angular momentum in the collapsing gas must be transferred to nearby material. Last, the observed magnetic fields of young stars require that most of the magnetic field lines initially present in the gas *not* be swept into the forming protostar.

41.1 Gravitational Instability: Jeans Instability

Consider first the simplest case: a nonrotating and unmagnetized gas. We recall from Chapter 35 the equations expressing conservation of mass and momentum in an unmagnetized fluid, and the equation for the gravitational potential:

$$\frac{\partial \rho}{\partial t} + \nabla \cdot (\rho \mathbf{v}) = 0 \quad , \tag{41.1}$$

$$\frac{\partial \mathbf{v}}{\partial t} + (\mathbf{v} \cdot \nabla)\mathbf{v} = -\frac{1}{\rho}\nabla p - \nabla \phi \tag{41.2}$$

$$\nabla^2 \phi = 4\pi G \rho \quad . \tag{41.3}$$

Suppose that there exists an equilibrium steady state solution $\rho_0(\mathbf{r})$, $\mathbf{v}_0(\mathbf{r})$, $p_0(\mathbf{r})$, $\phi_0(\mathbf{r})$ satisfying Eqs. (41.1 to 41.3) with $\partial \mathbf{v}_0/\partial t = \partial \rho_0/\partial t = 0$.

To determine the conditions under which this equilibrium solution is unstable to gravitational collapse, we introduce a small perturbation, denoted by subscript 1:

$$\mathbf{v} = \mathbf{v}_0 + \mathbf{v}_1 \quad , \tag{41.4}$$

$$\rho = \rho_0 + \rho_1 \quad , \tag{41.5}$$

$$p = p_0 + p_1 \quad , \tag{41.6}$$

$$\phi = \phi_0 + \phi_1 \quad . \tag{41.7}$$

We now linearize the equations, retaining only terms that are first-order in the perturbations. Conservation of mass and momentum, and Poisson's equation, give the equations that the perturbations must obey:

$$\frac{\partial \rho_1}{\partial t} + \mathbf{v}_0 \cdot \nabla \rho_1 + \mathbf{v}_1 \cdot \nabla \rho_0 = -\rho_1 \nabla \cdot \mathbf{v}_0 - \rho_0 \nabla \cdot \mathbf{v}_1 \quad , \tag{41.8}$$

$$\frac{\partial \mathbf{v}_1}{\partial t} + (\mathbf{v}_0 \cdot \nabla)\,\mathbf{v}_1 + (\mathbf{v}_1 \cdot \nabla)\,\mathbf{v}_0 = \frac{\rho_1}{\rho_0^2} \nabla p_0 - \frac{1}{\rho_0} \nabla p_1 - \nabla \phi_1 \, , \tag{41.9}$$

$$\nabla^2 \phi_1 = 4\pi G \rho_1 \quad . \tag{41.10}$$

Up to this point, the analysis is fully general (for $\mathbf{B} = 0$) and for an arbitrary equation of state. If we now consider an **isothermal** gas ($p = \rho c_s^2$), Eq. (41.9) simplifies to

$$\frac{\partial \mathbf{v}_1}{\partial t} + (\mathbf{v}_0 \cdot \nabla)\,\mathbf{v}_1 + (\mathbf{v}_1 \cdot \nabla)\,\mathbf{v}_0 = -c_s^2 \nabla \left(\frac{\rho_1}{\rho_0} \right) - \nabla \phi_1 \quad . \tag{41.11}$$

This gives us three equations (41.8, 41.10, and 41.11) for the three unknown functions (ρ_1, $\mathbf{v}_1$, and ϕ_1).

Jeans (1928) considered the problem of an initially uniform, stationary gas, with $\nabla \rho_0 = 0$, $\nabla \phi_0 = 0$, $\mathbf{v}_0 = 0$. Taking the divergence of Eq. (41.11), and using (41.8) and (41.10), one obtains

$$\frac{\partial^2 \rho_1}{\partial t^2} = c_s^2 \nabla^2 \rho_1 + (4\pi G \rho_0) \rho_1 \quad . \tag{41.12}$$

If we now consider plane-wave perturbations,

$$\rho_1 = const \times \exp\left[i \left(\mathbf{k} \cdot \mathbf{r} - \omega t \right) \right] \quad , \tag{41.13}$$

we obtain the **dispersion relation**

$$\omega^2 = k^2 c_s^2 - 4\pi G \rho_0 \quad . \tag{41.14}$$

Defining $k_\mathrm{J}^2 \equiv (4\pi G \rho_0)/c_s^2$, the dispersion relation becomes

$$\omega^2 = (k^2 - k_\mathrm{J}^2) c_s^2 \quad . \tag{41.15}$$

Therefore, ω is real if and only if $k \geq k_J$: if $k < k_J$, ω becomes imaginary, corresponding to exponential growth. The **Jeans instability** therefore occurs for wavelength

$$\lambda > \lambda_J \equiv \frac{2\pi}{k_J} = \left(\frac{\pi c_s^2}{G\rho_0}\right)^{1/2} , \tag{41.16}$$

and we define the **Jeans mass**:

$$\begin{aligned} M_J &\equiv \frac{4\pi}{3}\rho_0 \left(\frac{\lambda_J}{2}\right)^3 = \frac{1}{8}\left(\frac{\pi k T}{G\mu}\right)^{3/2} \frac{1}{\rho_0^{1/2}} \\ &= 0.32\, M_\odot \left(\frac{T}{10\,\mathrm{K}}\right)^{3/2} \left(\frac{m_\mathrm{H}}{\mu}\right)^{3/2} \left(\frac{10^6\,\mathrm{cm}^{-3}}{n_\mathrm{H}}\right)^{1/2} . \end{aligned} \tag{41.17}$$

It is gratifying that when we substitute densities and temperatures observed for quiescent dark clouds, we find a mass typical of stars! In the limit of $k \ll k_J$ (long wavelength), the exponentiation time or "growth time" is

$$\tau_J = \frac{1}{k_J c_s} = \frac{1}{\sqrt{4\pi G\rho_0}} = \frac{2.3\times 10^4\,\mathrm{yr}}{\sqrt{n_\mathrm{H}/10^6\,\mathrm{cm}^{-3}}} . \tag{41.18}$$

To understand the value of the growth time τ_J, note that a uniform pressureless sphere of initially stationary gas with density ρ_0 will collapse with all shells reaching the center simultaneously in a finite time known as the **free-fall time** (Spitzer 1978),

$$\tau_\mathrm{ff} = \left(\frac{3\pi}{32 G\rho_0}\right)^{1/2} = \frac{4.4\times 10^4\,\mathrm{yr}}{\sqrt{n_\mathrm{H}/10^6\,\mathrm{cm}^{-3}}} . \tag{41.19}$$

The free-fall time τ_ff is only slightly longer than the Jeans growth time τ_J.

The preceding analysis is elegant but, unfortunately, deeply flawed. The assumption that $\nabla\phi_0 = 0$ is completely unphysical – we *cannot* have $\nabla\phi_0 = 0$ everywhere, as this implies $\nabla^2\phi_0 = 0$ everywhere, and therefore $\rho_0 = 0$: we are discussing a vacuum. For this reason, Binney & Tremaine (2008) referred to the above derivation of M_J as "the Jeans swindle." However, for finite systems, rigorous analyses give instability criteria that are close to Jeans's, with critical masses for unstable growth that depend in detail on the geometry, but are close to the Jeans mass (41.17).

41.2★ Parker Instability

Jeans's analysis applies to a more-or-less uniform, stationary, unmagnetized cloud, but gravitational instability can arise in many geometries. Parker (1966) considered the equilibrium of a plane-parallel system with a magnetic field, with z being the coordinate normal to the plane. Let the gas have density $\rho_{\rm gas}(z)$ and one dimensional velocity dispersion c_s (arising from both thermal motions and "turbulence"), assumed to be independent of z, so that the effective pressure is $p(z) = \rho_{\rm gas}(z)c_s^2$. Let there be a magnetic field $\mathbf{B}(z) = B(z)\hat{\mathbf{x}}$ present. Suppose also that cosmic rays (trapped on the magnetic field lines) contribute a pressure $p_{\rm CR}(z)$.

To keep things simple, assume that the magnetic pressure and cosmic ray pressure are each proportional to the gas pressure:

$$\frac{B^2}{8\pi} = \alpha\, \rho_{\rm gas} c_s^2 \quad , \tag{41.20}$$

$$p_{\rm CR} = \beta\, \rho_{\rm gas} c_s^2 \quad . \tag{41.21}$$

From study of the motions of stars in the gravitational potential of the disk, the average midplane mass density $\rho_{\rm tot}(z=0) \approx 0.10\, M_\odot\, {\rm pc}^{-3} = 6.8 \times 10^{-24}\,{\rm g\,cm}^{-3}$ (Kuijken & Gilmore 1989). This mass density is dominated by stars. At the midplane, the vertical gravity vanishes: $\nabla\phi = 0$. Near the midplane of the disk, we can approximate the total mass density $\rho_{\rm tot} \approx const$ (provided we do not depart too far from the midplane) and integrate (41.10) to obtain

$$\nabla\phi \approx 4\pi G \rho_{\rm tot}\, z\, \hat{\mathbf{z}} \quad . \tag{41.22}$$

With this potential, the equation of momentum conservation becomes

$$0 = -\nabla p_{\rm gas} - \nabla p_{\rm CR} - \nabla\left(\frac{B^2}{8\pi}\right) + \frac{(\mathbf{B}\cdot\nabla)\mathbf{B}}{4\pi} - \rho_{\rm gas}\nabla\phi \tag{41.23}$$

$$= -(1+\alpha+\beta)c_s^2 \frac{d\rho_{\rm gas}}{dz} - 4\pi G \rho_{\rm tot}\rho_{\rm gas} z \quad , \tag{41.24}$$

$$\frac{d\ln\rho_{\rm gas}}{dz} = \frac{-4\pi G\rho_{\rm tot}}{(1+\alpha+\beta)c_s^2} z \quad , \tag{41.25}$$

$$\rho_{\rm gas} = \rho_{\rm gas,0}\exp[-(z/h)^2] \quad , \tag{41.26}$$

$$h \equiv \left[\frac{2(1+\alpha+\beta)c_s^2}{4\pi G\rho_{\rm tot}}\right]^{1/2} \tag{41.27}$$

$$= 140\,(1+\alpha+\beta)^{1/2}\left(\frac{c_s}{7.4\,{\rm km\,s}^{-1}}\right)\,{\rm pc} \quad . \tag{41.28}$$

Mast & Goldstein (1970) measured the radial velocities of 268 high-latitude H I clouds, and obtained a 1-D velocity dispersion $c_s = 7.4\,{\rm km\,s}^{-1}$; we adopt this value as representative. What values of α and β are appropriate? For typical values $n_{\rm H}(z=0) \approx 1\,{\rm cm}^{-3}$ and $B \approx 4\,\mu{\rm G}$, we find $\alpha = 0.56$.

The pressure due to cosmic rays is dominated by $\sim 0.1 - 1\,\mathrm{GeV}$ particles. The cosmic ray flux at these energies is uncertain (see §13.5), but chemical diagnostics in diffuse molecular clouds (including the abundance of H_3^+) favor a high cosmic ray ionization rate, consistent with the cosmic ray proton spectrum X3 in Fig. 13.5, for which $p_{\rm CR}(E > 1\,\mathrm{MeV}) = 1.22 \times 10^{-12}\,\mathrm{erg\,cm^{-3}}$. Thus we estimate $\beta = 1.06$. The scale height h and $\langle |z| \rangle = h/\sqrt{\pi}$ are then estimated to be

$$h \approx 225\,\mathrm{pc} \quad , \quad \langle |z| \rangle = \frac{h}{\sqrt{\pi}} \approx 130\,\mathrm{pc} \quad . \tag{41.29}$$

Observations of the inner Galaxy show $\langle |z| \rangle \approx 150\,\mathrm{pc}$ (Crovisier 1978; Malhotra 1995), while just beyond the solar circle (galactocentric radius $\sim 8.5\,\mathrm{kpc}$) the H I has $\langle |z| \rangle \approx 180\,\mathrm{pc}$ (Dickey et al. 2009). We note that our assumption of $\rho_{\rm tot}(z) = const$ will *over*estimate $\nabla\Phi$ and therefore will *under*estimate h and $\langle |z| \rangle$. We conclude that the observed vertical distribution of H I (averaged over large regions) is in agreement with this equilibrium model.

The equilibrium model has a significant fraction of the pressure contributed by cosmic rays. These cosmic rays are "tied" to magnetic field lines, but can stream parallel to field lines. In the equilibrium solution there is no streaming, because the magnetic field lines are perpendicular to the cosmic ray pressure gradient. However, Parker (1966) pointed out that this equilibrium was unstable: if the magnetic field lines were perturbed in the vertical direction, gas could flow *down* the field lines into "valleys," adding weight to the valley regions, while cosmic rays could flow *out* of the valleys, removing pressure support from the valleys. This is now known as the **Parker instability**. Parker (1966) considered initially plane-parallel structures with small perturbations of the form $f(z)e^{ik_x x}$, with $f(0) = 0$, and showed that an isothermal gas with $B \neq 0$ is always unstable to growth of perturbations of sufficiently long wavelength. The perturbations saturate at finite amplitude, with the gas concentrated in denser regions, and the magnetic field lines bulging upward between the density peaks. Giz & Shu (1993) estimated the growth times for a realistic gravitational potential, finding that the most rapidly growing mode has

$$\lambda_x \approx 500\,\mathrm{pc} \quad . \tag{41.30}$$

The growth time for the most rapidly growing mode is approximately equal to the time for a compressive wave in the gas to travel a distance $\lambda_x/2$, or

$$\tau \approx \frac{\lambda_x/2}{c_s} \approx 3 \times 10^7\,\mathrm{yr} \quad . \tag{41.31}$$

This time is short enough for the instability to grow as the ISM passes through a spiral density wave.

Mouschovias (1974) showed that the Parker instability evolves to a new equilibrium structure with concentrations of gas separated horizontally by a distance $\lambda_x/2$; the magnetic field lines are compressed in the gas concentrations, but bulge out between them. Giant H II regions in other galaxies are frequently seen to be located like "beads on a string" along spiral arms, with $\sim$ kpc separations. Mouschovias et al. (1974) proposed that the Parker instability is involved in the formation of the giant molecular cloud complexes that host giant H II regions.

41.3 Insights from the Virial Theorem

For a region in equilibrium, with uniform pressure p_0 and magnetic field $\mathbf{B}_0$ at the surface, the virial theorem (see §35.5) states that

$$0 = 2E_{\rm KE} + 3(\Pi - \Pi_0) + (E_{\rm mag} - E_{\rm mag,0}) + E_{\rm grav} \quad , \tag{41.32}$$

$$E_{\rm KE} \equiv \int \rho \frac{v^2}{2}\, dV \quad , \tag{41.33}$$

$$\Pi \equiv \int p\, dV \quad , \quad \Pi_0 = p_0 V \quad , \tag{41.34}$$

$$E_{\rm mag} \equiv \int \frac{B^2}{8\pi}\, dV \quad , \quad E_{\rm mag,0} \equiv \oint d\mathbf{S} \cdot \left[\mathbf{r}\frac{B^2}{8\pi} - \frac{\mathbf{B}(\mathbf{r}\cdot\mathbf{B})}{4\pi}\right] \quad , \tag{41.35}$$

$$E_{\rm grav} = -\frac{G}{2} \int dV_1 \int dV_2 \frac{\rho(\mathbf{r}_1)\rho(\mathbf{r}_2)}{|\mathbf{r}_1 - \mathbf{r}_2|} \quad . \tag{41.36}$$

We can use the virial theorem to determine conditions for instability. Let us consider two special cases.

41.3.1 Nonrotating Nonmagnetized Isothermal Core

Consider a spherical "core" with mass M and radius R, with external pressure p_0 at the surface. The gravitational energy can be written

$$E_{\rm grav} = -\frac{3}{5} a \frac{GM^2}{R} \quad , \tag{41.37}$$

where the dimensionless factor $a = 1$ for uniform mass density, and $a > 1$ if the density is centrally peaked. Mouschovias & Spitzer (1976) find $a \approx 1.67$ from numerical models of clouds on the verge of collapse. If the gas has 1-dimensional velocity dispersion $c_s = const$, then

$$\Pi = M c_s^2 \quad . \tag{41.38}$$

If the gas is in equilibrium with $v = 0$, then the virial theorem requires that

$$0 = 3Mc_s^2 - 4\pi p_0 R^3 - \frac{3}{5} a \frac{GM^2}{R} \quad . \tag{41.39}$$

The external pressure p_0 must be given by

$$p_0 = \frac{1}{4\pi R^3}\left[3Mc_s^2 - \frac{3}{5}a\frac{GM^2}{R}\right] \quad . \tag{41.40}$$

If p_0 is small, then the equilibrium has

$$R \approx \frac{aGM}{5c_s^2} \quad \text{for } p_0 \ll \frac{375c_s^8}{4\pi a^3 G^3 M^2} \quad . \tag{41.41}$$

For fixed M, the external pressure has a maximum allowed value $p_{\rm max}(M)$, obtained by finding the value of R for which the right-hand side of Eq. (41.40) is maximized. If $a = const$, the maximum possible pressure is

$$p_{0,\rm max}(M) = \frac{3^4 5^3}{4^5\pi}\frac{c_s^8}{a^3 G^3 M^2} = \frac{3.15}{a^3}\frac{c_s^8}{G^3 M^2} \approx 0.68\frac{c_s^8}{G^3 M^2} \quad . \tag{41.42}$$

We can now turn the argument around: for a given pressure p_0, the maximum core mass that can be in equilibrium is

$$M_{\rm BE}(p_0) = \frac{225}{32\sqrt{5\pi}}\frac{c_s^4}{(aG)^{3/2}}\frac{1}{\sqrt{p_0}} \tag{41.43}$$

$$= 0.26\left(\frac{T}{10\,{\rm K}}\right)^2\left(\frac{10^6\,{\rm cm^{-3}\,K}}{p_0/k}\right)^{1/2} M_\odot \quad . \tag{41.44}$$

$M_{\rm BE}$ in Eq. (41.43) is known as the **Bonnor-Ebert mass** (Bonnor 1956; Ebert 1957). The pressure-bounded isothermal sphere with $M = M_{\rm BE}$ has central density $\rho \approx 14\rho_0$, where $\rho_0 = p_0/c_s^2$ is the density at the surface.

The Bonnor-Ebert mass $M_{\rm BE}$ differs from the Jeans mass $M_{\rm J}$ only by a numerical constant of order unity: $M_{\rm BE} \approx 1.18 M_{\rm J}$. Only cores with $M > M_{\rm BE}$ are unstable to collapse. This would seem to explain the fact that the "typical" stellar mass is $\sim 1\,M_\odot$: only cores with $M > M_{\rm BE}$ will become stars.

We now consider the effects of magnetic fields.

41.3.2 Nonrotating Magnetized Isothermal Core

The virial theorem states that

$$0 = 3\left(\Pi - \Pi_0\right) + \left(E_{\rm mag} - E_{\rm mag,0}\right) + E_{\rm grav} \quad . \tag{41.45}$$

Let $B_{\rm rms}$ be the rms magnetic field within the clump. Then,

$$E_{\rm mag} = \frac{B_{\rm rms}^2}{8\pi}V \quad . \tag{41.46}$$

Assume that the magnetic field in the clump is poloidal (i.e., in polar coordinates, there is no azimuthal field: $B_\phi = 0$), and let

$$\Phi = \int_S \mathbf{B} \cdot d\mathbf{S} \quad , \tag{41.47}$$

where the integral is over a surface S bounded by the magnetic equator. If we approximate

$$B_{\rm rms} \approx \frac{\Phi}{\pi R^2} \quad , \tag{41.48}$$

then

$$E_{\rm mag} \approx \frac{4\pi}{3} R^3 \frac{(\Phi/\pi R^2)^2}{8\pi} = \frac{1}{6\pi^2} \frac{\Phi^2}{R} \quad , \tag{41.49}$$

where R is the radius of a sphere with volume V. Because the field within the core will be nonuniform, the actual magnetic energy will be somewhat larger than given by Eq. (41.49). The magnetic field will extend into the surrounding medium. We expect $E_{\rm mag,0}$ [the surface integral in Eq. (41.35)] to be proportional to $E_{\rm mag}$. Let

$$E_{\rm mag} - E_{\rm mag,0} = \frac{b}{6\pi^2} \frac{\Phi^2}{R} \quad . \tag{41.50}$$

Numerical models point to an effective value $b \approx 1.25$ for magnetized clumps on the verge of collapse (Mouschovias & Spitzer 1976). Because $(E_{\rm mag} - E_{\rm mag,0}) \propto R^{-1}$ and $E_{\rm grav} \propto R^{-1}$, the ratio $(E_{\rm mag} - E_{\rm mag,0})/E_{\rm grav}$ remains constant as the cloud expands or contracts. If

$$\frac{\Phi}{M} > \left(\frac{\Phi}{M}\right)_{\rm crit} = 3\pi\sqrt{\frac{2aG}{5b}} = 1.54 \times 10^{-3} \sqrt{\frac{a}{b}} \ {\rm gauss\,cm^2\,g^{-1}} \quad , \tag{41.51}$$

then magnetic pressure will prevent the clump from collapsing.

To be able to undergo gravitational collapse, the clump must have a flux-to-mass ratio Φ/M that is less than the critical flux-to-mass ratio:

$$\frac{\Phi}{M} < \left(\frac{\Phi}{M}\right)_{\rm crit} = 1.8 \times 10^{-3} \ {\rm gauss\,cm^2\,g^{-1}} \quad , \tag{41.52}$$

where we set the dimensionless factors $a \approx 1.67$ and $b \approx 1.25$. A core with $\Phi/M < (\Phi/M)_{\rm crit}$ is termed **magnetically supercritical**: the mass-to-flux ratio exceeds the critical value $(M/\Phi)_{\rm crit}$, and the magnetic field alone cannot prevent collapse.

In §32.10, we saw that observations suggest ordered magnetic fields in molecular clouds with the median magnetic field strength at a given density given by Eq.

(32.17): $B \approx 49\, n_4^{0.65}\,\mu$G for $0.03 \lesssim n_4 \lesssim 10^3$, where $n_4 \equiv n_{\rm H}/10^4\,{\rm cm}^{-3}$. For this estimate of the median magnetic field strength, a spherical region of mass M and uniform density would have

$$\frac{\Phi}{M} = 5.8 \times 10^{-3} n_4^{-0.02} \left(\frac{M}{M_\odot}\right)^{-1/3} \,{\rm gauss\,cm^2\,g^{-1}} \quad . \tag{41.53}$$

For $M = 1\,M_\odot$ and $n_{\rm H} = 10^4\,{\rm cm}^{-3}$, this is a factor of ~ 3 larger than $(\Phi/M)_{\rm crit}$ – thus, for the median magnetic field strength found by Crutcher et al. (2010), the magnetic pressure will prevent clump collapse. Note that the empirical relation $B \propto n^{0.65}$ found by Crutcher et al. is close to the $n^{2/3}$ behavior that would result from flux-freezing and homologous compression, corresponding to Φ/M being nearly independent of density in eq. (41.53).

For our idealized relations between density and clump size [Eq. (32.12)] and between density and magnetic field strength [Eq. (32.17)], we conclude that cores will be magnetically *sub*-critical, with the magnetic field able to resist gravity. Another way to see this is by comparing the estimated Alfvén speed with the 3-D velocity dispersion, Eq. (32.19), where the observations of Crutcher et al. (2010) appear to imply $(v_A)_{0.5}/\sigma_v \approx 0.75 n_4^{0.46}$, implying that the magnetic field dominates turbulence ($v_A > \sigma_v$) for $n_4 \gtrsim 1.6$.

Individual density peaks exhibit considerable scatter around the idealized relations, and therefore the above conclusions regarding the importance of magnetic fields will certainly not apply in all cases. Whether cores are generally magnetically subcritical or supercritical remains in dispute. Crutcher (2010) argued that the cores of dark clouds are generally supercritical. However, the frequent observation of strong $|B_\parallel|$ for $n_{\rm H} \gtrsim 10^4\,{\rm cm}^{-3}$ (Crutcher et al. 2010) – see above Figure 32.6 – appears to establish the dynamical importance of magnetic fields in at least some cases.

We know that gravitational collapse does occur, and that the resulting stars (see below) have Φ/M far smaller than given by Eq. (41.53) for $n_{\rm H} \lesssim 10^7\,{\rm cm}^{-3}$. The question now arises: How are cores managing to lower their values of Φ/M as they contract, so that they can ultimately collapse to form stars?

41.4 Magnetic Flux Problem: Ambipolar Diffusion

We have seen above (§35.4) that the magnetic field in the ISM often acts as though the field lines are "frozen" into the fluid. If flux-freezing continues to hold in a collapsing clump, Φ will be conserved.

We have already seen that for $M \approx 1\,M_\odot$ clumps to be able to collapse, something must act to reduce the flux by a factor ~ 3 – unless the magnetic field strengths in molecular clouds have been overestimated by an order of magnitude, the "flux-freezing" approximation must break down.

T Tauri stars have surface magnetic fields $2 \pm 1\,$kG (Johns-Krull 2007), corresponding to a flux per mass $\Phi/M \approx 3 \times 10^{-8}\,{\rm gauss\,cm^2\,g^{-1}}$. This is $\sim$5 orders

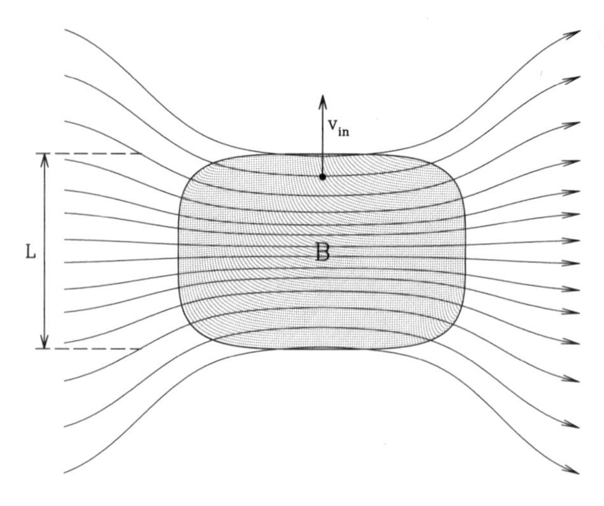

Figure 41.1 Geometry for ambipolar diffusion. B is the magnetic field strength at the cloud center; field lines are moving out of the cloud with velocity $\mathbf{v}_{\rm in}$ relative to the neutral gas.

of magnitude below the Φ/M estimated for a $\sim 1\,M_\odot$ clump in a cloud of density $n_{\rm H} \approx 10^4\,{\rm cm}^{-3}$. Therefore, nearly all of the magnetic flux that was initially present in this gas must escape before we are able to observe the T Tauri star.

The mechanism for this escape is presumably the process known as **ambipolar diffusion**. The magnetic field is coupled only to charged particles – electrons, ions, and charged dust grains. It remains a good approximation to assume the magnetic field is "frozen" into the plasma (i.e., the charged particles), but the plasma can **drift** relative to the neutral gas.

In a dense molecular cloud, the fractional ionization is extremely low (see Fig. 16.3), and the mass density in the plasma is almost negligible. If we ignore the inertia of the plasma, then the instantaneous velocity of the plasma is such that the $\mathbf{J} \times \mathbf{B}/c$ force per volume is exactly balanced by the force/volume resulting from collisions of charged particles with neutrals:

$$\frac{1}{c}\mathbf{J} \times \mathbf{B} = -n_i n_n \frac{m_i m_n}{m_i + m_n} \langle \sigma v \rangle_{\rm mt} (\mathbf{v}_i - \mathbf{v}_n) \quad , \qquad (41.54)$$

where $\langle \sigma v \rangle_{\rm mt}$ is the momentum transfer rate coefficient for ion-neutral scattering [see Eq. (2.39)]. If we assume that gradients are perpendicular to the magnetic field lines ($\mathbf{B} \cdot \nabla \to 0$), then

$$(\mathbf{v}_i - \mathbf{v}_n) = -\frac{(m_i + m_n)}{m_i m_n} \frac{1}{n_i n_n \langle \sigma v \rangle_{\rm mt}} \nabla \frac{B^2}{8\pi} \quad . \qquad (41.55)$$

Suppose that a clump has characteristic magnetic field strength B and characteristic dimension L (see Figure 41.1). The time scale for the magnetic field to slip out of the clump is

$$\tau_{\rm slip} = \frac{L/2}{|\mathbf{v}_i - \mathbf{v}_n|} = \frac{8\pi(L/2)^2}{B^2} n_i n_n \langle\sigma v\rangle_{\rm mt} \frac{m_i m_n}{(m_n + m_i)} \quad . \tag{41.56}$$

where we have approximated $|\nabla B^2| \approx B^2/(L/2)$. To evaluate this, suppose that (1) $L \approx 1.23 n_4^{-0.81}$ pc (Eq. 32.12), (2) the magnetic field strength is given by the observed relation $B \approx 49 n_4^{0.65}\,\mu$G for the median field strength [Eq. (32.17)], and (3) the fractional ionization $n_i/n_{\rm H} \approx 1 \times 10^{-7} n_4^{-1/2}$ (see Fig. 16.3). Then,

$$\tau_{\rm slip} \approx 7 \times 10^7 n_4^{-1.42}\,{\rm yr} \quad . \tag{41.57}$$

At densities $n_{\rm H} \gtrsim 10^5\,{\rm cm}^{-3}$, this time scale is short enough ($\lesssim 3 \times 10^6\,$yr) that ambipolar diffusion may be able to reduce the magnetic flux in a contracting clump. For $n_{\rm H} \gtrsim 2 \times 10^6\,{\rm cm}^{-3}$, the ambipolar diffusion time from Eq. (41.57) is shorter than the gravitational free-fall time $(3\pi/32G\rho)^{1/2} = 4.5 \times 10^5 n_4^{-0.5}\,$yr.

At very high densities, much of the free charge is located on dust grains. The ambipolar diffusion time is affected by the effect of grains on the ionization balance (see §14.8), and by the coupling of charged grains to both the magnetic field and the neutral gas. The dynamical evolution of dense cores is therefore sensitive to the distribution of grain sizes present; at these densities, the grain size distribution is expected to have been modified by coagulation and fragmentation, but the details are not yet clear.

41.5 Angular Momentum Problem

The discussion thus far has neglected rotation, but self-gravitating cores will generally have nonzero angular momentum. Consider a region of mass M, initial radius R_0, and angular momentum J, and neglect coefficients of order unity that depend on the density distribution in the cloud. If the core is uniformly rotating, the rotational kinetic energy

$$E_{\rm rot} \approx \frac{J^2}{(4/5)MR^2} \propto \frac{1}{R^2} \quad , \tag{41.58}$$

while the gravitational self-energy

$$|E_{\rm grav}| \approx \frac{6GM^2}{5R} \propto \frac{1}{R} \quad . \tag{41.59}$$

It is, therefore, clear that if the cloud contracts with $J = const$, the rotational kinetic energy will stop contraction at a radius

$$R_{\rm min} \approx \frac{(J/M)^2}{GM} = 8 \times 10^{15} \left(\frac{J/M}{10^{21}\,{\rm cm}^2\,{\rm s}^{-1}}\right)^2 \frac{M_\odot}{M}\,{\rm cm} \quad . \tag{41.60}$$

The specific angular momentum J/M will vary from core to core. At the solar circle, we are orbiting around the inner Galaxy with an angular frequency $\Omega \approx 225\,\mathrm{km\,s^{-1}}/8.5\,\mathrm{kpc} = 9 \times 10^{-16}\,\mathrm{rad\,s^{-1}}$; a $\sim 1\,\mathrm{pc}$ radius region rotating with this frequency would have specific angular momentum $J/M \approx 3 \times 10^{21}\,\mathrm{cm^2\,s^{-1}}$. A $\sim 0.1\,\mathrm{pc}$ radius cloud in solid body rotation with rotational velocity $v = 0.1\,\mathrm{km\,s^{-1}}$ at the "equator" would have specific angular momentum $J/M \approx 1 \times 10^{21}\,\mathrm{cm^2\,s^{-1}}$. It is clear that interstellar gas cannot contract to anything approaching the size of a star ($\sim 10^{11}\,\mathrm{cm}$) if its angular momentum is conserved. When the gravitational contraction is stopped by angular momentum, the rotationally supported core will take on a flattened disk geometry. Such **protostellar disks** are observed.

In order to overcome the "angular momentum barrier" and contract to protostellar dimensions, the core must transfer nearly all of its angular momentum to nearby material, either before or after it has become disk-like. There are two mechanisms that can do this – gravitational torques and magnetic torques. The collapsing gas may develop nonaxisymmetric density patterns so that the gravitational field can exert a torque on nearby matter – spiral density waves in a disk would be one example of this. If the density field in the collapsing region and in the gas around it is highly nonaxisymmetric, the gravitational torques could remove angular momentum from the clump on a time scale of order the dynamical time $1/(4\pi G\rho)^{1/2}$. However, the degree of nonaxisymmetry that will arise in the collapsing clump and its environs is uncertain.

The magnetic fields in molecular clouds are strong enough that magnetic torques due to bent magnetic field lines can be important. Consider a spherical surface of radius R bounding a rotating core. Let the angular momentum $\mathbf{J}$ be along the $\hat{\mathbf{z}}$ axis. The torque per area exerted at a point on the surface a distance r from the rotation axis is $\sim r B_{\parallel} B_{\perp}/8\pi$, where $B_{\parallel}$ and $B_{\perp}$ are the components of the magnetic field parallel and perpendicular to the surface [see Eq. (35.16)], and we have neglected trigonometric factors. Thus the magnetic braking time will be of order

$$t(\text{magnetic braking}) \approx \frac{J}{R^3 B^2} \quad , \tag{41.61}$$

if $\mathbf{B}$ has substantial components both normal to the surface *and* parallel (or antiparallel) to the local surface velocity. If the magnetic field is contributing to the support of the system against its self-gravity – which appears to be the case in typical molecular clouds and cloud cores (see §32.10) – the magnetic braking time can be of order the dynamical time. Detailed MHD simulations of realistic collapsing clumps are in their infancy, and it is not yet clear how the magnetic field strength varies as a clump collapses. However, if the magnetic field energy remains comparable in magnitude to the gravitational energy ($B^2 R^3 \sim GM^2/R$) and if the clump is rotationally supported [$J^2/MR^2 \sim GM^2/R \rightarrow J \sim (GM^3 R)^{1/2}$] then

$$t(\text{magnetic braking}) \sim \frac{J}{R^3 B^2} \sim \frac{JR}{GM^2} \sim \frac{1}{\sqrt{GM/R^3}} \quad , \tag{41.62}$$

i.e., the magnetic braking time can be of order the dynamical time $\sim 1/(G\rho)^{1/2}$ if the magnetic energy is comparable to the gravitational energy.

41.6 Accretion Disks

Eventually the collapsing clump becomes a rotationally supported disk with a density peak at its center. Dynamical processes in the disk then remove angular momentum from the inner parts of the disk and transfer this angular momentum to the outer parts of the disk or perhaps the ambient cloud; this allows the inner-disk material to fall inward and add to the growing star. The processes responsible for this momentum transfer are not yet well-understood. Ordinary molecular viscosity is ineffective, but momentum transport may occur as the result of "turbulent viscosity," or by magnetic torques. If the disk has not retained a significant ordered magnetic field, it can generate magnetic fields via the **magnetorotational instability**, or "**MRI**" (Balbus & Hawley 1991). The MRI requires a minimum electrical conductivity; whether the conductivity of the disk will be high enough for the MRI to operate is not yet known.

41.7 Radiation Pressure

A star-forming region emits radiation. Initially, the energy comes from accretion onto the protostar, with a luminosity

$$L_{\rm grav} \approx \frac{GM}{R}\dot{M} = 120\frac{M}{M_\odot}\frac{R_\odot}{R}\frac{\dot{M}}{10^{-6}\,M_\odot\,{\rm yr}^{-1}}\,L_\odot \quad . \tag{41.63}$$

Once the central density and temperature become high enough for nuclear fusion to begin, the luminosity comes from release of nuclear energy in conversion of first D, and then H, into He. As radiation streams away from the star, a force will be exerted on atoms, molecules, or dust particles that absorb or scatter the radiation. As seen in Fig. 23.12, the radiation pressure cross section for dust can be large. The ratio of radiation pressure force to gravitational attraction is

$$\frac{F_{\rm rad.pr.}}{F_{\rm grav}} = \frac{L\langle\sigma_{\rm rad.pr.}\rangle}{4\pi R^2 c}\frac{R^2}{1.4GMm_{\rm H}} \tag{41.64}$$

$$= \frac{L_\odot\langle\sigma_{\rm rad.pr.}\rangle}{5.6\pi G\,M_\odot m_{\rm H}c}\frac{L/M}{L_\odot/\,M_\odot} \tag{41.65}$$

$$= 0.0327\left(\frac{\langle\sigma_{\rm rad.pr.}\rangle}{10^{-21}\,{\rm cm}^2{\rm H}^{-1}}\right)\frac{L/M}{L_\odot/\,M_\odot} \quad . \tag{41.66}$$

If the radiation has a color temperature $10^4 \lesssim T \lesssim 10^5$ K, and the dust is similar to the dust in the local diffuse ISM, then $\langle\sigma_{\rm rad.pr.}\rangle \approx 10^{-21}\,{\rm cm^2H^{-1}}$ (see Fig. 23.12), and radiation pressure on the dust will exceed the gravitational attraction on the (dust+gas) if $L/M \gtrsim 30\,L_\odot/\,M_\odot$. For spherically symmetric flow, this would appear to prevent the formation of massive stars ($M \gtrsim 10\,M_\odot$), which will have $L/M \gtrsim 10^3\,L_\odot/\,M_\odot$ once nuclear burning commences. However, accretion is not a spherically symmetric process – accretion onto the star takes place via an optically thick disk. The radiation leaving the star is effectively collimated into a limited range of angles (presumably with a bipolar geometry); within these lobes, optically thin dusty gas is driven away, but near the equatorial plane the accreting matter is shielded from the short-wavelength radiation from the accreting stellar object.

Radiation pressure considerations also apply to galactic disks. If the luminosity surface density Σ_L is too large, dusty gas may be repelled from the disk. The ratio of radiation pressure force to gravitational force is just

$$\frac{F_{\rm rad.pr.}}{F_{\rm grav}} = 0.0327\frac{\langle\sigma_{\rm rad.pr.}\rangle}{10^{-21}\,{\rm cm^2H^{-1}}}\frac{\Sigma_L/\Sigma_M}{L_\odot/\,M_\odot}\quad, \tag{41.67}$$

where Σ_L is the disk luminosity per area, and Σ_M is the disk mass surface density. Therefore, for dust and a radiation spectrum such that $\langle\sigma_{\rm rad.pr.}\rangle \approx 10^{-21}\,{\rm cm^2H^{-1}}$, a disk with energetic star formation such that $\Sigma_L/\Sigma_M \gtrsim 30\,L_\odot/\,M_\odot$ can have a disk that is supported largely by radiation pressure (Thompson et al. 2005), and higher values of Σ_L/Σ_M could drive an outflow of gas and dust from the disk using radiation pressure alone (Murray et al. 2011). Such high values of Σ_L/Σ_M may result from a strong burst of star formation. Radiation pressure on dust may help account for the presence of fragile PAHs $\sim$4 kpc above the disk of the starburst galaxy M82 (see Plate 7).

Chapter Forty-two

Star Formation: Observations

Some theoretical considerations related to star formation were developed in Chapter 41. Here we present a few of the basic observational data pertaining to star formation. A thorough treatment of the field can be found in the excellent text by Stahler & Palla (2004).

Molecular clouds are found in a variety of sizes and densities. As discussed in Chapter 32 (see Table 32.2), the terminology that is used to refer to the substructure within them is rather arbitrary, and sometimes confusing, so we repeat it here.

Self-gravitating density peaks within an isolated "dark cloud" are usually referred to as **cores**. The cores have masses of order 0.3 to 10 $M_\odot$. Each core is likely to form a single star or a binary star.

In the case of giant molecular clouds (GMCs), the term **clump** is used to refer to self-gravitating regions with masses as large as $\sim 10^3\, M_\odot$. Clumps may or may not be forming stars; those that are, are termed **star-forming clumps**. Such clumps will generally contain a number of **cores**.

42.1 Collapse of Cores to form Stars

When a core becomes gravitationally unstable, it will begin to collapse. Exactly how this collapse proceeds is uncertain in detail, but we think we understand the overall outlines.

During the initial stages, radiative cooling in molecular lines is able to keep the gas cool. As a result, the gas pressure remains unimportant during this phase, and the matter moves inward nearly in free-fall. The velocities at this stage are not large,

$$v \lesssim \sqrt{\frac{GM_{\rm c}}{R_{\rm c}}} = \left(\frac{4\pi}{3}\right)^{1/3} G^{1/2} M^{1/3} \rho_{\rm c}^{1/6} \tag{42.1}$$

$$\approx 0.4 \left(\frac{M_{\rm c}}{M_\odot}\right)^{1/3} n_6^{1/6}\,{\rm km\,s^{-1}}, \tag{42.2}$$

where $M_{\rm c}$ and $\rho_{\rm c} = 1.4 n_{\rm H} m_{\rm H}$ are the mass and density of the core, and $n_{\rm H} = 10^6 n_6\,{\rm cm^{-3}}$.

Because the density is higher in the interior, the free-fall time $(3\pi/32G\rho)^{1/2}$ is shortest there, and the collapse proceeds in an "inside-out" mannner, with the center collapsing first, and the outer material later falling onto the central matter.

If cores had no angular momentum, and if magnetic fields were negligible, the collapse process would be relatively simple to understand and model. However, as we have seen earlier (§32.10), molecular clouds appear to have magnetic energies $E_{\rm mag}$ comparable to the kinetic energy $E_{\rm KE}$ (contributed mainly by the "turbulent" motions), and sufficient angular momentum to become dynamically important long before stellar densities are reached.

If the core is to collapse, the magnetic flux/mass ratio Φ/M must be less than the critical value $(\Phi/M)_{\rm crit}$. While low density cores appear to have too much magnetic flux to collapse [see Eq. (41.53)], the fact that stars *do* form implies that ambipolar diffusion – or perhaps some other process – is able to reduce Φ/M to below $(\Phi/M)_{\rm crit} = 1.8 \times 10^{-3}\,{\rm gauss\,cm^2\,g^{-1}}$, so that the magnetic field will not prevent gravitational collapse.

The infalling gas will generally have nonzero angular momentum, and (if it remains cold) the material will collapse to form a rotationally supported disk, with the material with the lowest specific angular momentum collected in a "protostar" at the center of the disk. Energy is dissipated as the infalling gas hits the disk. Angular momentum transport – due to the magneto-rotational instability (MRI) if the gas is sufficiently ionized, or due to gravitational torques or turbulent viscosity if the ionization is too low to support the MRI – will cause some material in the disk to move inward, with additional release of gravitational energy. The energy so released will heat the disk, and will be radiated away.

The dominant sources of energy are (1) the gravitational energy released as material is added to the protostar and as the prostar contracts, and (2) the energy released when the protostar is able to ignite fusion reactions to first "burn" deuterium, and then hydrogen. The protostar will have a significant luminosity, allowing it and the surrounding core to be observed as a luminous infrared source.

42.2 Class 0, I, II, and III Protostars

The spectrum of the radiated energy will depend on the amount of obscuration around the protostar. Let $R_{\rm disk}$ be some characteristic radius of the disk. The characteristic surface density of the disk is then

$$\Sigma_{\rm disk} = \frac{M_{\rm disk}}{\pi R_{\rm disk}^2} = 28\left(\frac{M_{\rm disk}}{0.1\,M_\odot}\right)\left(\frac{100\,{\rm AU}}{R_{\rm disk}}\right)^2 {\rm g\,cm^{-2}} \quad . \tag{42.3}$$

If the disk is not yet fully grown, there may be a significant amount of mass that is still infalling, and the column density along a sightline from the surface of the protostar to infinity,

$$\Sigma_{\rm infall} = \int_{R_\star}^{\infty} \rho dr \quad , \tag{42.4}$$

may be large, even for sightlines that do not intercept the disk itself. The spectrum of the energy radiated to infinity by the protostar will depend on $\Sigma_{\rm infall}$. Protostars and young stellar objects (YSOs) are conventionally divided into four different classes based on the overall shape of the infrared spectrum, characterized by a spectral index

$$\alpha \equiv \frac{d\log(\lambda F_\lambda)}{d\log\lambda} \quad , \tag{42.5}$$

or $\nu F_\nu \propto \nu^{-\alpha}$. Ground-based observations often employ atmospheric windows at 2.2 μm (K band) and 10 μm (N band), and low-mass ($M \lesssim 3\,M_\odot$) protostars may be classified by using the ratio of the observed flux densities in the K and N bands:

$$\alpha_{K,N} = \frac{\log[(\lambda F_\lambda)_{10\,\mu\rm m}] - \log[(\lambda F_\lambda)_{2.2\,\mu\rm m}]}{\log(10/2.2)} \quad . \tag{42.6}$$

- **Class 0** protostars are objects that are so heavily obscured that their spectra peak at $\lambda > 100\,\mu$m. For these sources, $\alpha_{K,N}$ is not a useful characteristic, because the source may be invisible at 2.2 μm, and at 10 μm there may be deep absorption by cold foreground silicate dust. Inward motions of the gas are sometimes revealed by asymmetric profiles of molecular emission lines. The lifetime of a Class 0 object is short, $\sim(1-3)\times 10^4$ yr (André et al. 2000).

- **Class I** protostars have $\alpha_{K,N} > 0$: there is more power being radiated near 10 μm than near 2 μm. Blackbodies with $T < 870$ K have $\alpha_{K,M} > 0$. Class I protostars are thought to have typical ages $\sim(1-2)\times 10^5$ yr.

- **Class II** YSOs have $-1.5 < \alpha_{K,M} < 0$. Blackbodies with $870 < T < 1540$ K have $-1.5 < \alpha_{K,M} < 0$. Class II YSOs correspond to classical T Tauri stars, which are pre-main-sequence stars, still undergoing gravitational contraction, with substantial accretion disks and accretion rates $\sim 10^{-6}\,M_\odot\,{\rm yr}^{-1}$.

- **Class III** protostars have $\alpha_{K,M} < -1.5$. Blackbodies with $T > 1540$ K have $\alpha_{K,M} < -1.5$. Class III YSOs correspond to "weak-lined" T Tauri stars, which are pre-main-sequence stars still undergoing gravitational contraction, but where the accretion disk is either weak or perhaps entirely absent.

The observed flux ratios are determined by the temperature of the emitting regions (star and inner disk) and by wavelength-dependent extinction. Classification of a given object as either Class I or Class II may depend on the source orientation. If we are viewing the object face-on, it may be classified as Class II, but an identical disk viewed edge-on could heavily redden the light reaching the observer, so that the object could be classified as Class I or even Class 0.

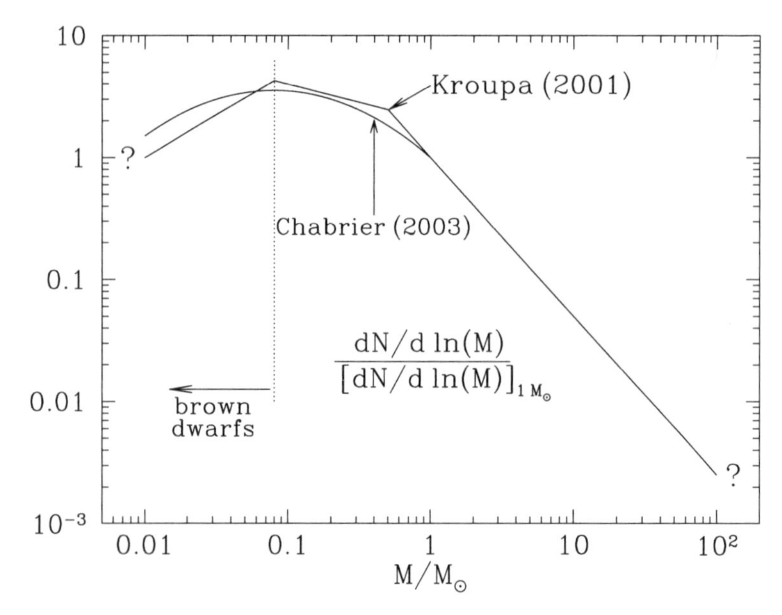

Figure 42.1 Stellar initial mass function (IMF) as estimated by Kroupa (2001) and Chabrier (2003). Shown is $dN/d\ln M$, the number of stars formed per logarithmic interval in stellar mass M, normalized to the value at $M = 1\,M_\odot$. Both Kroupa (2001) and Chabrier (2003) have $dN/d\ln M \propto M^{-1.3}$ for $M > 1\,M_\odot$.

In principle, $\alpha_{K,N}$ can also be increased by differential extinction due to cold foreground dust in the cloud, but this is not expected to be a big effect: with $A_K \approx 0.12 A_V$ and $A_N \approx 0.06 A_V$, foreground reddening in a cloud with $A_V \approx 10$ mag will only increase $\alpha_{K,N}$ by $\sim 0.043 A_V \approx 0.4$.

42.3 Initial Mass Function

If we sum over the stars formed in a large star-forming region (e.g., the Orion Nebula Cluster), we can discuss the distribution of initial stellar masses – the **initial mass function**, or **IMF**. Beginning with the pioneering work of Salpeter (1955), there have been many studies of the IMF in different regions of the Milky Way, and in other galaxies. There is no reason to think that the IMF should be universal, yet it shows remarkable uniformity from region to region. There may be systematic variations in the IMF depending on environmental conditions, but the variations are surprisingly small. It is difficult to determine the IMF at the high-mass end because massive stars are rare, and at the low-mass end because low-mass stars are faint. Nevertheless, for 0.01 to 50 $M_\odot$, there is reasonable agreement between different studies.

Figure 42.1 shows two recent estimates (Kroupa 2001; Chabrier 2003) for the

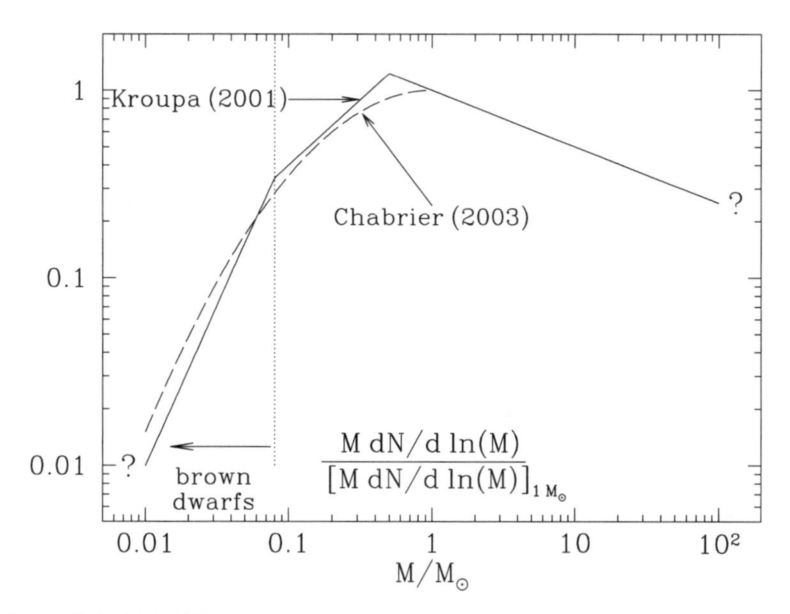

Figure 42.2 $MdN/d\ln M$, the mass formed per logarithmic interval in stellar mass M, for IMFs of Kroupa (2001) and Chabrier (2003), normalized to the value at $M = 1\,M_\odot$.

IMF in the disk of the Milky Way. For $M \gtrsim 1\,M_\odot$, the observations are consistent with a power law $dN/dM \propto M^{-2.3}$, very close to the slope $dN/dM \propto M^{-2.35}$ originally found in the pioneering study by Salpeter (1955). The Kroupa and Chabrier estimates for the IMF differ only in detail. While appreciable numbers of low-mass stars are formed, the mass per logarithmic mass interval peaks near ~ 0.5–$1\,M_\odot$. Table 42.1 provides some useful integral properties of the IMF. For example, for a total star formation rate $\dot{M}$, the rate of formation of $M > 8\,M_\odot$ stars is $\dot{M} \times 0.2118/19.14\,M_\odot$. If $M > 8\,M_\odot$ stars become Type II supernovae, then the Milky Way star formation rate $\sim 1.3\,M_\odot\,\mathrm{yr}^{-1}$ (see §42.4 below) corresponds to a Type II SN rate $0.014\,\mathrm{yr}^{-1} = 1/70\,\mathrm{yr}$.

Table 42.1 Some Properties of the Chabrier (2003) IMF[a]

Mass range ($M_\odot$)	mass/total mass	$\langle M\rangle/\,M_\odot$
0.01–0.08	0.0482	0.0379
0.08–1	0.3950	0.2830
1–8	0.3452	2.156
8–16	0.0749	10.96
16–100	0.1369	32.31
8–100	0.2118	19.14
0.01–100	1.0000	0.3521

[a] For lower and upper cutoffs of 0.01 and 100 $M_\odot$.

42.4 Star Formation Rates

What is the rate of star formation in the Milky Way galaxy?

Because massive stars are highly luminous and create H II regions that can be detected at large distances, we can use the observed number of high-mass stars (or the H II gas photoionized by them), together with theoretical estimates of stellar lifetimes, to estimate the rate at which massive stars are being formed in the Galaxy. The IMF can then be used to estimate the total rate at which stars are being formed. The difficulty is in counting the stars. Optical surveys (including observations of the bright Hα line from H II regions) are incomplete because of obscuration by interstellar dust. We must instead use long-wavelength tracers of massive stars.

[N II]205 μm emission is excited by collisions of N II with electrons in H II regions. Because N and H are ionized together, the ratio N II/H II depends only on the N abundance, which we think we know.[1] At low densities, the [N II]205 μm emissivity is proportional to the electron-proton recombination rate, and therefore (on average) proportional to the rate of hydrogen ionization. Bennett et al. (1994) used all-sky COBE observations of the [N II]205 μm line to estimate the total rate of photoionization of H in the Milky Way to be $Q_{0,\rm MW} \approx 3.5 \times 10^{53}\,{\rm s}^{-1}$. McKee & Williams (1997), taking into account the radial gradient in the N abundance, and correcting for suppression of [N II]205 μm emission at high densities, used the same COBE data to estimate $Q_{0,\rm MW} \approx 1.9 \times 10^{53}\,{\rm s}^{-1}$. The systematic uncertainties (including the assumed electron temperature and N abundance) are estimated to be $\sim 50\%$.

Free–free radio emission is negligibly affected by interstellar dust, and the free–free emissivity in H II regions is proportional to the proton-electron recombination rate. Murray & Rahman (2010) used observations by the *Wilkinson Microwave Anisotropy Probe (WMAP)* to determine the total rate of emission of ionizing photons in the Milky Way to be $Q_{0,\rm MW} = (3.2 \pm 0.5) \times 10^{53}\,{\rm s}^{-1}$, after correcting for the effects of absorption by dust. This appears to be the most reliable determination of the rate of emission of ionizing photons.

With estimates of the number of ionizing photons emitted over the lifetime of massive stars, and an assumed form for the IMF, Murray & Rahman (2010) use their measurement of $Q_{0,\rm MW}$ to estimate the the total rate of star formation in the Milky Way (averaged over the past ~ 3 Myr – the lifetime of early O-type stars) to be

$$\left(\frac{dM_\star}{dt}\right)_{\rm MW} = (1.3 \pm 0.2)\,M_\odot\,{\rm yr}^{-1} \quad ; \tag{42.7}$$

allowing for uncertainties in the IMF, the star formation rate should be in the range $0.9\,M_\odot\,{\rm yr}^{-1} < (dM_\star/dt)_{\rm MW} < 2.2\,M_\odot\,{\rm yr}^{-1}$.

Radio recombination lines (see §10.7) are another extinction-free tracer of ionized gas, with the additional benefit of kinematic information from the Doppler shifts, providing additional information to aid in distance estimation. Recombina-

[1] In H II regions ionized by very early-type stars, some of the N can be N III – see Table 15.2.

tion line surveys have detected a large population of previously unknown Galactic H II regions (Bania et al. 2010); future work will be able to provide an independent estimate for the total photoionization rate $Q_{\rm MW}$, and the implied Galactic star formation rate $(dM_\star/dt)_{\rm MW}$.

42.5 Schmidt-Kennicutt Law

The total mass of molecular gas in the Milky Way is $\sim 10^9\, M_\odot$ (see Table 1.2). If the typical density of this gas is $n_{\rm H} \approx 50\,{\rm cm}^{-3}$, with a free-fall time $(3\pi/32G\rho)^{1/2} \approx 6\times 10^6$ yr, the maximum rate at which stars could be made would be

$$\dot{M}_{\rm ff} = \frac{M_{\rm tot}}{\tau_{\rm ff}} \approx 200\, M_\odot\,{\rm yr}^{-1} \quad . \tag{42.8}$$

The actual star formation rate (see Eq. 42.7) is $\sim$2 orders of magnitude below this value, for two reasons: (1) most of the mass in GMCs is *not* undergoing free-fall collapse, and (2) even in regions that do collapse, only a fraction of the gas ends up in stars. One of the major challenges to ISM theory is to understand why the star formation rate has the observed value. This is a formidable problem that involves understanding the excitation and damping of the MHD turbulence in molecular clouds, transport of angular momentum out of contracting regions, ambipolar diffusion to remove magnetic flux from contracting regions, and the important effects of "feedback" – the effects of outflows and radiation from protostars and stars on the surrounding gas, either stimulating or suppressing further star formation.

Because star formation is the result of gravitational collapse, one expects the specific star formation rate (star formation rate per unit gas mass) to be larger in higher density regions. Schmidt (1959) proposed that the star formation rate per volume varied as a power of the local density ρ. The physics of star formation is complex, and the star formation rate will depend on physical properties other than gas density. Nevertheless, it has proven useful to examine the empirical relationship between observed star formation rate and gas density. Because the volume density ρ is difficult to determine from afar, Kennicutt (1998) examined the relationship between the global star formation in a galaxy and $\Sigma_{\rm gas,disk}$, the gas surface density averaged over the "optical disk" of the galaxy, finding that the star formation rate per unit area $\Sigma_{\rm SFR,disk}$ varied approximately as

$$\Sigma_{\rm SFR,disk} = (2.5\pm0.7)\times10^{-4}\left(\frac{\Sigma_{\rm gas,disk}}{M_\odot\,{\rm pc}^{-2}}\right)^{1.4\pm0.15} M_\odot\,{\rm kpc}^{-2}\,{\rm yr}^{-1} \quad . \tag{42.9}$$

This is often referred to as the **Schmidt-Kennicutt Law**. It is remarkable that this relation extends from the low gas surface densities of gas-poor spiral disks to the very high surface densities in the cores of luminous starburst galaxies. Because it relates gas surface densities and star formation rates averaged over the entire optical disk, it describes the *global* star formation rate.

For well-resolved nearby galaxies, it is possible to use *local* values of $\Sigma_{\rm SFR}$ and $\Sigma_{\rm gas}$. On 500 pc scales in the star-forming spiral galaxy M51a, the gas and star formation rate surface densities vary as (Kennicutt et al. 2007)

$$\Sigma_{\rm SFR,500} = (5.9\pm1.4)\times10^{-5}\left(\frac{\Sigma_{\rm gas,500}}{M_\odot\,{\rm pc}^{-2}}\right)^{1.56\pm0.04} M_\odot\,{\rm kpc}^{-2}\,{\rm yr}^{-1}. \quad (42.10)$$

Because of the nonlinear dependence of $\Sigma_{\rm SFR}$ on $\Sigma_{\rm gas}$, it is not surprising that the coefficient in the local relation Eq. (42.10) is smaller than in Eq. (42.9) using disk-averaged surface densities. Recall that $\Sigma_{\rm gas}$ in Eq. (42.9) is averaged over the optical disk of the galaxy, while $\Sigma_{\rm gas}$ in Eq. (42.10) is averaged over only 500 pc length scales. Because the gas surface density is very clumpy, and $\Sigma_{\rm SFR}$ increases more rapidly than linearly, we expect the prefactor in the "local" relation to be *smaller* than in the "global" relation.

If one considers molecular gas only, then one finds a tighter relationship. In M51, Kennicutt et al. (2007) found, on 500 pc scales,

$$\Sigma_{\rm SFR,500} = (1.7\pm0.4)\times10^{-4}\left(\frac{\Sigma_{\rm H_2,500}}{M_\odot\,{\rm pc}^{-2}}\right)^{1.37\pm0.03} M_\odot\,{\rm kpc}^{-2}\,{\rm yr}^{-1}. \quad (42.11)$$

This relation extends from $\Sigma_{\rm H_2,500} \approx 10\,M_\odot\,{\rm pc}^{-2}$ to $\sim 400\,M_\odot\,{\rm pc}^{-2}$ – a factor of ~ 160 in $\Sigma_{\rm SFR}$. For the local Milky Way value of $A_V/N_{\rm H}$, $\Sigma_{\rm H_2} = 10 - 400\,M_\odot\,{\rm pc}^{-2}$ corresponds to A_V from 0.48 to 20 mag.

We can try to obtain a qualitative understanding of the local relationship given by Eq. (42.11). Let us assume the H_2 to be in a uniform slab on the ~ 500 pc scales over which both $\Sigma_{\rm H_2}$ and $\Sigma_{\rm SFR}$ have been averaged in the observations leading to Eq. (42.11). Suppose that the vertical thickness of the molecular gas varies as $h \propto \Sigma_{\rm gas}^{\beta}$, Then $\rho \propto \Sigma_{\rm gas}^{1-\beta}$, the free-fall time $\tau_{\rm ff} \propto \Sigma_{\rm gas}^{(\beta-1)/2}$, and $\Sigma_{\rm gas}/\tau_{\rm ff} \propto \Sigma_{\rm gas}^{3/2-\beta/2}$. If the star formation rate/area is proportional to $\Sigma_{\rm gas}/\tau_{\rm ff}$, then the empirical relationship (42.11) is recovered if $\beta = 0.26 \pm 0.06$, so that $\rho \propto \Sigma^{0.74\pm0.06}$ – the internal density increases as $\Sigma_{\rm H_2}$ rises. These arguments are of course simplistic – in the Milky Way and in M51, GMCs are not uniform plane-parallel slabs. Recall that GMCs in the Milky Way tend to have $A_V \approx 10$ mag. Hence $\Sigma_{\rm H_2}$ running from 10 to 100 $M_\odot\,{\rm pc}^{-2}$ would correspond to the GMC covering factor increasing from 0.05 to 0.5. If individual clouds had a fixed rate of star formation per unit mass, we would have expected $\Sigma_{\rm SFR} \propto \Sigma_{\rm H_2}$ over this range, yet the observations indicate that $\Sigma_{\rm SFR}$ varies as $\Sigma_{\rm H_2}^{1.37}$. One interpretation would be that an increase in $\Sigma_{\rm SFR}$ leads to an increased pressure in the ISM, producing higher densities within GMCs, and higher SFR per unit mass. This trend appears to continue to the extraordinarily high surface density $\Sigma_{\rm H_2} \approx 400\,M_\odot\,{\rm pc}^{-2}$, which corresponds to $A_V \approx 20$ mag.

We are obviously far from understanding the process of star formation either on a star-by-star basis, or averaged over large regions within galaxies.

Much remains to be learned.

Appendix A

List of Symbols

Symbol	Definition
α	$\equiv e^2/\hbar c = 1/137.04$, fine-structure constant
α_A	thermal rate coefficient for case A recombination; see §14.2
α_B	thermal rate coefficient for case B recombination; see §14.2
α_N	polarizability of atom or molecule N; see Eq. (2.29)
β_ν	escape probability for photon of frequency ν; see §19.1
$\langle\bar{\beta}\rangle$	angle- and frequency-averaged escape probability; see §19.1
$\epsilon(\omega)$	complex dielectric function; see Eq. (22.7)
ϵ_1	$\equiv \mathrm{Re}(\epsilon)$
ϵ_2	$\equiv \mathrm{Im}(\epsilon)$
$\zeta_{\mathrm{CR},p}$	primary ionization rate for an H atom due to cosmic ray protons
ζ_{CR}	primary ionization rate due to cosmic rays
κ_ν	attenuation coefficient; see Eq. (7.9)
κ_ν	opacity = absorption cross section per unit mass; see §32.4
$\kappa(T)$	thermal conductivity; see §34.3
$\lambda_{\ell u}$	wavelength of photon emitted in transition $u \rightarrow \ell$
μ	electric dipole moment of molecule or grain
μ_{B}	Bohr magneton $\equiv e\hbar/2m_e c = 9.274 \times 10^{-21}\,\mathrm{erg\,gauss^{-1}}$
ν	frequency (oscillations per unit time)
ϕ_s	number of secondary ionizations per primary ionization; see §13.3
ρ	mass density
σ	Stefan-Boltzmann constant, $5.670 \times 10^{-5}\,\mathrm{erg\,s^{-1}\,cm^{-2}\,K^{-4}}$
σ	reaction or excitation/deexcitation cross section; see §2.1
σ	electrical conductivity; see Eqs. (22.8 and 35.35)
$\sigma_{\mathrm{abs}}(\lambda)$	dust absorption cross section per H nucleon
$\sigma_{\mathrm{sca}}(\lambda)$	dust scattering cross section per H nucleon
$\sigma_{\mathrm{ext}}(\lambda)$	$\equiv \sigma_{\mathrm{abs}} + \sigma_{\mathrm{sca}} =$ dust extinction cross section per H nucleon
$\sigma_{\mathrm{dust}}(\lambda)$	attenuation cross section per H nucleon; see Eq. (15.20)
$\sigma_{\mathrm{rad.pr.}}$	radiation pressure cross section per H nucleon; see §23.10.1
σ_v	velocity dispersion; see §6.5
Σ_{gr}	dust projected area per H nucleon; see Eq. (31.6)
Σ_{-21}	$\equiv \Sigma_{\mathrm{gr}}/10^{-21}\,\mathrm{cm^2 H^{-1}}$
τ_ν	optical depth at frequency ν; see Eq. (7.14)
χ	UV intensity at 1000Å relative to Habing (1968); see Eq. (12.5)
ω	angular frequency $2\pi\nu$ (radians per unit time)
ω_p	plasma frequency; see Eq. (11.2)
a_0	Bohr radius $\equiv \hbar^2/m_e e^2 = 5.292 \times 10^{-9}\,\mathrm{cm}$

Symbol	Definition
$a_{\rm eff}$	radius of equal-volume sphere; see Eq. (22.6)
Å	Angstrom $\equiv 10^{-8}$ cm
A_λ	extinction (in mag.) at wavelength λ
A_B	extinction (in mag.) at $B = 4400$ Å
A_V	extinction (in mag.) at $V = 5500$ Å
b	impact parameter; see §2.2.1
b_n	departure coefficient; see §3.8
B, $\mathbf{B}$	magnetic field strength, magnetic field vector
c	speed of light, $2.998 \times 10^{10}\,{\rm cm\,s^{-1}}$
$C_{\rm abs}(\lambda)$	absorption cross section at wavelength λ
$C_{\rm sca}(\lambda)$	absorption cross section at wavelength λ
$C_{\rm ext}(\lambda)$	$\equiv C_{\rm abs} + C_{\rm sca} =$ absorption cross section at wavelength λ
D_L	luminosity distance; see §8.2
DM	dispersion measure; see Eq. (11.12)
e	$\lvert$electron charge$\rvert$, 4.803×10^{-10} esu
E_j	energy of a level j
$E(B-V)$	$\equiv A_B - A_V$
EM	emission measure; see §10.5
$f_{\ell u}$	oscillator strength; see §6.3
$f(X;T)$	partition function per unit volume for species X; see Eq. (3.5)
F	total (electronic + nuclear) angular momentum quantum number; see §
F	flux = power per area; see §9.1
F_ν	flux density = power per area per unit frequency; see §8.2
$F_\star$	empirical depletion parameter; see Eq. (9.36)
FWHM	full width at half maximum; see Eq. (6.32)
$g_{\rm ff}$	Gaunt factor for free-free transitions; see §10.2
g_i	degeneracy of a level i; see §3.1
G	gravitational constant, $6.673 \times 10^{-8}\,{\rm erg\,cm\,g^{-2}}$
G_0	UV intensity (6–13.6 eV) relative to Habing (1968); see Eq. (12.6)
h	Planck's constant = 6.626×10^{-27} erg s
$\hbar$	$h/2\pi = 1.055 \times 10^{-27}$ erg s
H I	atomic hydrogen
H II	H^+, ionized hydrogen
I	nuclear angular momentum quantum number; see §4.6
I_ν	specific intensity = power per area per unit frequency per unit solid ang
j_ν	emissivity; see Eq. (7.9)
J	electronic angular momentum of an atom or ion
k	Boltzmann's constant, $1.381 \times 10^{-16}\,{\rm erg\,K^{-1}}$
k	$\equiv 2\pi/\lambda =$ wavenumber
K	degree Kelvin
λ	wavelength
L	total electronic angular momentum quantum number; see §4.5
$L_\odot$	solar luminosity, $3.826 \times 10^{33}\,{\rm erg\,s^{-1}}$

Symbol	Definition
$m_{\rm H}$	hydrogen mass, 1.674×10^{-24} g
$m(\omega)$	complex refractive index; see Eq. (11.4)
$M_{\rm H\,I}$	total mass of H I
$M_\odot$	solar mass, 1.989×10^{33} g
n_γ	photon occupation number; see Eq. (7.3)
n_e	electron density
$n_{\rm H}$	H nucleon density $= n({\rm H}) + n({\rm H}^+) + 2n({\rm H}_2)$
n_2	$n_{\rm H}/10^2\,{\rm cm}^{-3}$
n_3	$n_{\rm H}/10^3\,{\rm cm}^{-3}$
n_4	$n_{\rm H}/10^4\,{\rm cm}^{-3}$
n_6	$n_{\rm H}/10^6\,{\rm cm}^{-3}$
p	momentum of a particle
p	pressure of a fluid
pc	parsec $= 3.086 \times 10^{18}$ cm
Q_0	rate of emission of $h\nu > 13.60$ eV (H-ionizing) photons
$Q_{0,49}$	$Q_0/10^{49}\,{\rm s}^{-1}$
Q_1	rate of emission of $h\nu > 24.59$ eV (He-ionizing) photons
$Q_{0,{\rm MW}}$	rate of emission of $h\nu > 13.60$ eV photons by stars in the Galaxy
$Q_{\rm abs}$	$\equiv C_{\rm abs}/\pi a_{\rm eff}^2$ (absorption efficiency factor); see Eq. (22.6)
$Q_{\rm sca}$	$\equiv C_{\rm sca}/\pi a_{\rm eff}^2$ (scattering efficiency factor); see Eq. (22.6)
$Q_{\rm ext}$	$\equiv C_{\rm ext}/\pi a_{\rm eff}^2$ (extinction efficiency factor); see Eq. (22.6)
$R_{\rm gr}$	rate coefficient for H_2 formation on dust; see Eq. (31.8)
$R_{\rm S0}$	Strömgren radius in absence of dust; see Eq. (15.2)
$R_\odot$	solar radius, 6.960×10^{10} cm
R_V	$\equiv A_V/E(B-V)$; see Eq. (21.3)
RM	rotation measure; see Eq. (11.23)
S	total electronic spin quantum number; see §4.5
S_ν	source function in radiative transfer; see Eq. (7.16)
T	temperature
T_2	$T/10^2$ K
T_4	$T/10^4$ K
T_6	$T/10^6$ K
T_A	antenna temperature; see Eq. (7.6)
T_B	brightness temperature; see Eq. (7.5)
$T_{\rm exc}$	excitation temperature; see Eq. (7.8)
$T_{\rm spin}$	spin temperature; see Eq. (8.1)
u_ν	specific energy density; see Eq. (7.7)
W	equivalent width, dimensionless; see Eq. (9.3)
W_λ	equivalent width, wavelength; see Eq. (9.4)
W_v	equivalent width, velocity; see §9.1
x_e	$\equiv n_e/n_{\rm H}$, referred to as the fractional ionization;
z_{int}	internal partition function; see §3.1
Z	partition function; see §3.1

Appendix B

Physical Constants

m_e	9.10938×10^{-28} g	electron mass
$m_e c^2$	510.999 keV	
m_p	1.67262×10^{-24} g	proton mass
$m_p c^2$	938.272 MeV	
$m_{\rm H}$	1.67353×10^{-24} g	H mass
m_p/m_e	1836.15	proton/electron mass ratio
amu	1.66054×10^{-24} g	atomic mass unit
e	4.80320×10^{-10}esu	charge quantum
	1.60218×10^{-19}C	
h	6.62607×10^{-27} erg s	Planck's constant
$\hbar$	1.05457×10^{-27} erg s	$h/2\pi$
c	$2.99792458 \times 10^{10}$ cm s^{-1}	speed of light
$\alpha = e^2/\hbar c$	$1/137.036$	fine structure constant
$a_0 = \hbar^2/m_e e^2$	5.29177×10^{-9} cm	Bohr radius
$R_\infty = m_e c^2 \alpha^2/2hc$	109737 cm^{-1}	Rydberg constant
$hcR_\infty = m_e c^2 \alpha^2/2$	13.6057 eV	Rydberg
$\mu_B = e\hbar/2m_e c$	9.27401×10^{-21} ergG^{-1}	Bohr magneton
$\mu_N = e\hbar/2m_p c$	5.05078×10^{-24} ergG^{-1}	nuclear magneton
$r_e = e^2/m_e c^2 = \alpha^2 a_0$	2.81794×10^{-13} cm	classical electron radius
$\sigma_T = 8\pi r_e^2/3$	6.65246×10^{-25} cm^2	Thomson cross section
eV	1.60218×10^{-12} erg	electron-volt
hc	1.98645×10^{-16} erg cm	
	1.23984×10^{-4} eV cm	
	1.23984 eV μm	
hc/k	1.43878 K cm	
G	$6.6742 \pm 10 \times 10^{-8}$ erg cm g^{-2}	Gravitational constant
k	1.38065×10^{-16} erg K^{-1}	Boltzmann constant
$\sigma = \pi^2 k^4/60\hbar^3 c^2$	5.67040×10^{-5} erg s^{-1}cm^{-3} K^{-4}	Stefan-Boltzmann const.
$a = \pi^2 k^4/15(\hbar c)^3$	7.56577×10^{-15} erg cm^{-3} K^{-4}	radiation constant

Energy Conversion Factors

E (eV)	E/k (K)	E/hc (cm^{-1})	E (erg)	E (J/mole)
1	1.1605×10^4	8.0656×10^3	1.6022×10^{-12}	9.6485×10^4
8.6173×10^{-5}	1	0.69603	1.3807×10^{-16}	8.3144
1.2398×10^{-4}	1.4388	1	1.9865×10^{-16}	11.963

Appendix C

Summary of Radiative Processes

Definitions:

$$
\begin{aligned}
g_\ell, g_u &\equiv \text{degeneracy of level } \ell, u \\
f_{\ell u} \equiv f_{\ell\to u} &\equiv \text{oscillator strength for transition } \ell \to u \\
\sigma_{\ell u}(\nu) \equiv \sigma_{\ell\to u}(\nu) &\equiv \text{cross section for transition } \ell \to u \\
n_\gamma \equiv \frac{(I_\nu)_{pol}}{h\nu^3/c^2} &\equiv \text{photon occupation number \ \ (single polarization state)} \\
\langle n_\gamma\rangle \equiv \frac{\langle I_\nu\rangle}{2h\nu^3/c^2} &\equiv \frac{u_\nu}{8\pi h\nu^3/c^3} = \text{angle and polarization averaged } n_\gamma \\
n_\ell, n_u &\equiv \text{number density of atoms in level } \ell, u
\end{aligned}
$$

Identities:

$$
\begin{aligned}
g_\ell f_{\ell u} &\equiv -g_u f_{u\ell} \\
g_\ell B_{\ell u} &\equiv g_u B_{u\ell} \\
A_{u\ell} &= \frac{8\pi^2 e^2\nu^2}{m_e c^3}|f_{u\ell}| = \frac{8\pi^2 e^2\nu^2}{m_e c^3}\frac{g_\ell}{g_u}f_{\ell u} = \frac{8\pi^2 e^2}{m_e c\lambda^2}\frac{g_\ell}{g_u}f_{\ell u} \\
B_{u\ell} &= \frac{c^3}{8\pi h\nu^3}A_{u\ell} = \frac{\pi e^2}{m_e h\nu}\frac{g_\ell}{g_u}f_{\ell u} \\
\int_0^\infty \sigma_{\ell u}(\nu)d\nu &= \frac{g_u}{g_\ell}\frac{c^2}{8\pi\nu^2}A_{u\ell} = \frac{\pi e^2}{m_e c}f_{\ell u} \\
\sigma_{\ell u}(\nu) &= \frac{\pi e^2}{m_e c}f_{\ell u}\phi_\nu = \frac{\lambda^2}{8\pi}\frac{g_u}{g_\ell}A_{u\ell}\phi_\nu
\end{aligned}
$$

Blackbody Radiation:

$$
B_\nu(T) = \frac{2h\nu^3}{c^2}\frac{1}{e^{h\nu/kT}-1} \quad \to \quad \frac{2kT\nu^2}{c^2} \quad \text{for} \quad \frac{h\nu}{kT} \ll 1
$$

$$
n_\gamma = \frac{1}{e^{h\nu/kT}-1} \quad \to \quad \frac{kT}{h\nu} \quad \text{for} \quad \frac{h\nu}{kT} \ll 1
$$

$$
u_\nu = \frac{4\pi}{c}B_\nu \quad ; \quad u = \int_0^\infty u_\nu d\nu = aT^4 \quad ; \quad a = \frac{\pi^2}{15}\frac{k^4}{\hbar^3 c^3}
$$

Frequency distribution function: $(\int \phi_\nu d\nu = 1)$

$$\text{Gaussian}: \quad \phi_\nu = \frac{1}{(2\pi)^{1/2}} \left(\frac{c}{\nu_0 \sigma_v}\right) \exp\left[-\frac{1}{2}\left(\frac{(\nu - \nu_0)}{\nu_0}\frac{c}{\sigma_v}\right)^2\right]$$

$$\sigma_v = \text{one dimensional velocity dispersion} = \left(\frac{kT}{M}\right)^{1/2} = \frac{b}{\sqrt{2}}$$

$$\frac{(\text{FWHM})_\nu}{\nu_0} = \frac{(\text{FWHM})_\lambda}{\lambda_0} = \frac{(\text{FWHM})_v}{c} = \left(\frac{\sigma_v}{c}\right)\sqrt{8\ln 2}$$

$$\text{Lorentzian}: \quad \phi_\nu = \frac{4\gamma_{\ell u}}{16\pi^2(\nu - \nu_0)^2 + \gamma_{\ell u}^2} \quad , \quad \frac{(\text{FWHM})_\nu}{\nu_0} = \frac{\gamma_{\ell u}}{2\pi\nu_0}$$

$$\gamma_{\ell u} = \text{damping constant} = \sum_{k<\ell} A_{\ell k} + \sum_{k<u} A_{uk}$$

Interaction of atoms with radiation:

$$\text{Absorption } \ell \rightarrow u: \quad n_\ell B_{\ell u} u_\nu = \langle n_\gamma \rangle n_\ell \frac{g_u}{g_\ell} A_{u\ell}$$

$$\text{Spontaneous emission } u \rightarrow \ell: \quad n_u A_{u\ell}$$

$$\text{Stimulated emission } u \rightarrow \ell: \quad n_u B_{u\ell} u_\nu = \langle n_\gamma \rangle n_u A_{u\ell}$$

Radiative transfer:

$$dI_\nu = -I_\nu\, \kappa_\nu\, ds + j_\nu\, ds = \left[-I_\nu + \frac{j_\nu}{\kappa_\nu}\right] d\tau_\nu \quad \text{(neglecting scattering)}$$

$$\kappa_\nu = n_\ell\, \sigma_{\ell u}(\nu) - n_u\, |\sigma_{u\ell}(\nu)| = n_\ell \left(1 - \frac{n_u/g_u}{n_\ell/g_\ell}\right) \sigma_{\ell u}(\nu)$$

$$j_\nu = n_u\, A_{u\ell}\, h\nu\, \frac{1}{4\pi} \phi_\nu$$

$$\frac{j_\nu}{\kappa_\nu} = B_\nu(T_{\text{exc}}) \quad , \quad \text{where} \quad \frac{n_u}{n_\ell} = \frac{g_u}{g_\ell} e^{-h\nu/kT_{\text{exc}}}$$

Curve of Growth for a Maxwellian Velocity Distribution (see Chapter 9)

$$\tau_0 = \sqrt{\pi}\frac{e^2}{m_e c}\frac{N_l f_{lu}\lambda}{b} = 1.497\times 10^{-2}\frac{\mathrm{cm}^2}{\mathrm{s}}\frac{N_\ell f_{\ell u}\lambda}{b} \tag{C.1}$$

$$= 0.7580\left(\frac{N_\ell}{10^{13}\mathrm{cm}^{-2}}\right)\left(\frac{f_{\ell u}}{0.4164}\right)\left(\frac{\lambda}{1215.7\mathrm{Å}}\right)\left(\frac{10\mathrm{kms}^{-1}}{b}\right) \tag{C.2}$$

1. Optically thin ($\tau_0 \leq 1\,,\quad 0 < \text{error} < +2.4\%$):

$$W \equiv \frac{W_\lambda}{\lambda} = \sqrt{\pi}\,\frac{b}{c}\sum_{n=1}^{\infty}\frac{(-1)^{n-1}\tau_0^n}{n!\sqrt{n}} \approx \sqrt{\pi}\,\frac{b}{c}\frac{\tau_0}{\left[1+\tau_0/(2\sqrt{2})\right]} \tag{C.3}$$

$$= \frac{\pi e^2}{m_e c^2}\frac{N_\ell f_{\ell u}\lambda}{\left[1+\tau_0/(2\sqrt{2})\right]} = 8.85\times 10^{-13}\,\mathrm{cm}\,\frac{N_\ell f_{\ell u}\lambda}{\left[1+\tau_0/(2\sqrt{2})\right]} \tag{C.4}$$

2. Flat portion of curve of growth ($1 \leq \tau_0 \lesssim 10^3\,;\quad -8\% < \text{error} < +2.4\%$ for $1 < \tau_0 < 600$):

$$W \equiv \frac{W_\lambda}{\lambda} \approx \frac{2b}{c}\left[\ln(\tau_0/\ln 2)\right]^{1/2}$$

3. Damping ("square root") portion of curve of growth ($\tau_0 \gtrsim 10^4$):

$$W \equiv \frac{W_\lambda}{\lambda} \approx \frac{1}{c}\left(\frac{e^2}{m_e c}N_\ell f_{\ell u}\lambda^2\gamma_{u\ell}\right)^{1/2} \qquad \text{where } \gamma_{u\ell} \equiv \sum_k A_{uk} + \sum_k A_{\ell k}$$

$$= 6.02\times 10^{-3}\left(\frac{A_{u\ell}}{6.27\times 10^8\,\mathrm{s}^{-1}}\right)\left(\frac{\lambda_{u\ell}}{1215\,\mathrm{Å}}\right)^2\left(\frac{g_u/g_\ell}{3}\frac{\gamma_{u\ell}}{A_{u\ell}}\right)^{1/2}\left(\frac{N_\ell}{10^{20}\,\mathrm{cm}^{-2}}\right)^{1/2}$$

$$= 6.56\,\frac{b}{c}\left(\frac{\tau_0}{10^4}\right)^{1/2}\left(\frac{\lambda}{1215\,\mathrm{Å}}\frac{10\,\mathrm{km\,s}^{-1}{}^{-1}}{b}\frac{\gamma_{u\ell}}{6.27\times 10^8\,\mathrm{s}^{-1}}\right)^{1/2}$$

Transition from flat to square root portion of curve of growth occurs at $\tau_0 = \tau_{\mathrm{damp}}$, where

$$\frac{\tau_{\mathrm{damp}}}{\ln(\tau_{\mathrm{damp}}/\ln 2)} \approx \frac{4\sqrt{\pi}\,b}{\lambda_{u\ell}\gamma_{u\ell}} \approx 931\left(\frac{b}{10\,\mathrm{km\,s}^{-1}}\right)\left(\frac{1215\mathrm{Å}}{\lambda}\right)\left(\frac{6.27\times 10^8\,\mathrm{s}^{-1}}{\gamma_{ul}}\right)$$

or $\tau_{\mathrm{damp}} \approx 8400$ for Lyα with $b = 10\,\mathrm{km\,s}^{-1}$.

4. Flat and damping portions together can be approximated by:

$$W \equiv \frac{W_\lambda}{\lambda} \approx \left[\frac{4b^2}{c^2}\ln(\tau_0/\ln 2) + \frac{b}{c}\frac{\gamma_{u\ell}\lambda_{u\ell}}{c}\frac{(\tau_0 - 1.25393)}{\sqrt{\pi}}\right]^{1/2}$$

For Lyα with $b = 10\,\mathrm{km\,s}^{-1}$, error $< +4.9\%$ for $\tau_0 > 1.25393$.

1. **Free-free (≡ bremsstrahlung) emission coefficient j_ν** (see §§10.1, 10.2):

$$j_\nu = \frac{8}{3}\left(\frac{2\pi}{3}\right)^{1/2} \frac{e^6}{c^3 m_e^{3/2}(kT)^{1/2}} n_e n_i Z_i^2 g_{\rm ff,\it i} e^{-h\nu/kT} \tag{C.5}$$

$$\approx 5.444 \times 10^{-41} \frac{n_e n_i Z_i^2}{T_4^{0.5}} g_{\rm ff,\it i} e^{-h\nu/kT} \ {\rm erg\,cm^3\,sr^{-1}\,s^{-1}\,Hz^{-1}} \tag{C.6}$$

$$\approx 2.55 \times 10^{-40} \frac{n_e n_i Z_i^2}{T_4^{0.5}} \left[1 - 0.118 \ln\left(\frac{0.1 Z_i \nu_9}{T_4^{3/2}}\right)\right] \frac{\rm erg\,cm^3}{\rm sr\,s\,Hz} \quad \text{for } h\nu \ll kT \tag{C.7}$$

$$\approx 3.35 \times 10^{-40} n_e n_i Z_i^{1.88} T_4^{-0.32} \nu_9^{-0.12} \ {\rm erg\,cm^3\,sr^{-1}\,s^{-1}\,Hz^{-1}} \quad \text{for } h\nu \ll kT \quad , \tag{C.8}$$

where $\nu_9 \equiv \nu/\text{GHz}$, $T_4 \equiv T/10^4\,\text{K}$, and the Gaunt factor in (C.7, C.8) has been taken from the approximations (C.13, C.14).

2. **Free-free absorption coefficient $\kappa_\nu = j_\nu/B_\nu(T)$** (see §10.4):

$$\kappa_\nu = \frac{4}{3}\left(\frac{2\pi}{3}\right)^{1/2} \frac{e^6}{m_e^{3/2} c (kT)^{1/2} h\nu^3} \left[1 - e^{-h\nu/kT}\right] n_e n_i Z_i^2 g_{\rm ff,\it i} \tag{C.9}$$

$$\approx 1.771 \times 10^{-26} g_{\rm ff,\it i} \frac{n_e n_i Z_i^2}{T_4^{3/2} \nu_9^2} \ {\rm cm^5} \quad \text{for } h\nu \ll kT \tag{C.10}$$

$$\approx 8.31 \times 10^{-26} \left[1 - 0.118 \ln\left(\frac{Z_i \nu_9/10}{T_4^{3/2}}\right)\right] \frac{n_e n_i Z_i^2}{T_4^{3/2} \nu_9^2} \ {\rm cm^5} \quad \text{for } h\nu \ll kT \tag{C.11}$$

$$\approx 1.091 \times 10^{-25} n_e n_i Z_i^{1.88} T_4^{-1.32} \nu_9^{-2.10} \ {\rm cm^5} \quad \text{for } h\nu \ll kT \tag{C.12}$$

where $\nu_9 \equiv \nu/\text{GHz}$, $T_4 \equiv T/10^4\,\text{K}$, and the Gaunt factor in (C.11, C.12) has been taken from the approximations (C.13, C.14).

3. **Gaunt factor for free-free transitions $g_{\rm ff}$** (see §10.2) for $\nu_p \ll \nu \ll kT/h$:

$$g_{\rm ff} \approx \frac{\sqrt{3}}{\pi}\left[\ln \frac{(2kT)^{3/2}}{\pi Z_i e^2 m_e^{1/2} \nu} - \frac{5\gamma}{2}\right], \quad \gamma = 0.577216... = \text{Euler's constant}$$

$$= 4.691 \left[1 - 0.118 \ln\left(\frac{Z_i \nu_9/10}{T_4^{3/2}}\right)\right] \tag{C.13}$$

$$\approx 6.155\, \nu_9^{-0.12} T_4^{0.18} \quad (\pm 10\% \text{ for } 0.14 < \nu_9/T_4^{3/2} < 250) \tag{C.14}$$

4. **Dispersion relation for a cold plasma** (see Chapter 11):

$$k^2 c^2 = \omega^2 - \omega_p^2 \quad , \quad \text{where} \quad \omega_p^2 \equiv \frac{4\pi n_e e^2}{m_e} \tag{C.15}$$

$$v_{\rm phase} = \frac{\omega}{k} = c\left[1 - \frac{\omega_p^2}{\omega^2}\right]^{-1/2} \qquad v_{\rm group} = \frac{d\omega}{dk} = c\left[1 - \frac{\omega_p^2}{\omega^2}\right]^{1/2} . \tag{C.16}$$

Appendix D

Ionization Potentials (eV)

Element		I→II	II→III	III→IV	IV→V	V→VI	VI→VII	VII→VIII
1	H	13.5984						
2	He	24.5874	54.418					
3	Li	5.3917	75.640	122.454				
4	Be	9.3227	18.211	153.896	217.719			
5	B	8.2980	25.155	37.931	259.375	340.226		
6	C	11.2603	24.385	47.888	64.494	392.091	489.993	
7	N	14.5341	29.601	47.445	77.474	97.890	552.067	667.046
8	O	13.6181	35.121	54.936	77.414	113.899	138.119	739.327
9	F	17.4228	34.971	62.708	87.175	114.249	157.163	185.187
10	Ne	21.5645	40.963	63.423	97.190	126.247	157.934	207.271
11	Na	5.1391	47.286	71.620	98.936	138.404	172.23	208.504
12	Mg	7.6462	15.035	80.144	109.265	141.33	186.76	225.02
13	Al	5.9858	18.829	28.448	119.992	153.825	190.49	241.76
14	Si	8.1517	16.346	33.493	45.142	166.767	205.267	246.32
15	P	10.4867	19.769	30.203	51.444	65.025	220.430	263.57
16	S	10.3600	23.338	34.86	47.222	72.595	88.053	280.954
17	Cl	12.9676	23.814	39.80	53.24	67.68	96.94	114.201
18	Ar	15.7596	27.630	40.735	59.58	74.84	91.290	124.41
19	K	4.3407	31.625	45.806	60.913	82.66	99.4	117.6
20	Ca	6.1132	11.872	50.913	67.273	84.34	108.78	127.21
21	Sc	6.5615	12.800	24.757	73.489	91.95	110.68	137.99
22	Ti	6.8281	13.576	27.492	43.267	92.299	119.533	140.68
23	V	6.7462	14.618	29.311	46.709	65.282	128.125	150.72
24	Cr	6.7665	16.486	30.959	49.16	69.46	90.635	160.29
25	Mn	7.4340	15.640	33.668	51.2	72.41	95.604	119.203
26	Fe	7.9025	16.199	30.651	54.91	75.0	98.985	124.98
27	Co	7.8810	17.084	33.50	51.27	79.50	102.0	128.9
28	Ni	7.6399	18.169	35.187	54.92	76.06	108.0	132.0
29	Cu	7.7264	20.292	36.841	57.38	79.8	103.0	139.0
30	Zn	9.3942	17.964	39.723	59.573	82.6	108.0	133.9

Notes:

- Ionization potentials from Kramida et al. (2014, accessed 2015.07.09).
- The light line separates ions with $I < I_{\rm He}$ from ions with $I > I_{\rm He} = 24.6\,{\rm eV}$.
- Ions to right of the heavy line (with $I > I_{\rm He\,II} = 54.4\,{\rm eV}$) are not abundant in gas photoionized by O or B stars and are therefore indicative of photoionization by WR stars, PN nuclei, or collisional ionization in shocked gas.
- For elemental abundances, see Table 1.4.

Appendix E

Energy-Level Diagrams

This appendix provides energy-level diagrams for the more abundant elements (H, C, N, O, Ne, Mg, Al, Si, S, Ar) in ionization states with 3 to 9, or 11 to 15 electrons. The diagrams show the ground state and excited states with $E < 13.6\,\mathrm{eV}$.

- Hyperfine splitting (interaction with the nuclear magnetic moment) is not shown.
- Atoms or ions with 2 electrons (e.g., He I) or 10 electrons (e.g., Ne I) are not shown because they have no excited states below $13.6\,\mathrm{eV}$ (but see Fig. 14.3 for the radiative decay pathways from higher levels of He I that are populated by radiative recombination.)
- Fine-structure splitting is indicated (although not to scale). The one-electron system – H I – is a special case: different orbitals $n\ell$ with the same principal quantum number n (e.g., $2s$ and $2p$) have nearly identical energies, and spin-orbit splitting is negligible (e.g., the ${}^2\mathrm{P}^{\mathrm{o}}_{1/2}$ and ${}^2\mathrm{P}^{\mathrm{o}}_{3/2}$ levels formed from the $2p$ orbital have nearly the same energy). Therefore, for H we show a single energy level for each principal quantum number n.
- Heavy lines show resonance lines – permitted transitions to/from the ground state multiplet. Resonance lines are shown in absorption (i.e., as upward arrows), but of course are also observable as emission lines (e.g., H Lyman α, or C IV 1548,1551).
- Dashed lines show intercombination ("semiforbidden") lines out of the ground state multiplet. These have small f values but may be observable in absorption if the column density is sufficient.
- Downward solid lines show additional transitions that may be observable in emission. Because the lowest few energy levels of multielectron atoms typically have the same parity, most of the emission lines shown in these figures are forbidden transitions.
- Wavelengths are in Å, unless otherwise noted.
- All wavelengths are in vacuo.

1 electron

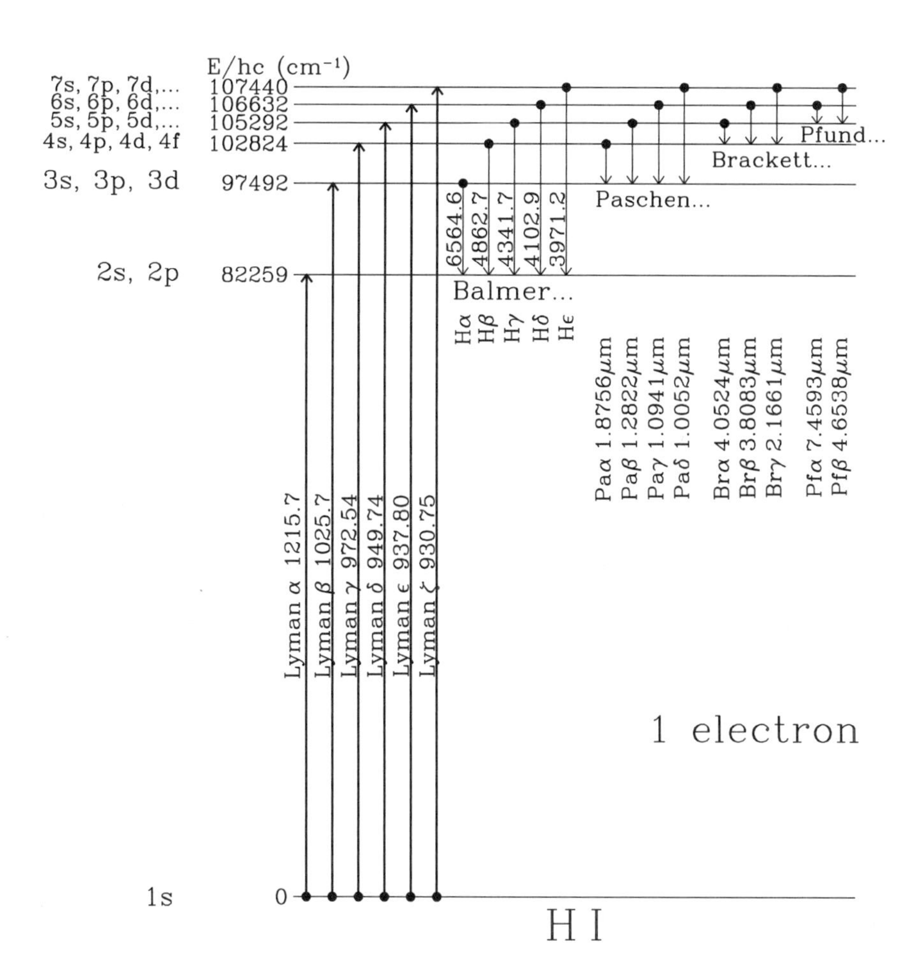

E/hc (cm⁻¹)
7s, 7p, 7d,... 107440
6s, 6p, 6d,... 106632
5s, 5p, 5d,... 105292
4s, 4p, 4d, 4f 102824
3s, 3p, 3d 97492
2s, 2p 82259
1s 0
Lyman α 1215.7
Lyman β 1025.7
Lyman γ 972.54
Lyman δ 949.74
Lyman ϵ 937.80
Lyman ζ 930.75
6564.6
4862.7
4341.7
4102.9
3971.2
Balmer...
Hα
Hβ
Hγ
Hδ
Hϵ
Paschen...
Paα 1.8756μm
Paβ 1.2822μm
Paγ 1.0941μm
Paδ 1.0052μm
Brackett...
Brα 4.0524μm
Brβ 3.8083μm
Brγ 2.1661μm
Pfund...
Pfα 7.4593μm
Pfβ 4.6538μm
1 electron
H I

3 electrons

(13.6 eV)/hc = 109692 cm⁻¹

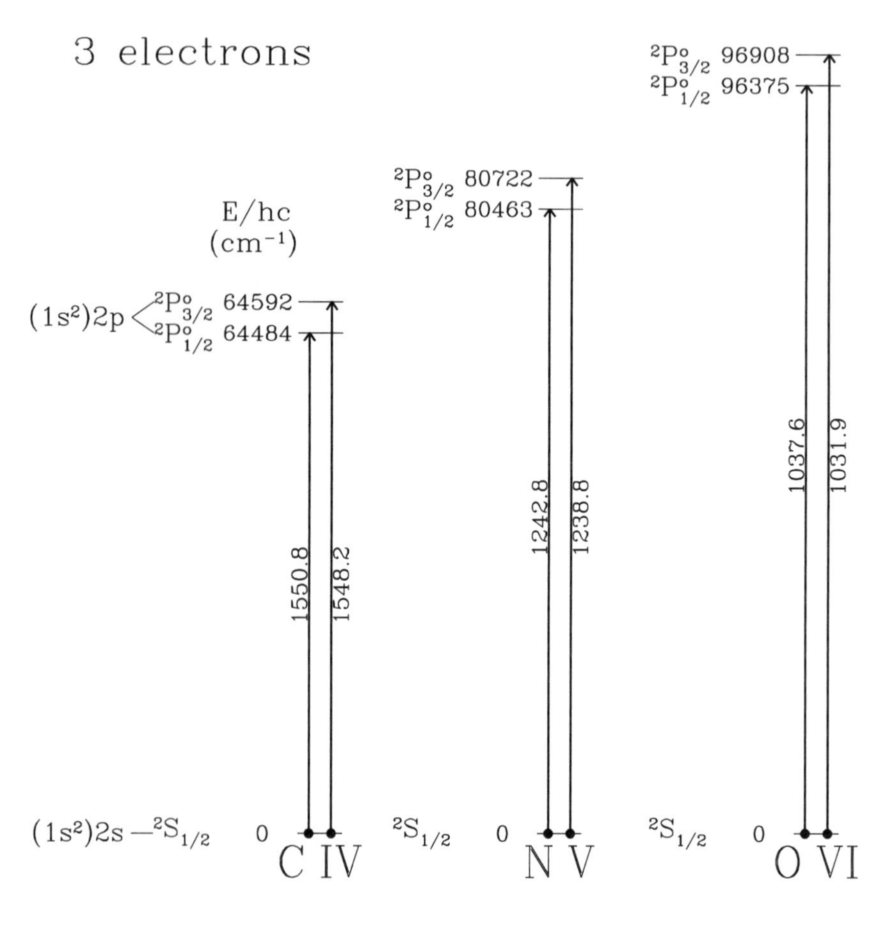
3 electrons
E/hc
(cm⁻¹)
(1s²)2p
²P°3/2 64592
²P°1/2 64484
(1s²)2s — ²S1/2 0
1550.8
1548.2
C IV
²P°3/2 80722
²P°1/2 80463
²S1/2 0
1242.8
1238.8
N V
²P°3/2 96908
²P°1/2 96375
²S1/2 0
1037.6
1031.9
O VI

4 electrons

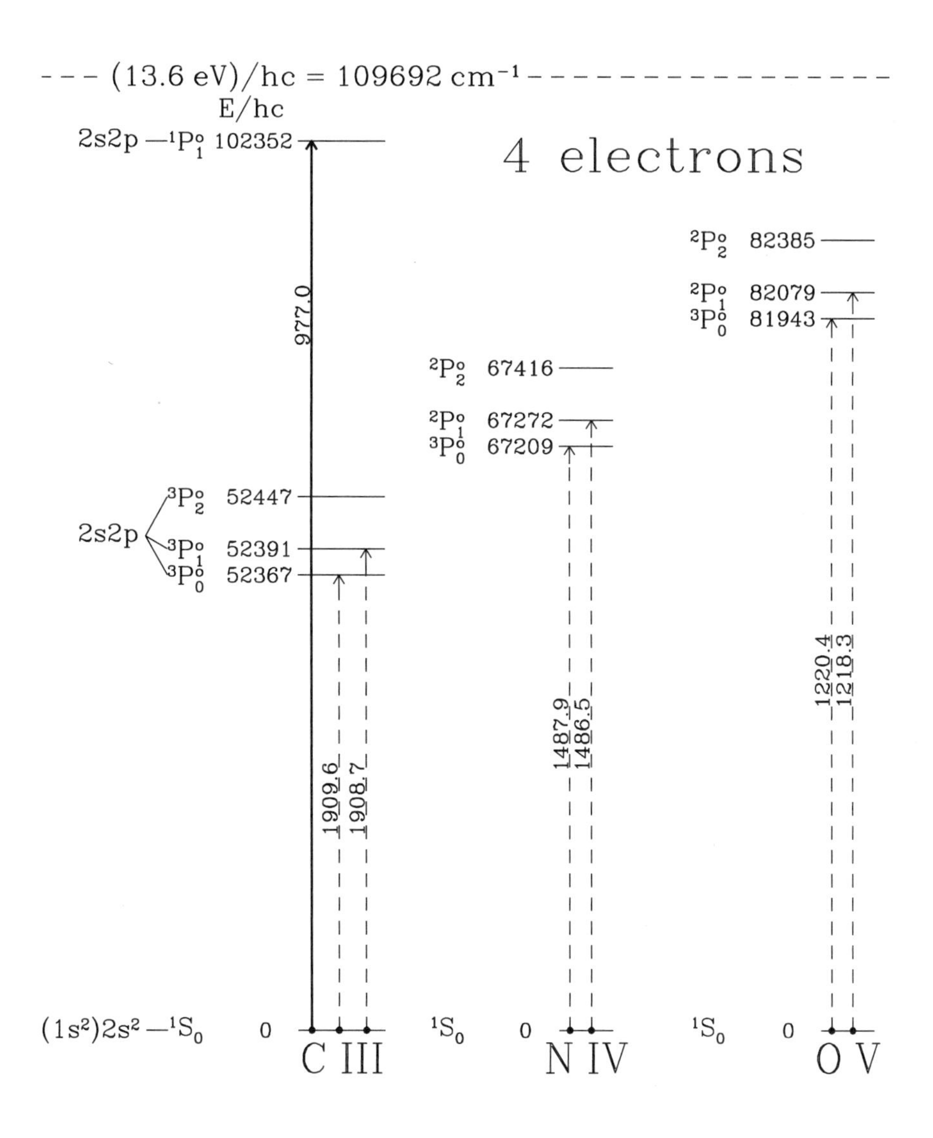

--- (13.6 eV)/hc = 109692 cm⁻¹ ---
E/hc
4 electrons
2s2p —¹P°₁ 102352
977.0
2s2p
³P°₂ 52447
³P°₁ 52391
³P°₀ 52367
1909.6
1908.7
(1s²)2s² —¹S₀ 0
C III
²P°₂ 67416
²P°₁ 67272
³P°₀ 67209
1487.9
1486.5
¹S₀ 0
N IV
²P°₂ 82385
²P°₁ 82079
³P°₀ 81943
1220.4
1218.3
¹S₀ 0
O V

5 electrons

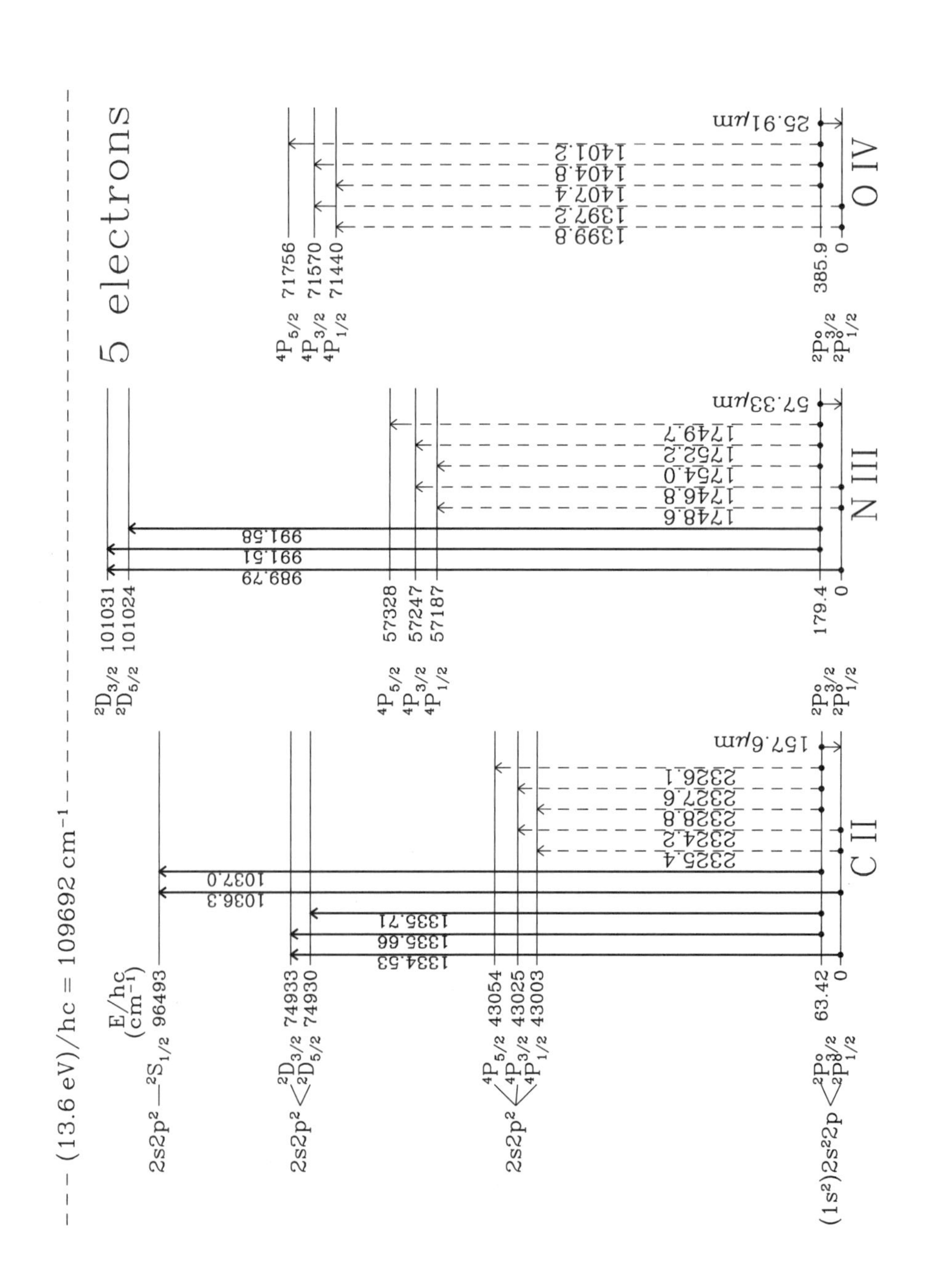
(13.6 eV)/hc = 109692 cm^{-1}
5 electrons
E/hc
(cm^{-1})
$2s2p^2$ — $^2S_{1/2}$ 96493
$2s2p^2$ $^2D_{3/2}$ 74933 $^2D_{5/2}$ 74930
$2s2p^2$ $^4P_{5/2}$ 43054 $^4P_{3/2}$ 43025 $^4P_{1/2}$ 43003
$(1s^2)2s^22p$ $^2P^o_{3/2}$ 63.42 $^2P^o_{1/2}$ 0
1334.53
1335.66
1335.71
1036.3
1037.0
2325.4
2324.2
2328.8
2327.6
2326.1
157.6μm
C II
$^2D_{3/2}$ 101031
$^2D_{5/2}$ 101024
$^4P_{5/2}$ 57328
$^4P_{3/2}$ 57247
$^4P_{1/2}$ 57187
$^2P^o_{3/2}$ 179.4
$^2P^o_{1/2}$ 0
989.79
991.51
991.58
1748.6
1746.8
1754.0
1752.2
1749.7
57.33μm
N III
$^4P_{5/2}$ 71756
$^4P_{3/2}$ 71570
$^4P_{1/2}$ 71440
$^2P^o_{3/2}$ 385.9
$^2P^o_{1/2}$ 0
1399.8
1397.2
1407.4
1404.8
1401.2
25.91μm
O IV

6 electrons

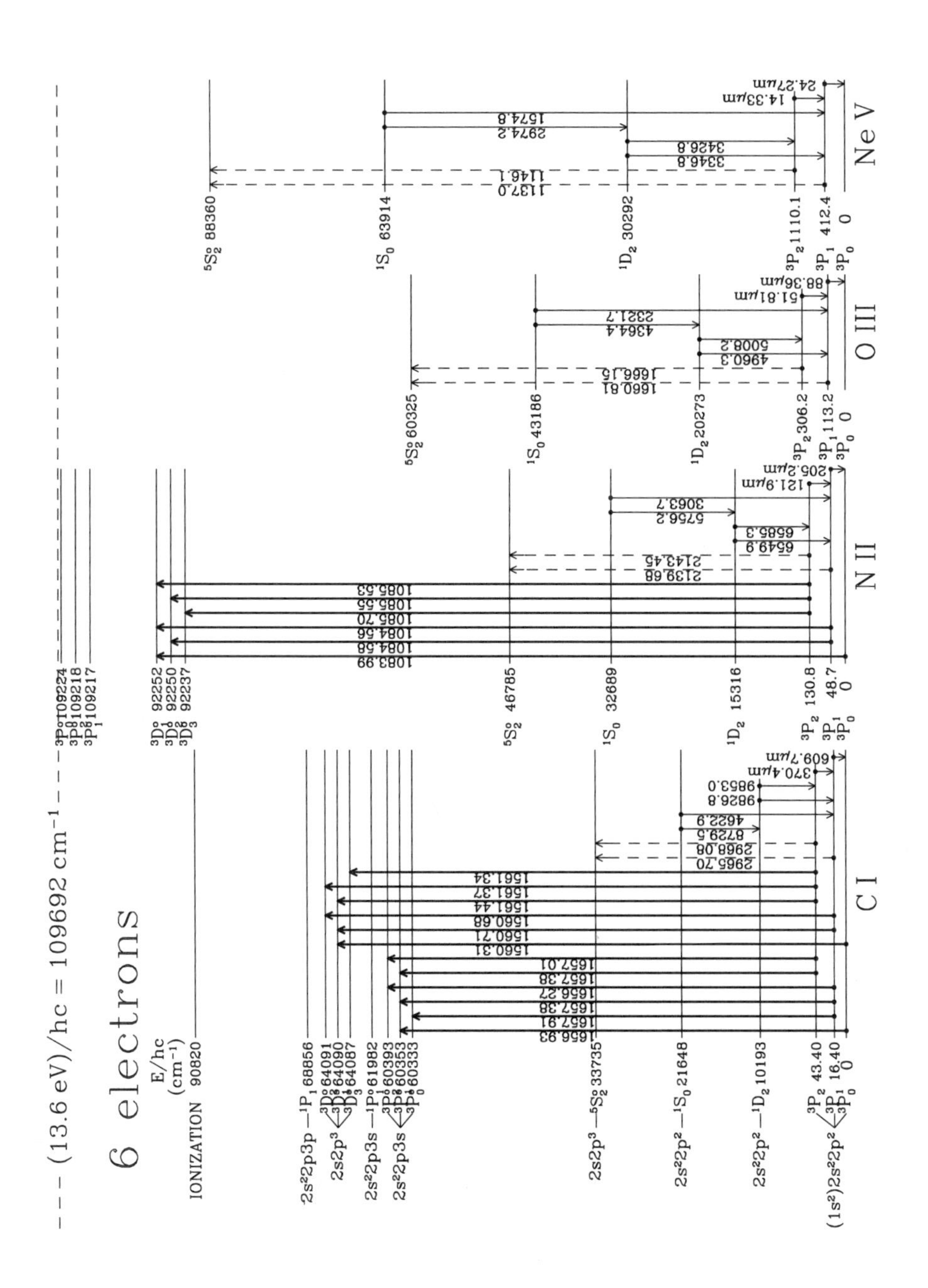

--- (13.6 eV)/hc = 109692 cm⁻¹ ---
6 electrons
E/hc
(cm⁻¹)
IONIZATION 90820
2s²2p3p —¹P₁ 68856
2s2p³ ³D°₂ 64091 ³D°₁ 64090 ³D°₃ 64087
2s²2p3s —¹P°₁ 61982
2s²2p3s ³P°₂ 60393 ³P°₁ 60353 ³P°₀ 60333
2s2p³ —⁵S°₂ 33735
2s²2p² —¹S₀ 21648
2s²2p² —¹D₂ 10193
(1s²)2s²2p² ³P₂ 43.40 ³P₁ 16.40 ³P₀ 0
1656.93 1657.91 1657.38 1656.27 1657.38 1657.01
1560.31 1560.71 1560.68 1561.44 1561.37 1561.34
2965.70 2968.08 8729.5 4622.9 9826.8 9853.0 370.4μm 609.7μm
C I
³P°₀ 109224 ³P°₂ 109218 ³P°₁ 109217
³D°₁ 92252 ³D°₂ 92250 ³D°₃ 92237
⁵S°₂ 46785
¹S₀ 32689
¹D₂ 15316
³P₂ 130.8 ³P₁ 48.7 ³P₀ 0
1083.99 1084.58 1084.56 1085.70 1085.55 1085.53
2139.68 2143.45 5756.2 3063.7 6549.9 6585.3 121.9μm 205.2μm
N II
⁵S°₂ 60325
¹S₀ 43186
¹D₂ 20273
³P₂ 306.2 ³P₁ 113.2 ³P₀ 0
1660.81 1666.15 4364.4 2321.7 4960.3 5008.2 51.81μm 88.36μm
O III
⁵S°₂ 88360
¹S₀ 63914
¹D₂ 30292
³P₂ 1110.1 ³P₁ 412.4 ³P₀ 0
1137.0 1146.1 2974.2 1574.8 3346.8 3426.8 14.33μm 24.27μm
Ne V

7 electrons

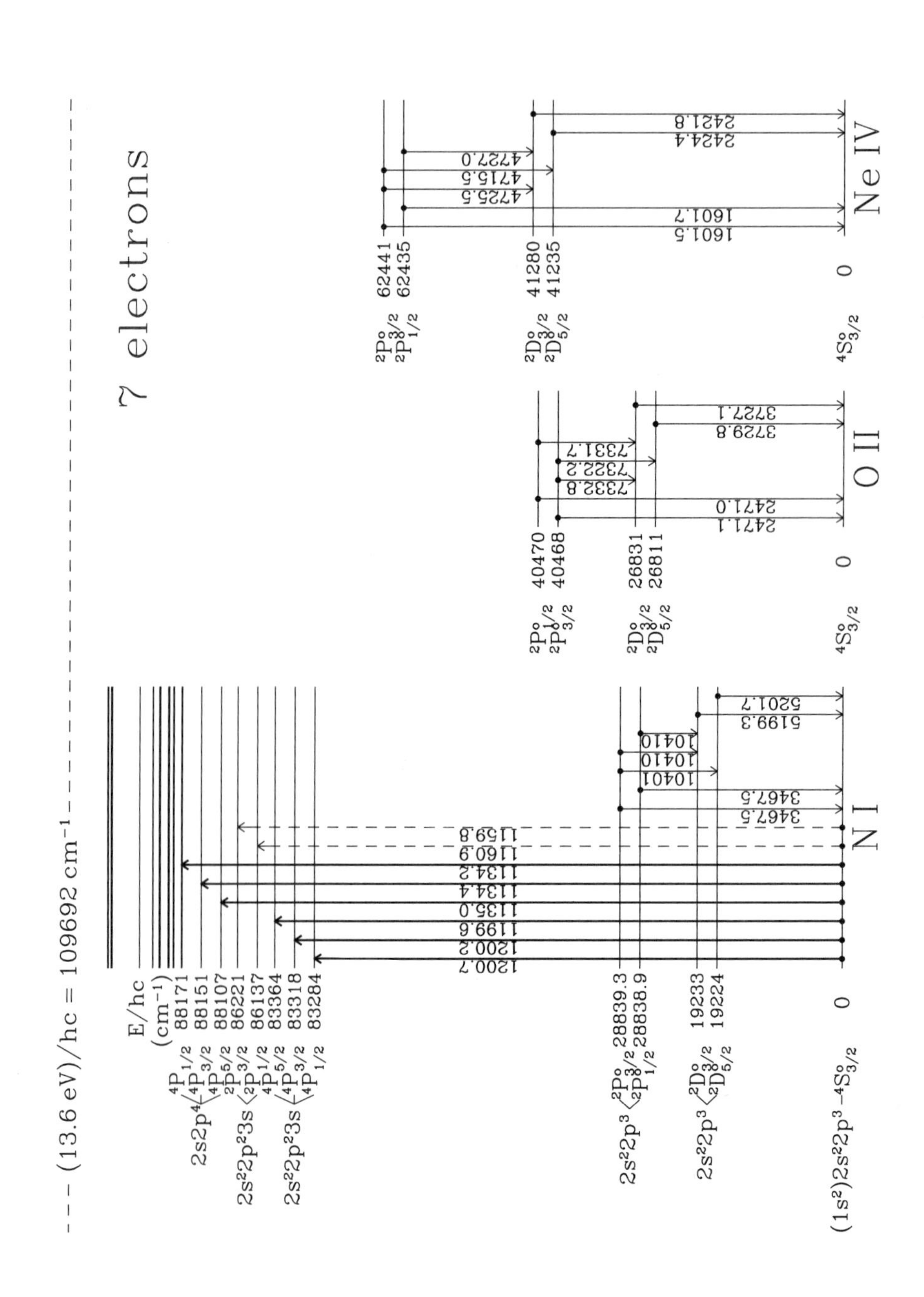

8 electrons

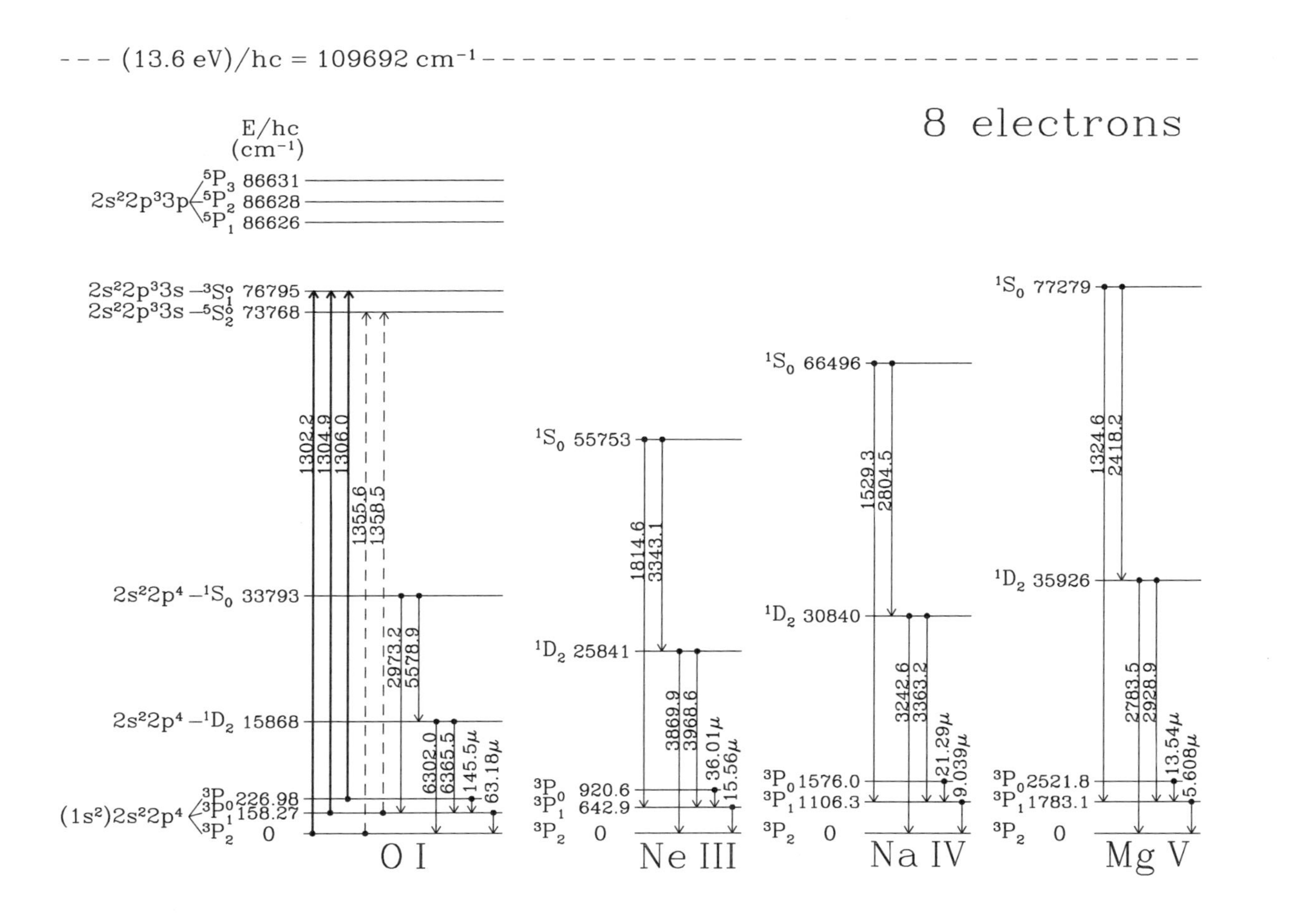

(13.6 eV)/hc = 109692 cm⁻¹
8 electrons
E/hc (cm⁻¹)
2s²2p³3p ⁵P₃ 86631
⁵P₂ 86628
⁵P₁ 86626
2s²2p³3s —³S°₁ 76795
2s²2p³3s —⁵S°₂ 73768
2s²2p⁴ —¹S₀ 33793
2s²2p⁴ —¹D₂ 15868
(1s²)2s²2p⁴ ³P₀ 226.98
³P₁ 158.27
³P₂ 0
1302.2
1304.9
1306.0
1355.6
1358.5
2973.2
5578.9
6302.0
6365.5
145.5μ
63.18μ
O I
¹S₀ 55753
¹D₂ 25841
³P₀ 920.6
³P₁ 642.9
³P₂ 0
1814.6
3343.1
3869.9
3968.6
36.01μ
15.56μ
Ne III
¹S₀ 66496
¹D₂ 30840
³P₀ 1576.0
³P₁ 1106.3
³P₂ 0
1529.3
2804.5
3242.6
3363.2
21.29μ
9.039μ
Na IV
¹S₀ 77279
¹D₂ 35926
³P₀ 2521.8
³P₁ 1783.1
³P₂ 0
1324.6
2418.2
2783.5
2928.9
13.54μ
5.608μ
Mg V

9 electrons

(13.6 eV)/hc = 109692 cm^{-1}

9 electrons

$(1s^2)2s^22p^5$ $^2P^o_{1/2}$, $^2P^o_{3/2}$

E/hc (cm^{-1})

Ne II: $^2P^o_{1/2}$ 780.42; $^2P^o_{3/2}$ 0; 12.81μm

Na III: $^2P^o_{1/2}$ 1366.3; $^2P^o_{3/2}$ 0; 7.319μm

Mg IV: $^2P^o_{1/2}$ 2228; $^2P^o_{3/2}$ 0; 4.488μm

Al V: $^2P^o_{1/2}$ 3442; $^2P^o_{3/2}$ 0; 2.879μm

Si VI: $^2P^o_{1/2}$ 5090; $^2P^o_{3/2}$ 0; 1.965μm

11 electrons

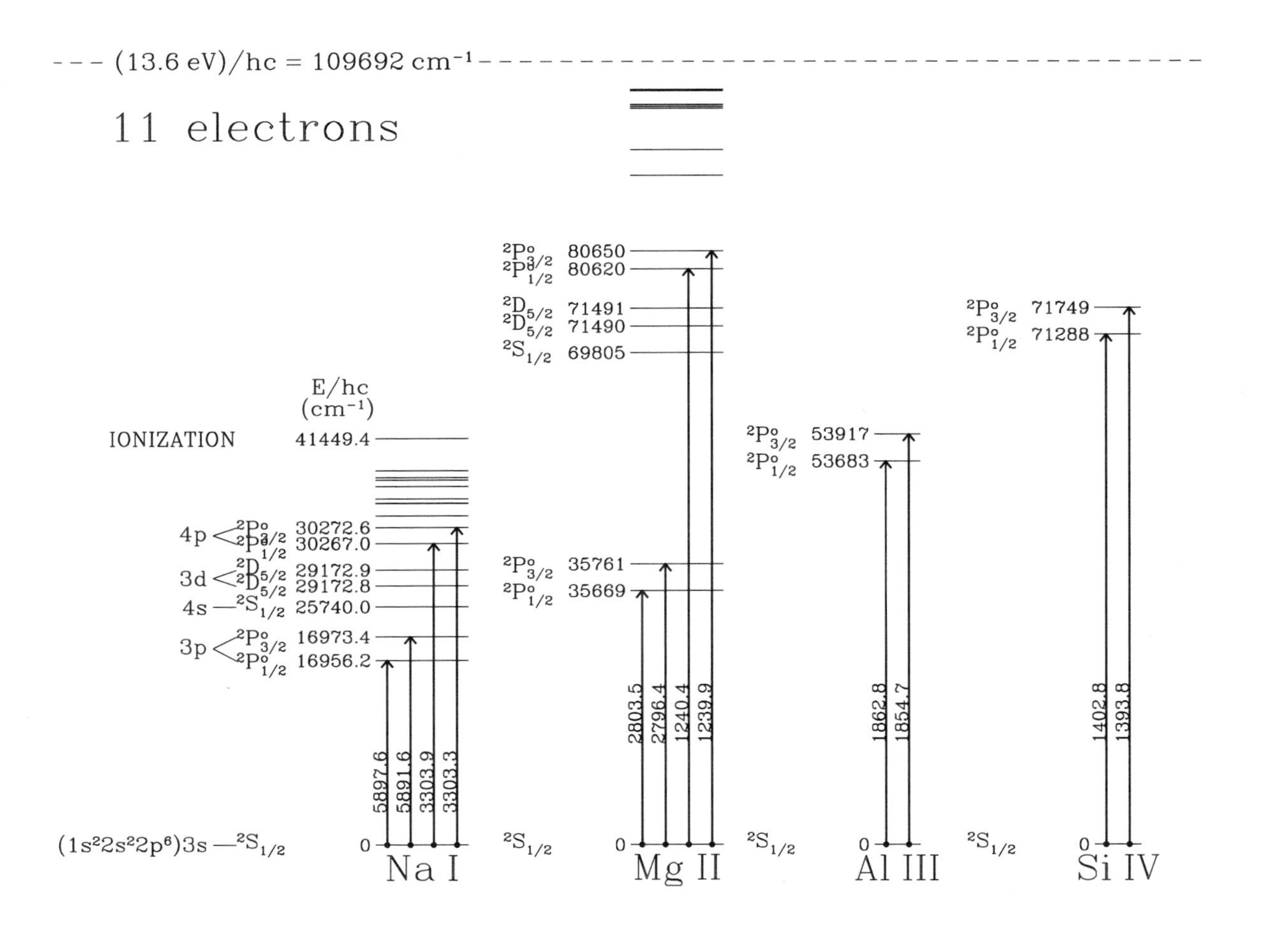

12 electrons

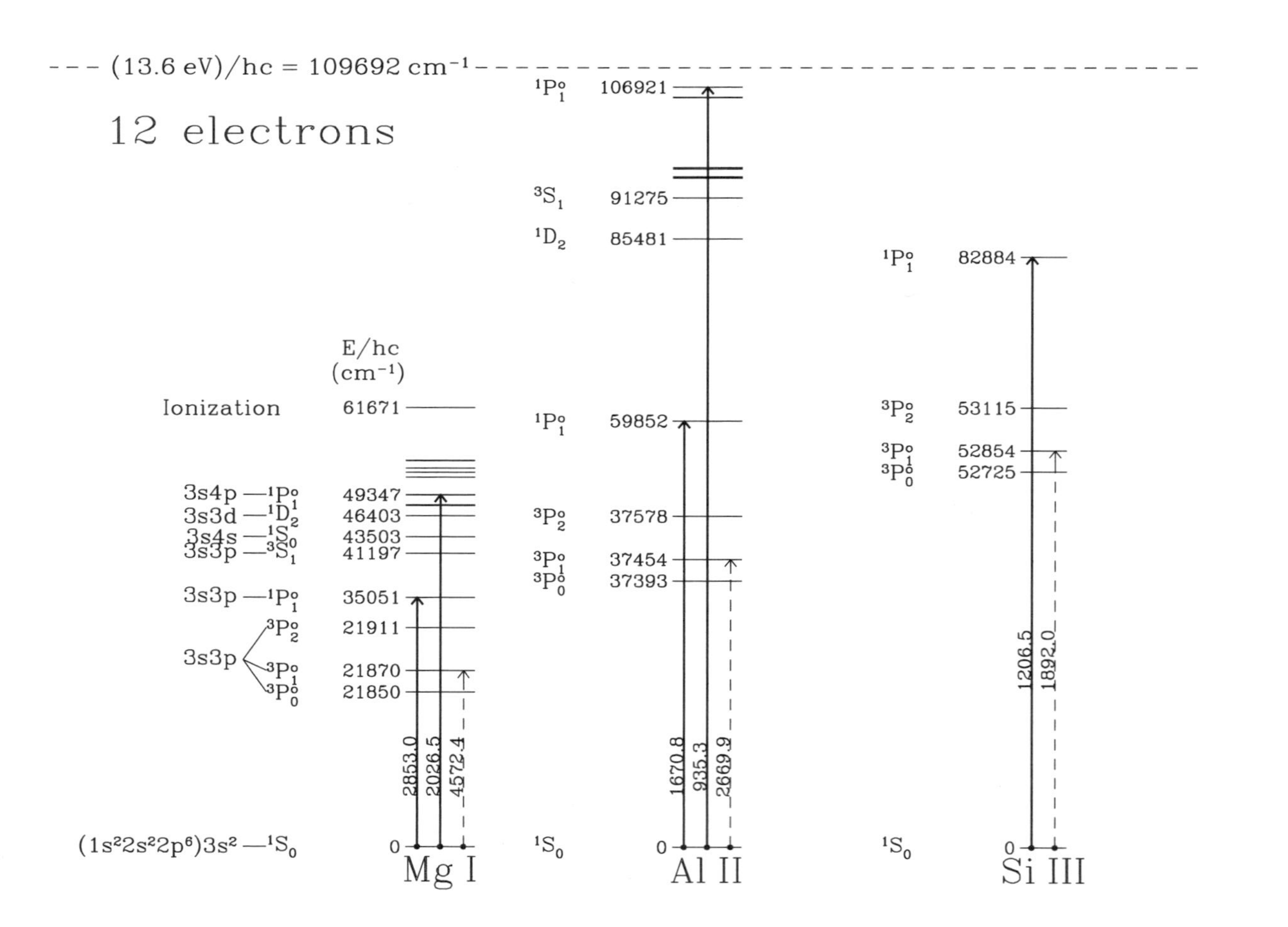

13 electrons

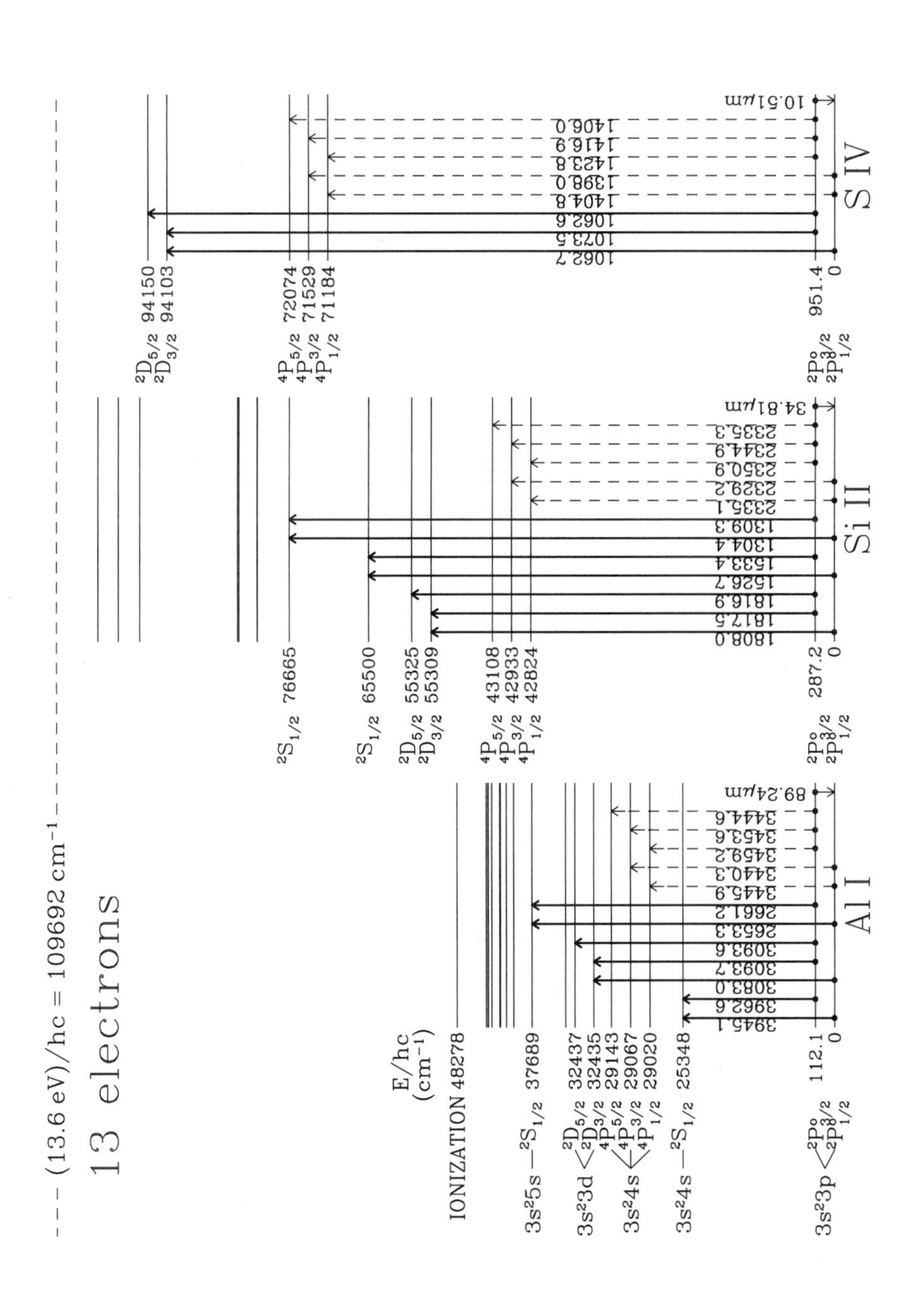

14 electrons

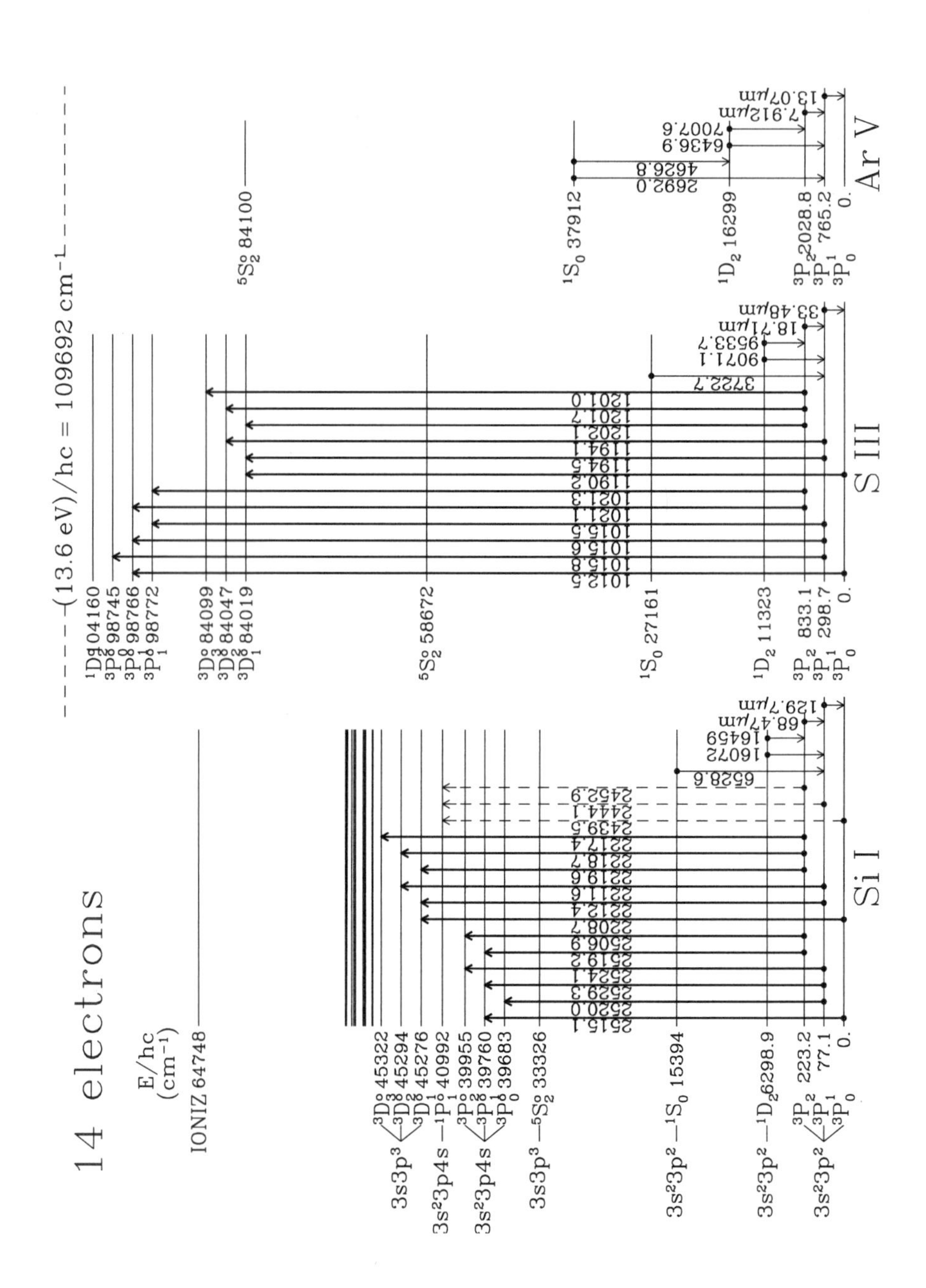

15 electrons

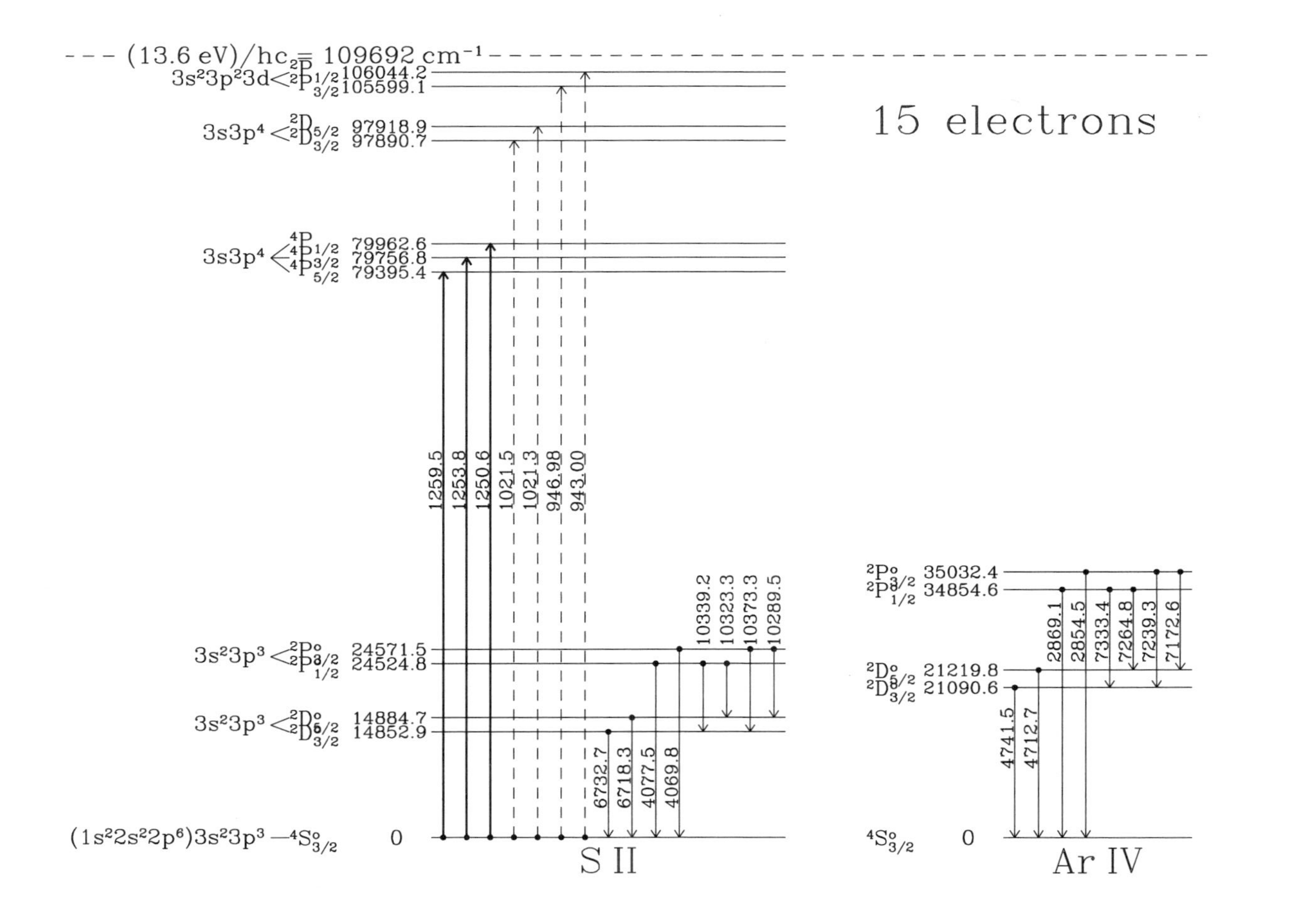

Appendix F

Collisional Rate Coefficients

Notation:

$$T_2 \equiv \frac{T}{10^2\,\mathrm{K}} \qquad ; \qquad T_4 \equiv \frac{T}{10^4\,\mathrm{K}} \quad .$$

Table F.1 Electron Collision Strengths for np^1 and np^5 Ions

	Ion	$\ell - u$	$\lambda_{u\ell}(\mu\mathrm{m})$	$\Omega_{u\ell}$	Note
$2s^2 2p$	C II	${}^2\mathrm{P}^{\mathrm{o}}_{1/2} - {}^2\mathrm{P}^{\mathrm{o}}_{3/2}$	157.7	$(1.55 + 1.25T_4)/(1 + 0.35T_4^{1.25})$	*a*
	N III		57.34	$1.21\, T_4^{0.151 + .056 \ln T_4}$	*b*
	O IV		25.91	$2.144\, T_4^{0.164 - 0.068 \ln T_4}$	*c*
$2s^2 2p^5$	Ne II	${}^2\mathrm{P}^{\mathrm{o}}_{3/2} - {}^2\mathrm{P}^{\mathrm{o}}_{1/2}$	12.815	$0.314\, T_4^{0.076 + 0.002 \ln T_4}$	*d*
	Mg IV		4.488	$0.342 + 0.00434 T_4$	*e*
$3s^2 3p$	Si II	${}^2\mathrm{P}^{\mathrm{o}}_{1/2} - {}^2\mathrm{P}^{\mathrm{o}}_{3/2}$	34.81	$4.45\, T_4^{-0.021 + 0.016 \ln T_4}$	*f*
	S IV		10.51	$8.54\, T_4^{-0.012 - 0.076 \ln T_4}$	*g*
	Ar VI		4.531	$5.90\, T_4^{0.203 - 0.146 \ln T_4}$	*h*
$3s^2 3p^5$	Ar II	${}^2\mathrm{P}^{\mathrm{o}}_{3/2} - {}^2\mathrm{P}^{\mathrm{o}}_{1/2}$	6.985	$2.93\, T_4^{0.084 - 0.014 \ln T_4}$	*i*
	Ca IV		3.207	$1.00\, T_4^{0.248 + 0.086 \ln T_4}$	*i*

a fit to Tayal (2008)
b fit to Stafford et al. (1994)
c fit to Tayal (2006*a*)
d fit to Griffin et al. (2001)
e Johnson & Kingston (1987)
f fit to Bautista et al. (2009)
g fit to Tayal (2000)
h fit to Saraph & Storey (1996)
i fit to Pelan & Berrington (1995)

Table F.2 Electron Collision Strengths for np^2 Ions

Ion	$\ell - u$	$\Omega_{u\ell}$	Note
C I	$^3P_0-^3P_1$	fitting formulae given in reference *a*	*a*
	$^3P_0-^3P_2$	fitting formulae given in reference *a*	*a*
	$^3P_1-^3P_2$	fitting formulae given in reference *a*	*a*
N II	$^3P_0-^3P_1$	$0.431\ T_4^{0.099+0.014 \ln T_4}$	*b*
"	$^3P_0-^3P_2$	$0.273\ T_4^{0.166+0.030 \ln T_4}$	*b*
"	$^3P_1-^3P_2$	$1.15\ T_4^{0.137+0.024 \ln T_4}$	*b*
"	$^3P_0-^1D_2$	$0.303\ T_4^{0.053+0.009 \ln T_4}$	*b*
"	$^3P_1-^1D_2$	$0.909\ T_4^{0.053+0.010 \ln T_4}$	*b*
"	$^3P_2-^1D_2$	$1.51\ T_4^{0.054+0.011 \ln T_4}$	*b*
"	$^3P_0-^1S_0$	$0.0352\ T_4^{0.066+0.018 \ln T_4}$	*b*
"	$^3P_1-^1S_0$	$0.105\ T_4^{0.070+0.021 \ln T_4}$	*b*
"	$^3P_2-^1S_0$	$0.176\ T_4^{0.065+0.017 \ln T_4}$	*b*
"	$^1D_2-^1S_0$	$0.806\ T_4^{-0.175-0.014 \ln T_4}$	*b*
O III	$^3P_0-^3P_1$	$0.522\ T_4^{0.033-0.009 \ln T_4}$	*c*
"	$^3P_0-^3P_2$	$0.257\ T_4^{0.081+0.017 \ln T_4}$	*c*
"	$^3P_1-^3P_2$	$1.23\ T_4^{0.053+0.007 \ln T_4}$	*c*
"	$^3P_J-^1D_2$	$0.243(2J+1)\ T_4^{0.120+0.031 \ln T_4}$	*c*
"	$^3P_J-^1S_0$	$0.0321(2J+1)\ T_4^{0.118+0.057 \ln T_4}$	*c*
"	$^1D_2-^1S_0$	$0.523\ T_4^{0.210-0.099 \ln T_4}$	*c*
Ne V	$^3P_0-^3P_1$	$1.408\ T_4^{-0.264-0.057 \ln T_4}$	*d*
"	$^3P_0-^3P_2$	$1.810\ T_4^{-0.444-0.060 \ln T_4}$	*d*
"	$^3P_1-^3P_2$	$5.832\ T_4^{-0.390-0.056 \ln T_4}$	*d*
"	$^3P_J-^1D_2$	$0.232(2J+1)\ T_4^{0.016+0.019 \ln T_4}$	*d*
"	$^3P_J-^1S_0$	$0.0273(2J+1)\ T_4^{0.027+0.042 \ln T_4}$	*d*
"	$^1D_2-^1S_0$	$5.832\ T_4^{-0.390-0.056 \ln T_4}$	*d*
S III	$^3P_0-^3P_1$	$3.98\ T_4^{-.227-0.100 \ln T_4}$	*e*
"	$^3P_0-^3P_2$	$1.31\ T_4^{-0.070+0.052 \ln T_4}$	*e*
"	$^3P_1-^3P_2$	$7.87\ T_4^{-0.171-0.033 \ln T_4}$	*e*
"	$^3P_0-^1D_2$	$0.773\ T_4^{-0.015+0.056 \ln T_4}$	*e*
"	$^3P_1-^1D_2$	$2.32\ T_4^{-0.015+0.057 \ln T_4}$	*e*
"	$^3P_2-^1D_2$	$3.86\ T_4^{-0.015+0.058 \ln T_4}$	*e*
"	$^3P_0-^1S_0$	$0.131\ T_4^{0.043+0.031 \ln T_4}$	*e*
"	$^3P_1-^1S_0$	$0.398\ T_4^{0.040+0.031 \ln T_4}$	*e*
"	$^3P_2-^1S_0$	$0.655\ T_4^{0.040+0.029 \ln T_4}$	*e*
"	$^1D_2-^1S_0$	$1.38 T_4^{0.140+0.093 \ln T_4}$	*e*
Ar V	$^3P_0-^3P_1$	0.257	*f*
"	$^3P_0-^3P_2$	0.320	*f*
"	$^3P_1-^3P_2$	1.04	*f*
"	$^3P_J-^1D_2$	$0.131(2J+1)\ T_4^{-0.092-0.150 \ln T_4}$	*f*
"	$^3P_J-^1S_0$	$0.413(2J+1)\ T_4^{-0.177+0.080 \ln T_4}$	*f*
"	$^1S_0-^1D_2$	$1.25\ T_4^{-0.017-0.009 \ln T_4}$	*f*

a Johnson et al. (1987)
b fit to Hudson & Bell (2005)
c fit to Aggarwal & Keenan (1999)
d fit to Lennon & Burke (1994)
e fit to Tayal & Gupta (1999)
f fit to Mendoza (1983)

Table F.3 Electron Collision Strengths for np^4 Ions

Ion	$\ell - u$	$\Omega_{u\ell}$	Note
O I	$^3P_2-^3P_1$	$0.0105\ T_4^{0.4861+0.0054\ln T_4}$	*a*
"	$^3P_2-^3P_0$	$0.00459\ T_4^{0.4507-0.0066\ln T_4}$	*a*
"	$^3P_1-^3P_0$	$0.00015\ T_4^{0.4709-0.1396\ln T_4}$	*a*
"	$^3P_J-^1D_2$	$0.0312(2J+1)\ T_4^{0.945-0.001\ln T_4}$	*b*
"	$^3P_J-^1S_0$	$0.00353(2J+1)\ T_4^{1.000-0.135\ln T_4}$	*b*
"	$^1D_2-^1S_0$	$0.0893\ T_4^{0.662-0.089\ln T_4}$	*b*
Ne III	$^3P_2-^3P_1$	$0.774\ T_4^{0.068-0.0556\ln T_4}$	*c*
"	$^3P_2-^3P_0$	$0.208\ T_4^{0.056-0.053\ln T_4}$	*c*
"	$^3P_1-^3P_0$	$0.244\ T_4^{0.086-0.058\ln T_4}$	*c*
"	$^3P_2-^1D_2$	$0.754\ T_4^{-0.011+0.004\ln T_4}$	*c*
"	$^3P_1-^1D_2$	$0.452T_4^{-0.010+0.004\ln T_4}$	*c*
"	$^3P_0-^1D_2$	$0.151\ T_4^{-0.010+0.003\ln T_4}$	*c*
"	$^3P_2-^1S_0$	$0.0840\ T_4^{0.029+0.015\ln T_4}$	*c*
"	$^3P_1-^1S_0$	$0.050\ T_4^{0.028+0.020\ln T_4}$	*c*
"	$^3P_0-^1S_0$	$0.0170\ T_4^{0.021+0.015\ln T_4}$	*c*
"	$^1D_2-^1S_0$	$0.269\ T_4^{0.055+0.034\ln T_4}$	*c*
Mg V	$^3P_2-^3P_1$	$0.929\ T_4^{0.130-0.041\ln T_4}$	*d*
"	$^3P_2-^3P_0$	$0.265\ T_4^{0.148-0.053\ln T_4}$	*d*
"	$^3P_1-^3P_0$	$0.317\ T_4^{0.102-0.040\ln T_4}$	*d*
"	$^3P_2-^1D_2$	$0.705\ T_4^{0.008-0.008\ln T_4}$	*d*
"	$^3P_1-^1D_2$	$0.448\ T_4^{0.005-0.012\ln T_4}$	*d*
"	$^3P_0-^1D_2$	$0.153\ T_4^{-0.002-0.011\ln T_4}$	*d*
"	$^3P_2-^1S_0$	$0.101\ T_4^{0.029-0.045\ln T_4}$	*d*
"	$^3P_1-^1S_0$	$0.0664\ T_4^{0.032-0.033\ln T_4}$	*d*
"	$^3P_0-^1S_0$	$0.0252\ T_4^{0.022-0.024\ln T_4}$	*d*
"	$^1D_2-^1S_0$	$0.175\ T_4^{0.105-0.033\ln T_4}$	*d*
Ar III	$^3P_2-^3P_1$	$4.04\ T_4^{0.031+0.002\ln T_4}$	*e*
"	$^3P_2-^3P_0$	$1.00\ T_4^{0.111-0.009\ln T_4}$	*e*
"	$^3P_1-^3P_0$	$1.42\ T_4^{-0.024-0.016\ln T_4}$	*e*
"	$^3P_2-^1D_2$	$2.94\ T_4^{0.011+0.014\ln T_4}$	*e*
"	$^3P_1-^1D_2$	$1.80\ T_4^{0.006+0.013\ln T_4}$	*e*
"	$^3P_0-^1D_2$	$0.602\ T_4^{0.004+0.014\ln T_4}$	*e*
"	$^3P_2-^1S_0$	$0.359\ T_4^{0.144+0.031\ln T_4}$	*e*
"	$^3P_1-^1S_0$	$0.220\ T_4^{0.004+0.014\ln T_4}$	*e*
"	$^3P_0-^1S_0$	$0.073\ T_4^{0.144+0.065\ln T_4}$	*e*
"	$^1D_2-^1S_0$	$1.18\ T_4^{0.128-0.028\ln T_4}$	*e*

a fit to Bell et al. (1998)
b fit to Zatsarinny & Tayal (2003)
c fit to Butler & Zeippen (1994)
d fit to Hudson et al. (2009)
e fit to Muñoz Burgos et al. (2009)

Table F.4 Electron Collision Strengths for $2s^2 2p^3$ Ions

Ion	$\ell - u$	$\Omega_{u\ell}$	Note
N I	$^4S^o_{3/2}-^2D^o_{5/2}$	$0.337\, T_4^{0.723-0.129\ln T_4}$	*a*
”	$^4S^o_{3/2}-^2D^o_{3/2}$	$0.224\, T_4^{0.726-0.125\ln T_4}$	*a*
”	$^2D^o_{5/2}-^2D^o_{3/2}$	$0.257\, T_4^{0.960-0.009\ln T_4}$	*a*
”	$^4S^o_{3/2}-^2P^o_{1/2}$	$0.055\, T_4^{0.759-0.140\ln T_4}$	*a*
”	$^4S^o_{3/2}-^2P^o_{3/2}$	$0.109\, T_4^{0.759-0.134\ln T_4}$	*a*
”	$^2D^o_{5/2}-^2P^o_{1/2}$	$0.139\, T_4^{0.559-0.009\ln T_4}$	*a*
”	$^2D^o_{5/2}-^2P^o_{3/2}$	$0.366\, T_4^{0.499+0.015\ln T_4}$	*a*
”	$^2D^o_{3/2}-^2P^o_{1/2}$	$0.141\, T_4^{0.476+0.030\ln T_4}$	*a*
”	$^2D^o_{3/2}-^2P^o_{3/2}$	$0.195\, T_4^{0.547+0.001\ln T_4}$	*a*
”	$^2P^o_{1/2}-^2P^o_{3/2}$	$0.123\, T_4^{0.738-0.059\ln T_4}$	*a*
O II	$^4S^o_{3/2}-^2D^o_{5/2}$	$0.803\, T_4^{0.023-0.008\ln T_4}$	*b*
”	$^4S^o_{3/2}-^2D^o_{3/2}$	$0.550\, T_4^{0.054-0.004\ln T_4}$	*b*
”	$^2D^o_{5/2}-^2D^o_{3/2}$	$1.434\, T_4^{-0.176+0.004\ln T_4}$	*b*
”	$^4S^o_{3/2}-^2P^o_{3/2}$	$0.140\, T_4^{0.025-0.006\ln T_4}$	*b*
”	$^4S^o_{3/2}-^2P^o_{1/2}$	$0.283\, T_4^{0.023-0.004\ln T_4}$	*b*
”	$^2D^o_{5/2}-^2P^o_{3/2}$	$0.349\, T_4^{0.060+0.052\ln T_4}$	*b*
”	$^2D^o_{5/2}-^2P^o_{1/2}$	$0.832\, T_4^{0.076+0.055\ln T_4}$	*b*
”	$^2D^o_{3/2}-^2P^o_{3/2}$	$0.326\, T_4^{0.063+0.052\ln T_4}$	*b*
”	$^2D^o_{3/2}-^2P^o_{1/2}$	$0.485\, T_4^{0.059+0.052\ln T_4}$	*b*
”	$^2P^o_{3/2}-^2P^o_{1/2}$	$0.322\, T_4^{0.019+0.037\ln T_4}$	*b*
Ne IV	$^4S^o_{3/2}-^2D^o_{5/2}$	$0.88\, T_4^{-0.080+0.007\ln T_4}$	*c*
”	$^4S^o_{3/2}-^2D^o_{3/2}$	$0.59\, T_4^{-0.091+0.012\ln T_4}$	*c*
”	$^4S^o_{3/2}-^2P^o_{1/2}$	$0.15\, T_4^{-0.006-0.005\ln T_4}$	*c*
”	$^4S^o_{3/2}-^2P^o_{3/2}$	$0.30\, T_4^{-0.003-0.005\ln T_4}$	*c*
”	$^2D^o_{5/2}-^2D^o_{3/2}$	$1.27\, T_4^{-0.013+0.017\ln T_4}$	*c*
”	$^2D^o_{5/2}-^2P^o_{1/2}$	$0.365\, T_4^{0.009+0.023\ln T_4}$	*c*
”	$^2D^o_{3/2}-^2P^o_{3/2}$	$0.90\, T_4^{0.007+0.035\ln T_4}$	*c*
”	$^2D^o_{3/2}-^2P^o_{1/2}$	$0.34\, T_4^{0.008+0.037\ln T_4}$	*c*
”	$^2D^o_{5/2}-^2P^o_{3/2}$	$0.51\, T_4^{-0.013+0.033\ln T_4}$	*c*
”	$^2P^o_{1/2}-^2P^o_{3/2}$	$0.35\, T_4^{0.099-0.014\ln T_4}$	*c*

a fit to Tayal (2006*b*)
b fit to Tayal (2007)
c fit to Ramsbottom et al. (1998)

Table F.5 Electron Collision Strengths for $3s^2 3p^3$ Ions

Ion	$\ell - u$	$\Omega_{u\ell}$	Note
S II	${}^4\mathrm{S}^\mathrm{o}_{3/2}{-}{}^2\mathrm{D}^\mathrm{o}_{3/2}$	$2.56T_4^{-0.071-0.023\ln T_4}$	*a*
”	${}^4\mathrm{S}^\mathrm{o}_{3/2}{-}{}^2\mathrm{D}^\mathrm{o}_{5/2}$	$3.83T_4^{-0.070-0.022\ln T_4}$	*a*
”	${}^4\mathrm{S}^\mathrm{o}_{3/2}{-}{}^2\mathrm{P}^\mathrm{o}_{1/2}$	$0.704T_4^{0.042+0.006\ln T_4}$	*a*
”	${}^4\mathrm{S}^\mathrm{o}_{3/2}{-}{}^2\mathrm{P}^\mathrm{o}_{5/2}$	$1.42T_4^{0.041-0.001\ln T_4}$	*a*
”	${}^2\mathrm{D}^\mathrm{o}_{3/2}{-}{}^2\mathrm{D}^\mathrm{o}_{5/2}$	$6.89T_4^{-0.103-0.022\ln T_4}$	*a*
”	${}^2\mathrm{D}^\mathrm{o}_{3/2}{-}{}^2\mathrm{P}^\mathrm{o}_{1/2}$	$1.47T_4^{0.014\ln T_4}$	*a*
”	${}^2\mathrm{D}^\mathrm{o}_{3/2}{-}{}^2\mathrm{P}^\mathrm{o}_{3/2}$	$2.39T_4^{-0.006}$	*a*
”	${}^2\mathrm{D}^\mathrm{o}_{5/2}{-}{}^2\mathrm{P}^\mathrm{o}_{1/2}$	$1.78T_4^{-0.012-0.006\ln T_4}$	*a*
”	${}^2\mathrm{D}^\mathrm{o}_{5/2}{-}{}^2\mathrm{P}^\mathrm{o}_{3/2}$	$4.06T_4^{0.005\ln T_4}$	*a*
”	${}^2\mathrm{P}^\mathrm{o}_{1/2}{-}{}^2\mathrm{P}^\mathrm{o}_{3/2}$	$1.80T_4^{0.032-0.001\ln T_4}$	*a*
Ar IV	${}^4\mathrm{S}^\mathrm{o}_{3/2}{-}{}^2\mathrm{D}^\mathrm{o}_{3/2}$	$0.762T_4^{0.012+0.023\ln T_4}$	*b*
”	${}^4\mathrm{S}^\mathrm{o}_{3/2}{-}{}^2\mathrm{D}^\mathrm{o}_{5/2}$	$1.144T_4^{0.012+0.022\ln T_4}$	*b*
”	${}^4\mathrm{S}^\mathrm{o}_{3/2}{-}{}^2\mathrm{P}^\mathrm{o}_{1/2}$	$0.393T_4^{0.014+0.182\ln T_4}$	*b*
”	${}^4\mathrm{S}^\mathrm{o}_{1/2}{-}{}^2\mathrm{P}^\mathrm{o}_{3/2}$	$0.785T_4^{0.013+0.183\ln T_4}$	*b*
”	${}^2\mathrm{D}^\mathrm{o}_{3/2}{-}{}^2\mathrm{D}^\mathrm{o}_{5/2}$	$7.055T_4^{-0.051-0.058\ln T_4}$	*b*
”	${}^2\mathrm{D}^\mathrm{o}_{3/2}{-}{}^2\mathrm{P}^\mathrm{o}_{1/2}$	$1.507T_4^{-0.005+0.043\ln T_4}$	*b*
”	${}^2\mathrm{D}^\mathrm{o}_{3/2}{-}{}^2\mathrm{P}^\mathrm{o}_{3/2}$	$2.139T_4^{0.021+0.010\ln T_4}$	*b*
”	${}^2\mathrm{D}^\mathrm{o}_{5/2}{-}{}^2\mathrm{P}^\mathrm{o}_{1/2}$	$1.533T_4^{0.025+0.005\ln T_4}$	*b*
”	${}^2\mathrm{D}^\mathrm{o}_{5/2}{-}{}^2\mathrm{P}^\mathrm{o}_{3/2}$	$3.939T_4^{0.004+0.032\ln T_4}$	*b*
”	${}^2\mathrm{P}^\mathrm{o}_{1/2}{-}{}^2\mathrm{P}^\mathrm{o}_{3/2}$	$2.065T_4^{0.425-0.052\ln T_4}$	*b*

a fit to Tayal & Zatsarinny (2010)
b fit to Ramsbottom et al. (1997)

Table F.6 Rate Coefficients for Fine-Structure Deexcitation by $X = H^+$, H, H_2, He

X	Ion	$\ell - u$	$k_{u\ell}(\mathrm{cm}^3\,\mathrm{s}^{-1})$	Note
H^+	C I	$^3P_0 - {}^3P_1$	$7.60\times 10^{-10} T_2^{0.464-0.042\ln T_2}$	*a*
H^+	C I	$^3P_0 - {}^3P_2$	$3.02\times 10^{-10} T_2^{1.217-0.126\ln T_2}$	*a*
H^+	C I	$^3P_1 - {}^3P_2$	$2.30\times 10^{-9} T_2^{0.919-0.111\ln T_2}$	*a*
H	C I	$^3P_0 - {}^3P_1$	$1.26\times 10^{-10} T_2^{0.115+0.057\ln T_2}$	*b*
H	C I	$^3P_0 - {}^3P_2$	$8.90\times 10^{-11} T_2^{0.228+0.046\ln T_2}$	*b*
H	C I	$^3P_1 - {}^3P_2$	$2.64\times 10^{-10} T_2^{0.231+0.046\ln T_2}$	*b*
H_2(para)	C I	$^3P_0 - {}^3P_1$	$0.67\times 10^{-10} T_2^{-0.085+0.102\ln T_2}$	*c*
H_2(ortho)	C I	$^3P_0 - {}^3P_1$	$0.71\times 10^{-10} T_2^{-0.004+0.049\ln T_2}$	*c*
H_2(para)	C I	$^3P_0 - {}^3P_2$	$0.86\times 10^{-10} T_2^{-0.010+0.048\ln T_2}$	*c*
H_2(ortho)	C I	$^3P_0 - {}^3P_2$	$0.69\times 10^{-10} T_2^{0.169+0.038\ln T_2}$	*c*
H_2(para)	C I	$^3P_1 - {}^3P_2$	$1.75\times 10^{-10} T_2^{0.072+0.064\ln T_2}$	*c*
H_2(ortho)	C I	$^3P_1 - {}^3P_2$	$1.48\times 10^{-10} T_2^{0.263+0.031\ln T_2}$	*c*
He	C I	$^3P_0 - {}^3P_1$	$1.76\times 10^{-11} T_2^{0.106-0.036\ln T_2}$	*d*
He	C I	$^3P_0 - {}^3P_2$	$4.44\times 10^{-11} T_2^{0.029}$	*d*
He	C I	$^3P_1 - {}^3P_2$	$8.33\times 10^{-11} T_2^{0.113+0.029\ln T_2}$	*d*
H	C II	$^2P^o_{3/2} - {}^2P^o_{1/2}$	$7.58\times 10^{-10}\, T_2^{0.128+0.009\ln T_2}$	*e*
H_2(para)	C II	$^2P^o_{3/2} - {}^2P^o_{1/2}$	$4.25\times 10^{-10}\, T_2^{0.124-0.018\ln T_2}$	*f*
H_2(ortho)	C II	$^2P^o_{3/2} - {}^2P^o_{1/2}$	$5.14\times 10^{-10}\, T_2^{0.095+0.023\ln T_2}$	*f*
He	C II	$^2P^o_{3/2} - {}^2P^o_{1/2}$	$0.38\times$ H rate	*g*
H	O I	$^3P_2 - {}^3P_1$	$3.57\times 10^{-10}\, T_2^{0.419-0.003\ln T_2}$	*b*
H	O I	$^3P_2 - {}^3P_0$	$3.19\times 10^{-10}\, T_2^{0.369-0.006\ln T_2}$	*b*
H	O I	$^3P_1 - {}^3P_0$	$4.34\times 10^{-10}\, T_2^{0.755-0.160\ln T_2}$	*b*
H_2(para)	O I	$^3P_2 - {}^3P_1$	$1.49\times 10^{-10} T_2^{0.369-0.026\ln T_2}$	*h*
H_2(ortho)	O I	$^3P_2 - {}^3P_1$	$1.37\times 10^{-10} T_2^{0.395-0.005\ln T_2}$	*h*
H_2(para)	O I	$^3P_2 - {}^3P_0$	$2.37\times 10^{-10} T_2^{0.255+0.016\ln T_2}$	*h*
H_2(ortho)	O I	$^3P_2 - {}^3P_0$	$2.23\times 10^{-10} T_2^{0.284+0.035\ln T_2}$	*h*
H_2(para)	O I	$^3P_1 - {}^3P_0$	$2.10\times 10^{-12}\, T_2^{1.117+0.070\ln T_2}$	*h*
H_2(ortho)	O I	$^3P_1 - {}^3P_0$	$3.00\times 10^{-12}\, T_2^{0.792+0.188\ln T_2}$	*h*
H	O I	$^1D_2 - {}^3P_J$	$1.21\times 10^{-13}(2J+1) T_4^{-0.045-0.078\ln T_4}$	*i*
H	Si II	$^2P^o_{3/2} - {}^2P^o_{1/2}$	$5.40\times 10^{-10} T_2^{0.152+0.034\ln T_2}$	*e*
H_2	Si II	$^2P^o_{3/2} - {}^2P^o_{1/2}$	$0.40\times$ H rate	*g*
He	Si II	$^2P^o_{3/2} - {}^2P^o_{1/2}$	$0.29\times$ H rate	*g*

a fit to Roueff & Le Bourlot (1990)
b fit to Abrahamsson et al. (2007)
c fit to Schroder et al. (1991)
d fit to Staemmler & Flower (1991)
e fit to Barinovs et al. (2005)
f fit to Flower & Launay (1977)
g assume $k_{u\ell} \propto \sqrt{\alpha_N/\mu}$
h fit to Jaquet et al. (1992)
i fit to Krems et al. (2006)

Table F.7 References for Rate Coefficients for Collisional Excitation of H_2

Collision		Reference
$H_2 + H$	experiment	Heidner & Kasper (1972)
	theory	Wrathmall & Flower (2006)
	theory	Wrathmall et al. (2007)
$H_2 + H_2$	experiment	Audibert et al. (1974)
	theory	Quéméner & Balakrishnan (2009)
$H_2 + e$	review	Tawara et al. (1990)

Appendix G

Semiclassical Atom

The absorption and emission properties of atoms are, of course, obtained from quantum-mechanical calculations. Nevertheless, it is instructive to note the close correspondence between quantum-mechanical results and a simple classical model.

Consider a charge q with mass m, moving in a "harmonic" potential:

$$U(r) = \frac{1}{2} m\omega_0^2 \left(x^2 + y^2 + z^2\right) \quad . \tag{G.1}$$

We want to include a "drag" force that represents the energy lost in electromagnetic radiation by the oscillating charge. Consider a drag force proportional to velocity:

$$\vec{F}_{\rm drag} = -\gamma m \vec{v} \quad , \tag{G.2}$$

where γ is a constant. The time-averaged energy loss to the drag force $-\gamma m\vec{v}$ is

$$\langle P \rangle = \gamma m \langle v^2 \rangle = \frac{1}{2} \gamma m \omega^2 |x_0|^2 \quad . \tag{G.3}$$

Classically, a charge undergoing simple harmonic motion $x = {\rm Re}[x_0 {\rm e}^{-i\omega t}]$, radiates a time-averaged power $P_{\rm rad} = \omega^4 q^2 |x_0|^2/(3c^3)$. Therefore, if we set

$$\gamma = \frac{2}{3} \frac{\omega_0^2 q^2}{mc^3} \quad , \tag{G.4}$$

the damping term will mimic the energy loss due to radiation.

Now, consider the response of the oscillator to an incident electromagnetic wave producing an electric field at the location of the atom:

$$\vec{E} = \hat{x}\, {\rm Re}\left[E_0 {\rm e}^{-i\omega t}\right] \quad . \tag{G.5}$$

The charged particle obeys the equation of motion

$$m\ddot{x} = -m\omega_0^2 x + qE_0 {\rm e}^{-i\omega t} - \gamma m \dot{x} \quad . \tag{G.6}$$

We seek a periodic solution $x_0 = {\rm e}^{-i\omega t}$; substituting this into (G.6) gives us

$$x_0 = \frac{qE_0}{m(\omega_0^2 - \omega^2) - im\omega\gamma} \quad . \tag{G.7}$$

Undergoing this periodic motion, the applied electric field E does work on the particle at the same rate (averaged over a cycle) that the particle loses energy to the damping force, given by Eq. (G.3). The incident electromagnetic wave has time-averaged energy density

$$\langle u \rangle = \frac{\langle E^2 \rangle}{8\pi} + \frac{\langle B^2 \rangle}{8\pi} = \frac{E_0^2}{8\pi} \quad . \tag{G.8}$$

The time-averaged rate of absorption of energy from the electromagnetic field can be written as the product of the energy flux $\langle u \rangle c$ and the absorption cross section $C_{\rm abs}$. The cross section for absorption of energy from the electromagnetic wave is

$$C_{\rm abs} = \frac{\langle P \rangle}{\langle u \rangle c} = \frac{4\pi q^2}{mc} \frac{\gamma \omega^2}{(\omega - \omega_0)^2 (\omega + \omega_0)^2 + \gamma^2 \omega^2} \tag{G.9}$$

$$= \frac{4\pi q^2}{mc} \frac{\gamma}{\omega_0^2 \left(\frac{\omega}{\omega_0} - \frac{\omega_0}{\omega} \right)^2 + \gamma^2} \quad . \tag{G.10}$$

This is called a **Drude absorption profile**. If $|\omega - \omega_0| \ll \omega_0$, then the profile can be approximated by a **Lorentz line profile**:

$$C_{\rm abs} \approx \frac{4\pi q^2}{mc} \frac{\gamma}{4(\omega - \omega_0)^2 + \gamma^2} \quad . \tag{G.11}$$

For comparison, quantum mechanics gives a Lorentz line profile

$$C_{\rm abs} \approx \frac{4\pi e^2}{m_e c} f_{\ell u} \frac{\gamma}{4(\omega - \omega_{\ell u})^2 + \gamma_{\ell u}^2} \quad , \tag{G.12}$$

which is identical to the classical result (G.11) except for the factor $f_{\ell u}$. Our classical harmonic oscillator has only one transition frequency. For the quantum atom, the electron moves in an anharmonic potential, and there are many transition frequencies, with the sum rule $\sum_u f_{\ell u} = 1$. What about the line width? The quantum damping parameter is

$$\gamma_{\ell u} = A_{u\ell} = \frac{2e^2 \omega_{\ell u}^2}{m_e c^3} \frac{g_\ell}{g_u} f_{\ell u} \quad . \tag{G.13}$$

Thus, we see that the quantum and classical results agree if

$$f_{\ell u} \leftrightarrow 1 \qquad \text{and} \qquad \frac{1}{3} \leftrightarrow \frac{g_\ell}{g_u} \quad . \tag{G.14}$$

We note that the $1s - 2p$ Lyman α transition of hydrogenic ions has $g_\ell / g_u = 1/3$. Therefore, the quantum mechanical result for the absorption cross section of atoms has a close correspondence to a simple classical damped harmonic oscillator.

Appendix H

Debye Length for a Plasma

It is often convenient to assume that the ions and electrons in a plasma are randomly located, but on large scales this cannot be correct, because a plasma must maintain overall electric neutrality – the positions of ions and electrons must be correlated.

Consider a proton–electron plasma, with average densities $n_{e0} = n_{p0}$ of electrons and protons. Choose a coordinate system centered on the position of a particular proton. The Coulomb potential contributed by the proton will result in an increased density of electrons (and a decreased density of other protons) in the neighborhood of our chosen proton, resulting in an excess of negative charge around the proton. Let $Q(r)$ be the expected charge within a region of radius r around the proton. Considering only the potential from the first proton, we can estimate the net charge density around it by assuming thermodynamic equilibrium:

$$n_e(r) = n_{e0}\mathrm{e}^{e^2/rkT} \quad , \quad n_p(r) = n_{p0}\mathrm{e}^{-e^2/rkT} \quad . \tag{H.1}$$

Because $n_{p0} = n_{e0}$, the net charge density is

$$\rho(r) = -en_{e0}\left[\mathrm{e}^{e^2/rkT} - \mathrm{e}^{-e^2/rkT}\right] \quad . \tag{H.2}$$

The net shielding charge within a radius R is given by

$$Q(R) = -en_{e0}\int_0^R 4\pi r^2 dr\left[\mathrm{e}^{e^2/rkT} - \mathrm{e}^{-e^2/rkT}\right] \quad . \tag{H.3}$$

Now suppose that $r \gg e^2/kT$. Then we expand the exponentials to obtain

$$Q(R) = -en_{e0}\int_0^R 4\pi r^2 dr\left[1 + \frac{e^2}{rkT} - \left(1 - \frac{e^2}{rkT}\right)\right] \tag{H.4}$$

$$= -4\pi en_{e0}\int_0^R \frac{2e^2}{rkT}r^2 dr = -4\pi en_{e0}\frac{e^2}{kT}R^2 \quad . \tag{H.5}$$

Shielding of the original proton charge by the induced negative charge density will be effective beyond a length scale L_{D} defined by $Q(L_{\mathrm{D}}) = -e$. This gives the **Debye length**:

$$L_{\mathrm{D}} = \left(\frac{kT}{4\pi n_e e^2}\right)^{1/2} = 690\, T_4^{1/2}(n_e/\,\mathrm{cm}^{-3})^{-1/2}\,\mathrm{cm} \quad . \tag{H.6}$$

Appendix I

Heuristic Model for Ion–Electron Inelastic Scattering

Consider an ion I^{+Z}, with net charge Ze, in an excited state u – the bound electrons are in an eigenstate of the potential $U(r)$ due to the nucleus and the other electrons. Let E_u and E_ℓ be the energy of the excited state and a lower level, and $E_{u\ell} \equiv E_u - E_\ell$.

If the potential $U(r)$ is now perturbed by $\delta U \gtrsim E_{u\ell}$, and the perturbation δU is maintained for a time $\sim h/E_{u\ell}$ and then removed, the electron wave function will have a substantial probability of ending up in a lower level ℓ.

The perturbation δU can be provided by a passing electron, with initial energy $\sim kT$, drawn into the Coulomb potential of the ion. If the electron passes within a distance $\sim a_0$ of the nucleus (where $a_0 \equiv \hbar^2/m_e e^2 = 5.292 \times 10^{-9}$ cm is the **Bohr radius**), then the perturbation due to the passing electron $\delta U \gtrsim E_{ul}$, and the duration of the perturbation will be of order the orbital period of the bound electron (if the initial energy $\sim kT \lesssim Ze^2/a_0$).

Let us then consider a very simple model: Suppose that an incident electron moves classically in a Coulomb potential $-Ze^2/r$ due to the ions, and will produce deexcitation of level u if and only if the distance of closest approach $r_{\rm min} \leq W a_0$, where we expect that the dimensionless number W will be of order unity. For this assumption, let us calculate the thermal rate coefficient.

Let v be the velocity of the incident electron at infinity. We want to relate $r_{\rm min}$ to v and the impact parameter b. Let $v_{\rm max}$ be the velocity at $r = r_{\rm min}$. Conservation of energy:

$$\frac{1}{2} m_e v_{\rm max}^2 = \frac{1}{2} m_e v^2 + \frac{Ze^2}{r_{\rm min}} \quad , \tag{I.1}$$

and conservation of angular momentum:

$$v_{\rm max} r_{\rm min} = v\, b \quad , \tag{I.2}$$

can be combined to obtain an equation for b in terms of $r_{\rm min}$ and v:

$$b = r_{\rm min} \left[1 + \frac{Ze^2/r_{\rm min}}{m_e v^2/2} \right]^{1/2} \quad . \tag{I.3}$$

According to our model, deexcitation will occur if $r_{\rm min} < Wa_0$, or $b < b_{\rm crit}(v)$, where

$$b_{\rm crit}(v) = Wa_0 \left[1 + \frac{Ze^2/Wa_0}{m_e v^2/2}\right]^{1/2} \quad . \tag{I.4}$$

With these assumptions, the cross section for collisional deexcitation is simply

$$\sigma_{u\ell}(v) = \pi b_{\rm crit}^2 = W^2 \pi a_0^2 \left[1 + \frac{Ze^2/Wa_0}{m_e v^2/2}\right] \quad . \tag{I.5}$$

The factor $[1+(Ze^2/Wa_0)/(m_e v^2/2)]$ in Eq. (I.5) represents the effects of "Coulomb focusing," making the cross section larger than the value $(\pi W^2 a_0^2)$ that it would have if the incident electron moved in a straight line.

Integrate (I.5) over a thermal velocity distribution:

$$\begin{aligned}
\langle\sigma v\rangle_{u\to\ell} &= \pi W^2 a_0^2 \left(\frac{m_e}{2\pi kT}\right)^{3/2} \int_0^\infty 4\pi v^2 dv\, {\rm e}^{-m_e v^2/2kT} v \left[1 + \frac{2Ze^2/Wa_0}{m_e v^2}\right] \\
&= \pi W^2 a_0^2 \left(\frac{8kT}{\pi m_e}\right)^{1/2} \left[1 + \frac{Ze^2}{Wa_0 kT}\right] \quad .
\end{aligned} \tag{I.6}$$

Note that

$$\frac{Ze^2}{a_0 kT} = \frac{15.78Z}{T_4} \quad . \tag{I.7}$$

For $T_4 \lesssim Z$, we can neglect the first term in the square bracket, and obtain the deexcitation rate coefficient:

$$\langle\sigma v\rangle_{u\to\ell} \approx \frac{h^2}{(2\pi m_e)^{3/2}} \frac{1}{(kT)^{1/2}} 2WZ \quad . \tag{I.8}$$

Note the similarity of the above Eq. (I.8) to Eq. (2.26) defining the collision strength $\Omega_{u\ell}$:

$$\langle\sigma v\rangle_{u\to\ell} = \frac{h^2}{(2\pi m_e)^{3/2}} \frac{1}{(kT)^{1/2}} \frac{\Omega_{u\ell}}{g_u} \quad . \tag{I.9}$$

This makes it clear why

- Electron-ion deexcitation rates vary approximately as $1/\sqrt{T}$ (i.e., collision strengths $\Omega_{u\ell}$ depend only weakly on T).
- "Collision strengths" $\Omega_{u\ell}$ are of order unity for electron–ion collisions.

Appendix J

Virial Theorem

The proof of the virial theorem is somewhat lengthy, and is provided here because it is not widely available to the student. We switch between vector ($\mathbf{r}$) and tensor (x_i) notation as convenient to clarify the presentation. Define

$$I \equiv \int \rho\, r^2\, dV \quad . \tag{J.1}$$

Then,

$$\begin{aligned}
\frac{dI}{dt} &= \int dV\, r^2 \frac{\partial \rho}{\partial t} + \oint d\mathbf{S} \cdot \left(\mathbf{v}\, \rho r^2\right) \\
&= -\int dV r^2 \nabla \cdot (\rho \mathbf{v}) + \int dV \nabla \cdot \left(\mathbf{v} \rho r^2\right) \\
&= \int dV \rho \mathbf{v} \cdot \nabla r^2 \\
&= \int dV \rho v_i \frac{\partial}{\partial x_i} x_j x_j \\
&= 2 \int dV \rho v_i x_j \frac{\partial}{\partial x_i} x_j = 2 \int dV \rho v_i x_j \delta_{ij} \\
&= 2 \int dV \rho v_i x_i \quad .
\end{aligned} \tag{J.2}$$

$$\begin{aligned}
\frac{1}{2}\frac{d^2 I}{dt^2} &= \int dV\, x_i \frac{\partial}{\partial t}(\rho v_i) + \int d\mathbf{S} \cdot \mathbf{v} \rho v_i x_i \\
&= \int dV x_i \frac{\partial}{\partial t}(\rho v_i) + \int dV \frac{\partial}{\partial x_j}(v_j \rho v_i x_i) \\
&= -\int dV x_i v_i \frac{\partial}{\partial x_j} \rho v_j + \int dV x_i \rho \frac{\partial}{\partial t} v_i + \int dV \frac{\partial}{\partial x_j}(v_j \rho v_i x_i) \\
&= \int dV \rho v_j \frac{\partial}{\partial x_j} x_i v_i + \int dV x_i \rho \frac{\partial}{\partial t} v_i \\
&= \int dV \left[\rho v_j v_i \delta_{ij} + \rho v_j x_i \frac{\partial}{\partial x_j} v_i + \rho x_i \frac{\partial}{\partial t} v_i\right] \\
&= \int dV \rho v^2 + \int dV \rho \mathbf{r} \cdot \left[(\mathbf{v} \cdot \nabla)\mathbf{v} + \frac{\partial}{\partial t}\mathbf{v}\right] \quad .
\end{aligned} \tag{J.3}$$

The first term on the right-hand side is just $2E_{\mathrm{KE}}$. The equation of momentum conservation (35.21) is now used to relate the second integral to the forces acting on the fluid (neglecting viscous stresses):

$$\int dV \rho \mathbf{r} \cdot \left[(\mathbf{v} \cdot \nabla)\mathbf{v} + \frac{\partial}{\partial t}\mathbf{v} \right] = Y_1 + Y_2 + Y_3 \ , \tag{J.4}$$

$$\begin{aligned} Y_1 &\equiv - \int dV \mathbf{r} \cdot \nabla \left(p + \frac{B^2}{8\pi} \right) \\ &= - \int dV \left\{ \nabla \left[\mathbf{r} \left(p + \frac{B^2}{8\pi} \right) \right] - \left(p + \frac{B^2}{8\pi} \right) \nabla \cdot \mathbf{r} \right\} \\ &= - \oint d\mathbf{S} \cdot \mathbf{r} \left(p + \frac{B^2}{8\pi} \right) + 3 \int dV \left(p + \frac{B^2}{8\pi} \right) \ , \end{aligned} \tag{J.5}$$

$$\begin{aligned} Y_2 &\equiv \frac{1}{4\pi} \int dV \mathbf{r} \cdot (\mathbf{B} \cdot \nabla) \mathbf{B} \\ &= \frac{1}{4\pi} \int dV x_i B_j \frac{\partial}{\partial x_j} B_i \\ &= \frac{1}{4\pi} \int dV \frac{\partial}{\partial x_j} (x_i B_i B_j) - \frac{1}{4\pi} \int dV B_i \frac{\partial}{\partial x_j} (x_i B_j) \\ &= \frac{1}{4\pi} \oint d\mathbf{S} \cdot (\mathbf{r} \cdot \mathbf{B}) \mathbf{B} - \frac{1}{4\pi} \int dV B_i B_j \frac{\partial}{\partial x_j} x_i \\ &= \frac{1}{4\pi} \oint d\mathbf{S} \cdot \mathbf{B} (\mathbf{r} \cdot \mathbf{B}) - 2 \int dV \frac{B^2}{8\pi} \ , \end{aligned} \tag{J.6}$$

$$Y_3 \equiv - \int dV \mathbf{r} \cdot \rho \nabla \Phi_{\rm grav} \tag{J.7}$$

where we have used $\nabla \cdot \mathbf{r} = (\partial/\partial x_i) x_i = 3$ and $(\partial/\partial x_j) B_j = 0$ to obtain (J.5) and (J.6). It can be shown that if the only source of $\Phi_{\rm grav}$ is mass *within* V, then

$$Y_3 = E_{\rm grav} = \frac{1}{2} \int dV_1 \int dV_2 \, G \, \frac{\rho(\mathbf{r}_1) \rho(\mathbf{r}_2)}{|\mathbf{r}_1 - \mathbf{r}_2|} \ . \tag{J.8}$$

This completes the proof of the virial theorem:

$$\frac{1}{2} \ddot{I} = 2E_{\rm KE} + 3(\Pi - \Pi_0) + (E_{\rm mag} - E_{\rm mag,0}) + E_{\rm grav} \ , \tag{J.9}$$

where

$$E_{\rm KE} \equiv \int \frac{\rho v^2}{2} dV \quad , \tag{J.10}$$

$$\Pi \equiv \int p\, dV \quad , \tag{J.11}$$

$$E_{\rm mag} \equiv \int \frac{B^2}{8\pi} dV \quad , \tag{J.12}$$

$$\Pi_0 = \frac{1}{3} \oint d\mathbf{S} \cdot \mathbf{r} p \qquad \to p_0 V \;\; \text{if } p = p_0 \text{ on S} \quad , \tag{J.13}$$

$$E_{\rm mag,0} \equiv \oint d\mathbf{S} \cdot \left[\mathbf{r} \frac{B^2}{8\pi} - \frac{\mathbf{B}\,(\mathbf{r} \cdot \mathbf{B})}{4\pi} \right] \qquad \to \frac{B_0^2}{8\pi} V \;\; \text{if } \mathbf{B} = \mathbf{B}_0 \text{ on S.} \tag{J.14}$$

Bibliography

Abazajian, K.N., Adelman-McCarthy, J.K., Agüeros, M.A., et al., 2009, "The Seventh Data Release of the Sloan Digital Sky Survey." *Ap. J. Suppl.*, **182**, 543

Abrahamsson, E., Krems, R.V., & Dalgarno, A., 2007, "Fine-Structure Excitation of O I and C I by Impact with Atomic Hydrogen." *Ap. J.*, **654**, 1171

Ackermann, M., Ajello, M., Atwood, W.B., et al., 2010, "Fermi LAT observations of cosmic-ray electrons from 7 GeV to 1 TeV." *Phys. Rev. D*, **82**, 092004

Aggarwal, K.M. & Keenan, F.P., 1999, "Excitation Rate Coefficients for Fine-Structure Transitions in O III." *Ap. J. Suppl.*, **123**, 311

Ali-Haïmoud, Y., Hirata, C.M., & Dickinson, C., 2009, "A refined model for spinning dust radiation." *M.N.R.A.S.*, **395**, 1055

Allen, C.W., 1973, *Astrophysical quantities, 3rd ed.* (London: University of London, Athlone Press)

Allers, K.N., Jaffe, D.T., Lacy, J.H., Draine, B.T., & Richter, M.J., 2005, "H_2 Pure Rotational Lines in the Orion Bar." *Ap. J.*, **630**, 368

Allison, A.C. & Dalgarno, A., 1969, "Spin Change in Collisions of Hydrogen Atoms." *Ap. J.*, **158**, 423

Altun, Z., Yumak, A., Badnell, N.R., Colgan, J., & Pindzola, M.S., 2004, "Dielectronic recombination data for dynamic finite-density plasmas. VI. The boron isoelectronic sequence." *Astr. Astrophys.*, **420**, 775

André, P., Ward-Thompson, D., & Barsony, M., 2000, "From Prestellar Cores to Protostars: the Initial Conditions of Star Formation." In "Protostars and Planets IV," (edited by V. Mannings, A.P. Boss, & S.S. Russell), 59–96 (Tucson: University of Arizona Press)

Angus, J.C. & Hayman, C.C., 1988, "Low-Pressure, Metastable Growth of Diamond and "Diamondlike" Phases." *Science*, **241**, 913

Arendt, R.G., Dwek, E., & Moseley, S.H., 1999, "Newly Synthesized Elements and Pristine Dust in the Cassiopeia A Supernova Remnant." *Ap. J.*, **521**, 234

Arendt, R.G., Odegard, N., Weiland, J.L., et al., 1998, "The COBE Diffuse Infrared Background Experiment Search for the Cosmic Infrared Background. III. Separation of Galactic Emission from the Infrared Sky Brightness." *Ap. J.*, **508**, 74

Armstrong, B., 1967, "Spectrum line profiles: The Voigt function." *J. Quant. Spectr. Rad. Trans.*, **7**, 61

Armstrong, J.W., Rickett, B.J., & Spangler, S.R., 1995, "Electron Density Power Spectrum in the Local Interstellar Medium." *Ap. J.*, **443**, 209

Asplund, M., Grevesse, N., Sauval, A.J., & Scott, P., 2009, "The chemical composition of the Sun." *Ann. Rev. Astr. Astrophys.*, **47**, 481

Audibert, M.M., Joffrin, C., & Ducuing, J., 1974, "Vibrational Relaxation of H_2 in the Range 500-40K." *Chem. Phys. Lett.*, **25**, 158

Axford, W.I., Leer, E., & Skadron, G., 1977, "The Acceleration of Cosmic Rays by Shock Waves." In "International Cosmic Ray Conference, 15th; Conference Papers, Vol. 11," 132–137 (Sofia: B'lgarska Akademiia na Naukite)

Badnell, N.R., 2006, "Radiative Recombination Data for Modeling Dynamic Finite-Density Plasmas." *Ap. J. Suppl.*, **167**, 334

Baines, M.J., Williams, I.P., & Asebiomo, A.S., 1965, "Resistance to the motion of a small sphere moving through a gas." *M.N.R.A.S.*, **130**, 63

Baker, J.G. & Menzel, D.H., 1938, "Physical Processes in Gaseous Nebulae. III. The Balmer Decrement." *Ap. J.*, **88**, 52

Balbus, S.A. & Hawley, J.F., 1991, "A powerful local shear instability in weakly magnetized disks. I - Linear analysis. II - Nonlinear evolution." *Ap. J.*, **376**, 214

Baldwin, J.A., Ferland, G.J., Martin, P.G., et al., 1991, "Physical Conditions in the Orion Nebula and an Assessment of its Helium Abundance." *Ap. J.*, **374**, 580

Baldwin, J.A., Phillips, M.M., & Terlevich, R., 1981, "Classification Parameters for the Emission-Line Spectra of Extragalactic Objects." *Publ. Astr. Soc. Pacific*, **93**, 5

Bania, T.M., Anderson, L.D., Balser, D.S., & Rood, R.T., 2010, "The Green Bank Telescope Galactic H II Region Discovery Survey." *Ap. J. Lett.*, **718**, L106

Barinovs, Ğ. & van Hemert, M.C., 2006, "CH^+ Radiative Association." *Ap. J.*, **636**, 923

Barinovs, Ğ., van Hemert, M.C., Krems, R., & Dalgarno, A., 2005, "Fine-Structure Excitation of C^+ and Si^+ by Atomic Hydrogen." *Ap. J.*, **620**, 537

Barnard, E.E., 1907, "On a Nebulous Groundwork

in the Constellation Taurus." *Ap. J.*, **25**, 218
Barnard, E.E., 1910, "On a Great Nebulous Region and the Question of Absorbing Matter in Space and the Transparency of the Nebulae." *Ap. J.*, **31**, 8
Bautista, M.A., Quinet, P., Palmeri, P., et al., 2009, "Radiative transition rates and collision strengths for Si II." *Astr. Astrophys.*, **508**, 1527
Bell, A.R., 1978, "The acceleration of cosmic rays in shock fronts. I." *M.N.R.A.S.*, **182**, 147
Bell, A.R., 2004, "Turbulent amplification of magnetic field and diffusive shock acceleration of cosmic rays." *M.N.R.A.S.*, **353**, 550
Bell, K.L., Berrington, K.A., & Thomas, M.R.J., 1998, "Electron impact excitation of the ground-state ^{3}P fine-structure levels in atomic oxygen." *M.N.R.A.S.*, **293**, L83
Bell, K.L., Gilbody, H.B., Hughes, J.G., Kingston, A.E., & Smith, F.J., 1983, "Recommended Data on the Electron Impact Ionization of Light Atoms and Ions." *J. Phys. Chem. Ref. Data*, **12**, 891
Benjamin, R.A., Skillman, E.D., & Smits, D.P., 1999, "Improving Predictions for Helium Emission Lines." *Ap. J.*, **514**, 307
Bennett, C.L., Fixsen, D.J., Hinshaw, G., et al., 1994, "Morphology of the interstellar cooling lines detected by COBE." *Ap. J.*, **434**, 587
Bergin, E.A. & Tafalla, M., 2007, "Cold Dark Clouds: The Initial Conditions for Star Formation." *Ann. Rev. Astr. Astrophys.*, **45**, 339
Bethe, H., 1933, *Handbuch der Physik*, vol. 24, PartI (Berlin: Springer)
Bhat, N.D.R., Cordes, J.M., Camilo, F., Nice, D.J., & Lorimer, D.R., 2004, "Multifrequency Observations of Radio Pulse Broadening and Constraints on Interstellar Electron Density Microstructure." *Ap. J.*, **605**, 759
Binney, J. & Merrifield, M., 1998, *Galactic Astronomy* (Princeton, NJ: Princeton Univ. Press), ISBN 0-691-00402-1
Binney, J. & Tremaine, S., 2008, *Galactic Dynamics, Second Edition* (Princeton, NJ: Princeton Univ. Press), ISBN 978-0-691-13026-2
Black, J.H. & van Dishoeck, E.F., 1991, "Electron densities and the excitation of CN in molecular clouds." *Ap. J. Lett.*, **369**, L9
Blandford, R.D. & Ostriker, J.P., 1978, "Particle Acceleration by Astrophysical Shocks." *Ap. J. Lett.*, **221**, L29
Blitz, L., Fukui, Y., Kawamura, A., et al., 2007, "Giant Molecular Clouds in Local Group Galaxies." In "Protostars and Planets V," (edited by B. Reipurth, D. Jewitt, & K. Keil), 81–96 (Tucson: University of Arizona Press)
Bloemen, J.B.G.M., 1987, "High-Energy Gamma Rays Probing Cosmic-Ray Spectral Differences throughout the Galaxy." *Ap. J. Lett.*, **317**, L15
Bloemen, J.B.G.M., Strong, A.W., Mayer-Hasselwander, H.A., et al., 1986, "The radial distribution of galactic gamma rays. III - The distribution of cosmic rays in the Galaxy and the CO-H_2 calibration." *Astr. Astrophys.*, **154**, 25
Blümer, J., Engel, R., & Hörandel, J.R., 2009, "Cosmic rays from the knee to the highest energies." *Prog. Part. Nucl. Phys.*, **63**, 293
Boehringer, H. & Hensler, G., 1989, "Metallicity-dependence of radiative cooling in optically thin, hot plasmas." *Astr. Astrophys.*, **215**, 147
Bohlin, R.C., Savage, B.D., & Drake, J.F., 1978, "A Survey of Interstellar H I from L-α Absorption Measurements. II." *Ap. J.*, **224**, 132
Bohren, C.F. & Huffman, D.R., 1983, *Absorption and Scattering of Light by Small Particles* (New York: Wiley)
Bonaldi, A., Ricciardi, S., Leach, S., et al., 2007, "WMAP 3-yr data with Correlated Component Analysis: anomalous emission and impact of component separation on the CMB power spectrum." *M.N.R.A.S.*, **382**, 1791
Bonilha, J.R.M., Ferch, R., Salpeter, E.E., Slater, G., & Noerdlinger, P.D., 1979, "Monte Carlo Calculations for Resonance Scattering with Absorption or Differential Expansion." *Ap. J.*, **233**, 649
Bonnor, W.B., 1956, "Boyle's Law and gravitational instability." *M.N.R.A.S.*, **116**, 351
Borkowski, K.J., Balbus, S.A., & Fristrom, C.C., 1990, "Radiative Magnetized Thermal Conduction Fronts." *Ap. J.*, **355**, 501
Borkowski, K.J., Reynolds, S.P., Green, D.A., et al., 2010, "Radioactive Scandium in the Youngest Galactic Supernova Remnant G1.9+0.3." *Ap. J. Lett.*, **724**, L161
Born, M. & Wolf, E., 1999, *Principles of Optics* (Cambridge: Cambridge Univ. Press), ISBN 0-521-64221
Boulanger, F. & Perault, M., 1988, "Diffuse Infrared Emission from the Galaxy. I - Solar Neighborhood." *Ap. J.*, **330**, 964
Bowen, D.V., Jenkins, E.B., Tripp, T.M., et al., 2008, "The Far Ultraviolet Spectroscopic Explorer Survey of O VI Absorption in the Disk of the Milky Way." *Ap. J. Suppl.*, **176**, 59
Bowen, I.S., 1934, "The Excitation of the Permitted O III Nebular Lines." *Publ. Astr. Soc. Pacific*, **46**, 146
Bowen, I.S., 1935, "The Spectrum and Composition of the Gaseous Nebulae." *Ap. J.*, **81**, 1
Bowen, I.S., 1947, "Excitation by Line Coincidence." *Publ. Astr. Soc. Pacific*, **59**, 196
Bradley, J., Dai, Z.R., Erni, R., et al., 2005, "An Astronomical 2175 Å Feature in Interplanetary Dust Particles." *Science*, **307**, 244
Bransden, B.H. & Joachain, C.J., 2003, *Physics of Atoms and Molecules, second edition* (New York: Prentice Hall), ISBN 0-582-35692-X
Brown, R.L. & Mathews, W.G., 1970, "Theoret-

ical Continuous Spectra of Gaseous Nebulae." *Ap. J.*, **160**, 939

Bryans, P., Kreckel, H., Roueff, E., Wakelam, V., & Savin, D.W., 2009, "Molecular Cloud Chemistry and the Importance of Dielectronic Recombination." *Ap. J.*, **694**, 286

Burgess, A., 1965, "Tables of hydrogenic photoionization cross-sections and recombination coefficients." *Mem. Royal Astr. Soc.*, **69**, 1

Butler, K. & Zeippen, C.J., 1994, "Atomic data from the IRON Project. V. Effective collision strengths for transitions in the ground configuration of oxygen-like ions." *Astr. Astrophys. Suppl.*, **108**, 1

Cami, J., Bernard-Salas, J., Peeters, E., & Malek, S.E., 2010, "Detection of C_{60} and C_{70} in a Young Planetary Nebula." *Science*, **329**, 1180

Cardelli, J.A., Clayton, G.C., & Mathis, J.S., 1989, "The Relationship between Infrared, Optical, and Ultraviolet Extinction." *Ap. J.*, **345**, 245

Cardelli, J.A., Mathis, J.S., Ebbets, D.C., & Savage, B.D., 1993, "Abundance of Interstellar Carbon toward Zeta Ophiuchi." *Ap. J. Lett.*, **402**, L17

Cardelli, J.A., Sofia, U.J., Savage, B.D., Keenan, F.P., & Dufton, P.L., 1994, "Interstellar Detection of the Intersystem Line Si II λ 2335 toward zeta Ophiuchi." *Ap. J. Lett.*, **420**, L29

Casassus, S., Dickinson, C., Cleary, K., et al., 2008, "Centimetre-wave continuum radiation from the ρ Ophiuchi molecular cloud." *M.N.R.A.S.*, **391**, 1075

Cesarsky, D., Lequeux, J., Abergel, A., et al., 1996, "Infrared spectrophotometry of NGC 7023 with ISOCAM." *Astr. Astrophys.*, **315**, L305

Chabrier, G., 2003, "The Galactic Disk Mass Function: Reconciliation of the Hubble Space Telescope and Nearby Determinations." *Ap. J. Lett.*, **586**, L133

Chandra, S., Kegel, W.H., & Varshalovich, D.A., 1984, "Einstein A-values for rotational transitions in the H_2O-molecule." *Astr. Astrophys. Suppl.*, **55**, 51

Chepurnov, A. & Lazarian, A., 2010, "Extending the Big Power Law in the Sky with Turbulence Spectra from Wisconsin Hα Mapper Data." *Ap. J.*, **710**, 853

Chernoff, D.F., 1987, "Magnetohydrodynamic Shocks in Molecular Clouds." *Ap. J.*, **312**, 143

Chernoff, D.F., McKee, C.F., & Hollenbach, D.J., 1982, "Molecular Shock Waves in the BN-KL Region of Orion." *Ap. J. Lett.*, **259**, L97

Chiar, J.E. & Tielens, A.G.G.M., 2001, "Circumstellar Carbonaceous Material Associated with Late-Type Dusty WC Wolf-Rayet Stars." *Ap. J. Lett.*, **550**, L207

Cioffi, D.F., McKee, C.F., & Bertschinger, E., 1988, "Dynamics of Radiative Supernova Remnants." *Ap. J.*, **334**, 252

Clayton, G.C., Anderson, C.M., Magalhaes, A.M., et al., 1992, "The First Spectropolarimetric Study of the Wavelength Dependence of Interstellar Polarization in the Ultraviolet." *Ap. J. Lett.*, **385**, L53

Colgan, J., Pindzola, M.S., & Badnell, N.R., 2004, "Dielectronic recombination data for dynamic finite-density plasmas. V: The lithium isoelectronic sequence." *Astr. Astrophys.*, **417**, 1183

Colgan, J., Pindzola, M.S., Whiteford, A.D., & Badnell, N.R., 2003, "Dielectronic recombination data for dynamic finite-density plasmas. III. The beryllium isoelectronic sequence." *Astr. Astrophys.*, **412**, 597

Cordes, J.M. & Lazio, T.J., 2003, "NE2001. I. A New Model for the Galactic Distribution of Free Electrons and Its Fluctuations." *arXiv*, **astro-ph/0207156v3**

Cordes, J.M., Weisberg, J.M., & Boriakoff, V., 1985, "Small-scale electron density turbulence in the interstellar medium." *Ap. J.*, **288**, 221

Cowie, L.L. & McKee, C.F., 1977, "The evaporation of spherical clouds in a hot gas. I - Classical and saturated mass loss rates." *Ap. J.*, **211**, 135

Cowlard, F.C. & Lewis, J.C., 1967, "Vitreous carbon — A new form of carbon." *J. Mat. Sci.*, **2**, 507

Cox, D.P. & Smith, B.W., 1974, "Large-Scale Effects of Supernova Remnants on the Galaxy: Generation and Maintenance of a Hot Network of Tunnels." *Ap. J. Lett.*, **189**, L105

Crinklaw, G., Federman, S.R., & Joseph, C.L., 1994, "The depletion of calcium in the interstellar medium." *Ap. J.*, **424**, 748

Crompton, R.W., Elford, M.T., & Robertson, A.G., 1970, "The momentum transfer cross section for electrons in helium derived from drift velocities at 77 K." *Australian Journal of Physics*, **23**, 667

Crompton, R.W., Gibson, D.K., & McIntosh, A.I., 1969, "The cross section for the $J = 0 \rightarrow 2$ rotational excitation of hydrogen by slow electrons." *Australian Journal of Physics*, **22**, 715

Crovisier, J., 1978, "Kinematics of Neutral Hydrogen Clouds in the Solar Vicinity from the Nançay 21-cm Absorption Survey." *Astr. Astrophys.*, **70**, 43

Crutcher, R.M., 1999, "Magnetic Fields in Molecular Clouds: Observations Confront Theory." *Ap. J.*, **520**, 706

Crutcher, R.M., 2004, "Observations of Magnetic Fields in Molecular Clouds." In "The Magnetized Interstellar Medium," (edited by B. Uyaniker, W. Reich, & R. Wielebinski), 123–132 (Katlenburg-Lindau: Copernicus GmbH)

Crutcher, R.M., 2010, "Role of Magnetic Fields in Star Formation." *Highlights of Astronomy*, **15**, 438

Crutcher, R.M., Wandelt, B., Heiles, C., Falgarone, E., & Troland, T.H., 2010, "Magnetic Fields in Interstellar Clouds from Zeeman Observations: Inference of Total Field Strengths by Bayesian Analysis." *Ap. J.*, **725**, 466

Dalgarno, A., 1976, "The interstellar molecules CH and CH^+." In "Atomic processes and applications," (edited by P.G. Burke), 109–132 (Amsterdam: North-Holland)

Dalgarno, A. & McCray, R.A., 1972, "Heating and Ionization of HI Regions." *Ann. Rev. Astr. Astrophys.*, **10**, 375

Dame, T.M., Hartmann, D., & Thaddeus, P., 2001, "The Milky Way in Molecular Clouds: A New Complete CO Survey." *Ap. J.*, **547**, 792

Dame, T.M., Ungerechts, H., Cohen, R.S., et al., 1987, "A composite CO survey of the entire Milky Way." *Ap. J.*, **322**, 706

Dartois, E., Muñoz Caro, G.M., Deboffle, D., & d'Hendecourt, L., 2004, "Diffuse interstellar medium organic polymers. Photoproduction of the 3.4, 6.85 and 7.25 μm features." *Astr. Astrophys.*, **423**, L33

Das, H.K., Voshchinnikov, N.V., & Il'in, V.B., 2010, "Interstellar extinction and polarization – A spheroidal dust grain approach perspective." *M.N.R.A.S.*, **404**, 265

Davies, R.D., Dickinson, C., Banday, A.J., et al., 2006, "A determination of the spectra of Galactic components observed by the Wilkinson Microwave Anisotropy Probe." *M.N.R.A.S.*, **370**, 1125

Davis, L.J. & Greenstein, J.L., 1951, "The Polarization of Starlight by Aligned Dust Grains." *Ap. J.*, **114**, 206

Day, K.L., 1979, "Mid-infrared optical properties of vapor-condensed magnesium silicates." *Ap. J.*, **234**, 158

de Oliveira-Costa, A., Devlin, M.J., Herbig, T., et al., 1998, "Mapping the Cosmic Microwave Background Anisotropy:Combined Analysis of QMAP Flights." *Ap. J. Lett.*, **509**, L77

de Oliveira-Costa, A., Kogut, A., Devlin, M.J., et al., 1997, "Galactic Microwave Emission at Degree Angular Scales." *Ap. J. Lett.*, **482**, L17

de Oliveira-Costa, A., Tegmark, M., Devlin, M.J., et al., 2000, "Galactic Contamination in the QMAP Experiment." *Ap. J. Lett.*, **542**, L5

de Oliveira-Costa, A., Tegmark, M., Gutierrez, C.M., et al., 1999, "Cross-Correlation of Tenerife Data with Galactic Templates-Evidence for Spinning Dust?" *Ap. J. Lett.*, **527**, L9

de Vries, B.L., Min, M., Waters, L.B.F.M., Blommaert, J.A.D.L., & Kemper, F., 2010, "Determining the forsterite abundance of the dust around asymptotic giant branch stars." *Astr. Astrophys.*, **516**, A86

Debye, P., 1909, "Der Lichtdruck auf Kugeln von beliebigem Material." *Annalen der Physik*, **335**, 57

Dere, K.P., Landi, E., Young, P.R., et al., 2009, "CHIANTI - an atomic database for emission lines. IX. Ionization rates, recombination rates, ionization equilibria for the elements hydrogen through zinc and updated atomic data." *Astr. Astrophys.*, **498**, 915

Desert, F.X., Boulanger, F., & Puget, J.L., 1990, "Interstellar dust models for extinction and emission." *Astr. Astrophys.*, **237**, 215

Dickey, J.M., Strasser, S., Gaensler, B.M., et al., 2009, "The Outer Disk of the Milky Way Seen in λ21 cm Absorption." *Ap. J.*, **693**, 1250

Dickey, J.M., Terzian, Y., & Salpeter, E.E., 1978, "Galactic neutral hydrogen emission-absorption observations from Arecibo." *Ap. J. Suppl.*, **36**, 77

Dickinson, C., Davies, R.D., Allison, J.R., et al., 2009, "Anomalous Microwave Emission from the H II Region RCW175." *Ap. J.*, **690**, 1585

Diehl, R. & Leising, M., 2009, "Gamma-Rays from Positron Annihilation." *arXiv*, **0906.1503v2**

Dobler, G., Draine, B., & Finkbeiner, D.P., 2009, "Constraining Spinning Dust Parameters with the WMAP Five-Year Data." *Ap. J.*, **699**, 1374

Dolginov, A.Z. & Mytrophanov, I.G., 1976, "Orientation of cosmic dust grains." *Ap. Sp. Sci.*, **43**, 291

Dong, R. & Draine, B.T., 2011, "Hα and Free-free Emission from the Warm Ionized Medium." *Ap. J.*, **727**, 35

Dopita, M.A. & Sutherland, R.S., 2000, "The Importance of Photoelectric Heating by Dust in Planetary Nebulae." *Ap. J.*, **539**, 742

Doty, S.D., Tidman, R., Shirley, Y., & Jackson, A., 2010, "Unbiased fitting of B335 dust continuum observations: approach and evidence for variation of grain properties with position." *M.N.R.A.S.*, **406**, 1190

Douglas, A.E. & Herzberg, G., 1941, "Note on CH^+ in Interstellar Space and in the Laboratory." *Ap. J.*, **94**, 381

Draine, B.T., 1978, "Photoelectric Heating of Interstellar Gas." *Ap. J. Suppl.*, **36**, 595

Draine, B.T., 1980, "Interstellar Shock Waves with Magnetic Precursors." *Ap. J.*, **241**, 1021

Draine, B.T., 1986, "Magnetohydrodynamic Shocks in Diffuse Clouds. III - The Line of Sight toward Zeta Ophiuchi." *Ap. J.*, **310**, 408

Draine, B.T., 1989*a*, "Interstellar extinction in the infrared." In "Infrared Spectroscopy in Astronomy, Proceedings of the 22nd Eslab Symposium, ESA SP-290," (edited by B. Kaldeich), 93–98 (Noordwijk: European Space Agency)

Draine, B.T., 1989*b*, "On the Interpretation of the λ 2175 Å Feature." In "IAU Symp. 135: In-

terstellar Dust," (edited by L. Allamandola & A. Tielens), 313–327 (Dordrecht: Kluwer)
Draine, B.T., 1995, "Grain Destruction in Interstellar Shock Waves." *Ap. Sp. Sci.*, **233**, 111
Draine, B.T., 1996, "Optical and Magnetic Properties of Dust Grains." In "*ASP Conf. Ser.* 97, Polarimetry of the Interstellar Medium," (edited by W.G. Roberge & D.C.B. Whittet), 16–28
Draine, B.T., 2003*a*, "Interstellar Dust Grains." *Ann. Rev. Astr. Astrophys.*, **41**, 241
Draine, B.T., 2003*b*, "Scattering by Interstellar Dust Grains. II. X-Rays." *Ap. J.*, **598**, 1026
Draine, B.T., 2004, "Astrophysics of Dust in Cold Clouds." In "The Cold Universe, Saas-Fee Advanced Course 32," (edited by D. Pfenniger & Y. Revaz), 213–303 (Berlin: Springer-Verlag), ISBN 3-540-40838-X
Draine, B.T., 2011, "On Radiation Pressure in Static, Dusty H II Regions." *Ap. J.*, **732**, 100
Draine, B.T. & Allaf-Akbari, K., 2006, "X-Ray Scattering by Nonspherical Grains. I. Oblate Spheroids." *Ap. J.*, **652**, 1318
Draine, B.T. & Anderson, N., 1985, "Temperature Fluctuations and Infrared emission from Interstellar Grains." *Ap. J.*, **292**, 494
Draine, B.T. & Bertoldi, F., 1996, "Structure of Stationary Photodissociation Fronts." *Ap. J.*, **468**, 269
Draine, B.T., Dale, D.A., Bendo, G., et al., 2007, "Dust Masses, PAH Abundances, and Starlight Intensities in the SINGS Galaxy Sample." *Ap. J.*, **663**, 866
Draine, B.T. & Flatau, P.J., 1994, "Discrete-dipole approximation for scattering calculations." *J. Opt. Soc. Am. A*, **11**, 1491
Draine, B.T. & Fraisse, A.A., 2009, "Polarized Far-Infrared and Submillimeter Emission from Interstellar Dust." *Ap. J.*, **696**, 1
Draine, B.T. & Katz, N., 1986*a*, "Magnetohydrodynamic Shocks in Diffuse Clouds. I - Chemical Processes." *Ap. J.*, **306**, 655
Draine, B.T. & Katz, N., 1986*b*, "Magnetohydrodynamic Shocks in Diffuse Clouds. II - Production of CH^+, OH, CH, and Other Species." *Ap. J.*, **310**, 392
Draine, B.T. & Lazarian, A., 1998*a*, "Diffuse Galactic Emission from Spinning Dust Grains." *Ap. J. Lett.*, **494**, L19
Draine, B.T. & Lazarian, A., 1998*b*, "Electric Dipole Radiation from Spinning Dust Grains." *Ap. J.*, **508**, 157
Draine, B.T. & Lee, H.M., 1984, "Optical Properties of Interstellar Graphite and Silicate Grains." *Ap. J.*, **285**, 89
Draine, B.T. & Li, A., 2001, "Infrared Emission from Interstellar Dust. I. Stochastic Heating of Small Grains." *Ap. J.*, **551**, 807
Draine, B.T. & Li, A., 2007, "Infrared Emission from Interstellar Dust. IV. The Silicate-Graphite-PAH Model in the Post-*Spitzer* Era." *Ap. J.*, **657**, 810
Draine, B.T. & Malhotra, S., 1993, "On Graphite and the 2175 Å Extinction Profile." *Ap. J.*, **414**, 632
Draine, B.T. & McKee, C.F., 1993, "Theory of interstellar shocks." *Ann. Rev. Astr. Astrophys.*, **31**, 373
Draine, B.T. & Roberge, W.G., 1982, "A Model for the Intense Molecular Line Emission from OMC-1." *Ap. J. Lett.*, **259**, L91
Draine, B.T. & Salpeter, E.E., 1979, "On the Physics of Dust Grains in Hot Gas." *Ap. J.*, **231**, 77
Draine, B.T. & Sutin, B., 1987, "Collisional Charging of Interstellar Grains." *Ap. J.*, **320**, 803
Draine, B.T. & Weingartner, J.C., 1996, "Radiative Torques on Interstellar Grains. I. Superthermal Spin-up." *Ap. J.*, **470**, 551
Draine, B.T. & Weingartner, J.C., 1997, "Radiative Torques on Interstellar Grains. II. Grain Alignment." *Ap. J.*, **480**, 633
Drake, G.W.F., 1986, "Spontaneous two-photon decay rates in hydrogenlike and heliumlike ions." *Phys. Rev. A*, **34**, 2871
Duley, W.W., Scott, A.D., Seahra, S., & Dadswell, G., 1998, "Integrated Absorbances in the 3.4 Micron CHn Band in Hydrogenated Amorphous Carbon." *Ap. J. Lett.*, **503**, L183
Dwek, E., Arendt, R.G., Bouchet, P., et al., 2010, "Five Years of Mid-infrared Evolution of the Remnant of SN 1987A: The Encounter Between the Blast Wave and the Dusty Equatorial Ring." *Ap. J.*, **722**, 425
Ebert, R., 1957, "Zur Instabilität kugelsymmetrischer Gasverteilungen. Mit 2 Textabbildungen." *Zeitschrift fur Astrophysik*, **42**, 263
Elitzur, M., 1992, "Astronomical masers." *Ann. Rev. Astr. Astrophys.*, **30**, 75
Elitzur, M. & Watson, W.D., 1978, "Formation of molecular CH^+ in interstellar shocks." *Ap. J. Lett.*, **222**, L141
Elmegreen, B.G., 2007, "On the Rapid Collapse and Evolution of Molecular Clouds." *Ap. J.*, **668**, 1064
Epstein, R.I., 1980, "The acceleration of interstellar grains and the composition of the cosmic rays." *M.N.R.A.S.*, **193**, 723
Esteban, C., Peimbert, M., García-Rojas, J., et al., 2004, "A reappraisal of the chemical composition of the Orion nebula based on Very Large Telescope echelle spectrophotometry." *M.N.R.A.S.*, **355**, 229
Falgarone, E., Troland, T.H., Crutcher, R.M., & Paubert, G., 2008, "CN Zeeman measurements in star formation regions." *Astr. Astrophys.*, **487**, 247
Federman, S.R., Sheffer, Y., Lambert, D.L., & Gilliland, R.L., 1993, "Detection of Boron, Cobalt, and Other Weak Interstellar Lines to-

ward Zeta Ophiuchi." *Ap. J. Lett.*, **413**, L51

Felli, M., Churchwell, E., Wilson, T.L., & Taylor, G.B., 1993, "The radio continuum morphology of the Orion Nebula - From 10 arcmin to 0.1 arcsec Resolution." *Astr. Astrophys. Suppl.*, **98**, 137

Ferch, J., Raith, W., & Schroeder, K., 1980, "Total cross section measurements for electron scattering from molecular hydrogen at very low energies." *J. Phys. B*, **13**, 1481

Fermi, E., 1949, "On the Origin of the Cosmic Radiation." *Phys. Rev.*, **75**, 1169

Fernández-Cerezo, S., Gutiérrez, C.M., Rebolo, R., et al., 2006, "Observations of the cosmic microwave background and galactic foregrounds at 12-17GHz with the COSMOSOMAS experiment." *M.N.R.A.S.*, **370**, 15

Fesen, R.A., Hammell, M.C., Morse, J., et al., 2006, "The Expansion Asymmetry and Age of the Cassiopeia A Supernova Remnant." *Ap. J.*, **645**, 283

Feuerbacher, B., Anderegg, M., Fitton, B., et al., 1972, "Photoemission from lunar surface fines and the lunar photoelectron sheath." In "Lunar and Planetary Science Conference Proceedings, Vol. 3," (edited by A. E. Metzger, J. I. Trombka, L. E. Peterson, R. C. Reedy, & J. R. Arnold), 2655–2963

Fiedler, R., Dennison, B., Johnston, K.J., Waltman, E.B., & Simon, R.S., 1994, "A Summary of Extreme Scattering Events and a Descriptive Model." *Ap. J.*, **430**, 581

Fiedler, R.L., Dennison, B., Johnston, K.J., & Hewish, A., 1987, "Extreme scattering events caused by compact structures in the interstellar medium." *Nature*, **326**, 675

Field, G.B., 1958, "Excitation of the Hydrogen 21-cm Line." *Proc. Inst. Radio Engineering*, **46**, 240

Field, G.B., 1959, "The Spin Temperature of Intergalactic Neutral Hydrogen." *Ap. J.*, **129**, 536

Field, G.B., Goldsmith, D.W., & Habing, H.J., 1969, "Cosmic-Ray Heating of the Interstellar Gas." *Ap. J. Lett.*, **155**, L149

Finkbeiner, D.P., 2003, "A Full-Sky Hα Template for Microwave Foreground Prediction." *Ap. J. Suppl.*, **146**, 407

Finkbeiner, D.P., 2004, "Microwave Interstellar Medium Emission Observed by the Wilkinson Microwave Anisotropy Probe." *Ap. J.*, **614**, 186

Finkbeiner, D.P., Langston, G.I., & Minter, A.H., 2004, "Microwave Interstellar Medium Emission in the Green Bank Galactic Plane Survey: Evidence for Spinning Dust." *Ap. J.*, **617**, 350

Finkbeiner, D.P., Schlegel, D.J., Frank, C., & Heiles, C., 2002, "Tentative Detection of Electric Dipole Emission from Rapidly Rotating Dust Grains." *Ap. J.*, **566**, 898

Fitzpatrick, E.L., 1999, "Correcting for the Effects of Interstellar Extinction." *Publ. Astr. Soc. Pacific*, **111**, 63

Fitzpatrick, E.L. & Massa, D., 1986, "An Analysis on the Shapes of Ultraviolet Extinction Curves. I - The 2175 A Bump." *Ap. J.*, **307**, 286

Fixsen, D.J., 2009, "The Temperature of the Cosmic Microwave Background." *Ap. J.*, **707**, 916

Fixsen, D.J., Bennett, C.L., & Mather, J.C., 1999, "COBE Far Infrared Absolute Spectrophotometer Observations of Galactic Lines." *Ap. J.*, **526**, 207

Fixsen, D.J. & Mather, J.C., 2002, "The Spectral Results of the Far-Infrared Absolute Spectrophotometer Instrument on COBE." *Ap. J.*, **581**, 817

Flower, D.R., 2001, "The rotational excitation of CO by H_2 ." *J. Phys. B*, **34**, 2731

Flower, D.R. & Launay, J.M., 1977, "Molecular collision processes. II - Excitation of the fine-structure transition of C^+ in collisions with H_2." *J. Phys. B*, **10**, 3673

Flower, D.R. & Launay, J.M., 1985, "Rate coefficients for the rotational excitation of CO by ortho- and para-H_2." *M.N.R.A.S.*, **214**, 271

Foing, B.H. & Ehrenfreund, P., 1994, "Detection of two interstellar absorption bands coincident with spectral features of C_{60}^+." *Nature*, **369**, 296

Folomeg, B., Rosmus, P., & Werner, H.J., 1987, "Vibration-rotation transition probabilities in CH^+ and CD^+." *Chem. Phys. Lett.*, **136**, 562

Forrey, R.C., Balakrishnan, N., Dalgarno, A., & Lepp, S., 1997, "Quantum Mechanical Calculations of Rotational Transitions in H-H_2 Collisions." *Ap. J.*, **489**, 1000

Fruchter, A., Krolik, J.H., & Rhoads, J.E., 2001, "X-Ray Destruction of Dust along the Line of Sight to γ-Ray Bursts." *Ap. J.*, **563**, 597

Furlanetto, S.R. & Furlanetto, M.R., 2007, "Spin-exchange rates in electron-hydrogen collisions." *M.N.R.A.S.*, **374**, 547

Galazutdinov, G.A., Han, I., & Krełowski, J., 2005, "Profiles of Very Weak Diffuse Interstellar Bands around 6440 Å." *Ap. J.*, **629**, 299

Galazutdinov, G.A., Musaev, F.A., Bondar, A.V., & Krełowski, J., 2003, "Very high resolution profiles of four diffuse interstellar bands." *M.N.R.A.S.*, **345**, 365

Garay, G., Moran, J.M., & Reid, M.J., 1987, "Compact Continuum Radio Sources in the Orion Nebula." *Ap. J.*, **314**, 535

Genzel, R. & Stutzki, J., 1989, "The Orion Molecular Cloud and star-forming region." *Ann. Rev. Astr. Astrophys.*, **27**, 41

Georgelin, Y.M. & Georgelin, Y.P., 1976, "The spiral structure of our Galaxy determined from H II regions." *Astr. Astrophys.*, **49**, 57

Gibb, E.L., Whittet, D.C.B., Boogert, A.C.A., & Tielens, A.G.G.M., 2004, "Interstellar Ice: The Infrared Space Observatory Legacy."

Ap. J. Suppl., **151**, 35

Gillett, F.C., Forrest, W.J., Merrill, K.M., Soifer, B.T., & Capps, R.W., 1975*a*, "The 8-13 Micron Spectra of Compact H II Regions." *Ap. J.*, **200**, 609

Gillett, F.C., Jones, T.W., Merrill, K.M., & Stein, W.A., 1975*b*, "Anisotropy of Constituents of Interstellar Grains." *Astr. Astrophys.*, **45**, 77

Gillmon, K., Shull, J.M., Tumlinson, J., & Danforth, C., 2006, "A FUSE Survey of Interstellar Molecular Hydrogen toward High-Latitude AGNs." *Ap. J.*, **636**, 891

Giz, A.T. & Shu, F.H., 1993, "Parker Instability in a Realistic Gravitational Field." *Ap. J.*, **404**, 185

Glover, S.C., Savin, D.W., & Jappsen, A.K., 2006, "Cosmological Implications of the Uncertainty in H^- Destruction Rate Coefficients." *Ap. J.*, **640**, 553

Godard, B., Falgarone, E., & Pineau Des Forêts, G., 2009, "Models of turbulent dissipation regions in the diffuse interstellar medium." *Astr. Astrophys.*, **495**, 847

Goicoechea, J.R. & Cernicharo, J., 2001, "Far-Infrared Detection of H_3O^+ in Sagittarius B2." *Ap. J. Lett.*, **554**, L213

Goldsmith, D.W., Habing, H.J., & Field, G.B., 1969, "Thermal Properties of Interstellar Gas Heated by Cosmic Rays." *Ap. J.*, **158**, 173

Gondhalekar, P.M., Phillips, A.P., & Wilson, R., 1980, "Observations of the interstellar ultraviolet radiation field from the S2/68 sky-survey telescope." *Astr. Astrophys.*, **85**, 272

Goodman, A.A., Barranco, J.A., Wilner, D.J., & Heyer, M.H., 1998, "Coherence in Dense Cores. II. The Transition to Coherence." *Ap. J.*, **504**, 223

Gorczyca, T.W., Kodituwakku, C.N., Korista, K.T., et al., 2003, "Assessment of the Fluorescence and Auger DataBase Used in Plasma Modeling." *Ap. J.*, **592**, 636

Gould, R.J., 1994, "Radiative Hyperfine Transitions." *Ap. J.*, **423**, 522

Gould, R.J. & Salpeter, E.E., 1963, "The Interstellar Abundance of the Hydrogen Molecule. I. Basic Processes." *Ap. J.*, **138**, 393

Griffin, D.C., Mitnik, D.M., & Badnell, N.R., 2001, "Electron-impact excitation of Ne^+." *J. Phys. B*, **34**, 4401

Gu, M.F., 2004, "Dielectronic Recombination Rate Coefficients of Na-like Ions from Mg II to Zn XX Forming Mg-like Systems." *Ap. J. Suppl.*, **153**, 389

Guhathakurta, P. & Draine, B.T., 1989, "Temperature Fluctuations in Interstellar Grains. I - Computational Method and Sublimation of Small grains." *Ap. J.*, **345**, 230

Gupta, R., Mukai, T., Vaidya, D.B., Sen, A.K., & Okada, Y., 2005, "Interstellar extinction by spheroidal dust grains." *Astr. Astrophys.*, **441**, 555

Habing, H.J., 1968, "The interstellar radiation density between 912 A and 2400 A." *Bull. Astr. Inst. Netherlands*, **19**, 421

Hall, J.S., 1949, "Observations of the Polarized Light from Stars." *Science*, **109**, 166

Hall, J.S. & Mikesell, A.H., 1949, "Observations of polarized light from stars." *A. J.*, **54**, 187

Han, J.L., Manchester, R.N., Lyne, A.G., Qiao, G.J., & van Straten, W., 2006, "Pulsar Rotation Measures and the Large-Scale Structure of the Galactic Magnetic Field." *Ap. J.*, **642**, 868

Haslam, C.G.T., Stoffel, H., Salter, C.J., & Wilson, W.E., 1982, "A 408 MHz all-sky continuum survey. II - The atlas of contour maps." *Astr. Astrophys. Suppl.*, **47**, 1

Heger, M.L., 1922, "Further study of the sodium lines in class B stars ; The spectra of certain class B stars in the regions 5630A-6680A and 3280A-3380A ; Note on the spectrum of γ Cassiopeiae between 5860A and 6600A." *Lick Observatory Bulletin*, **10**, 141

Heidner, R.F. & Kasper, V.V., 1972, "Experimental Rate Constant for $H+H_2(v=1) \rightarrow H+H_2(v=0)$." *Chem. Phys. Lett.*, **15**, 179

Heiles, C., 1989, "Magnetic Fields, Pressures, and Thermally Unstable Gas in Prominent H I Shells." *Ap. J.*, **336**, 808

Heiles, C., 2000, "9286 Stars: An Agglomeration of Stellar Polarization Catalogs." *A. J.*, **119**, 923

Heiles, C. & Crutcher, R., 2005, "Magnetic Fields in Diffuse HI and Molecular Clouds." In "Cosmic Magnetic Fields (Lecture Notes in Physics Vol. 664)," (edited by R. Wielebinski & R. Beck), 137–182 (Berlin: Springer Verlag)

Heiles, C., Goodman, A.A., McKee, C.F., & Zweibel, E.G., 1993, "Magnetic fields in star-forming regions – Observations." In "Protostars and Planets III," (edited by E. H. Levy & J. I. Lunine), 279–326 (Tucson: Univ. of Arizona Press)

Heiles, C. & Troland, T.H., 1982, "Measurements of Magnetic Field Strengths in the Vicinity of Orion." *Ap. J. Lett.*, **260**, L23

Heiles, C. & Troland, T.H., 2003, "The Millennium Arecibo 21 Centimeter Absorption-Line Survey. II. Properties of the Warm and Cold Neutral Media." *Ap. J.*, **586**, 1067

Heiles, C. & Troland, T.H., 2004, "The Millennium Arecibo 21 Centimeter Absorption-Line Survey. III. Techniques for Spectral Polarization and Results for Stokes V." *Ap. J. Suppl.*, **151**, 271

Helder, E.A., Vink, J., Bassa, C.G., et al., 2009, "Measuring the Cosmic-Ray Acceleration Efficiency of a Supernova Remnant." *Science*, **325**, 719

Henning, T., 2010, "Cosmic Silicates." *Ann. Rev. Astr. Astrophys.*, **48**, 21

Henry, R.C., Anderson, R.C., & Fastie, W.G., 1980, "Far-Ultraviolet Studies. VII - The Spectrum and Latitude Dependence of the Local Interstellar Radiation Field." *Ap. J.*, **239**, 859

Herbig, T., Lawrence, C.R., Readhead, A.C.S., & Gulkis, S., 1995, "A Measurement of the Sunyaev-Zel'dovich Effect in the Coma Cluster of Galaxies." *Ap. J. Lett.*, **449**, L5

Herrnstein, J.R., Moran, J.M., Greenhill, L.J., et al., 1999, "A geometric distance to the galaxy NGC 4258 from orbital motions in a nuclear gas disk." *Nature*, **400**, 539

Herschel, W., 1785, "On the Construction of the Heavens." *Philosophical Transactions Series I*, **75**, 213

Heyer, M.H. & Brunt, C.M., 2004, "The Universality of Turbulence in Galactic Molecular Clouds." *Ap. J. Lett.*, **615**, L45

Hillenbrand, L.A. & Hartmann, L.W., 1998, "A Preliminary Study of the Orion Nebula Cluster Structure and Dynamics." *Ap. J.*, **492**, 540

Hiltner, W.A., 1949*a*, "Polarization of Light from Distant Stars by Interstellar Medium." *Science*, **109**, 165

Hiltner, W.A., 1949*b*, "Polarization of Radiation from Distant Stars by the Interstellar Medium." *Nature*, **163**, 283

Ho, L.C., 2008, "Nuclear Activity in Nearby Galaxies." *Ann. Rev. Astr. Astrophys.*, **46**, 475

Hoang, T., Draine, B.T., & Lazarian, A., 2010, "Improving the Model of Emission from Spinning Dust: Effects of Grain Wobbling and Transient Spin-up." *Ap. J.*, **715**, 1462

Hoang, T. & Lazarian, A., 2009, "Grain Alignment Induced by Radiative Torques: Effects of Internal Relaxation of Energy and Complex Radiation Field." *Ap. J.*, **697**, 1316

Hoare, M.G., Kurtz, S.E., Lizano, S., Keto, E., & Hofner, P., 2007, "Ultracompact Hii Regions and the Early Lives of Massive Stars." In "Protostars and Planets V," (edited by B. Reipurth, D. Jewitt, & K. Keil), 181–196

Hobbs, L.M., York, D.G., Thorburn, J.A., et al., 2009, "Studies of the Diffuse Interstellar Bands. III. HD 183143." *Ap. J.*, **705**, 32

Hogg, D.W., 1999, "Distance measures in cosmology." *arXiv*, **astro-ph/9905116v4**

Hollenbach, D., Kaufman, M.J., Bergin, E.A., & Melnick, G.J., 2009, "Water, O_2, and Ice in Molecular Clouds." *Ap. J.*, **690**, 1497

Hollenbach, D. & Salpeter, E.E., 1971, "Surface Recombination of Hydrogen Molecules." *Ap. J.*, **163**, 155

Hollenbach, D.J. & Tielens, A.G.G.M., 1999, "Photodissociation regions in the interstellar medium of galaxies." *Revs. Modern Phys.*, **71**, 173

Houde, M., Dowell, C.D., Hildebrand, R.H., et al., 2004, "Tracing the Magnetic Field in Orion A." *Ap. J.*, **604**, 717

Huber, K.P. & Herzberg, G., 1979, *Molecular Spectra and Molecular Structure IV. Constants of Diatomic Molecules* (New York: Van Nostrand Reinhold)

Hudgins, D.M., Bauschlicher, Jr., C.W., & Allamandola, L.J., 2005, "Variations in the Peak Position of the 6.2 μm Interstellar Emission Feature: A Tracer of N in the Interstellar Polycyclic Aromatic Hydrocarbon Population." *Ap. J.*, **632**, 316

Hudson, C.E. & Bell, K.L., 2005, "Effective collision strengths for fine-structure transitions for the electron impact excitation of N II." *Astr. Astrophys.*, **430**, 725

Hudson, C.E., Ramsbottom, C.A., Norrington, P.H., & Scott, M.P., 2009, "Breit-Pauli R-matrix calculation of fine-structure effective collision strengths for the electron impact excitation of Mg V." *Astr. Astrophys.*, **494**, 729

Hughes, J.P., Gorenstein, P., & Fabricant, D., 1988, "Measurements of the Gas Temperature and Iron Abundance Distribution in the Coma Cluster." *Ap. J.*, **329**, 82

Hummer, D.G., 1988, "A Fast and Accurate Method for Evaluating the Nonrelativistic Free-Free Gaunt Factor for Hydrogenic Ions." *Ap. J.*, **327**, 477

Hummer, D.G. & Storey, P.J., 1987, "Recombination-line intensities for hydrogenic ions. I - Case B calculations for H I and He II." *M.N.R.A.S.*, **224**, 801

Hunter, S.D., Bertsch, D.L., Catelli, J.R., et al., 1997, "EGRET Observations of the Diffuse Gamma-Ray Emission from the Galactic Plane." *Ap. J.*, **481**, 205

Huss, G.R. & Draine, B.T., 2007, "What can presolar grains tell us about the solar nebula?" *Highlights of Astr.*, **14**, 353

Indriolo, N., Geballe, T.R., Oka, T., & McCall, B.J., 2007, "H_3^+ in Diffuse Interstellar Clouds: A Tracer for the Cosmic-Ray Ionization Rate." *Ap. J.*, **671**, 1736

Inoue, A.K. & Kamaya, H., 2010, "Intergalactic dust and its photoelectric heating." *Earth, Planets, and Space*, **62**, 69

Israel, F.P., 1978, "H II regions and CO clouds - The blister model." *Astr. Astrophys.*, **70**, 769

Jacobs, V.L. & Rozsnyai, B.F., 1986, "Multiple ionization and X-ray line emission resulting from inner-shell electron ionization." *Phys. Rev. A*, **34**, 216

Jaeger, C., Mutschke, H., Begemann, B., Dorschner, J., & Henning, T., 1994, "Steps toward interstellar silicate mineralogy. 1: Laboratory results of a silicate glass of mean cosmic composition." *Astr. Astrophys.*, **292**, 641

Jaquet, R., Staemmler, V., Smith, M.D., & Flower,

D.R., 1992, "Excitation of the fine-structure transitions of O(^3P_J) in collisions with ortho- and para-H_2." *Journal of Physics B Atomic Molecular Physics*, **25**, 285

Jeans, J.H., 1928, *Astronomy and cosmogony* (Cambridge, UK: Cambridge Univ. Press)

Jenkins, E.B., 2009, "A Unified Representation of Gas-Phase Element Depletions in the Interstellar Medium." *Ap. J.*, **700**, 1299

Jenkins, E.B. & Shaya, E.J., 1979, "A Survey of Interstellar C I - Insights on Carbon Abundances, UV Grain Albedos, and Pressures in the Interstellar Medium." *Ap. J.*, **231**, 55

Jenkins, E.B. & Tripp, T.M., 2011, "The Distribution of Thermal Pressures in the Diffuse, Cold Neutral Medium of Our Galaxy. II. An Expanded Survey of Interstellar C I Fine-structure Excitations." *Ap. J.*, **734**, 65

Jennings, D.E., Bragg, S.L., & Brault, J.W., 1984, "The v = 0 - 0 spectrum of H_2." *Ap. J. Lett.*, **282**, L85

Jenniskens, P. & Desert, F.X., 1994, "A survey of diffuse interstellar bands (3800-8680 A)." *Astr. Astrophys. Suppl.*, **106**, 39

Jenniskens, P., Mulas, G., Porceddu, I., & Benvenuti, P., 1997, "Diffuse interstellar bands near 9600Å: not due to C_{60}^+ yet." *Astr. Astrophys.*, **327**, 337

Johansson, S. & Letokhov, V.S., 2005, "Astrophysical Lasers in Optical Fe II Lines in Gas Condensations near ϵ Carinae." *Am. Inst. Phys. Conf. Ser.*, **770**, 399

Johns-Krull, C.M., 2007, "The Magnetic Fields of Classical T Tauri Stars." *Ap. J.*, **664**, 975

Johnson, C.T., Burke, P.G., & Kingston, A.E., 1987, "Electron scattering from the fine structure levels within the $1s^22s^22p^2$ ^3P ground state of C I." *Journal of Physics B Atomic Molecular Physics*, **20**, 2553

Johnson, C.T. & Kingston, A.E., 1987, "Electron excitation of the $1s^22s^22p^5$ $^2P_{3/2}$–$^2P_{1/2}$ transition and analysis of the $1s^22s2p^6$ nl resonances for Ne II and Mg IV." *J. Phys. B*, **20**, 5757

Jones, A.P. & D'Hendecourt, L.B., 2004, "Interstellar Nanodiamonds." In "*ASP Conf. Ser.* 309, Astrophysics of Dust," (edited by A.N. Witt, G.C. Clayton, & B.T. Draine), 589–602 (San Francisco, CA: ASP)

Jones, A.P., Duley, W.W., & Williams, D.A., 1990, "The structure and evolution of hydrogenated amorphous carbon grains and mantles in the interstellar medium." *Q.J.R.A.S.*, **31**, 567

Jones, R.V. & Spitzer, L.J., 1967, "Magnetic Alignment of Interstellar Grains." *Ap. J.*, **147**, 943

Jura, M., 1975, "Interstellar Clouds Containing Optically Thin H_2." *Ap. J.*, **197**, 575

Kaastra, J.S. & Mewe, R., 1993, "X-ray emission from thin plasmas. I - Multiple Auger ionisation and fluorescence processes for Be to Zn." *Astr. Astrophys. Suppl.*, **97**, 443

Kalemos, A., Mavridis, A., & Metropoulos, A., 1999, "An accurate description of the ground and excited states of CH." *J. Chem. Phys.*, **111**, 9536

Karzas, W.J. & Latter, R., 1961, "Electron Radiative Transitions in a Coulomb Field." *Ap. J. Suppl.*, **6**, 167

Kauffmann, G., Heckman, T.M., Tremonti, C., et al., 2003, "The host galaxies of active galactic nuclei." *M.N.R.A.S.*, **346**, 1055

Kelly, H.P., 1969, "Frequency-Dependent Polarizability of Atomic Oxygen Calculated by Many-Body Theory." *Phys. Rev.*, **182**, 84

Kemper, F., Vriend, W.J., & Tielens, A.G.G.M., 2004, "The Absence of Crystalline Silicates in the Diffuse Interstellar Medium." *Ap. J.*, **609**, 826

Kemper, F., Vriend, W.J., & Tielens, A.G.G.M., 2005, "Erratum: "The Absence of Crystalline Silicates in the Diffuse Interstellar Medium"." *Ap. J.*, **633**, 534

Kennicutt, Jr., R.C., 1998, "The Global Schmidt Law in Star-forming Galaxies." *Ap. J.*, **498**, 541

Kennicutt, Jr., R.C., Calzetti, D., Walter, F., et al., 2007, "Star Formation in NGC 5194 (M51a). II. The Spatially Resolved Star Formation Law." *Ap. J.*, **671**, 333

Kerr, T.H., Hibbins, R.E., Fossey, S.J., Miles, J.R., & Sarre, P.J., 1998, "Ultrafine Structure in the λ 5797 Diffuse Interstellar Absorption Band." *Ap. J.*, **495**, 941

Knacke, R.F., McCorkle, S., Puetter, R.C., & Erickson, E., 1985, "Interstellar dust spectra between 2.5 and 3.3 microns - A search for hydrated silicates." *A. J.*, **90**, 1828

Koda, J., Scoville, N., Sawada, T., et al., 2009, "Dynamically Driven Evolution of the Interstellar Medium in M51." *Ap. J. Lett.*, **700**, L132

Kogut, A., Banday, A.J., Bennett, C.L., et al., 1996, "Microwave Emission at High Galactic Latitudes in the Four-Year DMR Sky Maps." *Ap. J. Lett.*, **464**, L5

Koike, C. & Tsuchiyama, A., 1992, "Simulation and alteration for amorphous silicates with very broad bands in infrared spectra." *M.N.R.A.S.*, **255**, 248

Koo, B.C. & McKee, C.F., 1992, "Dynamics of Wind Bubbles and Superbubbles. I - Slow Winds and Fast Winds. II - Analytic theory." *Ap. J.*, **388**, 93

Kraetschmer, W. & Huffman, D.R., 1979, "Infrared extinction of heavy ion irradiated and amorphous olivine, with applications to interstellar dust." *Ap. Sp. Sci.*, **61**, 195

Kramers, H.A., 1923, "On the theory of X-ray absorption and of the continuous X-ray spectrum." *Phil. Mag.*, **48**, 836

Kramida, A.E., Ralchenko, Y., Reader, J., & NIST ASD Team, 2014, *NIST Atomic Spectra Database (version 5.2)* (Gaithersburg, MD: National Institute of Standards and Technology), http://physics.nist.gov/asd

Krems, R.V., Jamieson, M.J., & Dalgarno, A., 2006, "The ^{1}D–^{3}P Transitions in Atomic Oxygen Induced by Impact with Atomic Hydrogen." *Ap. J.*, **647**, 1531

Krolik, J.H., McKee, C.F., & Tarter, C.B., 1981, "Two-Phase Models of Quasar Emission Line Regions." *Ap. J.*, **249**, 422

Kroto, H.W. & Jura, M., 1992, "Circumstellar and interstellar fullerenes and their analogues." *Astr. Astrophys.*, **263**, 275

Kroupa, P., 2001, "On the variation of the initial mass function." *M.N.R.A.S.*, **322**, 231

Krymskii, G.F., 1977, "A regular mechanism for the acceleration of charged particles on the front of a shock wave." *Akademiia Nauk SSSR Doklady*, **234**, 1306

Kuijken, K. & Gilmore, G., 1989, "The Mass Distribution in the Galactic Disc - Part III - the Local Volume Mass Density." *M.N.R.A.S.*, **239**, 651

Kulsrud, R.M., 2005, *Plasma Astrophysics* (Princeton: Princeton Univ. Press)

Kwan, J., 1977, "On the Molecular Hydrogen Emission at the Orion Nebula." *Ap. J.*, **216**, 713

La Porta, L., Burigana, C., Reich, W., & Reich, P., 2008, "The impact of Galactic synchrotron emission on CMB anisotropy measurements. I. Angular power spectrum analysis of total intensity all-sky surveys." *Astr. Astrophys.*, **479**, 641

Ladjal, D., Barlow, M.J., Groenewegen, M.A.T., et al., 2010, "Herschel PACS and SPIRE imaging of CW Leo." *Astr. Astrophys.*, **518**, L141

Lai, S., Velusamy, T., Langer, W.D., & Kuiper, T.B.H., 2003, "The Physical and Chemical Status of Pre-Protostellar Core B68." *A. J.*, **126**, 311

Landau, L.D. & Lifshitz, E.M., 1972, *Quantum Mechanics, Non-relativistic Theory* (Oxford: Pergamon Press)

Landau, L.D. & Lifshitz, E.M., 2006, *Fluid Mechanics* (Amsterdam: Elsevier/Butterworth-Heinemann)

Landau, L.D., Lifshitz, E.M., & Pitaevskii, L.P., 1993, *Electrodynamics of Continuous Media* (Oxford: Pergamon Press)

Landi, E. & Landini, M., 1999, "Radiative losses of optically thin coronal plasmas." *Astr. Astrophys.*, **347**, 401

Larson, R.B., 1981, "Turbulence and star formation in molecular clouds." *M.N.R.A.S.*, **194**, 809

Lazarian, A. & Draine, B.T., 1999*a*, "Nuclear Spin Relaxation within Interstellar Grains." *Ap. J. Lett.*, **520**, L67

Lazarian, A. & Draine, B.T., 1999*b*, "Thermal Flipping and Thermal Trapping: New Elements in Grain Dynamics." *Ap. J. Lett.*, **516**, L37

Lazio, T.J.W., Waltman, E.B., Ghigo, F.D., et al., 2001, "A Dual-Frequency, Multiyear Monitoring Program of Compact Radio Sources." *Ap. J. Suppl.*, **136**, 265

Le Bourlot, J., Pineau des Forêts, G., & Flower, D.R., 1999, "The cooling of astrophysical media by H_2." *M.N.R.A.S.*, **305**, 802

Le Petit, F., Roueff, E., & Herbst, E., 2004, "H_3^+ and other species in the diffuse cloud towards ζ Persei: A new detailed model." *Astr. Astrophys.*, **417**, 993

Le Teuff, Y.H., Millar, T.J., & Markwick, A.J., 2000, "The UMIST database for astrochemistry 1999." *Astr. Astrophys. Suppl.*, **146**, 157

Lee, H., 2003, "Asymmetric Deviation of the Scattering Cross Section around Lyα by Atomic Hydrogen." *Ap. J.*, **594**, 637

Leitch, E.M., Readhead, A.C.S., Pearson, T.J., & Myers, S.T., 1997, "An Anomalous Component of Galactic Emission." *Ap. J. Lett.*, **486**, L23

Lennon, D.J. & Burke, V.M., 1994, "Atomic data from the IRON project. II. Effective collision strengths for infrared transitions in carbon-like ions." *Astr. Astrophys. Suppl.*, **103**, 273

Lennon, M.A., Bell, K.L., Gilbody, H.B., et al., 1988, "Recommended Data on the Electron Impact Ionization of Atoms and Ions: Fluorine to Nickel." *J. Phys. Chem. Ref. Data*, **17**, 1285

Lepp, S., 1992, "The Cosmic-Ray Ionization Rate." In "IAU Symp. 150: Astrochemistry of Cosmic Phenomena," 471–475

Li, A. & Draine, B.T., 2001, "On Ultrasmall Silicate Grains in the Diffuse Interstellar Medium." *Ap. J. Lett.*, **550**, L213

Liszt, H., 2001, "The spin temperature of warm interstellar H I." *Astr. Astrophys.*, **371**, 698

Lodders, K., 2003, "Solar System Abundances and Condensation Temperatures of the Elements." *Ap. J.*, **591**, 1220

Lombardi, M., Alves, J., & Lada, C.J., 2006, "2MASS wide field extinction maps. I. The Pipe nebula." *Astr. Astrophys.*, **454**, 781

MacPherson, G.J., Davis, A.M., & Zinner, E.K., 1995, "The distribution of aluminum-26 in the early Solar System - A reappraisal." *Meteoritics*, **30**, 365

Maddalena, R.J., Morris, M., Moscowitz, J., & Thaddeus, P., 1986, "The Large System of Molecular Clouds in Orion and Monoceros." *Ap. J.*, **303**, 375

Malhotra, S., 1995, "The Vertical Distribution and Kinematics of H i and Mass Models of the Galactic Disk." *Ap. J.*, **448**, 138

Mannheim, K. & Schlickeiser, R., 1994, "Interactions of cosmic ray nuclei." *Astr. Astrophys.*,

286, 983

Markova, N. & Puls, J., 2008, "Bright OB stars in the Galaxy. IV. Stellar and wind parameters of early to late B supergiants." *Astr. Astrophys.*, **478**, 823

Markova, N., Puls, J., Repolust, T., & Markov, H., 2004, "Bright OB stars in the Galaxy. I. Mass-loss and wind-momentum rates of O-type stars: A pure Hα analysis accounting for line-blanketing." *Astr. Astrophys.*, **413**, 693

Marlow, W.C., 1965, "Mean polarizability of excited molecular hydrogen." *Proc. Phys. Soc.*, **86**, 731

Maroulis, G., Makris, C., Xenides, D., & Karamanis, P., 2000, "Electric dipole and quadrupole moment and dipole polarizability of CS, SiO and SiS." *Molec. Phys.*, **98**, 481

Martin, D.C., Seibert, M., Neill, J.D., et al., 2007, "A turbulent wake as a tracer of 30,000 years of Mira's mass loss history." *Nature*, **448**, 780

Martin, P.G., 1971, "On interstellar grain alignment by a magnetic field." *M.N.R.A.S.*, **153**, 279

Martins, F., Schaerer, D., & Hillier, D.J., 2005, "A new calibration of stellar parameters of Galactic O stars." *Astr. Astrophys.*, **436**, 1049

Mast, J.W. & Goldstein, Jr., S.J., 1970, "The Motions of High-Latitude Hydrogen Clouds." *Ap. J.*, **159**, 319

Mathis, J.S., Mezger, P.G., & Panagia, N., 1983, "Interstellar radiation field and dust temperatures in the diffuse interstellar matter and in giant molecular clouds." *Astr. Astrophys.*, **128**, 212

Mathis, J.S., Rumpl, W., & Nordsieck, K.H., 1977, "The Size Distribution of Interstellar Grains." *Ap. J.*, **217**, 425

Mazzotta, P., Mazzitelli, G., Colafrancesco, S., & Vittorio, N., 1998, "Ionization balance for optically thin plasmas: Rate coefficients for all atoms and ions of the elements H to NI." *Astr. Astrophys. Suppl.*, **133**, 403

McCall, B.J., Drosback, M.M., Thorburn, J.A., et al., 2010, "Studies of the Diffuse Interstellar Bands. IV. The Nearly Perfect Correlation Between $\lambda\lambda$6196.0 and 6613.6." *Ap. J.*, **708**, 1628

McCall, B.J., Huneycutt, A.J., Saykally, R.J., et al., 2003, "An enhanced cosmic-ray flux towards ζ Persei inferred from a laboratory study of the $H_3^+ - e^-$ recombination rate." *Nature*, **422**, 500

McCall, B.J., Huneycutt, A.J., Saykally, R.J., et al., 2004, "Dissociative recombination of rotationally cold H_3^+." *Phys. Rev. A*, **70**, 052716

McKee, C.F., 1989, "Photoionization-Regulated Star Formation and the Structure of Molecular Clouds." *Ap. J.*, **345**, 782

McKee, C.F. & Ostriker, J.P., 1977, "A Theory of the Interstellar Medium - Three Components Regulated by Supernova Explosions in an Inhomogeneous Substrate." *Ap. J.*, **218**, 148

McKee, C.F. & Williams, J.P., 1997, "The Luminosity Function of OB Associations in the Galaxy." *Ap. J.*, **476**, 144

McKellar, A., 1940, "Evidence for the Molecular Origin of Some Hitherto Unidentified Interstellar Lines." *Publ. Astr. Soc. Pacific*, **52**, 187

Mendez, R.H., Kudritzki, R.P., Herrero, A., Husfeld, D., & Groth, H.G., 1988, "High resolution spectroscopy of central stars of planetary nebulae. I - Basic atmospheric parameters and their interpretation." *Astr. Astrophys.*, **190**, 113

Mendoza, C., 1983, "Recent advances in atomic calculations and experiments of interest in the study of planetary nebulae." In "Planetary Nebulae, IAU Symposium 103," (edited by D. R. Flower), 143–172

Meneguzzi, M., Audouze, J., & Reeves, H., 1971, "The production of the elements Li, Be, B by galactic cosmic rays in space and its relation with stellar observations." *Astr. Astrophys.*, **15**, 337

Mennella, V., Baratta, G.A., Esposito, A., Ferini, G., & Pendleton, Y.J., 2003, "The Effects of Ion Irradiation on the Evolution of the Carrier of the 3.4 Micron Interstellar Absorption Band." *Ap. J.*, **587**, 727

Mennella, V., Brucato, J.R., Colangeli, L., & Palumbo, P., 1999, "Activation of the 3.4 Micron Band in Carbon Grains by Exposure to Atomic Hydrogen." *Ap. J. Lett.*, **524**, L71

Menten, K.M., Reid, M.J., Forbrich, J., & Brunthaler, A., 2007, "The distance to the Orion Nebula." *Astr. Astrophys.*, **474**, 515

Menzel, D.H., 1968, "Oscillator Strengths for High-level Transitions in Hydrogen." *Nature*, **218**, 756

Merrill, P.W., 1934, "Unidentified Interstellar Lines." *Publ. Astr. Soc. Pacific*, **46**, 206

Meyer, J., Drury, L.O., & Ellison, D.C., 1997, "Galactic Cosmic Rays from Supernova Remnants. I. A Cosmic-Ray Composition Controlled by Volatility and Mass-to-Charge Ratio." *Ap. J.*, **487**, 182

Mie, G., 1908, "Beiträge zur Optik trüber Medien, speziell kolloidaler Metallösungen." *Annalen der Physik*, **330**, 377

Min, M., Waters, L.B.F.M., de Koter, A., et al., 2007, "The shape and composition of interstellar silicate grains." *Astr. Astrophys.*, **462**, 667

Mishchenko, M.I., Hovenier, J.W., & Travis, L.D. (eds.), 2000, *Light scattering by nonspherical particles: Theory, Measurements, and Applications* (San Diego: Academic Press)

Mitnik, D.M. & Badnell, N.R., 2004, "Dielectronic recombination data for dynamic finite-density plasmas. VIII. The nitrogen isoelectronic sequence." *Astr. Astrophys.*, **425**, 1153

Miville-Deschênes, M., Ysard, N., Lavabre, A., et al., 2008, "Separation of anomalous and synchrotron emissions using WMAP polarization data." *Astr. Astrophys.*, **490**, 1093

Möbius, E., Bzowski, M., Chalov, S., et al., 2004, "Synopsis of the interstellar He parameters from combined neutral gas, pickup ion and UV scattering observations and related consequences." *Astr. Astrophys.*, **426**, 897

Moos, G., Gahl, C., Fasel, R., Wolf, M., & Hertel, T., 2001, "Anisotropy of Quasiparticle Lifetimes and the Role of Disorder in Graphite from Ultrafast Time-Resolved Photoemission Spectroscopy." *Physical Review Letters*, **87**, 267402

Morton, D.C., 1975, "Interstellar Absorption Lines in the Spectrum of Zeta Ophiuchi." *Ap. J.*, **197**, 85

Morton, D.C., 2003, "Atomic Data for Resonance Absorption Lines. III. Wavelengths Longward of the Lyman Limit for the Elements Hydrogen to Gallium." *Ap. J. Suppl.*, **149**, 205

Moseley, J., Aberth, W., & Peterson, J.R., 1970, "H^+ + H^- Mutual Neutralization Cross Section Obtained with Superimposed Beams." *Phys. Rev. Lett.*, **24**, 435

Mouschovias, T.C., 1974, "Static Equilibria of the Interstellar Gas in the Presence of Magnetic and Gravitational Fields: Large-Scale Condensations." *Ap. J.*, **192**, 37

Mouschovias, T.C., Shu, F.H., & Woodward, P.R., 1974, "On the Formation of Interstellar Cloud Complexes, OB Associations and Giant H II Regions." *Astr. Astrophys.*, **33**, 73

Mouschovias, T.C. & Spitzer, Jr., L., 1976, "Note on the Collapse of Magnetic Interstellar Clouds." *Ap. J.*, **210**, 326

Muñoz Burgos, J.M., Loch, S.D., Ballance, C.P., & Boivin, R.F., 2009, "Electron-impact excitation of Ar^{2+}." *Astr. Astrophys.*, **500**, 1253

Murphy, E.J., Helou, G., Condon, J.J., et al., 2010, "The Detection of Anomalous Dust Emission in the Nearby Galaxy NGC 6946." *Ap. J. Lett.*, **709**, L108

Murray, N., Ménard, B., & Thompson, T.A., 2011, "Radiation Pressure from Massive Star Clusters as a Launching Mechanism for Supergalactic Winds." *Ap. J.*, **735**, 66

Murray, N. & Rahman, M., 2010, "Star Formation in Massive Clusters Via the Wilkinson Microwave Anisotropy Probe and the Spitzer Glimpse Survey." *Ap. J.*, **709**, 424

Nakanishi, H. & Sofue, Y., 2003, "Three-Dimensional Distribution of the ISM in the Milky Way Galaxy: I. The H I Disk." *Publ. Astr. Soc. Japan*, **55**, 191

Nakanishi, H. & Sofue, Y., 2006, "Three-Dimensional Distribution of the ISM in the Milky Way Galaxy: II. The Molecular Gas Disk." *Publ. Astr. Soc. Japan*, **58**, 847

Napiwotzki, R., 1999, "Spectroscopic investigation of old planetaries. IV. Model atmosphere analysis." *Astr. Astrophys.*, **350**, 101

Neogrády, P., Medvěd, M., Černušák, I., & Urban, M., 2002, "Benchmark calculations of some molecular properties of O_2, CN and other selected small radicals using the ROHF-CCSD(T) method." *Molec. Phys.*, **100**, 541

Neufeld, D.A., 1990, "The Transfer of Resonance-Line Radiation in Static Astrophysical Media." *Ap. J.*, **350**, 216

Nieten, C., Neininger, N., Guélin, M., et al., 2006, "Molecular gas in the Andromeda galaxy." *Astr. Astrophys.*, **453**, 459

Nota, T. & Katgert, P., 2010, "The large-scale magnetic field in the fourth Galactic Quadrant." *Astr. Astrophys.*, **513**, A65

Novak, G., Dotson, J.L., & Li, H., 2009, "Dispersion of Observed Position Angles of Submillimeter Polarization in Molecular Clouds." *Ap. J.*, **695**, 1362

Nussbaumer, H. & Storey, P.J., 1983, "Dielectronic recombination at low temperatures." *Astr. Astrophys.*, **126**, 75

Nussbaumer, H. & Storey, P.J., 1984, "Dielectronic recombination at low temperatures. II Recombination coefficients for lines of C, N, O." *Astr. Astrophys. Suppl.*, **56**, 293

Nussbaumer, H. & Storey, P.J., 1986, "Dielectronic recombination at low temperatures. III - Recombination coefficients for Mg, Al, Si." *Astr. Astrophys. Suppl.*, **64**, 545

O'Dell, C.R., 2001, "Structure of the Orion Nebula." *Publ. Astr. Soc. Pacific*, **113**, 29

O'Dell, C.R., Henney, W.J., Abel, N.P., Ferland, G.J., & Arthur, S.J., 2009, "The Three-Dimensional Dynamic Structure of the Inner Orion Nebula." *A. J.*, **137**, 367

O'Dell, C.R., Wen, Z., & Hu, X., 1993, "Discovery of New Objects in the Orion Nebula on HST Images - Shocks, Compact Sources, and Protoplanetary Disks." *Ap. J.*, **410**, 696

Okumura, A., Kamae, T., & for the Fermi LAT Collaboration, 2009, "Diffuse Gamma-ray Observations of the Orion Molecular Clouds." *arXiv*, **0912.3860**

Olofsson, J., Augereau, J., van Dishoeck, E.F., et al., 2009, "C2D Spitzer-IRS spectra of disks around T Tauri stars. IV. Crystalline silicates." *Astr. Astrophys.*, **507**, 327

Onaka, T., Yamamura, I., Tanabe, T., Roellig, T.L., & Yuen, L., 1996, "Detection of the Mid-Infrared Unidentified Bands in the Diffuse Galactic Emission by IRTS." *Publ. Astr. Soc. Japan*, **48**, L59

Osterbrock, D.E., 1961, "On Ambipolar Diffusion in H I Regions." *Ap. J.*, **134**, 270

Osterbrock, D.E., 1974, *Astrophysics of Gaseous*

Nebulae (San Francisco: W. H. Freeman and Co.)
Osterbrock, D.E., 1989, *Astrophysics of Gaseous Nebulae and Active Galactic Nuclei* (Mill Valley, CA: University Science Books)
Osterbrock, D.E. & Ferland, G.J., 2006, *Astrophysics of Gaseous Nebulae and Active Galactic Nuclei, 2nd edition* (Sausalito, CA: University Science Books), ISBN 1-891-38934-3
Palmeri, P., Mendoza, C., Kallman, T.R., Bautista, M.A., & Meléndez, M., 2003, "Modeling of iron K lines: Radiative and Auger decay data for Fe II-Fe IX." *Astr. Astrophys.*, **410**, 359
Parker, E.N., 1966, "The Dynamical State of the Interstellar Gas and Field." *Ap. J.*, **145**, 811
Pelan, J. & Berrington, K.A., 1995, "Atomic data from the IRON Project. IX. Electron excitation of the $^2P_{3/2}$-$^2P_{1/2}$ fine-structure transition in chlorine-like ions, from Ar II to Ni XII." *Astr. Astrophys. Suppl.*, **110**, 209
Pendleton, Y.J. & Allamandola, L.J., 2002, "The Organic Refractory Material in the Diffuse Interstellar Medium: Mid-Infrared Spectroscopic Constraints." *Ap. J. Suppl.*, **138**, 75
Petrosian, V., Silk, J., & Field, G.B., 1972, "A Simple Analytic Approximation for Dusty Strömgren Spheres." *Ap. J. Lett.*, **177**, L69
Pineau des Forets, G., Flower, D.R., Hartquist, T.W., & Dalgarno, A., 1986, "Theoretical studies of interstellar molecular shocks. III - The formation of CH^+ in diffuse clouds." *M.N.R.A.S.*, **220**, 801
Pogge, R.W., Owen, J.M., & Atwood, B., 1992, "Imaging Spectrophotometry of the Orion Nebula Core. I - Emission-Line Mapping and Physical Conditions." *Ap. J.*, **399**, 147
Prasad, S.S. & Tarafdar, S.P., 1983, "UV Radiation Field Inside Dense Clouds - Its Possible Existence and Chemical Implications." *Ap. J.*, **267**, 603
Puls, J., Markova, N., & Scuderi, S., 2008, "Stellar Winds from Massive Stars - What are the REAL Mass-Loss Rates?" In "*ASP Conf. Ser.* 388, Mass Loss from Stars and the Evolution of Stellar Clusters," (edited by A. de Koter, L.J. Smith, & L.B.F.M. Waters), 101–108
Purcell, E.M., 1969, "On the Absorption and Emission of Light by Interstellar Grains." *Ap. J.*, **158**, 433
Purcell, E.M., 1979, "Suprathermal Rotation of Interstellar Grains." *Ap. J.*, **231**, 404
Purcell, E.M. & Pennypacker, C.R., 1973, "Scattering and Absorption of Light by Nonspherical Dielectric Grains." *Ap. J.*, **186**, 705
Quéméner, G. & Balakrishnan, N., 2009, "Quantum calculations of H_2-H_2 collisions: From ultracold to thermal energies." *J. Chem. Phys.*, **130**, 114303
Rachford, B.L., Snow, T.P., Destree, J.D., et al., 2009, "Molecular Hydrogen in the Far Ultraviolet Spectroscopic Explorer Translucent Lines of Sight: The Full Sample." *Ap. J. Suppl.*, **180**, 125
Rachford, B.L., Snow, T.P., Tumlinson, J., et al., 2002, "A Far Ultraviolet Spectroscopic Explorer Survey of Interstellar Molecular Hydrogen in Translucent Clouds." *Ap. J.*, **577**, 221
Radhakrishnan, V., Murray, J.D., Lockhart, P., & Whittle, R.P.J., 1972, "The Parkes Survey of 21-cm Absorption in Discrete-Source Spectra. II. Galactic 21-cm 0bservations in the Direction of 35 Extragalactic Sources." *Ap. J. Suppl.*, **24**, 15
Ralchenko, Y., Kramida, A.E., Reader, J., & NIST ASD Team, 2010, *NIST Atomic Spectra Database (version 4.0.0)* (Gaithersburg, MD: National Institute of Standards and Technology), http://www.nist.gov/physlab/data/asd.cfm
Ramsbottom, C.A., Bell, K.L., & Keenan, F.P., 1997, "Effective collision strengths for fine-structure forbidden transitions among the $3s^2 3p^3$ levels of Ar IV." *M.N.R.A.S.*, **284**, 754
Ramsbottom, C.A., Bell, K.L., & Keenan, F.P., 1998, "Effective collision strengths for fine-structure forbidden transitions among the $2s^2 2p^3$ levels of Ne IV." *M.N.R.A.S.*, **293**, 233
Rathborne, J.M., Jackson, J.M., Chambers, E.T., et al., 2010, "The Early Stages of Star Formation in Infrared Dark Clouds: Characterizing the Core Dust Properties." *Ap. J.*, **715**, 310
Raymonda, J.W., Muenter, J.S., & Klemperer, W.A., 1970, "Electric Dipole Moment of SiO and GeO." *J. Chem. Phys.*, **52**, 3458
Rest, A., Foley, R.J., Gezari, S., et al., 2011, "Pushing the Boundaries of Conventional Core-Collapse Supernovae: The Extremely Energetic Supernova SN 2003ma." *Ap. J.*, **729**, 88
Reynolds, S.P., Borkowski, K.J., Green, D.A., et al., 2008, "The Youngest Galactic Supernova Remnant: G1.9+0.3." *Ap. J. Lett.*, **680**, L41
Rho, J., Kozasa, T., Reach, W.T., et al., 2008, "Freshly Formed Dust in the Cassiopeia A Supernova Remnant as Revealed by the Spitzer Space Telescope." *Ap. J.*, **673**, 271
Roberge, W.G. & Draine, B.T., 1990, "A New Class of Solutions for Interstellar Magnetohydrodynamic Shock Waves." *Ap. J.*, **350**, 700
Roberge, W.G., Jones, D., Lepp, S., & Dalgarno, A., 1991, "Interstellar Photodissociation and Photoionization Rates." *Ap. J. Suppl.*, **77**, 287
Robertson, J., 2003, "Electronic and atomic structure of diamond-like carbon." *Semiconductor Science Technology*, **18**, S12
Roche, P.F. & Aitken, D.K., 1984, "An investigation of the interstellar extinction. I - Towards dusty WC Wolf-Rayet stars." *M.N.R.A.S.*, **208**,

481
Roche, P.F. & Aitken, D.K., 1985, "An investigation of the interstellar extinction. II - Towards the mid-infrared sources in the Galactic centre." *M.N.R.A.S.*, **215**, 425
Rodgers, W. & Williams, A., 1974, "Integrated absorption of a spectral line with the Voigt profile." *J. Quant. Spectr. Rad. Trans.*, **14**, 319
Roueff, E. & Le Bourlot, J., 1990, "Excitation of forbidden C I fine structure transitions by protons." *Astr. Astrophys.*, **236**, 515
Sahai, R. & Chronopoulos, C.K., 2010, "The Astrosphere of the Asymptotic Giant Branch Star IRC+10216." *Ap. J. Lett.*, **711**, L53
Salem, M. & Brocklehurst, M., 1979, "A Table of Departure Coefficients from Thermodynamic Equilibrium (b_n Factors) for Hydrogenic Ions." *Ap. J. Suppl.*, **39**, 633
Salpeter, E.E., 1955, "The Luminosity Function and Stellar Evolution." *Ap. J.*, **121**, 161
Saraph, H.E. & Storey, P.J., 1996, "Atomic data from the IRON Project. XI. The $^2P_{1/2} - {}^2P_{3/2}$ fine-structure lines of Ar VI, K VII and Ca VIII." *Astr. Astrophys. Suppl.*, **115**, 151
Sarre, P.J., Miles, J.R., Kerr, T.H., et al., 1995, "Resolution of intrinsic fine structure in spectra of narrow diffuse interstellar bands." *M.N.R.A.S.*, **277**, L41
Savage, B.D., Cardelli, J.A., & Sofia, U.J., 1992, "Ultraviolet Observations of the Gas Phase Abundances in the Diffuse Clouds toward Zeta Ophiuchi at 3.5 kilometers per second Resolution." *Ap. J.*, **401**, 706
Scaife, A.M.M., Hurley-Walker, N., Green, D.A., et al., 2009, "AMI observations of Lynds dark nebulae: further evidence for anomalous cm-wave emission." *M.N.R.A.S.*, **400**, 1394
Schlegel, D.J., Finkbeiner, D.P., & Davis, M., 1998, "Maps of Dust Infrared Emission for Use in Estimation of Reddening and Cosmic Microwave Background Radiation Foregrounds." *Ap. J.*, **500**, 525
Schmidt, M., 1959, "The Rate of Star Formation." *Ap. J.*, **129**, 243
Schmutzler, T. & Tscharnuter, W.M., 1993, "Effective radiative cooling in optically thin plasmas." *Astr. Astrophys.*, **273**, 318
Schneider, I.F., Dulieu, O., Giusti-Suzor, A., & Roueff, E., 1994, "Dissociative Recombination of H_2^+ Molecular Ions in Hydrogen Plasmas between 20 K and 4000 K." *Ap. J.*, **424**, 983
Schroder, K., Staemmler, V., Smith, M.D., Flower, D.R., & Jaquet, R., 1991, "Excitation of the fine-structure transitions of C in collisions with ortho- and para-H_2." *J. Phys. B*, **24**, 2487
Schure, K.M., Kosenko, D., Kaastra, J.S., Keppens, R., & Vink, J., 2009, "A new radiative cooling curve based on an up-to-date plasma emission code." *Astr. Astrophys.*, **508**, 751
Scott, A. & Duley, W.W., 1996*a*, "The Decomposition of Hydrogenated Amorphous Carbon: A Connection with Polycyclic Aromatic Hydrocarbon Molecules." *Ap. J. Lett.*, **472**, L123
Scott, A. & Duley, W.W., 1996*b*, "Ultraviolet and Infrared Refractive Indices of Amorphous Silicates." *Ap. J. Suppl.*, **105**, 401
Sedov, L.I., 1959, *Similarity and Dimensional Methods in Mechanics* (New York: Academic Press, 1959)
Sellgren, K., Werner, M.W., Ingalls, J.G., et al., 2010, "C_{60} in Reflection Nebulae." *Ap. J. Lett.*, **722**, L54
Seon, K.I., Edelstein, J., Korpela, E., et al., 2011, "Observation of the Far-ultraviolet Continuum Background with SPEAR/FIMS." *Ap. J. Suppl.*, **196**, 15
Serkowski, K., 1973, "Interstellar Polarization (review)." In "IAU Symp. 52: Interstellar Dust and Related Topics," (edited by J.M. Greenberg & H.C. van de Hulst), 145–152
Seyfert, C.K., 1943, "Nuclear Emission in Spiral Nebulae." *Ap. J.*, **97**, 28
Shen, Y., Draine, B.T., & Johnson, E.T., 2008, "Modeling Porous Dust Grains with Ballistic Aggregates I. Methods and Basic Results." *Ap. J.*, **689**, 260
Shenoy, S.S., Whittet, D.C.B., Chiar, J.E., et al., 2003, "A Test Case for the Organic Refractory Model of Interstellar Dust." *Ap. J.*, **591**, 962
Shull, J.M., 1978, "H_2 Resonance Fluorescence with Lyman-α." *Ap. J.*, **224**, 841
Slavin, J.D., Jones, A.P., & Tielens, A.G.G.M., 2004, "Shock Processing of Large Grains in the Interstellar Medium." *Ap. J.*, **614**, 796
Smith, B., Sigurdsson, S., & Abel, T., 2008, "Metal cooling in simulations of cosmic structure formation." *M.N.R.A.S.*, **385**, 1443
Smith, J.D.T., Draine, B.T., Dale, D.A., et al., 2007, "The Mid-Infrared Spectrum of Star-Forming Galaxies: Global Properties of PAH Emission." *Ap. J.*, **656**, 770
Smith, L.J., Norris, R.P.F., & Crowther, P.A., 2002, "Realistic ionizing fluxes for young stellar populations from 0.05 to 2 $Z_\odot$." *M.N.R.A.S.*, **337**, 1309
Snell, R.L., Howe, J.E., Ashby, M.L.N., et al., 2000, "Water Abundance in Molecular Cloud Cores." *Ap. J. Lett.*, **539**, L101
Snowden, S.L., Freyberg, M.J., Plucinsky, P.P., et al., 1995, "First Maps of the Soft X-Ray Diffuse Background from the ROSAT XRT/PSPC All-Sky Survey." *Ap. J.*, **454**, 643
Snowden, S.L., McCammon, D., Burrows, D.N., & Mendenhall, J.A., 1994, "Analysis Procedures for ROSAT XRT/PSPC Observations of Extended Objects and the Diffuse Background." *Ap. J.*, **424**, 714
Sobolev, V.V., 1957, "The Diffusion of Lα Radiation in Nebulae and Stellar Envelopes." *Soviet Astronomy*, **1**, 678

Sofia, U.J., Lauroesch, J.T., Meyer, D.M., & Cartledge, S.I.B., 2004, "Interstellar Carbon in Translucent Sight Lines." *Ap. J.*, **605**, 272

Sofia, U.J. & Parvathi, V.S., 2010, "Carbon Abundances in Interstellar Gas and Dust." In "Cosmic Dust – Near and Far," (edited by T. Henning, E. Grün, & J. Steinacker), 236–242

Solomon, P.M., Rivolo, A.R., Barrett, J., & Yahil, A., 1987, "Mass, luminosity, and line width relations of Galactic molecular clouds." *Ap. J.*, **319**, 730

Spitzer, L., 1962, *Physics of Fully Ionized Gases, 2nd edition* (New York: Interscience)

Spitzer, L., 1976, "High-velocity interstellar clouds." *Comments Astrophys.*, **6**, 177

Spitzer, L., 1978, *Physical Processes in the Interstellar Medium* (New York: Wiley)

Staemmler, V. & Flower, D.R., 1991, "Excitation of the C($2p^2\ {}^3P_J$) fine structure states in collisions with He($1s^2\ {}^1S_0$)." *J. Phys. B*, **24**, 2343

Stafford, R.P., Bell, K.L., & Hibbert, A., 1994, "Electron impact excitation of N III - Fine-structure collision strengths and Maxwellian-averaged rate coefficients." *M.N.R.A.S.*, **266**, 715

Stahl, O., Wade, G., Petit, V., Stober, B., & Schanne, L., 2008, "Long-term monitoring of θ^1 Ori C: the spectroscopic orbit and an improved rotational period." *Astr. Astrophys.*, **487**, 323

Stahler, S.W. & Palla, F., 2004, *The Formation of Stars* (Weinheim: Wiley-VCH), ISBN 3-527-40559-3

Stancil, P.C., Schultz, D.R., Kimura, M., et al., 1999, "Charge transfer in collisions of O^+ with H and H^+ with O." *Astr. Astrophys. Suppl.*, **140**, 225

Stasińska, G. & Szczerba, R., 2001, "The temperature structure of dusty planetary nebulae." *Astr. Astrophys.*, **379**, 1024

Stecher, T.P. & Donn, B., 1965, "On Graphite and Interstellar Extinction." *Ap. J.*, **142**, 1681

Steel, T.M. & Duley, W.W., 1987, "A 217.5 nanometer Absorption Feature in the Spectrum of Small Silicate Particles." *Ap. J.*, **315**, 337

Stephens, J.R., Blanco, A., Bussoletti, E., et al., 1995, "Effect of composition on IR spectra of synthetic amorphous silicate cosmic dust analogues." *Plan. Sp. Sci.*, **43**, 1241

Stone, J.M., Ostriker, E.C., & Gammie, C.F., 1998, "Dissipation in Compressible Magnetohydrodynamic Turbulence." *Ap. J. Lett.*, **508**, L99

Storey, P.J., 1981, "Dielectronic recombination at nebular temperatures." *M.N.R.A.S.*, **195**, 27P

Strömgren, B., 1939, "The Physical State of Interstellar Hydrogen." *Ap. J.*, **89**, 526

Strong, A.W., Bennett, K., Bloemen, H., et al., 1994, "Diffuse continuum gamma rays from the Galaxy observed by COMPTEL." *Astr. Astrophys.*, **292**, 82

Strong, A.W., Moskalenko, I.V., & Ptuskin, V.S., 2007, "Cosmic-Ray Propagation and Interactions in the Galaxy." *Ann. Rev. Nucl. Part. Sci.*, **57**, 285

Sutherland, R.S. & Dopita, M.A., 1993, "Cooling Functions for Low-Density Astrophysical Plasmas." *Ap. J. Suppl.*, **88**, 253

Swings, P. & Rosenfeld, L., 1937, "Considerations Regarding Interstellar Molecules." *Ap. J.*, **86**, 483

Tammann, G.A., Loeffler, W., & Schroeder, A., 1994, "The Galactic supernova rate." *Ap. J. Suppl.*, **92**, 487

Tanaka, M., Matsumoto, T., Murakami, H., et al., 1996, "IRTS Observation of the Unidentified 3.3-Micron Band in the Diffuse Galactic Emission." *Publ. Astr. Soc. Japan*, **48**, L53

Tawara, H., Itikawa, Y., Nishimura, H., & Yoshino, M., 1990, "Cross Sections and Related Data for Electron Collisions with Hydrogen Molecules and Molecular Ions." *J. Phys. Chem. Ref. Data*, **19**, 617

Tayal, S.S., 2000, "Electron Collision Excitation of Fine-Structure Levels in S IV." *Ap. J.*, **530**, 1091

Tayal, S.S., 2006*a*, "Breit-Pauli R-Matrix Calculation for Electron Collision Rates in O IV." *Ap. J. Suppl.*, **166**, 634

Tayal, S.S., 2006*b*, "New Accurate Oscillator Strengths and Electron Excitation Collision Strengths for N I." *Ap. J. Suppl.*, **163**, 207

Tayal, S.S., 2007, "Oscillator Strengths and Electron Collision Rates for Fine-Structure Transitions in O II." *Ap. J. Suppl.*, **171**, 331

Tayal, S.S., 2008, "Electron impact excitation collision strength for transitions in C II." *Astr. Astrophys.*, **486**, 629

Tayal, S.S. & Gupta, G.P., 1999, "Collision Strengths for Electron Collision Excitation of Fine-Structure Levels in S III." *Ap. J.*, **526**, 544

Tayal, S.S. & Zatsarinny, O., 2010, "Breit-Pauli Transition Probabilities and Electron Excitation Collision Strengths for Singly Ionized Sulfur." *Ap. J. Suppl.*, **188**, 32

Taylor, G.I., 1950, "The formation of a blast wave by a very intense explosion 1: theoretical discussion." *Proc. Roy. Soc. London, A*, **201**, 159

Taylor, J.H. & Cordes, J.M., 1993, "Pulsar Distances and the Galactic Distribution of Free Electrons." *Ap. J.*, **411**, 674

Tenorio-Tagle, G., 1979, "The gas dynamics of H II regions. I - The champagne model." *Astr. Astrophys.*, **71**, 59

Thomas, M.A. & Humbertson, J.W., 1972, "The polarizability of helium." *J. Phys. B*, **5**, L229

Thompson, T.A., Quataert, E., & Murray, N., 2005, "Radiation Pressure-supported Starburst Disks and Active Galactic Nucleus Fueling." *Ap. J.*, **630**, 167

Tielens, A.G.G.M., 2008, "Interstellar Polycyclic Aromatic Hydrocarbon Molecules." *Ann. Rev. Astr. Astrophys.*, **46**, 289

Tielens, A.G.G.M., Meixner, M.M., van der Werf, P.P., et al., 1993, "Anatomy of the Photodissociation Region in the Orion Bar." *Science*, **262**, 86

Tommasin, S., Spinoglio, L., Malkan, M.A., et al., 2008, "Spitzer IRS High-Resolution Spectroscopy of the 12 μm Seyfert Galaxies. I. First Results." *Ap. J.*, **676**, 836

Townes, C.H. & Schawlow, A.L., 1975, *Microwave Spectroscopy.* (New York, NY (USA): Dover Publications)

Troland, T.H. & Crutcher, R.M., 2008, "Magnetic Fields in Dark Cloud Cores: Arecibo OH Zeeman Observations." *Ap. J.*, **680**, 457

Trumpler, R.J., 1930, "Absorption of Light in the Galactic System." *Publ. Astr. Soc. Pacific*, **42**, 214

van de Hulst, H.C., 1957, *Light Scattering by Small Particles* (New York: John Wiley & Sons)

van der Werf, P.P., Stutzki, J., Sternberg, A., & Krabbe, A., 1996, "Structure and chemistry of the Orion bar photon-dominated region." *Astr. Astrophys.*, **313**, 633

van Dishoeck, E.F., 1988, "Photodissociation and photoionization processes." In "Rate coefficients in astrochemistry," 49–72 (Dordrecht: Kluwer)

van Dishoeck, E.F. & Black, J.H., 1986, "Comprehensive models of diffuse interstellar clouds - Physical conditions and molecular abundances." *Ap. J. Suppl.*, **62**, 109

van Veelen, B., Langer, N., Vink, J., García-Segura, G., & van Marle, A.J., 2009, "The hydrodynamics of the supernova remnant Cassiopeia A. The influence of the progenitor evolution on the velocity structure and clumping." *Astr. Astrophys.*, **503**, 495

Veilleux, S. & Osterbrock, D.E., 1987, "Spectral Classification of Emission-Line Galaxies." *Ap. J. Suppl.*, **63**, 295

Verner, D.A., 1999, "Subroutine rrfit, version 4." http://www.pa.uky.edu/~verner/fortran.html

Verner, D.A. & Ferland, G.J., 1996, "Atomic Data for Astrophysics. I. Radiative Recombination Rates for H-like, He-like, Li-like, and Na-like Ions over a Broad Range of Temperature." *Ap. J. Suppl.*, **103**, 467

Verner, D.A., Ferland, G.J., Korista, K.T., & Yakovlev, D.G., 1996, "Atomic Data for Astrophysics. II. New Analytic Fits for Photoionization Cross Sections of Atoms and Ions." *Ap. J.*, **465**, 487

Verner, D.A. & Yakovlev, D.G., 1995, "Analytic fits for partial photoionization cross sections." *Astr. Astrophys. Suppl.*, **109**, 125

Vijh, U.P., Witt, A.N., & Gordon, K.D., 2005, "Blue Luminescence and the Presence of Small Polycyclic Aromatic Hydrocarbons in the Interstellar Medium." *Ap. J.*, **633**, 262

Visser, R., van Dishoeck, E.F., & Black, J.H., 2009, "The photodissociation and chemistry of CO isotopologues: applications to interstellar clouds and circumstellar disks." *Astr. Astrophys.*, **503**, 323

Voit, G.M., 1991, "Energy Deposition by X-Ray Photoelectrons into Interstellar Molecular Clouds." *Ap. J.*, **377**, 158

Voss, R., Diehl, R., Hartmann, D.H., & Kretschmer, K., 2008, "Population synthesis models for ^{26}Al production in starforming regions." *New Astronomy Review*, **52**, 436

Wakker, B.P., 2006, "A FUSE Survey of High-Latitude Galactic Molecular Hydrogen." *Ap. J. Suppl.*, **163**, 282

Wang, J.Z., Seo, E.S., Anraku, K., et al., 2002, "Measurement of Cosmic-Ray Hydrogen and Helium and Their Isotopic Composition with the BESS Experiment." *Ap. J.*, **564**, 244

Wang, W., Lang, M.G., Diehl, R., et al., 2009, "Spectral and intensity variations of Galactic ^{26}Al emission." *Astr. Astrophys.*, **496**, 713

Wannier, G.H., 1953, "Motion of Gaseous Ions in Strong Electric Fields." *Bell. system Tech. J.*, **32**, 170

Watson, C., Povich, M.S., Churchwell, E.B., et al., 2008, "Infrared Dust Bubbles: Probing the Detailed Structure and Young Massive Stellar Populations of Galactic H II Regions." *Ap. J.*, **681**, 1341

Watson, R.A., Rebolo, R., Rubiño-Martín, J.A., et al., 2005, "Detection of Anomalous Microwave Emission in the Perseus Molecular Cloud with the COSMOSOMAS Experiment." *Ap. J. Lett.*, **624**, L89

Waxman, E. & Draine, B.T., 2000, "Dust Sublimation by Gamma-ray Bursts and Its Implications." *Ap. J.*, **537**, 796

Weaver, R., McCray, R., Castor, J., Shapiro, P., & Moore, R., 1977, "Interstellar bubbles. II - Structure and evolution." *Ap. J.*, **218**, 377

Webber, W.R. & Lockwood, J.A., 2001, "Voyager and Pioneer spacecraft measurements of cosmic ray intensities in the outer heliosphere: Toward a new paradigm for understanding the global solar modulation process: 1. Minimum solar modulation (1987 and 1997)." *J. Geophys. Res.*, **106**, 29323

Webber, W.R. & Yushak, S.M., 1983, "A Measurement of the Energy Spectra and Relative Abundance of the Cosmic-Ray H and He Isotopes over a Broad Energy Range." *Ap. J.*, **275**, 391

Weibel, E.S., 1959, "Spontaneously Growing Transverse Waves in a Plasma Due to an Anisotropic Velocity Distribution." *Phys. Rev. Lett.*, **2**, 83

Weingartner, J.C. & Draine, B.T., 2001*a*, "Dust Grain-Size Distributions and Extinction in the Milky Way, Large Magellanic Cloud, and Small Magellanic Cloud." *Ap. J.*, **548**, 296

Weingartner, J.C. & Draine, B.T., 2001*b*, "Electron-Ion Recombination on Grains and Polycyclic Aromatic Hydrocarbons." *Ap. J.*, **563**, 842

Weingartner, J.C. & Draine, B.T., 2001*c*, "Forces on Dust Grains Exposed to Anisotropic Interstellar Radiation Fields." *Ap. J.*, **553**, 581

Weingartner, J.C. & Draine, B.T., 2001*d*, "Photoelectric Emission from Interstellar Dust: Grain Charging and Gas Heating." *Ap. J. Suppl.*, **134**, 263

Weingartner, J.C. & Draine, B.T., 2003, "Radiative Torques on Interstellar Grains. III. Dynamics with Thermal Relaxation." *Ap. J.*, **589**, 289

Weingartner, J.C., Draine, B.T., & Barr, D.K., 2006, "Photoelectric Emission from Dust Grains Exposed to Extreme Ultraviolet and X-Ray Radiation." *Ap. J.*, **645**, 1188

Welty, D.E. & Fowler, J.R., 1992, "Ultraviolet, Optical, and Infrared Observations of the High-Latitude Molecular Cloud toward HD 210121." *Ap. J.*, **393**, 193

Wen, Z. & O'Dell, C.R., 1995, "A Three-Dimensional Model of the Orion Nebula." *Ap. J.*, **438**, 784

Wesson, R., Liu, X.W., & Barlow, M.J., 2005, "The abundance discrepancy - recombination line versus forbidden line abundances for a northern sample of galactic planetary nebulae." *M.N.R.A.S.*, **362**, 424

Whitham, G.B., 1974, *Linear and Nonlinear Waves* (New York: Wiley)

Whittet, D.C.B., 2010, "Oxygen Depletion in the Interstellar Medium: Implications for Grain Models and the Distribution of Elemental Oxygen." *Ap. J.*, **710**, 1009

Whittet, D.C.B., Bode, M.F., Longmore, A.J., et al., 1988, "Infrared spectroscopy of dust in the Taurus dark clouds - Ice and silicates." *M.N.R.A.S.*, **233**, 321

Whittet, D.C.B., Boogert, A.C.A., Gerakines, P.A., et al., 1997, "Infrared Spectroscopy of Dust in the Diffuse Interstellar Medium toward Cygnus OB2 No. 12." *Ap. J.*, **490**, 729

Wiese, W.L., Smith, M.W., & Glennon, B.M., 1966, *Atomic Transition Probabilities. Volume I. Hydrogen Through Neon* (Washington, D.C.: U.S. Government Printing Office)

Williams, J.P. & McKee, C.F., 1997, "The Galactic Distribution of OB Associations in Molecular Clouds." *Ap. J.*, **476**, 166

Wilson, T.L., Filges, L., Codella, C., Reich, W., & Reich, P., 1997, "Kinematics and electron temperatures in the core of Orion A." *Astr. Astrophys.*, **327**, 1177

Witt, A.N., Gordon, K.D., Vijh, U.P., et al., 2006, "The Excitation of Extended Red Emission: New Constraints on Its Carrier from Hubble Space Telescope Observations of NGC 7023." *Ap. J.*, **636**, 303

Witt, A.N. & Vijh, U.P., 2004, "Extended Red Emission: Photoluminescence by Interstellar Nanoparticles." In "Astrophysics of Dust," , vol. 309 of *ASP Conf. Ser.* (edited by A.N. Witt, G.C. Clayton, & B.T. Draine), 115–139 (San Francisco, CA: ASP)

Wolff, M.J., Clayton, G.C., Kim, S.H., Martin, P.G., & Anderson, C.M., 1997, "Ultraviolet Interstellar Linear Polarization. III. Features." *Ap. J.*, **478**, 395

Wolfire, M.G., Hollenbach, D., & McKee, C.F., 2010, "The Dark Molecular Gas." *Ap. J.*, **716**, 1191

Wolfire, M.G., Hollenbach, D., McKee, C.F., Tielens, A.G.G.M., & Bakes, E.L.O., 1995, "The Neutral Atomic Phases of the Interstellar Medium." *Ap. J.*, **443**, 152

Wolfire, M.G., McKee, C.F., Hollenbach, D., & Tielens, A.G.G.M., 2003, "Neutral Atomic Phases of the Interstellar Medium in the Galaxy." *Ap. J.*, **587**, 278

Wolniewicz, L., Simbotin, I., & Dalgarno, A., 1998, "Quadrupole Transition Probabilities for the Excited Rovibrational States of H_2." *Ap. J. Suppl.*, **115**, 293

Wood, B.E., Müller, H., & Linsky, J.L., 2003, "Mass Loss Rates for Solar-like Stars Measured from Lyα Absorption." In "The Future of Cool-Star Astrophysics: 12th Cambridge Workshop on Cool Stars, Stellar Systems, and the Sun (2001 July 30 - August 3)," , vol. 12 (edited by A. Brown, G. M. Harper, & T. R. Ayres), 349–358

Woodall, J., Agúndez, M., Markwick-Kemper, A.J., & Millar, T.J., 2007, "The UMIST database for astrochemistry 2006." *Astr. Astrophys.*, **466**, 1197

Wooden, D.H., Harker, D.E., Woodward, C.E., et al., 1999, "Silicate Mineralogy of the Dust in the Inner Coma of Comet C/1995 01 (Hale-Bopp) Pre- and Postperihelion." *Ap. J.*, **517**, 1034

Wouthuysen, S.A., 1952, "On the excitation mechanism of the 21-cm (radio-frequency) interstellar hydrogen emission line." *A. J.*, **57**, 31

Wrathmall, S.A. & Flower, D.R., 2006, "A quantum-mechanical study of rotational transitions in H_2 induced by H." *J. Phys. B*, **39**, L249

Wrathmall, S.A., Gusdorf, A., & Flower, D.R., 2007, "The excitation of molecular hydrogen by atomic hydrogen in astrophysical media." *M.N.R.A.S.*, **382**, 133

Wright, E.L., Mather, J.C., Bennett, C.L., et al., 1991, "Preliminary Spectral Observations of the Galaxy with a 7 deg Beam by the Cosmic Background Explorer (COBE)." *Ap. J.*, **381**, 200

Yan, M., Sadeghpour, H.R., & Dalgarno, A., 1998, "Photoionization Cross Sections of He and H_2." *Ap. J.*, **496**, 1044

Yu, S., Drouin, B.J., Pearson, J.C., & Pickett, H.M., 2009, "Terahertz Spectroscopy and Global Analysis of H_3O^+." *Ap. J. Suppl.*, **180**, 119

Zatsarinny, O., Gorczyca, T.W., Korista, K., Badnell, N.R., & Savin, D.W., 2004*a*, "Dielectronic recombination data for dynamic finite-density plasmas. VII. The neon isoelectronic sequence." *Astr. Astrophys.*, **426**, 699

Zatsarinny, O., Gorczyca, T.W., Korista, K.T., Badnell, N.R., & Savin, D.W., 2003, "Dielectronic recombination data for dynamic finite-density plasmas. II. The oxygen isoelectronic sequence." *Astr. Astrophys.*, **412**, 587

Zatsarinny, O., Gorczyca, T.W., Korista, K.T., Badnell, N.R., & Savin, D.W., 2004*b*, "Dielectronic recombination data for dynamic finite-density plasmas. IV. The carbon isoelectronic sequence." *Astr. Astrophys.*, **417**, 1173

Zatsarinny, O. & Tayal, S.S., 2003, "Electron Collisional Excitation Rates for O I Using the B-Spline R-Matrix Approach." *Ap. J. Suppl.*, **148**, 575

Zeldovich, Y.B. & Raizer, Y.P., 1968, *Physics of Shock Waves and High-Temperature Hydrodynamic Phenomena, Vols. I & II* (New York: Academic Press)

Zhang, Y., Liu, X., Luo, S., Péquignot, D., & Barlow, M.J., 2005, "Integrated spectrum of the planetary nebula NGC 7027." *Astr. Astrophys.*, **442**, 249

Zhang, Y., Liu, X., Wesson, R., et al., 2004, "Electron temperatures and densities of planetary nebulae determined from the nebular hydrogen recombination spectrum and temperature and density variations." *M.N.R.A.S.*, **351**, 935

Zubko, V., Dwek, E., & Arendt, R.G., 2004, "Interstellar Dust Models Consistent with Extinction, Emission, and Abundance Constraints." *Ap. J. Suppl.*, **152**, 211

Zygelman, B., 2005, "Hyperfine Level-changing Collisions of Hydrogen Atoms and Tomography of the Dark Age Universe." *Ap. J.*, **622**, 1356

Index